Mathematics for Elementary School Teachers

FOURTH EDITION

Mathematics for Elementary School Teachers

Tom Bassarear
Keene State College

HOUGHTON MIFFLIN COMPANY   Boston   New York

This book is dedicated to my wife, Yvette, and my two children, Emily and Josh, who had to put up with my many absences from the family over the past four years. I'm back!

Publisher: *Richard Stratton*
Senior Sponsoring Editor: *Lynn Cox*
Senior Marketing Manager: *Jennifer Jones*
Marketing Associate: *Mary Legere*
Senior Development Editor: *Maria Morelli*
Editorial Assistant: *Laura Ricci*
Associate Project Editor: *Kristen Truncellito*
Art and Design Manager: *Gary Crespo*
Cover Design Manager: *Anne S. Katzeff*
Senior Photo Editor: *Jennifer Meyer Dare*
Composition Buyer: *Chuck Dutton*
New Title Project Manager: *Susan Brooks-Peltier*

Cover photograph: © *Jim Scherer/StockFood Creative/Getty Images*

PHOTO CREDITS:
Chapter 1: p.7, The Far Side" by Gary Larson © 1987 FarWorks, Inc. All Rights Reserved. The Far Side" and the Larson" signature are registered trademarks of FarWorks, Inc. Used with permission; p. 30: © North America Syndicate. **Chapter 7:** p. 464: PEANUTS © United Features Syndicate, Inc. **Chapter 8:** p. 499, (left) © Christi Carter/Grant Heilman Photography, (center) © L.Lee Rue III/Photo Researchers, (right) © Grant Heilman/Grant Heilman Photography; p. 502, (top left) Jonathan A. Meyers/JAM Photography, (top center) © Steve Jay Crise/CORBIS, (top right) © Ralph A. Clevenger/CORBIS, (bottom left) © Ed Eckstein/CORBIS, (bottom right) Michael Javorka/Getty Images; p. 503, Scala/Art Resource, NY; p. 507, Reprinted by permission of John L. Hart FLP, and Creators Syndicate, Inc.; p. 521, (left) Courtesy Kraft Foods, (center) © Bettmann/CORBIS, (right) © Bettmann/CORBIS; p. 530, Michael Javorka/Getty Images; p. 546, (top) © Bettmann/CORBIS, (bottom left) Jonathan A. Meyers/JAM Photography, (bottom right) © PR Inc./Photo Researchers; p. 560, (top left) © Bettmann/CORBIS, (top right) Scala/Art Resource, NY, (middle left) © Grant Heilman/Grant Heilman Photography, (middle right) © Runk Schoenberger/Grant Heilman Photography, (bottom left) © Grant Heilman/Grant Heilman Photography Frank Ward Photography, (bottom right) © Grant Heilman/Grant Heilman Photography © Roland and Sabrina Michaud. All Rights Reserved. John Hillelson Agency. **Chapter 9:** p. 582, © Roland and Sabrina Michaud. All Rights Reserved. John Hillelson Agency; p. 583, M.C. Escher's ceiling at Philips International. © 2006 The M.C. Escher Company—Holland. All Rights Reserved. www.mcescher.com; p. 593, M.C. Escher's Study of Regular Division of the Plane with Birds. © 2006 The M.C. Escher Company—Holland. All Rights Reserved. www.mcescher.com; p. 607, The Far Side" by Gary Larson © 1990 FarWorks, Inc. All Rights Reserved. The Far Side" and the Larson" signature are registered trademarks of FarWorks, Inc. Used with permission; p. 611, (left) © David Saunders and Marvin Harrell, Emporia University, KS, (right) © David Saunders and Marvin Harrell, Emporia University, KS; p. 618, © American Antiquarian Society; p. 619: (left) © Ron Wu/Index Stock Imagery, (right) AMNH Neg. 336663-Sun Mask. Photo by Arthur Singer, courtesy the Library, American Museum of Natural History; p. 620, (top) From SYMMETRY: A Unifying Concept, © 1994 by Istvan Hargittai and Magdolna Hargittai, published by Shelter Publications, Inc., P.O. Box 279, Bolinas, CA. Available in bookstores and Amazon.com. Reprinted with permission, (bottom) By Permission of the British Library, print 1007628.001; or 6810, 27v; p. 638, © Steve Rosenthal, Photographer; p. 633, (top left) © Tom Bassarear, (top right) © Tom Bassarear, (bottom left) © Tom Bassarear, (bottom right) © Tom Bassarear; p. 634, (top left) © Jonathan Holstein Collection, (top right) Indian Pine Quilt. Artist unidentified, Maine; 1880–1890. Cotton with cotton embroidery 86 x 82". Collection American Folk Art Museum, New York. Gift of Cyril Irwin Nelson in memory of his grandparents, Guerdon Stearns and Elinor Irwin (Chase) Holden, and in honor of his parents, Cyril Arthur and Elise Macy Nelson. 1982.22.1; (bottom) Photo from p. 10 of Quilts from Aunt Amy © 1999 by Mary Tendall Etherington and Connie Tesene, published by Martingale and Company.

Copyright © 2008 by Houghton Mifflin Company. All rights reserved.

No part of this work may be reproduced or transmitted in any form or by any means, electronic or mechanical, including photocopying and recording, or by any information storage or retrieval system without the prior written permission of Houghton Mifflin Company unless such copying is expressly permitted by federal copyright law. Address inquiries to College Permissions, Houghton Mifflin Company, 222 Berkeley Street, Boston, MA 02116-3764.

Printed in the U.S.A.

Library of Congress Control Number: 2006921991

Instructor's examination copy
   ISBN-10: 0-618-83320-X
   ISBN-13: 978-0-618-83320-7

For orders, use student text ISBNs
   ISBN-10: 0-618-76836-X
   ISBN-13: 978-0-618-76836-3

Contents

Preface   xvii

1 Foundations for Learning Mathematics   1

SECTION 1.1   Getting Comfortable with Mathematics   1

  Beliefs About and Attitudes Toward Mathematics   1
  What Is Mathematics?   3
  The NCTM Standards   4
  Using Mathematics   5

SECTION 1.2   Problem-Solving   6

INVESTIGATION 1.1   Pigs and Chickens   7

  Problem-Solving and Toolboxes   10
  Polya's Four Steps   11
  Why Emphasize Problem-Solving?   13

INVESTIGATION 1.2   How Much Will the Patio Cost?   15

SECTION 1.3   Patterns   16

INVESTIGATION 1.3   The Sum of the First 100 Numbers   17
INVESTIGATION 1.4   Pascal's Triangle   21

  The Power of Patterns   22

SECTION 1.4   Representation   23

INVESTIGATION 1.5   Coin Combinations   24

EXERCISES   26

SECTION 1.5   Reasoning and Proof   30

INVESTIGATION 1.6   Does Your Answer Make Sense?   30
INVESTIGATION 1.7   Making Generalizations   31

  Inductive Reasoning   32

INVESTIGATION 1.8   Inductive Thinking with Fractions   32
INVESTIGATION 1.9   Sequences, Patterns, Reasoning, and Proof   33

  Deductive Reasoning   36

INVESTIGATION 1.10   Deductive Reasoning and Venn Diagrams   40
INVESTIGATION 1.11   Why Does Odd + Odd Equal Even?   41

  Intuitive Reasoning   42

INVESTIGATION 1.12   The Nine Dots Problem   42

SECTION 1.6   Communication   44

  Learning by Communicating   45

INVESTIGATION 1.13   Communicating Patterns in a Magic Square   46
INVESTIGATION 1.14   Darts, Proof, and Communication   47

SECTION 1.7   Connections   48

  Making Connections Between the Problem and What Is Inside Your Head   48
  Connecting New Concepts to Old Concepts   49
  Making Connections Among Different Concepts   49
  Connecting Different Models for the Same Concept   50
  Connecting Conceptual and Procedural Knowledge   50
  Making Connections Between Mathematics and "Real Life" and Between
    Mathematics and Other Disciplines   51

INVESTIGATION 1.15 How Many Pieces of Wire?   51

EXERCISES   54

CHAPTER SUMMARY   58

CHAPTER 1 REVIEW EXERCISES   59

2 Fundamental Concepts   61

SECTION 2.1 Sets   62

    Sets as a Classification Tool   62

INVESTIGATION 2.1 Classifying Quadrilaterals   63

    Defining Sets   64
    Describing Sets   64

INVESTIGATION 2.2 Describing Sets   65

    Kinds of Sets   66
    Defining Subsets   66

INVESTIGATION 2.3 How Many Subsets?   67

    The Empty Set   69
    Equal and Equivalent Sets   69
    Venn Diagrams   70
    Operations on Sets   70
    Venn Diagrams and Relationships Between Sets   72

INVESTIGATION 2.4 Translating Among Representations   73

INVESTIGATION 2.5 Finding Information from Venn Diagrams   73

    Venn Diagrams as a Communication Tool   75

EXERCISES 2.1   76

SECTION 2.2 Algebraic Thinking   80

    Algebra as a Set of Rules and Procedures   81

INVESTIGATION 2.6 A Variable by Any Other Name Is Still a Variable   81

    The Equals Sign and Equivalence   83
    Algebra as the Study of Structures   83
    Algebra as the Study of Relationships Among Quantities   84

INVESTIGATION 2.7 Baby-sitting   86

INVESTIGATION 2.8 Choosing Between Functions   88

INVESTIGATION 2.9 Matching Graphs to Situations   90

INVESTIGATION 2.10 Developing "Graph Sense"   91

    Algebra as Generalized Arithmetic   91

INVESTIGATION 2.11 Looking for Generalizations   92

INVESTIGATION 2.12 How Many Dots?   94

EXERCISES 2.2   96

SECTION 2.3 Numeration   101

    Origins of Numbers and Counting   101
    Patterns in Counting   103
    The Egyptian Numeration System   104
    The Roman Numeration System   105
    The Babylonian Numeration System   106
    The Development of Base 10: The Hindu-Arabic System   109
    Advantages of Base 10   110
    Connecting Geometric and Numerical Representations   111

INVESTIGATION 2.13 Relative Magnitude of Numbers   113

    Units   114

INVESTIGATION 2.14 What If Our System Was Based on One Hand? **115**

INVESTIGATION 2.15 How Well Do You Understand Base 5? 116

EXERCISES 2.3 119

CHAPTER SUMMARY 122

CHAPTER 2 REVIEW EXERCISES 123

3 The Four Fundamental Operations of Arithmetic   127

SECTION 3.1 Understanding Addition 128

Addition 128
Contexts for Addition 128
A Pictorial Model for Addition 128
Number Lines: An Addition Model 130
Properties of Addition 131

INVESTIGATION 3.1 A Pattern in the Addition Table 133

INVESTIGATION 3.2 Mental Addition 134

Children's Strategies for Addition 136

INVESTIGATION 3.3 Children's Strategies for Adding Large Numbers 138

Algorithms 138
Adding Before Base 10 139
Investigating Addition Algorithms in Base 10 139

INVESTIGATION 3.4 An Alternative Algorithm 142

INVESTIGATION 3.5 Children's Mistakes 142

Estimation 143
Estimation Strategies for Addition 144

INVESTIGATION 3.6 What Was the Total Attendance? 145

A Refined Technique for Rounding 146

INVESTIGATION 3.7 Estimating by Making Compatible Numbers 147

INVESTIGATION 3.8 How Much Money Did Sandy Spend? 147

Number Sense and Operation Sense 148

INVESTIGATION 3.9 Number Sense with Addition 148

EXERCISES 3.1 150

SECTION 3.2 Understanding Subtraction 152

Contexts for Subtraction 152
A Pictorial Model for Subtraction 153
Subtraction with Number Lines 154
Properties of Subtraction 154

INVESTIGATION 3.10 Mental Subtraction 155

INVESTIGATION 3.11 Mental Subtraction in a Different Context 156

Children's Strategies for Subtraction 157

INVESTIGATION 3.12 Children's Strategies for Subtraction with Large Numbers 159

Understanding Subtraction in Base 10 159
Justifying the Standard Algorithm 160
An Alternative Algorithm 161

INVESTIGATION 3.13 An Algorithm Used in Many Other Countries 161

INVESTIGATION 3.14 Rough and Best Estimates with Subtraction 163

INVESTIGATION 3.15 Numbers Sense with Subtraction 164

Carrying and Borrowing 164
Reviewing Addition and Subtraction 165

EXERCISES 3.2 166

SECTION 3.3 Understanding Multiplication 168

    Multiplication 168
    Contexts for Multiplication 169
    A General Model for Multiplication 170
    Area Model for Multiplication 170
    Cartesian Product Model for Multiplication 171
    Changes in Units 171
    Properties of Multiplication 172
    The Multiplication Table 173

INVESTIGATION 3.16 A Pattern in the Multiplication Table 174

INVESTIGATION 3.17 Mental Multiplication 175

    Understanding Multiplication with Larger Numbers 176
    The Multiplication Algorithm in Base 10 178

INVESTIGATION 3.18 An Alternative Algorithm 181

INVESTIGATION 3.19 Developing Estimation Strategies for Multiplication 181

INVESTIGATION 3.20 Using Various Strategies in a Real-Life Multiplication Situation 183

INVESTIGATION 3.21 Number Sense with Multiplication 184

EXERCISES 3.3 186

SECTION 3.4 Understanding Division 189

    Contexts for Division 189
    Models for Division 190
    The Missing Factor Model of Division 192
    Division and Number Lines 192
    Properties of Division 193
    Division by Zero 193
    Division with Remainders 194

INVESTIGATION 3.22 Mental Division 195

    Division Algorithms 195

INVESTIGATION 3.23 Understanding Division Algorithms 196

INVESTIGATION 3.24 The Scaffolding Algorithm 198

INVESTIGATION 3.25 Estimates with Division 199

INVESTIGATION 3.26 Number Sense with Division 201

    Operation Sense 202

INVESTIGATION 3.27 Applying Models to a Real-life Situation 203

INVESTIGATION 3.28 Operation Sense 204

    Order of Operations 205
    Reviewing Multiplication and Division 205

EXERCISES 3.4 207

CHAPTER SUMMARY 211

CHAPTER 3 REVIEW EXERCISES 212

4 Number Theory 215

SECTION 4.1 Divisibility and Related Concepts 216

INVESTIGATION 4.1 Interesting Dates 216

    Odd and Even Numbers 218

INVESTIGATION 4.2 Patterns in Odd and Even Numbers 218

INVESTIGATION 4.3 Understanding Divisibility Relationships 221

INVESTIGATION 4.4 Determining the Truth of an Inverse Statement 222

    Divisibility Rules 223

INVESTIGATION 4.5 Understanding Why the Divisibility Rule for 3 Works 224

INVESTIGATION 4.6 Divisibility by 4 and 8 226

INVESTIGATION 4.7 Creating a Divisibility Rule for 12   228

EXERCISES 4.1   229

SECTION 4.2 Prime and Composite Numbers   232

    Determining Whether a Number Is Prime or Composite   233

INVESTIGATION 4.8 The Sieve of Eratosthenes   233

    Prime Factorization   235
    The Fundamental Theorem of Arithmetic   236
    A Largest Prime Number?   236

INVESTIGATION 4.9 Numbers with Personalities: Perfect and Other Numbers   237

EXERCISES 4.2   239

SECTION 4.3 Greatest Common Factor and Least Common Multiple   241

INVESTIGATION 4.10 Cutting Squares Using Number Theory Concepts   241

    The Greatest Common Factor   242

INVESTIGATION 4.11 Methods for Finding the GCF   243

    Least Common Multiple   245
    Strategies for Finding the LCM   245
    Using Prime Factorization and Exponents to Find the LCM   246

INVESTIGATION 4.12 Relationships Between the GCF and the LCM   248

    Yet Another Way to Find the LCM   249

INVESTIGATION 4.13 Going Deeper into the GCF and the LCM   249

    Relationships of Operations on the Set of Natural Numbers   250

EXERCISES 4.3   252

CHAPTER SUMMARY   253

CHAPTER 4 REVIEW EXERCISES   254

5 Extending the Number System   255

SECTION 5.1 Integers   256

    Integer Connections   256
    Representing Integers   257
    Operations with Integers   258
    Understanding Addition with Integers   258
    Absolute Value   259
    The Other Two Cases of Integer Addition   260
    Understanding Subtraction with Integers   261

INVESTIGATION 5.1 Subtraction with Integers   261

    Connecting Whole-Number Subtraction to Integer Subtraction   262
    Connecting Subtraction Contexts to Algorithms   262
    Understanding Multiplication with Integers   263

INVESTIGATION 5.2 The Product of a Positive and a Negative Number   263

    Understanding Division with Integers   264

EXERCISES 5.1   265

SECTION 5.2 Fractions and Rational Numbers   267

    Two Different Sets   268
    Fractions in History   268

INVESTIGATION 5.3 Rational Number Contexts: What Does 3/4 Mean?   269

INVESTIGATION 5.4 Fractions as Measure   271

INVESTIGATION 5.5 Rational Numbers and Multiple Representations   272

INVESTIGATION 5.6 Determining an Appropriate Representation   273

INVESTIGATION 5.7 How Many Blue Balloons?   274
    Equivalent Fractions   276
    Patterns and Order: Approaches to Procedures   277
    Addressing a Common Difficulty in Language   278
    Simplest Form   278

INVESTIGATION 5.8 Ordering Rational Numbers   279
    The Denseness of the Set of Rational Numbers   280

EXERCISES 5.2   281

SECTION 5.3 Understanding Operations with Fractions   284

    Addition of Rational Numbers   284

INVESTIGATION 5.9 Using Fraction Models to Understand Addition of Fractions   285
    Connecting Concepts to a Procedure for Adding Fractions   286
    Mixed Numbers and Improper Fractions   286

INVESTIGATION 5.10 Connecting Improper Fractions and Mixed Numbers   287
    Understanding Subtraction of Rational Numbers   288

INVESTIGATION 5.11 Estimation and Mental Arithmetic: Sums and Differences of Fractions   289

INVESTIGATION 5.12 Understanding Multiplication of Rational Numbers   292
    Connecting Multiplication of Fractions to the Multiplication Algorithm   294
    Justifying the Equality of Equivalent Fractions   296

INVESTIGATION 5.13 Division of Rational Numbers   296
    Making Sense of the Standard Division Algorithm   298

INVESTIGATION 5.14 Estimating Products and Quotients   299
    Applying Fraction Understandings to Nonroutine Problems   301

INVESTIGATION 5.15 When Did He Run Out of Gas?   301

INVESTIGATION 5.16 They've Lost Their Faculty!   302

EXERCISES 5.3   303

SECTION 5.4 Beyond Integers and Fractions: Decimals, Exponents, and Real Numbers   308

    Decimals   308
    Connecting Decimals and Integers   309

INVESTIGATION 5.17 Base 10 Blocks and Decimals   310
    Zero and Decimals   311

INVESTIGATION 5.18 When Two Decimals Are Equal   311

INVESTIGATION 5.19 When Is the Zero Necessary and When Is It Optional?   312

INVESTIGATION 5.20 Connecting Decimals and Fractions   313
    Decimals, Fractions, and Precision   314

INVESTIGATION 5.21 Ordering Decimals   315

INVESTIGATION 5.22 Rounding with Decimals   316

INVESTIGATION 5.23 Decimals and Language   317
    Operations with Decimals: Addition and Subtraction   318
    Multiplication with Decimals   318
    Division with Decimals   320

INVESTIGATION 5.24 Decimal Sense: Grocery Store Estimates   321

INVESTIGATION 5.25 Decimal Sense: How Much Will the Project Cost?   322

INVESTIGATION 5.26 How Much Did They Pay for Their Home?   322

INVESTIGATION 5.27 How Long Will She Run?   323

INVESTIGATION 5.28 Exponents and Bacteria   324

INVESTIGATION 5.29 Scientific Notation: How Far Is a Light-Year?   326
    Irrational Numbers   327

INVESTIGATION 5.30 Square Roots   328
    The Real-Number System   329
    Properties of Real Numbers   330

EXERCISES 5.4   331

CHAPTER SUMMARY   334

CHAPTER 5 REVIEW EXERCISES   335

6 Proportional Reasoning   337

    Additive Versus Mutiplicative Comparisons   337

SECTION 6.1 Ratio and Proportion   338

    The Unit Concept Matures   338
    Ratios, Rates, and Proportions   339
    Ratios, Rates, and Proportions in Mathematics and Real Life   339
    Delving Further into Ratios   340
    When Are Ratios Like and When Are They Unlike Fractions   341

INVESTIGATION 6.1 Unit Pricing—Is Bigger Always Cheaper?   342

INVESTIGATION 6.2 How Much Money Will the Trip Cost?   343

    Rates as Functions   344

INVESTIGATION 6.3 Reinterpreting Old Problems   345

INVESTIGATION 6.4 Using Estimation with Ratios   346

    Real-Life Applications   348

INVESTIGATION 6.5 Is the School on Target?   348

INVESTIGATION 6.6 Transferring Credits   349

INVESTIGATION 6.7 Stuck Behind a Truck   349

INVESTIGATION 6.8 From Raw Numbers to Rates   350

INVESTIGATION 6.9 How Much Does That Extra Light Cost?   352

    Dimensional Analysis Explained   352

EXERCISES 6.1   354

SECTION 6.2 Percents   357

    The Origin of Percent   357
    Uses of Percent   357

INVESTIGATION 6.10 Who's the Better Free-Throw Shooter?   358

INVESTIGATION 6.11 Understanding a Newspaper Article   359

INVESTIGATION 6.12 Buying a House   361

    Connections Between Percent and Other Mathematical Topics   363

INVESTIGATION 6.13 Sale?   364

    Percent Change   364

INVESTIGATION 6.14 What Is a Fair Raise?   365

INVESTIGATION 6.15 How Much Did the Bookstore Pay for the Textbook?   367

    Percents Less than 1 or Greater than 100   368

INVESTIGATION 6.16 The Copying Machine   368

INVESTIGATION 6.17 132% Increase?   369

    Interest   370

INVESTIGATION 6.18 Saving for College   371

INVESTIGATION 6.19 How Much Does That Credit Card Cost You?   373

EXERCISES 6.2   373

CHAPTER SUMMARY   377

CHAPTER 6 REVIEW EXERCISES   378

7 Uncertainty: Data and Chance 380

Data Interpretation and Chance in Society 381

SECTION 7.1 The Process of Collecting and Analyzing Data 382

Four Basic Components 382

INVESTIGATION 7.1 What Is Your Favorite Sport? 383
INVESTIGATION 7.2 How Many Siblings Do You Have? 386

Measures of Central Tendency 389
Deepening Our Understanding of Measures of Central Tendency 390

INVESTIGATION 7.3 Going Beyond a Computational Sense of Average 392

Another Interpretation of Mean 393

INVESTIGATION 7.4 How Many Peanuts Can You Hold in One Hand? 393
INVESTIGATION 7.5 How Long Does It Take Students to Finish the Final Exam? 397

Measures of Center Revisited 399
Dispersion, Variation, and Distributions 400
Exploring Data with Larger Numbers and Different Settings 403

INVESTIGATION 7.6 Videocassette Recorders 404

Interpreting Graphs 408

INVESTIGATION 7.7 Fatal Crashes 409
INVESTIGATION 7.8 Hitting the Books 411
EXERCISES 7.1 413

SECTION 7.2 Going Beyond the Basics 423

Comparing Two Sets of Data 423

INVESTIGATION 7.9 How Many More Peanuts Can Adults Hold Than Children? 423
INVESTIGATION 7.10 Scores on a Test 427
INVESTIGATION 7.11 Which Battery Do You Buy? 428

Another Tool to Understand Variation 430
Normal Distribution 430
Different Distributions 431

INVESTIGATION 7.12 Understanding Standard Deviation 431
INVESTIGATION 7.13 Analyzing Standardized Test Scores 435
INVESTIGATION 7.14 How Long Should the Tire Be Guaranteed? 436

Scatter Plots 437
Inferential Statistics 441

INVESTIGATION 7.15 Comparing Students in Three Countries 443

Weighted Average 445

INVESTIGATION 7.16 Grade Point Average 445
INVESTIGATION 7.17 What Does Amy Need to Bring Her GPA Up to 2.5? 446
EXERCISES 7.2 448

SECTION 7.3 Concepts Related to Chance 455

Preliminary Terms and Concepts 456
When All Outcomes Are Equally Likely 457
When All Outcomes Are Not Equally Likely 457
Theoretical and Experimental Probabilities 457

INVESTIGATION 7.18 Probability of Having 2 Boys and 2 Girls 458
INVESTIGATION 7.19 Probability of Having 3 Boys and 2 Girls 462
INVESTIGATION 7.20 Probability of Having at Least 1 Girl 463
INVESTIGATION 7.21 50-50 Chance of Passing 464
INVESTIGATION 7.22 What Is the Probability of Rolling a 7? 466
INVESTIGATION 7.23 What Is the Probability of Rolling a 13 with 3 Dice? 468
INVESTIGATION 7.24 "The Lady or the Tiger" 469

INVESTIGATION 7.25 Gumballs   470
    Fair Games   472
INVESTIGATION 7.26 Is This a Fair Game?   472
INVESTIGATION 7.27 What About This Game?   473
    Expected Value   474
INVESTIGATION 7.28 Insurance Rates   474
EXERCISES 7.3   477

SECTION 7.4 Counting and Chance   480
INVESTIGATION 7.29 How Many Ways to Take the Picture?   480
INVESTIGATION 7.30 How Many Different Election Outcomes?   482
INVESTIGATION 7.31 How Many Outcomes This Time?   484
INVESTIGATION 7.32 Pick a Card, Any Card!   486
INVESTIGATION 7.33 So You Think You're Going to Win the Lottery?   487
EXERCISES 7.4   488
CHAPTER SUMMARY   489
CHAPTER 7 REVIEW EXERCISES   494

8 Geometry as Shape   498

SECTION 8.1 Basic Ideas and Building Blocks   498
    Why Geometry? Because It's Fun!   499
INVESTIGATION 8.1 Playing Tetris   499
    Why Geometry? Because It Has Practical Applications   501
INVESTIGATION 8.2 Different Objects and Their Function   501
    Why Geometry? Because We See It in Art All Over the World   503
    Children   504
    Geometry and Spatial Thinking   505
    Back to the Greeks   506
    Basic Terms and Ideas   508
INVESTIGATION 8.3 Point, Line, and Plane   508
    Line Segments and Rays   510
    Relationships Between Lines   511
    Angles   512
    Naming Angles   513
    Measuring Angles   513
INVESTIGATION 8.4 Measuring Angles   515
    Classifying Angles   516
EXERCISES 8.1   518

SECTION 8.2 Two-Dimensional Figures   521
INVESTIGATION 8.5 Recreating Shapes from Memory   523
INVESTIGATION 8.6 All the Attributes   524
INVESTIGATION 8.7 Classifying Figures   526
    Polygons   529
    Triangles   529
INVESTIGATION 8.8 Why Triangles Are So Important   530
INVESTIGATION 8.9 Classifying Triangles   531
INVESTIGATION 8.10 Triangles and Venn Diagrams   533
    Triangle Properties   535
    Special Line Segments in Triangles   535
    Congruence   537

INVESTIGATION 8.11 Congruence with Triangles   538

    Quadrilaterals   539

INVESTIGATION 8.12 Quadrilaterals and Attributes   541

INVESTIGATION 8.13 Challenges   542

INVESTIGATION 8.14 Relationships Among Quadrilaterals   543

    Relationships Among Quadrilaterals   544
    Convex Polygons   545
    Other Polygons   545

INVESTIGATION 8.15 Sum of the Interior Angles of a Polygon   547

    Curved Figures   549
    Coordinate Geometry   550

INVESTIGATION 8.16 What Are My Coordinates?   550

INVESTIGATION 8.17 Understanding the Distance Formula   551

INVESTIGATION 8.18 The Opposite Sides of a Parallelogram Are Congruent   552

INVESTIGATION 8.19 Midpoints of Any Quadrilateral   553

EXERCISES 8.2   555

SECTION 8.3 Three-Dimensional Figures   559

INVESTIGATION 8.20 What Do You See?   561

INVESTIGATION 8.21 Connecting Polygons to Polyhedra   562

INVESTIGATION 8.22 Features of Three-Dimensional Objects   564

    Families of Polyhedra   565
    Pyramids   566

INVESTIGATION 8.23 Prisms and Pyramids   566

    Regular Polyhedra   567
    Relationships Among Polyhedra   568
    Connecting Two-Dimensional Representations to Three-Dimensional Objects   569

INVESTIGATION 8.24 Different Views of a Building   569

INVESTIGATION 8.25 Isometric Drawings   571

INVESTIGATION 8.26 Cross Sections   572

INVESTIGATION 8.27 Nets   573

    Cylinders, Cones, Spheres   573

EXERCISES 8.3   575

CHAPTER SUMMARY   577

CHAPTER 8 REVIEW EXERCISES   579

9 Geometry as Transforming Shapes   582

    Transformations   583

SECTION 9.1 Congruence Transformations   585

    Translations   585

INVESTIGATION 9.1 Understanding Translations   586

    Properties of Translations   587
    Reflections   588

INVESTIGATION 9.2 Understanding Reflections   588

INVESTIGATION 9.3 Understanding Rotations   589

INVESTIGATION 9.4 Understanding Translations, Reflections, and Rotations   591

    Combining Slides, Flips, and Turns   592

Contents xv

INVESTIGATION 9.5 Connecting Transformations 593

    Operations on Shapes and Operations on Figures 594
    Congruence 595

INVESTIGATION 9.6 Transformations and Art 596

EXERCISES 9.1 597

SECTION 9.2 Symmetry and Tessellations 603

    Reflection and Rotation Symmetry 604

INVESTIGATION 9.7 Reflection and Rotation Symmetry in Triangles 604

    Finding the Amount of Rotation in Rotation Symmetry 605

INVESTIGATION 9.8 Reflection and Rotation Symmetry in Quadrilaterals 606
INVESTIGATION 9.9 Reflection and Rotation Symmetry in Other Figures 607
INVESTIGATION 9.10 Letters of the Alphabet and Symmetry 608

    What Are Patterns? 609

INVESTIGATION 9.11 Patterns 609

    Patterns in Nature 611
    Symmetries and Patterns 611

INVESTIGATION 9.12 Symmetries of Strip Patterns 612

    The Seven Symmetries of Strip Patterns 614
    Wallpaper Patterns 615

INVESTIGATION 9.13 Analyzing Brick Patterns 615

    Symmetry Breaking 617
    Symmetry for Three-Dimensional Objects 618
    Tessellations 620

INVESTIGATION 9.14 Which Triangles Tessellate? 621
INVESTIGATION 9.15 Which Regular Polygons Tessellate? 622
INVESTIGATION 9.16 Tessellating Trapezoids 625
INVESTIGATION 9.17 More Tessellating Polygons 627
INVESTIGATION 9.18 Generating Pictures Through Transformations 629

EXERCISES 9.2 630

SECTION 9.3 Similarity 637

INVESTIGATION 9.19 Understanding Similarity 638
INVESTIGATION 9.20 Similarity Using an Artistic Perspective 640
INVESTIGATION 9.21 Using Coordinate Geometry to Understand Similarity 641

EXERCISES 9.3 642

CHAPTER SUMMARY 644

CHAPTER 9 REVIEW EXERCISES 647

10 Geometry as Measurement 649

SECTION 10.1 Systems of Measurement 650

    Development of Measurement Systems 650
    The Metric System 654
    Metric Length 654

INVESTIGATION 10.1 Developing Metric Sense 655

    Metric Volume 655
    Metric Mass 656
    Metric Temperature 656
    Time and Angles 656
    Becoming Comfortable with Metric Measurements 657

INVESTIGATION 10.2 Converting Among Units in the Metric System 657

Measurement of Other Quantities   658
Precision   658
Measurement as a Function   659

EXERCISES 10.1   660

SECTION 10.2 Perimeter and Area   663

Perimeter   664
Circumference and π 664

INVESTIGATION 10.3   What Is the Length of the Arc?   664

Area   665

INVESTIGATION 10.4   Converting Units of Area   669

Pythagorean Theorem   669

INVESTIGATION 10.5   Using the Pythagorean Theorem   670

INVESTIGATION 10.6   Understanding the Area Formula for Circles   671

INVESTIGATION 10.7   A 16-Inch Pizza Versus an 8-Inch Pizza   672

INVESTIGATION 10.8   How Big Is the Footprint?   673

INVESTIGATION 10.9   Making a Fence with Maximum Area   674

EXERCISES 10.2   675

SECTION 10.3   Surface Area and Volume   682

Understanding the Surface Area of Prisms   682
Understanding the Surface Area of Cylinders   684
Understanding the Surface Area of Pyramids   684
Understanding the Surface Area of Cones   685
Understanding the Surface Area of Spheres   686
Volume   686
Understanding the Volumes of Prisms   686
Understanding the Volumes of Cylinders   687
Understanding the Volumes of Pyramids   688
Understanding the Volumes of Cones   689
Understanding the Volume of Spheres   689

INVESTIGATION 10.10   Are Their Pictures Misleading?   691

Determining the Volumes of Irregularly Shaped Objects   691

INVESTIGATION 10.11   Finding the Volume of a Hollow Box   692

INVESTIGATION 10.12   Surface Area and Volume   693

EXERCISES 10.3   695

CHAPTER SUMMARY   699

CHAPTER 10 REVIEW EXERCISES   699

Appendix A   Principles and Standards for School Mathematics   A-1

Numbers and Operations   A-1
Algebra   A-1
Geometry   A-1
Measurement   A-2
Data Analysis and Probability   A-2
Problem Solving   A-2
Reasoning and Proof   A-2
Communication   A-3
Connections   A-3
Representation   A-3

ANSWERS TO SELECTED EXERCISES   A-4

INDEX   I-1

Preface

WHAT THE COURSE IS ABOUT

This course and this book are about developing and *retaining* the mathematical knowledge students will need as beginning mathematics teachers. I prefer to say that we are going to *uncover* the material rather than cover the material. The following analogy is useful. When archaeologists explore a site, they carefully uncover the site; as time goes on, they see more and more of the underlying structure. This is exactly what can and should happen in a mathematics course. When this happens, the students are more likely to *own* rather than *rent* the knowledge. When I ask my students to self-assess whether it feels like they are owning or renting, they know what I mean, and more importantly, as these words get inside them, they ask themselves this question as they are learning.

There are three ways to describe how the learning here is different from a traditional text.

Constructed versus Received Knowledge

When students are given problems that *require* them to grapple with important mathematical ideas, they learn those ideas more deeply than if they are simply presented with the concepts via lecture and then given problems for practice. Additionally, there is a need to shift the focus from one where students are studying mathematics to one where the students are also *doing* mathematics. That is, they are looking for patterns, making and testing predictions, making their own representations of a problem, inventing their own language and notation, etc. For example, in Exploration 8.9, when students try to define convex, the class discussion uncovers many different ways of seeing and expressing this idea and illustrates the need for clear and precise terminology in a way that students find empowering.

The textbook is written in an engaging and informal manner, engaging the students in an interplay between investigations and narrative. The Investigations are not called Examples, as they are in most textbooks, because of a different focus. The Investigations are not Examples of material just presented, but rather they are problems that have the student actively investigating important mathematical ideas. Virtually all recent research on the learning process supports this method.

The pencil icon, a popular feature with students, is a frequent reminder to the student to stop and think rather than just read on. Many instructors, at the beginning of the course, require the students to write in their journals when they see the pencil icon and to turn in their notes. This enables the instructor to better understand the students' thinking and enables the students to see firsthand the value of active learning. Many of my students continue this process throughout the course.

While the main text and *Explorations* manual were written to be used together, they can be used independently. The Explorations are cross-referenced in the textbook with a pentomino icon to indicate an Exploration that would precede that content in the textbook. Note that not every Exploration is cross-referenced in the text because there is not a one-to-one correspondence between the two, and some Explorations cut across multiple concepts.

Connected versus Fragmented Learning

The multifaceted nature of mathematical connections is elaborated in Chapter 1. In fact, understanding can be defined in terms of connections: The extent to which you understand a new idea can be seen by the quality and quantity of connections between that idea to what you already know. There are two ways in which connections are built into the structure of the text.

1. Mathematical connections

- Several themes—functions, breaking apart and putting together, conventions, proportional reasoning, unit, and multiplicity—pervade the whole book, that is, you find them in every chapter. These are elaborated later in the preface.
- The mathematical content is not isolated.
  You see this in every chapter. For example, Investigation 1.1 helps students see how the algebraic formula is closely connected to guess-test-revise; Investigation 1.15 is later connected both to fractions and to remainder. The four operations are constantly connected to each other in their development in Chapter 3. Then in Chapter 5, the connections between operations with fractions and operations with whole numbers are discussed; how decimals are connected to whole numbers and to fractions is also discussed. In Chapter 6, we look back at some problems done in Chapter 5 and see how they can now be solved more efficiently with the concepts of ratio. In Chapter 9, students see how our work with numbers and shapes is similar.
- All mathematical procedures are developed actively.
  In this way, the student knows not only how the procedure works but also why it works. For example, students understand why we move over when we multiply the second row in whole number multiplication; they realize that "carrying" and "borrowing" are essentially the same procedure; they understand that standard deviation is essentially the average distance from the average; they see that pi can be seen as how many times you can wrap any diameter around the circle.
- Section summaries and chapter summaries
  Here, ideas that have been developed separately are explicitly connected and woven together.
- Connection icon/boxed inserts
  This is one of the most popular features of the textbook with students! Relevant and interesting connections occur in boxed inserts:
  - History—considerations of interesting topics that are connected to concepts developed in the text
  - Mathematics—connections between the concept being addressed and other mathematical concepts
  - Language—etymology of terms and/or nuances of terms
  - Outside the Classroom—applications and uses of mathematical concepts and procedures in the business world, science, and everyday life

2. Connections to children's thinking

In this book you will see a focus on children's thinking, for two reasons. First, listening to the students' thinking is an essential part of good teaching. If students see this in their math courses, as well as the methods and student teaching courses, the habit of paying attention to students' thinking becomes part of how they view teaching. Second, when the student sees examples of children's thinking and connections between problems in this course and problems children solve, both the quality and quantity of the students' cognitive effort increase.

It is important for me to be very clear—I do not see this focus as watering down the content. In fact, I believe that the content of this book is more rigorous than traditional texts. I offer a reframing of how to measure the rigor of a course—do not measure the rigor by the rigor of the text, but by the rigor of the students' thinking at the end of the course. In Explorations, when I ask students to justify or prove their results (e.g., Explorations 1.1, 3.3, 5.10, 8.6, 10.6), my students are willing and able to do this work.

In the textbook, these connections to children's thinking occur when relevant. There is also an icon for Classroom Connection, which appears in the margins and in exercise problems that come directly from children's work.

The website for the book (described later in the preface) has an extensive cross-reference to children's problems—in videos, journal articles, books, and links to other websites. I use these sparingly in my class—keeping the primary focus on mathematical ideas. However, having students read an excerpt of an article or showing a snippet of a videotape has the outcome of my students spending more time on homework than they used to when they saw little connection between this course and the "real-life" of elementary teaching.

AUTHENTIC VERSUS CONTRIVED PROBLEMS

While most texts have many "real-life" problems, this text also differs in how those problems are made and presented.

Realistic versus pseudo-real-life problems

In Section 2.2, the question of paying a baby-sitter is explored. This is often portrayed as a linear function; e.g., if the rate is $8 per hour, $y = 8x$. However, in actuality, it is not a linear function but rather a step-wise function. Similarly, when determining the cost of carpeting a room, the solution path is often presented as dividing the area of the room by the cost per square yard; again, this is not how the cost is actually determined.

Realistic versus contrived problems

Many textbooks, under the guise of real-life problems, will create problems that no one in real-life solves, for example, determining the percentage of a quilt pattern that is each color. While this problem can have pedagogical value, it should not be offered as an example of a real-life problem. In this book, you will find many problems—that I have needed to solve, problems friends have had, problems children have had, problems I have read about—where the content fits with the content of this course.

You will also find problems in the text where you are asked to state the assumptions you make in order to solve the problem (e.g., Section 6.2 exercises 38–43). You will find problems that have the messiness of "real-life" problems: where the problem statement is ambiguous, where there is too little or too much information, or even contradictory information.

Integrated Big Content Ideas Throughout the Course

So many people think of elementary mathematics as chapter titles, and for a variety of reasons, I have chosen to live within a more traditional partitioning of mathematical content: fractions, decimals, percent, statistics, etc. However, a critically important outcome of this course has to be the students' realization of the underlying connectedness of mathematics. Below are several big ideas that I have tried consciously to weave into the whole textbook.

Functions   Functional relationships (the idea of a unique and predictable relationship between two sets or two variables) permeate mathematics and thus this book. Students will see that all mathematical operations are functions. In the operator construct of fractions, they use the fraction as a function machine, and I use function language in Investigation 5.5 (Determining an Appropriate Representation) to solve a fraction problem. Geometric transformations are functions, and geometric relationships (e.g., the relationship between the number of faces, the edges and vertices in polyhedra) embody functions, too. Students will explore the lack of a functional relationship between perimeter and area, and they will find that the act of measuring establishes a functional relationship between an attribute of an object and a number.

Composition-decomposition   The ability to understand most computation algorithms and to do mental arithmetic and estimation relies on the ability to decompose and recompose numbers. For example, 15×12 can be seen as 30×6 or as $15 \times 10 + 15 \times 2$. However, the processes of composition and decomposition do not stop with numbers. When transforming or measuring geometric figures we decompose and recompose them. Similarly, the problem-solving heuristic of breaking a problem into parts manifests this idea.

Conventions   Many students have no idea that some of the mathematics that they "learn" is invariant and some is convention. Making this distinction is an important idea that encourages students to look more carefully at what we ask them to learn and humanizes mathematics. For example, adding from right to left is a convention rather than mathematically required. On the other hand, in our numeration system, the order of numerals in a number is not a convention—$43 \neq 34$. Similarly, there are conventions in graphing and there are no-nos.

Proportional reasoning   Like functions, proportional reasoning shows up in almost every chapter. Proportional reasoning is critical to a deep understanding of fractions, of slope, of probability, of percent, of similarity, and so many other concepts. This idea shows up every year in the Exploration 7.2, Typical Person. As students discuss the graphs they have constructed, they grapple with why some graphs are valid and others are invalid. For example, the width of the bars and the height of the bars is a convention; however, squishing the bars together when the frequency for a value is zero is not allowable.

Unit   We refer to units when developing an understanding of numeration. One of the reasons that multiplication and division problems are generally more difficult than addition and subtraction problems is that in multiplication and division, the units are not the same. For example, if someone bicycles for 3 hours at 15 miles/hour, that person will travel 45 miles—the unit for the multiplicand is hours; for the multiplier, it is miles/hour; and for the product, it is miles. Three different units for each of the three numbers in the problem! We refer to unit fractions, and we need to distinguish between units and wholes when working with fractions. When we make a scale for a graph, we are selecting a unit. We need units when making things—for example, we refer to a unit in quilt designs and tessellation patterns. And yes, of course, we talk about units in measurement.

Multiplicity   This is the antithesis of the notion of mathematics as black and white, with most problems having one right answer, one best way to get to the answer, and one way to represent mathematical ideas. By the end of the course, if your instructor has you construct a portfolio, you should be able to write coherently about each of the following:

- Virtually all nontrivial mathematical problems can be solved in more than one way.
- Virtually every mathematical idea has multiple connections to other ideas.
- Most mathematical situations can be represented in multiple ways.

Develop Mathematical Thinking

I believe mathematical thinking is as much a mindset as it is a constellation of related "skills" and thus cannot be directly taught as much as encouraged to develop. When one can think mathematically, one can appropriately use a variety of problem-solving strategies; one can justify one's thinking, which involves being able to communicate; one looks for patterns and can arrive at hypotheses that one can check and revise and then ultimately make generalizations. For instance, in Exploration 4.4 (African Sand Drawings), as the students make drawings they look for patterns, make and check hypotheses, and come to realize that number theory ideas help to explain why different drawings take different numbers of turns to complete.

Algebraic thinking   I have chosen to de-emphasize the use of algebraic procedures—but not algebraic thinking—for two fundamental reasons. First, like it or not, many of our students in the course are algebra phobic. I take to heart a statement in Revitalizing Undergraduate Mathematics Education: "Teach the students we have, not the ones we wish we had." I clearly think algebra is an important tool, but this course is not the course to pursue this and all the other goals of the course. Second, I try to de-emphasize algebraic solutions to problems and emphasize other solution paths—e.g., guess-check-revise and draw a diagram—because solving equations algebraically is not a tool that elementary school students have, and we clearly want children to develop a wide repertoire of problem-solving tools. Thus, I believe the course should encourage the development of algebraic thinking, which is different from emphasizing the use of formulas to solve problems.

Problem-solving and communication   Problem-solving is not something we do after the students learn the concepts and procedures. Rather, problem-solving should permeate all of the Investigations and Explorations. Developing a repertoire of problem-solving tools is an important goal in the course, with an emphasis on being able to solve nonroutine and multistep problems. Therefore, this book has fewer exercises that are "just like" the examples. I believe that spending more time on fewer but richer problems results in students developing more mathematical power than doing lots of problems. Alfred North Whitehead's book *The Aims of Education* states this principle more eloquently than I ever could.

One of the most exciting pieces of feedback from instructors who have used this book is from one who said that for the first time students were actually talking about mathematical ideas. As a result of the repeated emphasis on active learning, discussing, justifying, and the like, many students come to see the crucial role of definitions and notation in mathematics. Many students find explaining and justifying their work and "explaining as if to a friend on the phone" to be very new and frustrating at first, but a consistent focus on communication bears rich fruit by the end of the course. As students struggle to explain an idea, a word like *prime* or *vertex* suddenly makes it much easier to describe the relationship they are trying to understand.

In the *Instructor's Resource Manual,* I discuss how learning logs and other devices help students to develop the ability to communicate their ideas and, more importantly, to see the value of being able to communicate mathematically. I have borrowed the notion of "first draft" from my English teacher colleagues. In class I will ask for someone's first draft of a definition or a concept, and throughout the text I encourage students to write and revise as a path to understanding.

ANNOTATED OVERVIEW OF THE COURSE

Let us take a walk through the book to help you get a more tangible (concrete) sense of the book and how the two volumes work together. Because of space limitations, I will highlight aspects of each chapter rather than discuss every section in detail.

Chapter 1: Foundations for Learning Mathematics

This chapter sets the tone for the whole book. The Explorations offer different kinds of problems to grapple with. In the Text, I take some time to explore important strands of mathematical thinking—problem-solving, communication, reasoning, patterns, representation, and connections. I present and discuss the NCTM Standards in more detail.

Chapter 2: Fundamental Concepts

When I first began teaching this course and was still trying to "cover" as much of the book as I could, one day I suddenly realized that what I really wanted from the material on sets and functions was for students to have an appreciation of the need for these ideas. This chapter lays that foundation. Section 2.1 gives students tools to be able to talk of sets and subsets and to be able to use a Venn diagram when the need arises in other chapters, for example, to understand the relationship between different sets of numbers. Similarly, the notion of function being a certain kind of relationship between two sets or two variables arises throughout the book. Any of the Explorations in Section 2.2 not only will have the students come to an understanding of the notion of functions, but also will get them started on developing mathematical thinking—looking for patterns, making and testing predictions, and making generalizations. The text's Investigations elaborate and refine this idea.

Numeration, discussed in Section 2.3, is a huge concept in elementary school mathematics. Exploration 2.8, Alphabitia, is one of the most powerful I have used. Most of my students report this to be the most significant learning of the semester and/or a turning point in the semester. The Exploration unlocks powerful understandings related to numeration, which the text supports by discussing the evolution of numeration systems over time.

Chapter 3: The Four Fundamental Operations of Arithmetic

These four operations are crucial for elementary school mathematics. Students come to realize that addition, subtraction, multiplication, and division are more than just plus, minus, times, and "gazinta." They see how the concepts of the operations, coupled with an understanding of base 10, enable them to understand how and why the procedures they have done by rote for years actually work. In Exploration 3.13, Egyptian duplation piques their interest. The Explorations and Investigations help them to make sense of both standard and nonstandard algorithms.

In addition to making sense of standard algorithms, I have extended the treatment of alternative algorithms to four Explorations in this chapter. My students have found these explorations to be both enlightening and fascinating.

Chapter 4: Number Theory

Number theory can be one of the most enjoyable chapters in the book. The African Sand Drawings exploration is one of my favorites because students find so many different patterns and can actually make and check hypotheses. Each of the Explorations in Chapter 4 has students grapple with core concepts of number theory in a connected way, and the text unpacks these concepts by investigating distinct aspects of them.

Chapters 5 and 6: Extending the Number System and Proportional Reasoning

I have chosen an unorthodox organization for Chapters 5 and 6 for conceptual and pedagogical reasons. The sets of integers, fractions, and decimals represent three historically significant extensions to the set of whole numbers. Given the limited time we have with the students, I find it makes more sense to emphasize rational numbers over integers and to connect the development of fraction concepts and operations in Chapter 5. Although the ratio construct is one of four major constructs of rational numbers, I find it to be more powerful to discuss ratio, proportion, and percents together in the context of Proportional Reasoning in Chapter 6.

In Exploration 5.6, students construct fraction manipulatives and then look for rules when ordering fractions, a critically important first step in seeing fractions as more than numerator and denominator. In Exploration 5.13 (Meanings of Operations with Fractions), having students represent problem situations with diagrams requires them to adapt their understanding of the four operations, developed in Chapter 3, to fraction situations. Having first constructed this concept through exploration, the student can approach an Investigation like 5.7 (Ordering Fractions), with a richer understanding of what it really means to say that one fraction is bigger than another.

Exploration 6.1 (Which Ramp Is Steeper?), although challenging, requires students to examine the relationship between length and height to predict steepness; from this they come to see slope as a measure of steepness. They also construct an understanding of proportions that enables them to work much more confidently with percents.

Chapter 7: Uncertainty: Data and Chance

The first two sections of this chapter (in both the text and Explorations) have been completely rewritten and organized. The current framework is much more aligned with my interpretation of recent research in statistics education and particularly the Guidelines for Assessment and Instruction in Statistics Education (GAISE) project report, which was endorsed by the American Statistical Association (http://www.amstat.org/education/gaise/). Students have responded enthusiastically to this new framework in pilots of the material in my and other instructors' classes. The students carefully walk through the stages of defining a question, collecting data, interpreting data, and then presenting data. Specific concepts new to the fourth edition include a discussion of scatter plots and more explicit attention to the notion of variation in the process of collecting and analyzing data.

Sections of the chapter that have reviewed well in the past have been retained. I am particularly excited that the Investigations with the concepts of mean and standard deviation have been so successful with students—as a result they can express these ideas conceptually instead of simply reporting the procedure.

Chapters 8 and 9: Geometry as Shape and Geometry as Transforming Shapes

I will confess that geometry is my favorite field in mathematics, and it pains me to find that so many of my students so dislike it. I have chosen to have three geometry chapters, each focusing on geometry from a different perspective. In Chapter 8, you have the option of introducing geometry through Explorations with Tangrams, Geoboards, or Pentominoes. This less formal, more concrete introduction allows students with bad memories of geometry to let down their guard, but at the same time do good mathematical explorations. The geometric transformations we explore in Chapter 9 can be one of the most interesting and exciting topics of the course. Quilts and tessellations both spark lots of interest and provoke good mathematical thinking. The text develops concepts and introduces terms that help students to refine understandings that emerge from the Explorations.

Chapter 10: Geometry as Measurement

This chapter addresses measurement from a conceptual framework (i.e., identify the attribute, determine a unit, and determine the amount in terms of the unit) and historical perspective. Both the Explorations and Investigations get the student to make sense of measurement procedures and to grapple with fundamental measurement ideas. Exploration 10.2 (How Tall?) generates lots of different solution paths and ideas, and lots of discussion about indirect measurement and precision. Exploration 10.9 (What Does π Mean?) has demystified π in the minds of many of my students and is a wonderful exercise in communication. Exploration 10.13 (Irregular Areas) requires students to apply notions of measuring area to a novel situation; every year students will hypothesize many different strategies, some of which are valid and some of which are not. The text looks at the larger notion of measurement, presents the major formulas in a helpful way, and illustrates different problem-solving paths.

Overall Features

Worthwhile mathematical tasks In developing and selecting Explorations and Investigations, I have searched for ones that will make all students feel challenged and successful. In developing the book, if I found that only some of the students engaged in an Investigation or Exploration, I either modified it or deleted it. One criteria, for me, of a good Investigation or Exploration is that there is more than one way to answer the question or problem that is posed.

Readability Reviewers of the text and students who have used the book have praised its readability. I have tried to be informal and engaging and yet not water down the ideas.

Exercises To paraphrase a colleague of mine, Neil Davidson, I look for problems that will require thinking. In the exercise sets that follow each section of the text chapters, while I do provide some traditional practice exercises (for which the answers are generally in the back of the book), I believe that grappling with nonroutine, multistep problems produces deeper understanding of mathematical ideas and greater problem-solving skills. I have also sought to include many exercises that are more like problems encountered in everyday life and work settings. You will find a number of problems that friends have reported struggling with or that have come up in my own life.

I need to explicitly mention something this book does not do that many other texts do: identify exercises by type. For example, denoting certain problems as "thinking critically" implies critical thinking is not necessary elsewhere. Labeling certain problems as communication problems implies that communication is only salient sometimes and marginalizes this aspect of mathematical thinking. Denoting certain problems as calculator problems reinforces the traditional role of the teacher deciding when to use and not to use calculators. Finally, "Just for Fun" implies that "real math" is hard work and not fun.

NCTM references Rather than simply quote from the NCTM's Principles and Standards for School Mathematics at the beginning of each chapter, I have sought to integrate the ideas from the Standards into the fabric of the course. Thus, there is some direct quoting, some paraphrasing, and some illustrating of my use of ideas from the Standards. In the *Instructor's Resource Manual*, I suggest various assessment practices that I believe embody the Assessment Standards more than the traditional daily homework, weekly quiz, and chapter test format.

Answers to selected exercises These are provided for student reference. The Instructor's Resource Manual contains complete solutions to all end-of-section exercises.

Features Within Each Chapter

What do you think? These provocative questions may be discussed in class, assigned as journal writings, or given as homework.

Chapter and section introductions These set the context for the work the students are to begin and connect the new ideas to previous ideas. I try to consistently offer a rationale for each of the topics. See, for example, how the introduction to Chapter 7 focuses on broader concepts of what it means to talk about data and chance.

Investigations Questions are posed and then the students are asked to "stop, think, and then read on." The pencil icon is a visual reminder to do so. Then there is a Discussion of the investigation, which generally involves several different solution paths. In some cases, the answers are given right there. In some cases, the answer is not given, thus giving you the option of giving the answer after the students have investigated the question or to use the question as a journal entry or as a homework problem.

Section summaries I have sought to make these more than simply a restating of what students have learned but rather a looking back at and weaving together of the ideas of the section and putting them in a broader context.

Chapter Summary and Chapter Review Exercises Each Chapter Summary begins by summarizing the big ideas of the chapter—those ideas that connect the different sections or connect ideas from this chapter with ideas from earlier chapters. The Chapter Summary also includes a complete list of key terms and basic concepts introduced in the chapter, with page references. Many students find this to be a useful reference. Students can then work the Chapter Review Exercises for extra practice with the chapter's major lessons.

Supplements

Geometer's Sketchpad

Instructor's Resource Manual with Complete Solutions This teacher-to-teacher resource manual serves as a guide to building an active, constructivist course. Part 1 includes five brief chapters. Chapter A contains an elaboration of the goals of the course, responses to Frequently Asked Questions about using the books, a suggested timetable, and suggestions for how to use manipulatives. Chapter B focuses on creating a positive learning environment, including a discussion of the physical environment, expectations, and norms. Chapter C presents some basics on cooperative learning. Chapter D examines some important differences among students that an instructor needs to think about, e.g., attitudes and beliefs, personal traits, gender, and ethnicity. Chapter E discusses assessment, a critical issue in a more innovative text such as this one.

Part 2 includes instructor's notes for each chapter of the Explorations. These notes include a brief chapter overview for the instructor, Intended Outcomes for each Exploration, an indication of time and materials needed, key Issues and Considerations that cover common difficulties/problems students may have, effective opening questions and other tips gleaned from my experience with the Explorations, and examples of students' work. Parts 1 and 2 are available online at the textbook website.

The *Instructor's Resource Manual* also contains complete solutions to all the end-of-section exercises from the text. As much as possible, I have sought to include different ways that students might solve the problems. Finally, a Printed Test Bank is included, supplying instructors with a copy of the test bank on the HM ClassPrep/HM Testing CD.

HM Testing CD-ROM This product combines an electronic version of the *Instructor's Resource Manual* with the testing and gradebook features of HM Testing. Staying true to the open-ended nature of the text's exercises, the majority of the test bank offers similar questions. However, also included are a handful of more traditional questions, which supply extra skill practice in an algorithmic format.

ETA Cuisenaire CD-ROM Houghton Mifflin has partnered with ETA Cuisenaire to provide a collection of engaging activities and ideas on CD-ROM. This CD-ROM includes video clips from actual classrooms and more than 450 activities that are searchable by concept, type of manipulative used, or grade level. Available to be packaged at a significant cost savings with new Bassarear textbooks, the Super Source CD-ROM will engage your students and prepare them for their future careers.

Manipulatives Kit

Companion Websites Visit the text specific website for students or instructors (college.hmco.com/pic/bassarear4e) and choose this textbook from the list provided to access additional resources:

- Over the life of edition, new Explorations that can be used with Geometer's Sketchpad
- Excel files that can be used with Explorations in Chapter 7
- Over 100 PowerPoint slides that accompany various Investigations and Explorations; these can be used as overheads or handouts
- A database of articles, videos, children's books, and case studies that is cross-referenced with the textbook
- Test Bank

Student Solutions Manual This student supplement contains complete solutions to those text exercises that have answers given at the back of the text.

Explorations The stepped-out activities in the Explorations manual make it easy for professors to incorporate hands-on learning. The Explorations encourage students to "see" math, discern patterns, solve problems, and make predictions, in some cases using manipulatives. When they begin teaching, students can use the activities as models in their own classrooms.

In the fourth edition of Explorations

There are more than twenty new explorations, including:

Chapter 1: Three new explorations

Chapter 2: A new exploration on equivalence and on relative magnitude

Chapter 3: Several explorations on number sense

Chapter 4: African Sand Drawings

Chapter 5: Linear representations of fractions, repeating decimals, and understanding decimal algorithms

Chapter 6: Using proportional reasoning to interpret data

Chapter 7: Eight new explorations

Many other explorations have been modified. Instructors who used previous editions can find on the website explorations no longer in the printed fourth edition textbook.

Acknowledgments

It is impossible to completely acknowledge all the people and materials that have influenced the construction of this textbook. I find that several conversations with and writings by Marty Simon, Deborah Schifter, and Deborah Ball have played over and over in my mind. For two years I worked on an NSF project called Mathematics for Tomorrow at Education Development Center. My work and conversations with Ellen Davidson and Jim Hammerman were very helpful.

I highly recommend books and research by the following authors, whose writings have enabled my practice to be more deeply connected to frameworks: John Clement, Paul Cobb, Robert Davis, Carol Dweck, Constance Kamii, Robert Karplus, Richard Lesh, Jack Lochhead, and Alan Schoenfeld.

There is not room to cite all the books that I read in preparing for this book. Certain of those books, though, were more influential than others, and I particularly recommend them. Over the course of preparing for the book, I paged through the NCTM Standards documents many times and found many of the books in the NCTM Navigation series very helpful. Other notable books include: *American Indian Design and Decoration* by Le Roy H. Appleton; *Capitalism & Arithmetic: The New Math of the 15th Century* by Frank Swetz; *From One to Zero: A Universal History of Numbers* by Georges Ifrah; *Introduction to Tessellations* by Dale Seymour & Jill Britton; *Mathematics* by David Bergamini (Ed.); *Metamorphosis: A Source Book of Mathematical Discovery* by Lorraine Mottershead; *Number Words and Number Symbols: A Cultural History of Numbers* by Karl Menninger; *Numbers: Their History and Meaning* by Graham Flegg; *Symmetry: A Unifying Concept* by Istvan & Magdolna Hargittai; *The History of Arithmetic* by Louis Charles Karpinski; and *The Language of Functions and Graphs* by the Shell Centre for Mathematical Education.

I certainly cannot claim that I invented all of the explorations, investigations, and problems in these two volumes. In many cases, I have made specific acknowledgments; if I missed someone, I apologize in advance. It is my hope that, with the increased amount of collaboration in mathematics education spurred by the NCTM and by advances in technology, teaching and writing will be much less solitary occupations than they have been in the past.

I would like to thank the following reviewers for their insights comments during the development of the fourth edition: Mark Bollman, Albion College (Michigan); Elaine Cohen, New Mexico State University–Las Cruces; Ruth Collins, Delaware Technical and Community College; Linda Herndon, Benedictine College (Kansas); Kathleen McDaniel, Buena Vista University (Iowa); Barbara Moses, Bowling Green State University (Ohio); Asmamaw Yimer, University of Wisconsin–Green Bay; and accuracy reviewer, David Meel, University of Wisconsin–Green Bay.

I would like to thank the many reviewers of previous editions noted below for their thoughtful and helpful comments throughout development. I would like to particularly thank Nadine S. Bezuk from San Diego University, whose critical comments on an earlier draft of my sections on rational numbers timed nicely with the disequilibrium I was experiencing at EDC and resulted in a book that is much stronger conceptually than it otherwise would have been.

Bernadette Antkoviak, Harrisburg Area Community College; Linda Beller, Brevard Community College; Julie J. Belock, Salem State College; Peter Berney, Yavapai College; Michael Bowling, Stephens College; Donald A. Buckeye, Eastern Michigan University; Doug Cashing, St. Bonaventure University; Forrest Coltharp, Pittsburg State University; Robert F. Cunningham, Trenton State College; Art Daniel, Macomb Community College; Mary J. DeYoung, Hope College; Charles Dietz, College of Southern Maryland; Maureen Dion, San Joaquin Delta Community College; Jerry Dwyer, University of Tennessee, Knoxville; Ronald Edwards, Westfield State University; Fred Ettline, College of Charleston; Larry Feldman, Indiana University of Pennsylvania; Merle Friel,

Humboldt State University; Karen Gaines, St. Louis Community College; Anita Goldner, Framingham State College; Elise Grabner, Slippery Rock University; William Haigh, Northern State University; Robert Hanson, Towson State University; J.B. Harkin, SUNY College at Brockport; Linda Herndon, Benedictine College; Susan K. Herring, Sonoma State University; Peter Incardone, New Jersey City University; Tess Jackson, Winthrop University; Loren P. Johnson, University of California, Santa Barbara; Isa S. Jubran, SUNY College at Cortland; Karla Karstens, University of Vermont; Lawrence L. Krajewski, Viterbo College; Mary Ann Byrne Lee, Mankato State University; Lois Linnan, Clarion University; John Long, University of Rhode Island; Vena Long, University of Missouri at Kansas City; Jane Ann McLaughlin, Trenton State College; Ronald J. Milne, Gashen College; Juan Molina, Austin Communicty College; Deborah Narang, University of Alaska, Anchorage; Kathy C. Nickell, College of DuPage; Sandra Powers, College of Charleston; Glenn Prigge, University of North Dakota; Dennis Raetzke, Rochester College; James E. Riley, Western Michigan University; Lew Romagnano, Metropolitan State College of Denver; Helen Salzberg, Rhode Island College; Dr. Connie S. Schrock, Emporia State University; Matt Seeley, Salish Kootenai College; Lauri Semarne; Jean M. Shaw, University of Mississippi; Jean Simutis, California State University, Hayward; Merriline Smith, California State Polytechnic University; Stephen P. Smith, Northern Michigan University; Larry Sowder, San Diego State University; Mary Teagarden, Mesa College; Gary Van Velsir, Anne Arundel Community College; Jeanine Vigerust, New Mexico State University; Clare Wagner, University of South Dakota; Tad Watanabe, Towson State University; Marvin S. Weingarden, Madonna University; J. Normon Wells, Georgia State University; Mary T. Williams, Francis Marion University; Andrew T. Wilson, Austin Peay State University; Mary Lou Witherspoon, Austin Peay State University; and Beverly Witman, Lorain County Community College.

Finally, I would like to thank the many people at Houghton Mifflin who have supported my desire to present to the mathematics community a book that I feel breaks new ground in some important ways. It is not as radical a departure as I had wanted but I understand that a book that looks too different will likely not sell and thus won't effect much change! I would like to express appreciation to the various members of the Houghton Mifflin team who have worked with me for the past two years to enable the fourth edition to be even stronger than the third edition: Richard Stratton (Publisher), Lynn Cox (Sponsoring Editor), Maria Morelli (Senior Development Editor), Laura Ricci (Editorial Assistant), Kristen Truncellito (Associate Project Editor), Jennifer Jones (Marketing Manager), Mary Legere (Marketing Associate), and Katie Huha (Senior Permissions Editor). Talking with friends who have published with other companies makes me appreciate, even more than I might otherwise, the spirit of teaming and collaboration that I have consistently found with my colleagues at Houghton. Two characteristics of mine—not being naturally good at paying attention to detail and constantly changing my ideas—can make for a book that is strong conceptually, but makes life more complicated for people involved in producing the book. I thank all of you for your patience.

Mathematics for Elementary School Teachers

CHAPTER 1

Foundations for Learning Mathematics

1.1 Getting Comfortable with Mathematics
1.2 Problem-Solving
1.3 Patterns
1.4 Representation
1.5 Reasoning and Proof
1.6 Communication
1.7 Connections

Knowing mathematics means being able to use it in purposeful ways. To learn mathematics, students must be engaged in exploring, conjecturing, and thinking rather than only in rote learning of rules and procedures. Mathematics learning is not a spectator sport. When students construct personal knowledge derived from meaningful experiences, they are much more likely to retain and use what they have learned. This fact underlies [the] teacher's new role in providing experiences that help students make sense of mathematics, to view and use it as a tool for reasoning and problem solving.[1]

NATIONAL COUNCIL OF TEACHERS OF MATHEMATICS

SECTION 1.1 GETTING COMFORTABLE WITH MATHEMATICS

Were you excited when you signed up for this course? Were you worried? Do you fondly remember your math lessons in elementary school, or have you been trying to forget them for all these years? You are at the beginning of a course in which you will reexamine elementary school mathematics to understand the underlying concepts better and to learn *why* mathematical procedures and formulas actually work. Your approach to this course depends on the attitudes and beliefs you bring to the classroom as a student; in subtle or not-so-subtle ways, you may pass these beliefs along when you enter the classroom as a teacher. So before we start working with mathematical concepts, let's look at our current conceptions of mathematics.

Beliefs About and Attitudes Toward Mathematics

This preliminary exercise is designed with two purposes in mind. First, it will help you examine and reflect on your beliefs and attitudes at the beginning of the course. Second, it will help you see a practical use of mathematics.

[1]NCTM, *Curriculum and Evaluation Standards for School Mathematics: Executive Summary* (Reston, VA: NCTM, 1989), p. 5.

Rate your attitudes Seven pairs of statements concerning attitudes toward and beliefs about mathematics are given in Table 1.1. Score your beliefs in the following manner:

- If you strongly agree with the statement in column 1, record a 1.
- If you agree with the statement in column 1 more than with the statement in column 2, record a 2.
- If you agree with the statement in column 2 more than with the statement in column 1, record a 3.
- If you strongly agree with the statement in column 2, record a 4.

Adaptive and maladaptive beliefs We will return to your responses to Table 1.1 soon. Meanwhile, let me say that in over twenty years of teaching mathematics in elementary school, middle school, high school, and college, I have worked with thousands of students. One thing (of many) these students have in common is that their beliefs about mathematics influence how they learn. With respect to their beliefs and attitudes, students generally fall into four basic categories:

- Some students work hard, confident in their problem-solving abilities.
- Some students work hard at memorizing everything the teacher says because they are fearful of getting a wrong answer.

TABLE 1.1

Column 1	Column 2
1. There will be many problems in this book that I won't be able to solve, even if I try really hard.	1. I believe that if I try really hard, I can solve virtually every problem in this book.
2. There is only one way to solve most "word" problems.	2. There is usually more than one way to solve most "word" problems.
3. The best way to learn is to memorize the different kinds of problems—rate problems, mixture problems, coin problems, etc.—and how to solve them.	3. The best way to learn is to make sure that I understand each step.
4. Some people have mathematical minds and some don't. Nothing they do can *really* make a difference.	4. Some students may have more aptitude for mathematics than others, but everyone can become competent in mathematics.
5. The teacher's job is to show us how to do problems and then give us similar problems to practice.	5. The teacher's job is more like that of a coach or guide—to help us develop the problem-solving tools we need.
6. A good test consists of problems that are just like the ones we have done in class.	6. A good test has problems at a variety of levels of difficulty, including some that are not just like the ones in the book.
7. I don't need to know all the ideas covered in this book because I'm going to teach younger children.	7. Even teachers of young children need to have a good understanding of the ideas in this book.

- Some students don't try very hard because they are convinced that mathematics is too difficult for them to understand.
- Some students don't try very hard because they don't think mathematics is relevant or useful.

The first group has what are called *adaptive beliefs* that help them approach math with a positive and confident attitude. The other three groups have different combinations of *maladaptive beliefs* that may keep them from thinking of learning as an evolving and enjoyable process. Many students find that they fit into more than one category or that the category they fall into depends on the teacher and the material.

Negative attitudes toward mathematics persist in many people. One article summarized a common attitude toward mathematics and mathematics teaching: "Mathematics—a set of facts, rules, and procedures—is a 'package' to be passively received. . . . Mathematics teachers are supposed to spend time explaining or 'covering' material from the textbook. . . . Teachers verify that students have received knowledge by checking the students' answers to make sure they are correct" (Martha Frank, "Problem Solving and Mathematical Beliefs," *Arithmetic Teacher*, 35(5), January 1988, pp. 32–34).

Stop and take stock of your reaction to this statement. To what extent do you agree with some or all of this statement?

That this view was held so widely by so many people was one of the reasons why the National Council of Teachers of Mathematics developed the many support materials that are available to teachers today, beginning with *The Curriculum and Evaluation Standards*, which were published in 1989.

In Table 1.1, the statements in column 1 indicate maladaptive beliefs, and the statements in column 2 indicate the corresponding adaptive beliefs. If you take the arithmetic average, or *mean*, of your scores (by adding up your scores and dividing by 7), you will get a number that we could call your belief index. If your belief index is less than 2, I would say that your beliefs are probably more maladaptive than adaptive. If you encounter difficulties in this course, it may be because some of your beliefs are hindering your ability to learn the material. If you do find this course frustrating, try to discuss your beliefs with your professor, with someone at a math center (if your college has one), or with a friend who is doing well in the course.

What Is Mathematics?

Many of my students ask, "Why do I have to learn all this mathematics in order to become an elementary school teacher?" I hope that by the end of the course, this book will have helped you to give a good answer to that question. A question that needs to be addressed before that question is "What is mathematics?" Think about this question for a minute and then read on. . . .

You may be surprised to learn that not all mathematicians give the same response to this question. *On the Shoulders of Giants: New Approaches to Numeracy*[3]

Outside the Classroom[2]

The pervasiveness of negative attitudes toward mathematics was powerfully illustrated in 1992 when Mattel introduced a new talking *Barbie* doll that said, "Math is tough." Now this may be true for many people, but having Barbie say it only reinforced that stereotypical perception of mathematics in the United States, especially among females. Mattel was persuaded to change Barbie's statement.

Language

Whenever you see the pencil icon, stop and think and briefly write your thoughts before reading on. Many of my students who took the time to think and write at these points (or at least to pause and think) have told me that it made a big difference in how much they learned.

---

[2]At appropriate places, you will see the boxes indicating a specific connection that I will be making—connections called MATHEMATICS, HISTORY, OUTSIDE THE CLASSROOM, LANGUAGE, and CLASSROOM CONNECTION. These notes will help develop the notion of the connectedness of mathematics.

[3]Lynn Arthur Steen, ed., *On the Shoulders of Giants: New Approaches to Numeracy* (Washington, DC: National Academy Press, 1990).

> ### ▪ Mathematics ▪
>
> Keith Devlin has written several fascinating and readable books on this subject,[4] one of which is a companion to a PBS series entitled *Life by the Numbers* (your college or local library probably has this book). The chapter titles for *Mathematics: The Science of Patterns* are "Counting," "Reasoning and Communicating," "Motion and Change," "Shape, Symmetry and Regularity," and "Position." Devlin discusses (among many other things) how mathematicians helped us to understand why leopards have spots and tigers have stripes, how mathematicians helped American ice skaters learn how to perform triple axel jumps, and how we use mathematics to measure the heights of mountains.

was written partly to help expand people's views of mathematics beyond the common stereotype of "mathematics is a bunch of formulas and rules for numbers." A group of mathematicians and mathematics educators brainstormed a number of possible themes for that book. In the end, it was agreed that the idea of *pattern* permeates all fields of mathematics. Five mathematicians were asked to write chapters on the following themes:

Dimension. In school, you have studied two- and three-dimensional shapes. Mathematicians have gone far beyond three dimensions for years. Recently, a field of mathematics has opened up the exploration of fractional dimensions. For example, the coastline of Britain (which can be modeled by a long, squiggly line) has been calculated to have a dimension of 1.26.

Quantity. This begins (with children) with the question "how many," for which the counting numbers (1, 2, 3, . . .) are appropriate; it moves in complexity to the question "how much," for which fractions and decimals were invented, and then to questions far more complex, for which other numbers and systems were invented.

Uncertainty. Questions of uncertainty permeate everyday life: How long will I live? What are my chances of getting a job after I graduate? What are the chances that my baby will be "normal"?

Shape. Humans' relationship with shape has a fascinating history—the shape of one's environment (desert, forest, mountain), what shape is best for packaging, the shapes that artists make, and the shapes we manufacture for quilts, clothing, and so on.

Change. We live in a world that is constantly changing. The development of computers enables us better to understand and manage change, whether it be the changing weather, the change in epidemics (such as AIDS), the change in populations (human and animal), or changes in the economy.

I cannot overstate the importance of getting you, as future elementary teachers, to expand your view of what mathematics is. Mathematics is far more than titles of courses and chapters in textbooks—whole numbers, fractions, decimals, percents, algebra, geometry, etc. These topics represent tools that are needed in order to answer important questions about dimension, quantity, uncertainty, shape, and change. "The numbers, lines, angles, shapes, dimensions, averages, probabilities, ratios, operations, cycles, and correlations that make up the world of mathematics enable people to make sense of a universe that otherwise might seem to be hopelessly complicated."[5] It is crucial that elementary teachers believe that mathematics is about fundamental ideas that are accessible to "regular" people and to understand these ideas so that *all* of their students will be at least competent and that *all* will see that mathematics is as important as other subjects.

The NCTM Standards

The National Council of Teachers of Mathematics (NCTM) published *The Curriculum and Evaluation Standards for School Mathematics* in 1989. This landmark book articulated, for the first time, why the nature of mathematics instruction in the United States has to change and described a much larger vision of what students ought to be learning in school about mathematics. The book presented detailed standards for elementary, middle, and high school mathematics.

---

[4]*Mathematics: The Science of Patterns* (New York: W. H. Freeman, 1996); *The Language of Mathematics: Making the Invisible Visible* (New York: W. H. Freeman, 1998); and *Life by the Numbers* (New York: Wiley, 1998).

[5]American Association for the Advancement of Science, *Benchmarks for Scientific Literacy* (New York: Oxford University Press, 1993), p. 25.

Two other important books followed: *Professional Standards for Teaching Mathematics* in 1991 and *Assessment Standards for School Mathematics* in 1995. *Principles and Standards for School Mathematics*, which incorporates and updates the three books just mentioned, was published in 2000. This book presents ten standards that run through mathematics from pre-kindergarten through 12th grade.

Standard 1: Number and Operation
Standard 2: Patterns, Functions, and Algebra
Standard 3: Geometry and Spatial Sense
Standard 4: Measurement
Standard 5: Data Analysis, Statistics, and Probability
Standard 6: Problem Solving
Standard 7: Reasoning and Proof
Standard 8: Communication
Standard 9: Connections
Standard 10: Representation

NCTM has organized each standard into three or four broad aspects which are described at a general level, and then articulated with examples at the pre-K to grade 2, grades 3–5, grades 6–8, and grades 9–12 level.

Chapters 2–10 in this textbook will engage you in the important ideas related to the first five standards for elementary school children. The last five standards permeate this textbook. Without them, all the concepts and formulas and procedures in the other areas are inert; that is, these process standards[7] make the content come to life. In most problems, you will be applying two or more of these process standards when successfully solving the problem. The rest of the first chapter and the Explorations for the first chapter have been designed to give you a sense of these five standards and their importance in the mathematics classroom. In the following sections, I ask that you read the NCTM summary of each of the process standards carefully. Assess the extent to which you understand what you are reading and the extent to which you "buy it." At the end of the section, go back and reread this summary to see whether my examples and elaborations have helped it to make more sense. None of the excellent elementary teachers I know applies these five standards "by the book." Rather, they make sure that these standards are alive in their lesson plans.

Using Mathematics

Let us now turn our attention to how people use mathematics in everyday situations and in work situations. Take a few minutes to jot down some instances in which you have used mathematics in your life and some instances in which you know that mathematics is used in different careers and work situations. Then read on. . . .

People use mathematics for various purposes, for example:

- To persuade a boss that our idea will make money
- To persuade a potential customer that our product will save money
- To predict—tomorrow's weather or who will win the election
- To make a personal decision—whether we can afford to buy a house

Classroom Connection

Deborah Meier, an award-winning principal of several elementary schools in the New York City area, wrote a book called *The Power of Their Ideas*.[6] One aspect of her leadership style was to ask, "So what?", which is the antithesis of the common anthem of "Just do it." I encourage my students *not* just to do it but to ask this question of me and of themselves. For example, I think it is important for you to examine your attitudes and beliefs toward mathematics. Do you buy this or have the last several pages simply been a blur of words? There are many wonderful teachers in this country, but there are also many teachers who "just do it." For them mathematics is just another subject to be "covered." Do you want your students to form this view of mathematics and carry it from your classroom? If not, then you need to ask yourself the "So what?" question frequently, and when you cannot give a satisfactory answer, talk with your instructor or go elsewhere—to the library, to the Web, to articles in *Teaching Children Mathematics*, to classrooms—and see for yourself the kinds of mathematical experiences that we want young children to have in school.

---

[6]Boston: Beacon Press, 1995.
[7]I will use the term *process standards* to refer to Standards 6–10. However, note that this is simply a phrase that enables me to communicate quickly. The longer version is "the standards that cut across all content areas in mathematics and that should be alive in virtually every lesson."

- To make a business decision—how much to charge for a new product or whether a new medicine (for example, a cure for AIDS) really works
- To help us understand how the world works (for example, why leopards have spots and tigers have stripes)

Solving problems In each of these examples, people are using mathematics as a tool for solving problems. To decide whether you can afford a new car, you have to collect data (on insurance, for example), add decimals, and work with percents (such as sales tax and interest on the loan). The mathematics you will use will help you solve the problem of whether to buy a car.

Think about this situation involving weather forecasters. In 1994 a hurricane brought severe rains to Georgia. Forecasters predicted that the Flint River would crest at 20 feet above flood level; the river actually crested at 13 feet above flood level, much to the relief of many residents. The forecasters used decimals, volume formulas, and conversions to determine the maximum volume of water that would be flowing. They based their final results on computer models of flooding rivers, and the computer models were based on data collected on previous flooding.

Now that we have discussed mathematics in general, we are ready to focus on developing problem-solving tools that can be used in a wide variety of situations.

SECTION 1.2 PROBLEM-SOLVING

WHAT DO YOU THINK?

- What problem-solving tools do you bring to this course?
- In addition to verifying your computation, how can you verify your answer to a problem?

In the following NCTM summary of the standard on problem-solving, I have added parenthetical comments that are meant to elaborate on the meaning and importance of each statement. As you read the summary, check the extent to which you understand each statement (both the NCTM's and mine).

Standard 6: Problem Solving
Instructional programs from prekindergarten through grade 12 enable all students to

- build new mathematical knowledge through problem-solving;
  [This is a major reason for a separate Explorations manual. In your explorations, you will encounter the ideas and concepts of the chapter firsthand, as opposed to simply being shown how to do it.]
- solve problems that arise in mathematics and in other contexts;
  [It is critical that you develop the knowledge to solve multistep, nonroutine problems, something we will elaborate on later in this chapter. Few problems outside school are "just like" the ones in the textbooks.]
- apply and adopt a variety of appropriate strategies to solve problems;
  [I will use the metaphor of a toolbox to develop this aspect.]
- monitor and reflect on the process of mathematical problem-solving.
  [Monitoring and reflecting are essential in order to "own" what you learn, as opposed to just "renting" this knowledge.]

NCTM *Principles and Standards for School Mathematics:*
(Reston, VA: NCTM, 2000, p. 52)

 We will use this icon to indicate that this is an appropriate place to cover this numbered Exploration in the separate Explorations manual, available with this text.

Section 1.2 / Problem-Solving 7

FIGURE 1.1

When you say **problem-solving** to most people, they think of an image something like Figure 1.1. Many of my students tell me that when they came into the course, their primary learning tool was memorization and their primary problem-solving tool was what they called "trial and error."

But problems need not be a source of dread. If you have done some of the explorations in the *Explorations* volume, you have already discovered some new tools for solving problems. In this section, we will examine some of the tools that are essential for solving multistep and nonroutine problems. Think of problems beyond the walls of the classroom that require mathematics. Generally, they are not one-step problems (such as simply dividing a by b), nor are they usually *just like* a problem you have solved before. We do our students a disservice if we lead them to believe that problem-solving is simply memorizing formulas and procedures and then test them on problems that are just like the ones we taught.

Before we do some problems, stop for a moment. What kinds of problem-solving tools do you bring to this course? Many people will find this exercise more productive if they think of actual problems they have had to solve, such as buying a car, saving for college, or deciding how much food and beverages to buy for a party. Stop and write down your thoughts before reading on. If possible, discuss your ideas with another student also.

Our first problem, though silly, is well known because it nicely illustrates a number of important problem-solving strategies.

INVESTIGATION 1.1

Pigs and Chickens

A farmer has a daughter who needs more practice in mathematics. One morning, the farmer looks out in the barnyard and sees a number of pigs and chickens. The farmer says to her daughter, "I count 24 heads and 80 feet. How many pigs and how many chickens are out there?"

Before reading ahead, work on the problem yourself or, better yet, with someone else. Close the book or cover the solution paths while you work on the problem.

Compare your answer to the solution paths below.

DISCUSSION

STRATEGY 1: *Use random trial and error*

One way to solve the problem might look like what you see in Figure 1.2.

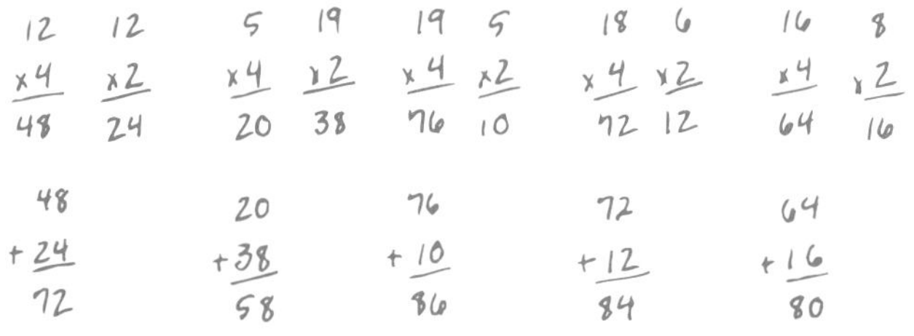

$$
\begin{array}{cccccccccc}
12 & 12 & 5 & 19 & 19 & 5 & 18 & 6 & 16 & 8 \\
\times 4 & \times 2 & \times 4 & \times 2 & \times 4 & \times 2 & \times 4 & \times 2 & \times 4 & \times 2 \\
\hline
48 & 24 & 20 & 38 & 76 & 10 & 72 & 12 & 64 & 16
\end{array}
$$

$$
\begin{array}{ccccc}
48 & 20 & 76 & 72 & 64 \\
+24 & +38 & +10 & +12 & +16 \\
\hline
72 & 58 & 86 & 84 & 80
\end{array}
$$

FIGURE 1.2

> **■ Language ■**
>
> The full description of this strategy is "Think, then guess, then check, then think, then revise (if necessary), and repeat this process until you get an answer that makes sense." A somewhat condensed description is think–guess–check–think–revise. I will refer to this strategy throughout the book simply as guess–check–revise, but I urge you not to let the strategy become mechanical.

Unfortunately, trial and error has a bad reputation in schools. The words *trial* and *error* do not sound very friendly. However, this strategy is often very appropriate. In fact, many advances in technology have been made by engineers and scientists who were guessing with the help of powerful computers using **what-if programs**. A what-if program is a logically structured guessing program. Informed trial and error, which I call **guess–check–revise**, is like a systematic what-if program. Random trial and error, which I call **grope-and-hope**, is what the student who wrote the solution in Figure 1.2 was doing. In this case, the student finally got the right answer. In many cases, though, grope-and-hope does not produce an answer, or if it does produce an answer, it is after many trials.

STRATEGY 2: Use guess–check–revise (with a table)

One major difference between this strategy and grope-and-hope is that we record our guesses (or hypotheses) in a table and look for patterns in that table. Such a strategy is a powerful new tool for many students because a table often reveals patterns. Look at Table 1.2. A key to "seeing" the patterns is to make a fourth column called "Difference." Do you see how this column helps? We will explore the notion of how seeing patterns can enhance our problem-solving ability in the next section.

TABLE 1.2

	Number of pigs	Number of chickens	Total number of feet	Difference	Thinking process
First guess	10	14	68		We need more feet, so the next guess needs to have more pigs.
Second guess	11	13	70	+2	Increasing the number of pigs by 1 adds 2 feet to the total. What if we add 2 more pigs?
Third guess	13	11	74	+4	Increasing the number of pigs by 2 adds 4 feet to the total. Because we need 6 more feet, let's increase the number of pigs by 3 in the next guess.
Fourth guess	16	8	80	+6	Yes!

From the table, we observe that if you add 1 pig (and subtract 1 chicken), you get 2 more feet. Similarly, if you add 2 pigs (and subtract 2 chickens), you get 4 more feet. Do you see why? Think before reading on. . . .

Because pigs have 2 more feet than chickens, each trade (substitute 1 pig for 1 chicken) will produce 2 more legs in the total number of feet. This observation would enable us to solve the problem in the second guess. Do you see how . . . ? After the first guess, we need 12 more feet to get to the desired 80 feet. Because each trade gives us 2 more feet, we need to increase the number of pigs by 6.

It is important to note that the guesses shown in Table 1.2 represent one of many variations of a guess–check–revise strategy.

STRATEGY 3: Make a diagram

Some people think in words, others in numbers, and still others in pictures. Sometimes making a diagram can lead to a solution to a problem. I stumbled across this approach one day as I was walking around the classroom listening

FIGURE 1.3

to students work on this problem in small groups. Figure 1.3 shows what one student had done. How do you think she had solved the problem? Write your thoughts before reading on. . . .

I asked her how she had solved the problem. She replied that she had made 24 chickens, which gave her 48 feet. Then she kept turning chickens into pigs (by adding 2 feet each time) until she had 80 feet! I was thrilled because she had represented the problem visually and had used reasoning instead of grope-and-hope. She was embarrassed because she felt she had not done it "mathematically." However, she had engaged in what I call mathematical thinking.

There are two aspects of this strategy that beg to be elaborated. First, it illustrates the notion of mathematical modeling, a very important aspect of mathematical problem-solving that will be discussed in more detail later. Simply stated, most complex mathematical problems are solved first by making a model of the problem—the model makes the problem simpler to see and thus to solve. The model enables us to do things that we could not do in the original situation. In this case, the model enables us to do what is biologically impossible but mathematically possible—we turn chickens into pigs until we get the right answer!

The other aspect of this problem is that, upon reflection, we realize its enormous potential. For example, what if the problem were 82 heads and 192 feet? No way you say? True, it would be tedious to draw 82 heads and then 2 feet underneath each head. This is exactly the power of mathematical thinking—you don't have to do all the drawing. Rather, the drawing stimulates the thinking that will lead to a solution. Think about what the diagram tells us, and then see whether you can solve the problem. . . .

If we drew 82 heads and then drew 2 feet below each head, that would tell us how many feet would be used by 82 chickens and how many feet we would still need. Having made the diagram for the simpler problem, we can do *that* for this problem without a diagram. That is, 82 chickens will use up 164 feet. Because $192 - 164 = 28$, we need 28 more feet—that is, 14 more pigs. So the answer is 14 pigs and 68 chickens. Check it out!

Furthermore, this solution connects nicely to an algebraic solution, as we shall see shortly. I have come to believe that for *many* students, the best way to understand the more abstract mathematical tools is first to solve problems using more concrete tools and then to see the connections between the concrete approach (in this case, a drawing or guess–check–revise) and the abstract approach (in this case, two equations).

STRATEGY 4: *Use algebra*

Because the range of abilities present among students taking this course is generally wide, it is likely that some of you fully understand the following algebraic strategy and some of you do not. Furthermore, many students enter this and other college math courses believing that the algebraic strategy is the *right* strategy, or at least the *best* strategy. Let's look at an algebraic solution and then see how it connects to other strategies and to the goals of this course.

Go back and review strategies 1 and 2. They both involved a total of 24 pigs and chickens. Can you explain in words why this is so? Think about this before reading on. . . .

Most students say something like "Because the total number of animals is 24" or "Well, 24 animals will have 24 heads." Therefore, if we say that

$p = $ the number of pigs

$c = $ the number of chickens

then the *number* of pigs plus the *number* of chickens will be 24. Hence, the first equation is

$$p + c = 24$$

Many students have difficulty coming up with the second equation. If this applies to you, look back at how we checked our guesses: We multiplied the number of pigs by 4 and the number of chickens by 2 and then added those two numbers to see how close that sum was to 80. In other words, we were doing the following:

4 × (The guess for number of pigs) + 2 × (the guess for number of chickens)
$\quad\quad\quad\quad (4 \times p) \quad\quad\quad\quad + \quad\quad\quad\quad (2 \times c)$

More conventionally, this would be written as

$$4p + 2c$$

Using guess–check–revise, we had the right answer when this sum was 80. Thus the second equation is

$$4p + 2c = 80$$

If you solve these two equations, using basic algebra, you will discover that $p = 16$ and $c = 8$.

The algebraic strategy in perspective   Students who do not do well with equations generally understand this algebraic solution better if they see it *after* they have used guess–check–revise. The reason for this is simple and is grounded in learning theory: *The teacher's explanation (or another student's) is far more meaningful if you can connect it to something you already know.* This is such an important learning principle that I want to elaborate. If I show you how to create the equation before you have attempted the problem on your own, my explanation is not likely to stick. In this case, your knowledge is "Teflon knowledge." The Teflon molecule was specially designed by chemists so that other materials tend not to stick to it. This is a wonderful property for frying pans to have, but not for human brains! To the extent that new ideas and concepts have **connections** to ideas and knowledge you already have, to that extent you are more likely to retain (that is, to **own** rather than **rent**) that new knowledge.

Algebra can be a very effective strategy for many problems. However, I offer the following cautionary notes with respect to the use of algebra in this course. First, the algebraic strategy is not more mathematical than the guess–check–revise strategy or drawing a diagram; it is simply more abstract. Second, it is better to use guess–check–revise and be confident of the process and of the answer than to use grope-and-hope to come up with an algebraic equation that you hope works out. Furthermore, guess–check–revise is a tool all of your future students should have, regardless of their age, whereas only a small percentage of your future students will have formal algebraic knowledge before grade 8 or 9.

> ■ *Language* ■
>
> Many authors define understanding in terms of connections. That is, you truly understand an idea only if it is well-connected to other ideas, and your depth of understanding is connected to how many connections you are making and the quality of those connections.

Problem-Solving and Toolboxes

An image of problem-solving I would like to develop in your mind is that of a **toolbox**. Imagine that your car breaks down and is towed to the garage, where a novice mechanic, right out of training school, is the first to look at it. The novice will probably try a few standard procedures: Insert the key to see what happens, check the battery connections, look for a loose wire, and so on. If none of those strategies work, the novice mechanic will be stumped and will have to summon the senior mechanic. The senior mechanic may try the same basic procedures and may solve the problem by *interpreting* the results. If this does not solve the problem, the mechanic will have to go to two toolboxes. The first toolbox is a physical one. The second toolbox is a mental one.

At the beginning of this course, many students are like the novice mechanic: When they encounter a problem, they have a limited repertoire of strategies. However, as the course develops, their toolbox grows in two ways. First, the number of tools grows. This is like the novice mechanic's learning to use the garage's diagnostic equipment. Second, their ability to use each tool also grows. This is analogous to the novice mechanic's learning how to use a voltmeter more skillfully.

Looking back   At this point, I want to introduce another process that most successful students have incorporated into their toolbox. They reflect on their work. That is, after solving a problem, they stop and examine their toolbox—what tools worked better and what new tools they used. Take a few minutes now to look at your notes on the strategies we discussed in Investigation 1.1 and then look at "4 Steps for Solving Problems" on the inside front cover of this book. What tools were used in the various solutions of the pigs-and-chickens problem? What made those tools work better? Then read on. . . .

My list includes

1. Draw a diagram—a picture is often worth a thousand words.
2. Guess–check–revise—starting here will help some students to develop the two equations.
3. Make a table—this increases the possibility of seeing more patterns.
4. Look for patterns—extending the table enables certain patterns to become more visible.
5. Develop an equation—a powerful tool, but also surrounded by quicksand for many.

Two final notes about Investigation 1.1:

1. The strategies discussed in Investigation 1.1 do not represent all the different ways in which the pigs-and-chickens problem has been solved. Because of space restrictions, only four were discussed.
2. Many students think that there is only one tool per problem. An important adaptive belief is that there are often several different ways (using different tools) to solve a problem and that the tools are often used in combination instead of separately.

Polya's Four Steps

George Polya developed a framework for problem-solving that breaks down problem-solving into four distinguishable steps. In 1945, he outlined these steps in a now-classic book called *How to Solve It*.

When you approach a problem—a math problem, a writing assignment, even a personal problem—if you think that you have to come up with an answer immediately and that there is only one "right" way to reach that answer, a solution may seem to be beyond your grasp. But if you break the problem down and creatively and mindfully approach each *step* of the problem, it generally becomes more manageable. Polya suggests that you first need to make sure you *understand the problem*. Once you understand the problem, you *devise a plan* for solving the problem. Then you *monitor your plan*; you check frequently to see whether it is productive or is going down a dead-end street. Finally, you *look back at your work*. This last step involves more than just checking your computation; for example, it includes making sure that your answer makes sense. For each of these four steps, there are specific strategies that we will explore in this chapter and that you will refine throughout this course.

Owning versus renting   Instead of just listing Polya's strategies, we are going to discover them by putting them into action. You will notice that I often ask you to stop, think, and write some notes. I really mean it! I have come to distinguish between those students who *own* what they learn and those who simply *rent* what they learn. Many students rent what they have learned just long enough to pass the test. However, within days or weeks of the final exam, it's gone, just like a video that has been returned to the store. One of the important differences between owners and renters is that those who own the knowledge tend to be *active readers*.

Think and then read on . . .   Throughout the book, I will often pose a question and ask you to "think and then read on. . . ." Rather than just look to the next paragraph and see the "answer," you will learn much more if you immediately cover up the next paragraph or close the book . . . stop . . . think . . . write down your thoughts . . . and then read on. The phrase "think and read on . . ." is there to remind you to read the book actively rather than passively. An *active reader* stops and thinks about the material just read and asks questions: Does this make sense? Have I had experiences like this? The active reader does the examples with pen or pencil, rather than just reading the author's description. I recommend that you keep a journal (your instructor may give you specific instructions). Keeping a journal has many benefits. Many people find that they learn more by writing down their responses to "think and read on . . ." rather than just stopping for a moment to reflect. Many students find that reading over their journals every week or at the end of each chapter provides new insights into the mathematics and into their own beliefs and attitudes about mathematics.

Using Polya's four steps   I encourage you to use Polya's four steps (on the inside front covers of this book and *Explorations*) in all of the following ways:

1. Use them as a guide when you get stuck.
2. Don't rent them, buy them. Buying them involves paraphrasing my language and adding new strategies that you and your classmates discover. For example, many of my students have added one whole step to help reduce anxiety: First take a deep breath and remind yourself to slow down!
3. Just as we did in Investigation 1.1, after you have successfully solved a problem, stop for a moment and reflect on the tools you used. As the course progresses, you should find that most problems involve the use of several strategies, and you should find that you use the tools more skillfully. For example, using "Make a diagram" skillfully involves deciding how precise your diagram needs to be, checking to see that the diagram illustrates the relevant given information, seeing whether the diagram can help you to paraphrase the problem or see it from a new perspective, and so on.

CLASSROOM CONNECTION

A colleague of mine was working through a word problem with her class one day and encouraging the students to think about what they were doing. Suddenly one of the students said, "But you don't need to do all this stuff you are teaching us; you just know the answer." She was stunned, and the ensuing discussion was informative. It turns out that many students believe that the difference between a student and a teacher is that the teacher just knows the answer or automatically knows how to get the answer. That is, teachers don't need such strategies as guess–check–revise, make a table, draw a diagram, look for patterns, etc. The truth is that we do! Virtually all of the most brilliant workers—whether they be engineers, scientists, businesspeople, carpenters, researchers, or entrepreneurs—approach complex problems by using the very tools that are being stressed in this text. Furthermore, even the top people in a field often make hypotheses that seem reasonable (to them and to their colleagues) but that turn out not to be true. So when you are working on these problems, please realize that the tools being discussed in this book are used by people in all kinds of situations.

Why Emphasize Problem-Solving?

Although Polya described his problem-solving strategies back in 1945, it was quite some time before they had a significant impact on the way mathematics was taught. One of the reasons is that the desired outcomes of mathematics instruction were defined too narrowly. If the stated goal of mathematics classes was to learn and practice the "right" techniques to answer textbook problems, chances are that students rarely saw how math applied to situations outside the classroom.

Another reason for the limited impact of Polya's work is that until recently, "problems" were generally defined too narrowly. For example, many of you learned how to do different kinds of problems—mixture problems, distance problems, percent problems, age problems, coin problems—separately but never realized that they have many principles in common. To use language from the NCTM, there has been too great a focus on single-step problems and routine problems. Consider the examples from the National Assessment of Educational Progress shown in Table 1.3.

TABLE 1.3

Problem	Percent correct Grade 11
1. Here are the ages of six children: 13, 10, 8, 5, 3, 3 What is the average age of these children?	72
2. Edith has an average (mean) score of 80 on five tests. What score does she need on the next test to raise her average to 81?	24

Source: Mary M. Lindquist, ed., *Results from the Fourth Mathematics Assessment of the National Assessment of Educational Programs* (Reston, VA: NCTM, 1989), pp. 30, 32.

To solve the first problem, one only has to remember the procedure for finding an average and then use it:

$$\frac{13 + 10 + 8 + 5 + 3 + 3}{6}$$

However, there is no simple formula for solving the second problem. Try to solve it on your own and then read on. . . .

To solve this one, you have to have a better understanding of what an average means. One approach is to see that if her average for 5 tests is 80, then her total score for the 5 tests is 400. If her average for the 6 tests is to be 81, then her total score for the 6 tests must be 486 (that is, 81×6). Because she had a total of 400 points after 5 tests and she needs a total of 486 points after 6 tests, she needs to get an 86 on the sixth test to raise her overall average to 81.

Many students still consider the second question to be a "trick" question unless the teacher has explicitly taught them how to solve that kind of problem. However, many employers note that problems that occur in work situations are rarely *just* like the ones in the book. What employers desperately need is more people who can solve the "trick" problems, because, as one wag put it, "life is a trick problem!"

The difference between traditional word problems and many real-life problems Table 1.4 lists differences between the word problems generally found in textbooks and real-life problems.

TABLE 1.4

Textbook word problems	Real-life problems
1. The problem is given.	1. Often, you have to figure out what the problem really is.
2. All the information you need to solve the problem is given.	2. You have to determine the information needed to solve the problem.
3. There is always enough information to solve the problem.	3. Sometimes you will find that there is not enough information to solve the problem.
4. There is no extraneous information.	4. Sometimes there is too much information, and you have to decide what information you need and what you don't.
5. The answer is in the back of the book, or the teacher tells you whether your answer is correct.	5. You, or your team, decides whether your answer is valid. Your job may depend on how well you can "check" your answer.
6. There is usually a right or best way to solve the problem.	6. There are usually many different ways to solve the problem.

When students undertake what some authors call more authentic problems, they come to realize that mathematics is more than just memorizing and using formulas, and they come to value their own thinking.

With respect to problem-solving, the NCTM has urged a turnabout. Traditionally, the teacher "taught" a new concept or skill, and *then* students did *problem-solving.* The NCTM has rejected this separation of teaching and problem-solving. That is, problem-solving involves more than just doing word problems to apply the concepts. In this book, we will investigate a wide variety of problems. By solving these problems, you will discover the meaning of the concepts and how to apply them and come to a much deeper understanding of and appreciation for the formulas or procedures you learn.

INVESTIGATION 1.2

How Much Will the Patio Cost?

Let's say you are building a patio in your back yard. You have decided to make the patio 12 feet by 8 feet. The local lumber store sells premade patio blocks, which measure 18 inches by 12 inches, for 75¢ each. How much will the patio cost?

Solve this problem on your own, taking time to understand the problem and consider how you might plan to solve it. Then compare your solution and strategies to the ones discussed below. . . .

DISCUSSION

STRATEGY 1: Make a diagram

Let us examine two kinds of diagrams that students often come up with. Because the blocks are $1\frac{1}{2}$ feet long and 1 foot wide, some students find a piece of

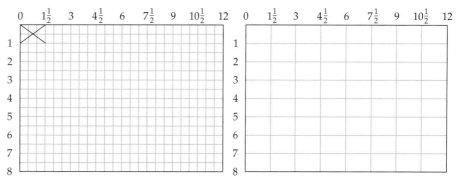

FIGURE 1.4   FIGURE 1.5

graph paper and let each square represent 1/2 foot, as shown in Figure 1.4. If you make one X on the grid for each patio block, you can determine the total number. What do you get?

Other students simply sketch the problem on a blank piece of paper as shown in Figure 1.5.

In either case, we can see that we will need 64 blocks, and 64 blocks times 75¢ per block is 4800¢ or $48.

STRATEGY 2: Divide

There is another way to solve the problem that is quicker but also requires more thinking at the beginning of the problem. If we divide the total area of the patio by the area of one block, the quotient will tell us how many blocks we need.

The total area of the patio is 12 feet × 8 feet = 96 square feet.

The area of each block is 1.5 feet × 1 foot = 1.5 square feet.

When we divide 96 square feet by 1.5 square feet, we get 64 blocks.

Some readers may be kicking themselves for not thinking of this strategy. In Chapter 3, you will find that one of the meanings of division is repeated subtraction. From one perspective, this problem is asking how many times we can subtract 1.5 from 96.

STRATEGY 3: *Use dimensional analysis*

We can treat the "labels" as algebraic terms. Just as numbers can "cancel," so too can labels. We will investigate canceling further in Chapter 5. Using *dimensional analysis*, we have

$$96 \text{ square feet} \div \frac{1.5 \text{ square feet}}{1 \text{ block}}$$

We can treat this as a division-of-fractions problem; thus we need to invert and multiply.

$$= 96 \text{ square feet} \times \frac{1 \text{ block}}{1.5 \text{ square feet}}$$

$$= 64 \text{ blocks}$$

Because square feet "cancel," the meaning of 64 is 64 blocks.

After Investigation 1.1, and also in the *Explorations*, you were asked to take some time to stop and reflect on what you learned from the activity. I have found that most of my best students intuitively do this after every problem that requires thinking on their part. I encourage you to do this on your own throughout the book. What problem-solving tools did you learn or refine in this investigation? Write them down before doing Investigation 1.3 in the next section. If possible, share your insights with another student.

SECTION 1.3 PATTERNS

WHAT DO YOU THINK?

- What do patterns have to do with learning mathematics?
- How does the ability to recognize and analyze patterns add to your problem-solving toolbox?

Patterns exist virtually everywhere in the world. Look at the floor, the walls, and the ceiling of most rooms, and you are sure to see patterns. Look at a wallpaper display in a store and see patterns. Virtually all clothing contains patterns. Look at an article of clothing that seems to have no patterns—a plain white blouse, a pair of pants. But now look at it through a magnifying glass; you will see patterns in the way the cloth was woven. Look at Islamic architecture and you will see more patterns than in probably any other kind of architecture. Mohammed, the founder of Islam, decreed that there be no statues or paintings of people in that religion. There are also patterns in weather that help us to make predictions.

Patterns are also an important part of how children organize the world: Nap comes after lunch; Monday we have art, Tuesday we have music; all dogs have tails. One of the goals of school mathematics is to build on children's curiosity to help them see patterns that underlie mathematical structure. For example, when we count, there is a repeating pattern in the ones place (0, 1, 2, 3, 4, 5, 6, 7, 8, 9. . .); when we count by 5, the ones place alternates

■ Outside the Classroom ■

In one sense, all patterns can be represented mathematically. This is not as radical a statement as you might think. Look at your television set. Most television sets still consist of 512 lines. The image from the transmitting station is coded by establishing a number (symbol) for each point on each of the 512 lines. Your television set then decodes that message and quickly puts the appropriate color in the appropriate spot, creating a picture.

CLASSROOM CONNECTION

A common exploration in preschool is to give the children a bunch of buttons and ask them what they see. The children will observe different colors, different shapes, different numbers of holes, different patterns with buttons, etc. Using this natural tendency to see in multiple ways and to look for patterns, we can help children develop the ability to form generalizations.

between 5 and 0; take any multiple of 9 and the sum of its digits will also be a multiple of nine (try 5986 × 9). In 1978, Mary Baratta-Lorton published a landmark book for elementary teachers called *Mathematics Their Way* (Menlo Park, CA: Addison-Wesley, 1976). She saw patterns as one of the unifying themes in elementary school mathematics. She said that

> [L]ooking for patterns trains the mind to search out and discover the similarities that bind seemingly unrelated information together in a whole. . . . A child who *expects* things to "make sense" *looks* for the *sense* in things and from this sense develops understanding. A child who does not see patterns often does not expect things to make sense and sees all events as discrete, separate, and unrelated. [Italics added.]

Patterns are a tool to help you learn mathematics, and mathematics often reveals seemingly hidden patterns that help us understand the world about us. Mathematics has been described by several writers as the science of pattern and order. Many important mathematical ideas have arisen from the recognition of patterns. However, to use a metaphor, patterns are not the pot of gold at the end of the rainbow; rather, patterns are the rainbow that leads to the pot of gold, which is the mathematical structure.

Now that we have discussed the prevalence of patterns in the natural world and the human-made world and the importance of patterns in the mathematics classroom, let me ask you to try to define the term *pattern*. That is, think of patterns and then consider what all these patterns have in common. What words, ideas, and images come to mind? Write down your thoughts and then read on. . . .

One aspect common to all patterns, whether geometric or numerical, is that there is repetition. When we examine this repetition, we find an underlying structure or organization. Understanding this structure enables us to make generalizations, which in turn enables the users of this knowledge to do more powerful mathematics, to design more powerful machines and equipment, or to make manufactured items more cheaply.

Recognizing patterns Being able to use patterns makes the guess–check–revise strategy much more powerful, as you saw in Investigation 1.1 (Pigs and Chickens). However, as many of you discovered, this is easier said than done. In order to use a pattern, you have to *recognize* it. As we found, making a table increases the likelihood of our recognizing a pattern. Recall also that we did not have just three columns in the pigs-and-chickens problem. *Extending* the table, we created a new column called "Difference" that helped to make the pattern more visible. We then *analyzed* the pattern to see that each time we changed a chicken into a pig, we added two feet. This enabled us to get the solution on the third guess.

Let us now investigate some problems that illustrate the pervasiveness and importance of patterns in mathematics.

INVESTIGATION 1.3 — The Sum of the First 100 Numbers

Legend has it that when Carl Gauss (1777–1855), one of the greatest mathematicians of all time, was in first grade, his teacher gave the class the following problem: Find the sum of the first 100 numbers. The teacher then sat down to grade papers, knowing that it would take most of the students a good half-hour to do this problem. Gauss came to the teacher's desk in 5 minutes with the answer. He had found a "large" pattern. Work on this problem and then read on. . . .

DISCUSSION

The goal of this investigation isn't to solve the problem as quickly as Gauss did but to understand why patterns are such a remarkable problem-solving tool and to realize that there are many different patterns, depending on how you approach the problem; it is much like seeing an interesting sculpture from different perspectives.

STRATEGY 1: Look for patterns that might make the problem smaller

Suppose we find the sum of the first ten numbers:

$$1 + 2 + 3 + 4 + 5 + 6 + 7 + 8 + 9 + 10 = 55$$

What is the sum of the second ten numbers?

$$11 + 12 + 13 + 14 + 15 + 16 + 17 + 18 + 19 + 20 = 155$$

What do you think will be the sum of the next ten numbers?

If you continue this pattern, you will get the answer relatively quickly.

STRATEGY 2: Start with a simpler problem and look for patterns

What if we look at the sums of $1 + 2$, $1 + 2 + 3$, $1 + 2 + 3 + 4$, etc., as shown in Table 1.5, and see what patterns emerge?

Do you see any patterns here? Think and then read on....

TABLE 1.5

The sequence	Sum
1 + 2	3
1 + 2 + 3	6
1 + 2 + 3 + 4	10
1 + 2 + 3 + 4 + 5	15

Actually, there are many patterns, and a crucial point about patterns and problem-solving is that not all patterns lead directly to a solution. An analogy to paths on a mountain is relevant: All paths lead somewhere; however, some paths are more helpful than others. Let us look at some of the patterns and then analyze one that helps us find the answer to the question.

Looking at differences   When we compare the sums of these sequences, we see that the difference between successive sums increases by 1 each time. For example, the difference between 6 and 3 is 3, the difference between 10 and 6 is 4, the difference between 15 and 10 is 5, and so on. Although this is a pattern, it's not very useful at this point because we can't use it to determine the sum of the first 100 numbers without knowing the sums of the first 98 and 99 numbers.

Comparing the number of terms and the sums   Another pattern emerges when, instead of looking at the relationship between numbers in the second column, we look at the problem from a different perspective. For example, what if we compare the number of terms in each sequence to the sum of the sequence (see Table 1.6). What do you see now?

TABLE 1.6

The sequence	Number of terms in the sequence	Sum
1 + 2	2	3
1 + 2 + 3	3	6
1 + 2 + 3 + 4	4	10
1 + 2 + 3 + 4 + 5	5	15

CLASSROOM CONNECTION

When looking at Table 1.7, some students ask, "How was I supposed to know to do that?" Researchers who looked at differences between "expert" problem-solvers and "novice" problem-solvers found that the experts were more likely to play around with a problem for a while if they didn't feel they had a good problem-solving strategy. There are many instances in which mathematical discoveries and solutions to problems have come from playing around with a problem. This idea of playfulness in mathematics is bizarre to many students. However, I seriously encourage you to think of it as a problem-solving tool. Furthermore, this adds a dash of creativity to problem-solving, which, in turn, makes it more enjoyable for many students.

CLASSROOM CONNECTION

"I know it but I can't explain it." Most of the time this statement indicates that the student's understanding is partial; that is, the student understands *what* but not *why*. One of the major goals of this course is to help future teachers across this bridge, from simply knowing what (to do) to also knowing why and being able to explain why.

Some students see that when there are 3 terms, the sum is 6, and 6 is 3 · 2; and when there are 5 terms, the sum is 15, and 15 is 5 · 3. One guess (or hypothesis) that can be made from this pattern is that when there are 7 terms in the sequence, the sum will be 7 · 4. Indeed, the sum of the first 7 numbers is 28! But how can we use this pattern to answer the question? Play with this for a bit before reading on. . . .

There are actually a couple of ways to use this pattern. One way is to make a new table and look at what is going on; Table 1.7 illustrates this.

TABLE 1.7

The sequence	Number of terms in the sequence	Sum	What we multiply the second column by to get the sum
1 + 2	2	3	$1\frac{1}{2}$
1 + 2 + 3	3	6	2
1 + 2 + 3 + 4	4	10	$2\frac{1}{2}$
1 + 2 + 3 + 4 + 5	5	15	3
1 + 2 + 3 + 4 + 5 + 6	6	21	$3\frac{1}{2}$
1 + 2 + 3 + 4 + 5 + 6 + 7	7	28	4

Some people can leap from this table to the first 100 numbers and say that we multiply the number of terms in the sequence, 100, by $50\frac{1}{2}$ because when the number of terms is an even number, what you multiply by is simply half of that number plus 1/2.

Some people don't like fractions and find that if they double the numbers in the second column, they get a table that is easier to work with (see Table 1.8). Some students can make the leap from looking at this table to realizing that the sum of the first 100 numbers will simply be 1/2 of 100 · 101. Do you see why this is so?

At the most general level, we can now say that the sum of the first n numbers is $\frac{1}{2}n(n + 1)$.

TABLE 1.8

The sequence	Number of terms in the sequence	Double the sum	The third column decomposed or broken down
1 + 2	2	6	2 · 3
1 + 2 + 3	3	12	3 · 4
1 + 2 + 3 + 4	4	20	4 · 5
1 + 2 + 3 + 4 + 5	5	30	5 · 6
1 + 2 + 3 + 4 + 5 + 6	6	42	6 · 7
1 + 2 + 3 + 4 + 5 + 6 + 7	7	56	7 · 8

STRATEGY 3: *Look for patterns in the whole string of numbers*

What happens if you add the first and last numbers? What does this lead to? Think and read on. . . .

$$1 + 2 + 3 + 4 + 5 + 6 + \cdots + 95 + 96 + 97 + 98 + 99 + 100$$

If we add 1 + 100 and then 2 + 99 and then 3 + 98 and so on, we always get 101. The question now is how many pairs adding up to 101 we have. Once we know that, we can multiply the number of pairs by 101. This is the large pattern that Gauss discovered! The diagram below, in which we connect the pairs of numbers, can help you to understand this strategy better.

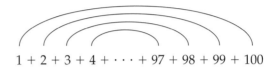

$$1 + 2 + 3 + 4 + \cdots + 97 + 98 + 99 + 100$$

This diagram illustrates having 50 sums of 101, that is 50 × 101.

STRATEGY 4: *Use a geometric representation*

FIGURE 1.6

Some students remember using manipulatives in elementary school, and manipulatives provide concrete models for solving problems. In this case, manipulatives in the form of square tiles can actually change the perspective from which one views the problem. Figure 1.6 shows the sum of the first five numbers represented with tiles. In this case, the student places the tiles on top of each other. If you see this as "half a rectangle," then you can see the connection between this figure and the formula $\frac{1}{2}n(n + 1)$. If you don't see the connection, copy this figure twice (either on paper or with tiles). Put the two copies together to make a rectangle. Because the area of a rectangle is found by multiplying the length times the width, and because each of these "triangles" has half the area of the rectangle, this is another way to solve the problem.

Classroom Connection

A point that I feel cannot be overemphasized is that for many problems from everyday life and business life, there is more than one way to solve the problem, just as there is usually more than one way to get from here to there—whether we are talking about paths up a mountain or routes from New York to Los Angeles. Many students have told me that before they took this course, they thought there was just *one right way* to do a problem, and so they never looked for patterns but instead looked for formulas or procedures. Once the students started looking for patterns, they found them everywhere, and over time they learned how to use their awareness of patterns more powerfully. We can refer to different ways to do a problem as different **solution paths**.

INVESTIGATION 1.4

Pascal's Triangle

The pattern below, called **Pascal's triangle**, is one of the most famous mathematical patterns. It was "discovered" by the mathematician Blaise Pascal (1623–1662).

This pattern occurs in many different fields of mathematics, as you will discover over the course of this book. What patterns do you see?

There are literally hundreds! Examine the triangle, play with it, and describe the patterns you see. You may choose to describe patterns with words, in a table or sequence, or by using mathematical notation.

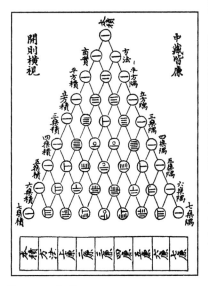

FIGURE 1.7

■ History ■

As you may have already noticed in other areas, a discovery is not necessarily named after the first person who discovered it. This is true of Pascal's triangle. Figure 1.7 comes from *The Precious Mirror of the Four Elements,* which was written in 1303 by Chu Shih-chieh, and he himself acknowledged that he had taken the triangle from a book written several hundred years earlier!

DISCUSSION

There are many patterns, and there are many ways to describe each pattern. Here are some patterns to get you started:

1. Both top diagonals contain only 1s. For example, start at the top and go down the left side. You have only 1s.
2. The counting numbers (1, 2, 3, 4, ...) are on the second diagonal.
3. There is no easy way to say the next one, so read it and then I will give an example: The sum of any sequence of counting numbers is found diagonally down from the last number counted. Look at 1, 2, 3, 4. The sum of those four numbers is 10, which is diagonally below 4. Now look at 1, 2, 3, 4, 5, the sum of which is 15, which is diagonally below 5!
4. It doesn't matter which way you go. The figure has symmetry such that each diagonal has a twin. For example, look at the second diagonal going down the right: 1, 2, 3, 4, 5, 6, 7, 8, ... Can you find its twin?

> **■ Mathematics ■**
>
> When I ask students at the end of my courses what they have learned about mathematics, many of them say that they have come to realize that mathematics is much more connected than they had thought. They saw more connections both among concepts (such as fractions, decimals, and percents) and among problems that they had solved. Do you see connections between Pascal's triangle and other problems in this chapter? Think and read on....
>
> If you didn't, look back at Investigation 1.3 and Exploration 1.1. Those sequences can be found within Pascal's triangle. Can you see them?

5. Each number (not equal to 1) in the triangle is the sum of the two numbers to its top right and its top left. Do you understand what I just wrote? Try it out, and then read on....

If you don't understand what I just said, look at this visual representation:

$$a \searrow \swarrow b$$
$$c$$

Choose any number c (not equal to 1) in the triangle. If we let a and b represent the two numbers immediately above c, then the value of c is equal to the value of $a + b$!

6. Add the numbers in each row in the triangle. What do you get? Do it now and then read on....

The sums of the rows make the following sequence: 1, 2, 4, 8, 16. This sequence can be described as the powers of 2. Can you guess the sum of the numbers in the 10th row? Try it and then read on....

In this case a table is helpful. See Table 1.9.

In other words, the sum of the numbers in the nth row can be obtained by raising 2 to the $(n - 1)$ power. If you have a calculator with an $\boxed{x^y}$ function, you can determine 2^9 by doing the following: 2 x^y 9 =; 512 will appear. More Explorations with Pascal's triangle can be found in Exploration 1.1. If you type "Pascal's triangle" into a search engine, you can explore it even further.

TABLE 1.9

Row	Sum
1	1
2	2
3	4
4	8
5	16
.	.
.	.
.	.
n	2^{n-1}

The Power of Patterns

Some of the greatest mathematicians in history made their mark by discovering and articulating patterns. The wonderful thing about patterns, though, is that they make mathematics more accessible to everyone. This was recognized by the authors of *Everybody Counts*,[8] who wrote,

> Virtually all young children like mathematics. They do mathematics naturally, discovering patterns and making conjectures based on observation. Natural curiosity is a powerful teacher, especially for mathematics. Unfortunately, as children become socialized by school and society, they begin to view mathematics as a rigid system of externally dictated rules governed by standards of accuracy, speed, and memory. Their view of mathematics shifts gradually from enthusiasm to apprehension, from confidence to fear. Eventually, most students leave mathematics under duress, convinced that only geniuses can learn it.

We explored several famous, powerful patterns in this section. You will see many of these patterns again, both in this book and beyond it. But more important, I hope you will see that stepping back from a problem can reveal patterns within it. Sometimes these patterns can help you devise a plan for solving the problem; at other times they emerge as you look back on your work.

[8]National Research Council, *Everybody Counts: A Report to the Nation on the Future of Mathematics Education* (Washington, DC: National Academy Press, 1989), pp. 53–54.

SECTION 1.4 REPRESENTATION

> **WHAT DO YOU THINK?**
>
> - Representation was not one of the standards in the 1989 NCTM Curriculum Standards. Why do you think it was made a separate standard in the 2000 Standards?
>
> - Why do you think I chose to put the Representation standard (Standard 10) after Patterns instead of at the end of the chapter?

Standard 10: Representation
Instructional programs from prekindergarten through grade 12 should enable all students to

- create and use representations to organize, record, and communicate mathematical ideas;
[When you can represent a situation or idea in multiple ways—tables, graphs, symbols, diagrams, words—your understanding of that situation or idea is much stronger than when you can represent the situation/idea in only one way.]

- select, apply, and translate among mathematical representations to solve problems;
[This relates to the toolbox metaphor and also to the notion of flexibility, as opposed to the mechanical application of rote procedures.]

- use representations to model and interpret physical, social, and mathematical phenomena.
[We will talk more about modeling later.]

(*Principles and Standards*, p. 67)

At the simplest level, a **representation** is a picture—a picture of your image of a problem. Representations can take many forms: diagrams, graphs, tables, sketches, equations, words, etc. One of the key ideas about representations is that most problems and most mathematical concepts can be represented in different ways. For example, there were several ways to represent the pigs-and-chickens problem from Investigation 1.1. Students commonly ask me which representation is "best," but this is just not a useful question. A more useful question is whether this representation fits the purposes before us. Representing the problem with 24 circles (representing 24 bodies) and then putting legs on the bodies is a very appropriate representation for this problem with children. This notion of multiple representations is also related to the standard on Connections, which will be discussed in Section 1.7. For example, some of my students *do* represent the problem with 24 circles. In this case, the next step is to *connect* this representation to other representations that are more useful with larger numbers, such as the table and the formula.

Just as many problems can be represented in multiple ways, this is also true for most important mathematical ideas. Let me use language to illustrate the idea of multiple representations. Think of the word *hot*. What does it mean? Write down the different meanings of *hot* before reading on. . . .

When I looked up this word in the dictionary, I found 14 different meanings. Following are phrases that illustrate some of these meanings: a hot forehead, hot peppers, a hot temper, hot for travel, a hot topic, a hot suspect, hot on the trail, I'm not so hot at math, a hot sports car.[9]

Now consider the word *fraction*. Let me be more specific: In how many different ways can you represent 3/4? Do this and then turn to page 50 in the textbook for a sneak preview of this important idea.

---

[9]*American Heritage Dictionary of the English Language* (Boston: Houghton Mifflin, 1992), p. 874.

CHAPTER 1 / Foundations for Learning Mathematics

Just as a literate person knows the multiple meanings of many words, a mathematically literate person knows multiple representations of important mathematical ideas.

I chose the next investigation because it illustrates several important aspects of representations.

INVESTIGATION 1.5 — Coin Combinations

You will find this and similar problems all over elementary school mathematics, and for good reason. As you will soon see, there is rich opportunity for mathematical development with these problems. The problem: How many different ways can you make 25 cents using dimes, nickels, and pennies? Solve this problem on your own before reading on. . . .

Language

Teachers traditionally use a phrase like "we will cover this much material in this course." I urge the word *uncover*. Think of an archaeological team that has discovered an ancient city. As they uncover the ruins carefully, the city emerges right before their eyes. So too when we look at different solution paths and different representations of a problem, new ideas and patterns emerge right before our eyes.

DISCUSSION

Do you see patterns in your work? Did you work systematically or just randomly? If you don't see patterns and/or you generated the combinations randomly, please go back and take a few minutes to do the problem over. This time think how you might generate the combinations by being systematic, and then look at your table. Patterns will help you to be more confident that you have all the combinations.

One of the neat things about this investigation is that there are multiple ways to represent the problem, some of which make it easier to see patterns. Furthermore, as shown with the pigs and chickens, there is more than one way to solve the problem, even if you are systematic. Let us **uncover** this problem now. First, look at the solution on the left, in which the combinations were generated randomly after the first two, which were combinations involving all of the same coin. Now look at the solution on the right. What do you think the person said to himself or herself when generating this table? Think before reading on. . . .

N	N	N	N	N		D	D	N	
25P						D	D	5P	
D	D	N				D	N	N	N
D	D	5P				D	N	N	5P
N	20P					D	N	10P	
N	N	N	10P			D	15P		
D	N	N	N			N	N	N	N	N
D	N	10P				N	N	N	N	5P
D	15P					N	N	N	10P	
N	N	N	N	5P		N	N	15P		
N	N	15P				N	20P			
D	N	N	5P			25P				

After the quarter, the person thought, how many ways would involve two dimes, then how many ways would involve one dime, then no dimes. Now describe patterns you see in this solution before reading on. . . .

There are many patterns; we will examine just a couple here. In the right-hand columns, the nickels form two triangles. Do you see them? Look at the 3rd, 4th, and 5th rows—you can draw a right triangle that encloses the nickels, with the right angle in the top left corner. That is, there are 3 nickels in the 4th row, 2 in the 5th row, and 1 in the 6th row. Then you skip a row, with no nickels, and then you have a similar, larger triangle. Another pattern is that the pennies go in a diagonal direction, increasing by 5 each time.

Below is another representation of the same solution path. Which representation do you prefer? To you, which representation shows the patterns better? Interestingly, some of my students prefer the one above because they like the letters, but others prefer the one below because seeing the numbers makes more sense to them.

Dimes	Nickels	Pennies
2	1	0
2	0	5
1	3	0
1	2	5
1	1	10
1	0	15
0	5	0
0	4	5
0	3	10
0	2	15
0	1	20
0	0	25

```
10 10 5
10 10 1 1 1 1 1
10 5  5 5
10 5  5 1 1 1 1 1
10 5  1 1 1 1 1 1 1 1 1 1
10 1  1 1 1 1 1 1 1 1 1 1 1 1 1
5  5  5 5 5
5  5  5 5 1 1 1 1 1
5  5  5 1 1 1 1 1 1 1 1 1
5  5  1 1 1 1 1 1 1 1 1 1 1 1 1
5  1  1 1 1 1 1 1 1 1 1 1 1 1 1 1 1 1 1 1
1  1  1 1 1 1 1 1 1 1 1 1 1 1 1 1 1 1 1 1 1 1 1 1
```

At the left is yet another representation for the dimes, nickels, and pennies. This one is more abstract in that you don't see each individual coin. Because of its abstractness, it will be harder for young children to make sense of than the previous representations. However, again partly because of its abstractness, certain patterns are easier to see. What patterns do you see in this table? Think before reading on. . . .

Once again there are many patterns, and we will discuss just two. The numbers of nickels form decreasing sequences: 1, 0, then 3, 2, 1, 0, then 5, 4, 3, 2, 1, 0. The lengths of the sequences are 2, 4, and 6. The numbers of pennies form increasing sequences: 0, 5 then 0, 5, 10, 15, then 0, 5, 10, 15, 20, 25. The lengths of these sequences are 2, 4, and 6 also.

As we noted at the beginning of this discussion, there is more than one way to be systematic. The solution at the left below and the table for that solution, at the right below, illustrate another solution path. Before we look at patterns in the table, take a minute to see if you can figure out the thinking process that generated this solution. . . .

In this case, the person began with combinations with 0 pennies, then combinations with 5 pennies, then 10 pennies, and so on. Now look for patterns in the solution path at the left and then in the table at the right. . . .

					Pennies	Nickels	Dimes
N	D	D			0	1	2
N	N	N	D		0	3	1
N	N	N	N	N	0	5	0
5P	D	D			5	0	2
5P	N	N	D		5	2	1
5P	N	N	N	N	5	4	0
10P	N	D			10	1	1
10P	N	N	N		10	3	0
15P	D				15	0	1
15P	N	N			15	2	0
20P	N				20	1	0
25P					25	0	0

There are 3 combinations with no pennies, 3 combinations with five pennies, 2 combinations with ten pennies, 2 combinations with fifteen pennies, 1 combination with twenty pennies, and 1 combination with twenty-five pennies. At the right, look at the nickels column. This time we have increasing sequences: 1, 3, 5, then 0, 2, 4, then 1, 3, then 0, 2, then 1, then 0. The sequences rotate from all odd numbers to all even numbers to all odd numbers to all even numbers to an odd number to an even number.

Summary

In the first half of Chapter 1, I have asked you to examine your beliefs and attitudes toward mathematics, I have introduced the NCTM Curriculum Standards, and we have investigated the standards on problem-solving, representation, and patterns. Before moving on, many instructors find it helpful for students to do some thinking and some problems that involve these themes. In the second half of Chapter 1, we will explore the remaining process standards.

EXERCISES

1. At a bicycle store, there were a bunch of bicycles and tricycles. If there were 32 seats and 72 wheels, how many bicycles and how many tricycles were there?

2. **a.** A farmer looks out into the barnyard and sees the pigs and the chickens. He says to his daughter, "I count 40 heads and 100 feet. How many pigs and how many chickens are out there?"

   b. What if the farmer saw 169 heads and 398 feet? How many pigs and how many chickens are out there?

3. A Martian farmer looks out into the barnyard and sees tribbles (which have four legs) and chalkas (which have seven legs). She says to her son, "I count 97 heads and 436 feet. How many tribbles and how many chalkas are out there?"

4. At a benefit concert, 600 tickets were sold and $1500 was raised. If there were $2 and $5 tickets, how many of each were sold?

5. For a certain event, 812 tickets were sold, for a total of $1912. If students paid $2 per ticket and nonstudents paid $3 per ticket, how many student tickets were sold?

6. Make up your own problem like the ones in Exercises 1–5.

7. If you have a bunch of 10¢ and 5¢ stamps, and you know that there are 20 stamps and their total value is $1.50, how many of each do you have?

8. This problem is adapted from the Figure This website, which has 80 rich problems that upper elementary and middle school children enjoy: Is there a combination of 15¢ and 33¢ stamps that can make $1.77? (Reprinted with permission from Figure This! (http://www.figurethis.org), copyright 2006 by the National Council of Teachers of Mathematics.)

9. Let's say you are building a patio in your back yard. You have decided to make the patio 12 feet by 9 feet. The local building supply store sells bricks that are 6 inches by 4 inches for 25¢ each. How much will the patio cost?

10. At the beginning of the month, Jack and Jill had $642 in a checking account. During the month, they wrote checks for $22, $53, and $55. They made a withdrawal of $50 from the automatic teller machine, and they made deposits of $142 and $100. How much money did they have at the end of the month? Solve this problem two different ways.

11. Sarah's diet allows 1500 calories per day. Thus far, she has had a glass of milk (90 calories), pancakes (150 calories), an apple (75 calories), a salad (150 calories), salad dressing (200 calories), and a piece of cake (350 calories). Can she have a steak (250 calories) and a salad with salad dressing for dinner? If not, what can she have? Solve the problem two different ways.

12. Sally works 40 hours a week and makes $6.85 an hour, but her kids are in child care for 32 hours a week and the day care center charges her $15 per day. If you deduct her child care expenses, how many dollars per hour does she actually make?

13. Martha planted 14 rows of apple trees, and each row had 21 trees. If each tree yielded an average of 250 apples, how many apples grew in her orchard?

14. A farmer needs to fence a rectangular piece of land. She wants the length of the field to be 80 feet longer than the width. If she has 1080 feet of fencing material, what should be the length and the width of the field?

15. Since its beginning, the U.S. Mint has produced over 288.7 billion pennies.
    a. What if we lined these pennies up. How long would the line be?
    b. The mint currently makes about 30 million pennies a day. How many is this per second? How many is this per year?

16. The Lewis and Clark expedition between 1804 and 1806 covered almost 8000 miles.
    a. If you traveled along their route, how much would you spend on gasoline?
    b. How many days would it take you?

17. The record for the longest migration is held by the arctic tern, which flies a round trip that can be as long as 20,000 miles per year, from the Arctic to the Antarctic and back. If the bird flies an average of 25 miles per hour and an average of 12 hours per day, how many days would it take for a one-way flight?

18. A family is planning a three-week vacation for which they will drive across the country. They have a van that gets 18 miles per gallon, and they have a sedan that gets 32 miles per gallon. How much more will they pay for gasoline if they take the van?
    a. First describe the assumptions you need to make in order to solve the problem.
    b. Solve the problem and show your work.
    c. What if the price of gas rose by 40¢ between the planning of the trip and the actual trip? How much more would the gas cost for the trip?

19. Joni is the owner of Red Oak Furniture, which is doing so well that the present staff is overworked. She has two choices: continue to pay overtime or hire a new person. She now has three employees, and they earn time-and-a-half for all time over 40 hours worked in a week. In the past three weeks, they have worked the number of hours shown. Should she hire a new employee?

	Week 1	Week 2	Week 3
Gerald	47	55	42
Nancy	43	51	48
Jose	56	51	42

20. a. Using each of the numbers 1–9 exactly once, fill in the blanks below:

   b. Find all solutions.

21. For each of the triangles below:
   (1) Describe as many patterns as you can find in the triangle.
   (2) Write the numbers that would be in the next row of the triangle, and explain how you arrived at your answer.

   a.　　　　1
   　　　2　　4
   　　3　　6　　9
   　　4　　8　　12　　16
   　　5　　10　　15　　20　　25

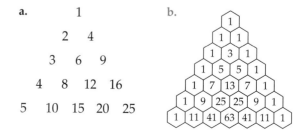

c. 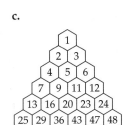   **d.**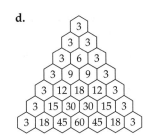

22. List the next term in each of the sequences below, and verbally describe the pattern that enabled you to predict the next term.
    a. □, △, □, □, △,
    b. ⌐, ⌊, ⌋,
    c. 1, 4, 7, 10,
    d. 2, 6, 18, 54,
    e. 1, 4, 9, 16,
    f. 1, 2, 4, 7, 11,
    g. 5, 1, 4, 2, 3, 3,
    h. 4, 5, 8, 13, 20, 29, 40,
    i. 1, 7, 21, 46, 85, 141,
    j. 1, 0, 2, 1, 2, 2, 4,
    k. 1, 3, 6, 10, 15, 21,
    l. 100, 90, 81, 73, 66,
    m. 81, 72, 63, 54, 45,

23. What patterns do you see in this sequence of numbers that are the squares of the first 16 numbers: 1, 4, 9, 16, 25, 36, 49, 64, 81, 100, 121, 144, 169, 196, 225, 256, ...

24. Continue the pattern in the three number sentences below, and then describe as many patterns as you can.
    $3 \times 3 - 1 = 8$
    $5 \times 5 - 1 = 24$
    $7 \times 7 - 1 = 48$

25. Examine the following phrase made famous by Fred Flintstone: y a b b a d a b b a d o o. If this pattern is repeated over and over, what letter will be in the 275th position?

26. Determine the following products: 67×67, 667×667.
    a. Describe the pattern you see.
    b. Predict the pattern of 6667×6667.
    c. This is an example of "what if" and "I wonder." What if the two numbers didn't have the same number of 6s? For example, is there any connection between what we have seen above and, for example, 667×67? Explore this question and report your findings.

27. Determine the following products: 1×1089, 2×1089, 3×1089.
    a. Describe the pattern(s) you see.
    b. Predict the next several products.
    c. Along the "what if" theme, will this pattern continue when you reach 10×1089 and further?

28. Determine the following products: 1×142857, 2×142857, 3×142857.
    a. Describe the pattern(s) you see.
    b. Do more computations until you can predict the next product.
    c. Along the "what if" theme, this pattern seems to break at 7×142857. Continue to find and write products of 142857 and 8, 9, 10, etc. What patterns do you see now?

29. Find a pattern in each of the following sets of equations and predict the next equation.
    a. $1^2 + 2^2 + 2^2 = 3^2$
       $2^2 + 3^2 + 6^2 = 7^2$
       $3^2 + 4^2 + 12^2 = 13^2$
    b. $1^3 + 2^3 = 3^2$
       $1^3 + 2^3 + 3^3 = 6^2$
       $1^3 + 2^3 + 3^3 + 4^3 = 10^2$
    c. $1 + 2 = 3$
       $4 + 5 + 6 = 7 + 8$
       $9 + 10 + 11 + 12 = 13 + 14 + 15$

30. a. In how many different ways can you make change for a quarter?
    b. In how many different ways can you make change for 50 cents?
    c. In how many different ways can you make change for one dollar without using pennies?

31. A special rubber ball is dropped from the top of a wall that is 16 meters high. Each time the ball bounces, it rises half as high as the distance it fell. The ball is caught when it bounces 1 meter high. How many times did the ball bounce?

32. a. On a digital clock, how many times in one day will the numbers be consecutive digits, as they are, for example, in 1:23?
    b. On a digital clock, which digit will appear most often in one day?

33. One of the most famous patterns in mathematics was discovered by the Italian mathematician Leonardo of Pisa (1170–1250), known to us as Fibonacci. He discovered this sequence, which we today call the Fibonacci sequence, while studying the birth rates of rabbits. He posed the following question: Suppose that a pair of rabbits produce a pair of baby rabbits every month, and that rabbits cannot reproduce until they are two months of age. How many pairs of rabbits will you have after one year (assuming, of course, that no rabbits die)? He found that if he listed the number of pairs of rabbits he had after each month, he had a very interesting sequence: 1, 1, 2, 3, 5, 8, 13, 21, 34, 55, 89, 144, 233, ...
    a. Add the first seven Fibonacci numbers. What do you notice?
       Add the first eight Fibonacci numbers. What do you notice?
       Try to describe this relationship both in words and using notation.
    b. Pick three consecutive terms. Multiply the first term and the third term. How does this compare to the square of the middle term? Describe this relationship using notation.

34. Palindromes are numbers whose value is the same backward as forward. For example, 1331 is a palindrome. Most children love explorations with palindromes

because they find the patterns fun and the investigations are so mathematically rich. How many palindromes can you find between 100 and 999?

35. Making palindromes from nonpalindromes is another interesting exploration. For example, take any number, reverse the digits, and add the two numbers together. Then determine how many steps it takes until a palindrome is reached. For example, $38 + 83 = 121$. Thus, we can say that 38 is a 1-step palindrome. On the other hand, 87 is a 4-step palindrome. Explore different 2-digit numbers. Can you describe those numbers that will be 2-step palindromes? 3-step palindromes? etc.?

36. Let's say there are 145 students in your class. If each student greeted and shook hands with every other student in the class, how many handshakes would there be?

37. This variation of the handshakes problem is on video 17 of *Teaching Math: A Video Library, K–4*, which many colleges and schools have: "If there are 20 children in a classroom and every child gives every child a valentine, how many valentines are distributed that day?" Does this problem have the same answer as the handshakes problem? Why or why not?

38. This problem was posed in the November 2001 "Problem Solvers" section of *Teaching Children Mathematics:* In the Animal Football League, the teams can score only 3-point field goals and 7-point touchdowns; no safeties or extra-credit points are allowed. In one contest, the Anteaters defeated the Bobcats 42 to 37. In how many different ways could each team have arrived at its final score?

   a. Do the problem yourself before reading on.
   b. Solve the problem using the approach described by this fifth grader and shown in the table: "I found the different ways that teams could arrive at the score by making a chart and checking each possibility. After subtracting the touchdown points from the total score, the difference would have to be divisible by 3 to be a possible combination."

	Touchdowns	Field Goals
1.	7	3
2.	14	6
3.	21	9
4.	28	12
5.	35	15
6.	42	18
7.		21
8.		24
9.		27
10.		30
11.		33
12.		36
13.		39
14.		42

This problem has several extensions:

   c. What scores are impossible to get?
   d. What is the greatest score that is impossible to get? Prove it.

39. This problem was posed in the April 2001 "Problem Solvers" section of *Teaching Children Mathematics:* Lori and Betty are pretending to have a garage sale. They found five 30¢ price tags and five 40¢ price tags left over from their mothers' last sale. If they use these price tags on their sale items, how many different amounts of money could they make from their sale?

   a. Solve the problem yourself.
   b. The two tables below illustrate one child's strategy. Continue this strategy to determine the answer. Describe this strategy as if to someone who missed class and doesn't understand it. Critique the strategy. That is, cite its strengths and weaknesses (or limitations).

30	30 + 40
30 + 30	30 + 30 + 40
30 + 30 + 30	30 + 30 + 30 + 40
30 + 30 + 30 + 30	30 + 30 + 30 + 30 + 40
30 + 30 + 30 + 30 + 30	30 + 30 + 30 + 30 + 30 + 40

   c. The table below shows another student's solution. Combine this strategy with being systematic to find all solutions.

Possibilities for 40¢

Possibilities for 30¢ +	0	1	2	3	4	5
0	0	40	80	120	160	200
1	30	70	110	150	190	230
2	60	100	140	180	220	260
3	90	130	170	210	250	290
4	120	160	200	240	280	320
5	150	190	230	270	310	350

   d. Use this table to find all the possibilities. Explain this table so that someone who doesn't understand it could use it.

40. This problem was explored in the October 1998 issue of *Teaching Children Mathematics*, pp. 105–107: South Main School has 618 students. Each student will have his or her picture taken next week by a photographer hired by the school. The photographer uses rolls of film having twenty-four exposures each. How many rolls of film will the photographer have to buy?

   Source: Reprinted with permission from *Teaching Children Mathematics,* October 1988, copyright 1988 by the National Council of Teachers of Mathematics.

SECTION 1.5 REASONING AND PROOF

WHAT DO YOU THINK?

- Do you think young children can do proofs?
- How can the classroom nurture intuitive reasoning?

Susan Jo Russell, who has worked extensively with elementary school children and teachers, states that "[M]athematical reasoning is about the development, justification, and use of mathematical generalizations."[10] Young children naturally make generalizations both in and out of the classroom. For example, they conclude that when the doorbell rings, someone is outside the door, and when they get into the car, they must be strapped in. Elementary school children regularly make *mathematical* generalizations. For example, they learn that when you add two numbers, you get the same amount when you "add forward as well as backwards" and that if you add two odd numbers together, you get an even number.

INVESTIGATION 1.6 Does Your Answer Make Sense?

This problem has a history that I will share after you solve it. All 261 fifth-graders in a school are going on a field trip, and each bus can carry 36 children and 4 adults. How many buses are needed? Do this problem before reading on. . . .

DISCUSSION

If you divide 261 by 36, you get 7.25. When this problem was given to seventh graders in one of the National Assessments of Educational Progress, a majority of children gave 7 as the answer. Why do think they did? . . . Many of them had been taught the rules of rounding mechanically, and thus they rounded 7.25 to 7. While the mathematical rule of rounding is to round 7.25 down to 7, the real-life application of this computation requires us to round 7.25 up to 8. When solving problems, it is always necessary not only to check your answer but your reasoning: Does the answer make sense? Does my reasoning make sense? The cartoon below illustrates this need quite comically.

With this example in mind, read what the NCTM has to say about reasoning. As with the other standards, check to see the extent to which

FIGURE 1.8*

[10] Susan Jo Russell, "Mathematical Reasoning in the Elementary Grades," in *Developing Mathematical Reasoning in Grades K–12*, ed. Lee V. Stiff, Reston, VA: NCTM, 1999, p. 1.
*© NAS. Reprinted with special permission of North American Syndicate.

these words make sense and the extent to which you buy them, and then read on. . . .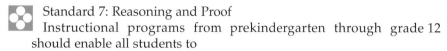

> Standard 7: Reasoning and Proof
> Instructional programs from prekindergarten through grade 12 should enable all students to
>
> - recognize reasoning and proof as fundamental aspects of mathematics; [This is crucial. You will see how children can and do "prove" conjectures, and you will come to understand the role of proof in the elementary classroom.]
> - make and investigate mathematical conjectures; [This is something children do naturally about the world. Much of this develops from looking for and making sense of patterns.]
> - develop and evaluate mathematical arguments and proofs; [Formal proofs don't happen until high school, but children can construct informal proofs.]
> - select and use various types of reasoning and methods of proof as appropriate. [We will examine several examples in this section.]
>
> (*Principles and Standards*, p. 56)

At the heart of this standard of Reasoning and Proof must be the belief that all mathematical ideas can make sense. I realize that many students come to college believing that making sense of mathematics is reserved only for the "smart" students; for the rest of the students, the best advice is "just do it." To me that is one of the maladaptive beliefs about mathematics.

INVESTIGATION 1.7 — Making Generalizations

Look at the following sequence of attribute blocks,[11] a manipulative commonly found in elementary classrooms. What patterns do you see? Write them down and then read on. . . .

DISCUSSION

When asked to describe the patterns, one child said, "there's triangle, triangle, circle, circle, square, square. . . . Then there's small, large, small, large, small, large. . . . Then there's thin, thick, thin, thick, thin, thick. . . . The fourth pattern

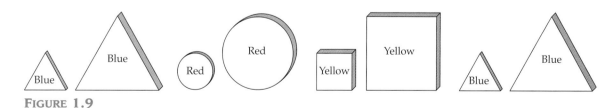

FIGURE 1.9

---

[11] National Council of Teachers of Mathematics, *Principles and Standards for School Mathematics*, Reston, VA: NCTM, 2000, p. 123.

is blue, blue, red, red, yellow, yellow." Such a problem is wonderful because there is more than one correct response. Another student showed the kind of reasoning we are discussing—the ability to generalize. She stated, "I think there are just two patterns. See, the shapes and colors are an AABBCC pattern." She proceeded to state that sizes and thin-thick are an ABABAB pattern. "So you really only have two different patterns." The first child was not wrong. However, the second child took the patterns to the next level of abstractness seeing that the shapes and colors both fit a more general description of the same pattern.

Children will encounter many different kinds of mathematical reasoning in their journey. Mathematical reasoning is often categorized as quantitative (using quantities) or qualitative (when quantities are not used). For example, reasoning that 49 + 36 = 50 + 35 involves quantitative reasoning, while determining that if you peel the label off a soup can the shape will be a rectangle involves qualitative reasoning. Many authors break mathematical reasoning down into more specific kinds, for example, numerical, proportional, algebraic, and spatial. Your competency at these kinds of reasoning will grow over the course of this book.

Doing mathematics involves discovery. The ability to see patterns and then make conjectures is an important part of developing mathematical reasoning. It is essential that children make, refine, and test conjectures in elementary school. Questions like: What do you think will happen next? What is the pattern? Is this true always? Sometimes?[12] are helpful and we will apply them in the following situation.

A broader way to describe the different kinds of reasoning is to talk about inductive reasoning, deductive reasoning, and intuitive reasoning. We will discuss these in turn. For most people, inductive reasoning in the most accessible form of reasoning, so we will begin there.

Inductive Reasoning

Inductive reasoning is a process of coming to a general conclusion from seeing patterns in specific examples and looking for the regularity in those patterns. This kind of reasoning is crucial in the child's construction of the world, and it involves making generalizations from seeing patterns in specific examples.

Let's examine a pattern you may not have seen in this context.

INVESTIGATION 1.8 Inductive Thinking with Fractions

Let's say you don't know how to add fractions with different denominators. Look at the three examples below. Can you see a pattern that would enable you to add other fractions? Think about this before reading on. . . .

$$\frac{1}{3} + \frac{1}{4} = \frac{7}{12}$$

$$\frac{1}{5} + \frac{1}{7} = \frac{12}{35}$$

$$\frac{1}{2} + \frac{1}{5} = \frac{7}{10}$$

---

[12] National Council of Teachers of Mathematics, *Principles and Standards for School Mathematics*, Reston, VA: NCTM, 2000, p. 57.

Classroom Connection

When I wrote this book, one of the criteria that I used when selecting problems for investigations and explorations was that the problems should be extendible. This gets to the "what if" and "I wonder" questions that young children naturally ask often. Unfortunately, this frequency seems to decline steadily as we get older. One of my goals in this course is that you resolve to reawaken this curiosity in mathematics. In this case, a "what if" question is "What if the numerators are not 1 but are the same, as in $\frac{2}{3} + \frac{2}{5} = ?$ This will be left as an exercise.

DISCUSSION

We can describe the pattern in words by saying that whenever you have two fractions with a 1 in the numerator, the numerator of the sum is found by adding the two denominators, and the denominator of the sum is determined by multiplying the two denominators. I wrote this description in everyday English. More formally, we would say that when you add two unit fractions whose denominators are relatively prime, the numerator of the sum is equal to the sum of the denominators of the two fractions, and the denominator of the sum is equal to the product of the denominators of the two fractions. Unit fractions are fractions whose numerators are 1, and two numbers that are relatively prime have no common factors other than 1. The pattern also works when the denominators are not relatively prime, but then the answer is not in simplest form; an example is

$$\frac{1}{4} + \frac{1}{6} = \frac{10}{24} = \frac{5}{12}.$$

The result can be expressed more succinctly with notation:

$$\frac{1}{x} + \frac{1}{y} = \frac{x+y}{xy}.$$

This apparently innate tendency to draw generalizations from specific examples needs careful attention in schools and is one of the many reasons why elementary teachers need a strong background in elementary mathematics. For example, consider explaining what an isosceles triangle is to young children. If all isosceles triangles shown to them have the base parallel to the bottom of the page, as do the first three triangles below, children will mistakenly not consider the triangle at the far right to be isosceles, although it is.

INVESTIGATION 1.9

Sequences, Patterns, Reasoning, and Proof

Do you recall the numerical sequence that we examined in Investigation 1.3? Young children can explore many similar sequences and then can determine the pattern and continue the sequence. Older students in elementary school can explore and figure out even more complex patterns. Let us examine three sequences here to understand how reasoning can help us go beyond simply noticing a pattern to understanding the structure of the sequence. For each of the sequences below, see whether you can determine the next number in the sequence, the 20th number (term) in the sequence, and then the nth term. Then read on. . . .

Sequence 1: 2, 5, 8, 11, 14, . . .
Sequence 2: 3, 6, 12, 24, 48, . . .
Sequence 3: 4, 9, 19, 39, 79, . . .

DISCUSSION

Sequence 1   Let us examine the first sequence: 2, 5, 8, 11, 14, ... It is not terribly difficult to determine that the 6th term will be 17. One way to determine the 20th term is simply to continue the sequence until you get to the 20th term.

This is a bit tedious, though you could program a spreadsheet to do the tedious part for you. With a bit of analysis, though, you could realize that you need to add 15 more 3s after the 5th term, and thus, the value of the 20th term is 14 + 45. If you saw this, great; if you didn't, look at the second row of Table 1.10, where I have not written the actual number but have broken down each term in the sequence so you can see *how it came to be*. Now, using the fact that multiplication is repeated addition, we can represent the terms more economically. This not only saves time but also begins to reveal the mathematical structure of the sequence. Do you see this? Do you understand the nth term now? If not, the last row might help. Here I have shown in bold the number that tells how many 3s. Notice that this number is always 1 less than the number representing the position of the term in the sequence. That is, the 4th term has three 3s, the 5th term has four 3s, etc. Thus, the nth term must have $(n - 1)$ 3s.

TABLE 1.10

1st	2nd	3rd	4th	5th	nth
2	5	8	11	14	
					$(n-1)$ times
2	2 + 3	2 + 3 + 3	2 + 3 + 3 + 3	2 + 3 + 3 + 3 + 3	2 + 3 + 3 + ⋯ + 3
2	2 + 3	2 + 2 · 3	2 + 3 · 3	2 + 4 · 3	2 + (n − 1) · 3
2	2 + 3	2 + **2** · 3	2 + **3** · 3	2 + **4** · 3	2 + (**n − 1**) · 3

> ■ *Language* ■
>
> We can describe arithmetic sequences in different ways. For example, we can say that the relationship between each term and the following term is that we add the same number each time. On the other hand, we can say that the difference between each pair of consecutive terms is constant.

Sequences like the one above, where the difference between each pair of consecutive terms is constant, are called **arithmetic sequences**. What is different is the starting number and the common difference. If we represent the starting number by a and the common difference by d, then we can state a rule for finding the nth term of any arithmetic sequence, as shown in the last column of Table 1.11.

TABLE 1.11

1st	2nd	3rd	4th	5th	nth
a	$a + d$	$a + 2 \cdot d$	$a + 3 \cdot d$	$a + 4 \cdot d$	$a + (n - 1) \cdot d$

Sequence 2 Now let's examine the second sequence: 3, 6, 12, 24, 48, ... You probably realized that this sequence is not "just like" the preceding one. If so, you are right. They are similar in that there is a pattern and there is a relationship between each term and the next. In this case, however, the relationship is that each term is twice, or double, the preceding term.

As before, in the second row of Table 1.12, I have broken down each term in the sequence so you can see how it came to be. Now, using the fact that exponentiation is repeated multiplication, we can represent the terms more economically. This not only saves time, but it also begins to reveal the mathematical structure of the sequence. Do you see this? Do you understand the nth term now?

Sequences like the one above, where the relationship between each term and the following term is that they always have the same ratio (in this case 2),

Table 1.12

1st	2nd	3rd	4th	5th	nth
3	6	12	24	48	
3	$3 \cdot 2$	$3 \cdot 2 \cdot 2$	$3 \cdot 2 \cdot 2 \cdot 2$	$3 \cdot 2 \cdot 2 \cdot 2 \cdot 2$	$3 \cdot \overbrace{2 \cdot 2 \cdots \cdot 2}^{(n-1) \text{ times}}$
3	$3 \cdot 2$	$3 \cdot 2^2$	$3 \cdot 2^3$	$3 \cdot 2^4$	$3 \cdot 2^{(n-1)}$

Table 1.13

1st	2nd	3rd	4th	5th	nth
a	$a \cdot r$	$a \cdot r^2$	$a \cdot r^3$	$a \cdot r^4$	$a \cdot r^{(n-1)}$

are called **geometric sequences**. What is different is the starting number and the common ratio, also called the unit (rate) of increase. If we represent the starting number by a and the common ratio by r, then we can state a rule for finding the nth term of any geometric sequence, as shown in the last column of Table 1.13.

Sequence 3 Finally, let us examine the third sequence: 4, 9, 19, 39, 79, ... Some students call this a hybrid sequence in that it is not "just like" either of the ones above. That is, the "bad news" is that we can't use either formula just developed. The "good news" is that we can use problem-solving tools and reasoning to determine the nth term. Try it yourself before reading on. . . .

One way of describing what repeats is to look at the relationship between each term and the following term. That is, what do you have to do to each term to produce the next term? One way of describing the operation is to say that you double each term and then add 1. We can represent this relationship between each term and the following term as $2n + 1$. This is nice, but it won't help us to determine the value of the 20th term or the nth term. Thus we have to look further. Here is where number sense and intuition come into play in mathematics. You might have realized that, after the first term, all of the terms end in 9. What if we wrote out a similar sequence—one in which each term is 1 more than the terms of our sequence. This has been done in the third row of Table 1.14. What do you see?

Table 1.14

	1st	2nd	3rd	4th	5th	nth
Original sequence	4	9	19	39	79	
Broken down	4	$4 \cdot 2 + 1$	$9 \cdot 2 + 1$	$19 \cdot 2 + 1$	$39 \cdot 2 + 1$	
Similar sequence	5	10	20	40	80	
Broken down	5	$5 \cdot 2$	$5 \cdot 4$	$5 \cdot 8$	$5 \cdot 16$	
Using exponents	5	$5 \cdot 2$	$5 \cdot 2^2$	$5 \cdot 2^3$	$5 \cdot 2^4$	$5 \cdot 2^{(n-1)}$
Original sequence	$5 - 1$	$(5 \cdot 2) - 1$	$(5 \cdot 2^2) - 1$	$5 \cdot 2^3 - 1$	$5 \cdot 2^4 - 1$	$5 \cdot 2^{(n-1)} - 1$

Either by breaking each number apart ($5, 5 \cdot 2, 5 \cdot 2 \cdot 2, 5 \cdot 2 \cdot 2 \cdot 2, 5 \cdot 2 \cdot 2 \cdot 2 \cdot 2$) or by realizing that this is a geometric sequence where $a = 5$ and $r = 2$, we can determine that the nth term of this similar sequence is $5 \cdot 2^{(n-1)}$. Then we can see that the nth term of the original sequence (4, 9, 19, 39, . . .) is simply 1 less; that is, $5 \cdot 2^{(n-1)} - 1$.

Deductive Reasoning

Let us begin our study of **deductive reasoning** with an amusing but true story. When Julia was three, she was taking a ride in the country with her mother. She looked out the window and saw a cow, but she had never seen a cow before. Thus, she exclaimed to her mother, "Look, mom, a big dog!" Do you understand Julia's mistake? Can you explain it? By the end of this section, you will be able to do this.

Julia had generalized, from her limited experience in the city, that all animals that have four legs and a tail are dogs. We can translate this sentence into if–then language as follows: If an animal has four legs and a tail, then it is a dog.

Statements of the form "if p, then q" are called **conditional statements**. The "if" part of a conditional is called the **hypothesis** of the implication, and the "then" part is called the **conclusion**.

All if–then statements have three interesting "relatives." In Julia's case, if we switched the if and then clauses, we would have a true statement: If it is a dog, then it has four legs and a tail.[13] These two statements (Julia's and the new one) are said to be *converses* of each other. A statement and its converse are often not both true. Consider some examples.

Statement	Converse
If it is a fruit, then it contains sugar.	If it contains sugar, then it is a fruit.
If it is a newborn baby, then it cries at night.	If it cries at night, then it is a newborn baby.
If it is cotton candy, then it has no protein.	If it has no protein, then it is cotton candy.

There are two other statements that have an interesting relationship to the original statement. Let us consider the original statement here to be the true one: If it is a dog, then it has four legs and a tail. We can write the **converse**, **inverse**, and **contrapositive** of any if–then statement. The second column of Table 1.15 shows the converse, inverse, and contrapositive in everyday English; the third column shows the statements in shorthand, where the hypothesis is denoted by p and the conclusion by q; and the fourth column shows the statements in their most succinct form. I hope you see that this last form enables us to grasp more easily the relationships among the four statements.

As we just saw, the converse of a true statement is not always true. What about the inverse and the contrapositive? What do you think?

---

[13]Here, we are assuming "typical" dogs. That is, we are ignoring three-legged dogs, which have lost a leg as a result of injury or disease.

Section 1.5 / Reasoning and Proof 37

> **Outside the Classroom**
>
> Most advertisers want consumers to believe the converse of a statement. That is, an advertiser might imply: If you drink Bleech beer, you will be cool. The converse of this statement is: if you want to be cool, drink Bleech beer.

TABLE 1.15

	Verbal	If–then	Most succinct representation
Statement	If it is a dog, then it has four legs and a tail.	If p, then q.	$p \rightarrow q$
Converse	If it has four legs and a tail, then it is a dog.	If q, then p.	$q \rightarrow p$
Inverse	If it is not a dog, then it doesn't have four legs and a tail.	If not p, then not q.	$\sim p \rightarrow \sim q$
Contrapositive	If it doesn't have four legs and a tail, then it is not a dog.	If not q, then not p.	$\sim q \rightarrow \sim p$

> **Language**
>
> It is a convention among mathematicians to represent statements with lower-case letters.

It turns out that a statement and its contrapositive are *logically equivalent*. That is, if a statement is true, the contrapositive of that statement is also true. Try this with the three examples described earlier. If you are reading this section actively, you will cover the right-hand column and see whether you can make the sentences yourself.

Statement	Contrapositive
If it is a fruit, then it contains sugar.	If it doesn't contain sugar, then it is not a fruit.
If it is a newborn baby, then it cries at night.	If it doesn't cry at night, then it is not a newborn baby.
If it is cotton candy, then it has no protein.	If it has protein, then it is not cotton candy.

Before you tune out because this is getting a bit formal, let me connect conditional statements to everyday life. Try to think of such examples. . . .

If you curse at the teacher, then you will get punished. The proverb, "The early bird gets the worm" can be restated as a conditional statement: If you are the early bird, then you will get the worm. In fact, many of our beliefs and actions are equivalent to if–then statements.

Deductive reasoning is a process of reaching a conclusion from one or more statements, called the hypotheses. An **argument** is a set of statements in which the last statement is called the conclusion and there is one or more hypotheses. The laws of deductive reasoning enable us to determine whether the reasoning in an argument is **valid** or **invalid**. Below are the four possibilities, stated in their general form at the left and with a mathematical example at the right. Case 1 and Case 2 show examples of valid arguments.

Case 1 In this case, we have a general statement that has been determined to be true, and we have a specific instance (example) of the first part of the statement. In this case, the conclusion is true. This law of logic is called the **Law of Detachment**. It is also called *affirming the hypothesis*.

If p, then q.	If a figure is a square, then it has four sides.
p is true.	This figure is a square.
Then q is true.	Therefore, this figure must have four sides.

Case 2 This law of logic is called **Modus Tollens**. It is also called *denying the conclusion*. Essentially, this is a restatement of what we saw earlier: The contrapositive of a true statement is also true. That is, if we know that $p \rightarrow q$, then we know that $\sim q \rightarrow \sim p$.

If p, then q.	If a figure is a square, then it has four sides.
q is not true.	This figure does not have four sides.
Then p is not true.	Therefore, this figure is not a square.

Case 3 and Case 4 show examples of two common mistakes (equating a statement and its converse or inverse) that lead to invalid arguments. In each of the cases, we can demonstrate the falsity of the conclusion with a counterexample.

Case 3 This kind of error in reasoning is called the *fallacy of affirming the consequence*.

If p, then q.	If a figure is a square, then it has four sides.
q is true.	This figure has four sides.
Then p is true.	Therefore, this figure is a square.

One counterexample is a trapezoid, which has four sides but is not a square.

Case 4 This kind of error in reasoning is called the *fallacy of denying the antecedent*.

If p, then q.	If a figure is a square, then it has four sides.
p is not true.	This figure is not a square.
Then q is not true.	Therefore, this figure does not have four sides.

One counterexample is a parallelogram, which is not a square but does have four sides.

Stop for a moment. Do you really understand what you just read or do you "sort of" understand it? Does it feel relevant? While you are not likely to teach these four cases, their value is in helping you to avoid logical errors that adults and children commonly make in their mathematical reasoning. Essentially, this branch of reasoning enables us to check the logic of our conclusions. That is, we make a conclusion that we think is valid. We can translate the statements (the given information that we already know to be true) into one of the four cases. If it fits case 1 or 2, then our conclusion is valid (as long as the preceding statements are true). If it fits case 3 or 4, then the conclusion is not valid, that is, it is not necessarily true.

An important kind of reasoning that is necessary for success in mathematics as well as many other fields is called *analytic thinking*. A common definition of analytic thinking is to break a complex idea into its component parts *and their relationships*. You will find this happening repeatedly throughout the book, whether it be the structure of base 10, the meanings of multiplication, the differ-

> ■ *History* ■
>
> The origin of formal logic can be traced back to the Greeks. Aristotle (384–322 B.C.) wrote a book that summarized principles of reasoning and laws of logic. Modern mathematics is built on laws of reasoning. As you will discover throughout this book, much of the mathematics that we use today and often take for granted is relatively recent. For example, the notation that we use dates back to the German mathematician Gottlob Frege (1848–1925).

ent ways that fractions can be interpreted, or understanding symmetry. Analytic thinking is an important tool in problem solving: take a complex problem and break it into parts (smaller pieces); then you put those pieces back together, which means you have to pay attention to how they are related to each other.

Venn diagrams Now let's examine conditional statements from another perspective. Recall our discussion about representations in Section 1.4. We have just seen different representations of conditional statements (if p, then q and $p \rightarrow q$). We can also represent conditional statements (for example, the four statements concerning dogs and four legs and a tail) with **Venn diagrams** (which we will investigate in more detail in Chapter 2). The Venn diagram in Figure 1.10 represents the valid relationship between all animals, those with 4 legs and a tail, and dogs. All animals that have four legs and a tail that are not dogs are inside the large circle but outside the dog circle. All other animals are inside the square but outside the circles. This seems rather formal and technical, but it is a good representation of what is going on inside Julia's head and the heads of other children as they are developing more and more sophisticated understandings of how animals are related. Later, other categories are developed, such as mammal, reptile, invertebrate, and so on.

The four Venn diagrams in Figure 1.11 illustrate the four statements in Table 1.15.

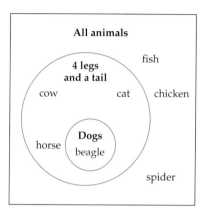

Figure 1.10

> ▪ *Language* ▪
>
> When using Venn diagrams in logic, many texts use the term *Euler diagram* in honor of the mathematician Leonard Euler (1707–1783), who popularized the use of Venn diagrams in logic.

Statement	Converse	Inverse	Contrapositive
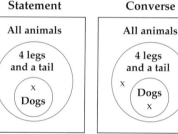		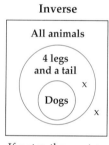	
If p, then q.	If q, then p.	If not p, then not q.	If not q, then not p.
$p \rightarrow q$	$q \rightarrow p$	$\sim p \rightarrow \sim q$	$\sim q \rightarrow \sim p$
If it is a dog, it has four legs.	If it has four legs, and a tail, it is a dog.	If it is not a dog, it doesn't have four legs and a tail.	If it doesn't have four legs and a tail, it is not a dog.

Figure 1.11

The first Venn diagram represents the true statement: If it is a dog, then it has four legs and a tail. Because all dogs have four legs and a tail, the small circle is entirely within the larger circle.

The second Venn diagram illustrates why the converse (If it has four legs and a tail, then it is a dog) is not necessarily true. There are two regions (denoted by the x's) that satisfy the statement "animals having four legs and a tail": inside the smaller circle and inside the larger circle but outside the smaller circle. This latter region is where the counterexamples come from: sheep, cows, horses, etc.

The third Venn diagram illustrates why the inverse (If it is not a dog, then it doesn't have four legs and a tail) is not necessarily true. There are two regions (denoted by the x's) that satisfy the statement "animals that are not dogs": inside the larger circle but outside the smaller circle and outside both circles. This former region is where the counterexamples come from. That is, sheep, cows, and horses are not dogs, but they do have four legs and a tail.

The fourth Venn diagram illustrates why the contrapositive (If it doesn't have four legs and a tail, then it is not a dog) is true. There is only one region for "doesn't have four legs and a tail"; outside both circles. And all animals in this region (such as chickens and fish) are definitely not dogs.

To the extent that you are always asking "does this make sense?" and looking to connect what you already know to what you are learning, your understanding will be deeper. One connection that students have told me is helpful is to realize that when representing the if–then hypothesis with a Venn diagram, the if part is the small circle and the then part is the larger circle. For example, "all humans are mammals" translates to the following conditional statement: If an animal is a human, then it is a mammal, which then translates to a Venn diagram where the mammal circle is the larger circle and the human circle is the smaller circle inside the large circle.

INVESTIGATION 1.10 — Deductive Reasoning and Venn Diagrams

Let us investigate two arguments to see how Venn diagrams can help determine the validity of an argument and to see how Venn diagrams connect to the other representation of logical arguments. For each argument, first represent the situation with a Venn diagram and determine whether the argument is valid or not valid. Then, determine whether the argument matches case 1, 2, 3, or 4. Do you get the same conclusion in both cases?

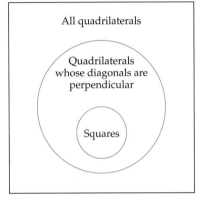

FIGURE 1.12

Argument 1:

If a quadrilateral is a square, then the diagonals are perpendicular.
The diagonals of this quadrilateral are perpendicular.
Therefore this quadrilateral is a square.

Argument 2:

If a number is divisible by 4, then it is also divisible by 2.
79 is not divisible by 2.
Therefore, 79 is not divisible by 4.

DISCUSSION

The first argument fits Case 3, and thus the conclusion is not valid.

If p, then q.
q is true.
Therefore, p is true.

It can be represented by the Venn diagram in Figure 1.12.
The first argument fits Case 2, and thus the conclusion is valid.

If p, then q.
q is not true.
Therefore, p is not true.

It can be represented by the Venn diagram in Figure 1.13.

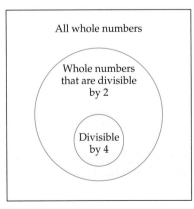

FIGURE 1.13

Biconditional statements There is one other concept from deductive reasoning that we will discuss in this chapter: the *biconditional* statement. A **biconditional statement** occurs when a statement and its converse are both true. This is especially relevant in mathematics: Most mathematical definitions are biconditional statements.

Consider the following two statements:

If two lines intersect to form right angles, then they are perpendicular.

If two lines are perpendicular, then they intersect to form right angles.

In this case, the statement and its converse are always both true. We can use "if and only if" language to combine them into the following biconditional statement, which is also the definition of perpendicular:

Two lines are perpendicular iff they intersect to form right angles.

The shorthand for "if and only if" is iff. So when you see iff in this book, realize that it is not a typographical error!

The following investigation, which represents a common discovery in first grade, will enable us to connect what many of you call common sense to formal deductive reasoning. This investigation also serves to illustrate that young children are often capable of proving their assertions.

INVESTIGATION 1.11 — Why Does Odd + Odd Equal Even?

Let us investigate the assertion that an odd number plus an odd number is an even number. How would you prove this assertion? Think before reading on. . . .

DISCUSSION

Many elementary children can offer a verbal proof that goes something like this: An odd number means that when you cut it in half, you have one left over. So if you add two odd numbers together, you will have two left over, so give one to each part and now the two parts are equal.

Some children prove the assertion spatially using Cuisenaire rods, another manipulative commonly found in elementary schools. All the children begin with an assertion: An odd number is a number that cannot be divided by two with no remainder, or an alternative formulation: A number is odd if it can't be cut into two equal whole numbers. Using Cuisenaire rods, the child picks a rod that represents an odd number, for example, 5. She then substitutes a 3 and a 2 rod for the 5 rod. She does the same for another odd number, for example, 9, substituting the 5 and 4 rods, for the 9 rod. The diagram below represents her argument: When I add these two odd numbers together, they make two equal halves, so the new number must be even.

FIGURE 1.14

Let us now investigate an algebraic proof that is a natural extension of the ones above (just as the two equations for solving the pigs-and-chickens problem was a natural extension of the trial and error approach). The general form for an odd number is $2a + 1$, where a can be any whole number. Do you see why? ...

We represent odd numbers algebraically this way because we know that $2a$ will *always* be an even number, no matter what the value of a is. Therefore, $2a + 1$ will always be odd, no matter what the value of a is.

The proof goes like this:

$$(2a + 1) + (2b + 1) = 2a + 2b + 2 = 2(a + b + 1).$$

Since I know that many students struggle to connect the algebraic notation to everyday English, let us do so here. Algebraically, we are saying that if we begin with any two different odd numbers, represented as $2a + 1$ and $2b + 1$, we can rearrange the terms to show that their sum can be represented as two times some amount. But any number that is two times some amount *must be* an even number!

This part of the proof can also be represented using the Law of Detachment:

If a number can be divided by two with no remainder, then it is an even number.

$2(a + b + 1)$ can be divided by two with no remainder.

Therefore $2(a + b + 1)$ is an even number.

Since we know that $2(a + b + 1)$ is an even number, we know that $(2a + 1) + (2b + 1)$ must be an even number because we have shown that they are mathematically equal. Therefore, we can conclude that an odd number plus an odd number is an even number.

Intuitive Reasoning

There is one other type of reasoning that occurs a lot in mathematics. **Intuitive reasoning** is not well understood, and we know very little about how to "teach" it. Let's look at a couple of examples.

INVESTIGATION 1.12

The Nine Dots Problem

Not all problem-solving involves computation and formulas, as this investigation shows.

Without lifting your pencil, can you go through all nine dots in Figure 1.15 with only four lines?

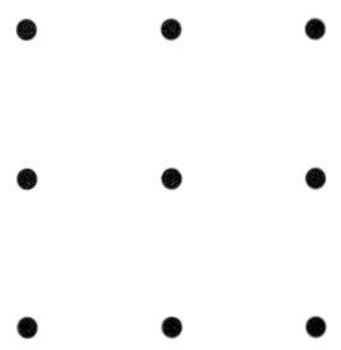

• • •

• • •

• • •

FIGURE 1.15

DISCUSSION

This is a very famous problem, which some of you may have already encountered because of its moral: This problem is impossible to solve as long as you "stay inside the box." That is, there is no possible way to go through all nine dots with four lines if you stay within the box. In order to solve the problem, you need to go "outside the box." If you haven't solved the problem yet, try to work with this hint. . . .

The solution to the problem can be seen on page 60. This idea of not getting stuck inside the box is crucial to good problem-solving. In many real-life problems, the solution to a problem requires that people think about the problem differently.

In this case, inductive reasoning is not very relevant. We don't have other examples before us from which we can generalize. Classic deductive reasoning is not applicable here. At some point, either out of desperation or after concluding that no solution within the square is possible, some people solve this puzzle by trying solutions that go outside the boundaries of the invisible square surrounding this set of dots. This is similar to figuring out the nth term of the 4, 9, 19, 39, 79, ... sequence by "feeling" its connectedness to the 5, 10, 20, 40, 80, ... sequence.

Let me give another example. Believe it or not, even though I am a mathematics teacher, I hated algebra in high school, mostly because I had poor teachers. The following kind of problem almost killed me in Algebra 2 because I couldn't solve it "the teacher's way," which then was "the right way" and the only way you could get credit. The actual problem (I have saved it!) was even more complex, but I have simplified the numbers here to illustrate the point.

The problem: A grocer has a barrel of peanuts that sell for $2 a pound and a barrel of cashews that sell for $5 a pound. If he wants to make 60 pounds of a mixture of the two that he will sell for $3 a pound, how much of each should he mix?

Now the classic way to approach this problem is to set up two equations in two unknowns and then solve. I will let p stand for the number of pounds of peanuts and c stand for the number of pounds of cashews (in my day we had to use x and y, even though p and c make it easier to remember which is which).

Equation 1: $p + c = 60$

Equation 2: $(2p + 5c) = 3 \cdot 60$

FIGURE 1.16

Try as I might, I just couldn't remember how to get Equation 2 for this kind of problem. However, I could get the answer using a way that "just came to me" and that I could not explain. I intuitively saw the problem in terms of ratios. What stood out for me was that $2 is only $1 away from $3 and that $5 is $2 away from $3. That is, 5 is twice as far from 3 as 2 is. Figure 1.16 represents this relationship visually.

Therefore, the correct answer must have twice as many pounds of peanuts as pounds of cashews. It took only a few rounds of "guess–check–revise" to determine that 40 pounds of peanuts and 20 pounds of cashews was the answer.

With practice, I was able to refine my method for even more complex problems. However, because I couldn't do it "the right way," I received no credit. I remember getting a 37 on one test. The only reason I passed the course is that the teacher graded on the curve. The starting guard on the basketball team was in the class; I almost always beat him, and whatever he got was a C. [Double standards for athletes have been and continue to be a common practice in many high schools and colleges.] The only reason I stayed in mathematics was that we moved, and my teacher in the new school was fascinated by my way and helped me to understand it and prove it. In fact, he encouraged me to submit it as an original piece of work in the science fair, which I won. Had my family not moved in the middle of that school year, I would probably not be a college mathematics teacher!

Even though intuition is not well understood, I believe it deserves mention in this course because many, if not most, mathematicians have experienced the power of intuition in their own mathematical work, and many important mathematical (and scientific) discoveries have been the result of what we call intuition. At the same time, as Blaise Pascal, a member of the Mathematics Hall of Fame, once said, "Chance favors the prepared mind." This statement has been interpreted by many mathematics educators to mean that intuition generally does not spring from a void. It generally comes because the person did have a lot of relevant information but wasn't making the connections at a conscious level. I find the analogy of islands in the ocean to be useful here. If you look at islands in an ocean from one perspective (aerial view), they are separate—that is, not connected. However, if you look at them from another perspective (topographical or relief map), they are like peaks of a single mountain range.

SECTION 1.6 COMMUNICATION

WHAT DO YOU THINK?

- Why does the NCTM encourage mathematics classrooms to provide numerous opportunities for communication?

Read the standard on Communication and consciously ask yourself whether the statements make sense. Are they meaningful? To what extent do you agree with them? Then read on. . . .

Standard 8: Communication
Instructional programs from prekindergarten through grade 12 should enable all students to

- organize and consolidate their mathematical thinking through communication;
  [Being able to communicate doesn't just come from thin air!]
- communicate their mathematical thinking coherently and clearly to peers, teachers, and others;
  [Most mathematical notation and vocabulary has been invented to make communication simpler.]
- analyze and evaluate the mathematical thinking and strategies of others;
  [Many of my students say they learn as much from their peers as they do from me.]
- use the language of mathematics to express mathematical ideas precisely.
  [The precision of mathematics ultimately makes it easier, not more difficult, to say what you see.]

(*Principles and Standards*, p. 60)

Albert Einstein is said to have remarked that "a description in plain language is a criterion of the degree of understanding that has been reached." Communication and understanding go hand in hand. "When students are challenged to think and reason about mathematics and to communicate the results of their thinking to others verbally or in writing, they are faced with the task of stating their ideas clearly and convincingly to an audience" (*Principles and Standards*, p. 85). Implicit in the previous sentence is that there are two distinct kinds of communication that you need to be aware of as you work in this course.

1. **Communicating with yourself about the problem**—that is, being able to make sense of your own strategies and solutions.

   If you go back (and I urge you to do so) and read "4 Steps for Solving Problems" on the inside front cover of this book and *Explorations*, you will realize that many of these steps are tools to help you communicate better with yourself about the problem.

2. Communicating with other people—sharing your observations and solutions and being able to understand others' observations and solutions.

>Look back on the investigations we have already done in this chapter and on the explorations you have done in *Explorations*. You should find that there were times when the words you used to communicate were not clear to others, or vice versa, and that better use of mathematical language could help facilitate communication, whether in describing your strategies for the handshakes problem, explaining assumptions you made about the coffee filter problem, or choosing language to describe patterns you saw in magic squares.

As we move to learning specific content in the following chapters, you will come to realize that one of the reasons for all the mathematical notation and language that sometimes seems to get thrown at you is that if you understand the notation and the terms (digit, place value, row, column, and so on), then two things are more likely. First, you can reduce ambiguity and confusion in mathematical conversations. Second, you are more likely to understand fully the mathematical structures being presented.

Learning by Communicating

There is another reason for emphasizing communication at the beginning of this course. Few people learn best in isolation. The National Training Laboratories in Bethel, Maine, have created the pyramid in Figure 1.17 to represent what we know about learning.

When we bring communication into the learning process—by having students work in small groups, by having students explain how they solved a problem, by having students justify the steps in their solution—the students are more likely to "own" the knowledge they gain.

Let us now examine an aspect of using patterns that was mentioned earlier: being able to describe patterns you see in such a way that others can understand what you are saying.

Retention Rates from Different Ways of Learning

Method	Average Retention Rates
Lecture	5%
Reading	10%
Audio-Visual	20%
Demonstration	30%
Discussion group	50%
Practice by doing	75%
Teach other/Immediate use of learning	90%

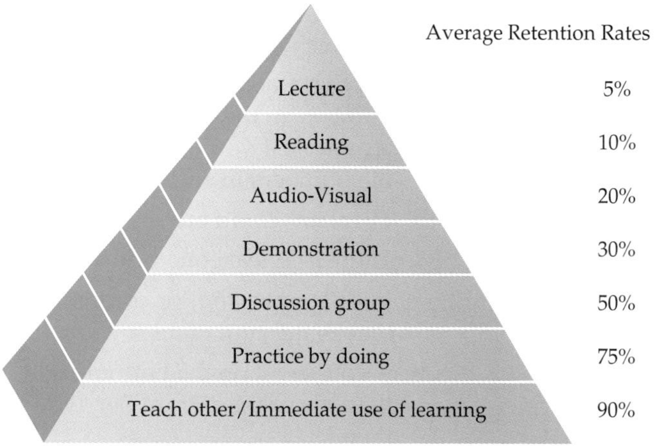

Source: Reprinted with permission from NTL Institute from "Retention Rates from Different Ways of Learning."

FIGURE 1.17

INVESTIGATION 1.13

13	3	2	16
8	10	11	5
12	6	7	9
1	15	14	4

Communicating Patterns in a Magic Square

Magic squares, triangles, and circles have fascinated human beings for thousands of years. Exploration 1.6 focuses on 3 × 3 magic squares. This investigation focuses on 4 × 4 magic squares and the importance of communication. The definition of a basic magic square is that when you add the numbers in any row, any column, or either diagonal, you get the same number. The square at the left was created by Albrecht Dürer in 1514 in his engraving *Melancholia*. He created a square in which the year appeared in the bottom row of the square! A quick examination of the square confirms that the magic sum for rows, columns, and diagonals is 34. Now take a closer look at this square. What observations do you make? What other patterns do you see in this magic square? Recall our earlier discussion of pattern—something where there is some repetition—for example, the sum of 34 in rows, columns, and diagonals. Observation will be defined as something interesting, but not repeating—for example, the 16 numbers in the square are consecutive numbers from 1 to 16. Write down your observations and patterns so that someone reading your description could see what you see. Each year, sometimes with a little prompting, my students fill two or more overhead transparencies with observations and patterns. Think and then read on. . . .

DISCUSSION

Read the descriptions below in two ways: First, do you understand my description? Second, would you word it differently? In my classes, when someone describes a not-simple pattern, I ask how many people understood that description—many, but nowhere near all, students raise their hands. Then I ask someone who used different language to describe the same pattern; again I ask how many understood this description. What we find is that two descriptions can be very different and both valid. Thus, it is not a matter of "which one is better" but rather realizing that just as there are usually different, but valid paths for solving a problem, so too are there usually different, but valid ways of describing what one sees in a problem.

- There are actually many more 34s. Here are six more! The sum of the four numbers in the center 2 × 2 square is 34; the sum of the four corner numbers is also 34; if you divide the 4 × 4 square into four 2 × 2 squares, the sum of the numbers in each of the 2 × 2 squares is also 34.
- There are 2 odd and 2 even numbers in each row and column.
- There is at least one prime number in each column.
- Middle rows and/or columns can be switched without affecting the square; that is, the sum of all rows, columns, and diagonals will still be 34.
- If you travel on the bottom-right to top-left diagonal, each number is 3 more than the previous number. Now look at the other diagonal. What do you see there?
- If you look at the middle two columns, each pair of numbers are consecutive numbers: 3 and 2, 10 and 11, 6 and 7, 15 and 14. Now look at the middle two rows. What do you see there?

- Square every number in the table. The sum of each row is 438 or 310, and the sum of these two numbers is 748. The sum of each column is 438 or 310, and the sum of these two numbers is 748.
- Add the first and second numbers in each row—the sum is either 16 or 18. Add the third and fourth numbers of each row—the sum is either 16 or 18. You can find a similar pattern by summing the top two numbers and then the bottom two numbers in each column; in this case, the sums alternate between 13 and 21.

INVESTIGATION 1.14

Darts, Proof, and Communication

Suppose you have a dart board like the one in Figure 1.18. You throw four darts, all of which land on the dart board. One of the questions I asked (in the exploration and of the fifth-graders I taught in the 1998–1999 school year) was what kinds of scores would be possible and what kinds of scores would be impossible. What do you think?

FIGURE 1.18

DISCUSSION

After a few minutes, one of the students, Erika, suddenly said, "Only even numbers are possible." I asked her how she came to that conclusion, and she said, "Well I know that an odd plus an odd is even and an odd plus an even is odd. [At this point, she held up four fingers to represent the four darts.] So the first two darts are odd and so when you add them, you have an even number. [She joined two of her fingers together to indicate the combined score from two darts.] Now this number (even) plus the next dart (odd) will make an odd number. [She now joined three of her fingers together to indicate the combined score from the first three darts.] Now this number (odd) plus the last dart (odd) will make an even number. So the only possible scores you can get are even numbers."

We can represent Erika's proof as shown in Figure 1.19.

$$(odd + odd) + odd + odd$$
$$(even + odd) + odd$$
$$odd + odd$$
$$even$$

FIGURE 1.19

Reflect on this problem for a moment, along with the standards we have discussed: problem-solving, patterns, representation, reasoning, and communication. Do you see them working interactively as opposed to separately? What aspects of each standard do you see in operation?

Erika chose to solve the whole problem in her head because she intuitively grasped the solution to the problem. I have presented her communication of her solution (verbally), but because not everyone learns and understands the same way, I have presented another way of representing the problem

(Figure 1.19). Other students solved the problem more inductively. That is, they saw patterns in their table that helped them to realize that odd + odd is even, and so on.

SECTION 1.7 CONNECTIONS

> **WHAT DO YOU THINK?**
>
> - How does making connections add to your toolbox?
> - What connections among concepts and problem-solving strategies have you seen thus far in the chapter?

Standard 9: Connections
Instructional programs from prekindergarten through grade 12 should enable all students to

- recognize and use connections among mathematical ideas;
  [Over the course of this book, you will come to see that there are rich relationships between most mathematical ideas and that their separateness (such as geometry being totally different from algebra, like English and Vietnamese) is an illusion.]
- understand how mathematical ideas interconnect and build on one another to produce a coherent whole;
  [When you buy this building-block metaphor, you take care to make sure that you understand the "big" ideas, which are like building blocks.]
- recognize and apply mathematics in contexts outside of mathematics.
  [There are many readable books, such as the ones mentioned before by Keith Devlin, that will amaze you!]

(*Principles and Standards*, p. 64)

Making *connections* is at the heart of the NCTM Standards. We have already noted connections between Pascal's triangle and Exploration 1.1 (The Handshakes Problem) and Investigation 1.3 (The Sum of the First 100 Numbers). We have also noted that the idea of connections is new to many students. This is addressed in the NCTM Curriculum Standards:

> The mathematics curriculum is generally viewed as consisting of several discrete strands. As a result, computation, geometry, measurement, and problem solving tend to be taught in isolation. It is important that children connect ideas both among and within areas of mathematics. Without such connections, children must learn and remember too many isolated concepts and skills rather than recognizing general principles relevant to several areas.

(*Curriculum Standards*, p. 32)

Let us examine some of the kinds of connections that we would like you to make in this course.

Making Connections Between the Problem and What Is Inside Your Head

Look back at the problems in this chapter that you solved on your own *and* that gave you some difficulty—that is, problems you had to do some thinking in order to solve, as opposed to those for which you immediately knew what to do and then did it. In these cases, you literally have to find some way to connect the words of the problem to a representation of the problem that will enable you then to solve the problem. Many of my students report that before this course, the teacher did the connecting for them. Because you are to become teachers, this is where you develop the ability to make such connections yourself.

Recall the example from the National Assessment of Educational Progress in Table 1.3. To solve the first problem, you simply had to add the numbers and divide by 6. However, to solve the second problem, you had to think about the information. In other words, you had to create in your head a *model* of the problem that made sense and that connected the relevant information to your mathematical knowledge. Developing the ability to connect the given information to your mathematical knowledge and to your problem-solving toolbox is one of the central objectives of this course. Some students have already developed this ability. However, every year, most of my students enter the course without these tools. Seeing those students' toolboxes develop and seeing their confidence grow is one of the many reasons why I cannot imagine being in any other profession!

Connecting New Concepts to Old Concepts

One important kind of connection is the connection between new ideas and something that is familiar to you. As we discussed in Section 1.2, if you are not able to make these connections, you may end up with "Teflon knowledge." Consider another example from the NAEP. Students were asked to select the decimal equivalent to 12 percent and then to select the decimal equivalent to .9 percent. The results are given in Table 1.16.

TABLE 1.16

	Percent correct	
	Grade 7	Grade 11
Which decimal is equivalent to 12 percent?	71	90
Which decimal is equivalent to .9 percent?	25	56

Under any circumstances, we would expect a smaller percentage of correct responses for the second question. However, the drop-off is enormous. For seventh-graders, the percentage who got the second question correct was barely one-third that for the first question. What connections could have helped these students? Try to answer this question yourself before reading on. . . .

First, they could have connected percents to rational numbers, reasoning that 12 percent means 12/100. Then they could have connected the fraction to a decimal: $12/100 = 0.12$. Following this reasoning for the second question, .9 percent means 0.9/100, which converts to 0.009. Alternatively, they could have applied the algorithm "Move the decimal point two places." In other words, .9 becomes .009. Of course, they had to realize that they had to put zeros to the left of .9 in order to be able to move two decimal places! We examine decimals and percents in more detail in Chapters 5 and 6.

Making Connections Among Different Concepts

Many, if not most, students have come to view mathematics as a collection of separate topics. Most mathematicians, however, see mathematics as a network of *interconnected* concepts, similar to a map of the highways in the United States, the freeways being equivalent to major connections and surface roads to smaller connections.

Looking at the four basic operations—addition, subtraction, multiplication, and division—we can make the following connections: Addition and subtraction are inverse operations, and multiplication can be seen as repeated addition. Likewise, multiplication and division are inverse operations, and division can be seen as repeated subtraction. We will investigate these concepts in Chapter 3. In Chapter 6, you will discover or rediscover that percents have close connections to ratios and to equivalent fractions. Many students see algebra and geometry as completely separate, whereas mathematicians see them as very connected.

Connecting Different Models for the Same Concept

In mathematics, many concepts can be represented in different ways. For example, consider the ways of representing 3/4 shown in Figure 1.20. One of them is not a valid representation of what we mean by 3/4. Do you understand why? Can you express that understanding in words?

When you can connect many different representations to the concept underlying all of them, you are well on your way to the kind of mathematical thinking encouraged by the NCTM. Many "good" problem-solvers are good not because they know more than others but because their knowledge is better connected.

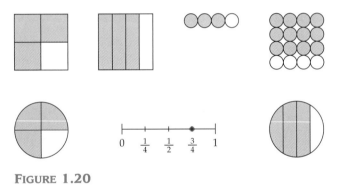

FIGURE 1.20

Connecting Conceptual and Procedural Knowledge

Most of you have probably learned a variety of standard procedures, also called **algorithms**. However, fewer of you are likely to know why they work. For example, although you can probably convert a mixed number to an improper fraction—for example, $4\frac{2}{3} = \frac{14}{3}$—do you know *why* we multiply the whole number by the denominator, add the numerator, and then put the whole thing over the denominator?

Similarly, do you know *why* we "move over" in whole-number multiplication?

$$\begin{array}{r} 47 \\ \times\ 35 \\ \hline 235 \\ 141 \\ \hline 1645 \end{array}$$

Outside the Classroom

Dorothy Wallace, a mathematics professor at Dartmouth College, developed a number of mathematics courses that she thought would be more interesting than the traditional courses to liberal arts students not majoring in math and science. One of these courses was Pattern, an interdisciplinary course on art and mathematics, which she co-taught with a local artist, Pippa Drew. In 1997, college professors from all over the country were invited to Dartmouth to get a sense of these courses. I was invited to join as a teacher-educator and facilitator. The group that I worked with consisted of 25 mathematics and art professors. They quickly came to see that they had far more in common than most of them had realized—both art and mathematics professors got excited about patterns. Finding similar patterns in different areas was exciting, and finding breaks in patterns was exciting as well.

Coming to understand why these and many other algorithms work will make you a more powerful problem-solver and a stronger teacher.

Making Connections Between Mathematics and "Real Life" and Between Mathematics and Other Disciplines

Mathematics enters into some of the most important decisions we make, both as individuals and as a society. Recall the explorations dealing with money. Think about buying tires, buying a house, investing your savings, comparison shopping, deciding how much wallpaper or paint to buy, building a fence, making a budget, figuring out a tip, choosing among credit cards—the list could go on for pages!

Think of how many political issues are connected to mathematics:

- The current and future effects of the federal deficit—How many people realize what a really big number 5 trillion is?
- The issue of income inequality—women earn an average of 70¢ for every dollar men earn.
- The cost of illiteracy and innumeracy to the United States—In 1990 it was estimated that American employers spent about $40 billion to train their employees in the basic reading, writing, and arithmetic skills they should have learned in school.

Unfortunately, all too many students do not see mathematics as connected. My greatest hope for this course is that it leads you to see that part of your job as a student is to make connections—to make sure you see how the new concepts and ideas connect to what you already know, to make sure you see connections within and between different concepts, and to make sure you see the connection between the mathematics in this course and mathematics beyond the classroom. If that happens, then our society will begin to be more mathematically literate rather than rank near the bottom internationally in mathematics performance.

The following investigation problem illustrates the importance of making relevant connections when solving problems.

INVESTIGATION 1.15 | How Many Pieces of Wire?

A jewelry artisan is making earring hoops. Each hoop requires a piece of wire that is $3\frac{3}{4}$ inches long. If the wire comes in 50-inch coils, how many $3\frac{3}{4}$ inch pieces can be made from one coil, and how much wire is wasted? Solve this problem on your own and then read on. . . .

But before you do, ask yourself, "Do I understand the problem? Does the problem's wording help me devise a plan for solving it? Once I have a solution, can I check it?" Perhaps most important, ask yourself, "What did thinking about *this* problem reveal to me about problem-solving in general?"

DISCUSSION

STRATEGY 1: Divide

Some people quickly realize that you can divide "to get the answer." If you use a calculator, it shows 13.333333.... If you use fractions, you get $13\frac{1}{3}$. Many people interpret these numbers to mean that you can get 13 pieces and you will have 1/3 inch wasted. Unfortunately, that is not correct. If you also got 1/3 inch as wasted, stop! Go back and see if you can figure out why 1/3 inch is not the correct answer and what the correct answer is. Then read on....

One of the reasons why I, like many other math teachers, stress the importance of labels is that they illustrate the meaning of what we are doing. The *meaning* of the quotient $\left(13\frac{1}{3}\right)$ is 13 whole hoops and 1/3 of a hoop. That is, 50 inches $\div$ $3\frac{3}{4}$ inches/hoop $= 13\frac{1}{3}$ hoops. The *meaning* of the fraction (1/3) is that what we have left would make (1/3) of a hoop. Because one whole piece is $3\frac{3}{4}$ inches long, 1/3 of a piece is 1/3 of $3\frac{3}{4}$; that is, $1\frac{1}{4}$ inches is wasted.

How would we check this answer? Think and then read on....

One way to check would be to multiply $3\frac{3}{4} \times 13$. Do you see why? This would tell us the length of the 13 whole pieces. If this number plus $1\frac{1}{4}$ equals 50, then our answers are correct. In fact, $3\frac{3}{4} \times 13 = 48\frac{3}{4}$ and $48\frac{3}{4} + 1\frac{1}{4} = 50$.

STRATEGY 2: "Act it out"

Some people understand the problem better when they draw a line to represent the 50-inch piece of wire. Draw a picture like that in Figure 1.21 to represent what happens when the artisan cuts off pieces $3\frac{3}{4}$ inches long until the wire is used up.

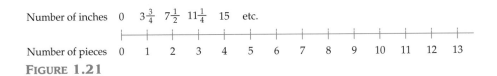

FIGURE 1.21

STRATEGY 3: Make a table

Refer to Table 1.17.

TABLE 1.17

Number of pieces	Number of inches	Thinking process
1	$3\frac{3}{4}$	
2	$7\frac{1}{2}$	
3	$11\frac{1}{4}$	
4	15	Using the concept of ratio and proportion, we can reason that if 4 pieces make 15 inches, then 12 pieces would make 45 inches.
12	45	So 1 more piece works.
13	$48\frac{3}{4}$	

The answer to the second part of the question is simply

$$50 - 48\tfrac{3}{4} = 1\tfrac{1}{4}$$

What kinds of connections do you see in this problem? Think and then read on. . . .

Connections Several of the kinds of connections mentioned at the beginning of Section 1.7 are evident in this problem.

- Connecting new concepts to old concepts: If the concept of division of whole numbers (investigated in Chapter 3, Section 4) is well understood, the student can immediately see this as a division problem, but one that involves fractions.
- Connecting different models for the same concept: In Chapter 3, Section 4, you will find that one model for division is repeated subtraction. Drawing the diagram of the wire and marking off $3\tfrac{3}{4}$-inch segments often triggers the connection between this problem and repeated subtraction, which then produces the insight "Oh, now I get it! I just have to divide 50 by $3\tfrac{3}{4}$!"
- Connecting conceptual and procedural knowledge: The procedure for dividing by fractions or decimals is not terribly complex. However, the result of the computation $\left(13\tfrac{1}{3}\right)$ is not quite the answer to the question.

Summary

We have finished our first pass over the NCTM's five process standards, and we have examined several specific aspects of those standards. I will argue that without them, the rules and formulas and procedures of mathematics are inert and lifeless, and that without them, *the mathematics classroom is not very interesting for most students.*

Many students find it helpful to keep a reflection notebook, also called a **journal** or a **learning log**. In fact, your instructor may make this part of the course. If you do take this time to examine your learning, you will own more of the ideas studied in this course and will be more likely to bring those ideas to the children you teach after you graduate.

With good instruction, where the five process standards are alive and well, students develop what the NCTM calls **mathematical power**.

Mathematical power includes the ability to explore, conjecture, and reason logically; to solve nonroutine problems; to communicate about and through mathematics; and to connect ideas within mathematics and between mathematics and other intellectual activity. Mathematical power also involves the development of personal self-confidence and a disposition to seek, evaluate, and use quantitative and spatial information in solving problems and making decisions. Students' flexibility, perseverance, interest, curiosity, and inventiveness also affect the realization of mathematical power.[14]

The last sentence brings us back to students' attitudes and beliefs—about mathematics, about learning, about teaching, and about themselves. In my doctoral dissertation (a study of how attitudes and beliefs influence the learning of mathematics), I wrote that affect (beliefs, attitudes, and emotions) influences both the *quantity* and the *quality* of the thinking you bring to the problem—that is, both how hard you try and how smart you try.

[14]NCTM, *Professional Standards for Teaching Mathematics* (Reston, VA: NCTM, 1991), p. 1.

I offer two metaphors. Each has been useful to a number of my students. Imagine driving in a car on a highway and suddenly encountering fog "thick as pea soup." What would you do? Of course, you would slow down. Now imagine you are reading a textbook and the words feel like a blur. Why not slow down? Or imagine working on a problem at the end of a chapter and it just doesn't make sense, but you plug along anyway. Why not slow down? Just as you can see much better, in fog, at 30 miles per hour than you can at 70 miles per hour, most people can make more sense of ideas and problems when they *slow down*.

Imagine you are going to visit someone in another city. The directions are not terribly clear, and you're not sure you made the right turn. What do you do? The reasonable course is to stop at the first opportunity (gas station or convenience store) and *ask for directions*. You might also stop and consult a map and your directions. Similarly, if you are reading an Investigation in the text or working on a problem and you feel lost, what do you do?

These metaphors deal with an underlying belief that mathematics is not just about "getting it" but also about "making sense." This practice of pausing frequently to ask whether the ideas or problems are making sense is crucial in order for you to own what you learn, as opposed to simply renting it until the test or the end of the course.

Exercises

1. Describe the mistake in inductive reasoning made by the child who says that the figure below is a diamond and "not" a square.

2. a. Along the lines of Investigation 1.8, make up a number of fraction addition problems where the two numerators are 2 (such as $\frac{2}{3} + \frac{2}{5}$). Describe the generalization you come up with in words and in mathematical notation.

   b. Make up several problems where the two numerators are 1 but the denominators have a common factor—for example, $\frac{1}{4} + \frac{1}{6}$. Describe the generalization you come up with in words and in mathematical notation.

3. Each one of the following conclusions has been made by children and adults. For which ones can you find a counter-example?

   a. If you double any number and add 1, the resulting number is always an odd number.

   b. If you triple any number and add 1, the resulting number is always an odd number.

   c. If you square a number, you get an even number.

   d. If a number is divisible by 2 and by 3, then it is divisible by 6.

   e. If a figure has four right angles, then it must be a rectangle.

4. Look at the clock below.

   a. How do this clock and a normal clock differ?

   b. What time is it?

   c. Draw the hour and minute hands in the positions they will occupy when it is 9:45.

   d. Why do you think the direction we call clockwise was selected?

5. At the writing of this book, many different companies were making television commercials to induce people to call collect using their company. One company advertised that you could call "anywhere anytime for just 10 cents a minute." Sounds pretty good. I pay an average of 10 cents a minute for long-distance calls placed from my home. How do they do it? Well, in small print at the

bottom of the picture, the reader learns that there is a $1.95 surcharge on every call. Describe the deal now.

6. Represent the statements below with Venn diagrams.
   a. All dogs like to be petted.
   b. Some Americans did not vote in the last election.
   c. Some students like mathematics.
   d. No cotton candy has protein.

7. Translate the following statements into if–then form.
   a. All fruit contains sugar.
   b. All dogs bark when the mail carrier comes to the door.
   c. No cats bark when the mail carrier comes to the door.
   d. All babies cry at night.
   e. In order to drive a car, you must pass a written test.

8. Write the inverse, converse, and contrapositive of each of these statements.
   a. If you exercise three times a week, then you will not get sick.
   b. If a number is a multiple of 10, then it is a multiple of 5.
   c. If Washington had not crossed the Delaware, then we would not have won independence from Britain.
   d. If you are a good elementary school teacher, then you know elementary mathematics well.

9. Determine whether the arguments below are valid or invalid.
   a. All polygons have angles.
      A circle has no angles.
      A circle is not a polygon.
   b. If you don't work hard, then you won't succeed.
      You work hard.
      Therefore, you will succeed.
   c. If you make an A on the midterm, you won't have to take the final.
      Jose did not take the final.
      Therefore, Jose made an A on the midterm.
   d. All blippies are orgs.
      All blippies are magas.
      Therefore, all orgs are magas.
   e. If I were rich, I would buy a cabin.
      I am not rich.
      Therefore, I have not bought a cabin.
   f. If a number is divisible by 8, then it is divisible by 4.
      x is not divisible by 8.
      Therefore, x is not divisible by 4.
   g. All comedians are funny.
      Shelly is not funny.
      Therefore, Shelly is not a comedian.

10. Using Venn diagrams, determine whether the arguments below are valid or invalid.
    a. All professional athletes are rich.
       All rock stars are rich.
       Therefore, some rock stars are professional athletes.
    b. All bunnies are skittish.
       Some skittish animals have fur.
       All cows have fur.
       Therefore, cows are skittish animals.
    c. All squares are rectangles.
       Some quadrilaterals are squares.
       Therefore, some quadrilaterals are rectangles.
    d. All rectangles are quadrilaterals.
       All quadrilaterals are polygons.
       Therefore, all rectangles are polygons.
    e. Michael Jordan drinks Gatorade.
       Michael Jordan has a lot of stamina.
       Therefore, if you drink Gatorade, then you will have a lot of stamina.

11. A boatman is to transport a fox, a goose, and a sack of corn across the river. There is room in his boat for only one of the three at a time. Furthermore, if the fox and the goose are left together, the fox will eat the goose. If the goose and the corn are left together, the goose will eat the corn. How can the boatman do the job?

12. The following two jar problems have been around for many years. They are contrived problems to promote reasoning. In these problems, a jar must be filled *completely* and emptied *completely*. For example, you cannot say, "Fill the big jar half full and fill the second jar half full" to get 4 gallons.
    a. You are given a 5-gallon pail and a 3-gallon pail, both of which are unmarked. You are asked to fetch 4 gallons of water from the well in one trip. How?
    b. You have three jugs of capacities 8, 5, and 3 gallons. The largest jug is full of water. The other two are empty. Your task is to redistribute the water so that you wind up with 2 gallons of water in the large jug and 3 gallons of water in each of the other jugs.

13. When Joe wakes up in the morning, the clock is blinking and says 3:30. His watch says 6:45. When did the power go off? What assumption(s) did you make in order to solve the problem?

14. Alice, Betty, and Carla traded hats and gloves. Alice wore Betty's hat and Carla's gloves. Write a paragraph explaining how you know whose hat and gloves Betty and Carla wore.

15. Look back at Investigation 1.14 with the dartboard. What if the scores were 1, 5, 10, 25?
    a. List all possible scores from 1 to 100.
    b. Select one score that is more than 4 and less than 100 that is impossible, and prove that it is impossible.

16. Look back at Investigation 1.14 with the dartboard. What if the scores were 1, 2, 4, 8?
    a. List all possible scores from 4 to 32.
    b. How many different ways could someone score 12?
    c. Prove that there are no more ways to score 12 than the ones you found in part (b).

17. Below is a 5 × 5 magic square.

11	18	25	2	9
10	12	19	21	3
4	6	13	20	22
23	5	7	14	16
17	24	1	8	15

a. Describe all the patterns you see in this magic square.

b. Describe one of the patterns in words, as though you were talking to someone on the phone.

18. The following magic figure originated in West Africa.

a. Describe the patterns you see in this magic figure.

b. Describe the relationship between this figure and the first magic square in Exploration 1.5.

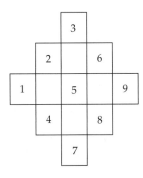

19. Examine the magic triangle below.

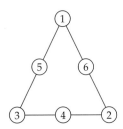

a. Describe all the patterns you notice.

b. Make a new magic triangle with different numbers.

c. What generalizations can you make about magic triangles of order 3—that is, those in which there are three numbers on each side of the triangle?

d. Make a magic triangle of order 4. Briefly describe the strategies and knowledge that enable you to make the three sums equal.

20. Below is a different kind of magic triangle.

a. Describe the patterns you see in this magic triangle.

b. Create a new magic triangle in which the nine numbers are not consecutive numbers.

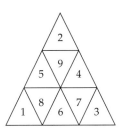

21. Leonard Euler, whom many consider to be the greatest mathematician of all time, created this magic square. Describe three patterns that you find in this magic square.

1	48	31	50	33	16	63	18
30	51	46	3	62	19	14	35
47	2	49	32	15	34	17	64
52	29	4	45	20	61	36	13
5	44	25	56	9	40	21	60
28	53	8	41	24	57	12	37
43	6	55	26	39	10	59	22
54	27	42	7	58	23	38	11

22. Benjamin Franklin created this magic square. Describe three patterns that you find in this magic square.

52	61	4	13	20	29	36	45
14	3	62	51	46	35	30	19
53	60	5	12	21	28	37	44
11	6	59	54	43	38	27	22
55	58	7	10	23	26	39	42
9	8	57	56	41	40	25	24
50	63	2	15	18	31	34	47
16	1	64	49	48	33	32	17

23. Suppose a family of 2 parents and their 10 children decides that each person will buy a gift for each member of the family for Christmas. How many gifts will be under the Christmas tree? Now suppose that the family decides that each parent and child will exchange gifts and that the 10 children will have a drawing in which each child chooses the name of one sibling to whom to give a gift. How many gifts will now be under the Christmas tree?

24. Write instructions for drawing a square, as though to someone who doesn't know the word *square* and has no formal training in mathematics.

25. Assume that someone is visiting you and has never used a phone. You are going to work, and this person is going to be home alone. Write directions to help this person make phone calls. Include all the different kinds of phone calls one might make.

26. a. How many $3\frac{2}{3}$-oz bottles of perfume can be filled from a jug containing 64 oz?
    b. How many ounces of perfume are left over per jug?
    c. Determine a jug size that is a whole number and that would produce no waste.

27. Find the two-digit number that, when added to its reverse, is the closest to 130. (*Note*: The reverse of 79, for example, is 97.) Show your work. If you got it on the first try, briefly explain your thinking. Otherwise, your successive guesses will suffice.

28. Two 2-digit numbers satisfy the following conditions:
    • The sum of the digits in each number is 10.
    • All four digits are different.
    • The sum of the numbers is 155.
    Determine the two numbers.

29. How many triangles can you find in the figure below?

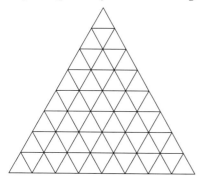

30. Variations of the following problem are common in elementary schools: See the 2001 *NCTM Yearbook*, p. 78, and *Teaching Children Mathematics*, February 1997, p. 326, for two discussions.
    a. A frog is climbing out of a well that is 8 feet deep. The frog can climb 4 feet per hour, but then it rests for an hour, during which time it slips back 2 feet. How long will it take for the frog to get out of the well?
    b. What if the well is 20 feet deep, the frog climbs 5 feet per hour, and the frog slips back 2 feet while resting?
    c. What if the well is 40 feet deep, the frog climbs 6 feet per hour, and the frog slips back 1 foot while resting?
    d. Describe patterns you see in the three problems.
    e. Solve the general case: A frog is in a well that is a feet deep, the frog climbs b feet per hour, and the frog slips back c feet while resting.

31. This problem appeared in *Teaching Children Mathematics* and was discussed by Kim Hartweg. "Aunt Katrina just moved into a new house and wants to brighten up the dining room wall by hanging six of her favorite decorative plates. Each plate is 8 inches in diameter. Aunt Katrina is having trouble deciding how to hang the plates in a straight line and evenly space them along her 104-inch wall.
    a. Draw a diagram to show the placement and spacing and explain how Aunt Katrina should hang her six decorative plates.
    b. What if one plate broke?
    c. What if she wanted to have the two end plates 12 inches from the end of the wall?

    Source: From *Teaching Children Mathematics*, December 2004/January 2005 issue, pp. 280–284, copyright 2005 by the National Council of Teachers of Mathematics.

32. In "Math Riddles: Helping Children Connect Words and Numbers," Carl Sherrill shows how children can learn many things while creating and solving riddles. Here are a few that my students created.
    a. The number of digits is the same as the number in the ones place. The sum of the digits is 10. All the digits are different. The digits are in increasing order.
    b. I am thinking of a 4-digit number. The sum of the digits is 27. The largest digit is used twice but not side by side. The smallest digit is in the largest place. The smallest digit is 1/3 as large as the largest digit.
    c. I am thinking of a 4-digit number. The sum of the digits is 21. Starting from left, the third digit is double the first and the fourth is double the second. All the digits are different. Add one more clue to produce a unique solution.
    d. I have 93 cents and 8 coins. I have the same number of quarters as pennies. Each type of coin is used. What are the coins?
    e. I have 7 coins. There are more nickels than quarters. I have twice as many dimes as nickels. How much money do I have?
    f. Make up your own riddle.

    Source: From "Math Riddles: Helping Children Connect Words and Numbers," *Teaching Children Mathematics,* March 2005, pp. 368–375, copyright 2005 by the National Council of Teachers of Mathematics.

33. This problem was posed in the September 1997 "Problem Solvers" section of *Teaching Children Mathematics*: How many rectangles appear in the figure below?

    a. One student invented her own notation to deal with the overlapping. After doing this for several cases, she concluded that she saw a pattern and didn't need to continue her diagram. Use her strategy, and then explain and justify why you don't need to continue the diagram for every size rectangle.

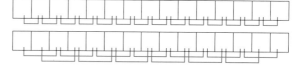

b. Another student used a combination of a sliding strategy and systematic thinking by referring to the rectangles by the position of the squares on each end. For example, the two 1 × 14 squares are listed below. Use this strategy to determine the answer, and describe your use of the strategy.

1 —————————————— 14
2 —————————————— 15

Source: Reprinted with permission from *Teaching Children Mathematics*, copyright © 1997 by the National Council of Teachers of Mathematics.

Chapter Summary

1. One's beliefs about mathematics can powerfully affect one's ability to do mathematics. Many students find that the big ideas of this chapter involve a major shift in the beliefs that they brought into the course.

2. George Polya's description of the four basic steps in solving problems is a useful framework, but it is the starting point, not the ending point, of your using it.

3. Many problems can be solved in a variety of ways. Some of the more common problem-solving strategies are guess–check–revise, make a diagram, develop an equation, make a table, look for patterns, connect the problem to a similar problem, break the problem into smaller pieces and act it out.

4. Solving a problem involves more than simply computation. The number is not the answer, and you are not finished unless the answer makes sense.

5. There is generally more than one way to represent a problem. Thus, it is important to construct a repertoire of representations in order to develop mathematical power and in order to understand other students' representations of problems.

6. Patterns permeate our lives as well as permeating mathematics. Recognizing patterns is the beginning of using them to understand mathematics and solve problems. You should be able to describe, extend, analyze, and create a wide variety of patterns.

7. Reasoning and proof are not just things to be learned by older students; they are also an essential part of elementary school mathematics.

8. There are three kinds of reasoning used in understanding mathematical ideas and solving mathematical problems: inductive, deductive, and intuitive reasoning.

9. Communication is an essential aspect of mathematics—being able to communicate with yourself as you learn mathematical ideas and solve problems, and being able to communicate with others. Tools to improve communication include diagrams, vocabulary, and symbols.

10. There are many aspects to being able to develop mathematical connections: making connections between the problem and what is inside your head, connecting new concepts to old concepts, making connections among different concepts, connecting different models for the same concept, connecting conceptual and procedural knowledge, and making connections between school mathematics and "real life" and between mathematics and other disciplines.

BASIC CONCEPTS

Section 1.1 Getting Comfortable with Mathematics

adaptive beliefs *2*   maladaptive beliefs *2*
NCTM standards *4*   process standards *5*

Section 1.2 Problem-Solving

problem-solving *6*
what-if program *8*
guess–check–revise *8*
grope-and-hope *8*
connections *10*
own *10*
rent *10*
toolbox *10*
reflect *10*

Polya's four steps *11*
    understand the problem *12*

devise a plan *12*
monitor your plan *12*
look back at your work *12*

Section 1.3 Patterns
solution path *20*
Pascal's triangle *21*

Section 1.4 Representation
representations *23*
uncover *24*

Section 1.5 Reasoning and Proof
inductive reasoning *32*
arithmetic sequence *34*
geometric sequence *35*
deductive reasoning *36*
conditional statement *36*

hypothesis *36*
conclusion *36*
converse, inverse, contrapositive *36*
valid argument *37*
invalid argument *37*
Law of Detachment (affirming the hypothesis) *38*
Modus Tollens (denying the conclusion) *38*
analytic thinking *38*
Venn diagram *39*
biconditional statement *41*
intuitive reasoning *42*

Section 1.7 Connections
model *49*
algorithms *50*
journals *53*
learning logs *53*
mathematical power *53*

CHAPTER 1 REVIEW EXERCISES

1. A school play charges $2 for students and $5 for adults. For the three days of the play, 458 tickets were sold and $1342 was raised. How many student tickets were sold?

2. A small factory makes three-legged stools and four-legged tables. Last month this factory used 100 legs to build 3 more stools than tables. How many stools did the factory make?

3. The Perez family is making a patio in the back yard. They are using patio blocks that measure 12 inches by 8 inches. If the patio is to be 16 feet by 12 feet and each block costs 75 cents, how much will the patio cost?

4. How many different ways can you make 50 cents if you have one quarter, five dimes, and ten nickels?

5. If a fence requires a post every 10 feet, how many posts are required for a fence that measures 100 by 100 feet?

6. A letter was posted that was covered with 10-cent stamps and 5-cent stamps. There were 12 stamps, and the total postage was 70 cents. How many of each stamp were on the letter?

7. Multiply several two-digit numbers by 99. Describe the pattern you see.

8. If the following pattern continues, what letter will be in the 243rd position?

    HIPHIPHOORAYHIPHIPHOORAY
    HIPHIPHOORAY . . .

9. Let's say you have a 4-gallon and a 9-gallon pail.
    a. How can you measure exactly 1 gallon of water?
    b. How can you measure exactly 3 gallons of water?

10. A caterpillar has been trapped in a glass that is 12 inches tall. The caterpillar can climb 3 inches per hour, but then it rests for an hour, during which time it slips back 1 inch. How long will it take for the caterpillar to get out?

11. Sunflower seeds are packed in packages each weighing $2\frac{1}{2}$ ounces. If there is a supply of 67 ounces of sunflower seeds, how many packages of seed can be made? How many ounces of sunflower seeds will be left over?

12. Suppose you have a dartboard with four concentric circles labeled 1, 2, 4, and 8. You can throw four darts, and the rule is that each dart must score.
    a. How many different ways could one get a score of 14?
    b. Are there any scores between 4 and 32 that are impossible?

13. Look at the 4 × 4 magic square below. Briefly list at least five observations or patterns you see in this magic square.

13	3	2	16
8	10	11	5
12	6	7	9
1	15	14	4

14. There are so many patterns in Pascal's triangle. This one is often called the hockey stick pattern because when the path is shaded in, it looks like a hockey stick. Two examples of this pattern are shown below. What relationship do you see among the numbers along the length of the stick and the number at the end of the stick?

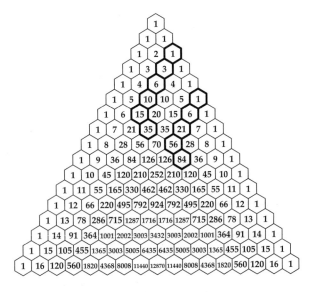

15. Represent each of the following statements with a Venn diagram.
    a. Some multiples of 7 are even numbers.
    b. All bicycles have two wheels.

16. Translate the following statements into if–then form.
    a. All meat contains protein.
    b. In order to pass the course, you must attend class.

17. Write the inverse, the converse, and the contrapositive of this statement: If you work hard, then you will succeed.

18. Determine whether the following argument is valid or invalid.

    No IMACs have a disk drive.
    This camera does not have a disk drive.
    Therefore, this camera is an IMAC.

19. Use a Venn diagram to determine whether the following argument is valid or invalid.

    If a penny has an Indian head on it, it was made before 1910.
    This penny was made before 1910.
    Therefore, this penny has an Indian head on it.

SOLUTION TO INVESTIGATION 1.12, P. 42

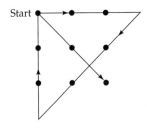

CHAPTER 2

Fundamental Concepts

2.1 Sets
2.2 Algebraic Thinking
2.3 Numeration

*I*n this chapter, we will explore three topics that occur throughout elementary school mathematics: sets, functions, and numeration. However, before we explore sets, we will touch on each of these three topics.

When we talk about a herd of deer, a bunch of carrots, an army of ants, a band of jackals, or a pack of lies, we are talking about a *set* of objects that have something in common. In mathematics, we use set language in many situations. For example, we talk about the set of even numbers or the set of common factors of 20 and 30. Many mathematical concepts and operations are defined in set language, and set language is often useful when we are talking about mathematical situations.

In many situations, both in daily life and in people's work lives, one set is *functionally* related to another set. The relationship may be simple—for example, the relationship between weight and postage—or it may be complex—for example, the relationship between income and tax.

The base 10 *numeration system*, which most people use without thinking about it, is one of the greatest inventions in human history. A deeper understanding of our numeration system increases a person's ability to make estimates and to solve more complex problems, to understand connections between whole numbers and decimals, to understand *why* the computation procedures work, and much more.

One of the foundations of the NCTM standards is the notion that mathematics ought to make sense to more people. One part of making sense of mathematics is understanding its vocabulary and symbols. Throughout this book, you will encounter many mathematical words and symbols, some of which will be familiar and some of which will be new to you. It is important for you to realize that these words (for example, *set*, *intersection*, *proportion*, and *parallelogram*) and symbols $\left(\text{e.g., } \cup, +, \%, \neq, \frac{a}{b}\right)$ were created to make it easier for people to discuss mathematical ideas. As you encounter new vocabulary and symbols, you might find it helpful to keep the following quotation in mind:

> Mathematics is often considered a difficult and mysterious science, because of the numerous symbols which it employs. . . . [T]he technical terms of any profession or trade are incomprehensible to those who have never been trained to use them. But

this is not because they are difficult in themselves. On the contrary they have invariably been introduced to make things easy.[1]

When you have finished this chapter, it might be useful to look back at the symbols introduced in the chapter. To what extent do the symbols make things easier? Is it important for the average person to know the concepts, or are they needed only by some people in some occupations? What do you think?

SECTION 2.1   SETS

WHAT DO YOU THINK?

- How do we use and apply set concepts in everyday life?
- Do we have to use circles to make Venn diagrams?

When mathematics educators discuss how much mathematics a teacher needs to know in order to teach well, there are two extreme positions. Those at one extreme argue that the content of this course should consist only of those concepts that the teachers will actually use with their future students. Those at the other extreme argue that the content of the course should be rigorous so that the teachers will know the mathematics at a much higher level.

My own position is somewhat in the middle. I believe that future teachers will generally be more motivated to learn concepts if they can see that their future students will work with these concepts too. I also feel that the teachers' knowledge needs to be at a higher level than they will teach at, for three reasons. First, if you have a strong understanding of the mathematical ideas, you can be more flexible in your lesson planning and in classroom discussions. Second, you will see more of the connections between concepts. Third, if you have a sense of where your students are headed, you can serve them better. That is, if an elementary teacher knows the concepts that middle school and high school teachers will build on, then the transition from elementary school mathematics to middle school and high school mathematics will not be as abrupt as it has been for all too many students.

Classroom Connection

Learning theorists tell us that a large part of the cognitive development of young children is driven by classification, and classification involves the idea of sets. For example, certain people belong in the set called "family." Most young children divide the human race into two sets: good people and bad people. Children grapple with the defining characteristics of sets; for example, there are many different-looking objects that are called chairs.

Sets as a Classification Tool

With this context, let us now look at why you might need to have some knowledge of various set concepts in order to teach more effectively at the elementary level. Children use set ideas in everyday life as they look for similarities and differences between sets; for example, they want to know why lions are in the cat family and wolves are in the dog family. Children also look for similarities and differences within sets; for example, they look within a set of blocks for the blocks that they can stack and the blocks that they can't.

Whether we realize it or not, we are classifying many times each day, and our lives are shaped by classifications that we and others have made.

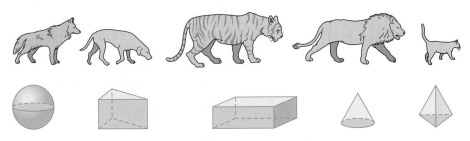

[1]*Introduction to Mathematics* (New York, 1911), pp. 59–69, cited in Robert Moritz, *On Mathematics* (New York: Dover Publications, 1914), p. 199.

INVESTIGATION 2.1

Classifying Quadrilaterals

Without classifying various objects and ideas, it would not be possible to have mathematics. This introductory investigation will help you to connect set language and concepts to other, more concrete mathematical ideas. Look at the eight shapes below. How can you classify these shapes into two groups so that each group has a common characteristic? Give a name to each group if you can. Work and then read on. . . .

DISCUSSION

There are many possible answers to this question. Let us examine some common ones.

One answer:
Those shapes in which all sides are equal

Those shapes in which not all sides are equal

Another answer:
Those shapes with at least one right angle

Those shapes with no right angles

Another answer:
Those shapes in which opposite sides are equal (parallelograms)

Those shapes in which opposite sides are not equal

Defining Sets

In Investigation 2.1, we began with a **set**, which is a collection of objects, and classified that set into smaller groups (*subsets*) having certain common features. In general, a **subset** is a set that is part of some other set. (A more precise definition will be given later.) Some subsets have names—for example, the subset of the set of shapes in Investigation 2.1 consisting of parallelograms. Other subsets are well defined but have no names; for example, there is no name for the set of geometric figures with at least one right angle.

We speak of individual objects in a given set as **members** or **elements** of the set. The symbol $\in$ means "is a member of." The symbol $\notin$ means "is not a member of." For example, if E is the set of even numbers, then $4 \in E$ but $3 \notin E$.

Describing Sets

There are three different ways to describe sets:

1. We can use words.
2. We can make a list.
3. We can use *set-builder notation*.

Words and lists In many cases, one of these representations is simpler or easier than the others. Let us examine some important mathematical sets and how we can describe them.

The first set of numbers that young children learn is called the set of **natural numbers**. We can use words to describe this set, or we can describe this set with a list.

N is the set of natural numbers or counting numbers.
$N = \{1, 2, 3, \ldots\}$

We use braces to indicate a set. The three dots are referred to as an ellipsis and are used to indicate that the established pattern continues indefinitely.

At some point, children realize that zero is also a number, and this leads to the next set: the set of **whole numbers** (W), which we can describe with words or with a list:

W is the set of natural numbers and zero.
$W = \{0, 1, 2, 3, \ldots\}$

Later, children become aware of negative numbers, so we have the set of **integers** (I):

$I = \{\ldots {}^-3, {}^-2, {}^-1, 0, 1, 2, 3, \ldots\}$

Set-builder notation Another important set is the set of **rational numbers** (Q), which we can describe in words:

Q is the set of all numbers that can be represented as the ratio of two integers as long as the denominator is not zero.

We cannot represent this set by making a list. Why is this? . . . In this case, and in many other cases, it is actually easier (as long as you understand the notation) to describe the set using what we call set-builder notation:

$$Q = \left\{\frac{a}{b} \mid a \in I \text{ and } b \in I, b \neq 0\right\}$$

This statement is read in English as "Q is the set of all numbers of the form $\frac{a}{b}$ such that a and b are both integers, but b is not equal to zero."

Set-builder notation always takes the form $\{x \mid x \text{ has a certain property}\}$. Although elementary school students do not use this notation to describe sets, it is helpful occasionally for you to see where your students will go. In high school, when students are working with more complex ideas, this notation makes communication easier.

Let us now apply these different ways of describing sets.

INVESTIGATION 2.2

Describing Sets

Consider the following set:

$$T = \{10, 20, 30, 40, 50, 60, 70, 80, 90, 100, 110, 120\}$$

Try to describe this set with words and with set-builder notation. What do you see as advantages and disadvantages of each of the three ways to describe this set?

If you have difficulty with this question, stop and think what might help. . . .

You might read the previous section again. You might write down the definitions and examples (or say the important ideas out loud to yourself). Or you might find that discussing this problem with a friend helps. The point here is to develop active learning habits rather than simply reading on to find the answer.

DISCUSSION

Verbal description:

T is the set of all multiples of 10 that are less than 130.

Set-builder notation:

$$T = \{x \mid x = 10n, 1 \leq n \leq 12, n \in N\}.$$

Many students initially have difficulty with set-builder notation. Therefore, let us examine the heart of the description.

This notation tells us that we are looking at all numbers that have the form "$10n$"—that is, multiples of 10. Because we are not talking about *all* multiples of 10, we have to let the reader know which multiples of 10 are in T. The mathematical phrase $1 \leq n \leq 12$ simply tells us that we are looking for multiples of 10 beginning with $10 \cdot 1$ and ending with $10 \cdot 12$. The last part of the description, "$n \in N$," simply lets us know that n must be a natural number.

The table below summarizes some of the advantages and disadvantages of each way of describing the set.

	Advantage	Disadvantage
Verbal description	Relatively simple	Subject to misinterpretation; in this case, some readers may not know whether 10 is included
List	No confusion	A bit impractical in some cases
Set-builder notation	Concise, clear	Can be imposing to a novice

Kinds of Sets

Finite and infinite sets In the course of this book, we will sometimes refer to *finite* and *infinite* sets. If the number of elements in a set is a whole number, that set is said to be **finite**. Some finite sets are small—for example, the set of Nobel Prize winners. Some finite sets are very large—for example, the grains of sand on all the beaches in the world. An **infinite set** has an unlimited number of members.

Consider the following infinite set. Does each way of describing the set make sense?

Verbal description E is the set of positive even numbers.
List $E = \{2, 4, 6, 8, \ldots\}$
Set-builder notation $E = \{x \mid x = 2n, n \in N\}$

Well-defined sets Thus far, each of the sets we have studied has been a **well-defined set**. That is, there is no ambiguity concerning whether a particular element is a member of the set. For example, 1 is not an element of the set of even numbers, and Virginia is not an element of the set called "states in New England." However, in everyday life, many sets are not well defined. For example, "the set of large numbers" is clearly not a well-defined set, whereas "the set of whole numbers greater than one billion" is well defined. Similarly, "the set of wealthy people" is not well defined. We can sharpen the description of this set by saying, for example, "the set of people whose net income is over $100,000." However, tax experts would still not consider this set to be well defined: If a law were enacted that imposed a tax on people whose net income was over $100,000, "net income" would need to be more precisely defined. Over the course of this book, we will frequently encounter problem situations in which part of the problem is that the problem is not well defined; in some cases, the set (of numbers or objects) in the problem is not well defined.

> ■ *Mathematics* ■
>
> Many students still falsely believe that one of the ways in which mathematics is different from other disciplines is that mathematics is "black and white"; for example, an answer is right or wrong, and there is one "best" way to solve a problem. With respect to sets, historically, a set is a set only if it is well defined. In some of the newer fields of mathematics, this "black and whiteness" is breaking down. In some fields, mathematicians speak of *fuzzy sets* and of the probability that a certain element is or is not in that set.

Defining Subsets

The notion of subset is critical because many practical applications of sets involve subsets. What do you think a subset is? Write your current definition of a subset and then read on. . . .

One informal definition of a subset is that a set is a subset of another set if it has fewer elements. Although this informal definition of subset makes sense, it presents problems in some situations. Thus, we give a more formal definition of subset:

A set X is a **subset** of a set Y if and only if every member of X is also a member of Y.

The symbol $\subseteq$ means "is a subset of." Thus we say $X \subseteq Y$. On the other hand, if a set X is not a subset of a set Y, we say $X \nsubseteq Y$.

There is actually another symbol that we can use when talking about subsets. This symbol ($\subset$) is used when we want to emphasize that the subset is a **proper subset**. A subset X is a proper subset of set Y if, and only if, the two sets are not equal *and* every member of X is also a member of Y. In the case of finite sets, this means that the proper subset has fewer elements than the given set.

If and only if We use the abbreviation **iff** to stand for "if and only if"; iff essentially means that the statement is true "backwards and forwards." For example, the definition of subset is equivalent to *both* statements below.

- If X is a subset of Y, then every member of X is also a member of Y.
- If every member of X is also a member of Y, then X is a subset of Y.

▪ Outside the Classroom ▪

Sets and subsets occur often in the business world. In fact, databases are closely connected to the idea of sets. In my office, I have a collection of articles on teaching mathematics. Let's say one article is entitled "Using Cooperative Learning to Teach Fractions." Physically, I have to store that article in one file folder—the folder called Cooperative Learning or the folder called Fractions or the folder called Elementary Math Methods. This presents a problem, because I wind up losing articles. I could make copies and have a copy in each folder, but that is costly and wasteful. With a database program, I can enter the article in a file and use the following descriptors: cooperative learning, fractions, and elementary math methods. When I want to review my resources on cooperative learning, I give a command to the computer saying, in the language of this section, "Look in the set of teaching resources and find the subset 'cooperative learning.'"

INVESTIGATION 2.3

How Many Subsets?

This investigation serves several purposes. It opens the idea of families of subsets, it shows how subsets might be useful, and it provides an opportunity to develop problem-solving tools.

Let's say that you and your friends decide to go out and get a large pizza. Let T represent the set of toppings that this restaurant offers:

T = {onions, sausage, mushrooms, peppers}

List all the possible different combinations of pizza that you could order, such as a mushroom and onion pizza. Then read on. . . .

DISCUSSION

The text will contain occasional reminders about the process of problem-solving. In this investigation, we focus on understanding the problem and looking back. Before you made your list, did you make sure you understood the question? After you make your list and before reading on, take a few moments to look back: How can you check your solution to make sure you didn't miss any combinations? . . .

TABLE 2.1

Onion	Sausage	Mushroom	Pepper
o	s	m	p
o, s	s, m	m, p	
o, m	s, p		
o, p	s, m, p		
o, s, m			
o, s, p			
o, m, p			
o, s, m, p			

STRATEGY 1: List all the combinations systematically

Begin with all combinations that involve onion, from the simplest to the most complex, and then do the same for each of the other toppings (see Table 2.1).

If you didn't use this strategy, do you understand it? If you cover up the table, can you reconstruct it? If you can't, then chances are you don't really understand this strategy.

There are many patterns within Table 2.1. How many can you see? How would you describe them so that someone who didn't see any patterns could see them from your description?

STRATEGY 2: List all the combinations using another system

Begin with all the ways to have one topping, then all the ways to have two toppings, etc. (see Table 2.2).

If you didn't use this strategy, do you understand it? If you cover up the table, can you reconstruct it? If you can't, then chances are you don't really understand this strategy.

> ■ **Mathematics** ■
>
> Do you remember Pascal's triangle from Chapter 1? It occurs in Tables 2.1 and 2.2. Can you find it?

TABLE 2.2

One of the four	Two of the four	Three of the four	All of the choices
o	o, s	o, s, m	o, s, m, p
s	o, m	o, s, p	
m	o, p	o, m, p	
p	s, m	s, m, p	
	s, p		
	m, p		

There are many patterns within Table 2.2. How many can you see? How would you describe them so that someone who didn't see any patterns could see them from your description?

One of the mistakes that students commonly make in such problems is to overlook a combination. There are patterns in both of the tables above that could help you to find all the combinations. Write down all the patterns you see and then read on. . . .

One pattern (in the "Three of the four" column) is that all of the four toppings are equally represented. Let's say you had somehow omitted the last combination: s, m, p. Looking over the three combinations you had found, you would notice that onion occurred *three* times but that sausage, mushroom, and pepper each occurred only *two* times. This observation itself names the combination you had missed: sausage, mushroom, and pepper.

Mathematically, there are 16 possible subsets of a set containing four elements. However, if you look at either of the tables above, there are only 15 subsets. What is the missing subset? If you haven't noticed it yet, stop and think before reading on. . . .

The Empty Set

The missing subset is plain pizza. How would we represent this subset using set notation? Recall that the subset called "mushroom pizza" can be represented as {m}, and the pizza with everything on it can be represented as {o, m, p, s}. Write your thoughts before reading on. . . .

One way is to put nothing inside the brackets { }. We also use the following symbol to represent a set that is empty: $\emptyset$. We use the terms **empty set** and **null set** interchangeably to mean the set with no elements.

This pizza problem also illustrates two mathematical statements that often baffle students and that will be left as exercises:

- Every set is a subset of itself.
- The empty set is a subset of every set.

Equal and Equivalent Sets

In many situations, the relationship between two sets is important. Let us examine two relationships that will come up occasionally.

Two sets are said to be **equal** iff they contain the same elements. For example, {1, 2, 3, 4, 5} = {5, 4, 3, 2, 1}.

Two sets are **equivalent** iff they have the same number of elements. More precisely, two sets are equivalent if their elements can be placed in a **one-to-one correspondence**. In such a correspondence, an element of either set is paired with exactly one element in the other set. We use the symbol ~ to designate set equivalence. For example, {United States, Canada, Mexico} ~ {1, 2, 3}.

This definition appears rather formal and intimidating to many students, who exclaim, "Why can't you just say that two sets are equivalent if they have the same number of elements?" The answer is that in most cases, this less formal description of equivalence works.

Equivalence and counting One reason for defining sets in terms of a one-to-one correspondence has to do with the difficulty young children have counting objects accurately. Watch young children trying to count the number of objects in a collection. Initially, they do not realize that each number has to match one and only one object, and so they can count a set several times and arrive at a different number each time! At some point, they realize (and you can see it in their pointing) that there must be a one-to-one correspondence between

> ■ *History* ■
>
> Earlier in this century, anthropologists discovered a tribe in New Guinea that did not have numbers for large amounts. For example, tribe members had no way of counting how many warriors went off to war. Without numbers, how do you think the tribe was able to determine how many warriors, if any, had been killed? The solution involved one-to-one correspondence. Before leaving, each warrior would place a large stone in the center of the village. When he returned, he would take one stone out of the pile. The number of stones remaining told the village how many warriors had not returned.

Venn Diagrams

Let us now explore applications of set theory. One way to represent sets is to use **Venn diagrams**, which are named after John Venn, the Englishman who invented these diagrams to illustrate ideas in logic. You have probably seen Venn diagrams used in other contexts, and we used them in Chapter 1.

Several years ago, I was teaching a graduate course in cooperative learning. One teacher explained how she had used the Venn diagram in Figure 2.1 to help her students understand the similarities and differences between butterflies and moths. In this Venn diagram, one region represents the set of moths' characteristics, another region represents the set of butterflies' characteristics, and the overlapping region represents the set of characteristics common to both.

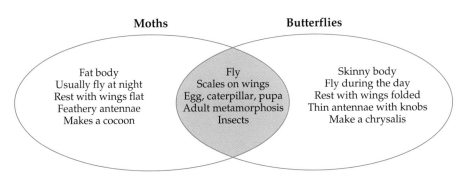

Figure 2.1

Operations on Sets

We will use Venn diagrams as we examine three operations on sets: intersection, union, and complement. Just as a doctor's operation consists of something the doctor does to a patient, a mathematical operation consists of something we do to a set of objects. In Chapter 3, we will examine the mathematical operations of addition, subtraction, multiplication, and division.

When we perform operations on sets of objects, it is often useful to refer to the set that consists of all the elements being considered as the **universal set**, or the **universe**, and to symbolize it as U. We represent U in the Venn diagram with a rectangle.

In the following discussions, we will let U be the set of students in a small class:

U = {Amy, Uri, Tia, Eli, Pam, Sue, Tom, Riki}

We begin with two subsets of U:

B = {students who have at least one brother}
S = {students who have at least one sister}

Figure 2.2 represents the students in this hypothetical class.

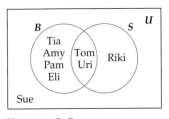

Figure 2.2

Before we discuss the various operations, does the diagram make sense to you? If you were in the class, where would your name appear? . . .

Intersection We can group this class of eight students into various subsets. How would you describe the subset consisting of Tom and Uri? Think and then read on. . . .

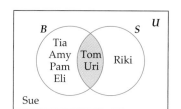
FIGURE 2.3

One way to describe this subset is "Those students who have at least one brother and at least one sister."

Mathematically, we call this subset the *intersection* of sets B and S. In mathematical language, we say that the **intersection** of two sets B and S consists of the set of all elements common to both B and S.

We represent the intersection of B and S by shading it (see Figure 2.3).

Using set-builder notation, we write

$$B \cap S = \{x \mid x \in B \text{ and } x \in S\}$$

The symbol $\cap$ is used to denote "intersection."

In Figure 2.1, the characteristics common to both moths and butterflies are listed in the intersection area. Connecting the concept of intersection to previous notation, we can say Tom $\in B \cap S$ (that is, Tom is a member of the set of students who have a brother and a sister), and we can say $(B \cap S) \subset U$ (that is, the set of students who have a brother and a sister is a subset of the entire class).

Union Let us examine another subset of the class: {Tia, Amy, Pam, Eli, Tom, Uri, Riki}.

How would you describe this subset in everyday English? Think and then read on. . . .

> ■ **Outside the Classroom** ■
>
> Can you see the connection between a mathematical intersection and a highway intersection?

There are actually several ways to describe this subset:

- Those students who have at least one brother or sister.
- Those students who have at least one sibling.
- Those students who are not an only child.

Mathematically, we describe this subset as the *union* of sets B and S. In mathematical language, we say that the **union** of two sets B and S consists of the set of all elements that are in set B *or* in set S *or* in both sets B and S.

We represent the union of B and S by shading it (see Figure 2.4).

Symbolically, we write

$$B \cup S = \{x \mid x \in B \text{ and/or } x \in S\}$$

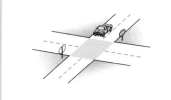
FIGURE 2.4

The symbol $\cup$ is used to denote "union."

Connecting the concept of union to previous notation, we can say Tom $\in B \cup S$, and we can also say $B \subset (B \cup S)$. Do you see why?

Complement Let us examine another subset of the class. Look back to Figure 2.2 and consider this subset of the class: {Tia, Amy, Pam, Eli, Sue}. How would you describe this subset in everyday English? Think and then read on. . . .

> ■ **Mathematics** ■
>
> If we denote the universal set as the set of integers, then the complement of the set of even numbers is the set of odd numbers. Do you see why?

One way to describe this subset is "Those students who have no sisters."

We describe this subset as the *complement* of set S. In mathematical language, the **complement** of set S consists of the set of all elements in U that are *not* in S.

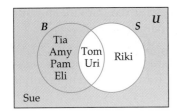

FIGURE 2.5

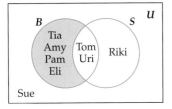

FIGURE 2.6

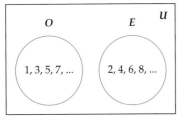

FIGURE 2.7

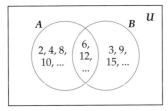

FIGURE 2.8

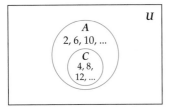

FIGURE 2.9

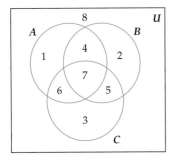

FIGURE 2.10

We represent the complement of S by shading it (see Figure 2.5).

In symbols, we write $\overline{S} = \{x \mid x \notin S\}$. We represent the complement of a set by placing a line over the set's letter.

Some people understand complement better if they think of the complement of S as "not S"—that is, all elements that are not in set S.

Subtraction Just as we have the operation of subtraction on whole numbers, we have the operation of subtraction on sets. Thinking of what subtraction means, what elements do you predict would be in B − S?

Verbally, we define set difference as the set of all elements that are in B that are not in S.

We represent B − S by shading it as shown at the left.

Symbolically, we write

$$B - S = \{x \mid x \in B \text{ and } x \notin S\}$$

Venn Diagrams and Relationships Between Sets

Venn diagrams are useful representations to help people more clearly see relationships among sets, for example, sets of numbers and sets of shapes. They are also a useful representation to help solve problems; recall our work with logic in Chapter 1.

When we are considering two sets, there are three ways in which they might be related.

(1) They can have nothing in common. In this case, we call them **disjoint sets**. Figure 2.7 illustrates this.

The set of odd and even numbers are disjoint.

$$O = \{1, 3, 5, 7, \ldots\}$$
$$E = \{0, 2, 4, 6, \ldots\}$$

(2) They can have some elements in common (Figure 2.8).

Multiples of 2 and 3 have some elements in common.

$$A = \{2, 4, 6, 8, 10, 12, \ldots\}$$
$$B = \{3, 6, 9, 12, 15, \ldots\}$$

(3) One set can be a subset of the other (Figure 2.9).

The multiples of 4 are a subset of the multiples of 2.

$$A = \{2, 4, 6, 8, 10, 12, \ldots\}$$
$$C = \{4, 8, 12, \ldots\}$$

When we are considering three sets, there are 7 possible regions where elements of the sets could lie. These regions are described in words below and in Figure 2.10.

Elements in region 1 are those elements that are in set A only.

Elements in region 2 are those elements that are in set B only.

Elements in region 3 are those elements that are in set C only.

Elements in region 4 are those elements that are in sets A and B but not C.

Elements in region 5 are those elements that are in sets B and C but not A.

Elements in region 6 are those elements that are in sets A and C but not B.
Elements in region 7 are those elements that are in sets A, B, and C.
Elements in region 8 are those elements that are in none of the 3 sets.

INVESTIGATION 2.4

Translating Among Representations

Consider the following situation:

U = the set of natural numbers from 1 to 30
A = the set of multiples of 2
B = the set of multiples of 3
C = the set of multiples of 5

Translate the following situations into notation:

A. The numbers that are multiples of 2 and 3
B. The numbers that are multiples of 2 and 3 but not 5
C. The numbers that are only multiples of 2, 3, and 5

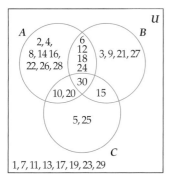

FIGURE 2.11

DISCUSSION

The numbers that are multiples of 2 and 3 can be represented as $A \cap B$. Here we are looking for what these two sets have in common.

The numbers that are multiples of 2 and 3 but not 5 can be represented as $(A \cap B) - C$. Here, we want to take away from $(A \cap B)$ those elements of set C.

The numbers that are only multiples of 2, 3, and 5 can be represented as $A \cap B \cap C$. Here, we are looking for numbers that are common to all three sets.

Throughout the book, we will use set terminology and/or set notation on those occasions where it will help to illustrate a concept or a problem.

INVESTIGATION 2.5

Finding Information from Venn Diagrams

The owners of Marksalot (a company that makes all sorts of markers) believe that healthy, happy employees are productive employees, so they had a consultant conduct an anonymous survey about employees' exercise habits, their smoking habits, and one particular aspect of their eating habits. Figure 2.12 represents the survey results.

U = the employees of Marksalot
E = the set of people who exercise regularly
S = the set of people who do not smoke at all
F = the set of employees who average at least five servings of fruits and vegetables each day, the amount recommended by the Food and Drug Administration

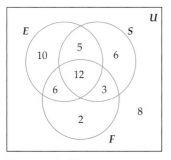

FIGURE 2.12

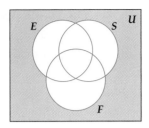

FIGURE 2.13

Answer the four questions below on your own and then check your answers. . . .

A. How many employees exercise regularly but don't average five daily servings of fruits and vegetables? Represent that subset visually and with symbols.

B. Describe the 8 employees who are outside all three circles (see Figure 2.13), first in everyday English and then with symbols.

C. Describe the following subset both in everyday English and visually:

$(E \cap F) \cup (E \cap S)$

D. Represent the results of the survey other than with a Venn diagram.

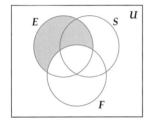

FIGURE 2.14

DISCUSSION

A. If we represent the question with symbols, we find that we want to look at the regions that are in E but not in F. There are 15 persons who match the subset "E but not F"—that is, $E \cap \overline{F}$. Visually, we have Figure 2.14.

B. In everyday English, these people do not exercise regularly, they do smoke, and they average less than five servings of fruits and vegetables a day.

In symbols, this set is $\overline{E \cup S \cup F}$. Because $E \cup S \cup F$ describes the subset of employees who fit into one or more of the healthy categories, the complement of this set describes the subset of employees who fit into none of the healthy categories. This is the subset that will probably require a disproportionate amount of medical services over their lifetimes. We will explore the concepts of probability and proportions in Chapters 6 and 7.

C. Many students find themselves intimidated by the apparent complexity of this problem. Looking at an analogous arithmetic problem may be useful. What if you were asked to find $(17 \times 34) + (30 \times 76) + (12 \times 56)$? Take a moment to think about how you might solve this problem and how this problem might be connected to finding $(E \cap F) \cup (E \cap S)$. . . .

Most students have little difficulty with the arithmetic problem. You know that you have to multiply the numbers inside the parentheses before you can add. Many mathematical operations, such as multiplication, are **binary operations**—operations that can be performed on only two elements at a time. If you had trouble with Question C, can you solve it now? Try to do so before reading on. . . .

Applying the idea of binary operations, we first find $(E \cap F)$ and $(E \cap S)$. Then we find the union of those two sets. See Figure 2.15.

One description of this set in everyday English is "Those employees who exercise regularly and who also don't smoke and/or who average at least five servings of fruit and vegetables each day." Some readers find this description to be awkward. Using the word *subset*, we can describe this group more clearly: We are talking about that subset of regular exercisers who also don't smoke and/or who eat at least 5 servings of fruits and vegetables each day.

If this explanation helped and you want to do more problems to see whether it does make sense, try some problems in the Exercises for which

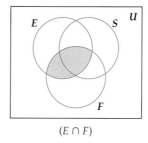 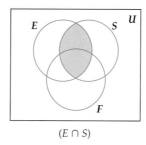 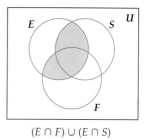

FIGURE 2.15

answers are at the back of the book. One note of caution: Being able to "get the answer" is not the same as "understanding the process." The larger goal of this book is for the mathematical concepts and operations to make sense.

D. Another way to represent the results of the survey is to list the numbers of employees in each of the eight different regions created by the Venn diagram:

- 33 people checked E
- 23 people checked F
- 26 people checked S
- 18 people checked E and F
- 17 people checked E and S
- 15 people checked F and S
- 12 people checked all 3
- 8 people checked none

We could also represent the results with a bar graph. If you don't see this possibility, look at Exercise 23.

Venn Diagrams as a Communication Tool

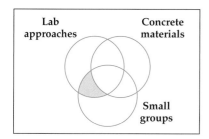

FIGURE 2.16
Source: Mathematics Resource Project, *Mathematics in Science and Society* (Palo Alto, CA: Creative Publications, 1977).

People other than mathematicians often use Venn diagrams in the same sense that "a picture is worth a thousand words." *Mathematics in Science and Society*, a book for elementary and middle school mathematics teachers, uses Venn diagrams to discuss teaching methodology. In the introduction to one of the chapters in the book, the authors make a teaching point: "This section is devoted to three valuable teacher tools: laboratory approaches, concrete materials and small groups of students. All three of these may—and quite often do—appear in the same lesson." The authors then use Venn diagrams to illustrate the kind of lesson they would be discussing. Try to describe, in everyday English, the kind of lesson represented in Figure 2.16. Then read on. . . .

They would be describing a lesson in which the students would be doing a math lab in small groups but would not be using concrete materials.

Summary

In this section, we have investigated those concepts from set theory that are most likely to occur in elementary mathematics. Although you are not likely to use these ideas in their formal sense, you will find that you encounter sets and subsets both in elementary school and in everyday life. Most sets in everyday life are finite, but many mathematical sets are infinite. Both in everyday life and in mathematics, there is confusion if a set is not well defined or clearly described. The notion of equivalent sets is one that early childhood educators see young children grapple with as they seek to make sense of relationships. Venn diagrams are used both in elementary school and in other places to describe relationships between various sets. These relationships involve the intersection, union, and complements of sets.

EXERCISES 2.1

1. Rewrite the following statements using mathematical symbols:
   a. 0 is not an element of the null set.
   b. 3 is not an element of set B.

2. Rewrite the following statements using mathematical symbols:
   a. The set D is not a subset of the set E.
   b. The set A is a subset of the set U.

3. Sets can be described in three ways: verbally, by listing the elements, or with set-builder notation. In each of the statements below, a set has been described verbally. Either describe the set in the other two ways or explain why it would be impossible to do so. Then explain which of the three descriptions you think would be most useful and why.
   a. The set of letters in the word *elementary*.
   b. The set of countries in Europe.
   c. The set of prime numbers less than 100.
   d. The set of fractions between 0 and 1.
   e. The set of students in your class.

4. Which of the sets in Exercise 3 are not well defined? Redefine those sets so that they are well defined.

5. Let U be the set of all colors and S be the subset {red, orange, yellow, green, blue, violet}.
   For the following questions, place the appropriate symbol in the blank: $\subset, \not\subset, \in, \notin$
   a. S __ U
   b. red __ U
   c. {magenta} __ U
   d. {green, blue} __ S

   Which of the following are true? Briefly explain your answer.
   e. $S \subseteq U$
   f. red $\subseteq U$
   g. gray $\in S$
   h. {green, blue} $\subseteq S$

6. Fill in the most appropriate symbol for each of the following: $\subset, \not\subset, \in, \notin$. Briefly justify your choice.
   a. 3 __ {1, 2, 3}
   b. {3} __ {1, 2, 3}
   c. {1} __ {{1}, {2}, {3}}
   d. {a} __ {a, b, c}
   e. {ab} __ {a, b, c, d}
   f. { } __ {1, 2, 3}

7. a. How many subsets does $A = \{p, i, c, k, l, e\}$ have?
   b. Can you make a generalization about the relationship between the number of elements in a set and the number of subsets? *Hint*: Look at Investigation 2.3.

8. Can you apply what you learned in Investigation 2.3 to answer the following question? There are 6 members on the Student Council. A committee consisting of 2 members is to be made. How many different committees are possible?

9. Justify the following statements, as though you were talking to a student who had not read this section.
   a. Every set is a subset of itself.
   b. The empty set is a subset of every set.

10. Make a Venn diagram with four circles. Use a compass or another device to make good circles. How many distinct regions are there in this diagram? What patterns do you notice in this diagram?

11. Find a Venn diagram from a newspaper, magazine, or elementary school mathematics book. Describe the Venn diagram in words.

12. Let $U = \{x \mid x \text{ is an American}\}$
    $F = \{x \mid x \text{ is a female}\}$
    $S = \{x \mid x \text{ is a smoker}\}$
    $P = \{x \mid x \text{ has a health problem}\}$
    a. Represent the following description with a Venn diagram and with symbols: An American nonsmoking healthy female.

b. Represent $F \cap (S \cup P)$ with a Venn diagram and in everyday English.

c. Represent $(F \cup P) \cap \overline{S}$ with a Venn diagram and in everyday English.

d. Convert the information from the first diagram into everyday English and into symbols.

e. Convert the information from the second diagram into everyday English and into symbols.

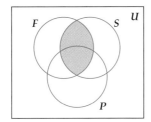

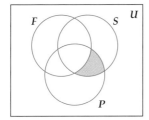

13. In the following Venn diagram,

    F = the set of students in the film club
    S = the set of students in the science club
    C = the set of students in the computer club

    Describe the following sets in English and in symbols.

    a.
    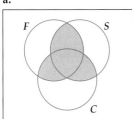

    b.
    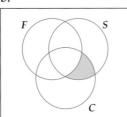

    Describe the following subsets in symbolic language and with a Venn diagram.

    c. Those people who are in the science and film clubs but not the computer club

    d. Those people who are in none of the three clubs

14. Use the sets below to answer parts (a) through (f):

    $U = \{1, 2, 3, 4, 5, 6, 7, 8, 9, 10, 11, 12, 13, 14, 15, 16, 17, 18, 19, 20\}$

    A: the numbers in U that divide 12 with no remainder
    B: the numbers in U that divide 15 with no remainder
    C: the numbers in U that divide 20 with no remainder

    a. Make a *clear* Venn diagram showing the sets U, A, B, and C.

    b. Represent the following subset with symbols: $\{7, 8, 9, 11, 13, 14, 16, 17, 18, 19\}$.

    c. Represent the following subset in everyday English and with symbols: $\{2, 4\}$.

    d. Represent the following subset in everyday English and with a diagram: $\overline{A \cap B}$.

    e. Represent the following subset in everyday English and with a diagram: $\overline{A \cup B}$.

    f. Represent the following subset with a diagram and with symbols: Those numbers that divide 12 or 15 with no remainder.

15. An elementary teacher has asked her students to place their names in the region that represents their answers to the question "Which pets live in your home?"

    D = the set of students who have at least one dog
    C = the set of students who have at least one cat
    O = the set of students who have a pet that is not a dog or a cat

    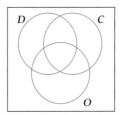

    Describe the following sets in everyday English and then with a diagram:

    a. $C \cap D$ b. $\overline{D \cup C}$ c. $C \cap O \cap D$

    Describe the following sets in symbols and then with a diagram:

    d. Students who have dogs but not cats

    e. Students who have at least one pet

    f. Students who have other pets but neither cats nor dogs

    Describe the following sets in everyday English and then with symbols:

    g.
    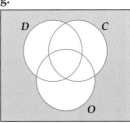

    h.
    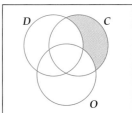

16. Why do you think we use circles instead of squares or triangles or other shapes when we make Venn diagrams? For example, is the representation below equivalent to the standard Venn diagram with three overlapping circles?

Explain your answer as though you were talking to a fellow student who does not understand.

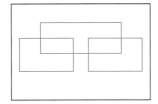

17. Recall the different kinds of sets of numbers described earlier: natural numbers, whole numbers, integers, and rational numbers. Make a Venn diagram to represent how these sets are related.

18. Pollsters often ask people's opinions. Politicians want to know how their position on an issue is viewed by particular constituencies—for example, by young voters, by African Americans, by women, by those who belong to the Sierra Club. Decisions about policies are often made on the basis of this polling information. Let's say that a public opinion survey was conducted to determine how much support there was for the president's policies. People were asked three questions:

   Do you support the president's economic policy?

   Do you support the president's foreign policy?

   Do you support the president's social policy?

   Let E, F, and S denote the sets of persons responding yes to the first, second, and third questions, respectively. The results of the survey are shown in the Venn diagram below. The numbers represent the percent of respondents.

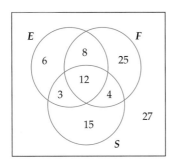

   a. What percent agree with his economic policy?
   b. What percent agree with just one of his policies?
   c. Describe the subset that would be represented by $F \cup (E \cap S)$, either in everyday English or by shading the appropriate portion of the diagram.
   d. If the president could make one single region of the Venn diagram larger (that is, make it have more members), which would it be? Why?

19. In a group of 120 students, 75 know how to use a Macintosh, 65 know how to use an IBM-compatible computer, and 20 do not know how to use either kind.

   a. How many students know how to use both kinds of computers?
   b. Are the sets in this problem well defined or not? Justify your response.

20. An advertising firm found that a certain ad that ran on both radio and TV was only heard on the radio by 21 percent of the people and was only seen on TV by 33 percent of the people. Just 10 percent of the population both heard the ad on the radio and saw it on TV.

   a. What percent of the people in the area has neither seen nor heard the ad?
   b. What percent of the people in the area only heard the ad on radio or only saw the ad on TV?

21. Refer to the discussion about the book *Mathematics in Science and Society* in the last subsection.

   a. What would be the focus of the authors' discussion following the Venn diagram below? Describe your answer in everyday English.

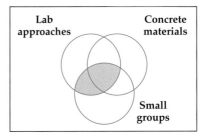

   b. The following Venn diagram appeared in the same book.

   Why do you think the authors used a question mark here?

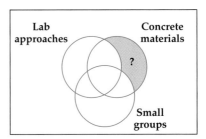

22. Make up a situation for which the following Venn diagram is appropriate.

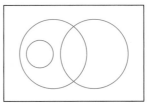

23. I was curious about the number of high-tech machines that my friends have, so I asked them if they had a cell phone, an iPod, and/or a DVD player. I have represented the results of my survey with a Venn diagram and with two different bar graphs.

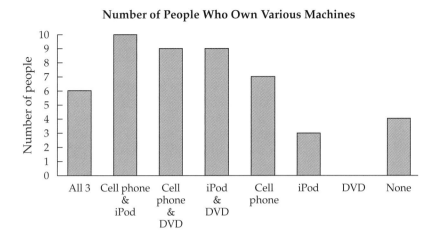

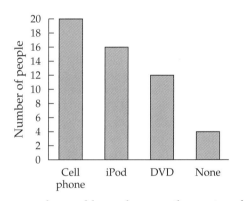
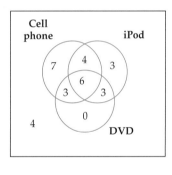

a. Which of the three graphs would you choose as the most useful? Why?
b. Jack thinks there is a mistake in the bar graph at the top. Assuming that the Venn diagram is accurate, do you agree with Jack? If you do, explain the mistake. If you don't, explain why, as though you were trying to convince Jack.
c. How many friends did I question?
d. From the Venn diagram, what conclusions could you draw about the relationships that characterize ownership of the three items?

24. Consider the following two subsets of U, the set of all people:

   I = the set of intelligent people
   S = the set of successful people

a. By yourself, answer the following question: How would you represent these two subsets with a Venn diagram?
b. Discuss your responses with other classmates. What did you learn from this activity? How did the Venn diagram contribute to this learning?

SECTION 2.2 ALGEBRAIC THINKING

WHAT DO YOU THINK?

- What does it mean to say that two sets are functionally related?
- What are some examples of functions in everyday life?
- Can we have functions without numbers?
- What is the reason for developing algebraic thinking in elementary school?

The NCTM repeatedly stresses that from the earliest grades, the curriculum should give students opportunities to focus on recognizing, describing, and extending patterns. Work with patterns helps them to see that "regularity is the essence of mathematics."[2] From these explorations, "algebra evolves as a general language to describe patterns in the world around us."[3] The NCTM states that elementary teachers should have sufficient familiarity with the "ideas of algebra"[4] so that their students will explore and develop, before they enter high school, the concepts and skills that are essential prerequisites to formal algebraic thinking. Students at the elementary school level can explore many kinds of relationships between two variables—for example, the relationship between the slope of a ramp and the distance traveled by a marble released at the top of the ramp. Understanding when relationships are functional and when they are not functional is one of the cornerstone ideas of algebra. However, algebraic thinking is more than just patterns and functional relationships. When I inform my students that they need to understand algebraic thinking, "hooray" is not the first reaction I hear! Thus, in order for you to want to learn the ideas in this section, I must answer, to your satisfaction, two related questions: (1) What do we mean by algebraic thinking? (2) What does algebraic thinking have to do with elementary school mathematics?

First, we need to address what algebra is and what it isn't. Algebra is not *just* a set of rules and procedures, though these rules and procedures are helpful when we are solving certain kinds of problems. Algebra is also a set of structures that make it easier for us to understand connections between certain kinds of problems and make it easier to solve problems. Algebra is also the study of relationships among variables—when variables are related in "nice" ways, we call that relationship a function. Finally, algebra can be seen as generalized arithmetic. We will explore each of these conceptions in turn, and I promise to present, in a basic way, the fundamental role or value of each conception and to show how this conception is manifested in the elementary classroom.[5]

Now, let me say a little bit more about what algebra isn't. Algebra is not a discrete subject that should be taught in ninth grade but rather a strand that should be integrated with other branches of mathematics, including geometry, statistics, and discrete mathematics. In fact, there is a trend in U.S. high schools to reconnect algebra and geometry rather than to teach them as separate courses.

Let us now explore and examine each of these conceptions of algebra in turn.

History

The word *algebra* first appears in "Hisab al-jabr walmuqabala" (*The Science of Reduction and Cancellations*) by Al-Khowarizmi around A.D. 825. It is regarded as the first book to have been written on algebra. Before the development of base 10 and modern notation, solving even very simple equations was difficult. His book was intended to be very practical; he introduced algebra as a means of finding quicker ways to solve problems that were part of everyday life at that time. Over time, *algebra* came to mean the study of equations.

---

[2] NCTM, *Curriculum and Evaluation Standards for School Mathematics* (Reston, VA: NCTM, 1989), p. 60.
[3] Elizabeth Phillips et al., *Patterns and Functions, Addenda Series. Grades 5–8* (Reston, VA: NCTM, 1991).
[4] Arthur F. Coxford, ed., *The Ideas of Algebra, K–12: 1988 Yearbook* (Reston, VA: NCTM, 1998).
[5] Several of the ideas in this section are adapted from "Conceptions of School Algebra and Uses of Variables" by Zalman Usiskin, published in *The Ideas of Algebra, K–12*, which was the 1988 Yearbook of the NCTM.

Algebra as a Set of Rules and Procedures

Think of something you do well now, such as skiing, playing softball, cooking, or knitting. Each of these areas has a set of rules and procedures that you need to learn in order to perform well. Now think of when you first learned to ski, to play softball, to cook, or to knit. You probably remember making mistakes and feeling overwhelmed at times, and you probably remember having fun too.

All this can be true in algebra as well. Now some of you might be thinking "Yeah, except for having fun." But skiing, playing softball, cooking, and knitting are not fun for everyone either. The point here is that learning the rules and procedures takes some time, and when most people understand the rules and procedures in a given situation or setting, they generally feel confidence and enjoyment. An important realization is that many of the rules and procedures in algebra are actually extensions of rules and procedures in arithmetic. As I stated in the Preface, seeing connections is a key theme of this book, and algebra has rich connections both to arithmetic and to geometry. We will explore some of these connections in more detail soon, but right now we are going to focus on one aspect of symbol manipulation in algebra that begins in elementary school.

INVESTIGATION 2.6

CLASSROOM CONNECTION

More than one high school teacher has reported a variation of the following response by a student: "If $x = 6$, what did you call it x for?"

■ *History* ■

The notation that we use to express algebraic ideas developed over a period of hundreds of years, and the driving force for inventing this notation and language was to make it easier to solve problems and communicate solutions and ideas. For example, in the 1500s, the expression $x^2 + 2x - 8$ would have been written as Zp 2Rm 8; R for an unknown, Z for zensus (an unknown squared), p for plus, and m for minus.

A Variable by Any Other Name Is Still a Variable

A crucial idea in algebraic thinking is that of the variable. In each of the following examples, the variables have a slightly different meaning. On a separate piece of paper, write down what each of the variables means, and then read on. . . .

$C = \pi d$

$5x = 30$

$\sin x = \cos x \cdot \tan x$

$1 = n \cdot (1/n)$

$y = kx$

DISCUSSION

Most people will call the first example a formula. C and d stand for circumference and diameter, whose values vary according to the circle; however, the value of π (3.14) does not vary. The second example is generally called an equation. Although x is the variable, in this case its value is 6. The third example is an example of an identity; it is true no matter what the value of x.

We generally refer to the fourth example as a property; in this case, n is used as a symbol to represent a property that is true for all numbers (except 0); the formal name of this property is the multiplicative inverse property. In the fifth example, the k is used to enable us to represent a family of functions in which the independent variable (x) and the dependent variable (y) are related in a certain way—in this case, a linear relationship.

Classroom Connection

This is a true story. A child comes home from the first day in first grade. His mother asks him what he learned, and he says, "I learned that five plus three equals eight." "That's great," says his mother, at which point the child asks, "Mommy, what does equals mean?"

Table 2.3

Example	What it is usually called	What the variables stand for
$C = \pi d$	Formula	C and d are concrete quantities. π is a constant.
$5x = 30$	Equation	x is the unknown.
$\sin x = \cos x \cdot \tan x$	Identity	x is an argument of a function.
$1 = n \cdot (1/n)$	Property	n is a symbol to represent a generalization.
$y = kx$	An equation of a function of direct variation	x is the independent variable, y is the dependent variable, and k is a constant.

Table 2.3 illustrates some of the important differences in these examples.

My sense is that some of the readers are now asking, "So what does this have to do with elementary school?" Lots! If children simply learn their addition and multiplication facts by rote, such as $6 + 5 = 11$ and $9 \cdot 7 = 63$, their ability to apply this knowledge is limited. The people who write curriculum materials know that, and thus you will also find problems like the following throughout elementary school textbooks:

Fill in the blank: $\square + 6 = 14$ and $\square \times 12 = 96$.

They do this because the goal is for students not just to know facts but also to see relationships between operations. Recall the conception of understanding, presented in Chapter 1, as the quantity and quality of connections that the learner has made. For example, in the first case, "something plus 6 equals 14" is translated as "that something is equal to 14 minus 6." Students need this flexibility, which comes from consistent application in classrooms of the process standards—problem-solving, reasoning, communication, connections, and representation.

There is another aspect of symbol manipulation that has to do with elementary school mathematics. In all of the five examples above, the equations can be manipulated. For example, we can change $C = \pi d$ into $\pi = \frac{C}{d}$. This is not just a manipulation but also a different representation of the idea. This representation highlights the fact that π is the ratio of the circumference of any circle to the diameter of that circle. During elementary school, children will learn lots of relationships that are expressed with algebraic language; for example, $A = LW$, $a + b = b + a$, an average (mean) is obtained by adding all the data and dividing by the number of data, etc.

Since the development of the ability to manipulate algebraic equations and expressions is one goal of high school mathematics, we will not spend much time on manipulation of algebraic expressions in this course, except to remind you that it is important for you to realize that algebra does have its own notation and language. Some of you are more comfortable and familiar with this language than others (think about the different approaches people took to Investigation 1.1 with the pigs and chickens). What is crucial in elementary school is that children develop the habit of making sense of manipulation of numbers. From the very beginning of school mathematics, children learn to manipulate mathematical symbols. During grade school, they learn to manipulate number sentences and to solve computation problems by breaking them down; for example, 8×6 is double 4×6—that is, $8 \times 6 = 2(4 \times 6)$. They also learn to use a variety of symbols: $=, +, -, \times, \div, >,$ and $<,$ to name just a few.

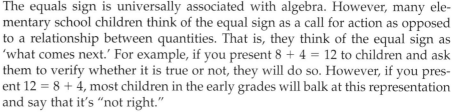

The Equals Sign and Equivalence

The equals sign is universally associated with algebra. However, many elementary school children think of the equal sign as a call for action as opposed to a relationship between quantities. That is, they think of the equal sign as 'what comes next.' For example, if you present 8 + 4 = 12 to children and ask them to verify whether it is true or not, they will do so. However, if you present 12 = 8 + 4, most children in the early grades will balk at this representation and say that it's "not right."

In learning to work with equals signs in more powerful ways, the larger idea is *equivalence,* which is one of the big ideas of mathematics. When you work with young children, your curriculum will provide various explorations to help students develop a richer understanding of the equal sign.

You will see the notion of equivalence throught this book. We discussed equivalence of different representations in Chapter 1. With respect to the equal sign signifying that the quantities on each side have the same value, recall Investigation 2.6. When we say that $\cos x \cdot \tan x = \sin x$, we are saying that $\cos x \cdot \tan x$ has the same value as $\sin x$. In the next chapter, we will see that children need to see the equivalence of 64 and 50 + 14 in order to subtract. In Chapter 5, we will examine equivalent fractions and the equivalence of fractional and decimal representations of the same amount.

Algebra as the Study of Structures

All mathematical structures can be described using algebraic notation. Many of these structures are at the heart of the elementary school curriculum:

- Properties of operations—closure, commutative, associative, identity and inverse, distributive, etc.
- Properties of numbers—for example, an "even" number can be divided into two equal halves, whereas an odd number cannot. The algebraic representation of an even number is $2n$, and the algebraic representation of an odd number is $2n + 1$ or $2n - 1$.
- Connections between operations, such as $a - b = a + {}^-b$

We will explore these structures in more detail in Chapters 3 through 6. Let me offer one example here. I have had the privilege of being in a classroom at the moment when a child has discovered that 2 + 9 (which is pretty hard for many first-graders) is "the same as" 9 + 2. This discovery is quickly generalized as "You can switch the numbers when adding." In this case, the student has discovered a very important structure in the system of whole numbers that we call the commutative property of addition. It is stated in its most concise form as $a + b = b + a$ for any whole numbers a and b. In elementary school, coming to understand structures often comes from "playing" with patterns and asking, "What do you see?" As the quote from Mary-Baratta Lorton in Chapter 1 indicates, if children come to expect mathematics to make sense and look for patterns, the transition from arithmetic to algebra will be so much smoother than it was for most readers of this book.

Is it crucial that second-graders can say and spell "commutative"? I don't think so. Is it crucial that they can apply this concept when adding and that they know it is not true when subtracting? Absolutely. Researchers have found

---

[6]Thomas Carpenter, Megan Frank, and Linda Levi, *Thinking Mathematically: Integrating Arithmetic and Algebra in Elementary School* (Portsmouth, NH: Heinemann, 2003), p. 12.

History

The word *function* was originally introduced by Leibniz, one of the inventors of calculus, to denote any quantity connected with a curve. Many famous mathematicians had a conception of functions that would be considered erroneous today. For example, Bernoulli regarded a function as any expression made up of a variable and some constants, and Euler regarded a function as any equation or formula involving variables and constants. The familiar notation $f(x)$ that we use today was introduced by Clairaut and Euler in the 1700s.

---

that the mistake shown below is common in second grade. In the following example, the student not only had trouble with place value and "borrowing" but also mistakenly assumed commutativity in subtraction. That is, the child assumed that $2 - 6 = 6 - 2$.

$$\begin{array}{r} 52 \\ -16 \\ \hline 44 \end{array}$$

In high school, students are expected to learn to do formal proofs and to know the properties in formal language. In elementary school, we want the students to continue to ask "why" and "what if" questions (Why doesn't the order matter when you add but it does when you subtract? What if you multiply? What if you are dealing with big numbers? Will it work with fractions too?) and we want them to have an understanding of basic structures and be able to apply this knowledge when solving problems. We will revisit this conception of algebra throughout the book, especially in Chapter 3 as we examine the operations of addition, subtraction, multiplication, and division with whole numbers.

Algebra as the Study of Relationships Among Quantities

"The concept of function is an important, unifying idea in mathematics. Functions which are special correspondences between the elements of two sets are common throughout the curriculum. In arithmetic, functions appear as the usual operations on numbers, where a pair of numbers corresponds to a single number, such as the sum of the pair; in algebra, functions are relationships between variables that represent numbers; in geometry, functions relate sets of points to their images under motions such as flips, slides, and turns; and in probability, they relate events to their likelihoods. The function concept also is important because it is a mathematical representation of many input–output situations found in the real world, including those that recently have arisen as a result of technological advances."

(*Curriculum Standards*, 1989, p. 154)

Many students come to college confusing functions with equations and then find college mathematics very difficult. Success at higher levels of mathematics rests on more than just being able to manipulate equations. In elementary school, children will not investigate functions and other algebraic ideas at a formal level. However, through their investigations, they do need to realize that relationships often exist between different variables (for example, the area of a rectangle varies according to the length and the width of the rectangle); that patterns and relationships can be represented in various ways (with words, symbols, graphs, and pictures); and that being able to communicate clearly the patterns they see is important.

If students engage in these kinds of investigations during elementary and middle school, then they will see many of the concepts that are developed at the formal level in high school as extensions of what they encountered earlier, not as something that is entirely new and foreign. We will now focus on the concept of a function, which is one of the concepts that underlies virtually all algebraic thinking.

A simple example of a function is the hourly wage. If you know that you are being paid at the rate of $8.00 per hour, you expect to receive $320 for working 40 hours, or $160 for working 20 hours, or $40 for working 5 hours. In other words, there is a clear, consistent relationship between the variable "hours

TABLE 2.4

Hours worked	Dollars earned
1	8
1.5	12
2	16
3	24
4	32
5	40
40	320

worked" and the variable "dollars earned." We can create a table of values for this function (see Table 2.4).

Formally, we define a **function** as a relationship between two sets in which each element of the first set, called the **domain** of the function, is matched with *exactly* one element of the second set, called the **range** of the function. In Exploration 2.3, you found functional relationships between other pairs of variables.

In the wage example, the first column (hours worked) represents the domain, and the second column (dollars earned) represents the range. Technically, what is in these columns is a subset of the domain and a subset of the range, respectively.

It is important to note that not all relationships between two sets are functions. For example, consider the relationship between the set of positive whole numbers and their square roots. In this case, if we ask, "What is the square root of 9?" the answer is +3 *and* −3. That is, each member of the set of positive whole numbers is matched with two square roots.

When we go outside the classroom to look for instances of functions in situations, we find that their value is often the predictability of the relationship. Consider the following questions:

- If I sell 50 computers, what will be my commission?
- If 75 students sign up for a course, how many sections will be offered?
- What are the consequences when you chew gum in Appleby High School?

If Jack sells 50 computers and receives $1000, whereas Jill sells 50 computers and receives only $700, then there is not a functional relationship between the number of computers sold and the commission.

If one department creates three sections when 75 students preregister and another department creates only two sections, then we say that at that college there is not a functional relationship between the number of students and the number of sections of a course.

If Cindy chews gum in one class and gets detention and chews gum in another class and receives no detention, then there is not a functional relationship between chewing gum and consequences.

Inputs, outputs, and function notation Many elementary teachers use a game called "What's my rule?" to introduce children to the idea of functions. The students give a number (the *input*); then the teacher performs an operation and gives them the number associated with that input (the *output*). The students then try to guess what the teacher is doing to the inputs. A game of "What's my rule?" is given below.

Teacher: What's my rule?
Student: 3
Teacher: 6
Student: 4
Teacher: 7
Student: I think I know the rule.
Teacher: What is your rule?
Student: You are adding 3. If I give you 10, you will say 13.
Teacher: That's right!

The teacher is adding 3. Algebraically, we would say $y = x + 3$. We can also use **function notation** to write the rule for adding 3 to any number: $f(x) = x + 3$. This mathematical sentence is read "f of x equals x plus 3." That is, $f(x)$ means "function of x."

Function notation is more precise because it emphasizes the relationship between the **input**, which is another name for a member of a function's domain,

CLASSROOM CONNECTION

I love this example from *Teaching Children Mathematics*, February 1997, p. 266. The pattern was "add two to the number." This was one student's response: "You say the first number, you don't say the next number, and you say the next number." Experienced teachers know that children will conceptualize and verbalize mathematical situations and problems in many different ways. Even though some ways are more efficient than others, the task is to begin with what the child is thinking and then move from there.

Input	Output
1	4
2	7
3	10

and the output, which is the corresponding member of its range. That is, the output is a function of the input. In this particular case of "What's my rule?" the output $f(x)$ is determined by adding 3 to the input x, and so we have

$$f(3) = 3 + 3 = 6$$
$$f(4) = 4 + 3 = 7$$
$$f(10) = 10 + 3 = 13$$

Let's play another round. What is the rule? Think and then read on. . . .

Using everyday English, we could say that the rule (or function) is to triple the input and add 1. Using function notation, we could say that $f(x) = 3x + 1$.

Elementary teachers often make a "function machine" to illustrate these situations. Function machines for the two situations just presented can be seen in Figure 2.17.

In this game, the domain and range do not have to be sets of numbers. Can you think of an example in which the domain and range are not numbers? Try to do so before reading on. . . .

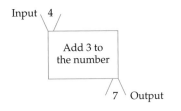
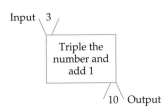

FIGURE 2.17

Here is one example from Margie Hoey, a second-grade teacher with whom I worked. One student placed some students in one group and some students in another group. The determining factor was whether the student was wearing blue or not. That is, the domain set consisted of the set of students in the class. The range set had two elements: "is wearing blue" and "is not wearing blue." For example,

Input	Output
Anastasia ⟶	Is wearing blue
Benito ⟶	
Cecilia ⤫	
Damon ⤫	
Emilia ⟶	Is not wearing blue

Do you see why this student's "rule" represents a function? Try to make up more rules that do not involve numbers.

INVESTIGATION 2.7 — Baby-sitting

Let us extend our understanding of functions by examining another relationship between the two variables "hours worked" and "dollars earned." Let's say Ellen baby-sits and charges $6 per hour. Describe how you would determine how much to pay Ellen. Determine whether the relationship between the two variables—hours and dollars—is a function. Whether you say it is or is not a function, can you justify your response? Then read on. . . .

TABLE 2.5

h	d
1	6
2	12
3	18
4	24
5	30
.	.
.	.
.	.

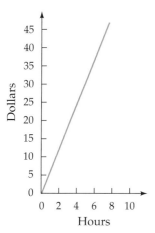

FIGURE 2.18

DISCUSSION

A common response to describing how to pay Ellen is to multiply the number of hours sat by 6. Let us use this situation to introduce several different ways to represent functions. Then we will examine more closely the relationship between time sat and dollars paid.

For purposes of simplicity, let us first look at different representations of this function using only whole-hour amounts.

Tables We can represent this function with a table in which h represents hours and d represents dollars (see Table 2.5). We can then use the table to determine how much money Ellen will make from baby-sitting.

Equations We can represent this function with an equation in which h represents hours and d represents dollars:

$$d = 6h$$

Graphs We can represent this function with a graph (see Figure 2.18).

In some cases, like this one, many people find it more convenient to refer to the input as the **independent variable** and to the output as the **dependent variable**. In this case, we say that the independent variable is the hours baby-sat and the dependent variable is the dollars earned. That is, the number of dollars a baby-sitter earns is dependent on the number of hours that he or she baby-sits.

Ordered pairs We can represent this function as a set of **ordered pairs** in which the first element of each ordered pair represents hours and the second element represents dollars:

$$B = \{(1, 6), (2, 12), (3, 18), (4, 24), \ldots\}$$

Mappings Finally, we can represent functions with arrow diagrams. Mathematicians often refer to this representation as a **mapping** of one set onto another set. We shall soon use this representation when we look at different kinds of functions.

h		d
1	→	6
2	→	12
3	→	18
4	→	24

A closer look at paying the baby-sitter At first glance, the question of how much to pay the baby-sitter is simple: Multiply the hours sat by 6. However, let us use the problem-solving strategy "act it out" to examine this problem more closely. For example, what if Ellen baby-sat from 7 to 11:15? How much would you pay her? Think before reading on. . . .

Some people say $24. Some people say $27—they round up to the nearest half-hour. In actuality, different people have different ways of determining how much to pay a baby-sitter. Let us examine the case of a couple, whom we will call the Alomars, who determine how much to pay their sitter in the following way: They round up the time to the nearest half-hour. We could now represent the Alomars' process for paying the baby-sitter in each of the ways we have just examined. Is the relationship between time sat and dollars earned a functional relationship in the case of the Alomars? How can you justify your answer? Think and then read on. . . .

Many students find that they need to represent the relationship to help them answer this question. Let us examine one representation more closely.

If you were to graph this relationship (using the Alomars' process), what would the graph look like? Try to make your own graph before reading on. . . .

The graph is shown on page 125. Take a look at the graph and see whether you can make sense of it before reading on.

If you are having trouble making sense of the graph, consider a few examples. Let's say the sitter sits for 2 hours and 50 minutes. Using the Alomars' process, we round this time to 3 hours and pay Ellen $18. As you can see, Ellen will receive $18 for all times between 2 hours and 31 minutes and 3 hours. So is this a function?

It is, because for any specific time (input), there is exactly one dollar amount (output) associated with that time.

The graph that represents this baby-sitting function is often confusing to students who see it for the first time. However, it is relatively common in real-world mathematics and is a member of the subset of functions called step functions.

Mathematical modeling As you have discovered in this section and in other mathematics courses, upon closer examination, a problem is often more complicated than it first appears. This is what makes "real-life" problems so challenging. The NCTM has stated that mathematical modeling should be an important part of mathematics education in school. Let us take some time to understand this concept and its importance. We will begin by examining the word *model*. A model of an object is not the object and is often a scaled-down version of the object. We speak of model airplanes, and architects and engineers build a model of a building or a bridge before constructing the actual object. In one sense, a model is a representation of the original object. There are two important features of many models. First, the model contains many of the properties of the original object. Second, the model can be manipulated and studied to help us better understand the (usually larger and more complex) object. A mathematical model is a mathematical structure that approximates the features of a situation. A mathematical model can be an equation (or a set of equations), a graph, or some other kind of diagram.

In this case, we have seen that the original straight-line graph turns out not to be a useful or accurate model of the baby-sitting situation. The step graph is a more accurate representation. This process of examining a situation and then developing a model that accurately represents that situation is called mathematical modeling. Constructing and interpreting mathematical models is one of the more important uses of mathematics in the real world. Over the course of this book, you will have many opportunities to construct and interpret mathematical models.

> ■ *Outside the Classroom* ■
>
> An important aspect of modeling was seen in Investigation 1.1 when some students model the problem by drawing circles to represent all the animals and two lines to represent chickens. The problem is solved by doing what is biologically impossible but mathematically possible—turning chickens into pigs (by adding two legs to each animal until we have a total of 48 legs). An important aspect of a model is that it makes it easier to "play" with the problem in a way that we could not do with the actual situation.

INVESTIGATION 2.8 — Choosing Between Functions

Jackie has been promoted to chief salesperson of Southside Computers. She has two choices as to how she will be paid. She can receive $50 for every computer system she sells, or she can receive $250 a week plus $25 for every computer system she sells. Looking at the past year, she finds that she has averaged 9.3 sales per week, and she figures that she can do a little better this year. What would you recommend? Work on this problem for a bit and then read on. . . .

DISCUSSION

You may want to start by plugging in a few numbers—that is, using guess–check–revise. Do you see that if Jackie sells only a *few* computers, she will be better off with the second plan but that the first plan is better if she sells a *lot* of computers? This realization leads to a specific mathematical question: At what point will the two plans give her the same money? Once she can answer that question, she can base her decision on whether she believes she will sell more or less than that amount. Below we see three very different strategies for solving this subproblem: guess–check–revise, use algebra, and make a graph. If you already solved the problem using one strategy, try to use the other strategies now and then read on. . . .

STRATEGY 1: Guess–check–and revise

This strategy is illustrated in Table 2.6.

TABLE 2.6

Sales	Plan A	Plan B	Reflection/Analysis
0	0	250	
2	100	300	Plan A goes up 100; plan B goes up 50. Plan A is still $200 below plan B. Each time you increase sales by 2 computers, plan A increases $50 *more* than plan B does.
10	500	500	

What does the last row mean? Think and read on. . . .

It means that Jackie makes the same amount under both plans when she sells 10 computers a month. If she thinks she will sell more than 10, then plan A is better for her.

STRATEGY 2: Use algebra

Let y = Jackie's weekly salary.
Let x = the number of computer systems she sells.

The equation that represents plan A is

$$y = 50x$$

The equation that represents plan B is

$$y = 250 + 25x$$

Do you understand how both of these equations were constructed? We have two equations in two unknowns that we can solve:

$$50x = 250 + 25x$$
$$25x = 250$$
$$x = 10$$

STRATEGY 3: Make a graph

Where the two lines cross will show the point at which her earnings under both plans will be equal.

History

René Descartes (1596–1650) invented the Cartesian coordinate system (named after him) to illustrate such relationships on graphs. Legend has it that he got the idea from watching a fly crawl across a ceiling. He realized that a coordinate system could represent the fly's location with one symbol—that is, the ordered pair.

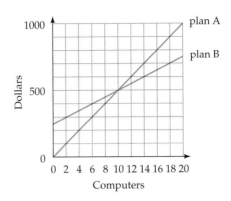

FIGURE 2.19

The two lines cross at the point (10, 500) (see Figure 2.19). In other words, the ordered pair (10, 500) is an element of both graphs.

Many of the investigations and explorations deal with the question of answering questions about change. Recall the discussion of *On the Shoulders of Giants* in Chapter 1—understanding change is regarded by many as one of the most important ways in which mathematics is used. For example, manufacturers test different shapes and materials for cups and then measure the temperature of the coffee to see how fast it cools; scientists measure the effectiveness of a new medicine by measuring how fast it kills the harmful bacteria. Some common elementary school experiments with change include measuring the weight of apple slices over time to measure dehydration and measuring the growth of plants to compare different conditions—more/less sunlight, more/less water, good/poor soil. Up to now, the investigations in this section have all been quantitative—that is, you have used numbers. In the next two investigations, we will examine various situations involving change, but we will move away from numbers deliberately. Doing so will make the algebraic ideas come more to the foreground.

INVESTIGATION 2.9

Matching Graphs to Situations

Below are descriptions of three runners in a race and three graphs. Match each description to the correct graph and explain your choice. The independent variable is time, and the dependent variable is distance from the start.

Alex started slowly, then ran a bit faster, and then ran even faster at the end of the race.

Manuel started quickly but then tired and slowed down a bit and then slowed down even more at the end.

Ragib started quickly, stopped to tie his shoe, and then ran even faster than before.

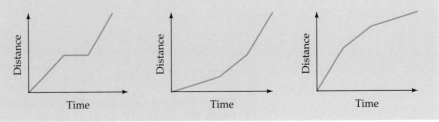

DISCUSSION

A slower speed has a smaller slope than a faster speed. Do you see why? How would you explain this to someone who doesn't? Let us use numbers to illustrate this idea. If someone goes 4 feet in 1 second, then in terms of graphing, we move 1 unit to the right (time) and 4 units up (distance). If someone goes 8 feet in 1 second, we move 1 unit to the right and 8 units up. Stopping means that the runner gets no closer to the end and so is represented by zero slope (horizontal line). Thus the second graph matches Alex—less steep than the others at the beginning and then steeper and steeper. The third graph matches Manuel—steeper than Alex at the beginning and then less and less steep. The first graph matches Ragib—steep at the beginning, horizontal in the middle to show that he stopped, and then even steeper than before.

INVESTIGATION 2.10

2.5

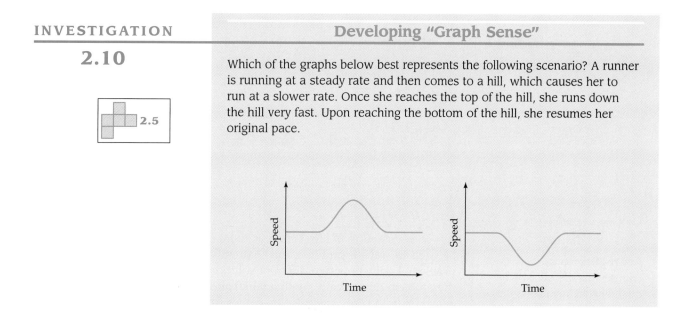

Developing "Graph Sense"

Which of the graphs below best represents the following scenario? A runner is running at a steady rate and then comes to a hill, which causes her to run at a slower rate. Once she reaches the top of the hill, she runs down the hill very fast. Upon reaching the bottom of the hill, she resumes her original pace.

DISCUSSION

Many students pick the first graph because that is what the race looks like—level, then a hill, and then level again. But the graph is a picture not of the layout of the race but, rather, of the person's speed. A steady speed means that the speed is not changing and is thus represented by a horizontal line. When the runner slows down the slope of the graph decreases; when she speeds up, the slope of the graph increases. Thus the second graph is correct.

Algebra as Generalized Arithmetic

This conception of algebra is related to algebraic structures and to functions, but having you see the connectedness between algebra and arithmetic is so important that we will explore it in another conception. In Investigation 2.11, we open the discussion of this conception of algebra.

INVESTIGATION 2.11

In chapter 1, we discussed the value of being able to recognize and use patterns in solving problems. Exploring patterns in elementary school is an important part of developing algebraic reasoning. However, it is important to state that patterns alone do not necessarily develop algebraic thinking. Algebraic thinking develops as the students learn to analyze various patterns (numerical and visual) in order to make and test predictions and finally to make generalizations. The next investigation offers rich possibilities for developing algebraic reasoning by investigating patterns.

Looking for Generalizations

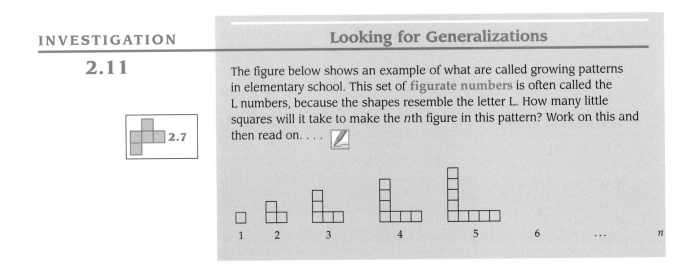

2.7

The figure below shows an example of what are called growing patterns in elementary school. This set of **figurate numbers** is often called the L numbers, because the shapes resemble the letter L. How many little squares will it take to make the nth figure in this pattern? Work on this and then read on. . . .

DISCUSSION

As we discussed in the first chapter, there are often different strategies that can be used to solve a problem and that embody different representations of the problem. I will show two different solution paths that come from representing the problem numerically (with a table) and geometrically (by taking the figure apart). In both cases, the solution comes from seeing a pattern (in the numbers and in the shapes).

STRATEGY 1: Make a table and look for patterns

Figure number	1	2	3	4	5	6	7
Number of squares	1	3	5	7	9	11	13

Looking at the relationship between the corresponding numbers in the first and second sets (input, output), many people will quickly see that the output number is always 1 less than double the input number. Thus the number of squares in the nth L number is simply $2n - 1$. The third row in the table below illustrates this relationship very clearly.

History

Few readers realize that representing the value of the nth triangular number in the form $n(n + 1)/2$ is a very recent development. Modern notation is generally assumed to have originated in the 1500s, and much of it did not crystallize in its current form until the nineteenth century. Thus, before 1500, this relationship could have been stated only in words: To find the value of any triangular number, multiply the number of the term that you are dealing with by one more than that number and then divide that result by 2. I don't know about you, but $n(n + 1)/2$ seems a lot easier to me!

Figure number	1	2	3	4	5	6	n
Number of squares	1	3	5	7	9	11	
Number of squares	1	$2 \cdot 2 - 1$	$2 \cdot 3 - 1$	$2 \cdot 4 - 1$	$2 \cdot 5 - 1$	$2 \cdot 6 - 1$	$2 \cdot n - 1$

STRATEGY 2: Break it apart

Each of the L shapes can be broken down (decomposed) into two "arms" and a base (or "arm connector").

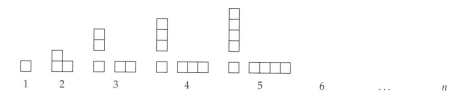

The base is always a single square. In each case, the arms are the same length, and the length is 1 less than the figure number. That is, the length of each of the arms of the 5th figure is 4 (1 less than 5). Thus the nth figure will have two arms, each of whose length is $(n - 1)$, plus one square that represents the base. Thus the number of squares of the nth figure will be equal to $(n - 1) + (n - 1) + 1$, which is mathematically equivalent to $2n - 1$.

STRATEGY 3: Break it apart another way

We could also have broken the shape apart this way: a bottom and a top.

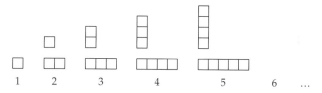

The bottom begins with one square and grows by one each time. The top begins with the second figure and also grows by one each time. The number of squares in the bottom always matches the figure number. That is, the fourth figure has a base of 4, the fifth figure has a base of 5, and so the nth figure will have a base of n. The number of squares in the top is always one less than the number of squares in the bottom. Thus, the nth figure will have a bottom of n and a top of $(n - 1)$. Adding the two together, we find that the total number of squares is $n + (n - 1)$, or $2n - 1$.

One of the characteristics of a worthwhile mathematical task is that it can be extended; in other words, "there is more than meets the eye!" Let's say we kept a cumulative total of the number of squares as we went along. What would be the total number of squares needed to make n of these L numbers? Think and then read on. . . .

Figure number	1	2	3	4	5	$\cdots$	n
Number of squares	1	3	5	7	9		
Sum	1	4	9	16	25		

In this case, it is not too difficult to deduce that the relationship between each figure number and the sum of that many L numbers is n^2; that is, $2^2 = 4$, $3^2 = 9$, $4^2 = 16$, $5^2 = 25$, etc. An interesting question here is "Why do you think this happens?" Please cover up the figure below before thinking about this question. Then read on. . . .

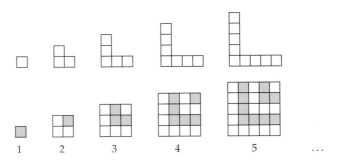

I have added dotted lines to the figures in the second row so that you can see how each L number needs a square in order to "fill out" to become a square itself, and that square that is needed just happens to be equal to the size of the square made by the previous L numbers put together. The arrows help you to see the progression.

Now, it is beyond the reach of elementary students to prove formally that $1 + 3 + 5 + \cdots + (2n - 1) = n^2$. However, it is not beyond the reach of elementary students to see squares in these shapes. In fact, one of the signs that your elementary classroom is functioning well would be that some of the students are able to make this geometric connection. This goes back to the Baratta-Lorton quote from Chapter 1: "Looking for patterns trains the mind to search out and discover the similarities that bind seemingly unrelated information together in a whole. . . . A child who *expects* things to 'make sense' *looks* for the *sense* in things and from this sense develops understanding. A child who does not see patterns often does not expect things to make sense and sees all events as discrete, separate, and unrelated" (italics added).

INVESTIGATION

2.12

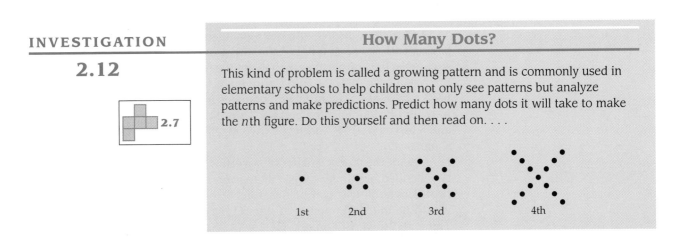

How Many Dots?

This kind of problem is called a growing pattern and is commonly used in elementary schools to help children not only see patterns but analyze patterns and make predictions. Predict how many dots it will take to make the *n*th figure. Do this yourself and then read on. . . .

DISCUSSION

I was fascinated by the discussion of this problem at a conference I attended because of the many different ways it was seen by children. Let us examine three of those ways.

STRATEGY 1: Seeing two lines

One group of children saw two lines. Each line consists of an odd number of dots and grows by 2 each time.

Figure number	1	2	3	4	5	n
Number of dots in each line	1	3	5	7	9	
Total number of dots	1	$3 \cdot 2 - 1$	$5 \cdot 2 - 1$	$7 \cdot 2 - 1$	$9 \cdot 2 - 1$	

The biggest conceptual problem here is how to connect the figure number to the number of dots in each line. The connection is that the number of dots in each line is always one less than double the figure number: $2n - 1$. Thus the total number of dots is $(2n - 1) + (2n - 1) - 1 = 4n - 3$. Do you see why we have to subtract 1 at the end?

STRATEGY 2: How many pairs of dots?

Another group of children saw the problem in terms of a center and then pairs of dots. That is, the first figure is just the center. The second figure is the center plus 2 pairs of dots. The third figure is the center plus 4 pairs of dots.

Figure number	1	2	3	4	5	n
Number of pairs of dots	0	2	4	6	8	
Total number of dots	1	$2 \cdot 2 + 1$	$4 \cdot 2 + 1$	$6 \cdot 2 + 1$	$8 \cdot 2 + 1$	

As in the previous strategy, the challenge now is how to connect the figure number to the number of pairs. In this case, we see that the number of pairs is always double the previous figure number. That is, in the *third* figure, the number of pairs is 4, which is 2 times 2. In the *fourth* figure, the number of pairs of dots is 6, which is 2 times 3.

STRATEGY 3: How much does it grow by?

Another group of children saw that the number of dots increases by 4 each time.

Figure number	1	2	3	4	5	6	7	n
Number of dots	1	5	9	13	17	21	25	
Number of dots	1	$4 + 1$	$2 \cdot 4 + 1$	$3 \cdot 4 + 1$	$4 \cdot 4 + 1$	$5 \cdot 4 + 1$	$6 \cdot 4 + 1$	

From this observation, there are still multiple ways to determine the nth case. One is by trial and error. Another is by using the observation that the pattern grows by 4 each time and representing the amount this way, such as $4 + 1$, $2 \cdot 4 + 1$, $3 \cdot 4 + 1$, etc. From this point, we can see that the multiplier of 4 is simply one less than the figure number. Thus the nth case will have $(n - 1)4 + 1$ dots, which simplifies to $4n - 4 + 1 = 4n - 3$.

Yet another way to use the observation of increasing by 4 each time is to remember that in a straight line, the slope is simply the rate of change. In this case, the rate of change is 4, and so the slope is 4. Thus, the equation is $y = 4x + b$. Now it is a matter of trial and error to determine that $b = -3$. That is, $y = 4x - 3$ is the number of dots in any figure.

Summary

In this section, we have explored the notion of algebraic thinking. I hope that when someone mentions the importance of algebraic thinking in elementary school, you will have a sense of what this means and will believe that it is important for this kind of thinking to begin in elementary school, not in middle or high school. What all four conceptions have in common is *sense making*, looking for patterns, and the five process standards. Just as we take off the training wheels when children have become confident at riding a bicycle, a crucial role of elementary teachers is to help students move beyond reasoning with actual numbers to more general reasoning—about relationships, about quantity, and about what symbols mean. In elementary school, it is crucial that students become skilled at seeing and communicating patterns; this in turn helps them to see relationships, which then lead to generalizations, which point to important mathematical structures and ideas. That is why the role of elementary teachers in working with patterns is so important for the children's future success in mathematics.

There are two other points that bear emphasizing. The notion of multiple representations was introduced in Chapter 1, and you saw its importance in this section in the many different ways to represent functions. As you saw in Investigations 2.11 and 2.12, different representations can lead to the same solution. It is just as important for children to become fluent at translating between representations. The notion of representations is related to the idea of modeling; in many problem situations, we make a model of the situation in order to solve the problem.

The other point is connections. In order to solve many problems, the solver needs to connect the problem to a diagram or table used to represent the problem. It is also important that you, and your future students, realize that arithmetic and algebra are connected. This is so that algebra will be seen not as a totally new subject but as one in which many of the rules are simply rules from arithmetic stated in a more general form. Realization of the connections between algebra and geometry is also crucial. Virtually all algebraic ideas can be represented geometrically, and virtually all geometric ideas can be represented algebraically.

EXERCISES 2.2

1. The table below shows the process by which a college department determines how many sections of a course to offer.

Number of students	Number of sections
Less than 6	Course not offered this semester
6–29	1
30–59	2
60–89	3
90–119	4

   a. Make a graph to represent this process.

   b. Is the relationship between the number of students and the number of sections a function? Justify your answer.

2. Let's say you have children, you hire a baby-sitter frequently, and you pay the baby-sitter $6 per hour. Describe how *you* would pay the baby-sitter, first in words, then with a graph. In order to receive full credit, the reader must be able to pay the baby-sitter the same amount that *you* would pay, based on your descriptions.

3. Let's say the entire freshman class of Wannago High School is going on a field trip. The number of buses

needed is a function of the number of people going. To keep matters simple, let's say that the school bus company has buses that can hold 36 passengers and that there will be two adults per bus.

  a. How many buses will be needed for 232 students?
  b. Represent this function graphically.

4. At the time this book was written, the postage on a letter was determined by the following set of rules:

   The first ounce costs 39¢.

   Each additional ounce, or fraction thereof, costs 24¢.

   a. How much would it cost to mail a letter weighing 6.3 ounces?
   b. Draw a graph of this function.

5. Archaeologists can estimate a person's height from the length of the femur (thigh bone). The formula for doing so is

   $$H = 2.3L + 61.4$$

   where all measurements are in centimeters. Let's say an archaeologist has found a femur of a human, and the bone is 45 centimeters long. How tall do we think the person was at the time of death?

6. In Investigation 2.3 in Section 2.1, we found that a set containing 4 elements has a total of 16 subsets.

   a. Fill in the blanks in the table that follows.
   b. Describe in words the relationship between the number of elements in a set and the number of subsets. Refer to the table.
   c. Describe the relationship with an equation.

Number of elements in the set	Number of subsets
1	?
2	?
3	?
4	16
5	?

7. Have you ever ridden on a bicycle where the saddle was too high or too low? The height of a saddle can be determined by adjusting it until the rider says, "It feels right at this height," but not all people guess right. Bicycle manufacturers have found that for the average person, the saddle height can be determined by the following formula: $h = 1.08i$, where h represents the saddle height and i represents the inseam.

   a. How tall should the saddle height be when you ride a bicycle?
   b. If you have a bicycle, measure the saddle height. If it is different from the number derived by applying the formula above, change it and see whether the height is more comfortable now.

8. Let's say a beautician charges $15 per haircut. She also determines that overhead not associated with number of haircuts (rent, utilities, equipment) is $150 per week. Consider the relationship between number of haircuts and profit. Is this a proportional function? That is, does double number of haircuts mean that profit is doubled?

9. John says that Investigation 1.1 (Pigs and Chickens) does not represent a function because two different inputs can have the same output. For example, (7, 17) and (0, 31) both map to the same number of feet. Do you agree with John? Why or why not?

10. There is a relationship between the successive squares of triangular numbers. This relationship can be expressed in everyday English as follows: "Select any triangular number and square it. Then take the previous triangular number and square it. Now take the difference between the two numbers."

    a. Express this relationship in notation.
    b. Is this relationship a function? Justify your response.

11. Jack and Jill run a catering business. Their fee for catering banquets is $150 plus $4 per person.

    a. Express this relationship with a table.
    b. Express this relationship with an equation.
    c. Express this relationship with a graph. What would be a realistic domain for the graph?
    d. Jack and Jill catered a banquet for which they charged $462. How many people attended?
    e. Their goal is to gross $2000 a month. How many jobs do they need to average in order to make their goal? (There is no single right answer.)

12. My parents used to live in Park Rapids, Minnesota. When I visited them, I flew to Minneapolis and rented a car. It is about 160 miles to Park Rapids. The last time I visited them, I spent three days with them, counting my arrival and departure days. The car rental agency offered two options: $45 a day plus 30¢ per mile or $185 for the week. Which do you recommend?

13. Let's say you have a choice between two checking plans. With the first plan, you will be charged $2 per month plus 15¢ per check. With the second plan, you will be charged $5 per month regardless of how many checks you write. If you write only a few checks, the first plan is clearly cheaper. Similarly, if you write 200 checks a month, the second plan is clearly cheaper.

    a. Determine the number x that will enable you to make the following statement: If you write 1 to x checks per month, choose the first plan. If you write more than x checks per month, choose the second plan.
    b. Determine the number x using a method different from the one you used in part (a).
    c. Which of the plans are functions: both, only one, or neither? Justify your response.

14. There is a functional relationship between the frequency of cricket chirps and the temperature. To find the temperature, the rule is to count the number of chirps in one minute, divide by 4, and add 40.
    a. What temperature does 50 chirps per minute correspond to?
    b. Without calculating, determine whether 100 chirps per minute will correspond to twice the temperature of 50 chirps per minute. That is, does chirping at twice the rate mean that the temperature is twice as high? Explain your reasoning.
    c. Translate the rule into an equation.
    d. Draw a graph to represent this relationship.

15. A convenience store advertised two phone cards. Which card would you buy? Why?
    Card 1: 3.9¢ a minute with 37¢ connection charge for each call.
    Card 2: 7.9¢ a minute with no connection charge.

16. Consider the graph below, which shows the depreciation of a copy machine over 5 years.
    a. What is the value of the machine when it is 2 years old?
    b. By how much does the machine decrease in value in 1 year?
    c. Predict when the value of the copy machine will be zero.

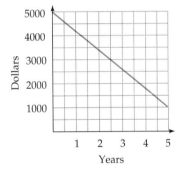

17. I can recall trying to explain to my children that thunder is the sound that lightning makes. It took some time for them to understand this because there is often a delay of many seconds between seeing the lightning and hearing the thunder. Lightning travels at 186,000 miles per second, whereas thunder travels at approximately 750 miles per hour.
    Let's say that you see a flash of lightning and 5 seconds later you hear the thunder. How far away did the lightning strike?

18. Look at the growing pattern below, in which each figure resembles the letter C. How many squares will it take to make the nth figure in this pattern?

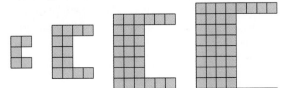

19. Look at the growing pattern below. How many squares will it take to make the nth figure in this pattern?

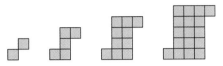

20. Look at the growing pattern below. How many squares will it take to make the nth figure in this pattern?

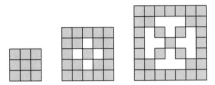

21. What if we made a pattern by joining pentagons? If the length of each side of the pentagon is 1 unit, then what would be the perimeter of n pentagons joined together? *Note:* The perimeter is the distance around the outside of the figure.

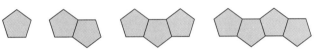

22. Variations of this problem have been found in many elementary schools. Begin with a wooden cube whose dimensions are 1 cm × 1 cm × 1 cm. Glue another cube to this cube. This new rod will be 1 cm longer than the first. Now, let's say we use a rubber stamp with a design, such as a happy face, and that stamp will cover a 1 cm by 1 cm square. Thus it would take 6 stamps to stamp the first cube and 10 stamps for the second figure. How many stamps would it take to cover a rod that was n centimeters in length?

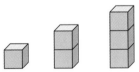

23. a. What is the surface area of a tower of cubes like the one shown at the left below?
    b. What is the surface area of a tower of cubes like the one shown at the right below?

24. This problem is taken from the February 1997 issue of *Teaching Children Mathematics*. The class had been studying patterns, and the teacher had assigned the children to create a numerical pattern. Liz made the following problem: Draw the pattern of circles that form a quadrilateral on the top and two triangles on the bottom. The center circle should be a vertex common to all three shapes. Starting at the top and moving down the page, fill in eight counting numbers in order. Any number may be used as the starting number.

a. Do this for several different sets of numbers. What do you see?

b. Compute the sum of the 4 numbers that make the quadrilateral, the sum of the numbers that make each triangle. Now add these three sums. Can you prove that the ones place of this sum will always be 4?

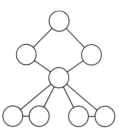

Source: Reprinted with permission from *Teaching Children Mathematics*, copyright © 1997 by the National Council of Teachers of Mathematics.

25. Here is another problem that has appeared in many places. Eight adults and two children need to cross a river in a small boat that can hold one adult or one or two children.
    a. How many one-way trips will it take for all of them to cross the river?
    b. How many trips will it take for 2 children and x adults?
    c. How many trips for 8 adults and 3 children?
    d. How many trips for 8 adults and y children?
    e. How many trips for x adults and y children?

26. Variations of this problem have also appeared in many places. Jack and Jill decide to start a rumor that there will be no school on the following Monday.
    a. If each of them tells 1 person, and each of those people tells exactly 1 other person, and so on, how long will it take until all 450 students in the school hear the rumor? Assume that the rate for telling the next person is 1 hour. That is, initially Jack and Jill know; after 1 hour, 2 more people know, ...
    b. What if each person tells exactly 2 people? How long will it take now?

27. Which of the graphs below best matches the situation of a child swinging on a swing, where the independent variable is time, and the dependent variable is height?

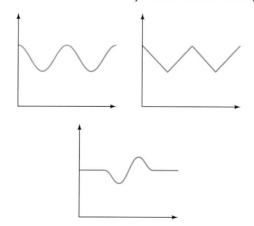

28. Which of the graphs below best matches the situation of a child swinging on a swing, where the independent variable is time, and the dependent variable is speed?

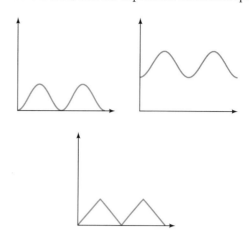

29. Sketch a graph to show this situation: A train pulls into a station, drops off some passengers and picks up new ones, and then leaves the station. The independent variable is time, and the dependent variable is speed.

30. Sketch a graph for the following situation: A person begins to walk home from the grocery store. Halfway home she realizes that she forgot something and returns to the store. Then she jogs all the way home. The independent variable is time, and the dependent variable is distance from home.

31. A family of four loves popcorn and has made a large batch to eat while they watch a movie. Sketch a graph to show the amount of popcorn in each person's bowl over time. Dad eats the popcorn steadily. Mom eats steadily but at a slower rate. Emily eats at about the same rate as Dad but refills her bowl when it is half empty. Josh eats slowly and eats only half of his popcorn.

32. Sketch a graph for this situation: Pierre is driving on the highway at a steady rate. He passes a car and then resumes his previous speed. Then he sees a car pulled over for speeding and slows down for a while, before later resuming his original speed.

33. Sketch a graph for the speed of the space shuttle from the time of lift off until its maximum speed.

34. Below is a graph of someone's drive home from work. Write the story.

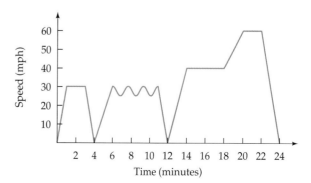

35. Below is a graph of a 400-meter race with three racers, A, B, and C. Write the story.

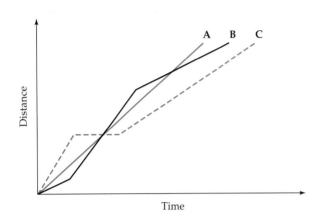

36. Below are several graphs. Write a story for each graph.

a. 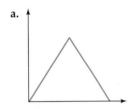 b.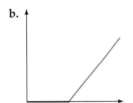

c. 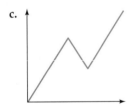 d.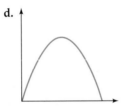

e. 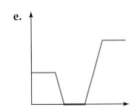 f.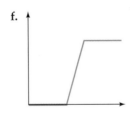

37. In virtually every school, someone raises the U.S. flag every day. Each of the graphs below represents a picture of the flag being raised. Translate each of the graphs into words. Which graph do you think is most realistic? Explain your reasoning.

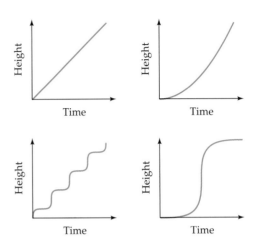

38. Go to the Web or to journals to find an example in pre-K–2, grades 3–5, or grades 4–6 of an exploration that develops the notion of algebraic thinking.

  a. Briefly describe the activity/exploration.

  b. Briefly summarize the important mathematical knowledge that can come out of this activity.

  c. Name one of the process standards that you feel will be developed during this activity, and briefly justify your choice.

39. Find a graph from a newspaper or magazine that represents a functional relationship between two variables.

  a. Explain why the relationship is functional.

  b. Describe the relationship in words, as though you were talking on the phone to someone who cannot see the graph.

  c. Make up and answer one question based on data in the graph.

40. Describe a real-life scenario involving a functional relationship between two variables.

SECTION 2.3 NUMERATION

WHAT DO YOU THINK?

- Why and when did humans invent numbers?
- Why do many mathematicians regard the invention of zero as one of the most important developments in the entire history of mathematics?

"Through and through the world is infested with quantity: to talk sense is to talk quantities. It is no use saying the nation is large—how large? It is no use saying that radium is scarce—how scarce? You cannot evade quantity. You may fly to poetry and music, and quantity and number will face you in your rhythms and your octaves."[7]

Numbers permeate our everyday lives. In Chapter 3, we will examine how we operate with numbers (addition, subtraction, multiplication, and division) and how we use exact numbers and estimates. In this section, we will lay the foundation for that work by examining the numeration system we use. Most adults don't think twice about our numeration system. Yet it represents one of the greatest advances in human history, and its widespread use is surprisingly recent—less than 500 years. If you look at Standard 1, you can see that the NCTM stresses the importance of students not just understanding how to count but also understanding how our numeration system works. Rather than read about why our numeration system is so powerful and so much "better" than other systems, you will appreciate our system and understand its properties better if you participate in the reenactment of the invention of numeration systems, which is the aim of Exploration 2.8: Alphabitia.

Origins of Numbers and Counting

History Did you know that the base 10 numeration system you use every day is called the Hindu-Arabic numeration system because it was invented by the Hindus and transmitted to the West by the Arabs? Did you know that this system has been in widespread use in the West for only 500 years? Did you know that people had to invent counting?

The earliest systems must have been quite simple, probably tallies. The oldest archaeological evidence of such thinking is a wolf bone over 30,000 years old, discovered in the former Czechoslovakia (see Figure 2.20). On the bone are 55 notches in two rows, divided into groups of five. We can only guess what the notches represent—how many animals the hunter had killed or how many people there were in the tribe.

Other anthropologists have discovered how shepherds, throughout the ages, have been able to keep track of their sheep without using numbers to count them. Each morning as the sheep left the pen, the shepherds made a notch on a piece of wood or on some other object. In the evening, when the sheep returned, they would again make a notch for each sheep. Looking at the two tallies, they could quickly see whether any sheep were missing. Anthropologists also have discovered several tribes in the twentieth century that did not have any counting systems!

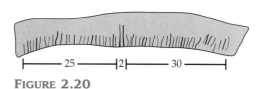

FIGURE 2.20

---

[7]Alfred North Whitehead, cited in Julian Weissglass, *Exploring Elementary Mathematics* (Dubuque, IA: Kendall Hunt Publishing, 1990), p. 100.

The beginnings of what we call civilization were laid when humans made the transition from being hunter-gatherers to being farmers—that is, from a nomadic life to a settled life. Archaeologists generally agree that this transition took place almost simultaneously in many parts of the world some ten to twelve thousand years ago. It was probably during this transition that the need for more sophisticated numeration systems developed. For example, a tribe need kill only a few animals, but one crop of corn will yield many hundreds of ears of corn.

The invention of numeration systems was not as simple as you might think. The ancient Sumerian words for one, two, and three were the words for man, woman, and many. The Aranda tribe in Australia used the word *ninta* for one and *tara* for two. Their words for three and four were *tara-ma-ninta* and *tara-ma-tara*.

Requirements for counting In order to have a counting system, people first needed to realize that the number of objects is independent of the objects themselves. Look at Figure 2.21. What do you see?

FIGURE 2.21

There are three objects in each of the sets. However, the number three is an abstraction that represents an amount. Archaeologists have found that people didn't always understand this. For example, the Thimshians, a tribe in British Columbia, had seven sets of words in their language for each number they knew, depending on whether the word referred to (1) animals and flat objects, (2) time and round objects, (3) humans, (4) trees and long objects, (5) canoes, (6) measures, and (7) miscellaneous objects. Whereas we would say three people, three beavers, three days, and so on, they would use a different word for "three" in each case. There is evidence in some of our own words of such origins. For example, we have many different nouns denoting two: a pair, a couple, a brace (of pheasants), a span (of horses), a duet.

Having a counting system also requires that we recognize a one-to-one correspondence between two equivalent sets: the set of objects we are counting and the set of numbers we are using to count them. (Notice the use of set language as we develop the concepts related to counting.) I watched both of my children miscount objects for some time, either counting too many or too few, because they had not yet realized that they needed to say the next number each time they touched the next object. It took some time for them to realize that each object represented the next number, as shown in Figure 2.22—that is, that there is a one-to-one correspondence between the set of objects and the set of numbers.

FIGURE 2.22

There is another aspect of counting that needs to be noted also. Most people think of numbers in terms of counting discrete objects. However, this is only one of the two major contexts in which numbers occur. For example, in Figure 2.23, there are 3 balls, there are 3 ounces of water in the jar, and the length

FIGURE 2.23

of the line is 3 centimeters. In the first case, the 3 tells us how many objects we have. However, in the two latter cases, the number tells how many of the units we have. In this example, the units are ounces and centimeters.

Working with numbers that represent discrete amounts is more concrete than working with numbers that represent measures. Recall the Thimshian language. Virtually all of the seven categories involved discrete sets. Ascribing numbers to continuous objects or events, such as length, time, or area, is more abstract.

Note that we distinguish between **number**, which refers to the amount (being counted or measured), and **numeral**, which refers to the symbol(s) used to designate the amount.

Patterns in Counting

As humans developed names for numbers larger than the number of fingers on one or two hands, the names for the larger numbers were often combinations of names for smaller numbers. Can you see the patterns and fill in the blanks for the three systems in Table 2.7 below? What patterns did you see? Did you find any surprises after you checked your answers on page 126?

People who have investigated the development of numeration systems, from prehistoric tallies to the Hindu-Arabic system, have discovered that many of the numeration systems had patterns, both in the symbols and in the

TABLE 2.7

Number	Greenland Eskimos	Aztecs	Luo of Kenya
1	atauseq	ce	achiel
2	machdlug	ome	ariyo
3	pinasut	yey	adek
4	sisasmat	naui	angwen
5	tadlimat	maculli	abich
6	achfineq-atauseq (other hand one)	chica-ce	ab-achiel
7	achfineq-machdlug	chic-ome	ab-ariyo
8	_____	chicu-ey	_____
9	achfineq-sisasmat	chic-naui	_____
10	qulit (first foot)	matlacti	apar
11	achqaneq-atauseq (first foot one)	matlacti-on-ce	apar-achiel
12	_____	_____	apar-ariyo
13	_____	_____	_____
15	achfechsaneq (other foot)	caxtulli	_____
16	_____	_____	_____
20	inuk navdlucho (a man ended)	cem-pouali	piero-ariyo

words, around the amounts we call 5 and 10. However, a surprising number of systems also show patterns around 2, 20, and 60. For example, the French word for eighty, *quatre-vingts*, literally means "four twenties."

As time went on, people developed increasingly elaborate numeration systems so that they could have words and symbols for larger and larger amounts. We will examine three different numeration systems—Egyptian, Roman, and Babylonian—before we examine our own base 10 system. Essentially, though, all numeration systems are simply a set of rules used with a set of symbols so that each amount can be represented by a unique element or combination of elements from that set of symbols! Do you see the connection to functions?

The Egyptian Numeration System

The earliest known written numbers are from the Egypt of about 5000 years ago. The Egyptians made their "paper" from a water plant called papyrus that grew in the marshes. They found that if they cut this plant into thin strips, placed the strips very close together, then placed another layer crosswise, and finally let it dry, they could write on the substance that resulted. Our word *paper* derives from their word for papyrus.

Symbols in the Egyptian system   The Egyptians developed a numeration system that combined picture symbols (hieroglyphics) with tally marks to represent numbers. Table 2.8 gives the primary symbols in the Egyptian system.

TABLE 2.8

1,000,000	100,000	10,000	1000	100	10	1
Astonished person	Polliwog or burbot fish	Pointing finger	Lotus flower	Scroll	Heelbone	Staff, stroke

The Egyptians could represent amounts using combinations of these basic numerals. To deepen your understanding of how the Egyptian system worked, use it to answer the following questions. Check your answers on page 126.

What is the value of each of the following numbers?

1.
2.

How would the Egyptians represent the following amounts?

3. 1202
4. 304

Working with the Egyptian system   Take a few minutes to think about the following questions. Write your thoughts before reading on. If possible, compare your responses with those of a classmate. . . .

1. What do you notice about the Egyptian system? Do you see any patterns?
2. What similarities do you see between this and the more primitive systems we have discussed?
3. What limitations or disadvantages do you find in this system?
4. Try to describe the rules for this system, as though you were talking to someone on the phone.

The Egyptian numeration system resembles many earlier counting systems in that it uses tallies and pictures. In this sense, it is called an *additive system*. Do you see why?

Look at the way this system represents the amount 2312. In one sense, the Egyptians saw this amount as $1000 + 1000 + 100 + 100 + 100 + 10 + 1 + 1$ and wrote it as 𒐚𒐚𓏺𓏺𓏺𓎆𓏽. In an **additive system**, the value of a number is literally the sum of the numerals.

However, this system represents a powerful advance: The Egyptians created a new numeral for every *power of 10*. Do you understand this?

They had a numeral for the amount 1. To represent amounts between 1 and 10, they simply repeated the numeral. For the amount 10, they created a new numeral. All amounts between 10 and 100 can now be expressed using combinations of these two numerals. For the amount 100, they created a new numeral, and so on.

These amounts for which they created numerals are called **powers of ten**. Recall, from your work with exponents from algebra, that we can express 10 as 10^1 and 1 as 10^0. Thus we can express the value of each of the Egyptian numerals as a power of 10:

1,000,000	100,000	10,000	1000	100	10	1
$10 \cdot 10 \cdot 10 \cdot 10 \cdot 10 \cdot 10$	$10 \cdot 10 \cdot 10 \cdot 10 \cdot 10$	$10 \cdot 10 \cdot 10 \cdot 10$	$10 \cdot 10 \cdot 10$	$10 \cdot 10$	10	1
10^6	10^5	10^4	10^3	10^2	10^1	10^0

The Egyptian system was not a full-blown base 10 system, as we shall find out, but it was a remarkable achievement for its time. Egyptian rulers could represent very large numbers. One of the primary limitations of this system, as we shall see in Chapter 3, was that computation was extremely cumbersome. It was so difficult, in fact, that the few who could compute enjoyed very high status in the society.

The Roman Numeration System

The Roman system is of historical importance because it was the numeration system used in Europe from the time of the Roman Empire until after the Renaissance. In fact, several remote areas of Europe continued to use it well into the twentieth century. Some film makers still list the copyright year of their films in Roman numerals.

Symbols in the Roman system   Table 2.9 below gives the primary symbols used by early Romans and later Romans.

TABLE 2.9

Amount	Early Roman	Later Roman
1	I	I
5	V or Λ	V
10	X	X
50	↓	L
100	⊙	C
500	⊂ and ⊃	D
1000	① or ∞	M

Again, to deepen your understanding of how the Roman system worked, use it to answer the following questions. Check your answers on page 126.

What is the value of each of the following numbers (using the numerals of the Later Roman system)?

1. MXI **2.** MDXCVII

The X precedes the C in MDXCVII above. Notice how this placement affects the value. We will discuss this aspect of the Roman system later.

How do you think the Romans would represent the following amounts?

3. 1102 **4.** 319

Working with the Roman system   Take a few minutes to think about the following questions. Write your thoughts before reading on. If possible, compare your responses with those of a classmate. . . .

1. What do you notice about the Roman system? Do you see any patterns?
2. What similarities do you see between this system and the Egyptian system?
3. What limitations or disadvantages do you find in this system?
4. Try to describe the rules for the Roman system, as though you were talking to someone on the phone.

Like the Egyptians, the Romans created new numerals with each power of 10, that is, 1, 10, 100, 1000, etc. However, the Romans also created new numerals at "halfway" amounts—that is, 5, 50, 500, etc. Why do you think they did this? Think and then read on. . . .

This invention reduced some of the repetitiveness that encumbered the Egyptian system. For example, 55 is not XXXXXIIII but LV.

Basically, the Roman system, like the Egyptian system, was an additive system. However, the Later Roman system introduced a *subtractive* aspect. For example, IV can be seen as "one before five." This invention further reduced the length of many large numbers. For example, when writing the amount 444, users of the Later Roman system no longer had to write CCCCXXXXIIII, but rather could write . . . well, what do you think?

Check your answer on page 126.

As in the Egyptian system, computation in the Roman system was complicated and cumbersome, and neither system had anything resembling our zero.

The Babylonian Numeration System

The Babylonian numeration system is a refinement of a system developed by the Sumerians several thousand years ago. Both the Sumerian and Babylonian empires were located in the region occupied by modern Iraq. The Sumerians did not have papyrus, but clay was abundant. Thus they kept records by writing on clay tablets with a pointed stick called a stylus. Thousands of clay tablets with their writing and numbers have survived to the present time; the earliest of these tablets were written almost 5000 years ago.

Symbols in the Babylonian system   Because the Babylonians had to make their numbers by pressing into clay instead of writing on papyrus, their symbols could not be as "fancy" as the Egyptian symbols. In fact, this physical constraint was probably the impetus for the advances that their system represents.

They had only two symbols, an upright wedge that symbolized "one" and a sideways wedge that symbolized "ten." In fact, the Babylonian writing system is called *cuneiform*, which means "wedge-shaped."

Amount	Symbol
1	▼
10	◄

Amounts could be expressed using combinations of these numerals; for example, 23 was written as ◄◄▼▼▼.

However, being restricted to two numerals creates a problem with large amounts. The Babylonians' solution to this problem was to choose the amount 60 as an important number. Unlike the Egyptians and the Romans, they did not create a new numeral for this amount. Rather, they decided that they would have a new *place*. For example, the amount 73 was represented as ▼ ◄▼▼▼. That is, the ▼ at the left represented 60 and the ◄▼▼▼ to the right represented 13. In other words, they "saw" 73 as 60 + 13.

Similarly, ▼▼▼ ◄▼▼ was seen as six 60s plus 12, or 372.
▼▼▼

Thus, we consider the Babylonian system to be a **positional system** because the value of a numeral depends on its position (place) in the number.

As with the other two systems, use the following questions to reinforce your understanding of the Babylonian system. Check your answers on page 126.

What is the value of each of the following numbers?

1. ▼▼ ◄▼▼   2. ▼▼▼ ◄◄◄▼▼   3. ▼ ◄▼ ◄◄▼
              ▼▼ ◄

How do you think the Babylonians would represent the following amounts?

4. 1202   5. 304

Working with the Babylonian system Take a few minutes to think about the following questions. Write your thoughts before reading on. If possible, compare your responses with those of a classmate. . . .

1. What do you notice about the Babylonian system? Do you see any patterns?
2. What similarities do you see between this system and the Egyptian and Roman systems?
3. What limitations or disadvantages do you find in the Babylonian system?
4. Try to describe the rules for this system, as though you were talking to someone on the phone.

Place value All three systems have vestiges of tallies for the first four numbers: IIII, ||||, ▼▼▼▼. The Romans invented a new numeral for 5. The Egyptians and Babylonians waited until 10 to create a new numeral. To represent larger amounts, the Babylonians invented the idea of the value of a numeral being a function of its place in the number. This is the earliest occurrence of the concept of **place value** in recorded history. With this idea of place value, they could represent any amount using only two numerals.

■ *History* ■

We are not sure why the Babylonians chose 60, although many hypotheses have been set forth. One is that 60 divides 360 evenly, and the Sumerians may have originally thought that there were 360 days in a year. Another possibility is that because many numbers divide 60 evenly (1, 2, 3, 4, 5, 6, 10, 12, 15, 20, 30, and 60), this choice made calculations with fractions much easier. Their choice of 60 affects our lives today, for it was they who divided a circle into 360 parts and divided an hour into 60 minutes.

We can understand the value of their system by examining their numbers with expanded notation. Look at the following Babylonian number:

𒐖 𒎙𒐈 𒌋𒐕

Because the 𒌋𒐕 occurs in the first (or rightmost) place, its value is simply the sum of the values of the numerals—that is, $10 + 10 + 1 = 21$. However, the value of the 𒎙𒐈 in the second place is determined by multiplying the face value of the numerals by 60—that is, $60 \cdot 23$. The value of the 𒐖 in the third place is determined by multiplying the face value of the numerals by 60^2—that is, $60^2 \cdot 2$.

The value of this amount is

$$60^2 \cdot 2 + 60 \cdot 23 + 21 = 7200 + 1380 + 21 = 8601$$

Thus, in order to understand the Babylonian system, you have to look at the face value of the numerals *and* the place of the numerals in the number. Therefore, although the Babylonian system still has additive aspects (for example, 5 = 𒐙), the value of a number is no longer determined simply by adding the values of the numerals. One must take into account the place of each numeral in the number.

The Babylonian system is more sophisticated than the Egyptian and Roman systems. However, there were some "glitches" associated with this invention. Before you read on, imagine that you were a Babylonian. What problems could you see with this new idea of place value?

What if there were nothing in a place? For example, how could the Babylonians represent the amount 3624? Try this and then read on. . . .

The need for a zero If we represent this amount from the Babylonian perspective, we note that $60^2 = 3600$. Thus the Babylonians saw 3624 as $3600 + 24$. They would use 𒐕 in the third place to represent 3600, and they would use 𒎙𒐘 in the first place to represent 24, but the second place is empty. Thus, if they wrote 𒐕 𒎙𒐘, how was the reader to know that this was not $60 + 24 = 84$? Again, try to imagine yourself as a Babylonian. How might you solve this problem? Think and read on. . . . If possible, compare your ideas with those of a classmate.

Archaeologists have found evidence that later Babylonian writers experimented with several different notations to represent an empty place. A Babylonian mathematical table from about 300 B.C. contains a new symbol 𒑊 that acts like a zero. Using this convention, they could represent 3624 as

𒐕 𒑊 𒎙𒐕𒐕
    𒐖

The slightly sideways wedges indicate that the second place is empty, and thus we can unambiguously interpret this numeral as representing

$$60^2 + 0 + 24 = 3624$$

This later Babylonian system is thus considered by many scholars to be the first place value system[8] because the value of every symbol depends on its place in the number and there is a symbol to designate when a place is empty.

In the exercises, you will find that the Mayans in Central America also created a numeration system with a zero between A.D. 300 and A.D. 900. As you may have discovered in your work as an Alphabitian in Exploration 2.8,

---

[8]Georges Ifrah, *From One to Zero: A Universal History of Numbers*, p. 373. Translated from the French, 1985.

creating a system with a base and with a symbol to act as a place holder is not a simple task. This is why it took humans so long to create such systems.

The Development of Base 10: The Hindu-Arabic System

We have now seen how the Egyptian and Roman systems represented important advances over tallying, allowing people to count and represent large amounts with numerals. We have seen how the Babylonians struggled with the concept of the value of a numeral depending on its place in the number. You are now in a better position to understand and appreciate the structure of the base 10 numeration system we use today.

We can still only guess at early stages of this system because we have no early archaeological evidence, as we have for the Egyptian and Babylonian systems. However, at some point, someone or some group discovered that all numbers could be represented using combinations of just ten symbols. One of these symbols, zero, constituted a tremendous leap in abstraction, because it represented nothing and also represented an empty place. You have seen how the Babylonians struggled with this dilemma for many hundreds of years.

Archaeologists tell us that the Hindu numeration system emerged around A.D. 600 and that the earliest known zero is found inscribed on the wall of a temple in central India dated A.D. 870. By A.D. 800, news of this system came to Baghdad, which had been founded in A.D. 762 by followers of Mohammed. Leonardo of Pisa, whom you encountered in Chapter 1, traveled throughout the Mediterranean and the Middle East, where he first heard of the new system. In his book *Liber Abaci* (translated as *Book of Computations*), published in 1202, he argued the merits of this new system, but it took some time for the Hindu-Arabic system to replace the Roman system that almost everyone in Europe used. In fact, a thirteenth-century law forbade bankers of Florence to use these Hindu-Arabic numerals, because it was felt that forgeries were easier in the new system than in the Roman system. However (as you will discover in Chapter 3), one of the biggest advantages of this new system was ease of computation, and it was not long before the merchants realized how much easier computing was in this new system.

Figure 2.24 on the next page traces the development of the ten numerals that make up our numeration system.[10] Just as the ancient Egyptians used papyrus and the Sumerians used clay tablets, the people in ancient India used dried palm leaves to write with. Some historians believe that the Indians could have joined up these strokes, so that $=$ became Z and $\equiv$ became $\gtrless$.

The development of numeration systems from the most primitive (tally) to the most efficient (base 10) has taken tens of thousands of years. Although the base 10 system is the one you grew up with, it is also the most abstract of the systems and possibly the most difficult initially for children. The next section describes the essential mathematical structure of the Hindu-Arabic numeration system. However, rather than just reading this section, stop first and reflect on what you have learned thus far in your own investigations. Imagine describing this system to a Babylonian, Egyptian, or Roman who has suddenly been transported into our time. Imagine that this person can understand English and has a counting table like Table 2.10 but is still struggling to make sense of this system. How would you describe the essential features of this system to that person? Do so before reading on. . . .

> **History**
>
> In 1229 the City Council of Florence, Italy, passed a law forbidding the use of base 10 numbers when entering records of money in account books. Numbers had to be written out.[9]

---

[9]Robert Kalpan, *The Nothing That Is* (New York: Oxford University Press, 1999), p. 102.
[10]Karl Menninger, *Number Words and Number Symbols: A Cultural History of Numbers*, trans. Paul Broneer (Cambridge, MA: M.I.T. Press, 1969), p. 418.

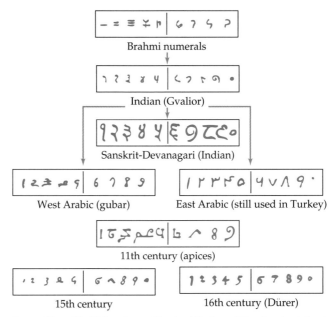

FIGURE 2.24

Source: From Karl Menninger, *Number Words and Number Symbols: A Cultural History of Numbers,* 1969, p. 418. Reprinted by permission of MIT Press.

Advantages of Base 10

Our base 10 numeration system has several characteristics that make it so powerful.

No tallies   The base 10 system has no vestiges of tallies. Any amount can be expressed using only 10 **digits**: 0, 1, 2, 3, 4, 5, 6, 7, 8, and 9. In fact, the word *digit* literally means "finger."

TABLE 2.10

Egyptian	Roman	Babylonian	Hindu-Arabic
|	I	▼	1
||	II	▼▼	2
|||	III	▼▼▼	3
||||	IV	▼▼▼▼	4
|||||	V	▼▼▼▼▼	5
||||||	VI	▼▼▼▼▼▼	6
|||||||	VII	▼▼▼▼▼▼▼	7
||||||||	VIII	▼▼▼▼▼▼▼▼	8
|||||||||	IX	▼▼▼▼▼▼▼▼▼	9
∩	X	❮	10
∩∩	XX	❮❮	20
∩∩∩∩∩	L	❮❮❮❮❮	50
∩∩∩∩∩∩	LX	▼	60
୭	C	▼❮❮❮❮	100

Decimal system   The base 10 system is a **decimal** system, because it is based on groupings (powers) of 10. The value of each successive place to the left is 10 times the value of the previous place:

100,000   10,000   1000   100   10   1

Ten ones make one ten.
Ten tens make one hundred.
Ten hundreds make one thousand.
Ten thousands make ten thousand.

Expanded form   When we represent a number by decomposing it into the sum of the values from each place, we are using **expanded form**. There are different variations of expanded form. For example, all of the expressions below emphasize the structure of the number, 234—some more simply and some using exponents.

$$200 + 30 + 4$$
$$= 2 \cdot 100 + 3 \cdot 10 + 4 \cdot 1$$
$$= 2 \cdot 10^2 + 3 \cdot 10^1 + 4 \cdot 10^0$$

Note: $10^1 = 10$ and $10^0 = 1$.

Expanded form is an important tool for two reasons. First, it is one of the many tools that mathematicians use to represent the process of decomposing numbers into smaller amounts to help us better understand the concepts and procedures that we use. Second, in Chapter 3, we will use expanded form to understand why the procedures we use to add, subtract, multiply, and divide actually work.

The concept of zero   The fourth reason why the base 10 system is powerful is the invention of the concept of zero, represented by the symbol 0. "The invention of zero marks one of the most important developments in the whole history of mathematics."[11] This is the feature that moves us beyond the Babylonian system. Recall the Babylonians' attempts to deal with the confusion when a place was empty. It was the genius of some person or persons in ancient India to develop this idea, which made for the most efficient system of representing amounts and also made computation much, much easier. One of the most difficult aspects of this system is that the symbol 0 has two related meanings: In one sense, it works just like any other digit (it can be seen as the number 0), and at the same time, it also acts as a place holder. It takes young schoolchildren several years to understand this fully and accept it. I recall my five-year-old daughter saying that zero meant "nothing" and insisting that it was not a number!

Connecting Geometric and Numerical Representations

Let me now ask a question that will help you to assess your understanding of these concepts and to extend your understanding: What do you think the fifth place in our **base 10 manipulatives** looks like? I ask this question of my

---

■ *History* ■

A French writer in the 1400s referred to zero as nothing but "a sign which creates confusion and difficulties." Some writers referred to the new Hindu-Arabic system as having nine digits and then a zero. Some even ridiculed this new system, thinking that it would go away once people came to their senses and went back to Roman numerals: "Just as the rag doll wanted to be an eagle, the donkey a lion, and the monkey a queen, the *cifra* [zero] put on airs and pretended to be a digit."[12]

---

[11]H.A. Freebury, *A History of Mathematics* (New York: Macmillan, 1958), p. 170.
[12]Karl Menninger, *Number Words and Number Symbols: A Cultural History of Numbers* (Cambridge, MA: The MIT Press, 1969), p. 422.

students, and many are baffled by the question and need a hint. If you find yourself baffled, read the following.

If you mentally step back and look at this system, you will notice that the first place is represented by a unit, which we will call a "small cube"; the second place is represented by a long; the third place is represented by a flat; and the fourth place is represented by a block (a "big cube").

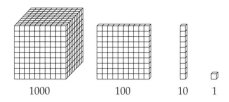

If you were to draw a picture of the fifth place, what would it look like? Think about this and read on. . . .

Rather than give a direct answer to the question, we will lead up to it. As we look at each place, we see some amazing patterns. Let us start at the beginning, with units.

- Ten ones make one ten. In a physical sense, ten units become one long.
- Then ten tens become one hundred. In a physical sense, ten longs become one flat.
- Then ten hundreds become one thousand. In a physical sense, ten flats become a "big cube."

Continuing this pattern, we see that ten thousands become one ten-thousand. In a physical sense, we can represent ten thousands as a "big long." There are other representations; for example, we could have represented 10,000 as a pile of ten "big cubes," as shown in Figure 2.25.

Both representations show a value of 10,000. Mathematically, we say that the former representation is more powerful because it makes a better connection. One of the reasons it makes a better connection is that it fits a larger pattern that you will soon see.

If we continue the pattern in this way, we can represent ten 10,000s (big longs) as one 100,000, and this amount can be represented visually as a "big flat," as shown in Figure 2.26.

Table 2.11 below shows the name of each place, the amount it represents, and its shape.

Do you see now why the representation of 10,000 as a "big long" is more powerful than the other representations?

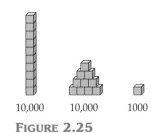

Figure 2.25

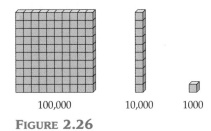

Figure 2.26

Table 2.11

Word	Million	Hundred thousand	Ten thousand	Thousand	Hundred	Ten	One
Exponential representation	10^6	10^5	10^4	10^3	10^2	10^1	10^0
Symbol	1,000,000	100,000	10,000	1000	100	10	1
Shape	huge cube	big flat	big long	big cube	flat	long	small cube

Think about the terms, look at the physical shape of each place, and look at the way we write numbers. All three representations are closely connected.

- The name changes every three terms—thousand, million, billion, etc.
- The shape changes every three terms—cube, long, flat.
- We place commas after every three terms, e.g., 345,234,186.

Is this a coincidence, or is this the reason why we separate large numbers by groups of three?

Many readers at this point are often lulled into a sense of complacency—"This is pretty basic and simple." However, it is *not* simple to most young children, many of whom struggle to learn to count and compute in base 10. Because part of the struggle is caused by teachers who know only the hows and not the whys of the system, we need to look more carefully at the basic ideas of this system. Since you know it so well, we will do this via a back door.

INVESTIGATION 2.13

Relative Magnitude of Numbers

How much is a million? How much bigger is a billion than a million? Our modern society deals with large numbers all the time.

> Politicians talk about a war costings $100 billion a year.
> The federal deficit is more than $8 trillion at the writing of this book.
> The closest star to us is about 24,600,000,000,000 miles from earth.
> The cleanup from Hurricane Katrina will involve the removal of 500 million cubic yards of debris.
> More than 25 million people have died from AIDS. Africa has 12 million of AIDS orphans.

Let's investigate an example commonly used in elementary classrooms. If a large paper clip is about 2 inches long, how long would a chain of 2 million of those paper clips be?

CLASSROOM CONNECTION

There are many children's books along the theme of children's natural questions about large numbers. David Schwartz's *How Much is a Million?*[13] is a classic. My favorite is *Big Numbers in Pictures That Show Just How Big They Are*[14]. The author begins with 1 pea and then goes up one place value at a time. When you get to 1 billion, the peas have completely filled a two-story house and are spilling into the yard.

DISCUSSION

Well, it would be 2 million inches, but can you "feel" 2 million inches? I can't. If we divide 2 million by 12 (the number of inches in a foot), we get 166,667 feet. That's still not a number most people can sense. However, if we divide that number by 5280 (the number of feet in a mile), we get about 32 miles. Most people can reference 32 miles. We either know a distance that is in that range or maybe half that range, so it would be a round trip.

Now how long would a chain of 1 billion such paper clips be? Think before reading on. Make your own estimate.

It would be 32,000 miles, because a billion is a thousand million. How far is 32,000 miles?

---

[13] David M. Schwartz, *How Much Is a Million?* (New York: Lothrop, Lee, & Shepard Books, 1985).
[14] Edward Packard, *Big Numbers: And Pictures That Show Just How Big They Are!* (Brookfield, CT: Millbrook Press, 2000).

Classroom Connection

Children in the early grades love to show you how high they can count. Once they get to thousands and millions, a common questions is "what is the largest number?" When my son asked this question and came to understand there is no biggest number, I told him that a googol, which is 1 with a 100 zeros, was the biggest number that had a name. He was so excited with this information and wowed his friends for weeks!

How long would a chain of 1 trillion paper clips be?

32,000,000 miles. The distance around the earth is about 24,000 miles. The moon is about $\frac{1}{4}$ million miles from earth, so one round trip would be $\frac{1}{2}$ million and two round trips would be 1 million, so 64 round trips would be 32 million miles.

1 million paper clips	32 miles	A bit longer than a marathon
1 billion paper clips	32,000 miles	More than the distance around the earth
1 trillion paper clips	32,000,000 miles	64 round trips to the moon

Were you surprised at how much bigger a billion is than a million; how much bigger a trillion is than a billion? Most people are. When we are not able to have a reference for thinking about large amounts, they literally swim in our heads. They lose their reality. If we could really sense $8 trillion dollars, we would clamor to reduce the debt. If we could really sense the number 40 million people living with AIDS, then we would likely take action.

Units

There is an old parable that says a journey of 1000 miles begins with a single step. The same can be said for counting. We always begin with 1. However, unlike the phrase "a rose is a rose is a rose," a 1 is not always the same. For example, a 1 in the millions place represents 1 million. This is the power of our numeration system, but it is very abstract. Europeans knew about this new Hindu-Arabic system for hundreds of years before it finally replaced the Roman system, which had been used for well over 1000 years. The idea that a numeral can represent different amounts depending on its place in a number involves the idea of unitizing. We will speak more of unitizing in Chapter 3 because it is one of the big accomplishments of the early grades.

At this point, we will elaborate this idea of unit because it will recur throughout this book. The first place in our system is called the ones place, the singles place, and the units place. By the time children come to school, they are generally very comfortable with the idea of counting, that is, one unit at a time. Over time they can count higher and higher. There are many conceptual jumps required in learning mathematics during one's school years, for example, the jump from whole numbers to fractions or from arithmetic to algebra.

One of the early jumps is from **simple units** to **composite units.** When counting objects, one is our simple unit. When asked to count a pile of objects, for example, 240 pennies, children will count one at a time. However, if they lose their count, they have to start all over. Some children realize that they can put the pennies into piles of 10. Now if they lose count, they can go back and count by tens, for example, 10, 20, 30, 40, etc. In this case, 10 is a composite unit, that is, it is composed of a number of smaller units. Some children can see that 1 pile is also 10 pennies. To be able to hold these two amounts simultaneously is a challenge for young children, and it is an essential milestone along the way.

We have composite units everywhere: 100 is equivalent to ten 10s, 1000 is equivalent to ten 100s. In fact, our language shows this: some people will say thirty-four hundred for 3400. We talk about 1 dozen eggs, a case of soda (24 cans), a pair, a pound (16 ounces). When we say that we will need 6 dozen eggs for a pancake breakfast fundraiser, we can see 6 dozen and we also know that this is 72 individual eggs.

Throughout the book, we will stop and take note when we are expanding our understanding of this idea of units.

INVESTIGATION 2.14

What If Our System Was Based on One Hand?

The people who developed base 10 decided to base it on two hands. What if they had decided to base it on one hand? That is, what if one-zero (10) had come not after we counted two hands but after we counted one hand? Our counting would look like this: 1, 2, 3, 4, 10, ... The manipulatives (see the figure below) would have the same basic shape as the ones you grew up with, but the long would be only half as long, and the flat would not be one-half as big but one-fourth as big. (Why that is true will be left as an exercise.)

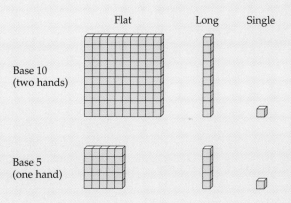

Because the structures of this new base are the same as in the system you grew up with, the counting follows the same rules. The table below shows the beginnings of the new system. If you find yourself struggling, make your own set of manipulatives (cut from graph paper) and represent each number manipulatively: 1 single, 2 singles, 3 singles, 4 singles.... The next number represents one hand and will now be called one-zero, or 10, because this is what a long will be in this system. The system continues all the way to 44. What comes next? Think carefully. This base has the same rules and characteristics as the base you grew up with, and it could have been the system you did grow up with if a different decision had been made almost 2000 years ago!

1	2	3	4	10
11	12	13	14	20
21	22	23	24	30
31	32	33	34	40
41	42	43	44	?

DISCUSSION

First of all, counting simply means adding 1 each time, so the next number after 44 is 44 + 1. In *this* base, a place is full after 4. Thus the ones place is full, and we "move" the 10 ones into the next place because we can repackage (regroup) 10 singles into 1 long. However, the longs place is also full once we get one more long. Thus we repackage (regroup) 10 longs into 1 flat, and we now have 1 flat, 0 longs, and 0 singles. That is, 100 is the next number after 44 in this base.

If you don't see this, here are two different strategies to help you make more sense of it. First, use manipulatives. In the first figure at the left, we see 44 (4 longs and 4 singles) plus 1. In the second figure, we see those 10 singles have become a long. In the third figure, we see the 10 longs have become a flat. We now have 1 flat, 0 longs, and 0 singles—that is, 100.

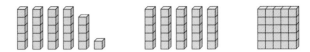

Classroom Connection

Jerome Bruner wrote that children learn mathematical ideas best if they begin at the concrete level and then move at their own rate to the symbolic level. I find this true of most adults. The most basic level is the level of manipulatives—the actual blocks themselves. If your instructor does not have base 5 blocks, you can make your own set by cutting flats, longs, and singles from graph paper. Many students find that actually making and moving the manipulatives helps them grasp the ideas more readily. At some point, many students find that they no longer need the manipulatives but that jumping all the way to using just numbers is hard. Bruner knew this and articulated a middle level that is a pictorial level. At this level, you don't need the manipulatives but find pictures of them useful. I have come to believe that most adults learn new mathematics concepts more deeply if they first experience these concepts at the concrete level and proceed from the concrete level to the visual level and finally to the symbolic level *at their own pace* and by connecting the representations at each level.

INVESTIGATION 2.15

How Well Do You Understand Base 5?

One way to assess your understanding of this new base is to figure out what numbers come after and before given numbers. Answer the questions below before reading on. Feel free to do the problems at the concrete, visual, or symbolic level. . . .

What number comes after 234 in this base? What number comes after 1024?
What number comes before 210? What number comes before 3040?

DISCUSSION

At the symbolic level, one might reason the first problem like this: After 234, we have the ones place, so the ones place will now be zero, the longs place will have one more, so it will be 4, and the flats place will still be 2.

At the visual level, one can see that the next number means 2 flats, 3 longs, 4 singles, and 1 more single. The diagram at the left is a more concrete picture of 234 + 1 (you can see each single on the long and flat); the digram at the right is a more abstract picture of 234 + 1.

In either picture, the number after 234 will look like 2 flats, 3 longs, and a handful of singles that must be exchanged for 1 long. Thus we now have 2 flats, 4 longs, and 0 singles—that is, 240.

To determine what number comes after 1024, let us stop and reflect for a moment. The manipulatives are like training wheels. They are useful to help the ideas develop, but eventually they need to come off. Having to represent problems visually becomes more tedious, especially as the numbers get larger, and thus the learner needs to internalize the basic ideas. In this case, the ones place is full—10 ones become one more long. Now we have 3 longs. No other places are affected, so the next number is 1030.

What number comes before 210? The answer comes quickly from the picture. Do you see how?

The number before 210 is 1 less than 210. You can see this by regrouping the long into 10 singles and taking one away, so that we now have 2 flats, 0 longs, and 4 singles—that is, 204. Another way to see it is to cover up one of the singles on the long above and realize that this represents the breaking up of that long into singles.

We can also do this by guess–test–revise. That is, logic says that it should be 204. Now check it out—it's easier to count forward than backward. That is, the anwer is 204 if the next number after 204 is 210. Add a single to 204, and we fill the singles place and add one to the longs place, which indeed is 210.

What comes before 3040? You could represent this as 3 big cubes, 0 flats, 4 longs, and 0 singles. In order to take 1 away from 3040, you would have to exchange 1 long for 10 singles and take away one single. You now still have 3 big cubes and still have 0 flats, but now you have 3 longs, and 4 singles. Thus the number before 3040 is 3034. At a more abstract level, as you internalize the properties of a base system, you simply know that the number before 3040 must have a 4 in the ones place (just as you "know" that the number before 7090 in base 10 must end in a 9). Since you are going backwards, you know that you have one less long. You also "know" that the flats and next place are not affected in this case, and thus the answer is 3034.

Either in explorations or in class, your instructor may have you work with other bases so that your understanding of the fundamental ideas of the base and place value become deeper and deeper. Having explored other bases, let us now revisit the fundamental ideas of base 10. Any base—base 2, base 5, base 10, base 12, whatever—will have the following fundamental characteristics.

1. Any base has only one-zero symbols. With those one-zero symbols, we can represent any amount; think of *Toy Story*—"to infinity and beyond!"
2. The value of each place is one-zero times the previous place.
3. Each place can contain only one symbol. When a place is full, we "move" to the next place by regrouping (exchanging) 10 of the given place to 1 of the next higher place.
4. The value of a numeral depends on its place in the number.
5. The value of a number is determined by multiplying each numeral by its place value and then adding these sums.
6. Zero represents an empty place and 0 represents an actual amount, with a place on the number line.

Summary

Almost certainly, base 10 has become the universal counting system today because we have 10 fingers on both hands. We also could have selected base 5—fingers on one hand—or base 20—digits on fingers and toes, not an uncommon choice when we look at historical records. (As you will find out in Chapter 3, computation would be much easier for most people if our species had one more finger on each hand, in which case we would have used base 12.)

We have explored different counting systems to give you an appreciation of the significance of base 10 and its abstractness—it took humans many thousands of years finally to invent such a powerful numeration system. In the course of working on the explorations, you have come to appreciate the importance of mathematical vocabulary, including the terms *digit* and *place* (ones place, tens place, etc.). You have realized that with the concepts of base and place value, and a symbol to represent "nothing," we can represent any amount using only a few digits (in base 10, we use 0, 1, 2, 3, 4, 5, 6, 7, 8, and 9). You have been introduced, through expanded form, to the tool of decomposing a number into its constituent parts. This notion of breaking an object or idea into its component parts is an essential tool in all scientific disciplines.

In this section, you have probably gained a deeper appreciation of both the power and the abstractness of our base 10 numeration system. To use an old phrase, I wish I had a dollar for every student who has said something like "Wow, no wonder it's hard for little kids to learn how to count; I never thought of it [our system] that way before." In Chapter 3, we will carefully examine the meanings of the four basic operations of addition, subtraction, multiplication, and division. In that chapter, you will come to understand better the value of expanded form. Then we will learn how the algorithms, which most adults take for granted, really work. By the end of Chapter 3, your appreciation of the Hindu-Arabic system should be even greater!

EXERCISES 2.3

1. **a.** Fill in the blanks in these counting systems.
   b. What patterns do you see in these systems?
   c. Describe the "rules" of this system as though you were talking on the phone to a friend who missed the class when this system was discussed.

Base 10	Maya	Luli of Paraguay	South American
1	hun	alapea	tey
2	ca	tamop	cayupa
3	ox	tamlip	toazumba
4	can	lokep	cajesa
5	ho	lokep moile alapea (four and one) or is alapea (one hand)	teente
6	uac	lokep moile tamep	teyente-tey
7	uuc	?	teyente-cayapa
8	uaxac	lokep moile lokep	?
9	bolon	lokep moile lokep alapea	teyente-cajesa
10	lahun	is yaoum (both hands)	caya-ente
11	buluc	is yaoum moile alapea (hands and one)	caya-ente-tey
12	lah-ca	is yaoum moile tamep	?
13	ox-lahun	?	?
15	ho-lahun	?	toazumba-ente
16	?	?	?
20	hunkal	is eln yaoum (hands, feet)	?
21	?	?	?
22	?	?	?

2. Find the base 10 equivalent of each of these numerals.
   a. 𓏺𓏺𓏺∩∩∩|
   b. 𓆼𓆼𓊖𓊖𓊖∩||
   c. MDCLXVI
   d. MDXIX
   e. CIX
   f. ≪≺ ≺≺ ▼▼
   g. ▼▼ ≺▼▼▼
   h. ≺≺▼▼▼

3. Represent these base 10 amounts in Egyptian, Roman, and Babylonian symbols.
   a. 312 **b.** 1206 **c.** 6000
   d. 10,000 **e.** 123,456

4. An amusing exercise is to convert English words into Roman values. For example, the English word LID would be worth 50 + 1 + 500 = 551.
   a. What is the value of MIX?
   b. What is the most valuable English word made up only of letters in the Roman numeration system?
   c. What is the most valuable English word that you can find, if we allow any English word but determine its value by adding only those letters that have values in the Roman system?

5. Represent the following base 10 numbers by sketching the base 10 manipulatives.
   a. 345 **b.** 2001

6. Represent the following base 10 numbers in expanded form.
   a. 345 **b.** 2001 **c.** 10,101

7. Rewrite each of the following in standard form.
   a. $4 \cdot 10^3 + 8 \cdot 10^2 + 5 \cdot 10 + 9$
   b. $3 \cdot 10^4 + 2 \cdot 10^2 + 4 \cdot 10$
   c. $7 \cdot 10^5 + 5 \cdot 10^4 + 3$

8. Represent the following numbers by sketching what they would look like using manipulatives.
   a. 304_5 **b.** 1001_2 **c.** $23b_{16}$

9. **a.** What is the largest four digit number in base 8?
   b. What is the largest three digit number in base 16?

10. The symbols developed by the ancient Greeks are visually quite fascinating and remind me of the children's game called hangman. The basic symbols for amounts up to 10,000 are given below.

    1 5 10 50 100 500 1,000 5,000 10,000
    | Γ Δ ⌐ H ⌐ X ⌐ M

    Translate the following Greek numbers into base 10:
    a. ⌐ΔΔΓ|| **b.** H⌐ΔΔΓ|

    Translate the following base 10 numbers into Greek:
    c. 347 **d.** 5555

11. An ancient Chinese mathematician named Sun-Tsu who lived in the first century A.D. described the use of calculating rods (made of bamboo) for representing numbers. The digits for 1 to 9 are represented as follows:

    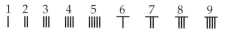

    However, when the digits from 1 to 9 appear in the tens column, they are represented as follows:

    1 2 3 4 5 6 7 8 9
    — = ≡ ≣ ≣ ⊥ ⊥ ⊥ ⊥

Thereafter, every time you move over one place, you change from one form to the other.

Thus the number 4763 would be represented as 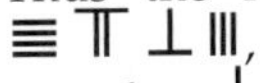, and the amount 8888 would be represented as 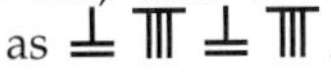.

Translate the following numbers into base 10:

a. 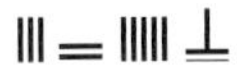 **b.** 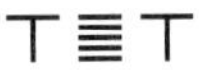

Represent the following numbers in the Chinese system:

c. 346 **d.** 12,345

12. One of the most impressive of the ancient numeration systems comes from the Mayans of Central America. Their numeration system was developed around 400 or 300 B.C. They used only three symbols: dot, line, and oval. The oval functions in a manner very similar to our zero. They also wrote their numbers vertically.

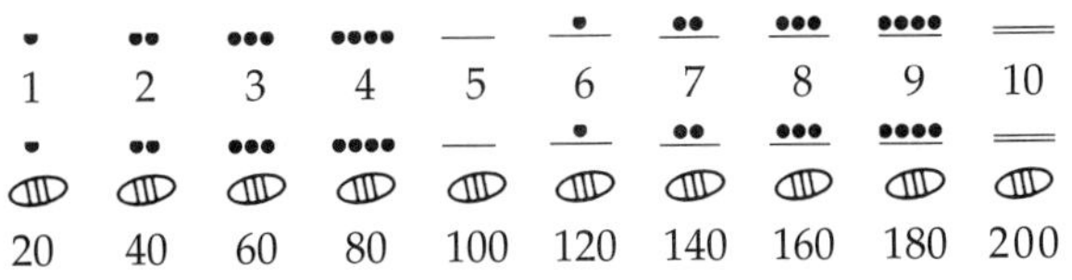
1
2
3
4
5
6
7
8
9
10
20
40
60
80
100
120
140
160
180
200

Many mathematics historians credit the Mayans with being the first civilization to develop a numeration system with a fully functioning zero. The values of their places are 1, 20^2, 20^3, 20^4, etc. They actually had two number systems, one for common counting and one for calendar counting. We will explore the common counting system, since it is a pure base 20 system. Below are three base 10 amounts and how they would be written in the Mayan system.

46 =	300 =	417 =
(2 × 20)	(15 × 20)	(1 × 400)
+ 6	+ 0	+ (0 × 20)
		+ 17

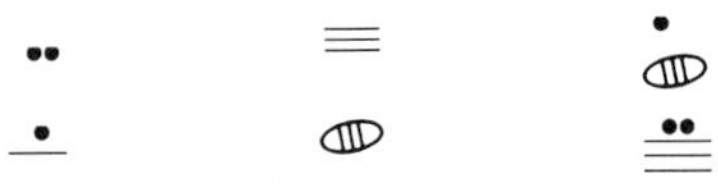

Translate the following Mayan numbers into base 10:

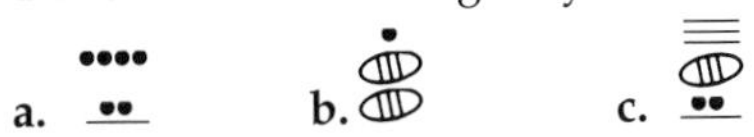
a.
b.
c.

Write the following base 10 numbers in Mayan:

d. 245 **e.** 500 **f.** 2813

13. The developers of Braille decided on the notation shown below for numbers in Braille. I tried for some time to figure out the logic behind this code until I read that the symbols for the digits are simply the symbols for the first 10 letters of the Braille alphabet! That is, the symbol for 1 is the same as the symbol for *a* and so on! Each digit has its unique shape of raised dots on a 2 × 3 grid as shown below. The convention is to use a backwards L to let the reader know that a number is coming. The symbol for a comma is shown also.

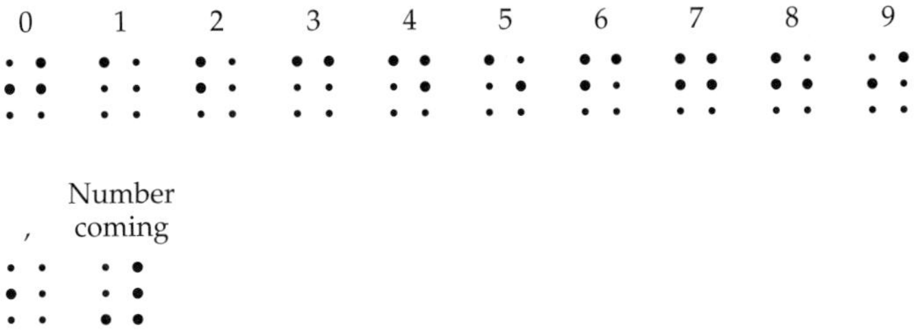
0
1
2
3
4
5
6
7
8
9
,
Number coming

Here is how the amount 245,608 would appear in Braille:

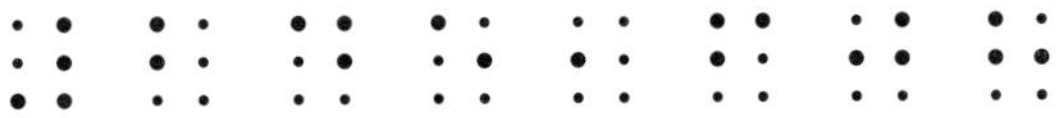

Convert these numbers to Braille.

a. 599 **b.** 3,603,200

Convert these Braille symbols into base 10 symbols.

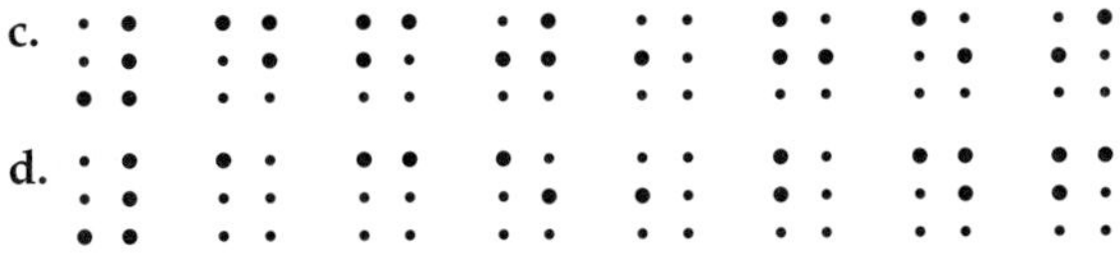
c.
d.

14. In the problems below, enter the first number into your calculator. Then describe what number and operation you would enter so that the calculator would display the following number. For example, if you entered 2345 and then the next number were 1875, you would enter "−500 =" on the calculator.

a. 3456	**b.** 37779	**c.** 123456
3056	30009	203456
d. 30405	**e.** 346734	**f.** 445566
34445	1	334455

g. Make up your own problem.

In Exercises 15 through 26, recall your examination of different bases in the explorations.

15. Tell what comes after:

a. 34_5	**b.** 1011_2	**c.** 99_{16}	**d.** 7099_{16}
e. 101_2	**f.** 111_{12}	**g.** 124_5	**h.** 405_6

16. Tell what comes before:

a. 1010_5	**b.** 340_5	**c.** 100_{16}	**d.** 1110_2
e. 1010_2	**f.** 110_{16}	**g.** 120_4	**h.** 60_7

17. Convert the following numbers into base 10.

a. 41_5	**b.** 55_6	**c.** 210_5	**d.** 2104_5
e. 101_5	**f.** 1111_6	**g.** 303_4	**h.** 606_7
i. 1101_2	**j.** 10001_2	**k.** 99_{16}	**l.** 909_{16}

18. Convert the following numbers from base 10 into the designated base.

   a. $44_{10} = ?_5$ b. $152_{10} = ?_5$ c. $92_{10} = ?_2$
   d. $206_{10} = ?_2$ e. $72_{10} = ?_{12}$ f. $402_{10} = ?_{16}$
   g. $44_{10} = ?_6$ h. $1252_{10} = ?_6$ i. $144_{10} = ?_{16}$
   j. $100_{10} = ?_5$ k. $99_{10} = ?_{16}$ l. $1052_{10} = ?_5$
   m. $2{,}500{,}000_{10} = ?_5$ n. $2{,}500{,}000_{10} = ?_2$

19. In what base does $25 + 25 = 51$?

20. Tell what base makes the following statement true: $23_{10} = 25_x$

21. For what base x is this statement true? $598_{10} = 734_x$

22. If $44_x = 28_{10}$, how many base x candy bars will fit into a box holding 110_5 candy bars?

23. Although the base 10 long is twice as long as the base 5 long, the base 10 flat is 4 times as large as the base 5 flat. Why is this so? That is, given that 10 is twice as much as 5, why isn't the base 10 flat also twice as big as the base 5 flat?

24. Explain the following conversion problems as though you were talking to a struggling student.

   a. $234_{10} = ?_5$ b. $405_8 = ?_{10}$

25. We speak of so many children having little or no understanding of place value. What does "place value" mean to you? Define the concept in your own words. This involves giving some meaningful explanation of what "place value" means mathematically.

26. We have discussed the development of more sophisticated numeration systems in class, culminating in the Hindu-Arabic base 10 system that we currently use. Describe the advantages of the Hindu-Arabic system over the Roman system.

27. The first edition of this book was published in 1997, which is in the twentieth century, not the nineteenth century. Can you explain why?

28. When we write large numbers, why do we insert a comma every three numbers instead of every two or four numbers?

29. You are Zirkle, from Zordon, and you have just finished your trip to the solar system. Your planet has managed to develop interplanetary travel without ever developing a sophisticated number system. You have examined three different bases (base 2, base 5, and base 6), and you are preparing your recommendations for your home planet. Summarize the pros and cons of each base, and then tell which base you recommend that your planet adopt. *Note:* Your species has no fingers, only flippers. Therefore, none of the bases has an advantage with respect to your anatomy.

   a. Summarize the pros and cons of each base in a table like the one below. *Note:* Entries in the cells need to demonstrate that you understand the advantage or disadvantage/limitation. That is, an entry should state *what* the advantage or disadvantage is and also give a brief *explanation* of the advantage or disadvantage.

Base	+	−
2		
5		
6		

   b. Write your final recommendation: an essay of at least one paragraph (more than two sentences). You may refer to your chart, so that you don't have to be redundant.

30. In some dairy states, a kind of base 6 is used for shipping milk. Six cartons of milk make 1 box; 6 boxes make 1 crate, 6 crates make 1 flat, and 6 flats make 1 pallet. Thus, if someone said they had filled 2413, that would mean 2 flats, 4 crates, 6 boxes, and 3 cartons.

   a. How many cartons of milk would this be in base 10?
   b. Which of the six characteristics of base 10 (on pages 109–110) does this system possess? Justify your responses.

31. NASA keeps track of time using the following system:

   Hours:Minutes:Seconds:Hundreths of a Second

   Thus 4:2:5:6 would mean that the program is 4 hours, 2 minutes, 5.06 seconds from launch.

   a. Let's say that a launch was temporarily stopped at 4:2:5:6 and then later stopped at 3:1:7:2. How much time had elapsed between the two stops?
   b. Which of the six characteristics of base 10 (on pages 109–110) does this system possess? Justify your responses.

32. One common mistake that young children make when counting is illustrated below: twenty-six, twenty-seven, twenty-eight, twenty-nine, twenty-ten, twenty-eleven, etc.

   Can you describe the nature of the child's mistake? That is, what is the child not understanding or what misconception of counting causes the mistake? Do you see this mistake differently now than you would have before this course? If so, please explain.

33. Another common mistake that young children make when counting is illustrated below: twenty-six, twenty-seven, twenty-eight, twenty-nine, thirty-one, thirty-two, etc.

   Can you describe the nature of the child's mistake? That is, what is the child not understanding or what misconception of counting causes the mistake? Do you see this mistake differently now than you would have before this course? If so, please explain.

34. In the October 1993 issue of the *Arithmetic Teacher*, Susan Bohan asked her elementary school class to make changes in the Egyptian system to make it easier to use. Below is the work of several children. In

each case, describe which, if any, of the six characteristics of base 10 (on page 123) the refined systems possess. Justify your responses.

a. In a slight modification of Mary's system, she put dots on each symbol to represent how many of those symbols were represented. For example, 70 was shortened from ∩∩∩∩∩∩∩ to ∩̈. Here are two examples of amounts using Mary's system:

b. Three boys, Mark, Jim, and Bob, made up new symbols for 3–9, shown below. They kept | for 1 and || for 2.

They then paired these symbols with the Egyptian numerals. For example, 70 was shortened from ∩∩ ∩∩∩ to ⌐J. They decided to have both a vertical and a horizontal aspect to their system—the new symbols (for how many of each amount) were written above the Egyptian symbols instead of next to them. Here are two examples:

5348   ⋆ △ □ ⌐J
       ⋎ ⊙ ∩ |

9105   ⧠ | ⋆
       ⋎ ⊙ |

c. After some discussion of the previous system, someone suggested that the Egyptian symbols weren't necessary anymore as long as everyone knew what each place stood for. The students realized that they did need a symbol for an empty place. They choose Z for 0.

Source: Reprinted with permission from *Arithmetic Teacher* copyright © 1993 by the National Council of Teachers of Mathematics.

35. This question came from a fifth-grader: Is 5 the middle number between 0 and 10? What do you think?

36. I have seen problems like these used in elementary classrooms to help students develop a deeper understanding of the relationship between places.

   a. How many hundreds are in 2600?
   b. How many hundreds are in 54,000?
   c. How many tens are in 250?
   d. How many tens are in 4500?
   e. What is the more common way of saying thirty-four hundred?
   f. What number is equivalent to 345 tens?

37. If it were possible to make a line of people that went around the world, how many people would it take?

38. a. How long is 1 million seconds? Pick a unit that we can all sense. For example, minutes is not a tangible unit because 17,000 minutes is not an amount that most people can relate to.
    b. How long is 1 billion seconds?

39. a. If you laid 1 million dollar bills end to end, how long would that line be?
    b. How long would 1 billion dollar bills be?

Chapter Summary

1. Many mathematical concepts and problems involve various kinds of sets and subsets of those sets.

2. The primary sets of numbers that children work with in elementary school are the sets of natural numbers, whole numbers, integers, and rational numbers. The relationships among them can be shown with a Venn diagram.

3. It is important for elementary school children to develop "algebraic thinking." There are several fundamental conceptions of algebra—algebra as a set of rules and procedures, algebra as the study of structures, algebra as the study of relationships among quantities, and algebra as generalized arithmetic.

4. When two sets or variables are said to be functionally related, this tells us that the relationship between those two sets or variables is such that knowing an element of one set (or the value of the independent variable) enables us to make predictions.

5. Algebraic ideas and relationships can be expressed in many different ways—verbally, with tables, with ordered pairs, with formulas, and with graphs. It is important to understand these ways and how to translate (move) easily among them.

6. When children count, they need to realize that there is a one-to-one relationship between the elements in the set they are counting and the set of natural numbers.

7. Our current (Hindu-Arabic) numeration system evolved over thousands of years, and people from

many cultures developed ideas that contributed to our system.

8. Our base 10 system has six important characteristics: (1) It has only 10 symbols, with which we can represent any amount. (2) The value of each place is 10 times that of the previous place. (3) Each place can contain only one symbol. (4) The value of a numeral depends on its place in the number. (5) The value of a number is determined by multiplying each numeral by its place value and then adding these sums. (6) Zero represents both an empty place and an actual amount.

BASIC CONCEPTS

Section 2.1 Sets

set 64
subset 64
iff 67
Venn diagrams 70
binary operations 74

member, element 64
proper subset 67
universal set, or the universe 70

Ways to describe sets
words, list 64
set-builder notation 64

Important mathematical sets
natural numbers (N) 64
whole numbers (W) 64
integers (I) 64
rational numbers (Q) 64

Kinds of sets
finite and infinite sets 66
well-defined set 66
empty set, null set 69
disjoint sets 72

Relationships between sets
equal 69
one-to-one correspondence 69
equivalent 69

Operations on sets
intersection 71
complement 71

union 71
subtraction 72

Section 2.2 Algebraic Thinking

conceptions of algebra 80
set of rules and procedures 81
study of relationship among quantities 84
function notation 85
output 86
mathematical model 88

generalized arithmetic 80
equivalence 83
study of structures 83
domain, range 85
function 85
input 85
independent variable, dependent variable 87
figurate numbers 92

Ways to represent functions:
table 87
ordered pairs 87
mapping 87

graph 87
equation 87

Section 2.3 Numeration

Origins of numbers and counting 101

Kinds of numeration systems
Egyptian 104
Babylonian 106

Roman 105
Hindu-Arabic 109

Characteristics of numeration systems
numeral 103
additive 105
place value 107
base 10 109
decimal 111
base 10 manipulatives: unit, long, flat, block 111
simple units 114
composite units 114

number 103
powers of 10 105
positional 107
digit 110
expanded form 111

CHAPTER 2 REVIEW EXERCISES

1. Describe the set of powers of 10 using:
   a. Set-builder notation
   b. List notation

2. Is "the set of countries in the world" a well-defined set or not? Justify your response.

3. Fill in the most appropriate symbol—$\in$, $\notin$, $\subset$, or $\not\subset$—for each of the following questions.
   a. 3   {1, 2, 3}
   b. {1, 2}   {1, 2, 3}
   c. $\emptyset$ {1, 2, 3}
   d. {3, 4}   {1, 2, 3}

4. Rewrite the following statements using mathematical symbols.
   a. The set D is not a subset of the set E.
   b. 0 is not an element of the null set.

5. Let
   U = the set of whole numbers between 0 and 30
   A = {1, 3, 5, 7, 9, 11, 13, 15, 17, 19, 21, 23, 25, 27, 29}
   B = {0, 5, 10, 15, 20, 25}
   a. Represent this system with a Venn diagram.
   b. List the elements of $A \cap B$.
   c. Describe in words the elements of $\overline{A}$.

6. Let
   U = the set of college students at your college
   E = students at your college who are planning to teach elementary school
   C = students who have cars on campus
   S = students who are out-of-state
   a. Draw a Venn diagram to represent this situation: $E \cap (C \cup S)$.
   b. Draw a Venn diagram to represent this situation: Those students at your college who are in-state and who have cars on campus.

7. In a group of 100 people, 70 have a DVD player, 60 have a CD player, and 20 have neither. How many have both?

8. What is the difference between saying that two sets are equal and saying that they are equivalent?

9. A function is described by the formula $f(x) = 3x - 2$.
   a. Draw a table, sketch a graph, or draw an arrow diagram to represent this function.
   b. Find $f(6)$.

10. The function that converts American dollars to Transylvanian rubles is given by the following hookup: M 3, then D 20, then A 13. Use the hookup to convert $100 into rubles.

11. Determine how many squares it will take to make the nth figure in this growing pattern:

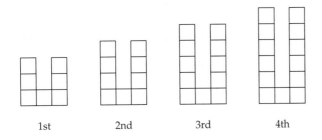

12. Determine how many squares it will take to make the nth figure in this growing pattern:

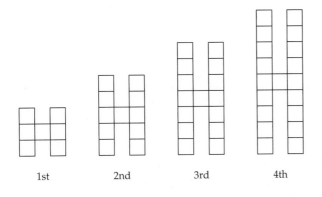

13. The first five rectangular numbers are 2, 6, 12, 20, 30.
    a. What is the 10th rectangular number?
    b. Write a rule for finding the nth rectangular number.

14. The first 7 items in a sequence are: 10, 14, 18, 22, 26, and 30.
    a. What is the 10th number in the sequence?
    b. Write a rule for finding the nth number in this sequence.

15. A cellular phone company offers two plans:
    a. $30 a month for 300 anytime minutes plus 10 cents a minute for additional minutes
    b. $40 a month for unlimited anytime minutes
    Fred is not sure which plan he should pick. Tell him the conditions under which each plan would be better for him.

16. Which of the following graphs best matches the situation of the temperature of a cup of coffee that is initially very hot and then sits on a desk? Explain your choice, and also explain why the other choices are not valid representations of the situation.

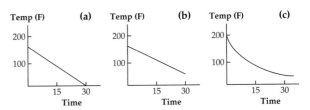

17. Sketch a graph for the following situation. He walked steadily for awhile and then began climbing a hill. As he walked up the hill, he walked more and more slowly until he reached the top of the hill. He walked more and more quickly as he walked down the hill and then maintained a steady pace once he made it to the bottom of the hill.

18. Write a story for which this graph is an accurate representation. Select and label the independent and dependent variables, for example, time and distance.

19. Write the following amounts in the Egyptian, (late) Roman, and Babylonian systems.
    a. 47   b. 95   c. 203   d. 3210

20. Indicate what number comes after
    a. 404_5 b. 1244_5 c. 1001_2
21. Indicate what number comes before
    a. 4320_5 b. 30040_5 c. 1100_2
22. Convert the number to base 10: $243_5 = ?_{10}$
23. Would you rather have $\$200_{10}$ or $\$1000_5$?
24. Explain why 321_5 and 321_6 are not the same.
25. If 10 is twice the value of 5, why doesn't a base 10 flat have twice the value of a base 5 flat?
26. The value of the fifth place in base 10 is ____ times the value of the fifth place in base 5.
27. Many students say that a major learning from Section 2.2 is that 10 is not just ten, but rather "one-zero." What does "one-zero" mean? Write your answer so that it works for any base.
28. In any base, the first place looks like a little cube, the second place looks like a long, the third place looks like a flat, etc. What does the seventh place look like?
29. Express the following number in expanded form: 2,068. Note that there are three variations of standard form—any of these is fine here.
30. Describe, in your own words, the primary characteristics that distinguish base 10 from earlier, non-base systems—for example, the Egyptian and Roman systems.

ANSWERS TO QUESTIONS IN TEXT:

Paying the baby-sitter: graph, p. 88

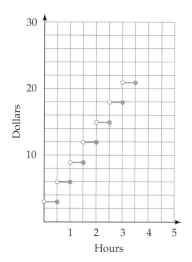

Patterns in counting: missing words in Table 2.7, p. 103

Number	Greenland Eskimos	Aztecs	Luo of Kenya
8	achfineq-pinasut		ab-adek
9			ab-angwen
12	achqaneq-machdlug	matlacti-on-ome	
13	achqaneq-pinasut	matlacti-on-yey	apar-adek
15			apar-abich
16	achfechsaneq-atauseq	caxtulli-on-ce	apar-ab-chiel

The Egyptian numeration system: symbols, p. 104

1. 1323   2. 143,022   3. [symbols]   4. [symbols]

The Roman numeration system: symbols, p. 106

1. 1011   2. 1597   3. MCII   4. CCCXIX

Working with the system

The Later Roman symbol for 444 was CDXLIV.

The Babylonian numeration system: places, p. 107

1. $120 + 12 = 132$
2. $5 \times 60 + 42 = 342$
3. $1 \times 60^2 + 11 \times 60 + 21 = 3600 + 660 + 21 = 4281$
4. 1202 would be seen as $20 \times 60 + 2 =$ [symbols]
5. 304 would be seen as five 60s plus 4, or [symbols]

CHAPTER 3

The Four Fundamental Operations of Arithmetic

3.1 Understanding Addition
3.2 Understanding Subtraction
3.3 Understanding Multiplication
3.4 Understanding Division

History

I cannot overemphasize how recent most of our everyday mathematics actually is. During the Middle Ages, many mathematicians considered an operation to be any mathematical technique or procedure that was considered important. In fact, especially when the Hindu-Arabic numeration system was still new in Europe, many mathematicians considered numeration to be an operation. It has only been in the last century that mathematicians have linked the concept of operation to the concept of function.[1]

Learning how to add, subtract, multiply, and divide—first with whole numbers, then with negative numbers, next with fractions, and finally with decimals—and then applying this knowledge to problems makes up the lion's share of elementary school mathematics. This area is also the one in which there is the most controversy, not unlike the controversy over phonics vs. whole language in language arts. Without getting into the controversy, I want to frame our work by taking some time to explore certain aspects that lead to competency with these operations. Our focus here will be understanding that

- The operations have multiple meanings vs. one single meaning—for example, subtraction is more than simply "take-away."
- These meanings are often connected—for example, one meaning of multiplication is repeated addition.
- There are many algorithms (standard procedures) for each operation; the one you learned for each operation is simply one of many.
- Knowledge of base 10 and place value is vitally important to see how these algorithms work.
- Practice and memorization are important, but practice without understanding is a waste of time.

[1]Frank J. Swetz, *Capitalism and Arithmetic: The New Math of the 15th Century* (La Salle, IL: Open Court, 1987), p. 181.

SECTION 3.1 UNDERSTANDING ADDITION

Addition

> **WHAT DO YOU THINK?**
>
> - How many words, aside from *combine*, can you think of that describe addition?
> - Not having place value, how might the Romans have added, for example, XXXVIII + XXVI?

Children encounter addition even before they have mastered counting. They naturally combine objects and want to know how many. For example, there are 4 people in our family, and 2 guests have come for dinner. How many people will be eating?

Examine the following addition problems and then write your responses to the following questions in your own words. Pretend that you are a young child who has not yet learned addition. Obviously, finding the answers to the problems will be easy for you. Thus you should focus on the following questions before reading on. . . .

1. What action words describe what is happening in these and other addition problems? One action word is *combining*. What others can you think of?
2. Other than "they all involve addition," can you think of other ways in which all four problems are alike? In what ways are some of the problems different from each other?

Four addition problems

1. Andy has 3 marbles, and his older sister Bella gives him 5 more. How many does he have now?
2. Keesha and José each drank 6 ounces of orange juice. How much juice did they drink in all?
3. Linnea has 4 feet of yellow ribbon and 3 feet of red ribbon. How many feet of ribbon does she have?
4. Josh has 4 red trucks and 2 blue trucks. How many trucks does he have altogether?

Contexts for Addition

Problems 1 and 4 are easier for most children, because the child can see the actual marbles and trucks and count 1, 2, 3, 4, 5, 6, 7, 8 marbles and 1, 2, 3, 4, 5, 6 trucks. Problems 2 and 3 are more abstract in that the child cannot actually see 6 ounces or 4 feet. The child might represent the problem with concrete objects, such as 4 buttons for the yellow ribbon and 3 buttons for the red ribbon. These problems represent two basic contexts in which we operate on numbers. Some numbers represent **discrete** amounts, and some numbers represent measured (**continuous**) amounts.

A Pictorial Model for Addition

One of the themes of this book is the power of "multiple representations"—that there is generally more than one way in which we can represent most mathematical concepts and problems. If we draw diagrams to represent the four addition problems, they seem on the surface to be quite different, except for the marbles and trucks problems (see Figure 3.1).

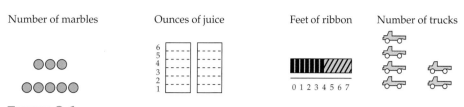

FIGURE 3.1

History

Many people do not realize how recent much of our mathematical notation is. The symbols we use for addition and subtraction, + and −, first appeared in Germany in the late 1400s. Historical records indicate that these symbols were first used to indicate sacks that were surplus or minus in weight. Michael Stiffe (mid-1500s) was the first person to use + and − as symbols of operation as well as of abbreviation. In 1631, William Oughtred first used the letter *x* to represent multiplication. Italian merchants introduced the symbol for division (÷) in the 1400s to indicate a half. For example, they wrote 4 ÷ to indicate $4\frac{1}{2}$. The equals sign first appeared in the late 1500s in a book by Robert Recorde. It was more than 100 years later that the use of this sign became widespread.

In each of these cases, however, we are joining two sets or we are increasing a set. We can highlight the similarities among the problems with the representations in Figure 3.2.

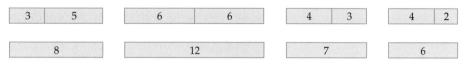

FIGURE 3.2

In general, we can represent any addition problem $a + b = c$ as shown in Figure 3.3.

FIGURE 3.3

Do you see any advantages of this general model over having no model at all? If you do, what advantages do you see? Think and then read on. . . .

1. This model captures the way in which *all* addition problems are similar— that is, joining and combining two amounts to make a larger amount.
2. This model is also related to the notion of parts and wholes, an abstraction that is important in the development of whole-number ideas and in understanding other mathematical ideas, like fractions.
3. This model also works well whether the elements to be combined are discrete objects or measurements.
4. Perhaps the most important advantage of the pictorial model is that we can define all four operations in this context, as we shall soon see. This reveals the essential connectedness of the four operations. When students see this connectedness and the connection between numbers and place value, they are likely to be more successful with nonroutine and multistep problems.

Did your list include these advantages? The same ideas in different words? Other advantages?

One researcher who has worked extensively with young children and looked at how children come to make sense of numbers and operations on numbers tells us that "(p)robably the major conceptual achievement of the early school years is the interpretation of numbers in terms of part and whole relationships."[2] It is important to note that understanding part–whole relationships with whole numbers allows numbers to be interpreted simultaneously as

---

[2] Lauren Resnick, "A Developmental Theory of Number Understanding," in *The Development of Mathematical Thinking,* ed. Herbert Ginsburg (New York: Academic Press, 1983), p. 114.

positions on the mental number line and as compositions of other numbers. For example, 18 is the number after 17 and before 19, but 18 can also be seen as $10 + 8$, $9 + 9$, $20 - 2$, and so on. Understanding that a number can be composed (put together) and decomposed (broken into parts) is essential for being able to work confidently with the four operations. This notion of composing and decomposing is one of the big ideas of elementary mathematics, and we will come back to it repeatedly throughout the text.

There are several classical stages in children's understanding of addition. At the most basic level, the child counts to determine the sum. For example, in the first problem on p. 128, many young children would answer the question by putting the marbles on the floor and then counting the two groups. Interestingly, the traditional definition of addition is in set language, and children intuitively begin with this model (see Figure 3.4).

FIGURE 3.4

Let $A =$ the set of Andy's marbles, and let $B =$ the set of Bella's marbles.

Then $n(A) = a$, the number of marbles Andy has, and $n(B) = b$, the number of marbles Bella has.

We then define **addition** in the following way:

If A and B are disjoint sets, containing a and b elements, respectively, then $a + b = n(A \cup B)$.

In other words, the **sum** of a and b is equal to the number of elements in the union of the two sets. Do you see why the sets have to be disjoint?

The numbers a and b are called **addends**.

In the next stage of development, the child "counts on." That is, the child begins with the first number and counts however many more the second number represents. A child solving Problem 1 in this manner would say, "4, 5, 6, 7, 8," probably keeping track of the second addend (5) with fingers. At the next level, the child realizes that it is possible to begin with the larger number (i.e., $3 + 5 = 5 + 3$) and counts, "6, 7, 8." Finally, the child simply knows that $3 + 5 = 8$.

Number Lines: An Addition Model

The number line plays an important role throughout arithmetic, but research shows that American schoolchildren tend to have trouble understanding and using this model. Therefore, we will take a little time to investigate the number line further as a tool for representing (and solving) addition problems. We will come back to the number line when we examine the other operations. Before you read on, try to think of physical examples of number lines in everyday life. . . .

Number lines are found on rulers, clocks, graphs, and thermometers. A number line can be constructed by taking a line (not necessarily a straight line) and marking off two points: zero (the origin) and one. The distance from 0 to the point 1 is called the unit segment, and the distance between all consecutive whole numbers is the same. Although number lines are most commonly used to represent length, they may be used to model all kinds of problems. For example, we could use a number line to indicate time, with each unit representing one unit of time—day, minute, year, etc.

How could we have represented the ribbon problem with a number line? Write down your thoughts before reading on. . . .

In the diagram at the left in Figure 3.5, we first draw an arrow (in this case representing the length of the ribbon) 4 units long. We draw another arrow 3 units long and connect the two arrows. The arrow at the top represents the combined length of the two shorter arrows.

Figure 3.5

In the diagram at the right, we start at the point on the line representing the length of the first ribbon and then draw an arrow 3 units long (representing the second ribbon). The location where the arrow ends tells us the combined length of the two ribbons.

Both diagrams represent 4 + 3 on the number line, although the one on the left more closely resembles the actual laying of the two ribbons end to end.

Properties of Addition

Another aspect of mathematical power comes from understanding the properties of the various operations. As usual, this is best grasped via exploration rather than presentation. Look at Table 3.1, the addition table for base 10.

Table 3.1

	0	1	2	3	4	5	6	7	8	9
0	0	1	2	3	4	5	6	7	8	9
1	1	2	3	4	5	6	7	8	9	10
2	2	3	4	5	6	7	8	9	10	11
3	3	4	5	6	7	8	9	10	11	12
4	4	5	6	7	8	9	10	11	12	13
5	5	6	7	8	9	10	11	12	13	14
6	6	7	8	9	10	11	12	13	14	15
7	7	8	9	10	11	12	13	14	15	16
8	8	9	10	11	12	13	14	15	16	17
9	9	10	11	12	13	14	15	16	17	18

■ *Mathematics* ■

If you did Exploration 2.8, recall how strange the Alphabitian system was to you. What observations made you more comfortable with adding? Can you think of analogous observations that might make the learning of base 10 addition facts easier for young children?

Sometimes students in this course forget how new addition is to young children. When young children start to learn about adding, Table 3.1 has been the traditional method of representing the 100 "addition facts" that they have to learn. It can look imposing to some children! However, understanding some properties of addition and base 10 can unlock its potential as a learning tool. Take a few moments to think about what addition means and to look at the addition table. What do you observe (insights or patterns) that might make learning the addition facts easier for children? Think before reading on. . . .

One observation (in children's language) is that "adding zero doesn't change your answer." Mathematicians call this the identity property of addition. It is represented in symbols as follows:

$$a + 0 = 0 + a = a$$

Children discover that when you add 1 to any number, you get the next number from counting. In other words, they connect addition and counting.

We also find that when we add any two numbers, we get the same sum regardless of the order in which we added; that is,

$$a + b = b + a$$

This commutative property of addition is one that all children come to recognize. I have been fortunate enough to be with a few children when they made this discovery. They were as excited as Edison probably was when he finally invented the light bulb. Their faces just shone when they shared with me the discovery that you get the same number either way!

This discovery occurs in different ways. Some students, more naturally curious than others, simply wonder whether the order matters; they discover that it doesn't. For other students, the discovery is of a visual nature. That is, if we draw a diagonal line from the top left corner to the bottom right corner of Table 3.1 and then fold the table in half along that line, we find that the two halves of the table are identical. Mathematically, we speak of the symmetric nature of the addition table.

There is another discovery, called "bridging with 10," that makes learning addition easier. For example, if you ask a child, "What is 7 + 5?" and then ask how the child got the answer, many children will say something like, "7 + 3 is 10 and 2 more is 12." What happens is that students quickly learn the combinations that make a long (that is, 10) when using base 10 manipulatives. Many will intuitively decompose 7 + 5 into 7 + 3 + 2.

This leads to a third property, called the associative property of addition. For example, 7 + (3 + 2) = (7 + 3) + 2. Formally, we say

$$(a + b) + c = a + (b + c)$$

One last property of addition often seems almost trivially obvious; it is called the closure property of addition. Stated in everyday language, the closure property means that the sum of any two whole numbers is a unique whole number. There are two parts to this property: (1) uniqueness (the sum will always be the same number) and (2) existence (the sum will always be a whole number).

From a more formal perspective, not all sets are closed under addition. For example, consider the set of odd whole numbers {1, 3, 5, …}. The sum of two odd numbers is not in the set of odd numbers; thus we say that the set of odd numbers is not closed under addition. In higher mathematics, there are many different kinds of sets, and it is often important to know whether the set is closed under the various operations.

CLASSROOM CONNECTION

For most college students, the closure property evokes one of two reactions: either "so what" or "here we go again, making something obvious look complicated." Let me turn the tables on you. How might this property be related to a question a child might actually ask? Think and read on. . . .

Children will literally ask closure questions: "Will *any* two numbers make a whole number?"

INVESTIGATION 3.1

6	7
7	8

A Pattern in the Addition Table

As you will come to appreciate by the end of this course, patterns can help make the learning of mathematics easier and more interesting. In the addition table, if you look at any 2 × 2 **matrix** (that is, a rectangular array of numbers or other symbols), the sums of the numbers in each of the two diagonals are equal. For example, in the matrix to the left, 6 + 8 = 7 + 7. Can you justify this pattern mathematically? Work on it and then read on. . . .

DISCUSSION

Description 1

At a verbal level, one could justify this pattern by saying that in any 2 × 2 matrix in the table, the two numbers in one diagonal are always identical and the other two numbers are always 1 less and 1 more than this number. Therefore, the sum of the two other numbers will "cancel out" so that you get the "same" sum in either case. If you had a hard time following the previous two sentences, you are not alone, although I have heard second graders explain this pattern in language as elaborate as that above.

Description 2

We can use some notation to make the description less ambiguous. Noting that the value of each number increases by 1 each time that we move across (or down) the table, we can let x represent the number in the top left corner of the diagonal. Thus, in relation to x, the values of the other three numbers are

x	$x + 1$
$x + 1$	$x + 2$

It is now a simple algebraic exercise to demonstrate that the sum of each diagonal is $2x + 2$.

Description 3

Yet other students will say that the sums of the diagonals are equal because "it's the same numbers in both cases." What do you think such a student might be seeing? Think before reading on. . . .

Let's say that we are looking at the matrix formed by the intersection of the 2 row and the 3 row and the 4 column and the 5 column. The numbers are 6, 7 and 7, 8. However, if we represent the numbers by their origin, we have

	4	5
2...	2 + 4	3 + 4
3...	2 + 5	3 + 5

Mathematics

We have defined addition in the context of whole numbers. In later chapters, we will examine how to extend/modify this definition so that it will fit other numbers—for example,

$$\frac{3}{4} + \frac{2}{5} \qquad {}^-4 + {}^-5$$

$$4.342 + 5.6 \qquad x + 22 = 37$$

Can you see that the simplistic notion of addition as joining and combining doesn't fit all of these situations?

The sum of the top-left-to-bottom-right diagonal is $(2 + 4) + (3 + 5)$. However, because of the commutative and associative properties, this sum is equal to $(2 + 5) + (3 + 4)$. In other words, we are indeed using the same numbers!

We can now generalize this cell by saying that the matrix formed by the intersection of the *a* row and the *b* row and the *c* column and the *d* column is

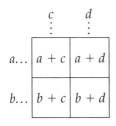

and

$$a + c + b + d = a + d + b + c$$

Our first work with addition in base 10 will be doing some addition problems mentally, for a couple of reasons. First, this will require you to think carefully about how your knowledge of place value applies to adding numbers. Second, much of our use of arithmetic does not involve pencil and paper or calculators, but rather mental computation—when estimating or when it is quicker to do a computation or part of a computation in our head than with a pencil or calculators.

INVESTIGATION 3.2

Mental Addition

Do the following five computations in your head. Briefly note the strategies you used, and try to give names to them.

Note: One mental tool all students have is being able to visualize the standard algorithm in their heads. For example, for the first problem, you could say: "9 + 7 = 16, carry the 1, then 5 + 3 = 8 + the carried 1 makes 9; the answer is 96." However, because you already know that method, I ask you not to use it here but to try others. There are actually quicker ways to do this problem in your head than using the traditional algorithm. See whether you can discover any of them.

1. 39 + 57 **2.** 68 + 35 **3.** 66 + 19 **4.** 545 + 228 **5.** 186 + 125

The right answer to each of these sums isn't really what's at issue. Rather, these strategies should serve as a jumping-off point for understanding mental addition more clearly.

DISCUSSION

Leading digit One strategy that works nicely with most addition problems is **leading digit**. Some people refer to it as **front end** because we add the "front" of the numbers first. For example, we can look at Problem 1 as 80 + 16 = 96; that is, we add the numbers in the tens column (3 + 5), add the numbers in the ones column (9 + 7), and then add 80 + 16 to get 96.

This strategy can be used with larger numbers. For example, try this strategy with Problem 4 and then read on. . . .

700 + 60 + 13, or 773

The leading-digit method is generally faster than doing the algorithm in your head. For example, doing Problem 4 in your head using the standard algorithm looks like this: 5 + 8 = 13, carry the 1; 4 + 2 + the carried 1 is 7, so that's 73; 5 + 2 = 7, so it's 773.

Each of the three representations below highlights a different aspect of this method. The expanded form at the left shows the number taken apart by place value and makes it easier to see the two addition problems (called *partial sums*) within the problem. The manipulatives make it easier to see why the expanded form works: Because addition is ultimately combining the two amounts, we can add the longs first and add the singles second and then add those two amounts. The representation at the right borders on a proof. First, we break the numbers apart in expanded form. Formally, it would take several applications of associative and commutative laws to move from the second to the third line.

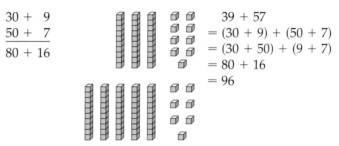

$$39 + 57$$
$$= (30 + 9) + (50 + 7)$$
$$= (30 + 50) + (9 + 7)$$
$$= 80 + 16$$
$$= 96$$

Try some of the other problems using the leading-digit method.

Compensation Another powerful mental math strategy is called **compensation**. Look at Problem 1 to see how this strategy works. Instead of trying to add 39 + 57, we could have added 40 + 56. Do you see how we transform 39 + 57 into 40 + 56?

By adding 1 to the first number and subtracting 1 from the second number, we have created a new problem that is much easier to solve mentally. Why is this? Mathematically, we used the following transformation:

$$39 + 57 = (40 - 1) + 57 = 40 + (57 - 1) = 40 + 56$$

Which other problems lend themselves to this strategy?

Break and bridge We can use the **break and bridge** strategy in Problem 2. To add 68 + 35, break 35 into 30 + 5 and do the following:

$$(68 + 30) + 5 = 98 + 5 = 103$$

Representing this strategy on a number line makes it easier to understand for many students.

We add 30 first (68 to 98) and add 5 next (98 to 103)

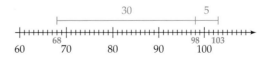

Which other problems lend themselves to this strategy?

Compatible numbers Another powerful strategy is creating **compatible numbers**. This often involves seeing pairs of digits whose sum is 10. Just as children quickly learn the pairs of digits whose sum makes a long (for example, 1 and 9, 2 and 8), most people can sense which combinations of numbers will make multiples of 100. Thus, in Problem 5, 186 + 125, we can see that the 180 and the 120 combine to make 300, and then we have 11 (6 + 5) more.

Do you see how place value and base 10 are used in this strategy? If you can see Problem 5 as 18 tens and 12 tens, then you have 30 tens—that is, 300. Look at Problem 2. What ways can you see to get multiples of 10 in that problem?

The compatible numbers strategy is especially useful when adding several numbers mentally. Try it with the following problem and then read on: 16 + 35 + 28 + 15.

We can obtain the exact sum by adding the 35 and the 15 together and the 28 and the 16 together. Next, we add the results:

50 + 44 = 94

Choosing a strategy   Which strategy you use is often a matter of preference. For example, Problem 2 (68 + 35) may be done mentally in at least four different ways, each of which is the easiest way for *some* students. Although in general no one way is inherently superior to another, it is better to be proficient at several methods than at just one.

Leading digit: 68 + 35 = 60 + 30 + 8 + 5
Compensation: 68 + 35 = 70 + 33
Break and bridge: 68 + 35 = 68 + 30 + 5
Compatible numbers: 68 + 35 = 65 + 35 + 3

Justifying strategies   How do these strategies use base 10, expanded form, and the commutative and associative properties? Think and then read on. . . .

Let's look at the compatible numbers strategy for Problem 2 in detail:

The action	Justification
68 + 35 = (65 + 3) + 35	Substitution
= 65 + (3 + 35)	Associative
= 65 + (35 + 3)	Commutative
= (65 + 35) + 3	Associative
= 100 + 3	Addition
= 103	Addition

Children's Strategies for Addition

How do children learn to add? You will explore the teaching of addition in your methods course. Here we will focus on the mathematics underlying common stages in the child's development of computation with addition. The discussion of each strategy will be modeled with appropriate representations.

Imagine that you are a child and you haven't yet learned the standard procedure for adding. How might you add 48 + 26? Here are several ways that we have found. Additional ways appear in the explorations and exercises.

Add up by 10s

Just as you read earlier that one stage in young children's development of addition is to add up, some children begin multidigit addition in the same way.

They add up by 10s from 48. What they say is, "48, 58, 68, 74." Some children will struggle a bit from 68 to 74. They use 10 as a bridge, and say, "48, 58, 68, plus 2 is 70, plus 4 more is 74." Many children will use a number line to explain this strategy.

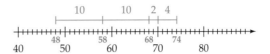

Break the second number apart

Another variation of adding up is to begin with the first number and break the second number into its place value parts. That is, they add the 10s and then the 1s.

$$48 + 20 = 68$$
$$68 + 6 = 74$$

A number line is a useful representation of this strategy.

Use partial sums

A very common approach that comes closer to one of the standard algorithms is to add, from left to right. In the beginning, children often write the **partial sums** (which are the sums of each place) and then move to the more conventional "carrying." The diagram below illustrates this approach with manipulatives. Constance Kamii and many others who have explored what is called a constructivist approach to learning arithmetic have found that when children are not just shown how to add, the vast majority of children will actually add from the largest to the smallest place. This is very interesting because we read from left to right. Personally, I have always added (and subtracted) from left to right. It always made more sense to me, and it is faster. Note that this method is essentially the leading-digit method that many people use when adding mentally.

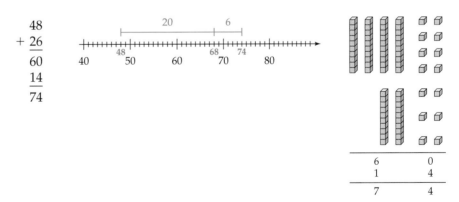

Use money and compensate

Depending on the problem, some children will use other frameworks, often money. For example, because 48 and 26 are close to a half-dollar and a quarter, some children will add 50 and 25 and then compensate, since 48 is 2 less than 50 and 26 is 1 more than 25.

$$50 + 25 - 2 + 1$$

This strategy, though it is not applying base 10 in this problem, is actually quite powerful and can be connected to base 10 through discussion.

Now that we have examined some different children's algorithms for smaller numbers, let us examine what they do with larger numbers.

INVESTIGATION 3.3

Children's Strategies for Adding Large Numbers

What if the numbers were bigger—for example, 368 + 574? Look back on the approaches described above. Can you adapt any of them to this problem? Try to do so before reading on. . . .

DISCUSSION

The "break the second number apart" strategy applies to larger numbers.

$$368 + 500 = 868$$
$$868 + 70 = 938$$
$$938 + 4 = 942$$

The "partial sums" strategy also applies.

$$300 + 500 = 800$$
$$60 + 70 = 130$$
$$8 + 4 = 12$$
$$800 + 130 = 930 \quad \text{and} \quad 930 + 12 = 942$$

The "partial sums" strategy can be modified to "keep each sum in its proper place," which is shown at the right.

```
   368          368
 + 574        + 574
  ----         ----
   800            8
   130           13
    12           12
```

Now that you have done some addition problems mentally and have seen children's approaches, and now that you have examined how place value knowledge enables those methods to work, let us look at one of the standard algorithms for adding.

Algorithms

A major goal of elementary school is to have students become computationally fluent. This means developing efficient algorithms for each operation. An **algorithm** is a single, clearly described method that works in all cases. To master algorithms for all the operations is a long and challenging process, and

■ History ■

An Arabic scholar named Al-Khowarizmi wrote many books on mathematics and astronomy. In A.D. 825, he published a book that explained the use of the Hindu numerals, including the standard procedures for the fundamental operations. The Latin translation of his name is Algorismus. This is where our word *algorithm* comes from.

many of my students remember how difficult certain algorithms were ("borrowing" in subtraction and all the steps of long division), and my recent teaching of fourth grade has reminded me how hard multidigit multiplication is to many students. The focus of the investigations in this book and of the *Explorations* volume is to develop your understanding of these operations so that you know not only how to compute but also why these procedures work.

The first arithmetic book in Europe was published in Treviso, Italy, in 1478. It described the *new* base 10 system and many different procedures for adding, subtracting, multiplying, and dividing. Because base 10 was so new to Europe at that time, many different algorithms were in use; over hundreds of years, certain algorithms "won." Thus the algorithms that you learned to add, subtract, multiply, and divide are not *the* algorithms but simply four of many. Furthermore, they are not universally used today. Some of you, in different parts of the country, learned different algorithms, and school children in different parts of the world learn very different algorithms for some of the operations, especially subtraction. So let us examine this notion of algorithm.

Adding Before Base 10

Before we begin examining addition in base 10, take a moment to think about how people added before base 10 was invented. Imagine that you were a Roman. How might you add these two numbers?

$$\begin{array}{r} \text{XXXVII} \\ + \quad \text{XLVI} \\ \hline \end{array}$$

As you found in Section 2.3, Romans couldn't just add the VII to the VI and "carry" the III.

We are not certain how the Romans actually added. There is evidence that the Babylonians did calculations in the sand with pebbles and inscribed the results on clay tablets. In fact, our word *calculate* comes from *calculus*, which means "small stone." This pebble method may have been a precursor to the abacus, which has been used extensively in Asia. The term *carrying* may have also come from carrying the pebbles from one column to the next. We do know that computation was rather cumbersome and that few people in those days could do what we call basic arithmetic. They were accorded high status in the society because of their knowledge!

CLASSROOM CONNECTION

The National Assessment of Education Progress (NAEP) has found that third-graders' success in computation decreases considerably when they move from two-digit problems to three-digit problems. Further research indicates that many students are memorizing rather than understanding the process.

Investigating Addition Algorithms in Base 10

Let us look at one of the more difficult of the three-digit addition problems, one in which we encounter zeros. Many of my students find that even though procedurally we treat zero "just like any other number," they feel just a bit uncomfortable when they run into zeros. Therefore, an analysis of such a problem is often helpful.

Add 267 + 133 as you normally would. Can you explain the "whys" of each step? Try to do so before reading on. You may or may not wish to use base 10 blocks.

In the context of our base 10 numeration system, we are combining 2 hundreds, 6 tens, and 7 ones with 1 hundred, 3 tens, and 3 ones. If the student understands the process of combining and regrouping, then this problem is *not* substantially more difficult than one with smaller numbers; it is only longer.

140 CHAPTER 3 / The Four Fundamental Operations of Arithmetic

Using manipulatives	Using words
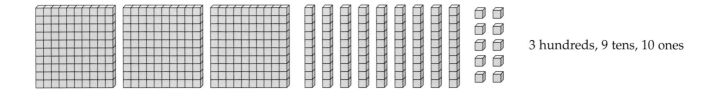	2 hundreds, 6 tens, 7 ones + 1 hundred, 3 tens, 3 ones

Recalling what addition means, when we literally combine the two sets, we have 3 hundreds, 9 tens, and 10 ones.

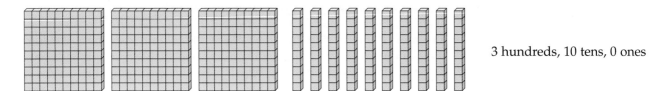

3 hundreds, 9 tens, 10 ones

However, this is not a valid answer. Thus we have to regroup. After we "trade" the 10 ones for 1 ten, we have 3 hundreds, 10 tens, and 0 ones.

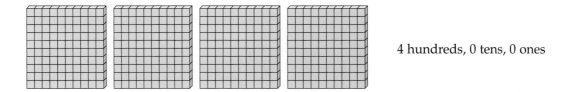

3 hundreds, 10 tens, 0 ones

Because we now have 10 tens (once again not a valid representation in base 10), we trade them for another hundred, which gives us 4 hundreds, no tens, and no ones—that is, 400.

4 hundreds, 0 tens, 0 ones

A commonly used algorithm   With this manipulative representation, you can now better understand the why of one common algorithm for addition (the one you probably learned in grade school). Using this algorithm, we go from right to left (that is, we begin with the ones place).

$7 + 3 = 10$; place the 0 in the ones place and put the 1 above the tens place; because $7 + 3$ is equivalent to 1 ten and 0 ones.

$$\begin{array}{r} \overset{1}{2}67 \\ 133 \\ \hline 0 \end{array}$$

$1 + 6 + 3 = 10$; place the 0 in the tens place and put the 1 above the hundreds place.

$$\begin{array}{r} \overset{11}{2}67 \\ 133 \\ \hline 00 \end{array}$$

$1 + 2 + 1 = 4$; place the 4 in the hundreds place. The sum is 400.

$$\begin{array}{r} \overset{11}{2}67 \\ 133 \\ \hline 400 \end{array}$$

We can also examine this procedure from a more formal perspective. Representing the problem in expanded form enables us to prove why it works. Knowing that addition can be seen as combining, we can represent our problem in the following way.

Statement	Justification
$267 + 133$	
$= (2 \cdot 100 + 6 \cdot 10 + 7) + (1 \cdot 100 + 3 \cdot 10 + 3)$	Expanded form
$= (2 \cdot 100 + 1 \cdot 100) + (6 \cdot 10 + 3 \cdot 10) + (7 + 3)$	Commutative and associative properties
$= (2 + 1)100 + (6 + 3)10 + (7 + 3)$	Distributive property
$= 3(100) + 9(10) + 10$	Addition
$= 3(100) + (9 + 1)10$	Distributive property
$= 3(100) + 10(10)$	Addition
$= 3(100) + 100$	Multiplication
$= (3 + 1)100$	Distributive property
$= 4(100)$	Addition
$= 400$	Multiplication

Seeing how numbers can be *composed* and *decomposed* makes it possible to understand the algorithm deeply. To understand the addition algorithm, we decompose the number using expanded form; we can then see how those different parts can be reconfigured—composed—to make our new whole, that is, the sum. In one sense, the manipulatives let you see the parts (expanded form) and the whole at the same time.

142 CHAPTER 3 / The Four Fundamental Operations of Arithmetic

INVESTIGATION

3.4

History

The earliest book on arithmetic is the papyrus of Ahmes, who was an Egyptian scribe in about 1700 B.C. It is believed that the book is a compilation of what was known at the time, some of which goes back to 3000 B.C. The title of the papyrus was "Directions for Obtaining the Knowledge of All Dark Things." The first arithmetic book printed in the West was printed in Treviso, Italy, in 1478. The first one in the United States was printed in 1729; it was written by Isaac Greenwood, a professor of mathematics at Harvard College.[3]

An Alternative Algorithm

As mentioned earlier, when base 10 was invented, many different algorithms for each operation were invented. That is, the algorithm that you learned is not the only one. Children often learn different algorithms in some countries. One algorithm that is a favorite of children is called the **lattice algorithm**. Observe the example below and see if you can figure out how it works and why it works.

```
   6 4 5
 + 7 2 8
  1/3 0/6 1/3
  1/3/7/3
```

DISCUSSION

It works this way. First, you write the problem. Below each place draw a square.
   Second, draw diagonal lines through each square that extend below the square.
   Third, write the result of each partial sum in the box.
   Last, add diagonally. If the sum in any diagonal addition is greater than 10, "carry" just as you do with the standard algorithm.
   If we represent the problem in terms of the place value of each part of the lattice, we see that the lattice "herds" each digit to the proper place.

```
   6 4 5     O = Ones
 + 7 2 8     T = Tens
  Th/H/T     H = Hundreds
  /H/T/O     Th = Thousands
```

INVESTIGATION

3.5

Children's Mistakes

The problem below illustrates a common mistake made by many children as they learn to add. Understanding how a child might make that mistake and then going back to look at what lack of knowledge of place value, of the operation, or of properties of that operation contributed to this mistake is useful. Look at the problem. What error on the part of the child might have resulted in this wrong answer? Make up a similar problem that the child might also do wrong, and use the child's thinking to solve that problem. Then explain what knowledge the child seems to be lacking.
   The problem: 38 + 4 = 78

DISCUSSION

In this case, it is likely that the child lined up the numbers incorrectly:

```
    4
 + 38
 ----
   78
```

[3]Louis Charles Karpinski, *The History of Arithmetic* (New York: Russell & Russell, 1965), p.81.

CLASSROOM CONNECTION

A friend of mine, David Sobel, was talking about mathematics with his six-year-old daughter, Tara. They were doing double-digit addition. David had just shown Tara that 20 + 20 = 40. Tara thought for a moment and then proudly announced that 50 + 50 must be 70. When David asked how she had got that answer, she said, "When you add the same numbers that have a zero at the end, you just skip ten!" In so many mathematical situations, there are many ways to interpret the data. It takes time and practice to improve your ability to analyze and interpret patterns. It is important to note that both David and I were encouraged, not discouraged, by Tara's thinking. Why was this?

CLASSROOM CONNECTION

The following story illustrates the need for number sense. Stopping at a dinosaur skeleton the museum tour guide said that the skeleton was 65 million and 10 years old. Someone asked how he knew this. He replied, "well it was 65 million years old when I began working here, and I've been here 10 years."

Giving other problems where the addends do not all have the same number of places will almost surely result in the wrong answer. For example, given 45 + 3, this child would likely get the answer 75. Given 234 + 42, the child would likely get 654. In this case, the child has not "owned" the notion of place value. Probably, part of the difficulty is not knowing expanded form (for example, that 38 means 30 + 8—that is, 3 tens and 8 ones).

Estimation

The availability of inexpensive calculators has profoundly changed the way we compute. Most people now rely on calculators when exact answers are needed. However, estimating skills are still very important in cases where an exact answer is not needed and to check the reasonableness of results obtained on the calculator. Estimation, in turn, requires good mental arithmetic skills, which come from an understanding of the nature of the operations, a firm understanding of place value, and the ability to use various properties (for example, the associative and distributive properties).

The NCTM Standards not only emphasize the need for students to understand how to estimate but also reinforce the importance of their being more actively involved in the construction of knowledge, as opposed to simply receiving procedures from teachers. Thus, the standards note that students need to "develop, analyze, and explain procedures for computation and techniques for estimation" (*Curriculum Standards*, p. 94).

When are numbers estimates and not exact numbers? Before we examine some methods of estimation, we need to understand when numbers represent estimates and when they represent exact numbers. All of the following numbers are estimates or approximations. What ways can you see to group them according to why they are estimates? For example, the age of a dinosaur bone is an estimate because present dating methods do not enable us to get an exact number; in other words, the exact age is unknown.

- A certain dinosaur bone is 65 million years old.
- The number of hungry children in the United States is 16 million.
- The area of the Sahara Desert is 3,320,000 square miles.
- The mean July temperature in Tucson, Arizona, is 86 degrees.
- Jane lives 55 miles from the nearest airport.
- My office is 12 feet by 9 feet.

Numbers are estimates when

1. The exact value is unknown—for example, predictions and numbers that are too large or difficult to determine.
2. The value is not constant—for example, population and barometric air pressure.
3. There are limitations in measurement—for example, the length of a desk and the area of a pond.

Note that these are not the only instances when a number is actually an estimate, but rather three common categories. The purpose here is not to memorize categories, but to realize that many numbers we see around us are actually estimates.

Rounding Just as many numbers in everyday life are estimates rather than exact numbers, many numbers are rounded. For example,

- It took Jackie 10 hours to get from Boston to Buffalo.
- Anna gets 34 miles per gallon with her new car.
- Rosie put 2100 miles on her car last month.
- Fred paid $18,000 for his new car.
- The population of Sacramento, California, is 370,000.

The preceding examples bring another question to mind: Why do we round? Stop for a minute to think of your own response. Then read on. . . .

Below are two of many possible reasons:

1. Rounding makes comprehension easier. Which of the following sentences would you prefer to read in a newspaper? Why?

   The school budget for the 34,168 students was $153,167,458.

   The school budget for the 34,000 students was $153 million.

2. Rounding makes computation easier. Let's say you are trying to determine the cost of buying new textbooks for your fifth-grade class. There are 23 students, and the textbooks cost $19.75. If you wanted to get a sense of the cost, you would probably round 19.75 to 20 and multiply 23 by 20.

When do we use estimation and when do we use exact computation? Before you read this section, write down your thoughts about this question. Then read on. . . .

Following are several examples of when people generally estimate:

- Making a budget—cost of college, cost of food per month
- Determining the cost of a trip or vacation—ski trip, camping trip, trip to Europe
- Deciding which to buy—whether to buy a new car or a used car
- Determining time—how long to get to . . .
- Determining whether we have enough money—being at the grocery store when short on cash
- Deciding how much the tip should be (at a restaurant)
- Determining how long the paper or project will take

Estimation methods As you will find in this chapter, we do not estimate in the same way in all situations. The method(s) we use to estimate depend partly on how close the estimate has to be and on whether we want to over- or underestimate deliberately (for example, underestimating when budgeting can cause serious problems, so budgets are sometimes deliberately overestimated). Finally, people usually prefer certain methods to other methods. In many cases, different methods will produce estimates that are nearly identical, and in some cases, certain methods are easier than other methods.

Estimation Strategies for Addition

Here we will analyze some estimation problems to understand better the application of base 10 concepts and mental math strategies.

Classroom Connection

Every year in my middle school mathematics methods course, my students develop a series of questions about a topic, and then they go into the schools and ask actual students these questions in order to learn more about what the students know and what they don't know. Every year my students make up estimation problems, and every year my students are quite stunned to find that many of the middle school students have few or no estimation and mental math strategies in their repertoire. This is all the more amazing in view of the fact that in their everyday lives, people require estimates more often than exact answers.

INVESTIGATION 3.6

What Was the Total Attendance?

In each of the following estimation problems, first obtain a rough estimate (5–10 seconds), then obtain the best estimate you can, and then read on. Use mental arithmetic strategies in your estimates.

A. Approximately what was the total attendance for the following three football games at Tiger Stadium?   75,145   34,135   55,124

DISCUSSION

Remember that one of the main goals of the investigations is for you to develop a repertoire of strategies. Check your estimates against the ones obtained below with different strategies. Do you find you like certain strategies better than others? Are there any that you find you have a hard time understanding? In this situation, there are several estimation strategies that work well:

Leading digit: $7 + 3 + 5 = 15$; that is, 150,000
The leading-digit method will always give you an estimate that is lower than the actual sum. Why is this?

Refined leading digit: $150,000 + (5 + 4 + 5 = 14$—that is, 14,000)
$= 150,000 + 14,000 = 164,000$

Rounding: $80,000 + 30,000 + 60,000 = 170,000$

Compatible numbers: $75 + 35 + 55 = 110 + 55 = 165$, which represents 165,000

B. Approximately what was the total attendance for three baseball games at Wrigley Field in Chicago? Make your own rough estimate and then a refined estimate before reading on. . . .

32,425   31,456   34,234

DISCUSSION

Are you getting quicker at estimating? Are you finding that you are applying the mental addition techniques? Each of the four methods above could have been used here. You may well have come up with another strategy called *clustering*, because all three numbers are relatively close together. In this case, a very quick, rough estimate would be

$30,000 \times 3 = 90,000$

If we use, for example, a refined leading digit strategy, we can get $90,000 + 7000$, and looking at the 425, 456, and 234, we can see that this is about 1000. Thus a more refined estimate is about 98,000.

Looking back There are two points to keep in mind when estimating and doing mental mathematics:

1. The method you use is often partially determined by the problem itself. If you want only a rough estimate, you might use leading digit or rounding. If you want a more refined estimate, you might use compatible numbers. If you want to make sure you have enough money, you might round everything up so that the estimated sum is definitely greater than the actual sum.
2. It is not so much a matter of there being right or wrong ways to estimate or there being better or poorer methods. There is a certain amount of latitude for preferences, just as some tennis players will like a certain type of racket and others will like another type of racket. However, if you have a large repertoire of estimating and mental math techniques in your toolbox, you will be more skillful.

A Refined Technique for Rounding

Most textbooks offer the following guidelines for rounding:

If the appropriate digit is less than 5, round down. Otherwise round up.

For example, if we estimate 35 + 57, we round to 40 + 60 = 100. There is a problem with this method, though. Suppose we want to estimate the sum of the following numbers: 35 + 42 + 76 + 45 + 35 + 85

If we have a problem in which many of the numbers end in a 5, this method will give an estimate that is too high.

A more refined set of guidelines for rounding makes the following modification:

If the digit to the left of the 5 is odd, round up.

If the digit to the left of the 5 is even, round down.

When we encounter problems with columns of addition, it is likely that the numbers of evens and odds will be close, thus canceling each other out. Let us see how this affects the estimate with the preceding problem:

Problem	Standard rounding	Refined rounding
35	40	40
42	40	40
76	80	80
45	50	40
35	40	40
85	90	80
318	340	320

If we use this refined method, our estimate will generally be closer to the actual amount. Why won't it always be closer?

Section 3.1 / Understanding Addition 147

INVESTIGATION
3.7

Estimating by Making Compatible Numbers

Using compatible numbers is an effective strategy to estimate the following sums. Try this on your own and then check below....

A.	38	B.	23
	72		359
	89		177
	65		675
	27		162
			315

DISCUSSION

A. In this case, a quick glance shows us that 38 and 65 will make a sum close to 100, and so will 72 + 27. If we see this, our estimate of 200 + 89 = 289 is quite close to the actual answer of 291.

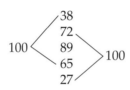

B. Using compatible numbers, we can estimate 200 + 500 + 1000 = 1700.

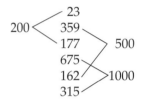

INVESTIGATION
3.8

How Much Money Did Sandy Spend?

Sandy wrote checks for the following amounts. Approximately how much did she just spend?

$173	Credit card
$97	Electricity
$63	Gas credit card
$212	Insurance
$4	Parking ticket
$36	Phone

A. As before, first make a very quick, rough estimate.

B. See how close you can get to the exact answer, but without using pencil-and-paper or a calculator.

DISCUSSION

A. To find a rough estimate, we could use the leading digit method; this would give us 3 hundreds, but then we would have to remember the 3 hundreds while adding the tens. A quicker way to make a rough estimate is to use compatible numbers: 173 + 212 make about 400; 97 and 4 make another 100, and then 63 + 36 make another 100. Our rough estimate is $600.

B. There are many ways to obtain a more accurate estimate. One way might begin by rounding:

170 + 100 + 60 + 210 + 40

Then we could add these rounded numbers to obtain an estimate of $580.

Again, it is not a matter of which strategies are better or right, but of having a sense of how accurate an estimate we need and then being able to employ a strategy that can get us close to the actual answer.

What would be a scenario in which Sandy would be satisfied with a rough estimate? What would be a scenario in which Sandy would want her estimate to be "pretty" close to the actual amount?

Looking back   Before we move on to subtraction, take a few moments to look back over the strategies for mental addition and estimation that we have explored. Do you find that you prefer some over others? Do you see that some strategies work better with some numbers than with others? Identify those places in which these strategies use properties of addition or important base 10 structures.

Number Sense and Operation Sense

This chapter is more than simply learning how to use the various algorithms efficiently and accurately. In this course, we are moving from merely understanding *how* to also understanding *why*. A big part of this transformation involves developing what is called number sense and operation sense. Essentially this means common sense when working with numbers and operations on numbers. Various authors have articulated many characteristics of number and operation sense. For the purpose of our introductory work, I offer the following six.

Number and operation sense involves the ability:

1. To take numbers apart and put them together.
2. To move fluently among different representations.
3. To recognize when one representation is more useful than another.
4. To perform mental computation and estimation flexibly.
5. To determine whether an answer is reasonable.
6. To understand the effect of different operations on numbers.

In each section of this chapter, we will investigate problems that will help to further develop your number sense. It is important to note that number and operation sense is being developed throughout; however, we will use some problems to focus explicitly on number and operation sense.

INVESTIGATION

3.9

Number Sense with Addition

A. I'm thinking of 3 numbers whose sum is greater than 70.
Tell whether each of the following must be true, might be true, can't be true:

A. All three numbers are greater than 20.

B. If two of the numbers are less than 20, the other must be greater than 20.

DISCUSSION

A. There are generally two ways to proceed. One is to try to explain why it *must* be true. Another is to find a counterexample to demonstrate that the statement is not true. In this case, we can see that $10 + 30 + 40 = 80$. The sum is greater than 70 but only two of the three numbers are greater than 20.

B. Here we can proceed logically. What if the two numbers less than 20 are 19? Then their sum is 38 and the third number must be at least 33. Do you see why? If the two numbers are less than 19, then the third number will have to be more than 33. Therefore, this statement is true.

> B. Take no more than 5 to 10 seconds to determine whether this answer is reasonable. That is, make your determination, using your number sense, without doing any pencil and paper work or trying to add the numbers mentally.

$$\begin{array}{r} 6{,}563 \\ +4{,}448 \\ \underline{7{,}203} \\ 17{,}214 \end{array}$$

DISCUSSION

If we look at the leading digits, we see that they add up to 17 (meaning 17,000). A very quick look at the rest of the amounts lets us quickly see that the answer must be greater than 17,214, and thus this answer is not reasonable. If you look closely, you can see that the correct answer is 18,214. This is a classic addition mistake that can easily be caught by quickly scanning the problem!

> C. Place $>$, $<$, or $=$ in the circle. Quickly look at the numbers on the left and right of the circle. Determine which symbol is appropriate.
> $463 + 924 + 723 \bigcirc 842 + 646 + 658$

DISCUSSION

The leading digit method is useful again: $4 + 9 + 7 = 20$ while $8 + 6 + 6 = 20$. So we have to look in the next place. Rounding, we have $60 + 20 + 20 = 100$. On the right we quickly see that the sum of these numbers $(40 + 40 + 60)$ is greater than 100, so the sum on the right is greater.

Summary

Many of my students report that this part of the course is very exciting. After all the hard work with Alphabitia and/or bases, they see the fruits of that work in this section, and they realize that computation is more complex than they realized. If they have had some experiences with young children, they make such comments as "Now I understand why some children find it really hard to add big numbers."

Let us review some important ideas from this section and take this time to connect addition to place value explicitly.

Addition means more than just sum or add. Analysis of various story problems show that many words describe addition: *combine, join, extend, increase*. The "combine" description helps us to see that from one perspective, addition can be seen as "fast counting," and indeed most children initially solve addition problems (4 + 3) by combining the two amounts and then counting.

There are two models: discrete and continuous. The discrete is more concrete to children (the concept of 3 balls is easier to "see" than 3 inches).

There are many different methods that can be used to solve addition problems with both small and large numbers

$$8 + 7 = 8 + (2 + 5) = (8 + 2) + 5 = 10 + 5 = 15 \quad \text{(bridging)}$$
$$= (7 + 1) + 7 = 7 + (1 + 7) = 7 + (7 + 1) = (7 + 7) + 1 \quad \text{(doubles)}$$

The properties are not just words but also useful ideas. The commutative and associative properties were both used in the cases above.

To do addition efficiently, we need to connect our place value knowledge of base 10: knowing how to take apart numbers and being able to do so flexibly.

EXERCISES 3.1

1. Refer to the base 10 addition table. If we look at any of the diagonals going in a top-left-to-lower-right direction, the value of each successive term increases by 2 each time. Explain this pattern.

2. In many textbooks, students are encouraged to learn their fact families. This is the fact family for 10: 1 + 9, 2 + 8, 3 + 7, 4 + 6, 5 + 5, 6 + 4, 7 + 3, 8 + 2, 9 + 1. Transform each fact pair to an ordered pair and graph the pairs on graph paper. Why does the line look the way it does?

3. For each number line problem below, identify the computation it models and briefly justify your answer.

    a.
    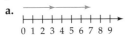
    b.
    0 1 2 3 4 5 6 7 8 9

    c.
    0 1 2 3 4 5 6 7 8 9

4. Through illustrations, demonstrate how to solve these problems with manipulatives.

    a. 76
       +47

    b. 524
       +268

5. Below is an addition algorithm from an old text. Explain why it works.

    ```
     36
    +48
     14
      7
     84
    ```

6. How does this addition algorithm work? Imagine someone asking, "How do you know where to put the 1 and the 3 in the 13?"

    ```
     368
    +574
      12
      13
       8
    ```

7. Pretend you are a young child again. Which algorithm do you think would be easiest to learn first: the standard algorithm, the partial sums algorithm, or the lattice algorithm? Explain your reasoning.

8. Finish the following addition problem, which has been done in expanded form.

    $345 = 3$ hundreds $+ 4$ tens $+ 5$ ones
    $+268 = 2$ hundreds $+ 6$ tens $+ 8$ ones

9. *Classroom Connection* The following computations represent mistakes that children commonly make in learning to add. In each case, figure out how the child could have gotten the answer and describe the cause of the child's error.

    a. 10 + 100 + 20 + 40 + 101 = 901
    b. 20 + 2 + 14 + 9 = 36
    c. 128 + 133 = 2511

10. *Classroom Connection* Identify and explain the errors in each of these computations, all of which have occurred in real classrooms.

    a. 36
       +28

        91

    b. 36
       + 28

        514

    c. 365
       +287

        742

11. Determine the following sums mentally. Briefly describe how you obtained the sum.

   a. 47
      +55

   b. 69
      +77

   c. 88
      +77

   d. 56
      +19

   e. 577
      +126

   f. 735
      +248

   g. 269
      + 87

   h. 75
      36
      187
      65

   i. 24
      53
      387

   j. 583
      +423

12. A traveling saleswoman is adding up her miles this week. How many miles did she travel? Determine the answer mentally.

   165   345
   78   142
   57

13. Estimate the following sums. Briefly describe how you obtained your estimate.

   a. 473
      345
      +355

   b. 6963
      3286
      +7147

   c. 88,865
      32,565
      +23,784

   d. 536,455
      94,352
      +659,346

   e. 473
      345
      134
      565
      943
      +355

   f. 69,655
      32,438
      90,432
      24,684
      79,833
      +71,347

   g. 234,345,343
      325,345,689
      587,247,678
      56,345,865
      +234,764,874

14. From each of the following lists, select three numbers whose sum will be closest to the target sum. Briefly explain your reasoning—how did you determine those three numbers?

	Numbers				Target
a.	37	83	56	74	150
b.	73	32	94	53	200
c.	24	39	47	97	170
d.	115	175	164	153	400
e.	185	372	153	274	650
f.	352	423	439	583	1200

15. Taking no more than 5 to 10 seconds, determine whether to place <, >, or = in the circle to make the number sentence true. That is, make your determination, using your number sense, without doing any pencil and paper work or trying to add the numbers mentally 426 + 315 + 408 ◯ 278 + 495 + 387

16. Place the digits 1, 2, 3, 6, 7, and 8 in the boxes to obtain

   a. The greatest sum
   b. The least sum

17. Choose among the digits 1, 2, 3, 4, 5, 6, 7, 8, and 9 to make the sum equal 500. You may use each digit only once. How many different ways can you make 500?

18. How many different ways can you make this a true statement, using each of the whole numbers 1 through 9 only once?

   For example, 152
                368
                479
                ___
                999

19. Determine the values of N and P that make this a true statement. Explain how you arrived at the answer.

   N
   N
   N
   $+ P$
   ___
   PN

20. Determine values for a, b, and c that will make the number sentence correct.

   abc
   +cba
   ___
   625

21. a. 1101_2
       $+111_2$

    b. 2004_5
       $+2441_5$

    c. 55_6
       $+55_6$

    d. 992_{16}
       $+265_{16}$

22. Solve the following addition problems with the lattice algorithm:

    a. 5568_{10}
       $+2745_{10}$

    b. 322_5
       $+234_5$

    c. 764_8
       $+215_8$

23. Determine the missing numbers that will complete the following addition problem.

    $4\square 4_6 + \square 35_6 = \square003_6$

24. a. Find the base for which the following statement is true: 35 + 25 = 62.

    b. Find the base for which the following statement is true: 135 + 444 = 601.

    c. Determine the base(s) in which $31_a = 25_b$.

SECTION 3.2 UNDERSTANDING SUBTRACTION

WHAT DO YOU THINK?

- Why do most current elementary texts discourage the term "borrow"?
- What other contexts for subtraction are there beside "take away"?

Now let us investigate the operation of subtraction, which presents some interesting issues for young children. We will connect the pictorial model developed for addition to subtraction, but first let's look at some questions to make sure you are actively involved.

Assume that you are a young child who has not yet learned subtraction. How might you solve these problems—using manipulatives or diagrams, or in your head using common sense and other mathematical knowledge? Purposely "forgetting" what you already know about subtraction will help you to see beyond the mechanics of the operation to its underlying meaning—and that's what this investigation is about. Note your thoughts before reading on. . . .

Five subtraction problems

1. Joe had 7 marbles. He lost 2 in a game. How many does he have left?
2. Billy has 23 marbles and Yaka has 32. How many more does Yaka have?
3. At the beginning of the week, the hospital had 32 ounces of insulin. During the week, 14 ounces of insulin were used. How much insulin did the hospital have at the end of the week?
4. Alicia was 45 inches tall at the beginning of the first grade and 53 inches at the end of the third grade. How many inches did she grow during this time?
5. The Jones farm has 35 pigs and 18 chickens. How many more pigs are there than chickens?

Now go back to these five problems and answer the following questions:

- What action words describe what is happening in these four problems?
- In what ways are the problems different? In what ways are they similar?
- What does subtraction mean? For example, what words come to mind when you think of subtraction?

CLASSROOM CONNECTION

Children often have trouble seeing the comparison model as subtraction. Do you see why? Think before reading on. . . .

Part of the reason is that many students will determine the answer by counting up from the smaller number. For example, when determining how many inches Alicia grew, many children see this as "45 plus what makes 53?"—that is, 45 + ? = 53. For this reason, some teachers refer to the comparison model as the *missing-addend model*.

Even with larger numbers, many children prefer to count up rather than to subtract. They will subtract if they are forced to, but it frequently does not jibe with their common sense. Recall that one of the lines in the NCTM Standards is that students should believe that mathematics makes sense.

Contexts for Subtraction

One way in which the problems are different is connected to the two contexts we discussed for addition—that is, whether we are subtracting discrete sets or measured amounts.

Problems 1, 2, and 5 are discrete problems.
Problems 3 and 4 are measurement (continuous) problems.

These problems differ in another way, and this difference is connected to the idea of parts and wholes that we find in all four operations. In **"take-away" problems**, a part is taken away from a whole. In **"comparison" problems**, two amounts are being compared. In some cases, the comparison is between two subsets of a large set (two parts of a whole); in other cases, the comparison is between two sets (wholes). Can you determine which of the five problems on the previous page are take-away problems and which are comparison problems? Think and then read on. . . .

Problems 1 and 3 are take-away problems, whereas the other three problems fit the comparison model. Figure 3.6 shows the subtraction problem $7 - 2$ in the two contexts.

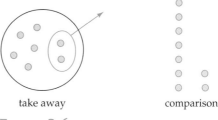

take away   comparison

FIGURE 3.6

A Pictorial Model for Subtraction

Now look at all five subtraction problems, and put aside the categories (discrete measurement, take away, and comparison). Is it possible, as we did with addition, to represent all of these problems with a single model? Write your thoughts and then read on. . . .

One way to express the commonality is that in all cases, we have a large amount and two smaller amounts whose sum is equal to the larger number. Recall the pictorial model for addition: If we invert that diagram, we have a pictorial representation of a general model for subtraction (see Figure 3.7).

FIGURE 3.7

It is important to note that *whole* and *part* refer to the numerical values rather than the contexts. For example, in Problem 2, we are comparing two wholes and saying that one whole contains 9 more marbles than the other.

We define **subtraction** formally in the following manner:

$$c - b = a \text{ iff } a + b = c$$

That is,

The difference between two numbers c and b is a if and only if c is the sum of a and b.

In mathematical language, c is called the **minuend**, b the **subtrahend**, and a the **difference**.

This model also highlights the connections between addition and subtraction problems (see Figure 3.8):

$3 + 5 = 8$
$5 + 3 = 8$
$8 - 3 = 5$
$8 - 5 = 3$

FIGURE 3.8

Referring to the composition and decomposition notion, we can join 3 and 5 to make 8, and we can decompose 8 into 5 and 3. (We used this awareness of

---

CLASSROOM CONNECTION

I recall reading, several years ago, a paper about children with mathematical learning disabilities that speaks to the need for learners to see the connectedness of mathematical ideas. The author was called in to diagnose a third-grader who was having difficulties with mathematics. The author discovered that the child did not see addition and subtraction as related, was not able to see the operations as parts and wholes, and was not able to use the commutative property of addition. That is, she saw the four number sentences on the right as four unrelated number sentences. It was no wonder that she was having difficulty remembering her addition and subtraction facts!

the ability to break amounts into convenient parts when we examined mental computation and estimation.)

Subtraction with Number Lines

Many measurement situations, such as time and distance, naturally lend themselves to a number line representation. Let's use an example to explore this model. Joanne had 8 feet of rope and used 3 feet to stake a tent. How much rope does she have left? Represent this problem with a number line and then read on. . . .

As with addition, we can use a number line to solve the problem in either of two ways (see Figure 3.9).

FIGURE 3.9

Do both ways make sense? Does one way feel more comfortable to you? As with the addition number lines, the left diagram more closely resembles the physical representation of the problem.

Properties of Subtraction

Think back to the properties of addition—identity, commutativity, associativity, and closure. Do those same properties hold for subtraction? Think and then read on. . . .

Some students think that subtraction has an identity property, because if we take away zero from a number, its value does not change; that is, $a - 0 = a$. This is true; however, if we reverse the order, the result is not true—that is, $0 - a \neq a$. Therefore, we generally say that the operation of subtraction does not have an identity property. In higher mathematics, we say that the operation of subtraction has a right-identity but not a left-identity. After examining a few cases, you can see that the operation of subtraction does not possess the commutative property or the associative property. The commutative property is not immediately understood by children. I recall one first-grader arguing that $3 - 5$ was 0 because "you can't have less than nothing" and another arguing that it was 2 because "you just turn them around." Finally, the operation of subtraction is not closed for the set of natural numbers because the difference of two numbers can be a negative number.

Just as we did with addition, we will begin our study of subtraction algorithms with mental subtraction. You know subtraction well, so having to do problems mentally will require you to think about the operation and how you apply your place value knowledge to subtract. A reminder: Because the commutative and associative properties do not hold for subtraction, we need to be more careful. Thus representations of the problems are even more important for subtraction than for addition.

INVESTIGATION 3.10

Mental Subtraction

Do the following computations in your head. Briefly note the strategies you used, and try to give names to them.

Note: One mental tool all students have is being able to visualize the standard algorithm in their heads. For example, for the first problem, you could say, "Cross out the 6, and now subtract $15 - 8 = 7$ and then $5 - 2 = 3$; the answer is 37." However, because you already know that method, I ask you not to use it here but to try others instead.

1. $65 - 28$ **2.** $62 - 29$ **3.** $184 - 125$ **4.** $132 - 36$ **5.** $1000 - 648$

CLASSROOM CONNECTION

One of the ways that I assess how well a student understands the various models for subtraction and the relationship between addition and subtraction is by seeing the extent to which a student can develop a repertoire of mental strategies for subtraction.

DISCUSSION

As with addition, there are a variety of subtraction strategies. As you read the discussion below, reflect on the strategies. Do they make sense to you? If not, ask yourself why. Is it because you don't understand how the strategy works? Is it because you don't feel your mental subtraction skills are strong enough to use it? Again, it's not so much a matter of right or better strategies as it is of having a repertoire. Nearly all of the strategies work better with certain kinds of numbers than with others.

In Problem 1, one alternative is to **add up**—that is, to ask how we get from 28 to 65. One student's actual thinking could look like this: "$28 + 30 = 58$, then add 7 to get to 65. So the answer is $30 + 7 = 37$." The add-up strategy is nicely illustrated with a number line.

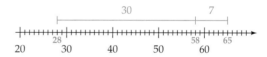

Another strategy for Problem 1 is to find the nearest ten and add up: "$28 + 2 = 30$, then we need 35 more to get to 65, and $35 + 2$ is 37."

Some students understand this problem better with a **number line**, as in Figure 3.10. In one sense, we are going from 28 to 30 to 65.

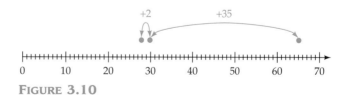

FIGURE 3.10

In Problem 2, compensate: Add 1 to both numbers. This transforms the difficult $62 - 29$ to the relatively simple $63 - 30$.

Some students understand the compensation strategy better by using a number line (see Figure 3.11). On a number line, $62 - 29$ can be interpreted as how long is the road from 29 to 62? If we move the beginning and the ending of the road 1 unit, we haven't changed the length of the road.

Thus 62 − 29 and 63 − 30 have the same difference.

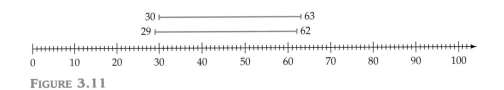

FIGURE 3.11

In Problem 3, add up: 125 + **60** = 185 and then take off **1**. The answer is 60 − 1 = 59.

In Problem 4, try the compensation strategy: 132 − 36 is close to 136 − 36 = **100**, but we must take away **4** to get 96. In other words, we added 4 to 132, but then we had to remember to take it off again.

Problem 4 can also be solved by using compatible numbers: 36 + **64** = 100, and then **32** more to get to 132, so the answer is 64 + 32 = 96.

In Problem 5, use compatible numbers: 48 and 52 are compatible numbers; that is, their sum is 100. Therefore, 648 + **52** makes 700. Then we add **300** to get to 1000. The answer is 352.

Problem 5 can also be solved by adding up: 648 + **300** = 948. We need **52** more. The difference is 300 + 52 = 352.

Do you understand all of the strategies discussed here? Try to make a problem in which one strategy would be more appropriate than another. Most students find that they retain these strategies if they make up some problems on their own and practice.

INVESTIGATION 3.11   Mental Subtraction in a Different Context

Mental math with non-base 10 numbers: You left home at 3:25 and got to the meeting at 6:15. How long did it take you to get to the meeting? Try it and then read on. . . .

DISCUSSION

Even with non-base 10 numbers, adding up is an effective strategy: 3:25 + 35 minutes makes 4 o'clock, 2 more hours gets you to 6 o'clock, and then 15 more minutes. In other words, you are adding 2 hours + 35 minutes + 15 minutes = 2 hours + 50 minutes.

Once again a number line is instructive (see Figure 3.12).

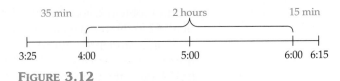

FIGURE 3.12

Alternatively, you might reason that 3:25 to 6:25 is 3 hours and you need to "back off" 10 minutes; that is, the answer is 2 hours and 50 minutes.

Children's Strategies for Subtraction

Just as we examined algorithms that children invent for larger and larger addition problems, let us examine algorithms they invent for subtraction. Below are three very different methods that children have invented for subtracting larger numbers. As you explore algorithms that children and adults have invented, you will come to appreciate both the complexity of subtraction (more children struggle with subtraction than addition) and how this process is made easier by connecting addition to subtraction and by understanding how place value works. The problem is $64 - 27$. You can find a discussion of this problem on page 454 in the April 1999 issue of *Teaching Children Mathematics*. Variations on the same theme appear in the May 1999 issue (p. 517) and the February 2003 issue (p. 349) of the same journal.

Break the second number apart   Essentially, this approach breaks apart the 27 and subtracts in pieces.

$$64 - 7 = 57$$
$$57 - 20 = 37$$

Do you see connection to any addition strategies? Think before reading on. . . .

> ▪ *Mathematics* ▪
>
> As you can see, on many occasions a problem can be interpreted and thus represented in many different ways.

This method is similar to adding up, where we begin with one number, break the second number into parts (expanded form), and add one place at a time. Here, we begin with the minuend and then subtract one place at a time.

As with addition, different representations illustrate and illuminate the problem in different ways. The manipulatives (below left) show that we are taking away 2 longs and 7 singles. We can take them away, one at a time. On the number line (below middle), we move 7 units, which brings us to 57, and then we move 20 units, which brings us to 37. We can also represent the process numerically (below right).

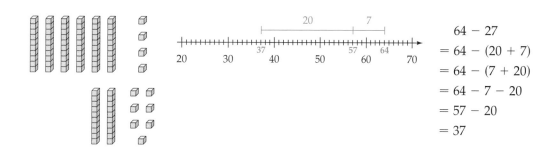

Using negative numbers   I remember reading in 1986 that a second-grader invented this strategy, and I was skeptical. Then I saw a second-grader invent it in a classroom where I had been working with the teacher. Since then I have read of its invention in many places. Apparently, it is rather common! You can find a discussion of this problem in the October 1991 issue of *The Arithmetic Teacher* (pp. 20–21) and in the February 2003 issue of *Teaching Children Mathematics* (p. 351).

It works like this: Subtract each place, beginning at the left. When the number in the subtrahend is greater than the number in the minuend, don't say, "I can't take 7 from 4," but rather "I take 7 from 4, which gives −3." Now combine the two differences: $40 - 3 = 37$.

$$\begin{array}{r} 64 \\ -27 \\ \hline \end{array} \qquad \begin{array}{l} 60 - 20 = 40 \\ 4 - 7 = -3 \\ 40 - 3 = 37 \end{array}$$

The representation (below left) shows the problem in expanded form. The representation (below right) illustrates why the method works.

$$\begin{array}{r} 60 + 4 \\ -20 + 7 \\ \hline 40 - 3 \end{array} \qquad \begin{aligned} (60 + 4) &- (20 + 7) \\ &= (60 - 20) + (4 - 7) \\ &= 40 + -3 \\ &= 37 \end{aligned}$$

Adding up   This method is very common, especially if the context of the problem is not take away but rather comparison or missing addend. You can find a discussion of this problem in the February 2003 issue of *Teaching Children Mathematics* (p. 348).

$$\begin{array}{r} 64 \\ -27 \\ \hline \end{array} \qquad \begin{array}{l} 27 + 30 = 57 \\ 57 + 7 = 64 \end{array}$$

The answer is $30 + 7 = 37$.

When asked to think out loud, the strategy sounds like this: "27 + what gets closest to 64? $27 + 30 = 57$. How many more to get to 64? $57 + 7 = 64$, so the answer is $30 + 7 = 37$."

Note that a common variation when the ones place in the difference is close to 10 is to overshoot the number and then come back. For example, "$27 + 40 = 67$. Now I need to take away 3 to get to 64, so the answer is $40 - 3 = 37$."

A number line nicely illustrates this strategy.

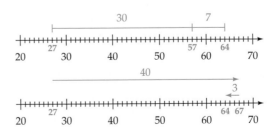

INVESTIGATION 3.12

Children's Strategies for Subtraction with Large Numbers

What if the numbers were bigger, for example 832 − 367? Look back on the approaches described above. How can you adapt them to this problem? Try to do so before reading on. . . .

DISCUSSION

"Break the second number apart" strategy extends but is now more cumbersome: Begin with the minuend and subtract each place.

```
  832      832      825      765
 −367       −7      −60     −300
          ─────    ─────    ─────
           825      765      465
```

Using negative numbers extends quite well. In fact, many of my students actually find this strategy easier than the algorithm they grew up with.

```
  832      800       30       2
 −367     −300      −60      −7
          ─────    ─────    ─────
           500      −30      −5

          500 − 30 = 470; 470 − 5 = 465
```

Adding up also becomes a bit cumbersome, but interestingly, some students feel quite comfortable with it. This is what they write:

```
  832
 −367
```

367 + 5 = 372 + 60 = 432 + 400 = 832. The answer is
5 + 60 + 400 = 465.

Note again that some children add up from left to right, and their solution path looks like this:

367 + 400 = 767; 767 + 60 = 827; 827 + 5 = 832. The answer is
5 + 60 + 400 = 465.

Understanding Subtraction in Base 10

Let us now examine some standard and nonstandard algorithms for subtraction. Again, recall the Alphabitian problems; for a child, subtracting 32 − 14 can be as overwhelming and confusing as many students originally found DA − BC in the Alphabitian system from *Explorations*.

Researchers have found that when subtraction problems have zeros, the success rate for most third-graders goes down drastically. However, this need not be so. Let us examine the following problem: 300 − 148. Although there is no single procedure that all students use to solve this problem (unless they are

■ *History* ■

When nothing remains, put down a small circle so that the place be not empty, but the circle must occupy it, so that the number of places will not be diminished when the place is empty and the second be mistaken for the first.[4]

This quote comes from a text written by Al-Khowarizmi in A.D. 825 as he tried to explain the procedure for subtracting in the new (Hindu) numeration system. Do you understand what he is saying? What function does the circle serve?

---

[4]Karl Menninger, *Number Words and Number Symbols: A Cultural History of Numbers* (Cambridge, MA: MIT Press, 1969), p. 413.

forced to by their teacher), we will examine here how this problem might be solved going from left to right.

Using manipulatives	Using words	Using symbols
	We need to regroup 300 so that we can "take away" 148.	300 −148
	We can trade 1 of the 3 hundreds for 10 tens, giving us 2 hundreds and 10 tens, which has the same value as 3 hundreds.	$\overset{2\ 1}{\cancel{3}00}$ −148
	Next, we can trade 1 of the tens for 10 ones. We now have 2 hundreds, 9 tens, and 10 ones, which still has the same value as 3 hundreds.	$\overset{2\ 9\ 1}{\cancel{3}\cancel{0}0}$ −148
	Now we can take away 1 hundred, 4 tens, and 8 ones.	$\overset{2\ 9\ 1}{\cancel{3}\cancel{0}0}$ −148 152

Classroom Connection

It is important to reemphasize that "standard" algorithms are not the "right" ones, or even the "best" ones, but rather the ones that, for various reasons, have become most widespread. Many educators believe that more harm than good is done by forcing all students to learn the "standard" algorithms.

Justifying the Standard Algorithm

It is beyond the scope of this book to prove formally every algorithm, procedure, and theorem. What is essential is that elementary students and teachers understand the algorithms that they use—that they understand the whys of the algorithm. Does the explanation above help you to understand the left-to-right subtraction procedure?

Some students find that representing the problem in expanded form is helpful. A key to understanding this algorithm is to understand the equivalence of 3 hundreds and 2 hundreds, 9 tens, and 10 ones. Although they look different, and the numerals are different (300 versus 29^{1}0), the amounts are equal. Another key to understanding this algorithm is to know why we needed to do the regrouping. This is easier to see at the physical level; without these regroupings, we cannot literally "take away" 1 flat, 4 longs, and 8 units.

Section 3.2 / Understanding Subtraction **161**

3 hundreds + 0 tens + 0 ones →   2 hundreds + 10 tens + 0 ones →   2 hundreds + 9 tens + 10 ones
−1 hundred + 4 tens + 8 ones   −1 hundred + 4 tens + 8 ones   −1 hundred + 4 tens + 8 ones

Cannot subtract the tens or   Cannot subtract the ones   Subtraction possible
the ones

An Alternative Algorithm

INVESTIGATION

3.13

An Algorithm Used in Many Other Countries

In my research on children's algorithms and algorithms in other parts of the world, I found more variation for subtraction than for any other operation. I also found that many people who grew up in other countries learned a very different algorithm. We will examine this alternative algorithm (which we will call the European algorithm) for several reasons: First, it reinforces the notion that mathematics is actually as creative an activity as art—very far from black and white, cut and dried. Second, like Alphabitia and working in bases, it requires you to really think about place value. Third, why it works is very different from why the previous algorithm works.

Below are two examples of its use and what a person using this algorithm would say.

For the first problem shown below, "You can't take 8 from 4, so you put a 1 next to the 4, cross out the 6, and put a 7. Now subtract: $14 - 8 = 6$, $8 - 7 = 1, 9 - 3 = 6$."

For the second problem, "You can't take 8 from 3, so you put a 1 next to the 3, cross out the 5, and put a 6. Now, you can't take 6 from 2, so you put a 1 next to the 2, cross out the 1, and put a 2. Now subtract: $13 - 8 = 5$, $12 - 6 = 6, 6 - 2 = 4$."

$$
\begin{array}{cccc}
984 & 98\overset{1}{4} & 623 & \overset{1\,1}{6}2\overset{}{3} \\
-368 & -3\underset{7}{6}8 & -158 & -\underset{2\,6}{1}\underset{}{5}8 \\
& 616 & & 465
\end{array}
$$

DISCUSSION

What do you think—pretty weird? Well, our method is just as weird to them. Actually, this method was used in many parts of the United States in the twentieth century and was replaced by the algorithm most of us grew up with as a consequence of discussions by mathematics educators. Now that you have seen how this works, can you explain why it works? That is, can you justify it, or prove it?

As I mentioned earlier, the justification for this algorithm is very different from the justification for the previous algorithm, in which the "borrowing" didn't change the value of the minuend, but only its representation. Here, however, the value of both the minuend and the subtrahend *do* change.

The number line is a very useful representation to help us understand why this algorithm works. The first number line shows 984 and 368. The difference (616) is simply how far apart the numbers are on the number line.

```
|————————————————————————————————————————|
368                                     984
```

Before you look at the second number line, recall that this algorithm does indeed change the value of both the minuend and the subtrahend. What is the new value of each of these in this problem? Think before reading on. . . .

The value of the minuend 984 increases to 994, ($9\overset{1}{8}4 = 980 + 14$), and the value of the subtrahend 368 increases to 378. Do you see something?

Indeed, the value of both numbers increases by 10. Because the value of both numbers is increased by the same amount, the difference (distance) between them remains the same.

The discussion of these two algorithms is a nice place to introduce a framework for examining algorithms that was presented by Hyman Bass in the February 2003 issue of *Teaching Children Mathematics*. Bass proposes that we examine five qualities of an algorithm:

- Accuracy (or reliability). The algorithm should always produce a correct answer.
- Generality. The algorithm applies to all instances of the problem, or class.
- Efficiency (or complexity). This refers to whether the cost (the time, effort, difficulty, or resources) of executing the algorithm is reasonably low compared to the input side of the problem.
- Ease of accurate use (vs. proneness to error). The algorithm can be used reasonably easily and does not lead to a high frequency of error in use.
- Transparency (vs. opacity). What the steps of the algorithm mean mathematically, and why they advance toward the problem solution, is clearly visible.

Let us use this framework to examine the traditional algorithm, the European algorithm, and the negative numbers algorithm.

Accuracy. Used correctly, all of these algorithms will always produce a correct answer.

Generality. All the algorithms will work with any subtraction problem.

Efficiency. The traditional algorithm doesn't work exactly the same in all cases—for example, when multiple "borrowing" is needed and when there are zeros in the minuend. The European algorithm also requires some additional thought if there are 9s in the subtrahend. The negative numbers algorithm works identically in all situations.

Ease of accurate use. As noted in the discussion above, in certain situations children are more likely to make errors with the traditional and European algorithms. The negative numbers strategy requires stronger computation skills because of the use of negative numbers—for example, $2 - 8 = -6$.

Transparency. The traditional algorithm is more transparent, which is a primary reason why it supplanted the "European" algorithm in the United States in the twentieth century. It is easier for children to understand the "why" of the traditional (you don't change the value) than the "why" of the European (because both increase by the same amount, the difference is still the same). The negative numbers strategy is also transparent.

INVESTIGATION 3.14

Rough and Best Estimates with Subtraction

Now let us apply these skills to problems where we are likely to estimate with subtraction. Remembering that one of the main goals of the investigations is to develop a repertoire of strategies, check your estimates against the ones discussed below. In each of the problems below, first obtain a rough estimate (5 to 10 seconds). Then obtain the best estimate you can. Do you find that you like certain strategies better than others? Are there any that you find you have a hard time understanding?

A. In 1995, Acstead State College had an enrollment of 4234, whereas Milburn College had an enrollment of 3475. How many more students did Acstead have than Milburn?

$$\begin{array}{r} 4234 \\ -3475 \end{array}$$

B. The attendance at the Yankees game was 73,468, whereas the attendance at the Mets game was 46,743. How much greater was the attendance at the Yankees game?

$$\begin{array}{r} 73{,}468 \\ -46{,}743 \end{array}$$

DISCUSSION

A. You can get a quick estimate by rounding: $4200 - 3500 = 700$

To get a more accurate estimate, you could use adding up:

$3475 + \mathbf{800} = 4275$, then back off **40** to get 4235. The estimate is $800 - 40 = 760$.

Do these strategies make sense to you? Did you do something different? How would you describe a scenario in which someone would be satisfied with a rough estimate? How would you describe a scenario in which someone would want the estimate to be pretty close to the actual amount?

B. Leading digit: $7 - 4 = 3$, giving an estimate of about 30,000.
Rounding: $70{,}000 - 50{,}000 = 20{,}000$.

Rounding and adding up: $47{,}000 + 3000$ makes 50,000. Adding up to 73,000 gives 23,000. Our estimate is $3000 + 23{,}000 = 26{,}000$.

■ *Outside the Classroom* ■

In Part B of Investigation 3.14, a Yankee fan would be more likely to do the first estimate, a Mets fan the second estimate, and a baseball announcer or executive the third estimate.

Looking back Take a few moments to look back over the strategies for mental subtraction and estimation that we have explored. Do you find that you prefer some over others? Can you connect these strategies to different models or contexts of subtraction—for example, take-away, comparison, missing addend, number line? Can you explain which strategies make powerful use of structures of our base 10 numeration system—for example, the leading digit strategy? Do you see that some strategies work with some numbers better than with others?

INVESTIGATION 3.15

Numbers Sense with Subtraction

A. Fill in the blanks to make this subtraction problem.

$$\begin{array}{r} \overline{}\,\overline{}\,\overline{}\,\overline{} \\ -4263 \\ \hline 3159 \end{array}$$

DISCUSSION

At first glance, some students panic. "There is no way I can do this problem!" However, if we stop for a moment and remember the relationship between addition and subtraction, we can see that we can quickly determine the missing number by simply adding 3159 + 4263!

B. Take no more than 5 to 10 seconds to determine whether this answer is reasonable. That is, make your determination, using number sense, without doing any pencil and paper work.

$$\begin{array}{r} 6,542 \\ -2,847 \\ \hline 3,124 \end{array}$$

DISCUSSION

We have learned that there are different contexts for algorithm and different algorithms for subtracting, and that addition and subtraction are inverse operations. For most people the simplest way to proceed here is to add 3124 and 2847 and see if the answer is close to 6542. A quick estimation of the leading two digits shows 3100 + 2900 = 6000, so the answer is not true.

Another way of looking is to inspect the places. Even with regrouping, the digit in the ones place must be a 5; it cannot be a 4. So, from this perspective, the answer also cannot be true.

Carrying and Borrowing

We have noted that we often need to regroup when adding and subtracting. However, using the traditional algorithms we use different words—we use *carry* for addition and *borrow* for subtraction. The following problem illustrates this point.

(14 ones → 1 ten and 4 ones)   (1 ten and 4 ones → 14 ones)

$$\begin{array}{r} \overset{1}{} \\ 36 \\ +28 \\ \hline 64 \end{array} \qquad \begin{array}{r} \overset{5\;1}{\cancel{6}4} \\ -36 \\ \hline 28 \end{array}$$

In the addition problem, because 6 + 8 > 9, we put the 4 in the ones place and "carry" the 1 to the tens place. In the subtraction problem, in order to subtract in each place, we need to "borrow" a 1 from the tens place and move it to the ones place whose value is now 14.

Most mathematics educators, including me, will argue that we do our children a disservice by using the words *carry* and *borrow*. Rather than my simply telling you why, take a minute to write down what is similar or the same about both processes and what is different. Then read on. . . .

What is the same about both is that an amount is moved from one place to another. What is the same about both is that this amount always represents a 10-for-1 exchange (10 ones for 1 ten or 1 ten for 10 ones).

What is different is the direction of the exchange. When adding, we move the 10 from the smaller place to a 1 in the larger place (right to left). When subtracting, we move the 1 from the larger place to a 10 in the smaller place. One of my students, describing the two processes as trading, said that addition is trading up and subtraction is trading down.

Because of the use of these two different words, many children and adults alike do not realize that these processes are virtually identical. Thus most textbooks now use the word *regrouping* instead of the words *carrying* and *borrowing*. This is a very big point, because many, many children are judged to have a learning disability not so much because they are slower or less mathematically inclined, but rather because if something doesn't make sense, it is harder for them to remember. When we use the same words for the same process, and when children see how closely connected addition and subtraction are, fewer children struggle with subtraction.

Reviewing Addition and Subtraction

We have now examined addition and subtraction rather carefully. In what ways do you see similarities between the two operations? In what ways do you see differences? Think and then read on. . . .

One way in which the two processes are alike is illustrated with the part–whole diagram used to describe each operation. These representations help us to see connections between addition and subtraction. In one sense, addition consists of adding two parts to make a whole. In one sense, subtraction consists of having a whole and a part and needing to find the value of the other part. You are becoming more comfortable with the idea of composition and decomposition, which appears repeatedly throughout mathematics. In one sense, addition is an act of composition, putting together two parts to make a new whole. In one sense, subtraction is an act of decomposition, breaking a whole into a given part and a part we need to find.

We see another similarity between the two operations when we watch children develop methods for subtraction; it involves the "missing addend" concept. That is, the problem $c - a$ can be seen as $a + ? = c$.

We saw a related similarity in children's strategies. Just as some children add large numbers by "adding up," some children subtract larger numbers by "subtracting down."

Earlier in this section, subtraction was formally defined as $c - b = a$ iff $a + b = c$. The negative numbers strategy that some children invent brings us to another way of defining subtraction, which we will examine further in Chapter 5 when we examine negative numbers. That is, we can define subtraction as adding the inverse: $a - b = a + -b$.

A very important way in which the two operations are different is that the commutative and associative properties hold for addition but not for subtraction. Translated into children's language, "You have to be much more careful when subtracting than when adding, because in subtraction the order does matter."

For example,

$$7 + 2 = 2 + 7 \quad \text{but} \quad 7 - 2 \neq 2 - 7.$$

Similarly,

$(7 + 5) + 2 = 12 + 2 = 14$ and $7 + (5 + 2) = 7 + 7 = 14$, but
$(7 - 5) - 2 = 2 - 2 = 0$ but $7 - (5 - 2) = 7 - 3 = 4$

Exercises 3.2

1. Make up a subtraction story problem for each of the following contexts. Briefly *explain* why the story problem is an example of the particular model.
   a. Take-away
   b. Missing-addend
   c. Comparison

2. a. Explain why the operation of subtraction is not commutative.
   b. Explain why the operation of subtraction is not associative.

3. For each number line problem below, identify the computation it models and briefly justify your answer.
   a.
   b.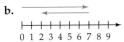
   c. (number line 0–9)

4. Represent the following problems on a number line. Explain each problem as though you were talking to someone who is not taking this class.
   a. $5 + 4$
   b. $8 - 3$

5. Which model of subtraction best illustrates each of the problems below:
   a. Reena had 25 quilts and sold 10 of them at the show. How many does she have left?
   b. The hall holds 100 people. Currently 65 tickets have been sold. How many more tickets can be sold?
   c. The sixth grade has sold 58 raffle tickets and the fifth grade has sold only 45. How many more has the sixth grade sold?

6. Through illustrations, demonstrate how to solve these problems with manipulatives.
   a. 524
      −268
   b. 600
      −345

7. *Classroom Connection* As you might expect, there are many more ways in which children make mistakes in subtracting than in adding. What is the child's mistake in this problem? What aspect of subtraction or place value is the child not applying or applying incorrectly?

   a. 76
      −48
      ——
      32
   b. 76
      −48
      ——
      22
   c. 76
      −48
      ——
      38

   d. 70
      −48
      ——
      38
   e. 70
      −48
      ——
      32
   f. 700
      −482
      ——
      228

   g. ¹ ¹ 364
      − 79
      ——
      395
   h. 32
      −29
      ——
      17

8. Examine the problem at the right.

    ³ ¹ ¹
    4 0 0
    −2 3 6
    ————
    1 7 4

   a. Explain, in mathematical terms, what the student did wrong. Where are the problems in the student's thinking?
   b. How would you help the student if the quote below represented her explanation:
      "You can't take 6 from 0 and you can't take 3 from 0, so you have to borrow from the four. Cross out 4 and write 3, cross out the zeros and write 10."

9. *Classroom Connection* Following are two children's solutions to $73 - 39$, taken from the December 2001 issue of *Teaching Children Mathematics* (p. 231). Study each child's work, and then describe how that child would find the answer to $65 - 28$. Then explain why that method works.

   a.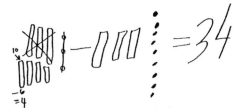

   (a) Samantha's solution

   b.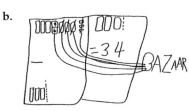

   (b) Alice's solution

   Source: Reprinted with permission from *Teaching Children Mathematics*, copyright © 2001 by the National Council of Teachers of Mathematics.

10. *Classroom Connection* Below are three methods, described in the February 2003 issue of *Teaching Children Mathematics*, that were invented by different children. In each case, study the child's work.

Then solve 904 − 367 using the child's method, and explain why that method works.

a. 872   872 − 300 = 572
   −345   572 − 40 = 532
                532 − 5 = 527

b. 872   345 + 7 = 352
   −345   352 + 20 = 372
                372 + 500 = 872

c. 872   872 − 340 = 532
   −345   532 − 5 = 527

11. *Classroom Connection* As I mentioned earlier, I have always subtracted from left to right. Here is how I solve 837 − 375.

   837   8 − 3 = 5, now look to the right: 3 < 7, so put a 4 in hundreds place
  −375   13 − 7 = 6, now look to the right: 7 > 5, put the 6 in the tens place
   462   7 − 5 = 2, put the 2 in the ones place

As in Problem 10, solve 904 − 367, this time using "my" method, and explain why that method works.

12. A student's work on a subtraction problem is shown here. Explain the numerals in the top row, 3, 9, and 1, as though you were talking to a third-grader who is having difficulty with the process.

   3 9 1
   4̸ 0̸ 0̸
  − 1 3 5
   2 6 5

13. Determine the following differences mentally. Briefly describe how you obtained your estimate.

a. 87   b. 70   c. 82   d. 500   e. 502
   −29   −23   −34   −134   −206

f. 473   g. 625   h. 808   i. 506   j. 4000
   −258   −475   −707   − 29   − 555

In Exercises 14–17, determine the exact answer mentally.

14. Peter bought a house for $116,000 and sold it for $145,000. How much profit did he make?

15. I left my house at 3:34 P.M. and arrived at 6:15 P.M. How long did the trip take me?

16. Washington College has 4132 full-time students, whereas Lincoln College has 2824 students. How much larger is Washington than Lincoln?

17. A fund-raiser has a goal of $55,000. So far, $34,854 has been collected. How much more is needed?

18. Place the digits 1, 2, 3, 6, 7, and 8 in the boxes to obtain
    a. The greatest difference
    b. The least difference

19. Choose among the digits 1, 2, 3, 4, 5, 6, 7, 8, and 9 to make the difference 234. You can use each digit only once. How many different ways can you make 234?

20. With three boys on a large scale, it read 170 pounds. When Adam stepped off, the scale read 115 pounds. When Ben stepped off, the scale read 65 pounds. What is the weight of each boy?

21. A mule and a horse were carrying some bales of cloth. The mule said to the horse, "If you give me one of your bales, I shall carry as many as you." "If you give me one of yours," replied the horse, "I will be carrying twice as many as you." How many bales was each animal carrying?

22. By estimating, quickly determine if check 238 will bounce. The beginning balance is $510. Sketch the process by which you estimated.

Check No.	Amount ($)	Balance
233	143	
234	97	
235	65	
236	212	
237	8	
238	45	

23. Estimate the following differences. Briefly describe how you obtained the difference.

a. 4473   b. 65,963   c. 73,463
   −2355   −29,147   −28,543

d. 43,433   e. 413,082   f. 320,283
   −16,328   −285,876   −184,438

g. 3,133,543 h. 71,234,033
   −1,903,253   −32,753,962

24. From each of the following lists, select two numbers whose difference will be closest to the target difference.

	Numbers			Target
a.	315	475	764	300
b.	185	372	953	650
c.	382	723	793	350

25. This problem requires some nice thinking. Find *a*, *b*, *c*, and *d* that will make the subtraction problem work. Is there more than one solution? Why or why not?

a. 6ab   b. 6xx
   −c8b   −1y8
    1da   47y

26. One of my students asked me this question after she had read the text: "Why do we have addition and multiplication tables, but not subtraction and division tables?" Write your response to her question.

27. Respond to the following as though you were talking to a principal who is familiar with the NCTM standards. You want to convince the principal to hire you. The principal has just said that she has only recently realized that the words *borrowing* and *carrying* are no longer used by effective math teachers. Rather, such teachers use alternative terms like *renaming* or *regrouping*. The principal asks whether you agree with the change in language and, if so, why.

28. Find the differences in the given base.

   a. 41_6
      -22_6

   b. 401_6
      -222_6

   c. 742_8
      -364_8

   d. 3040_8
      $- 507_8$

29. Determine the missing numbers that will complete the following subtraction problem.

   $3\_2\_\_{}_5 - 441_5 = 20\_4_5$

30. Joanna was driving from Fresno to Bakersfield. Because it was a business meeting, she would be reimbursed for her mileage. When she left Fresno, she noted that her odometer said 67,324. Unfortunately, she forgot to check her odometer in Bakersfield. However, on the way home, she saw a sign that said she was 24 miles from Fresno. At that point, her odometer said 67,564. She now had enough information to determine the distance from Fresno to Bakersfield. How far is it?

SECTION 3.3 UNDERSTANDING MULTIPLICATION

WHAT DO YOU THINK?

- What other meanings are there for multiplication beyond "repeated addition"?
- Why do we "move over" when we multiply?
- Are there more even or more odd numbers in the multiplication table?

Do you remember learning about multiplication in elementary school? If so, what do you remember? When I ask my own students this question, most of their answers have to do with trying to learn the multiplication table. Admittedly, multiplication is more complex than addition and subtraction. However, students who develop a good understanding of the structures of multiplication and its connections with the other operations report being amazed to find that the multiplication facts are easier to learn and that multiplying with larger numbers can actually make sense!

As before, we will first examine the several meanings of multiplication and different models for representing them. Examination of patterns and properties will provide a foundation for understanding how the algorithms work. Then we will examine the connections among the various operations. This, in turn, paves the way for a strong operation sense, which enables you to solve the nonroutine and multistep problems that arise in everyday and work settings.

Multiplication

Many elementary methods textbooks group addition and subtraction in one chapter and then multiplication and division in another chapter. This is not simply because children learn addition and subtraction first but is also because multiplication and division are more complex and abstract concepts. In fact, many of my students report that although they learned *more* about addition and subtraction in this course, they learned quite a bit of *new* material about multiplication and division.

Mathematics starts to crumble for many students when they get to multiplication, especially learning (or memorizing) the multiplication tables. In fact, many adults do not know all the single-digit multiplication facts, such as 8 × 7. Given how available calculators are, this is not a huge problem, although it is sometimes embarrassing. However, more problematic is the fact that few adults understand the different meanings of multiplication, and this lack of knowledge limits their ability to solve real-world problems. One of the goals of the following examination of multiplication is for you to understand the different meanings of multiplication and why it is useful to know these different meanings.

As with our introduction to addition and subtraction, assume that you are a young child who has not yet learned multiplication. How might you

CLASSROOM CONNECTION

Many young children encounter the concept of multiplication even before school; for example, if the tooth fairy gives me fifty cents for each tooth, how much will I get for all my teeth?

solve these problems—using manipulatives or diagrams, or using common sense and other mathematical knowledge? Note your thoughts before reading on. . . .

Four multiplication problems

1. One piece of candy costs 4¢. How much would 3 pieces cost?
2. If Jackie walks at 4 mph for 3 hours, how far will she have walked?
3. A carpet measures 4 feet by 3 feet. What is the area of the carpet?
4. Carla has 4 blouses and 3 skirts. How many combinations can she wear? (Assume that all possible combinations go together!)

Contexts for Multiplication

Now go back to these four problems and answer the following questions:

- In what ways are the problems different? In what ways are they similar?
- What does multiplication mean? For example, what words come to mind when you think of multiplication?

If we represent the four problems with diagrams, as in Figure 3.13, the differences appear to be tremendous, even though all four problems can be solved as 3 times 4.

Problem 1 is like the discrete sets we encountered with addition and subtraction. It is also literally **repeated addition**; it is solved by adding 4 + 4 + 4.

Problem 2 involves a certain number of measured units and can be represented with a number line that shows repeated addition of the measures.

Problem 3 involves measures, but it involves repeated addition less than it involves counting the number of new units. This problem illustrates the area context of multiplication.

Problem 4 involves discrete objects, but making sense of it relies not on repeated addition but on putting combinations in an array. In set language, what we did in this problem was to examine two sets and look at all possible ways

 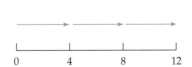 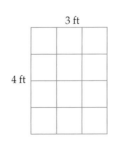 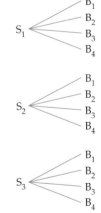

FIGURE 3.13

of pairing the elements of those sets. We will see later in this section that we can represent this context with the *Cartesian product model of multiplication*.

A General Model for Multiplication

As you can see, multiplication is a more complex concept than addition and subtraction, and it is more than simply repeated addition. Can you describe the four problems in such a way that they all have something in common, as we did for addition and subtraction? Work on this question and then read on. . . .

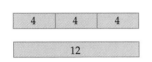

FIGURE 3.14

Our general model for addition can be extended to multiplication. Figure 3.14 shows one diagram for 3 · 4. Do you see the resemblance between the addition and multiplication models? As with addition and subtraction, the *general model of multiplication* can be cast in part–whole language: The product (the whole) is built from parts that are equal in size or amount.

Traditionally, multiplication is represented as

$$n \text{ times } a = \overbrace{a + a + a + \cdots + a}^{n \text{ times}}$$

We generally use a dot to denote multiplication (since x is used in algebra to denote a variable). If $n \cdot a = b$, then n is called the **multiplier**, a is called the **multiplicand**, and b is called the **product**. Furthermore, n and a can be said to be **factors** of b; b is a **multiple** of both n and a.

Using boxes, we can represent a **general model for multiplication** (Figure 3.15):

n times a = the amount a added n times

That is, a is the value of whatever is in each box.

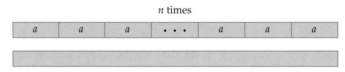

FIGURE 3.15

The general model shown in Figure 3.15 works well for Problems 1 and 2. However, the last two problems do not fit the model as well. We can force-fit them, but repeated addition is not the essence of those models. Therefore, we look for other models that can represent these contexts better.

The need for this is not readily apparent when we are working with whole numbers. Thus, a sneak preview: In Chapter 5, we will see that the repeated addition model of multiplication just doesn't work well for 3/4 times 2/3; that is, adding three-fourths two-thirds of a time simply doesn't make sense as repeated addition. Let us now examine more closely the area model and the Cartesian product model of multiplication.

FIGURE 3.16

Area Model for Multiplication

The **area model for multiplication** is an important one that we will see throughout the book. Figure 3.16 is a discrete representation for 4 · 3.

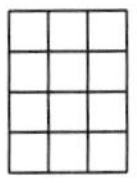

FIGURE 3.17

If we change each dot to a square, as in Figure 3.17, many students can better see the connection between the repeated addition model and the area model.

As you may recall, the formula for the area of a rectangle is length times width. Thus any multiplication problem $a \cdot b$ can be represented as a rectangle whose length is a units and whose width is b units. We will use this context in the next section when we examine the multiplication algorithm and in Chapter 5 when we examine multiplication of fractions and decimals.

Cartesian Product Model for Multiplication

As we saw in Problem 4, the Cartesian product of any two sets A and B is the set consisting of *all* possible ways of combining elements of the first set with elements of the second set. Using more formal language, we say that the Cartesian product of any two sets A and B consists of all possible ordered pairs such that the first element is from set A and the second element is from set B. In mathematical notation, we write

$$A \times B = \{(a, b) \mid a \in A \text{ and } b \in B\}$$

In Problem 4, the Cartesian product of the set S of skirts and the set B of blouses is

$$\{(S_1, B_1), (S_1, B_2), (S_1, B_3), (S_1, B_4), (S_2, B_1), (S_2, B_2), (S_2, B_3), (S_2, B_4),$$
$$(S_3, B_1), (S_3, B_2), (S_3, B_3), (S_3, B_4)\}$$

The definition of multiplication as Cartesian product is similar to our definition of addition using set language:

If $n(A) = a$ represents the number of elements in set A, and if $n(B) = b$ represents the number of elements in set B, then the number of elements in the Cartesian product of sets A and B is equal to the product of a and b.

In notation,

$$a \cdot b = n(A \times B)$$

That is, the value of a times b is equal to the number of elements in the Cartesian product of set A and set B.

There are many real-life situations in which the Cartesian product is a useful problem-solving tool. In Chapter 7, we will find that this concept can help us examine probability situations.

> ■ *Language* ■
>
> Both representations in Figures 3.16 and 3.17 can be seen as rectangular arrays.

Changes in Units

There is an important difference between multiplication and addition (or subtraction). When we add or subtract two amounts, the labels do not change: oranges plus oranges = oranges. However, this is not true with multiplication. Look back to the four multiplication problems presented at the beginning of this section. Do you see this?

- We are multiplying 3 *pieces* by 4 *cents per piece* and getting 12 *cents*.
- We are multiplying 4 *miles per hour* by 3 *hours* and getting 12 *miles*.
- We are multiplying 4 *feet* by 3 *feet* and getting 12 *square feet*.
- We are multiplying 4 *blouses* by 3 *skirts* and getting 12 *outfits*.

This matter of changing units is a crucial one that will develop over the course of the book.

Properties of Multiplication

Which of the properties that hold for addition also hold for multiplication? Write your thoughts and then read on. . . .

When adding, we found that "zero doesn't do anything." However, when we multiply any number by zero, the product is zero. This property of multiplication, which is not intuitively obvious to many children, is known as the **zero property of multiplication**:

$$a \cdot 0 = 0 \cdot a = 0$$

Many children are either intrigued or confused by the fact that when adding, "zero doesn't change your answer," but that when multiplying, "one doesn't change your answer." Stated in everyday language, "when we multiply any number by 1, we get the same number." This is called the **identity property of multiplication**:

$$a \cdot 1 = 1 \cdot a = a$$

Just as with addition, when we multiply any two numbers, we get the same product regardless of the order; that is, $a \cdot b = b \cdot a$. This property is known as the **commutative property of multiplication**.

Just as with addition, grouping doesn't matter when we multiply several numbers. For example, $(5 \cdot 4) \cdot 7 = 5 \cdot (4 \cdot 7) = 140$, although the problem is much easier to do mentally in the first way. Why is this? This grouping property is known as the **associative property of multiplication**:

$$(a \cdot b) \cdot c = a \cdot (b \cdot c)$$

There is another property, called the **distributive property**, that connects multiplication to the operations of addition and subtraction. Consider the following problem given to third-grade students: How many cans of soda are in three 24-can cases? When students are given freedom to invent their own solutions to these problems rather than "discover" the teacher's way, they will solve this problem in a variety of ways, three of which are discussed below.

Some will simply add $24 + 24 + 24$.

Others will represent the problem with base 10 blocks, as in Figure 3.18. They will count 6 tens and 12 ones, convert to 7 tens and 2 ones, and give the answer of 72.

> ■ *Mathematics* ■
>
> For many years, I happily told my students that multiplication is commutative. However, I have since participated in several discussions about this notion, which have helped me to see another layer of why multiplication is more complex than addition. In a strictly numerical sense, multiplication is commutative. However, when we add context, which we generally do with children, we have to realize that the context does not commute. Let me use the following stories to illustrate this.
>
> Let's say two children have birthday parties, and one child has 7 party bags with 8 candies in each, and the other has 8 party bags with 7 candies in each. Numerically, both problems yield 56 candies, but the two problems don't look alike to many children. They are different situations.
>
> Look at another example. One person walks 3 miles/hour for 4 hours while another person walks 4 miles/hour for 3 hours. These are two different problems, even though they have the same outcome of 12 miles.

> ■ *Mathematics* ■
>
> The distributive property is a powerful one that we will encounter later in this chapter when we learn why the multiplication algorithm works. We will also see it in Chapter 5 when we examine multiplication with fractions and decimals, and it occurs throughout algebra. It is also widely misunderstood; hence many students do not have this property in their toolbox. This is like a mechanic's toolbox not containing a crescent wrench!

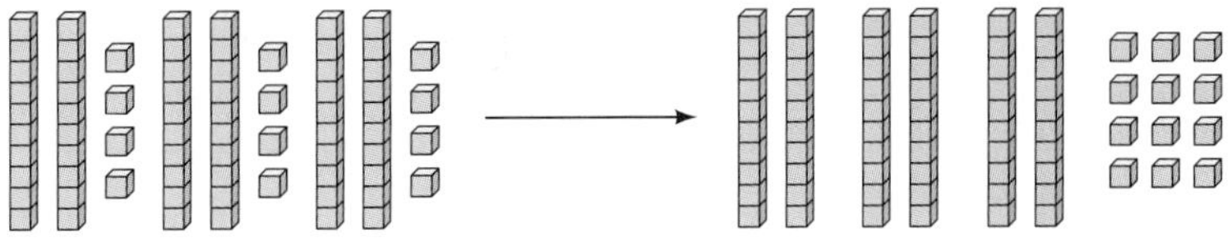

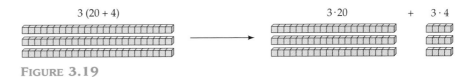

FIGURE 3.18

In this case, the students are using the distributive property of multiplication over addition:

$$a(b + c) = ab + ac$$

They have transformed $3 \cdot 24$ into $3 \cdot (20 + 4) = 3 \cdot 20 + 3 \cdot 4 = 60 + 12 = 72$. That is, 3 groups of 24 can be decomposed by breaking the 24 into 2 tens and 4 ones, which is equivalent to 3 groups of 20 and 3 groups of 4 (see Figure 3.19).

FIGURE 3.19

Others will solve the problem with money: 24¢ is 1 penny less than a quarter; do this three times and you will have 3 quarters take away 3 pennies—that is, $75 - 3 = 72$. In this case, the students are using the distributive property of multiplication over subtraction:

$$a(b - c) = ab - ac$$

The students have transformed $3 \cdot 24$ into $3 \cdot (25 - 1) = 3 \cdot 25 - 3 \cdot 1 = 75 - 3 = 72$

The Multiplication Table

We discussed patterns in the multiplication table in Exploration 3.8. Let us further examine the multiplication table to see how patterns can help children learn the table's 100 "multiplication facts." What patterns do you see in Table 3.2? Write your observations and then read on. . . .

TABLE 3.2

	1	2	3	4	5	6	7	8	9	10
1	1	2	3	4	5	6	7	8	9	10
2	2	4	6	8	10	12	14	16	18	20
3	3	6	9	12	15	18	21	24	27	30
4	4	8	12	16	20	24	28	32	36	40
5	5	10	15	20	25	30	35	40	45	50
6	6	12	18	24	30	36	42	48	54	60
7	7	14	21	28	35	42	49	56	63	70
8	8	16	24	32	40	48	56	64	72	80
9	9	18	27	36	45	54	63	72	81	90
10	10	20	30	40	50	60	70	80	90	100

INVESTIGATION 3.16

A Pattern in the Multiplication Table

Let us examine one pattern. A student noted that every odd number in the table is "surrounded" by even numbers but that the reverse is not the case. Let us investigate the first observation. Why do you think every odd number in the table is surrounded by even numbers? Work on this problem before reading on. . . .

DISCUSSION

A key to unlocking this mystery comes from the following generalizations: The product of two even numbers is always an even number, the product of two odd numbers is always an odd number, and the product of an odd number and an even number is always an even number. We will explore odd and even numbers further in Chapter 4. If you didn't make much progress on this question before, try to use this information and see whether you can explain this phenomenon now. Then read on. . . .

Let us look first at a specific case and then at a general case. The diagram below shows 35 (from 7 · 5) and the numbers that surround it in the table.

		6	7	8
		.	.	.
		.	.	.
		.	.	.
4	...	24	28	32
5	...	30	35	40
6	...	36	42	48

More generally, we can write

		Even	Odd	Even
		.	.	.
		.	.	.
		.	.	.
Even	...	Even	Even	Even
Odd	...	Even	Odd	Even
Even	...	Even	Even	Even

Thus we see that each of the eight numbers surrounding an odd number in a multiplication table is either the product of two even numbers or the product of an odd number and an even number. Thus none of them can be an odd number.

Using this pattern I will present two ways in which this knowledge can help with multiplication facts. First, a brief reflection on Investigation 3.16 yields the deduction that 3/4 of the multiplication facts are even numbers; that is, only one in four multiplication facts is an odd number. Second, nowhere in the multiplication table do we have two odd numbers in a row. Thus, for example, if I know that 7 · 7 is 49, then 8 · 7 can't be 55 or 57. The generalizations about the products of odd and even numbers would help, for example, a student who wasn't sure whether 7 · 7 was 48 or 49.

Ultimately, the most important outcome of looking for and examining patterns is that the student comes to see that mathematics is more than just a bunch of facts and rules and that these "relationships" not only are sometimes intriguing but also make us more mathematically powerful.

Now let us apply some of this knowledge about multiplication to a real-life setting.

INVESTIGATION 3.17

3.9

Mental Multiplication

Just as we did with addition and subtraction, we will begin our exploration of computation with multiplication by doing some mental multiplication. Many of my students are surprised to find that with some thought and practice, they can do many more multiplication problems mentally than they would have thought possible. Doing so successfully generally requires us to apply the various properties of multiplication, often creatively!

Do the following computations in your head. Briefly note the strategies you used, and try to give names to them.

1. 64×5 **2.** 16×25 **3.** 15×12 **4.** 849×2 **5.** 60×30

As you read these strategies, once again monitor your own thinking. Do you understand how the strategies work? If not, do you simply need to reread the discussion, or would it be helpful to make up and do some similar problems, or would it be more helpful to practice with a friend?

DISCUSSION

In Problem 1, use **multiples of 10** as a reference point:

$$64 \times 5 = \frac{1}{2} \text{ of } 64 \times 10 = \frac{1}{2} \text{ of } 640 = 320$$

In Problem 2, the **halve and double** method gives

$$16 \times 25 = 8 \times 50 = 400$$

Do you see the use of the commutative and associative properties here?

Pattern 3 is a very interesting problem because it lends itself naturally to many different solution paths. Students solve it by using the distributive property in two different ways. Some solve it by breaking 15×12 into ten 15s and two 15s. Others break 15×12 into ten 12s and five 12s. Some will cut the 12 in half and use the associative property $2 \times (6 \times 15)$. Yet others will apply double and halve: $15 \times 12 = 30 \times 6$.

In Problem 4, try compatible numbers in conjunction with the distributive property:

$$849 \times 2 = (850 - 1) \times 2 = 1700 - 2 = 1698$$

For Problem 5, use multiples of 10:

$$60 \times 30 = 6 \times 3 \times 10 \times 10 = 18 \times 100$$

What people often think in this case is "18 with two zeros," or 1800.

Because looking back or reflecting is such an important learning tool, and because so few students bring that tool into this course, I am risking overemphasizing it. Now that you have examined various mental multiplication strategies, do you see advantages and limitations of the various strategies? Do you understand the names given to the strategies, or can you come up with names that you like better?

Understanding Multiplication with Larger Numbers

Before we examine how and why the standard multiplication algorithm works, a couple of comments are in order. Over the years, I have come to appreciate that this is perhaps the most difficult algorithm for my students to understand, and it is also the hardest algorithm for children to master. Thus I urge you to read the material very carefully and thoughtfully. This jump from multiplication facts to multiplying multidigit numbers is a big one, and we will take it in steps.

The big jump for the children in learning how to multiply is to apply their knowledge of base 10 and place value and to let go of strategies that work well for addition. For example, 23×4 can be found in many ways, as shown below. Some ways use base 10; some ways use powerful tools, and some ways do not. Some of these ways, like the first three, help the children to see the close connection between addition and multiplication. The way at the far right illustrates a more powerful, efficient way to determine the answer. Your methods course will teach you how to move children from ways like those at the left to ways like those at the right. For your mathematical development, it is important that you understand the representations and approaches below and that you are able to see how to break the problem apart, using base 10 and place value ideas.

Add	Add pairs	Count the 10s Add the 3s	Take apart (1)	Take apart (2)	Take apart (3)
1					
23		23	$25 \times 4 - 2 \times 4$	10×4	20×4
23		23		10×4	3×4
23	23 ⟩ 46	23		3×4	
23	23	23			
——	⟩ 92	——			
92	23 ⟩ 46	80	$100 - 8$	$40 + 40 + 12$	$80 + 12$
	23	12			
		92			

One starting point for developing an understanding of the standard algorithm is to go beyond the basic multiplication facts, one step at a time. For example, how might you solve 16×7 if you didn't know the algorithm? Do so before reading on. . . . ✎ When this problem is given to children, they solve it in many ways. Children naturally invent the distributive property by applying multiplication as repeated addition to their multiplication facts. For example, 16×7 can be broken into $(8 \times 7) + (8 \times 7)$ or

$(10 \times 7) + (6 \times 7)$. Both are valid. However, over time, we want the children to understand that breaking the numbers apart, using base 10, generally proves more powerful. Thus, this decomposition is ultimately more powerful: $16 \times 7 = (10 \times 7) + (6 \times 7)$.

Toward that end, the problems become larger and larger. For example, the children can construct the answer to 35×14 with the following string, where they begin with what they know and build up:

3.12

$3 \times 4 = 12$

$30 \times 4 = 120$ because they have studied the pattern when multiplying by 10

$5 \times 4 = 20$

$35 \times 4 = 140$ because 35×4 is equivalent to thirty 4s plus five 4s

$35 \times 10 = 350$ multiplying by 10

$35 \times 14 = 490$ because 35×14 is equivalent to four 35s plus ten 35s

These problems, which are called string problems or cluster problems, serve multiple purposes. (See Exploration 3.12 for more problems like these.) They help to lay the conceptual foundation for understanding the standard algorithm. However, they also provide good practice in applying the distributive property, which is essential to understanding the algorithm.

Another step in the development is to examine problems in context. For example, we can frame the problem 23×12 as: How much will it cost to blacktop a playground that measures 23 meters by 12 meters. The natural representation of this problem, a rectangle, virtually *requires* the children to move toward the more powerful representation of multiplication as an array. Below you can see various representations of the problem and then some of the solution paths that come out of these representations.

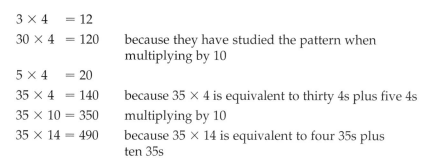

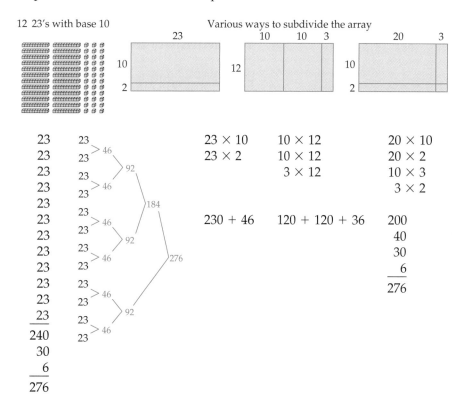

Let's discuss the three rectangular representations shown above. First, make sure that you understand how all three represent 23 × 12. All three rectangles employ the powerful, but often only partially understood, distributive property.

The first rectangle illustrates one way of cutting the rectangle apart. That is, twelve 23s can be broken into ten 23s and two 23s.

The second rectangle illustrates one way of cutting the rectangle apart. That is, twenty-three 12s can be broken into ten 12s plus ten more 12s plus three more 12s.

The third rectangle illustrates a way of cutting the rectangle apart that connects to the standard algorithm. This way can be elicited by asking the children to fill the space with the least number of manipulatives. This can be done by filling the top left section with flats, the bottom right section with singles, and the other two sections with longs.

Some very powerful insights have come both from children and from teachers when they were doing and studying these problems. Problems like these help both children and teachers to internalize (to "own") a more powerful representation of multiplication (as an array) and to see more powerful ways of breaking apart multiplication problems (using the distributive property). They help teachers to see that although multiplication can be seen as repeated addition, if that is *all* you see multiplication as, then your students can achieve only limited understanding. Teachers also realize that we are still looking at parts and wholes. However, parts and wholes with addition and subtraction are conceptually much simpler. For example, in adding two numbers (say 68 + 43), almost every way of taking apart the numbers will connect the closest place value (70 − 2 or 40 + 3). However, when we are looking at multi-digit multiplication, there are many ways to take apart the numbers, rather than just one or two ways.

Now let us examine the standard multiplication algorithm in a very systematic way.

Classroom Connection

One of the teachers I worked with was so stunned as a result of our work with multiplication that she exclaimed, "I will never tell my students that multiplication is just repeated addition again!"

The Multiplication Algorithm in Base 10

3.13

From this section and Explorations 3.12 and 3.13, you now have a better understanding of different models of multiplication and an understanding of the importance of the distributive property. We will now examine the standard multiplication algorithm, which is perhaps the most difficult of the algorithms to understand at the conceptual level. It rests on an understanding that any multiplication problem can be represented as a rectangle and on an understanding of the distributive property. We will examine this algorithm with 56 · 34.

The rectangular model of the product   Without base 10, we are left with an imposing problem! Figure 3.20 shows 34 rows of 56 units. The answer is there, but who wants to count them all?

However, if we apply our knowledge of base 10, we can find alternatives to counting each single unit.

Using base 10 to model the product   One first step is to see what the problem would look like if we represented it using our knowledge of base 10 (see Figure 3.21). That is, we would have 56 (5 tens and 6 ones) 34 times. This makes our job only slightly easier. Who wants to add 56 thirty-four times?

Once again, our knowledge of base 10 can make the counting process less tedious.

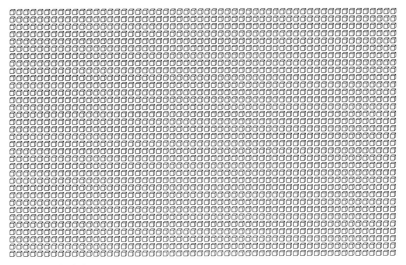

FIGURE 3.20

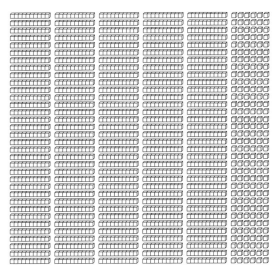

FIGURE 3.21

Using our knowledge that 10 ones make a ten, and that 10 tens make a hundred, we can turn units into longs and longs into flats (see Figure 3.22). Do you see the connection between these processes and the following diagram? Do you see that this diagram is another representation of 56 × 34? Do you see connections between this diagram and how you would compute 56 × 34?

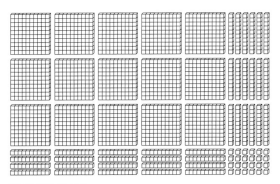

FIGURE 3.22

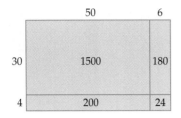

Counting the new groups When we represent the problem using the rectangular model of multiplication, we see that the standard algorithm systematically does the regrouping for us. Let us first look at how we can quickly count the amount in each of the four regions in Figure 3.22. Can you use your knowledge of multiplication to do so? Try to do this on your own before reading on. . . .

- The units region consists of $4 \cdot 6 = 24$.
- The longs region to the left of the units region consists of $4 \cdot 5 = 20$ longs, which have a value of 200.
- The longs region above the units region consists of $3 \cdot 6 = 18$ longs, which have a value of 180.
- The flats region consists of $3 \cdot 5 = 15$ flats, which have a value of 1500.

Adding these four regions, we have

```
1500
 180
 200
  24
----
1904
```

Thus, when we use the algorithm to determine the product $56 \cdot 34$, we determine four products (in this case, $4 \cdot 6$, $4 \cdot 5$, $3 \cdot 6$, and $3 \cdot 5$), and we know what to do with the products. These products are called **partial products**.

Do you see how the four regions in Figure 3.22, the four sums in the addition problem above, and the four partial products are related to one another?

When we represent the problem in expanded form, we can see how the four partial products connect to the algorithm.

```
      50 +  6              56
   ×  30 +  4            × 34
   ----------            ----
     200 + 24             224
1500 + 180               1680
                         ----
                         1904
```

The standard multiplication algorithm Now look at the algorithm in its most concise form. Can you justify each step of this algorithm?

When we move up to larger multiplication problems, the manipulatives and visual representations become more cumbersome. However, if the students understand the process, they can apply that understanding to the larger problems. This understanding is also the foundation for skillful estimation. For example, when we represent 324 times 6 in expanded form, we have three partial products. The value of the partial product $3 \cdot 6$ (which represents 300 times 6) is much larger than the value of the other two partial products (120 and 24). Therefore, 300 times 6 (1800) is a quick and reasonable estimate for this problem.

```
    324             300 + 20 + 4
  ×   6           ×            6
                  ---------------
                   1800 + 120 + 24
```

As with addition and subtraction, we come to understand the algorithm by decomposing the numbers (using expanded form) and looking at how the ideas of base and place value help us to determine the total product. Recall again the statement by Lauren Resnick: "Probably the major conceptual achievement of the early school years is the interpretation of numbers in terms of part and whole relationships." When one understands what the operations mean and has the tools (expanded form, place value, and so forth), then one can make full sense of the operations, and that forms the basis for what the NCTM standards have called "mathematical power."

INVESTIGATION 3.18

An Alternative Algorithm

The lattice algorithm that you learned for addition can be modified for multiplication. Observe the example below and see if you can figure out how it works and why it works.

DISCUSSION

It works this way.

First, write the first number horizontally and the second number vertically.

Second, make a box that has the same dimensions as your problem. In this case, since we have a 2-digit by 2-digit problem, we make a 2 × 2 box.

Third, as with lattice addition, we draw diagonal lines through each square that extend below the square.

Fourth, we write the result of each partial product in the appropriate box.

Last, as with lattice addition, add diagonally. If the sum in any diagonal addition is greater than 10, "carry" just as you do with the standard algorithm.

The "why" of this algorithm is very similar to the why of lattice addition. The diagram at the right shows the place value of each digit inside the lattice. As with lattice addition, the diagonals keep each digit with others of the same place value.

Th = Thousands
H = Hundreds
T = Tens
O = Ones

INVESTIGATION 3.19

Developing Estimation Strategies for Multiplication

In each of the problems below, first obtain a rough estimate; then obtain the best estimate you can before reading on.

A. Chip rode 78 miles last week. At this rate, how many miles will he ride this year?

DISCUSSION

STRATEGY 1: Find a lower bound and an upper bound for the answer

Because we want to estimate the product of 78 and 52, if we round both numbers down to the nearest 10, we have 70 × 50 = 3500. Similarly, if we round both numbers up to the nearest 10, we have 80 × 60 = 4800. The actual product of 78 and 52 lies between these two numbers. The estimate from rounding down is called a **lower bound**, and the estimate from rounding up is called an **upper bound**. Using this technique, we can quickly say that Chip will ride between 3500 and 4800 miles this year.

STRATEGY 2: Round one number up and round the other number down

Thus 78 × 52 becomes 80 × 50 = 4000.

STRATEGY 3: Use expanded form and estimate the sum of the four partial products

```
         70 +  8
         50 +  2
        ─────────
        140 + 16
      3500 + 400
```

One thought process might go like this: "70 × 50 = 3500, 50 × 8 is 400, and 70 × 2 is more than 100. Thus the answer will be 3500 plus more than 500, let's say 4050."

Determine the actual answer. How close were the different estimates?

> **B.** Jane plans to start graduate school in September. She figures that she can save $345 per month for the next 9 months. How much will she have saved? Work on this and then read on. . . .

DISCUSSION

Use the distributive property:

$$345 \times 9 = (345 \times 10) - (345 \times 1) = 3450 - 345 \approx 3100$$

> **C.** There were 47,752 Americans killed or missing in Vietnam. The number of Americans killed or missing in World War II was about 6 times that number. Approximately how many Americans were killed or missing in World War II? Work on this and then read on. . . .
>
> $$\begin{array}{r} 47{,}752 \\ \times 6 \\ \hline \end{array}$$

DISCUSSION

We can get a rough estimate using rounding:

$$50{,}000 \times 6 = 300{,}000$$

We can get a more refined estimate by rounding and using mental math (double and halve):

$$48 \times 6 = 96 \times 3 = 288, \text{ or } 288,000$$

INVESTIGATION 3.20

Using Various Strategies in a Real-life Multiplication Situation

A warehouse has 50 bays (places to stack pallets). Four pallets can be stacked in each bay, each pallet can hold 24 cartons, and each carton holds 12 boxes. How many boxes can the warehouse contain? If each box sells for $4, what is the value of the merchandise in a full warehouse? Work on this problem yourself before reading on. . . .

DISCUSSION

As with many of our problems, there are several strategies that will lead to a solution. Many of my students tell me that they can understand the problem when they read my solutions but that they have trouble coming up with a strategy on their own. Therefore, I urge you not to read the discussions of the investigations until after you have attempted the problem (on your own or with a friend), and when you do read the discussion, read "actively."

STRATEGY 1: Draw a diagram

There are many possible ways to represent this problem with a diagram. The diagram in Figure 3.23 is not "the right diagram" but rather an example of a useful diagram. If you didn't solve the problem or if you didn't solve it with a diagram, take a few moments to look at this diagram. Does it help? How? Does it connect to or stimulate your thinking about what operation(s) might be involved?

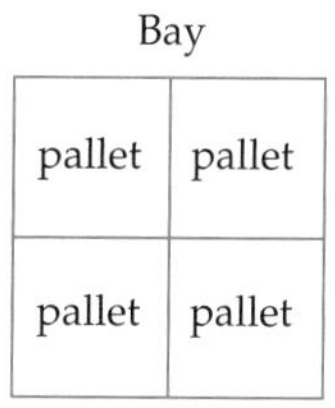

FIGURE 3.23

STRATEGY 2: Use smaller numbers

What if there were 3 bays, 4 pallets in each bay, 2 cartons in each pallet, and 5 boxes in each carton?

Many researchers have found that if they give two problems that are mathematically identical, but one of which has big or messy numbers, the success rates can be dramatically different. Do the smaller numbers help you to see the problem more clearly so that you can deduce what operation(s) to use? Some students find that this strategy makes it easier to draw a diagram that helps them figure out what to do.

STRATEGY 3: Use dimensional analysis

Recall this tool, which was introduced in Investigation 1.2 (How Much Will the Patio Cost?). Using dimensional analysis, you cancel the larger units to concentrate on the aspect of the problem you want to solve. In this problem, we have 50 bays. However, we don't want to know the amount in terms of bays, we want to know it in terms of a smaller unit—boxes.

We can multiply 50 bays by 4 pallets per bay. Using dimensional analysis, this looks like

$$50 \text{ bays} \cdot \frac{4 \text{ pallets}}{\text{bay}} = 200 \text{ pallets}$$

That is, if we use pallets as our unit of measurement, we have 200 pallets. Recall the discussion of *changes in units* on page 166. Because we do not want the answer in terms of pallets, we continue to change units:

$$200 \text{ pallets} \cdot \frac{24 \text{ cartons}}{\text{pallets}} = 4800 \text{ cartons}$$

Translating to an even smaller unit, we find that

$$4800 \text{ cartons} \cdot \frac{12 \text{ boxes}}{\text{carton}} = 57{,}600 \text{ boxes}$$

The value of the merchandise is

$$57{,}600 \text{ boxes} \cdot \frac{\$4}{\text{box}} = \$230{,}400$$

A student more confident with using dimensional analysis can do the entire problem in one step:

$$50 \text{ bays} \cdot \frac{4 \text{ pallets}}{\text{bay}} \cdot \frac{24 \text{ cartons}}{\text{pallet}} \cdot \frac{12 \text{ boxes}}{\text{carton}} \cdot \frac{\$4}{\text{box}} = \$230{,}400$$

I recall being shown this tool in a high school chemistry class, and I have used it in all kinds of school and nonschool problems since then. If this tool is new to you, does dimensional analysis make sense to you now? If it does not, you might want to work on this with a classmate or with your instructor.

INVESTIGATION 3.21

Number Sense with Multiplication

3.14

A. Without computing determine whether this number sentence is true or false.
68 × 4 = (34 × 2) + (34 × 2)

DISCUSSION

If we add 34 groups of 2 and 34 more groups of 2, we get 68 groups of 2, but 68 × 4 means 68 groups of 4. Therefore, this problem is false. It is also a common mistake made by children and adults, because both numbers at the right are half of the corresponding numbers at the left, so there is that sense of being equal.

B. This problem is similar to puzzles archaeologist have to solve; the circles represent places where they cannot read the numbers. We have recovered these fragments of this multiplication problem. What was the problem and what is the answer?

```
    O6
×  4O
   32O
  OOO
  OOO4
```

DISCUSSION

This problem requires some analysis. Since the product has a 4 in the ones place, a first step is to ask 6 times what gives a 4 in the ones place? Many people stop with the 4 because $6 \times 4 = 24$. However, if we are careful, we also see that $6 \times 9 = 54$. So now what do we do? What number in the tens place of the multiplicand will get us 324? That is 96×4, 86×4, 76×4, etc. None of these work. What about 96×9? Way too big. Try smaller 46×9? 36×9? This works. Therefore, we have now solved part of the puzzle:

$$\begin{array}{r} 36 \\ \times\ 49 \\ \hline 324 \\ OOO \\ \hline OOO4 \end{array}$$

Actually, at this point, it is a simple matter to complete the problem to determine the rest of the missing places, and you can do this on your own.

C. Will the result of 43×87 be in the hundreds, thousands, or ten thousands? Explain your choice.

DISCUSSION

If we use our estimation and rounding skills, we quickly see that $40 \times 90 = 3600$, so the answer is in the thousands.

D. Is this answer reasonable? Why or why not?

$$\begin{array}{r} 84 \\ \times\ 66 \\ \hline 4824 \end{array}$$

DISCUSSION

Once we realize that there are four partial products, we can quickly resist the intuitive appeal of $4 \times 6 = 24$ and $8 \times 6 = 48$. When children transer all of what they have learned from addition to multiplication, this is a common answer. Do you see why? With addition, you add each place, so if you multiply each place you get 4824.

Looking back Before we move on to division, take a few moments to look back over the strategies for mental multiplication and estimation that we have explored. Do you find that you prefer some over others? Can you identify those places in which these strategies use properties of multiplication (for example, the distributive property)? Can you see which strategies make use of the multiplication algorithm (for example, partial products)? Can you see which strategies make use of base 10 structures (for example, expanded notation)? Do you see that some strategies work with some numbers better than with others?

EXERCISES 3.3

1. Make up story problems and diagrams to illustrate 5 · 3 in the four different contexts shown in Figure 3.13.

2. Write a realistic story to represent each problem below.
   a. 12 · 4   b. 35 · 3   c. 5 · (3 + 7)

3. Do you think we could speak of a distributive property of addition over multiplication? How would you determine whether it held or not?

4. For each of the following questions, refer to the base 10 multiplication table.
   a. The top-left-to-lower-right diagonal divides the table into three sets: diagonal, upper part, and lower part. What do you notice about the upper part and the lower part? What property is illustrated here?
   b. Look at the top-right-to-lower-left diagonal. Describe the patterns you see. Justify the patterns.
   c. The products of the diagonals of any 2 by 2 matrix are equal. Explain why.
   d. Look at the 9s column in the multiplication table. The sum of the digits is always 9. Can you explain why?
   e. Look at the 9s column in the multiplication table. As we go down the column, the ones digit decreases by 1 each time and the tens digit increases by 1 each time. Can you explain why?

5. Let's say you forgot what 9 · 7 is. How would you figure out the product, using other multiplication facts? There are many, many different ways!

6. Fingers are a set of manipulatives that most people have, and they can be used to help students remember their multiplication facts for 9. This is how the method works. Suppose you want to know 9 · 7. Bend down finger number 7. The number of fingers to the left of the bent finger gives you the value of the tens place of the product, and the number of fingers to the right of the bent finger gives you the value of the ones place of the product. Why does this work?

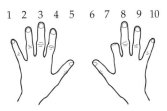

7. Before the Russian Revolution in 1917, most of the people in Russian villages still used the Roman system and a multiplication algorithm known today as the Russian peasant algorithm. It works by simultaneously doubling one of the numbers and halving the other. Fractions are simply disregarded; for example, half of 19 is $9\frac{1}{2}$—drop the $\frac{1}{2}$. We find the product by first looking in the "Halve" column. Whenever you see an odd number in that column, circle that row. To find the answer, add the circled numbers in the "Double" column. The example below shows how it works.

Problem: 25 · 19.

Double	Halve
(25)	· 19)
(50)	· 9)
100	· 4
200	· 2
(400)	· 1)

Circling the appropriate numbers, we add:

  25
  50
 400

 475

Solve the following problems using the Russian peasant algorithm.
   a. 25 · 17   b. 48 · 39   c. 120 · 42
   d. Explain why this algorithm works.

8. Another ancient method of multiplication is shown below with the problem 102 · 96.

   Find the average: 99
   Take half the difference of the numbers: 3
   Look up the square of 99: 9801
   Look up the square of 3: 9
   Subtract and you have your answer: 9792

   a. Use this method to determine the product of 36 and 54.
   b. Explain why it works.

9. a. Write directions for multiplying 46 · 37 using the lattice algorithm.
   b. Explain why this method works.
   c. What mathematics do you have to know in order to be able to use this method?
   d. Discuss some advantages and disadvantages of this algorithm.

10. Through illustrations, demonstrate how to solve these problems with manipulatives.
    a. 25   b. 43   c. 29
       × 7   ×12   ×15

11. Determine the following products in two different ways other than using the standard algorithm or the lattice algorithm.
    a. 42   b. 41
       ×15   ×39

12. Draw a diagram to represent 34 · 28. Explain how to obtain the product from the diagram.

13. Look at this multiplication problem, solved using the standard algorithm. In this problem, we multiply 2 · 6 and carry the 1. What is it that we are carrying? What is the mathematical meaning of this carrying?

    $$\begin{array}{r} 42 \\ \times 36 \\ \hline 252 \\ 126 \\ \hline 1512 \end{array}$$

14. *Classroom Connection* Below are three alternative representations, using partial products, of a two-digit by two-digit multiplication problem. These are sometimes called developmental algorithms, because teachers have used them to help children see the four partial products. One child wrote, "I like this (way) better because it allows you to give your brain a rest plus it's faster...!" ("Alternative Algorithms," *Teaching Children Mathematics*, April 2001). In each case, explain where the numbers come from, why they are placed where they are, and the value of this particular representation.

    a. $\begin{array}{r} 72 \\ \times 34 \\ \hline 8 \\ 280 \\ 60 \\ 2100 \\ \hline 2448 \end{array}$ b. $\begin{array}{r} 72 \\ \times 34 \\ \hline 2100 \\ 280 \\ 60 \\ 8 \\ \hline 2448 \end{array}$ c. $\begin{array}{r} 72 \\ \times 34 \\ \hline 8 \\ 28 \\ 6 \\ 21 \\ \hline 2448 \end{array}$

15. In each of the following, use the information from the computed problem to determine the answer mentally.

    $$\begin{array}{r} 587 \\ \times 345 \\ \hline 2935 \\ 23480 \\ 176100 \\ \hline 202{,}515 \end{array}$$

    a. $587 \times 40 = ?$
    b. $587 \times 45 = ?$
    c. $587 \times 305 = ?$
    d. $587 \times 300 = ?$

16. Will the product of 347×43 be in the thousands or ten thousands? Explain your choice.

17. a. Someone determined 25×12 by adding (25×10) and (25×2). What property were they using?
    b. Someone determined $8 \times 7 \times 5$ mentally by multiplying $8 \times 5 \times 7$. What properties did they use?

18. *Classroom Connection* On page 454 in the April 1999 issue of *Teaching Children Mathematics*, a teacher shares a method invented by one of her students to keep track of the regroupings in multidigit multiplication. Explain this child's method. That is, why does she do what she does?

    $$\begin{array}{r} 236 \\ \times 57 \\ \hline 24 \\ 1412 \\ 13 \\ 10500 \\ \hline 13452 \end{array}$$

    Source: Reprinted with permission from Teaching Children Mathematics, copyright © 1999 by the National Council of Teachers of Mathematics.

19. *Classroom Connection* Identify and explain the errors in the problems below, all of which have occurred in real classrooms.

    a. $\begin{array}{r} 46 \\ \times 28 \\ \hline 404 \\ 101 \\ \hline 1414 \end{array}$ b. $\begin{array}{r} 46 \\ \times 28 \\ \hline 3248 \\ 812 \\ \hline 4060 \end{array}$ c. $\begin{array}{r} 46 \\ \times 28 \\ \hline 368 \\ 92 \\ \hline 460 \end{array}$ d. $\begin{array}{r} 46 \\ \times 28 \\ \hline 92 \\ 368 \\ \hline 3772 \end{array}$

20. *Classroom Connection* A child multiplied 57×4 and got 288. How might the child have done this? What ideas from place value and/or multiplication has the child misapplied?

21. *Classroom Connection* When children work to develop methods of multiplying large numbers, the following conceptual error is very common. For example, a child will conclude that 34×26 will be equal to $(30 \times 20) + (4 \times 6)$. Why doesn't this work?

22. Determine the products mentally.

    a. $\begin{array}{r}16\\ \times 16\end{array}$ b. $\begin{array}{r}16\\ \times 25\end{array}$ c. $\begin{array}{r}66\\ \times 5\end{array}$ d. $\begin{array}{r}849\\ \times 2\end{array}$ e. $\begin{array}{r}60\\ \times 30\end{array}$ f. $\begin{array}{r}450\\ \times 20\end{array}$

    g. $\begin{array}{r}35\\ \times 12\end{array}$ h. $\begin{array}{r}65\\ \times 42\end{array}$ i. $\begin{array}{r}736\\ \times 4\end{array}$ j. $\begin{array}{r}35\\ \times 16\end{array}$ k. $\begin{array}{r}632\\ \times 4\end{array}$

In Exercises 23 and 24, determine the answer mentally.

23. A concert hall has 36 seats in a row, and there are 25 rows. How many people can be seated?

24. A camping show rents booth space for $1.50 per square foot. How much would it cost to rent a 10 foot by 15 foot space?

25. Estimate as closely as you can.
    a. 41×68 b. 56×74 c. 62×83
    d. 57×38 e. $34{,}345 \times 48$ f. 417×23

26. By estimating, determine which of the following is wrong. Explain your reasoning.
    a. $3312 \times 13 = 43{,}056$
    b. $23 \times 874 = 30{,}102$
    c. $563 \times 86 = 48{,}418$

27. Which two numbers will have a product closest to 2000?
    43   34   65   83   111

28. Which is a better estimate of 468×9: 420×10 or 460×10?

In Exercises 29–31, first determine a rough estimate in 5 to 10 seconds. Then determine your best estimate, which means no pencil and paper and less than 30 seconds or so. (Otherwise, we defeat the purpose of estimating!) Then determine the exact answer.

29. It takes the average person 6 seconds to read this sentence. In that time, 24 people will be born on the earth. At this rate, how many people are born each year? How long will it take for the number of births worldwide to equal the number of people in your state?

30. A student lives in Hanover, which is 62 miles from her school. How many miles per semester does she travel if she drives to school twice a week? How much does she spend for gas? Assume a 15-week semester.

31. Certain fund-raisers have a goal of raising $55,000 in 1 month. After 5 days, the amount that has been raised is $9345. Are they on track?

32. If a dripping faucet wastes water at the rate of 75 ounces per day, how many gallons does it waste per year?

33. People at the copy center at Keene State College told me that they used 1200 reams of paper last semester at my college. Departments are charged 3¢ per copy when they use the copy center. If departments use their own copy machines, which they do for small jobs, they are charged 5¢ per copy.
    a. How many pages were run off at the copy center?
    b. How much did the college pay to the copy center?
    c. How much did the college save by having a copy center?

34. Take your pulse for 15 seconds. On the basis of this measurement, how many times would you say it has beaten since you got up this morning? This week? Since you were born?

35. How many Cheerios are in a box? First, define the problem better. Devise a strategy to estimate the number; then devise another strategy. Are your results close? Which strategy do you think is more accurate?

36. What base 10 concepts and/or properties are we making use of when we employ the leading-digit method of estimation?

37. An article in the January 1995 issue of *Teaching Children Mathematics* (p. 269) reported the following statistic: Every 57 minutes an under-age drinker is involved in a traffic fatality.
    a. Estimate the number in 1 year.
    b. What additional information would you like to have to clarify the matter?

38. Is this answer reasonable? Why or why not?

    $$\begin{array}{r} 408 \\ \times\ 304 \\ \hline 12,032 \end{array}$$

39. Which of the following might represent the cube of 156?
    a. 3,796,416   b. 4,251,528
    c. 3,944,312   d. 3,581,577

40. Select any two-digit number and multiply it by 99. Repeat this process for two or more two-digit numbers. Look for patterns that would enable you to compute the product of any two-digit number and 99 in your head.
    a. Describe how you can determine the product of any two-digit number and 99.
    b. Describe how you can determine the product of any two-digit number and 999.

41. Determine the value of *a* that will make the multiplication problem correct.

    $$\begin{array}{r} a6 \\ \times\ 3a \\ \hline 1564 \end{array}$$

42. Place the digits 0, 1, 3, 5, 7 in the boxes to obtain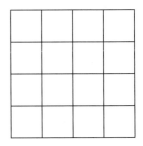
    a. The greatest product
    b. The least product

43. Let's say you have an inexpensive calculator that has only an eight-digit or nine-digit display, and thus you cannot get the answers to these computations directly from your calculator. Determine the products by a means other than doing them out longhand, assuming that you have only these resources: your knowledge of multiplication and base 10 and a calculator with an eight- or nine-digit display. Justify your method; that is, why will it give you the desired product?
    a. $999,999,999 \cdot 56$
    b. $987,654,321 \cdot 9$
    c. $11,111,111,111 \cdot 45$
    d. $34,000 \cdot 56,000$
    e. $(399,999)^2$

44. Take your house number or room number and double it, add 5, multiply by 50, add your age, and subtract 250. What do you get? Can you explain why this trick works?

45. The square on the left is a multiplication magic square. How is a multiplication magic square similar to an addition magic square? How is it different? Can you make another multiplication magic square?

14	39	6	74
111	4	26	21
26	21	111	4
6	74	14	39

46. *Classroom Connection* This problem was described in the October 1993 issue of *The Arithmetic Teacher*, ("Becca's Investigation," pp. 78–81). A child, Rebecca, had made an interesting discovery. Two very different-looking problems gave the same product. That is, 36×8 and 48×6 both yielded 288. She noted that in both cases she had a 48 and a 24. She wondered if there were other cases where two different numbers yielded the same product. What do you think? How many pairs can you find where we are multiplying a one-digit by a one-digit number?

    Source: Reprinted with permission from *The Arithmetic Teacher*, copyright © 1993 by the National Council of Teachers of Mathematics.

47. Without doing anything on paper, answer the following question: When we multiply 83 by some number we get a number that ends in 4.

a. What can you tell about the missing multiplier?
b. Can you determine the second number if the product is between 5000 and 6000?
c. Can you determine the second number if the product is less than 2000?

48. This problem is similar to puzzles archaeologists have to solve. We have recovered these fragments of this multiplication problem. What was the problem and what is the answer?

$$\begin{array}{r} 486 \\ \times\ \ 60 \\ \hline 002 \\ 000 \\ \hline 0000 \end{array} \qquad \begin{array}{r} 486 \\ \times\ 62 \\ \hline 972 \end{array} \qquad \begin{array}{r} 486 \\ \times\ 67 \\ \hline \end{array}$$

49. For certain kinds of two-digit numbers, there is a shortcut to find their product.

$$25 \cdot 25 = 625 \qquad 22 \cdot 28 = 616$$
$$37 \cdot 33 = 1221 \qquad 76 \cdot 74 = 5624$$

a. Determine the kinds of numbers for which this shortcut works.
b. Describe the procedure in English, as though to someone on the phone.
c. Explain why it works.

50. On first sight, 34×23 and $(3x + 4)(2x + 3)$ are very different problems. However, change the representation and the similarities are striking.

a. Describe the similarities between the two computations. *Hint:* Represent 34×23 as $(30 + 4)(20 + 3)$.

$$\begin{array}{r} 34 \\ \times 23 \\ \hline 102 \\ 68 \end{array} \qquad \begin{array}{r} 3x + 4 \\ \times 2x + 3 \\ \hline 9x + 12 \\ 6x^2 + 8x \end{array}$$

b. Now explain why the famous FOIL algorithm (First, Outer, Inner, Last) works.

51. Determine the product of the following Alphabitian problem: $BC \times AB$

52. Make a base 5 multiplication table. Use it to determine the following products:

a. $\begin{array}{r}43_5 \\ \times\ 12_5\end{array}$ b. $\begin{array}{r}204_5 \\ \times\ 32_5\end{array}$

53. Determine what numbers go in the blank spaces and finish the problem.

$$\begin{array}{r} 2\ 0\ 4\ \square_5 \\ \square\ \ 4_5 \\ \hline \square\ \square\ 3\ 3\ 2 \\ \square\ 1\ 4\ 1 \end{array}$$

SECTION 3.4 UNDERSTANDING DIVISION

WHAT DO YOU THINK?

- Do you think our multiplication and division algorithms would work with the Roman or Egyptian numeration system?
- When we do long division, we "bring down" the next number. What does *bring down* mean?

Division is probably the least understood of the four basic operations. In workshops with elementary teachers, I have found that although most of them can perform division computations, many of them find word problems requiring division to be baffling. Many of the classic difficulties people have with division—understanding the long-division algorithm, knowing how to deal with remainders, and knowing when to divide and when to use other operations—go back to a lack of understanding of what division means. To develop that understanding, I ask you to examine the following division problems and then answer the questions below in your own words.

Contexts for Division

Assume that you are a young child who has not yet learned division. How might you solve these problems, using manipulatives or diagrams, or using common sense and other mathematical knowledge? Note your thoughts before reading on....

3.15 *Four division problems*

1. Applewood Elementary School has just bought 24 Apple computers for its 4 fifth-grade classes. How many computers will each classroom get?
2. Jeannie is making popsicles. If she has 24 ounces of juice and each popsicle takes 4 ounces, how many popsicles can she make?
3. Melissa, Vanessa, Corissa, and Valerie bought a bolt of cloth that is 24 yards long. If they share it equally, how many yards of cloth does each person get?
4. Carlos has 24 apples with which to make apple pies. If it takes 4 apples per pie, how many pies can he bake?

Now go back to these four problems and answer the following questions:

- In what ways are the problems different? In what ways are they similar?
- What does division mean? For example, what words come to mind when you think of division?

As with the other operations, some of these problems involve discrete objects (Problems 1 and 4), whereas other problems involve measured (continuous) amounts (Problems 2 and 3).

Another difference emerges when we solve these problems as young children do. Left to themselves, most children will solve Problems 1 and 3 in a way that is very different from the way they will solve Problems 2 and 4. Do you see why?

Models for Division

Let us solve Problems 1 and 4 simultaneously to understand two different models of division.

To solve the first problem, many children will give one computer to each class, dealing them out one by one: one for our class, one for your class, and so on. Some students use multiplication facts or guess-and-check and start out bigger: two for our class, two for your class, and so on. They then keep dealing until all the computers have been distributed.

Visually, we have

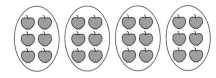

Each classroom gets 6 Apple computers.

This model is called the **partitioning model** because we solve the problem by first setting up the appropriate number of groups (partitions), which we then fill.

To solve the fourth problem, most children do something very different. They make groups of 4 (each representing one pie) until there are no more apples left.

Visually, we have

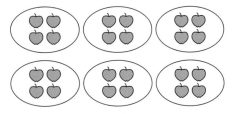

We have enough apples to make 6 pies.

This model is called the **repeated subtraction model** because we repeatedly subtract the given amount until we can do so no more. Some books call this the measurement model of division.

Now let us look at how the models are related. First we need some terminology.

Division terminology In both cases, the problem would be written as

$$24 \div 4 = 6 \text{ or as } 4\overline{)24}^{\,6}$$

The division of the number b by the nonzero number n is formally defined as

$$b \div n = a \text{ iff } a \cdot n = b$$

The number b is called the **dividend**, n is called the **divisor**, and a is called the **quotient**.

Comparing the two division models With this terminology, we can compare the two models in Table 3.3. What patterns or connections between the two models do you notice? Think and then read on. . . .

TABLE 3.3

	Dividend		Divisor	Quotient
Partitioning	24 computers	divided by	4 classes gives	6 computers per class
Repeated subtraction	24 apples	divided by	4 apples per pie gives	6 pies

We can also represent both division contexts with the general part–whole box model we have developed for the other three operations. At the most general level, we have

	Dividend	Divisor	Quotient
Partitioning	Whole	Number of parts (groups)	Number in each part (group)
Repeated subtraction	Whole	Number in each part (group)	Number of parts (groups)

Stop for a moment to reflect on these models. Does this representation connect with the problems we have done here and the problems you did in your explorations? How are these two models of division alike, and how are they different?

We can represent the models visually:

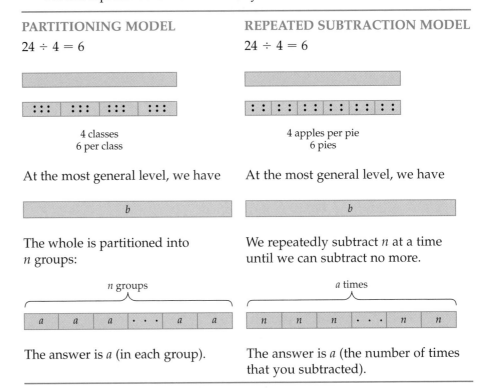

The Missing Factor Model of Division

Going back to the formal definition of division, $b \div n = a$ iff $a \cdot n = b$, we can understand another model for division.

Just as some children interpret and solve some subtraction problems by finding the missing addend, some children interpret and solve some division problems by finding the missing factor. For example, consider the two problems we have just analyzed. Some children will solve either or both of these problems by asking themselves: 4 times what is equal to 24?

This model for division, called the **missing-factor model** for division, gives us another tool for working with division, one that we will use often in this and future chapters.

Division and Number Lines

Although the number line is not commonly used to represent division problems, it can be used. What problem is represented in the number line in Figure 3.24? Does the number line fit both models or just one model? Think and then read on. . . .

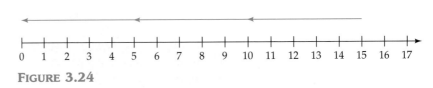

FIGURE 3.24

The problem is $15 \div 5 = 3$. The number line fits only the repeated subtraction model. In other words, we cannot use the number line to model a partitioning problem. Why is this?

Properties of Division

Of the properties we have investigated for the other operations, identity, commutativity, associativity, and closure, which do you think hold for division? Think and then read on. . . .

If you have been making connections among the operations, you probably have realized that addition and multiplication are alike in that each operation has an identity and that the commutative and associative properties hold for those operations. You saw that those properties did not hold for subtraction, although subtraction does have a right-identity. Therefore, you might be inclined to say the same for division. Division does have a right-identity; that is, $a \div 1 = a$ for all numbers. The commutative and associative properties do not hold for division. Similarly, the set of whole numbers is not closed under division. Why is this?

The definition of division enables us to see the connection between division and multiplication, and the models of division enable us to understand how division with smaller numbers works. As you may have already realized, just as we noted earlier that multiplication is more complicated conceptually than addition and subtraction, so is division. There are two complexities that pose problems for most children: division by zero and division with remainders. Let us examine them now.

Division by Zero

Our definition for division implies that "you can't divide by zero." Can you explain why? Think about this and then read on. . . .

There are several ways in which we can investigate this problem. We can use inductive reasoning: Make a table and look for patterns. Do you understand Table 3.4? What pattern do you see that can help you to explain why "you can't divide by zero"?

■ Mathematics ■

"*USS Yorktown* dead in water after dividing by zero"
In September 1997 the computer system aboard the *USS Yorktown* shut down the ship. The problem was caused by someone who erroneously entered a zero in the wrong place, which caused the computer to divide by zero, which it cannot do. The computer system crash resulted in shutting down the ship's engines and the ship was dead in the water for three hours.

■ History ■

"Black holes result from God dividing the universe by zero."
Anonymous

TABLE 3.4

Computation	Dividend	Divisor	Quotient
$5 \div 1$	5	1	5
$5 \div 0.1$	5	0.1	50
$5 \div 0.01$	5	0.01	500
$5 \div 0.001$	5	0.001	5000
$5 \div 0.00000001$	5	0.00000001	?

As the divisor becomes smaller and smaller, the quotient becomes larger and larger.

We can also approach the problem of dividing by zero by seeing whether either context fits the model. What if we had 24 computers and we wanted to distribute them to zero classrooms? How many computers would each classroom get? That doesn't make any sense. Let's try repeated subtraction. What if we had 24 apples and each pie took zero apples? How many pies could we bake? This doesn't make sense either. From a practical (contextual) level, division by zero just doesn't make sense. Thus we say that division by zero is undefined.

A third approach to this problem is to assume that it is possible; this is called **indirect proof**. If division by zero were possible, then a nonzero x divided by zero would be equal to some number k (that is, $x \div 0 = k$). However, if this is true, then according to our definition of division, $k \cdot 0 = x$, but this is not possible because we know that $k \cdot 0 = 0$.

Division with Remainders

3.18

There is a wonderful children's book called *The Doorbell Rang*. In this story, the mother has a plate of cookies that are to be shared equally among the children. Before the children eat the cookies, however, the doorbell rings and another person comes. Now they have to figure out how many cookies each person gets, and then another person comes, and another, and. . . .

Let's say there are 12 cookies and 3 children—each child gets 4 cookies. What if another child comes?—now each child gets 3 cookies. What if another child comes?—now each child gets 2 cookies, and there are two left over. We can now say that $12 \div 5$ is 2 with 2 cookies left over. As we saw above, the set of whole numbers is not closed under division; in everyday English, this means that when we divide one whole number by another, the answer is not always a whole number. However, the *division algorithm* assures us that a solution involving whole numbers will exist for any whole-number division problem, other than dividing by zero.

The **division algorithm** states that for any whole numbers a and b, where $b \neq 0$, there are absolutely going to be two numbers, which we will designate as q and r, that enable us to know the answer to $a \div b$. That is,

$a = bq + r$

There is one other stipulation, that r be a whole number less than b. Do you see why?

Without this stipulation, we lose the uniqueness of r and q. For example, if we divide 33 by 4, we could say $33 \div 4 = 8$ R 1 or $33 \div 4 = 7$ R 5.

Because communication is one of the NCTM process standards, I want to express the division algorithm in its most concise form:

$\forall\, a, b \in W, (b \neq 0) \exists\, q, r \in W \ni a = b \cdot q + r, 0 \leq r < b$

A direct translation in English is as follows: For any whole numbers a and b, except that b cannot be equal to zero, there exist two whole numbers q and r such that a is equal to b times q plus r, where r is less than b and could possibly be equal to zero.

Translated into everyday English, this is equivalent to the following: For any two whole numbers a and b (as long as b is not equal to zero), we can find two whole numbers q and r that make this equation true ($a = bq + r$), and r must be greater than or equal to zero but less than b.

```
      6
   _____
 8 ) 5742
     48
     ___
      9
```

Again, some readers will say, "So what?" There are two responses to this question. The first is that many aspects of our numeration system and operations on numbers that seem so obvious to you as adults are neither obvious nor simple to young children. Second, understanding some of the formalities of mathematics will help you to recognize and address problems that children often have. For example, when faced with division problems with remainders, some children actually wonder whether there *is* a solution to the problem. And some children, whose multiplication facts are poor, make mistakes when dividing that cause huge problems. For example, consider the example at the left. In this case, the student does not own the understanding that r must be less than b and thus is doomed to failure—and to more frustration. Exploration 3.18 also addresses the idea of remainders when dividing.

INVESTIGATION 3.22

Mental Division

Mental division is new territory for most of my students. Even the division idea of **canceling zeros** is one that many students feel apprehensive about before this course. Therefore, examine each of the problems below carefully. How might you determine the exact answer, applying what you know about division? Briefly note the strategies you used, and try to give names to them.

1. $20\overline{)6000}$ 2. $400\overline{)20{,}000}$ 3. $8\overline{)152}$ 4. $5\overline{)345}$

DISCUSSION

In Problem 1, we will examine two different ways to use **canceling**.
One method of canceling transforms the problem from 6000/20 to (60 · 100)/20, which can then be simplified to 3 · 100 because 20 divides both 60 and 100 without remainder.

$$\frac{6000}{20} = \frac{60 \cdot 100}{20} = \frac{\overset{3}{\cancel{60}} \cdot 100}{\underset{1}{\cancel{20}}} = 300$$

Another way to cancel in Problem 1 is to transform it from 6000/20 to 600/2, which is equal to 300.

$$\frac{6000}{20} = \frac{600 \cdot 10}{2 \cdot 10}$$

In this case, we are actually applying the idea of **equivalent fractions**, which we will examine in Chapter 5. That is, we are dividing both numerator and denominator by the same amount, 10.

Can you use either of these strategies to solve Problem 2? If so, which?
Problem 3 lends itself to compatible numbers; for example, $160 \div 8 = 20$. We can use the distributive property to get an exact answer, since

$$\frac{152}{8} = \frac{160 - 8}{8} = 20 - 1 = 19$$

We can use compatible numbers in Problem 4: $350 \div 5 = 70$.
We can also use the distributive property in Problem 4:

$$\frac{345}{5} = \frac{350 - 5}{5} = 70 - 1 = 69$$

In Problem 4, we can also use the idea of equivalent fractions and multiply both numerator and denominator by 2. The resulting fraction, 690/10, easily simplifies to 69.

Division Algorithms

At the beginning of each course, I invite my students to write math autobiographies in which I ask them to tell me math-related memories, both positive and negative, from early childhood through high school. A common sentence in these autobiographies goes something like "I liked mathematics until long division." To this day, many adults think that understanding why the standard division algorithm works is like understanding the structure of quantum physics—that is, something beyond the ability of "ordinary people." They know what to do, but not why. For example, faced with 252 divided by 4, most people say something like "4 gazinta 25 6 times," but "gazinta" is a mystery. The purpose of the next Investigation is to convince you that "gazinta" is not as esoteric as quantum physics.

INVESTIGATION 3.23

3.20

Understanding Division Algorithms

Solve the two division problems as though you didn't know any algorithms; you may use manipulatives or draw diagrams or use reasoning. Then read on. . . .

A. Warren has 252 guests coming to his wedding. Each table holds 4 guests. How many tables will he need?

B. Mickey has 252 marbles that he wants to distribute equally into 4 piles. How many marbles are in each pile?

DISCUSSION

A. If you represent these problems with base 10 blocks, you may have found that they weren't very useful for Problem A because it essentially asks how many groups of 4 there are in 252. Some students solve this problem by using the missing-factor model: 4 times what makes 252?

Guess	Result	Thinking
50 tables	200 people	Too low, try a bigger number, let's say 60 tables.
60 tables	240 people	We need 3 more tables for the 12 remaining people.
63 tables	252 people	

B. The base 10 blocks are more useful for Problem B. In this case, Mickey's problem is to divide (distribute) 252 into four groups equally. Using base 10 manipulatives, he can figure out how to do this most efficiently. Because he cannot give 100 to each person, he has to trade the 2 hundreds (flats) for 20 longs. He still has 252, but physically he has 25 longs and 2 units.

Now he can place 6 longs in each group. The corresponding symbolic step is shown at the left.

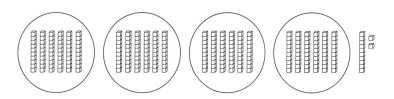

He is left with 1 long and 2 units, representing 12; thus he trades the long for 10 units. This lets him place 3 ones in each group. The corresponding symbolic step in shown at the left.

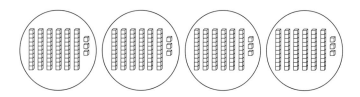

Looking back Write your response to the following questions before reading on. . . .

1. What is happening when we say "gazinta"?
2. What is happening when we multiply?
3. What is happening when we subtract after we multiply?
4. What is happening when we bring down the next number?

Let us look at the answers to these questions.

1. When we say "gazinta," we are simply determining the first step in distributing the amount in the dividend into equal groups. For example, in the problem 36)864, when we say that 36 "doesn't go into 8," mathematically, this means that when we divide 864 into 36 equal groups, we do not have enough to have 1 hundred in each group, and therefore we have to look at the next place. When we say that 36 goes into 86 two times, mathematically this means that when we distribute 86 tens into 36 groups of equal size, we have enough for 2 tens in each group, and thus we place the 2 over the 6, that is, in the tens place.

$$\begin{array}{r} 2 \\ 36\overline{)864} \end{array}$$

2. When we multiply, we determine how much has been distributed, or, in other words, how much of our original amount has been used up. In the case of 36)864, we find that we have used up 720.

$$\begin{array}{r} 2 \\ 36\overline{)864} \\ 72 \end{array}$$

3. We subtract after we multiply so that we can find out how much of the dividend is left after the first distribution. In this case, we have 14 tens left over. That is, these tens cannot be distributed equally into 36 equal-size groups, and so they have to be traded in for ones.

$$\begin{array}{r} 2 \\ 36\overline{)864} \\ \underline{72} \\ 14 \end{array}$$

■ *History* ■

The common division algorithm evolved from a method called a *danda*, which is derived from the same root as the word giving. The "giving" part is referred to by most people as the "bringing down" part. That is, when you have finished subtracting, you "bring down" the next digit in the dividend (the number you are dividing). From another perspective, you are giving that number to the remainder. The earliest version of this algorithm dates to the fourteenth century.

4. When we bring down the next number, we are determining how much we have to distribute. In this case, we have the 14 tens and the 4 ones that we bring down—that is, a total of 144—to distribute equally into 36 groups.

$$\begin{array}{r} 24 \\ 36\overline{)864} \\ \underline{72} \\ 144 \\ \underline{144} \end{array}$$

It is important to emphasize that although our numeration system was invented over a thousand years ago, the division algorithm that you learned in elementary school is only one of many algorithms for division that were invented over this time.

Even when a teacher helps children to construct the standard division algorithm, some children have difficulty. Part of this difficulty often stems from not knowing their times tables very well, but part of it stems simply from not being able to internalize all the seemingly different steps. As one child once said, "It's weird. If we're doing division, why are we multiplying and subtracting?"

INVESTIGATION 3.24

The Scaffolding Algorithm

Another algorithm, which has enjoyed favor in different parts of the world over the centuries, is called the **scaffolding algorithm**.

$$\begin{array}{r} 699 \text{ R } 162 \\ 6\overline{)4356} \\ \underline{36} \\ 75 \\ \underline{54} \\ 216 \\ \underline{54} \\ 162 \end{array} \qquad \begin{array}{r} 6 \\ 20 \\ 200 \\ 500 \\ 6\overline{)4356} \\ \underline{3000} \\ 1356 \\ \underline{1200} \\ 156 \\ \underline{120} \\ 36 \\ \underline{36} \end{array}$$

DISCUSSION

Let us examine how it works. Consider the problem $6\overline{)4356}$. A child with poor multiplication facts will often have trouble with the first step. If the child chooses wrong, such as $6 \times 6 = 36$, then the whole problem is doomed. See the work shown above. The alternative approach is to look at the problem from a different vantage point. Instead of asking how many 6s are in 43 or $6 \times$ what is close to 43, we ask the child to look at the problem in context. For example, if a company produces 4356 bottles of soda each day, how many six-packs would this be? Then ask the child to estimate. Let's say the estimate is 500. The child determines how much 500 six-packs would be: $500 \times 6 = 3000$. This guess is recorded above the problem (see the work shown above), and then what remains after this guess is determined: Subtracting from 4356, we have 1356 left. Now the question is: How many six-packs would this be? Using the knowledge from the previous guess, 200 is a good guess, and 200 × 6 = 1200. Subtracting again, this gives us 156 left. Let's say the child guesses 20; this uses up 120 and leaves 36, at which point the child might guess 6.

Once again, it is not that any one algorithm is "best," but rather that different people and different countries have preferred different algorithms over time. In many cases, good arguments can be made for several algorithms.

INVESTIGATION 3.25

Estimates with Division

Now let us apply these mental division skills to problems in which we are likely to use division to estimate. In each of the problems below, obtain an estimate before reading on. . . .

Do you find that you like certain strategies better than others? Are there any that you have a hard time understanding? Can you describe a scenario in which a rough estimate would be adequate? A scenario in which a more refined estimate would be more appropriate?

A. Mr. & Mrs. Smith pay $9100 a year for their son's college. Translate this into a monthly payment that they can put into their budget.

DISCUSSION

Using the missing-factor model, ask yourself, "12 times what is closest to 91?" 12 × 7 = 84, 12 × 8 = 96. Because 91 is just about in the middle (of 84 and 96), it is reasonable to conclude that $12 \times 7\frac{1}{2}$ will be close to 91, and so our estimate is $7\frac{1}{2}$ hundred a month, or, more conventionally, $750 per month.

B. Pierre just bought a new van, and he wants to see what kind of mileage he gets. He filled up with gas after going 489 miles, and the car took 19 gallons of gas. Estimate the mileage.

DISCUSSION

One strategy is to round both the divisor and the dividend up.

$\frac{489}{19}$ will be close to $\frac{500}{20}$, and this is equivalent to $\frac{50}{2}$, and so he is getting about 25 mpg.

This strategy is interesting because it is not identical to the one used in multiplication, where we get more accurate estimates if we round one number up and the other down. In the case of division, we obtain more accurate estimates if we round both numbers up or both numbers down to manageable numbers. Can you explain why? Work on this for a while and then read on. . . .

Hint: It might be helpful to make up some problems and check this out. For example, try 3750 ÷ 13. Decrease both: to 3600 and 12, which are compatible numbers. Increase both: to 3900 and 13. Then decrease one and increase the other—for example, 3750 to 4000 and 13 to 10. Compare the results. Make up some other problems. Can you use ideas developed thus far to help you?

> **C.** A large business ran 5432 copies in the last 7 business days. How many copies is it averaging per day?

DISCUSSION

A rough estimate could be gotten by making compatible numbers, increasing the 5432, and solving the problem. For example, 5600 ÷ 7 gives an estimate of 800.
A more refined strategy would be to use multiplication:

$$7 \times 700 = 4900$$
$$7 \times 800 = 5600$$

Because 5432 is closer to 5600 than it is to 4900, a rough estimate would be just under 800.

Do any of these strategies reflect the work you did in Exploration 3.20 with the risograph machine?

> **D.** Employees put 25¢ in a can for each cup of coffee they drink. Last week the can contained $29.50. How many cups of coffee were drunk?

DISCUSSION

STRATEGY 1: *Solve the problem in dollars*

That is, 25¢ per cup means 4 cups per dollar.
  You can estimate $30 × 4 cups per dollar = 120 cups.

  Using this strategy, you could actually use mental math and get the exact answer in two ways:

$$29.50 \times 4 = (29 \times 4) + \left(\frac{1}{2} \times 4\right) = 116 + 2 = 118$$

or

$$29.50 \times 4 = (30 \times 4) - \left(\frac{1}{2} \times 4\right) = 120 - 2 = 118$$

STRATEGY 2: *Solve the problem in cents*

That is, $29.50 is about $30, which is equivalent to 3000¢.

$$\frac{3000}{25} = \frac{2500 + 500}{25} = 100 + 20 = 120$$

Looking back Take a few moments to look back over the strategies for mental division and estimation that we have explored. Have you added to your repertoire? Do you understand how the distributive property and expanded notation are applied? Do you see that some strategies work with some numbers better than with others? Can you connect these strategies to different models or contexts of division—for example, missing factor or repeated subtraction?

INVESTIGATION 3.26

3.21

Number Sense with Division

A. What is the value of the divisor that makes this division problem?

$$?\overline{)557} \quad 69\,R\,7$$

DISCUSSION

To solve this problem, we use several aspects of our knowledge. One solution path is to see that 69 × 10 is 690, so the divisor is less than 10 but not by much. One's first guess would either be 9 or 8. When we try 8, we find that 560/8 does equal 69 R 7. Another approach is to look at the remainder and see that the divisor must be either 8 or 9 in order to have a remainder of 7.

B. Take no more than 5 to 10 seconds to determine whether this answer is reasonable. That is, make your determination using number sense, without doing any pencil and paper work.

$$62\overline{)4368} \quad 704$$

DISCUSSION

Most people find it easier to translate this to a multiplication problem: 704 × 62. We quickly see that 700 × 60 = 4200. Looking at this problem as equal groups, we are looking for 62 groups of 704. Since 60 groups of 700 is already 4200, we see that 62 groups of 700 (and thus 702) will be quite a bit higher than 4200. You could do the 'gazinta' here: 62 gazinta 436 seven times and 62 × 7 = 434, which you subtract from 436 to get 2. You can quickly see that the next digit in the divisor is now a 0. Then you need to bring down the next number. This becomes a lot to keep in one's head. Some people can do this, but one of the points of the course is that there is no need to make math hard when it can be made simpler, when one uses number sense and operation sense.

C. Determine the three missing digits:

4☐☐2 ÷ 8☐ = 48

How can you apply your understanding of multiplication and division to solve this problem? What problem-solving strategies seem more helpful? Work on this yourself and then read on. . . .

DISCUSSION

One of the key themes in this book is the idea of multiple representations, which is like looking at an object from different perspectives—from above, from below, at ground level, from the side, and so on. Sometimes, one representation sheds more light on a problem than another. If you were stuck, look at the following representation of the problem. Does this help?

$$\frac{4☐☐2}{8☐} = 48$$

If you are still stuck, try using a multiplication representation:

8☐ · 48 = 4☐☐2

What does this tell you now about the ones digit of the first number? Think and then read on. . . .

It must be a 4 or a 9. Do you see why? Test your intuition—which one do you think it is?

Now you can finish the problem. When you have come up with a solution, don't forget to look back at your work. What did this investigation tell you about operations? How did your problem-solving toolbox—looking for patterns, working backwards, multiple representations, and so on—grow as a result of your doing this problem?

Operation Sense

3.22

Now that we have examined each of the four operations, let us explore some problems in which we can apply our knowledge. Someone who knows which operation to perform in a problem is said to have good *operation sense*, a term that represents an important skill but is hard to define. Operation sense is more than just knowing what operation to use in what situation. If we want mathematics problems to model real-life problems, then, as you have seen in this book, many problems will not be routine, one-step problems. In the context of solving more complex problems, students with good operation sense have the following "knowledge":

- They can see the relationships among the operations—for example, multiplication is the inverse of division.
- They can apply their understanding of the properties of the operations—for example, they can use the distributive property.
- The diagrams that they draw connect the problem to the models of the operations.

- They have a sense of the effects of each operation—for example, the larger the divisor, the smaller the quotient.

One of the goals of Investigations 3.9, 3.15, 3.21, and 3.26 and Explorations 3.4, 3.7, 3.14, 3.21, and 3.22 is for you to come to appreciate the need for operation sense and to develop this ability. That is, the successful completion of these problems requires you to see and use various connections between these four operations.

INVESTIGATION 3.27

Applying Models to a Real-life Situation

A teacher is going to do a project with the two fifth-grade classes in the school. One class has 21 students, and the other has 15 students. She has decided that all of these students will be divided into 12 groups and that each group will need 72 inches of string. She has a roll of string that is 882 inches long. Does she have enough string, or will she need to buy more? Do this problem yourself without using a calculator before reading on. . . .

DISCUSSION

You can use either multiplication or division to solve this problem. Using multiplication, $72 \cdot 12 = 864$ tells you that you will have enough string for each group, with 18 inches of string to spare. If you use division, the problem can be interpreted either as a partitioning problem or as a repeated subtraction problem. Do you see why?

- If you divide 882 inches by 72 inches per group, this says that you are seeing the problem in terms of repeated subtraction. What does your answer of 12 with a remainder of 18 tell you?
- If you divide 882 inches by 12 groups, this says that you are seeing the problem in terms of partitioning. What does your answer of 73 with a remainder of 6 tell you?

In the repeated subtraction model, the 12 says that if you repeatedly cut 72 inches, you can do this 12 times, and you will have 18 inches of string left over. In the partitioning model, the 73 says that if each group is to have an equal amount of string, you have enough string to make 12 pieces each of which is 73 inches long and you will have 6 inches left over. In either case, you have to understand what your quotient means, and you have to think about what the remainder means.

There are two other aspects of this problem that are worthy of emphasis.

First, although the mathematical answer is that the teacher has enough string, the teacher might actually conclude from her computations that she needs more string. Why is this?

What if the students make mistakes? Because $72 \cdot 12 = 864$, the teacher has only 18 inches of string in reserve. Thus, although the mathematical answer is that the teacher has enough string, the real-life answer would depend on the nature of the project. For example, if the students were to cut the pieces into specified lengths and I anticipated that some groups would mismeasure the lengths, I would not feel confident that I had enough string.

Second, compare this question to the following problem: A teacher is going to have her students do a project in which each group of students will need 72 inches of string. If the teacher has 882 inches of string, how many groups can she have in her class?

This problem is a one-step, routine problem, the kind the NCTM says should be deemphasized. The student simply has to decide whether to add, subtract, multiply, or divide; if the student just guesses, there is a 25 percent chance of being right. This is what many students do. Larry Sowder[5] examined children's strategies for solving word problems and reported these kinds of strategies: "Look at the numbers; they will tell you which operation to use. Try all the operations and choose the most reasonable answer. Look for key words or phrases to tell which operation to use."

Furthermore, the actual problem has been sanitized so that the student simply has to use the numbers given. In the first version of the problem, the number of students in each class (21 and 15) is *irrelevant to the question being asked*. These numbers were given not to try to trick you, but because in real-life settings, we generally have lots of information. The first part of the problem is to decide whether we have enough information to solve the problem and to decide what information is relevant to the question we are asking.

INVESTIGATION 3.28

Operation Sense

A. Select the correct operations to make the sentence true. You can use an operation more than once.

6 O 5 O 3 = 10

DISCUSSION

A few rounds of guess-test-revise generally convince most students that addition and subtraction are not possible. If we explore multiplication and division, trying multiplication 6 × 5 reveals that ÷ 3 will produce an answer of 10.

B. Make a number sentence that will produce the desired answer once the unknown amounts are known.

Germaine bought **A** CDs for **B** dollars each. He sold the whole lot for **C** dollars. How much profit did he make?

DISCUSSION

If we think about it, we can see the repeated addition in a group of CDs with the same price. In selling, he is taking the CDs away. The tricky part is realizing that, assuming he is making a profit, the C comes first. That is, the number sentence is C − (A × B).

C. Is this answer reasonable or not?

There were 35 students in a class and each student paid $24 for a field trip. The total cost was about $700.

---

[5]Larry Sowder, "Children's Solutions of Story Problems," *Journal of Mathematical Behavior*, 7 (1988), pp. 227–228.

DISCUSSION

If we just use leading digit, we could believe this answer because we have 30 × 20. However, if we use our mental addition to see that 35 × 20 is 700, we realize that this answer is not reasonable.

Order of Operations

Complex computations that use more than one operation pose a potential problem. To see what I mean, do this problem with pencil and paper and then on your calculator: 3 · 4 − 8 ÷ 2. What did you come up with? Stop, think, and then read on. . . .

Unless you have a very inexpensive or antique calculator, you will get 8. However, if you do each operation in the order in which it appears in the problem, you will get 2. Why did the calculator give a different answer?

The fact that we can get two different answers from the same problem has led to rules called the order of operations. These rules tell you the order in which to perform operations so that each expression can have only one value. Most calculators are programmed to obey the order of operations.

What did the calculator actually do to get 8? Write the process in words and see whether a friend can use your explanation to see how to get 8. Then read on. . . .

In the absence of parentheses, multiply 3 times 4 to get 12, then divide 8 by 2 to get 4. Now subtract 4 from 12 to get 8. You may remember order of operations from school as "Please Excuse My Dear Aunt Sally," which is a mnemonic to remind you of the order in which a computation is done: Start inside the *p*arentheses, then do *m*ultiplication and *d*ivision as they occur from left to right, and finally do *a*ddition and *s*ubtraction as they occur from left to right. (The *e* from "Excuse" refers to exponents, which we will discuss in Chapter 5.)

If we insert parentheses so that more people would understand this problem, it would read as: (3 · 4) − (8 ÷ 2). How many different answers can you get depending on different placements of parentheses?

Reviewing Multiplication and Division

In Sections 3.3 and 3.4, we have examined multiplication, division, and connections among all four operations.

We have found that multiplication can be represented as repeated addition, as the area enclosed by the multiplier and the multiplicand, or as the number of combinations (Cartesian product). Similarly, division problems can be seen as partitioning, performing repeated subtraction, or finding a missing factor.

Visual models help us to see connections between multiplication and division. In one sense, multiplication consists of adding two or more equal-size parts to make a whole. Similarly, division can be seen as decomposing a whole into parts that are all the same size.

You have learned that there are certain properties that hold for multiplication of whole numbers but not for division: commutativity, associativity, and closure. Just as we have the identity property of addition, we have the identity property of multiplication. Thus the numbers 0 and 1 have special importance, and they also have a rich history, as you will find in Chapter 4. We also learned of a property that connects operations: the distributive property. This property is a key to understanding how the multiplication algorithm works.

We have also examined the various connections among the four operations, including order of operations, and applied our understanding of those connections so that our operation sense becomes more powerful.

A key goal of this section is for you to better understand how these four operations are related to one another. What similarities do you see among the operations? What connections do you see among them? Take a few minutes to draw a diagram that shows the relationships among the four operations. Then read on. . . .

Addition and subtraction are inverse operations. Multiplication and division are inverse operations.

One model of multiplication is repeated addition, and one model of division is repeated subtraction. We have a missing-addend model for subtraction, and we have a missing-factor model for division.

The set of whole numbers is closed under the operations of addition and multiplication and is not closed under the operations of subtraction and division. The commutative and associative properties hold for addition and multiplication, but not for subtraction and division. We have identity properties for addition and multiplication; because subtraction and division are not commutative, we have only right-identities for these operations. We can speak of the distributive property of multiplication over addition and of the distributive property of multiplication over subtraction.

With addition and subtraction, the units in each part are always identical; for example, 3 *sheep* plus 5 *sheep* equals 8 *sheep*. However, with multiplication and division, the units are different; for example, 3 *miles per hour* times 4 *hours* yields a product of 12 *miles*.

In many addition and multiplication situations, we are joining or combining parts to make a whole. In many subtraction and division situations, we are taking a whole apart.

These statements represent a verbal description of the operations' similarities and connections. One of many visual representations of these similarities and connections is shown in Figure 3.25.

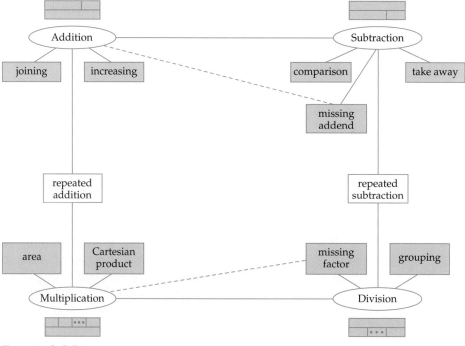

FIGURE 3.25

We have seen that the standard algorithms are related to some of the operations' contexts and models. These algorithms rely on our base 10 system. Again, it is important to emphasize that these standard algorithms are not the only ones, nor are they necessarily the best ones.

Units and measurement   In Chapter 2 we launched the idea of *unit* as a "big idea" of mathematics. In this chapter, you have seen how the notion of unit is critical in many areas. When we add and subtract, the units remain the same. However, when we multiply or divide two numbers, the units of the two numbers and the units of the answer are not the same. Dimensional analysis is a strategy that keeps track of units as we move through a complex series of computations where the units are changing.

In order to use computation procedures confidently, the child must be able to move flexibly among different units. For example, in addition, when "carrying," 12 ones are equivalent to 1 ten and 2 ones.

In subtraction, when borrowing, 1 hundred is equal to 10 tens. When multiply 2-digit numbers, the product is often a 3-digit number.

When adding 2-digit numbers, the sum is usually a 2-digit number.

35 can be seen as 3 tens and 5 ones or 35 ones.

The year 1900 can be seen as 1 thousand and 9 hundreds or as 19 hundreds.

Sometimes the units are discrete: 3 balloons.

Sometimes the units are rates: 3 miles per hour.

When estimating, we often use units, some of which are invented. For example, one way to get an estimate of the height of a tree is to have a friend stand at the base. If the friend is about 6 feet tall and, using your fingers as unit, you determine that the tree is about 7 of your friends tall, your estimate is that the tree is about 40 feet tall. In this case, the distance between your fingers is a simple unit and 6 feet is the composite unit.

EXERCISES 3.4

1. Four children want to share 168 jelly beans. How many does each get? Which model of division is operative here? Model the problem with a diagram.

2. a. Make up a division story problem for the partitioning model. Make a diagram and explain why this problem is an example of the partitioning model.

   b. Make up a division story problem for the repeated subtraction model. Make a diagram and explain why this problem is an example of the repeated subtraction model.

3. For each number line problem below, identify the computation it models and briefly justify your answer.

   a. 0 1 2 3 4 5 6 7 8 9

   b. 0 1 2 3 4 5 6 7 8 9

4. Represent the following problems on a number line. Explain each problem as though you were talking to someone who was not taking this class.

   a. $6 \cdot 3$ b. $12 \div 3$

5. Solve this problem with manipulatives and justify your process: $4\overline{)652}$

6. Solve these problems by a means other than the standard or the scaffolding algorithm.

   a. $641 \div 25$ b. $1842 \div 68$

7. In each of the following, use the information from the computed problem to determine the answer mentally.

   $$\begin{array}{r} 742 \\ 36\overline{)26{,}712} \\ 25{,}200 \\ \hline 1\,510 \\ 1\,440 \\ \hline 72 \\ 72 \end{array}$$

   a. $36 \times 400 = ?$

   b. $742 \times 36 = ?$

   c. $25{,}200 \div 360 = ?$

8. *Classroom Connection* Identify and explain the errors below, all of which have occurred in real classrooms.

   a. $$\begin{array}{r} 54 \\ 7\overline{)3528} \\ 35 \\ \hline 28 \\ 28 \end{array}$$

   b. $$\begin{array}{r} 540 \\ 7\overline{)3528} \\ 35 \\ \hline 28 \\ 28 \end{array}$$

   c. $$\begin{array}{r} 79\ R\ 26 \\ 8\overline{)658} \\ 56 \\ \hline 98 \\ 72 \end{array}$$

9. Determine the quotients mentally.

   a. $\dfrac{6000}{20}$ b. $\dfrac{4800}{60}$ c. $\dfrac{10{,}000}{200}$

   d. $\dfrac{9000}{30}$ e. $\dfrac{3600}{90}$ f. $\dfrac{20{,}000}{400}$

10. Estimate as closely as you can.

    a. $8\overline{)5432}$ b. $7\overline{)29{,}123}$ c. $20\overline{)342{,}354}$

11. By estimating, determine which of the following equations is wrong.

    a. $75{,}296 \div 16 = 4706$
    b. $34{,}272 \div 48 = 714$
    c. $38{,}844 \div 498 = 780$

12. Which of the following would be a better estimate of 225/9: 250/10 or 230/10? Explain your answer.

13. A computer company has 3765 copies of a hot-selling program and wants to ship the same number to each of its 18 distributors. Approximately how many programs does each distributor get? Francie estimated by rounding 3765 to 4000 and 18 to 20 and then divided 4000 by 20 in her head; her estimate was 200. However, Nadine says that this way of estimating isn't good because, if you are going to round, you should round one number up and one number down. Nadine rounded 18 up to 20 and 3765 down to 3600, and so her estimate was $3600 \div 20 = 180$. With whom do you agree? Why?

In Exercises 14–24, do each of the following. First, determine a rough estimate in 5 to 10 seconds. Then determine your best estimate. ("Best estimate" means no pencil and paper and less than 30 seconds or so. Otherwise, we defeat the purpose of estimating!) Then determine the exact answer.

14. Several years ago, the flat roof over my kitchen was leaking. My friend at the hardware store told me of a new product that would fix my roof. Each can of roof sealant would cover 100 square feet and cost $26. My roof was 23 feet 8 inches long and 15 feet 6 inches wide. How many cans of roof sealant did I need to buy?

15. Mindy rode her bike 3 times last week. At the beginning of the week, the odometer read 302.4, and at the end of the week it read 315.2. At this rate, how many miles will she ride this year?

16. Mr. and Mrs. Smith will pay $8400 this year for their daughter's college tuition. Approximately how much is this per month?

17. Jackie Adams bought a piece of property for $28,543 and sold it for $73,463. Approximately how much profit did she make?

18. Sandy wants to get an idea of the cost of lumber for a project. She has determined that she needs a total of about 450 feet of boards, and the cost is 23¢ per foot. Approximately how much will they cost?

19. Two college students are riding across the United States on bicycles. They are averaging 112 miles per day.

    a. If it is 2987 miles from San Francisco to New York, will they complete the trip in 1 month?
    b. Approximately how far would they travel in 30 days if they averaged 167 miles per day?

20. We bought a new van in June 1990. In December 1991, we had logged 18,345 miles. Our plan was to keep the van until 100,000 miles. At the rate at which we were putting miles on the van, approximately when would the van hit 100,000 miles?

21. The Alvarez family has decided that camping will be more fun if they buy the following items. Approximately how much will the family spend?

Lantern	$35.95
New stove	$59.99
Four new air mattresses at	$14.95 each
A screen house to put over the picnic table	$57.50
A new set of camping pots and pans	$24.95

22. A soft drink manufacturer produces 3240 cans in an 8-hour day. Cans are packaged 24 to a case. How many cases are produced each week? each month?

23. A machine makes 400 candies a minute. At this rate, how long will it take this machine to make 2000 boxes, each of which contains 24 candies?

24. Julie is working out on a computerized ski machine. She starts at 12:05 and will exercise for 30 minutes. Her goal is to burn 400 calories. At 12:17, the machine says that she has burned 148 calories. At this rate, will she reach her goal?

25. Joe had planned to sell 10 pencils at 20¢ each. However, his little brother broke two of the pencils. If he wants to make the same amount of money, how much should he charge for the pencils now?

26. Orange soda comes in six-packs that are packed four to a carton. To make sure that each of the 247 students in the sixth grade will have a soda for the class picnic, how many cartons should we order? How could you figure this out mentally?

27. Janice has decided to buy a computer for her son. So far, she has $450 in the bank. The computer costs $1234. If she can afford $15 per week, when will she have the money?

28. Melanie recorded a tape of her guitar playing. It cost her $1200 to record the master tape, and it costs $2.50 to make each copy. If she sells the tapes for $10 each, how many tapes must she sell to break even?

29. a. How many buses are needed if each bus can take 25 passengers and there are 334 students?
    b. How many buses are needed if we also place two adults on each bus?

30. Victor read 8 books in 34 weeks. He wanted to figure out about how long he spent reading each book, so he divided. Victor got 4 remainder 2 as an answer and is confused about what the 2 represents. How would you help Victor understand the problem?

31. Wei is having a party. She is expecting 120 people and wants enough juice so that each person can have two glasses. If the juice comes in 32-ounce containers, how many cans of juice should she buy? If each can costs 89 cents, how much will the juice for her party cost?

32. Using a calculator, Ralph multiplied by 10 when he should have divided by 10. The display read 300. What should the correct answer be?

33. This problem is over two thousand years old! A dog is chasing a rabbit that has a 150-foot head start. If the dog runs 9 feet for every 7 feet jumped by the rabbit, in how many leaps will the dog catch up with the rabbit?

34. Mathematics abounds in children's literature. Consider this age-old children's rhyme:
    As I was going to St. Ives
    I met a man with seven wives.
    Every wife had seven sacks;
    Every sack had seven cats;
    Every cat had seven kits.
    Kits, cats, sacks, and wives:
    How many were going to St. Ives?

35. A factory can make 250 gizmos every hour. If the factory operates 24 hours per day, how many gizmos can they make in one year?

36. Mars is 34 million miles away. Presently, spaceships go about 16,000 miles per hour. How long would it take a spaceship to get to Mars?

37. In 2004 the U.S. federal debt was $8,200,000,000,000. Let's say that the decision was made to reduce the debt by $10 million per day. How long would it take to pay off the debt at this rate?

38. When someone turns 21, how many days old is that person?

39. Given the increasingly fast advances in technology, it is not far-fetched to think that someday, every person in the United States could have an Internet account. Imagine that 275 million Americans each had an Internet account and that an Internet directory was available. If we printed that directory, how many pages would it contain?

40. How many people are listed in the phone book in the town or city in which your college or university is located? First, describe and justify your plan for answering this question.

41. In 1993, a newspaper reported that nearly 3,000,000 crimes occur each year in U.S. schools. What do you think was meant by crimes? Do you believe this figure? Do you think it is too small? too big? Why?

42. A newspaper story said that on any given day, 7 percent of all Americans eat at McDonald's. Is this true or another case of journalistic exaggeration? Write your report.
    One fact: There are approximately 9000 McDonald's restaurants.

43. Find an article from a newspaper or magazine that demands from the reader knowledge of numeration, basic operations, or estimation. Describe the mathematical knowledge you used in analyzing the article.

44. Determine which number goes where: 7   90   20
    An adult dolphin is about ___ feet long and has ___ teeth. Dolphins live about ___ years.

45. A student of a teacher with whom I was working made up the following problem: Determine which number goes where: 30   48   1961   20   20,000
    The Berlin wall was set up in ____. The wall was approximately ___ miles long or ____ kilometers long. More than _____ people fled before the wall was erected. The wall was about ____ feet tall.

46. Place the digits 2, 4, 7, and 9 in the boxes below to obtain:
    a. The greatest quotient   b. The least quotient
    □)□□□

47. If you divide a number that is larger than 75 by a number that is less than 3, what can you tell about the relative value of the quotient? Explain your choice.
    a. It must be greater than 25.
    b. It must be less than 25.
    c. It will be between 3 and 75.
    d. There is not enough information to conclude any of the above.

48. When 96 is divided by x, a quotient of $y \neq 1$ and a remainder of 5 result. Find x and y.

49. Find the five-digit number that has the following pattern: If you put a 1 after it, the number is three times as large as it would be if you had put a 1 before it.

50. Solve the following division problem and represent the remainder as a fraction in lowest terms. *Note:* The problem must be done in base 5.
    a. $4_5 \overline{)22202_5}$ b. $13_5 \overline{)3203_5}$

51. Beatrice has decided to move to Borneo tomorrow! She has $\$1432_5$ in one account and $\$304_5$ in another account. Before she leaves, she must pay back $\$314_5$ to Peter. Last week she babysat for 24_5 hours at the rate of $\$3_5$ an hour. What will be her net worth after she closes her bank accounts, repays Peter, and gets her money for babysitting?

52. A kindergarten teacher has 110_5 children in her class. She has enough gumdrops so that each child can have 4_5 gumdrops. On the way to class, she dropped half of the gumdrops, and when she gets to school, she finds that half of the students are absent. How many gumdrops will each child get?

53. In what base is the following problem true?

$$5_x \overline{)3350_x} \quad \text{421 R 1}$$

54. Determine the numbers in the empty boxes.

a.

a	b	a + b	a · b
65		131	
	72		86976

b.

a	b	a + b	a ÷ b
153			17
		20	19

c.

a	b	a − b	a · b
		20	1056

d.

a	b	a · b	a ÷ b
		20	5
	37		925

55. Find the missing digits.
  a. $74 \cdot 8\square = 6\square\square 4$
  b. $\square\square 4 \cdot 64\square = 489{,}346$
  c. $18 \cdot (\square\square 4 + 29\square) = 10{,}116$
  d. $(\square 3) \cdot (\square) \cdot (\square 7) = 1547$
  e. $1530 = \square\square \cdot 48 + 42$

56. In the following puzzle, each letter has a value between 0 and 9. Find the value of each symbol so that the following computations are all accurate.

```
   A         C        B       F       A        B
  +A        +C       −E      ×D      ×A       ×F
  ---      ---      ---     ---     ---      ---
   B        DE        B       F       G       HF

   C                  J               K
  ×A              H)DB             A)HD
  ---
  DC
```

57. For each of the following, determine possible whole numbers a, b, and c for which the statement is true.
  a. $(a \div b) \div c = a \div (b \div c)$
  b. $a(b - c) = ab - ac$
  c. $(a - b) - c = a - (b - c)$

58. A popular kind of problem in many books is the broken calculator problem.
  a. If the 7 key is broken, how would you determine $66 + 77$ on your calculator?
  b. If the 7 key is broken, how would you determine 23×70 on your calculator?
  c. If the 5 key is broken, how would you determine $260/5$ on your calculator?
  d. If the 6 key is broken, how would you determine 26×44 on your calculator?
  e. The − key is broken. How can you find $834 - 653$.

59. In each of the problems below, select the correct operations by a means other than random trial and error. Briefly explain the reasoning behind your guesses.
  a. $14(48 \text{ O } 32) = 224$
  b. $(462 \text{ O } 243) \text{ O } 15 = 47$
  c. $4 \text{ O } (8 \text{ O } 5) = 52$
  d. $56 \text{ O } 42 \text{ O } 352 = 2000$

60. In each of the cases below, the actual number have been replaced by variables. Write the number sense that would correctly determine the answer.
  a. The students in the three fifth grade classes saved A cans, B cans, and C cans, respectively. If they receive D cents per can, how much will they earn?
  b. Emily has just gone to college. Her cell phone will cost A dollars per month and her iTunes account is B dollars per month. How much money will she spend during the C months of the college year?
  c. Marissa has A dollars in her bank account. She babysat last week for a neighbor for B hours at a rate of C dollars per hour. She bought a cell phone for D dollars and spent E dollars on a dress. How much money does she have now?

61. Determine which of these answers are reasonable. You have about 5 seconds to make your determination. That is, make your determination, using number sense, without doing any pencil and paper work or trying to subtract the numbers mentally.
  a. There were 35 students in a class and each student paid $28 for a field trip. The total cost was about $600.
  b. A bookstore chain bought 45,000 copies of a new book and has 652 stores. Each store will have about 70 books.

Chapter Summary

1. Many students have said that really understanding base 10 and the four operations, was, for them, the beginning of a new attitude toward mathematics. We will continue to examine new and important mathematical ideas throughout this book, but the foundation for much of elementary mathematics has now been laid.
2. Each operation has multiple meanings.
3. Many algorithms have been developed to enable us to compute more efficiently.
4. The standard algorithm for each operation does not connect equally well to each meaning of the operation.
5. Being able to make sense of algorithms requires:
   - The ability to apply base 10 and place value concepts
   - The ability to compose and decompose the numbers (for example, to use expanded form)
6. Patterns enable us to understand the operations more deeply.
7. In many real-life problems, the answer depends on knowing how to interpret one's computation.
8. Being able to perform mental math and to estimate requires
   - The ability to apply base 10 and place value concepts
   - The ability to compose and decompose the numbers (for example, to use expanded form)
   - The ability to apply properties of the operations, especially the commutative, associative, and distributive properties
9. Numbers in real-life settings are sometimes exact, sometimes rounded, and sometimes estimates.
10. In real-life problem-solving, one needs to know when to find an exact answer and when to find an estimate.
11. Real-life problem solvers need to know whether their estimates are reasonable.
12. People may use rounded numbers rather than exact numbers for a variety of reasons.

BASIC CONCEPTS

Section 3.1 Understanding Addition

Addition terminology:
- sum *130*
- addition *130*
- addends *130*

Addition contexts:
- discrete *128*
- continuous *128*

Addition models:
- pictorial *128*
- number line *130*
- tables *131*

Addition properties:
- identity *132*
- commutativity *132*
- associativity *132*
- closure *132*

Algorithms for addition:
- common *140*
- lattice *142*

Other terminology:
- matrix *133*
- algorithm *138*
- decompose *141*
- compose *141*

Section 3.2 Understanding Subtraction

Subtraction contexts:
- take away *152*
- comparison *152*
- missing addend *152*

Subtraction terminology:
- subtraction *153*
- minuend *153*
- subtrahend *153*
- difference *153*

Subtraction models:
- pictorial *153*
- number lines *154*

Algorithms for subtraction:
- standard *160*
- alternative *161*
- carry, borrow *164*

Section 3.3 Understanding Multiplication

Multiplication contexts and models:
- repeated addition *169*
- area *170*
- general model *170*
- Cartesian product *171*

Multiplication terminology:
- multiplier *170*
- multiplicand *170*
- product *170*
- factor *170*
- multiple *170*

Multiplication properties:
- zero *172*
- identity *172*
- commutativity *172*
- associativity *172*
- distributivity *172*
- distributive property of multiplication over addition *173*
- distributive property of multiplication over subtraction *173*

Algorithms for multiplication:
- standard *180*
- partial products *180*
- lattice *181*

Section 3.4 Understanding Division

Division contexts and models:
  partitioning **190**   missing-factor **192**
  repeated subtraction **190** number line **192**

Division terminology:
  dividend **190**   divisor **190**
  quotient **190**

Division concepts:
  division by zero **193**   division algorithm **194**

Other terminology:
  indirect proof **193**   operation sense **202**
  order of operations **205**

Algorithms for division:
  standard **196**
  scaffolding **198**

Number sense and operation sense:

Number sense
  with addition **148**   with subtraction **164**
  with multiplication **184**   with division **201**

Operation sense **202**

Mental Math and Estimation

Strategies for mental computation and estimation:
(Note that some of these strategies can be used both for mental computation and for estimation, some strategies are applicable to more than one operation, and some are alternative ways to solve the same problem.)

mental addition **134**	leading-digit
break and bridge **135**	(front end) **134**
compatible	compensation **135**
numbers **135**	add up **136**, **155**
multiples of 10 **136**, **175**	partial sums **137**
number line **137**, **155**	rounding **144**, **146**
clustering **145**	halve and double **175**
expanded form **182**	lower bound **182**
partial products **182**	upper bound **182**
canceling **195**	canceling zeros **195**
equivalent fractions **195**	missing-factor
	model **199**

CHAPTER 3 REVIEW EXERCISES

1. State the problem that is represented in each case below:

   a.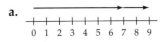

   b.

   c.

   d.

2. Represent the following problem on a number line: 3(4 + 2)

3. An alternative algorithm for adding whole numbers is shown below. Make up and solve a few problems using this algorithm. Explain why it works, as though to a parent who has learned only the traditional right-to-left algorithm.

   ```
      832
     +549
     1300
       70
       11
     1381
   ```

4. Place the digits 3, 4, 5, 7, 9 in the boxes to obtain the smallest difference. Briefly explain your reasoning.

   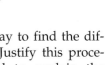

5. The figure below shows one way to find the difference between 800 and 126. Justify this procedure. You're not being asked to explain the procedure (that is, *how* to get the answer of 674) but, rather, to explain *why* this procedure works.

   $$\begin{array}{r} {\scriptstyle 7\ 9} \\ {\scriptstyle \cancel{1}\ \cancel{1}} \\ \cancel{8}\cancel{0}0 \\ -126 \\ \hline 674 \end{array}$$

6. Make up two subtraction story problems, one illustrating the take-away model and one illustrating the comparison or missing-addend model.

7. It has been said that we should not use the terms "borrowing" and "carrying" because they are the same thing. Describe what it is about them that is "the same" mathematically. You need to go deeper than just saying that they both involve regrouping.

8. Using a sheet of graph paper, cut out a piece that measures 37 × 24 and explain how to get the product *just from the picture* and your understanding of

multiplication. Assume you do not know the algorithm, but you do know base 10, place value, and what multiplication means.

9. Why do we move over in the second row when we multiply? Use the example below. Specifically, what is the mathematical reason for moving the 74 over instead of putting it below the 4 and 8?

$$\begin{array}{r} 37 \\ \times 24 \\ \hline 148 \\ 74 \\ \hline 888 \end{array}$$

10. Imagine a parent of a fourth-grader asking you, "Why is 13 × 25 equal to (10 × 25) + (3 × 25)?" Write a short paragraph (2–3 sentences) that would help the parent to understand *why*. Note that you are not explaining why we have children break numbers apart. You are simply helping the parent to understand why 13 × 25 = (10 × 25) + (3 × 25).

11. Place the digits 1, 2, 4, 5, and 8 in the boxes to obtain the greatest product. Describe your first two guesses and the thinking strategies behind those guesses. Points for this question will be based on the quality of the two guesses *and* on your justification of them. That is, random trial and error gets no points.

12. Pete is in fourth grade, and his group has been exploring multiplication. They came up with a method that works, using place value: Multiply the numbers in the ones place and then multiply the numbers in the tens place, just like we do addition. But it doesn't work. Why not?

$$\begin{array}{r} 34 \\ \times 26 \\ \hline 600 \\ 24 \\ \hline 624 \end{array}$$

13. We can find the sum of 29 and 17 mentally by transforming the problem into 30 + 16. However, if we transform 29 × 17 into 30 × 16, it doesn't work. Why not?

14. Look at the multiplication table in the text. If you take any 2 × 2 box in the table and multiply the diagonals, the products of the diagonals are equal. Why? Here is one example, shown at the right: 12 × 20 = 240 = 15 × 16.

12	15
16	20

For Exercises 15–19, find the answer using methods other than random trial and error. Briefly explain your solution path—that is, the thinking behind your work.

15. Find *a* and *b* if *a* + *b* = 46 and *a* × *b* = 480.

16. How could you figure the remainder to 73,932,500 ÷ 97665? That is, 73,932,500 ÷ 97665 = *x* remainder *y*. Find *y*.

17. I am thinking of a number that has a remainder of 16 when divided by 75. What is that number? There is more than one answer.

18. Using a calculator, Li multiplied by 5 when he should have divided by 5. The display read 300. What should the correct answer be?

19. Find the missing digits: (□9) · (□) · (□1) = 7821

20. When we divide, what is going on *mathematically* when we multiply? For example, here we multiply 5 by 8 and later we multiply 5 by 6.

$$\begin{array}{r} 86 \\ 5\overline{)432} \\ \underline{40} \\ 32 \\ \underline{30} \\ 2 \end{array}$$

21. Make up two division story problems, one illustrating the partitioning model and one illustrating the repeated-subtraction model.

22. Do the following computations *in your head*, using a strategy other than the standard pencil-and-paper algorithm. Describe the "how" *and* the "why" of the strategy you used in each case. In each case, 2–4 lines should be sufficient.

   a. 329 + 567
   b. 34 + 48 + 77 + 66
   c. 91 − 39
   d. 502 − 206
   e. 600 − 246
   f. 35 × 19
   g. 632 × 4
   h. 29 × 12
   i. 40,000 ÷ 80

23. Explain how you can obtain the answers to these problems easily in your head using properties:

   a. 36 + 82 + 64
   b. 13 × 19 + 7 × 19
   c. 1592 ÷ 8

For problems 24–28, obtain your best estimate. Note that "estimate" means two things: (1) all computation is in your head, and (2) what you can do in 10–15 seconds. Then briefly explain how you got your estimate.

24. a. 3684 + 2853 + 6241 + 8312
    b. 44,268 − 28,843

c. 468×9

d. $48{,}412 \div 14$

25. Cletha bought a house for $216,250 and sold it for $345,300. How much profit did she make?

26. A student lives in Hanover, which is 62 miles, each way, from school. How many miles per semester does she travel if she drives to school twice a week for 14 weeks?

27. Pierre just bought a new van, and he wants to see what kind of mileage he gets. He filled up with gas after going 489 miles, and the car took 19 gallons of gas. Estimate the mileage.

28. I had planned a spaghetti dinner for my friends. I needed pasta ($1.53), sauce ($2.37), French bread ($1.99), garlic cloves ($.52), parmesan cheese ($3.50), a head of lettuce ($1.19), a pepper ($.89), and salad dressing ($1.20). About how much did this meal cost me?

29. To estimate 638×42, we can use the strategy of rounding one number up and the other down. Explain why 660×40 will give a closer estimate than will 640×40.

30. If a person exercises an average of 3 hours per week from age 18 to age 70, how many days has she or he exercised over this time?

31. The fuel tank in a car holds 20 gallons of gasoline. The car has a fuel efficiency rating of 18 miles per gallon. If the gas tank is full when the family begins, and their destination is 1200 miles away, how many times will they have to stop and fill the gas tank to get to their destination?

32. The copy center at Keene State College told me that it used 1200 reams of paper last semester. It charges departments 3¢ per copy.

    a. How many pages were run off at the copy center?

    b. How much did the college pay to the copy center?

    c. If the departments use their own copy machines, which they do for small jobs, they are charged 5¢ per copy. How much did the college save by having a copy center that semester?

33. Perform the following computations in base 5.

    a. 3442_5 b. 4000_5 c. 44_5 d. $5\overline{)20202}_5$
       $+2333_5$ -1234_5 $\times 32_5$

34. In what base does $13 \times 3 = 42$?

35. Find the base for which the following statement is true: $135 + 444 = 601$

36. Find the base for which the following statement is true: $103_{10} = 1213_x$

37. Find the missing digits:

    a. 3 8   b. 6 3   c. ab
       + 2 8   - 2 8 $\times 3a$
       -------   -------   -------
       7 9 3 2   2 7 1 9   1564

CHAPTER 4

Number Theory

4.1 Divisibility and Related Concepts

4.2 Prime and Composite Numbers

4.3 Greatest Common Factor and Least Common Multiple

History

The Pythagoreans were a brotherhood who believed that the whole universe could be explained using only natural numbers. They were not simply mathematicians, though; they believed that all areas of life, including science, music, and philosophy, could be explained in terms of numbers. For example, if the ratio of the lengths of two otherwise identical strings is 2:1, the notes are exactly one octave apart.

The Pythagoreans were a secretive society. Because none of their accomplishments were attributed to individual members, we cannot say who discovered what; thus, all of their discoveries are simply attributed to Pythagoras. We also know that at least 28 women were classified as Pythagoreans and that Pythagoras' wife, Theono, was a teacher in his school.[1]

Around the fourth century B.C. in Greece, a profound shift in the focus of mathematics took place—a shift from simply wanting to know *how* to wanting to know *why*. Until that time, numbers were primarily used for practical purposes. Probably the most recognizable example of this changed focus is the Pythagorean theorem: $a^2 + b^2 = c^2$. Long before the Pythagoreans, other people had known of this relationship. However, the Pythagoreans and other early Greek mathematicians were not content to know that this relationship held for certain right triangles; they proved that it *must* hold for *any* right triangle. The field of number theory was born during this time.

Number theory is a branch of mathematics that focuses primarily on natural numbers and their relationships.[2]

Number theory offers many rich opportunities for explorations that are interesting, enjoyable, and useful. These explorations have payoffs in problem-solving, in understanding and developing other mathematical concepts, in illustrating the beauty of mathematics, and [in] understanding the human aspects of the historical development of numbers. (*Curriculum Standards*, p. 91)

A curious reader might ask why future elementary teachers should study number theory. There are several reasons. Many number theory concepts, including factors and multiples, prime and composite numbers, and divisibility relationships, are firmly embedded in the elementary school curriculum.

Many students at all levels find numbers and number patterns fascinating. Being able to see patterns and to make and test predictions is an important part of mathematical thinking, making number theory fertile ground for mathematically rich investigations. For thousands of years, people throughout the world have been fascinated by patterns in numbers and relationships between different sets of numbers. In this chapter, we will explore some of the more interesting and famous ones. Remember Carl Gauss, from Chapter 1, who

---

[1]Cited in Lancelot Hogben, *Mathematics for the Millions* (New York: Norton, 1937), p. 190.
[2]Technically, number theory deals with the larger set of integers; recall our discussion in Chapter 2 of various sets of numbers: natural numbers (*N*), whole numbers (*W*), and integers (*I*). Pedagogically, however, it works better to focus our investigations on the set of natural numbers, which is also called the set of counting numbers.

found a way to add the first 100 natural numbers so quickly? As an adult, he once said, "Mathematics is the queen of the sciences, and arithmetic is the queen of mathematics." By arithmetic he meant number theory.

Number theory connects with other mathematical concepts and fields. Some of the topics that number theory relates to are fractions (least common multiple and greatest common factor), probability (patterns in counting different combinations), decimals (patterns in repeating as opposed to nonrepeating decimals), algebra (modular arithmetic), and geometry (string art, star patterns).

SECTION 4.1 DIVISIBILITY AND RELATED CONCEPTS

WHAT DO YOU THINK?

- What do the terms *odd* and *even* mean?
- How can you tell whether one number is divisible by another without dividing?
- What connections do you see between divisibility and decomposition?

In this section, we will investigate the basic concepts that will help you to understand the various structures of natural numbers. In Chapter 3, we examined two different ways in which numbers can be decomposed: additive and multiplicative. For example, 12 can be decomposed into $10 + 2$, and it can be decomposed into $4 \cdot 3$. Depending on circumstances, either decomposition might be relevant. In this chapter, we will explore multiplicative decompositions of natural numbers. As you will discover, being able to decompose numbers and knowing which decomposition is relevant are themes that recur throughout elementary mathematics.

INVESTIGATION 4.1

Interesting Dates

We will begin our investigations with a playful question. Many mathematics teachers and mathematicians would smile when writing down December 8, 2096. Do you see why? Think before reading on. . . .

If we were to write this date in shorthand, we would write 12/8/96.

A. Describe the relationship among these three numbers in words.
B. Determine how many instances of this pattern will occur in the years 2000–2019.

DISCUSSION

A. One way of describing the relationship among the three numbers is to say that the third number is the product of the first two: 96 is the product of 12 and 8.

There are several mathematical concepts and terms that come out of this relationship: *factor*, *multiple*, *divisible*, *divisor*, and *divides*. These terms are all related. Therefore, we will define one term formally and state how the other terms are related to it.

If a and b are two whole numbers ($b \neq 0$) and there is a third natural number c such that $a \cdot c = b$, then we say that a **divides** b, and we write $a \mid b$.

When one number does not divide another number, we write $a \nmid b$.

- If $a \mid b$, then a is a **factor** of b.
- If $a \mid b$, then b is a **multiple** of a.
- If $a \mid b$, then b is **divisible** by a.
- If $a \mid b$, then a is a **divisor** of b.

B. Let us now explore the second question. There are no such dates in 2000. In 2001, there is only one such date: 1/1/01. Before you look at the table below, make your own table and then read on. . . .

Note: Table 4.1 is not the only possible table.

TABLE 4.1

1/1/01					
1/2/02	2/1/02				
1/3/03	3/1/03				
1/4/04	2/2/04	4/1/04			
1/5/05	5/1/05				
1/6/06	2/3/06	3/2/06	6/1/06		
1/7/07	7/1/07				
1/8/08	2/4/08	4/2/08	8/1/08		
1/9/09	3/3/09	9/1/09			
1/10/10	2/5/10	5/2/10	10/1/10		
1/11/11	11/1/11				
1/12/12	2/6/12	3/4/12	4/3/12	6/2/12	12/1/12
1/13/13					
1/14/14	2/7/14	7/2/14			
1/15/15	3/5/15	5/3/15			
1/16/16	2/8/16	4/4/16	8/2/16		
1/17/17					
1/18/18	2/9/18	3/6/18	6/3/18	9/2/18	
1/19/19					

In one sense, we are done—we have found all 53 possible dates. However, there is still a lot of interesting mathematics to be gleaned from looking for and examining patterns and relationships. Take a few minutes to examine the table above. What patterns and relationships do you observe in the table?

One pattern is that many dates have a twin. For example, the dates in the first and second columns, the dates in the third and fourth columns, and the dates in the fifth and sixth columns are twins. Someone who sees this connection early can finish the problem more quickly—sort of a "2 for 1" deal. Mathematically, we can talk about symmetry in our set.

We can also see that some dates do not have twins. For example, 1/7/07 has a twin, but 1/13/13 doesn't. Why not?

If we stop for a moment, we realize that 13/1/13 is not a valid date.

Some valid dates have no twins: 2/2/04, 3/3/09, 4/4/16. In this case, the numbers 4, 9, and 16 are square numbers.

Odd and Even Numbers

One of the first ways in which children encounter number theory is with odd numbers and even numbers. Let me introduce this discussion with a story. One morning my five-year-old son, Josh, asked me at the breakfast table if 44 was an odd or an even number; at that time, my 45th birthday was a few days away. I asked him what he thought, and he said he thought it was even. I asked him what he thought an even number was. Before you read his response, think about this question yourself. Then read on. . . .

Josh's response was that even numbers were fair. I asked him what he meant by fair. His response was, "Well, it's fair if two teams can get the same amount."

Take a few moments to write your own definition of odd and even. Given our explorations in Alphabitia, in *Explorations* can you make a definition that will work for that system too?

Below are several different ways in which even numbers have been defined. Note that the notion of *even* is closely tied to the notion of decomposition. That is, if a number can be decomposed in the following way, then it is even.

A number is an even number if:

- It has 2 as a factor.
- It is divisible by 2.
- Its ones digit is 2, 4, 6, 8, or 0. (This characteristic does not hold for all bases; do you see why?)
- It can be represented as the sum of two equal numbers.
- There exists another natural number a such that $x = 2a$.

Take a minute to write the analogous ways in which odd numbers can be defined.

Please don't skip over this. You will use this thinking in the next investigation.

■ Mathematics ■

The concept of evenness can be constructed in additive and multiplicative terms. For example, evenness can be thought of in additive terms: x is an even number if $a + a = x$; that is, if it is the sum of two equal numbers. Evenness can also be thought of in multiplicative terms: x is an even number if $2a = x$—that is, if x is the double of some number.

INVESTIGATION 4.2 Patterns in Odd and Even Numbers

In this investigation, we will examine and discover some of the many patterns that emerge when odd and even numbers are combined. There are several reasons for investigating odd and even numbers. One is that many people find it fun; I have seen the excitement on young children's faces as they discover these patterns. Another reason is that in order to explain these patterns, one has to look more closely at the concepts. Finally, we move from noticing and describing a pattern to being able to explain *why* it is true.

Begin the investigation by thinking about the following questions and noting your responses before reading on. . . .

A. What do you notice about adding two odd numbers? two even numbers? an odd number and an even number?

B. Can you explain the pattern you see in the sum of two odd numbers?

CLASSROOM CONNECTION

It is important to emphasize that most concepts in the elementary school curriculum can be represented concretely and that most, but not all, students need to experience these concepts concretely in order to understand them deeply. This is true with the concepts of factor, multiple, and divisible.

Cuisenaire rods are commonly used manipulatives in elementary schools that work nicely with many elementary mathematics concepts. They come in ten different lengths, each length being a different color (see Figure 4.1). The length of each of the rods is a multiple of the length of the shortest rod.

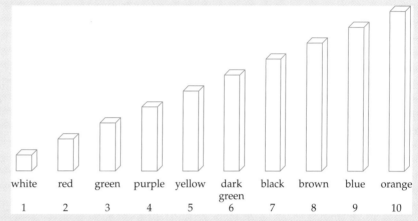

FIGURE 4.1

Children will spontaneously build "trains" with Cuisenaire rods. To see whether one number is a factor of another number, we can begin by making a large train using Cuisenaire rods of only one color. For example, consider a 20 train (two orange rods). A child can readily test whether 4 is a factor of 20 by seeing whether the large train can be made using only purple (4) rods (see Figure 4.2).

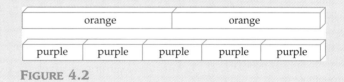

FIGURE 4.2

DISCUSSION

A. The patterns for adding odd and even numbers can most succinctly be represented as

   odd + odd = even
   even + even = even
   odd + even = odd

B. Now that we have seen *what* the pattern is, how might we explain *why* it holds for all odd and even numbers? What tools do you have that might enable you to show *why* this is true?

STRATEGY 1: Make a drawing

Let us begin by examining the question from a geometric perspective. See Figure 4.3, which represents any odd number. Why is this figure a valid representation of any odd number?

FIGURE 4.3

Think of two Cuisenaire rods, one of which is one unit longer than the other. Any odd number can be represented in this manner. Can you connect any of your definitions to this diagram? How?

If we think of *even* as meaning able to be separated into two groups that have the same amount, then *odd* must mean that we can't do that; in other words, one of the rods will be one unit longer. What if we combine two different odd numbers? Draw that diagram. What do you see? At this point, make another attempt to explain why the sum of two odd numbers must be an even number. You may find that you can refine or extend your ideas from your first explanations, or you may find that the picture and explanation above are different from what you had written.

When we build a concrete model to show the combination of two odd numbers, the first odd number has one unit left over and the second odd number has one unit left over, as in Figure 4.4. When we combine the two numbers, the two units become a pair.

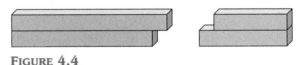

FIGURE 4.4

There are two ways in which this model connects with our idea of *even*. Some students might say that now there are no units left over, and thus we have an even number. Other students might refer to Figure 4.4 and, using language from our work with base 10 blocks, say that when we combine the two numbers, we have two sticks of equal length; thus, it is an even number.

STRATEGY 2: Use algebra

Let us now examine an algebraic explanation for the fact that the sum of two odd numbers must be an even number. Because one of the characteristics of *all* even numbers is that they are divisible by 2, we can say that $2n$ represents *any* even number as long as n represents a natural number. Because one of the characteristics of *all* odd numbers is that they are not divisible by 2, we can say that $2n + 1$ represents any odd number. Why is this? Try to use these ideas to show with symbols why the sum of any two odd numbers is an even number. Then read on. . . .

Statement	Reason
$(2n + 1) + (2m + 1)$	This represents the sum of *any* two odd numbers, where m and n represent *any* natural numbers.
$= 2n + 2m + 2$	We can change the original expression to this expression because of the associative and commutative properties.
$= 2(n + m + 1)$	We can change the previous expression to this expression because of the distributive property.

■ History ■

In a fascinating book with the not so fascinating title of *The Penguin Dictionary of Curious and Interesting Numbers*, David Wells describes the values that many people associated with the first several natural numbers. The Pythagoreans considered 2 and all even numbers to be female, and they considered 3 and all odd numbers to be male. Female numbers were considered to be human and earthy, whereas male numbers were considered to be divine and heavenly. Not surprisingly, they associated the number 5 with marriage, because it is the unity of the first male and female numbers. Some readers may be angered at the Greeks' characterization of male and female. I think it is important for people to realize that mathematics is more than just numbers and that mathematics has helped to shape our cultural values in many ways.

However, this form means that this number, $2(n + m + 1)$, has to be an even number. Why?

See whether you can prove the other two statements (even + even = even and odd + even = odd) in a similar fashion.

In Chapter 2, we stated that many relationships between two sets in mathematics are functional relationships. Is this relationship we have just examined (that is, whether the sum of two numbers will be odd or even) a functional relationship? Why or why not? Think and read on. . . .

Function language If we cast this notion of the sum of two numbers being odd or even in function language, we can see that the input set consists of all ordered pairs of natural numbers; that is,

Input set = $\{(a, b) \mid a, b \in N\}$

Similarly, the output set consists of two elements, even and odd; that is,

Output set = {even, odd}

We can illustrate this with a mapping diagram (see Figure 4.5):

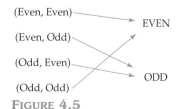

FIGURE 4.5

We see that every element of the input set is connected to exactly one element of the output set; for example, even + even always produces an even number. Therefore, this relationship is a functional one.

INVESTIGATION 4.3 — Understanding Divisibility Relationships

There are a host of theorems that concern divisibility. We will examine two here, partly because we will need them later in the section and partly because this investigation provides practice in the kind of mathematical thinking originally developed by the early Greeks—that is, advancing from demonstrating that a statement is true in a particular case to proving that the statement is true for all numbers—what we call proving the general case.

A. We know that $6 \mid 42$ and $6 \mid 72$. Does it necessarily follow that 6 divides their sum—that is, does 6 divide $(42 + 72)$? Before doing any computation, what do you think?

DISCUSSION

In this particular case, it is true; 6 does divide $42 + 72$, because $6 \cdot 19 = 114$.

B. In fact, pick any three natural numbers a, b, and c for which $a \mid b$ and $a \mid c$. Then $a \mid (b + c)$. Can you prove this?

◼ Outside the Classroom ◼

The question in Investigation 4.3 may seem to be silly to some people, but science is full of examples of products or medicines or mathematical principles that worked in many test cases but were later found not to work in all cases. In some instances, it was a matter of "Aw, shucks, we were wrong," but in other cases, it was a matter of lives being lost. Therefore, this ability to be able to prove that a hypothesis either is true in only some cases or is true in all cases is an important ability in many areas of life.

Let me restate this general question in English and in mathematical notation:

English: If a number divides two numbers, will it necessarily divide their sum?

Notation: If $a\,|\,b$ and $a\,|\,c$, is it always true that $a\,|\,(b+c)$?

How might we prove this? Work on this question and then read on. . . .

DISCUSSION

One proof uses the definition of divisibility and looks like this:

If $a\,|\,b$, then there is a natural number x such that $ax = b$.

Similarly, if $a\,|\,c$, then there is a natural number y such that $ay = c$.

The proof consists of transforming $b + c$ to show that it must be divisible by a:

$b + c = ax + ay$ Here we are substituting ax for b and ay for c.

$ = a(x + y)$ This is true because of the distributive property.

We now have a valid proof. Do you see why? Think and then read on. . . .

We have proved the general case, because if $b + c$ can be expressed as the product of a and some natural number, then by definition $(b + c)$ is divisible by a. Therefore, we have proved that if $a\,|\,b$ and $a\,|\,c$, then $a\,|\,(b+c)$.

Two theorems similar to this one, which will be left unproved here are:

If $a\,|\,b$ and $a\,|\,c$, then $a\,|\,(b-c)$

If $a\,|\,b$ and $a\,|\,c$ and $a\,|\,d$, then $a\,|\,(b+c+d)$

INVESTIGATION 4.4 Determining the Truth of an Inverse Statement

Suppose we turn things around and consider the inverse statement: What if a number a divides neither of two larger numbers? Can we say that it will never divide the sum of those two numbers? Think and then read on. . . .

This question can be stated in English and in notation as follows:

English: If a does not divide b and a does not divide c, then will a not divide their sum?

Notation: If $a \nmid b$ and $a \nmid c$, then $a \nmid (b+c)$?

DISCUSSION

Although this hypothesis seems reasonable, we can show that it is invalid by using a **counterexample**, an example that proves a hypothesis to be false.

For example, consider $a = 3$, $b = 7$, and $c = 2$.

Although $3 \nmid 7$ and $3 \nmid 2$, $3\,|\,(7 + 2)$.

Thus it is not true that if $a \nmid b$ and $a \nmid c$, then $a \nmid (b+c)$.

Note: A mathematical statement is considered to be true only if it is true 100 percent of the time. If there is even one exception (a counterexample), then the statement is considered to be mathematically false.

Divisibility Rules

Before the widespread use of calculators, knowing divisibility rules was quite useful. For example, simplifying 42/54 is much easier for a student who immediately sees that both numbers are divisible by 6. Think back to Investigation 4.1. Are there any dates in the year 2096 that work? Work on this and then read on. . . .

There are several ways to answer this question. One is simply random guess–check–revise. Another way is to be systematic: $1 \cdot ? = 96$, $2 \cdot ? = 96$, $3 \cdot ? = 96$, and so forth. Another way is first to find all the factors of 96. We will explore the second strategy now and the third strategy in the next section.

There are four dates in 2096 that work: 4/24/96, 6/16/96, 8/12/96, and 12/8/96. If we use a systematic approach, we are essentially asking, in each case, does this number divide 96, and if so, is the result a valid date? For example, does 3 divide 96? You can answer this by dividing. However, if you know the divisibility rule for 3, you immediately know that 3 does indeed divide 96.

Several of the divisibility rules are quite simple. Let us examine them first.

Divisibility by 2   When will a number be divisible by 2? Another way of asking this question is: If you think of (the set of) all numbers that are divisible by 2, what do they have in common?

What they have in common is that they are all even numbers. Thus the divisibility rule for 2 is

   A number is divisible by 2 iff it is an even number.

An equivalent statement is

   A number is divisible by 2 iff its ones digit is a 0, 2, 4, 6, or 8.

Note that this statement holds for base 10. Divisibility rules in other bases will be left as exercises.

Divisibility by 5   This rule is simple in base 10:

   A number is divisible by 5 iff its ones digit is a 0 or a 5.

Do you see why this rule is so easy in base 10? Think and then read on. . . .

Think back to base 10 manipulatives. When we count by 5s, we are counting by half longs.

Divisibility by 10   A number is divisible by 10 iff the ones digit is 0.

These three divisibility rules are generally the easiest to remember. Do you think it is a coincidence that 2 and 5 both divide 10?

Divisibility by 3   Make up several numbers and determine whether they are divisible by 3. Look at the numbers that are divisible by 3. Can you see any patterns that would enable you to determine that a number is or is not divisible by 3 without having to divide? If you don't, consider this statement: "Divisible by 3" is analogous to an exclusive club. That is, every number that is divisible by 3 has some characteristic that every number not divisible by 3 does not have. Look again at your two sets of numbers (those that are divisible by 3 and those that are not); make up more numbers and

---

CLASSROOM CONNECTION

Because this book does not emphasize the more formal aspects of mathematics, we will generally be concerned with justifying the if–then statements that we encounter, rather than formally proving them. In the *NCTM Principles and Standards*, we find the term *justify* over and over again. If students in elementary and middle grades become used to justifying their conclusions, then the step to logical proofs is not a huge one. However, we will occasionally present proofs because this more formal aspect of mathematics becomes increasingly important in secondary and college mathematics. It is my assumption that helping children develop the intuitive and inductive parts of the mind is the best foundation for the more formal and deductive work that comes later.

CLASSROOM CONNECTION

Most elementary teachers have the students do skip counting before multiplication. The set of multiples of 5 is the first such set that most children learn because it is the simplest set of multiples in base 10.

place them in the appropriate set, for this is a great chance to develop the part of the problem-solving toolbox dealing with looking for patterns. Then read on. . . .

INVESTIGATION 4.5

Understanding Why the Divisibility Rule for 3 Works

Formally, we say that

A number is divisible by 3 iff the sum of the digits of the number is divisible by 3.

For example, 3|567 because 3|(5 + 6 + 7). Let us use the divisibility relationships that we explored earlier to explain *why* the divisibility rule for 3 works. Recall that one of the main purposes of this course is to learn not just the whats of mathematics but also the whys.

Pick a few numbers that are divisible by 3 and see whether you can tell what the sum of the digits has to do with divisibility by 3. For example, 243 is divisible by 3. Why? Explore this for a while and then read on. . . .

DISCUSSION

Begin with manipulatives We will first work through a concrete representation and then move to the more general case. If we represent 243 with base 10 manipulatives, we have Figure 4.6.

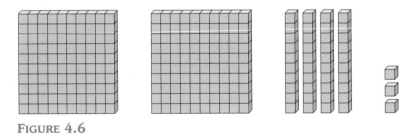

FIGURE 4.6

In one sense, we can look at this representation as three distinct groups: flats, longs, and units. The key to understanding why this rule works lies in the fact that if $a|b$ and $a|c$, then $a|(b + c)$. We will use an unproved extension of that theorem: If $a|b$, $a|c$, and $a|d$, then $a|(b + c + d)$ for all natural numbers a, b, c, and d. Do you see what this theorem has to do with showing why 243 is divisible by 3? Think and then read on. . . .

If we can break down our number into various components so that each component is divisible by 3, then the number will be divisible by 3. What can we do to the flats and the longs to make them divisible by 3? Think and then read on. . . .

If we cut one unit from each flat and each long, we still have the same amount—that is, 243 (see Figure 4.7). But now the 99 blocks and the 9 blocks are divisible by 3.

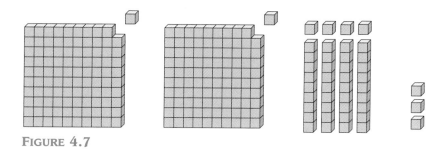

FIGURE 4.7

We simply move the "extra" units created in cutting the flats and longs and put them with the original 3 units (see Figure 4.8). Now we have 9 ones, and 9 is divisible by 3.

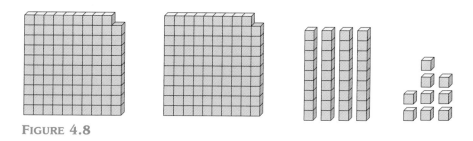

FIGURE 4.8

Now use expanded notation   We can use expanded notation now to show what we have done and to understand better why the divisibility rule for 3 works.

Statement	Reason
$243 = 200 + 40 + 3$	We have rewritten the amount using expanded notation.
$= (100 + 100) + (10 + 10 + 10 + 10) + 3$	We have substituted equivalent amounts (for example, $100 + 100$ for 200).
$= (99 + 1 + 99 + 1) + (9 + 1 + 9 + 1 + 9 + 1 + 9 + 1) + 3$	100 and 10 are not divisible by 3, but if we take 1 away from each 100 and each 10, what remains *is* divisible by 3.
$= 2(99) + 4(9) + (2 + 4 + 3)$	We have used the commutative and associative properties to rearrange what we had.

Finally, use the theorem   We can now use the theorem: If $a\,|\,b, a\,|\,c$, and $a\,|\,d$, then $a\,|\,(b + c + d)$. That is, if we can show that 3 divides each of the three expressions above [2(99), 4(9), and (2 + 4 + 3)], then 3 must divide the sum 243; that is, 243 is divisible by 3! Let us examine each of the three expressions in turn.

We know that $3\,|\,2(99)$ because $3\,|\,99$. This knowledge comes from another (unproved) theorem that states that if $a\,|\,b$, then $a\,|\,bc$ for all natural numbers a, b, and c. This can be stated in English as follows: If a number divides another number, then it divides any multiple of that number.

Similarly, $3\,|\,4(9)$ because $3\,|\,9$.

Therefore, demonstrating that $3\,|\,243$ rests upon knowing that $3\,|\,(2 + 4 + 3)$. However, $2 + 4 + 3$ is simply the sum of the digits of 243!

Generalizing the justification This does not constitute a proof of the divisibility rule for 3 but is, rather, an examination into the structure of the rule. You may want to do the above kind of analysis with a few other numbers to better understand why it works. If we had represented 243 as *abc*, we would have a proof of the divisibility rule for 3 for any three-digit number. That is,

$$abc = 100a + 10b + c = 99a + a + 9b + b + c = 99a + 9b + (a + b + c)$$

Divisibility by 9 As you discovered when you worked with the multiplication table in Exploration 3.6, the 3 family and the 9 family are related. So too are their divisibility rules. Before reading the rule below, see whether you can guess what it is. You can approach this task in two different ways: Choose several numbers and multiply them by 9 or make up a bunch of numbers and divide them by 9. The first approach creates one set; the second approach creates two sets, divisible by 9 and not divisible by 9. Here is another chance to apply your problem-solving skills (looking for patterns) and content knowledge (looking to see whether you can adapt the divisibility rule for 3). In either case, see what you can come up with before reading on. . . .

A number is divisible by 9 iff the sum of its digits is divisible by 9.

The justification of this rule is left as an exercise.

INVESTIGATION 4.6

Divisibility by 4 and 8

Just as the divisibility rules for 3 and 9 are similar, so too are the divisibility rules for 2, 4, and 8.

A. As before, choose several numbers and determine which are divisible by 4. Then find patterns that will lead you to be able to predict whether a number is divisible by 4 without dividing. Then read on. . . .

B. Choose several numbers and determine which are divisible by 8. Find patterns to help you predict whether a number is divisible by 8. Then read on. . . .

DISCUSSION

A. The divisibility rule for 4 is not terribly obvious. Let me offer a hint before presenting the rule. Consider the number 532. The test for divisibility by 4 centers on whether $4 \mid 32$. Similarly, with the number 123,456, the test for divisibility by 4 centers on whether $4 \mid 56$.

Now try to express this rule in English as though you were explaining it to someone on the phone. Then read on. . . .

There are several valid ways to express the rule. One looks like this:

A number is divisible by 4 iff the number represented by the last two digits is divisible by 4.

By the last two digits, I mean the ones digit and the tens digit.
Before we examine why this rule works, you may want to try it with a few numbers.

The key to understanding how this rule works is to realize that no matter how many hundreds we have, *that* part of the number will always be divisible by 4. This is true because $4\mid 100$, and therefore any multiple of 100 will also be divisible by 4.

Proof for three-digit numbers:

Statement	Reason
$abc = 100a + (10b + c)$	Rewrite using expanded form.
$4\mid 100a$	Because $4\mid 100$, $4\mid 100a$.
If $4\mid(10b + c)$, then $4\mid abc$	Because of the theorem, if $a\mid b$ and $a\mid c$, then $a\mid(b + c)$.

B. Before stating the rule for divisibility by 8, let me give you a hint: The rule for 8 is similar to the rule for 4. Think and then read on. . . .

To help you understand the rule rather than just memorizing it, consider the number 2328 and whether it is divisible by 8. This is another case in which expanded notation is helpful:

$$2328 = 2000 + 300 + 20 + 8$$

Will any of the parts be divisible by 8? Think and read on. . . .

Because $8\mid 1000$, any multiple of 1000 will be divisible by 8, so $8\mid 2000$. However, 8 does not divide 100, so we need to look at the last *three* digits to determine divisibility by 8. That is, if $8\mid 328$, then the original number, 2328, will be divisible by 8.

The divisibility rule for 8 uses language similar to the rule for 4:

A number is divisible by 8 iff the number represented by the last three digits is divisible by 8.

Some students have noted that although 8 does not divide 100, 8 does divide any even multiple of 100. In other words, $8\mid 200$, $8\mid 400$, and so forth. Thus we can say that if the digit representing the hundreds place is even, we need look at only the last two digits. What if the digit representing the hundreds place is odd; is there a pattern that might enable us to determine divisibility by 8 by looking at only the last two digits. What do you think? This question will be left as an exercise.

Divisibility by 6 The divisibility rule for 6 has a different flavor from the other rules. How might you determine whether a large number is divisible by 6 without actually dividing? Think and then read on. . . .

A number n is divisible by 6 iff n is divisible by 2 and by 3.

This can also be stated as

$6\mid n$ iff $2\mid n$ and $3\mid n$

Why do you think this rule must be true?

Let us now investigate how we might apply this divisibility rule to a larger number.

INVESTIGATION 4.7

Creating a Divisibility Rule for 12

What about a divisibility rule for 12? Can we use what we know, or do we just have to divide (longhand or with a calculator)? Think and then read on. . . .

DISCUSSION

When I have asked my classes this, I generally get different hypotheses. Let us examine two of them. Some students say that if a number is divisible by 2 and by 6, then it must be divisible by 12. Other students say that if a number is divisible by 3 and by 4, then it must be divisible by 12. Is this one of those cases in which both answers are correct, or is only one correct? Think and then read on. . . .

In this case, the first hypothesis is not always true. Can you see why? For example, $2 \mid 18$ and $6 \mid 18$ but $12 \nmid 18$. If you thought this first hypothesis was correct, can you now see why it isn't? If you knew that it wasn't correct, can you explain why? In either case, take a few minutes to think about how you would explain to someone why this hypothesis is not true. Then read on. . . .

This hypothesis doesn't work because "divisible by 2" is a redundant condition. If a number is divisible by 6, then it *must* be divisible by 2. Thus, saying that a number is divisible by 2 and divisible by 6 is mathematically equivalent to saying simply that it is divisible by 6.

Revisiting Venn diagrams The preceding sentence makes great sense to some students and "sort of" sense to others. If you are in the latter set, the old adage that a picture is worth a thousand words may be relevant. Let us examine some Venn diagrams and see how they can contribute to our understanding of divisibility rules.

To say that if a number is divisible by 6, it must be divisible by 2 means that the set of numbers divisible by 6 is a subset of the set of numbers divisible by 2, as shown in the figure at the left in Figure 4.9. When we look at divisibility by 3 and by 4, however, we see that they are not disjoint sets (like the sets of even and odd numbers), but rather overlapping sets. The diagram at the right in Figure 4.9 illustrates the relationship between the numbers divisible by 3 and the numbers divisible by 4. The intersection of these two sets is the set of numbers divisible by 12.

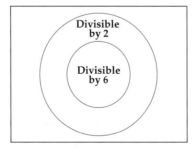
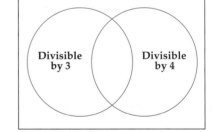

FIGURE 4.9

Thus we can say that a number is divisible by 12 iff it is divisible by 3 and by 4.

Summary

The ideas of divisibility, divides, factor, and multiple are highly interconnected. As we examined patterns in dates and patterns in even and odd numbers, we saw how the concepts of factor, multiple, and divisibility enabled us to understand better the problems we were investigating. Which term we use often depends on our perspective. For example, we might say that 6 divides 24, that 24 is divisible by 6, that 6 is a factor of 24, that 24 is a multiple of 6, or that 6 is a divisor of 24. This notion of divisibility is useful because the relationship between two numbers is often important in explaining patterns and making predictions. We also used divisibility ideas to develop and explain the various divisibility rules. We also found that in order to understand why divisibility rules work, it was helpful to understand some basic principles and terminology of deductive reasoning: the inverse, converse, and contrapositive of a statement.

The notion of divisibility is highly connected to another important idea that you have seen several times in this book already: decomposition. Just as an understanding of computation algorithms in Chapter 3 depended on decomposing the numbers, so too does an understanding of divisibility require that one be comfortable with taking numbers apart and putting them together—that is, decomposing and composing them. As we saw in Chapter 3, we can decompose numbers additively or multiplicatively; for example, we can see 6 as $3 + 3$ or as $2 \cdot 3$. In this section, we found that divisibility relates to multiplicative decompositions.

EXERCISES 4.1

1. Write at least two different definitions of *odd number*. When would it be preferable to use each definition?

2. Earlier in this section, we examined the relationships between adding odd numbers and adding even numbers. This question asks you to examine the relationships between multiplying odd numbers and multiplying even numbers.
   a. Explain why the product of an odd number and an odd number is an odd number. *Hint:* Look at the discussion of why the sum of an even number of odd numbers is an even number.
   b. How would you describe the rule for determining whether the product of many natural numbers is even or odd? For example, is $2 \cdot 3 \cdot 4 \cdot 5 \cdot 6$ an odd number or an even number?

3. a. Find four different odd numbers whose sum is 20.
   b. How many combinations of four different odd numbers whose sum is 30 can you find? Describe your strategies.

4. a. What can you say about two numbers if their sum is even and their product is odd? Justify your answer.
   b. Let's say we have three numbers whose sum is even. Must the product of the three numbers be even or odd, or are both possible? Justify your answer.

5. Is 443_7 an even or odd number? Explain your response.

6. There are some interesting connections between the set of odd and even numbers and the sets of figurate numbers you encountered in Exploration 2.6.
   a. The table below is generated by finding the sum of consecutive odd numbers. Can you explain why the sum of consecutive odd numbers generates the set S of square numbers, $S = \{1, 4, 9, 16, 25, 36, \ldots\}$?

   $$1$$
   $$1 + 3 = 4$$
   $$1 + 3 + 5 = 9$$
   $$1 + 3 + 5 + 7 = 16$$

   b. Another way to generate the set of square numbers is shown below. Can you explain how the following sequence leads to the set of square numbers?

   $$1$$
   $$1 + 2 + 1$$
   $$1 + 2 + 3 + 2 + 1$$
   $$1 + 2 + 3 + 4 + 3 + 2 + 1$$
   $$1 + 2 + 3 + 4 + 5 + 4 + 3 + 2 + 1$$

7. I'm thinking of two numbers. Determine the numbers if:
   a. Their sum is 45 and their difference is 15.
   b. Their sum is 48 and their quotient is 3.
   c. Their sum is 61, and if the digits in each of the numbers are reversed, their sum is 115.

8. a. Find the following products: 37 · 3 and then 37 · 6. What do you predict will be the product of 37 · 9? Describe, in words, the pattern that you see.
   b. Now predict how long you think the pattern will continue and explain your prediction.
   c. Some students believe that the pattern breaks when we get to 37 · 30, and others don't. What do you think? Why?

9. Any four-digit palindrome (e.g., 1331; 2772) is divisible by 11! Why?

10. a. The total cost of a used car is $4,346. Can this be paid in 12 monthly payments?
    b. A warehouse has received 12,468 copies of a new computer game. Can they ship the same amount to all 12 stores?
    c. There are 425 students in the school and 15 homerooms. Can each homeroom have the same number of students?

11. In order to make the most accurate calendar, the following system was devised. Leap years occur every 4 years except for years ending in 00, for example, 2000. In this case, the year is a leap year if it is divisible by 400. Which of the following years are leap years:
    a. 1776   b. 1800   c. 2100   d. 2424

12. In Investigation 4.1, we explored one set of dates.
    a. Consider the following dates: 2/3/45 and 6/7/89. First, write a general description for all dates that will be members of this set. Second, determine how many instances of this pattern will occur in this century.
    b. Create your own set of dates that have a number theory connection. First, write a general description for all dates that will be members of this set. Second, determine how many instances of this pattern will occur in this century.

13. This problem has been around for many years. Can you determine the ages of Alfie's children from analyzing the following conversation?

    Alfie: Can you guess the ages of my three children? The product of their ages is 36.

    Bernice: That's not enough information. Give me another hint.

    Alfie: The sum of their ages is the same as our street address.

    Bernice: That's still not enough information. Give me another hint.

    Alfie: My oldest child is a girl.

    Bernice: Now I can figure it out.

Note: Problems 14 and 15 can consist of a rote practice of the divisibility rules or they can be a wonderful opportunity for thinking. *Before* checking each number for the specific divisibility rule, try to predict the answer. For example, does 123,456 "feel" like the sum of the digits will be divisible by 3? Or in Exercise 15(b), before checking to see whether 4|98, does it "feel" like it will or it won't? If your intuition is often right, can you explain why? In many cases, what is behind a good intuition is a sense of patterns.

14. Use divisibility rules to answer the following.
    a. Which of the following numbers divide 123,456? 3, 4, 6, 8, 9
    b. Which of the following numbers divide 2,345,678? 3, 4, 6, 8, 9

15. Test the numbers below for divisibility by 2, 3, 4, 5, 6, 8, 9, and 12.
    a. 222,444   b. 213,498   c. 987,987,987

16. a. Is 199,999 divisible by 9?
    b. Is 939,393,939 divisible by 8?
    c. Is 78,888,888 divisible by 8?
    d. Is 26,052 divisible by 13?

17. Find the digits that will make these statements true:
    a. 9|35_6   b. 3|5_45   c. 3|54_5   d. 144|467,_ _2

18. Write a definition for *divides* in your own words. Then have someone who is not in this course read your definition and explain to you, on the basis of your definition, what *divides* means.

19. The children's book *A Remainder of One* tells the story of a bug who is constantly left out. When the army of bugs marches in rows of 2, he is the only one left out. When they march in rows of 3, he is still the only one left out. When they march in rows of 4, he is still the only one left out. Only when they march in rows of 5 does he fit in. How many bugs are in this army?

    Elinor J. Pinczes, *A Remainder of One*, illustrated by Bonnie Mackain, Boston: Houghton Mifflin, 1995.

20. One aspect of parenting that has nearly driven me crazy is that kids lose things. My children love to play card games. However, they lose cards, so we have several decks of cards, none of which is complete. This situation reminded me of a problem that I have seen in many guises: We have assembled quite a pile of cards and are playing a game in which all the cards are dealt out before playing. We have made the following amazing discovery:

    If two people are playing, one card will be left over.
    If three people are playing, two cards will be left over.
    If four people are playing, three cards will be left over.
    If five people are playing, four cards will be left over.
    If six people are playing, five cards will be left over.
    If seven people are playing, no cards will be left over.
    How many cards are in this deck?

21. This problem from China is almost 2000 years old: Find a number that when divided by 3 gives a remainder of 1, when divided by 5 gives a remainder of 4, and when divided by 7 gives a remainder of 2.

22. Joe wrote the divisibility rule for 4 in the following way: A number is divisible by 4 if the last two digits are divisible by 4. Is his expression accurate? Is it clear? If you like it, explain why. If you think it is not accurate or not clear, explain what is inaccurate or vague.

23. We learned earlier that if the digit representing the hundreds place is even, then a number is divisible by 8 iff the number represented by the last two digits is divisible by 8. Finish and justify the second part of this modification of the divisibility rule for 8: If the digit representing the hundreds place is an odd number, then a number is divisible by 8 iff _____.

24. Prove the divisibility rule for 9 for any three-digit number.

25. Some textbooks present two other divisibility rules. Examine each of the rules below to make sure you can use them. Then explain why they work.
    a. A number is divisible by 7 if the number obtained by subtracting double the ones digit from the number represented by the remaining digits is divisible by 7.
    b. A whole number is divisible by 11 if the sum of the digits representing even powers of 10 minus the sum of the digits representing odd powers of ten is also divisible by 11.

26. Devise and justify a divisibility rule for
    a. 18
    b. 24
    c. 25
    d. 2 in base 5
    e. 4 in base 5
    f. 3 in base 12

27. Classify each of the following as true or false, assuming that a, b, and c are integers. If the statement is true, briefly explain why it is true. If the statement is false, give a counterexample. Next rewrite it so that it is true, and then justify the new statement.
    a. If $a|b$ and $a \nmid c$, then $a|bc$.
    b. If a number is divisible by 3, then every digit of that number is divisible by 3.
    c. If $c|ab$, then $c|a$ or $c|b$.
    d. If $4|a$ and $6|a$, then $24|a$.
    e. If $12|a$, then $6|a$.
    f. If $d|a$, then $d|a^2$.

28. Each of the conjectures below is either true or false. For each, if you think it is true, try to prove that it is true or justify why you think it is true. If you think it is false, provide a counterexample.
    a. The sum of 3 consecutive numbers is divisible by 3.
    b. The product of any 3 consecutive numbers is divisible by 6.
    c. The product of any 4 consecutive numbers is divisible by 24.
    d. The square of any odd number can be represented in the form $8n + 1$, where $n \in W$.

29. a. What can you say about the divisibility of the sum of three consecutive numbers? In other words, are there any numbers that will always divide this sum, regardless of the three numbers?
    b. What about the sum of any four consecutive numbers?
    c. What about the sum of any five consecutive numbers?

30. For 6 consecutive years, Joan's age was divisible by her granddaughter's age. What were their ages during that period?

31. a. Let's say you have a bunch of 6¢ and 5¢ stamps and a package for which the postage is $1.93. Can you make $1.93 with only 5¢ and 6¢ stamps?
    b. What if you have only 5¢ stamps and 6¢ stamps and the postage is $1.80? How many different ways can you make the postage?

32. a. What is the units digit of 3^{900}? *Hint:* Make a table and look for a pattern.
    b. What is the remainder when 3^{123} is divided by 5?
    c. What is the remainder when 3^{900} is divided by 2?

33. Look at the following:
    $$17 | 10^2 - 7^2$$
    $$35 | 30^2 - 5^2$$
    a. Create another example that fits this pattern.
    b. Can you generalize this pattern? That is, can you express the relationship in such a way that someone else could use your directions to create more examples?
    c. Can you prove your generalization?

34. At the end of this section, we used Venn diagrams to represent some divisibility relationships.
    a. Represent with a Venn diagram the relationship between the set of numbers divisible by 4 and the set of numbers divisible by 8.
    b. Represent with a Venn diagram the relationship between the set of numbers divisible by 3 and the set of numbers divisible by 9.
    c. Represent with a Venn diagram the relationship between the set of numbers divisible by 5 and the set of numbers divisible by 10.
    d. Represent with a Venn diagram the relationship between the set of numbers divisible by 5 and the set of numbers divisible by 25.
    e. Represent with a Venn diagram the relationship between the set of numbers divisible by 2, the set of numbers divisible by 4, the set of numbers divisible by 8, and the set of numbers divisible by 16.

35. Can you tell whether the following sum will be an odd or an even number without actually adding the numbers? Play with some other examples and then summarize and justify whatever hypotheses/rules you come up with.
    $$21 + 56 + 37 + 23 + 49 + 20 + 73 + 34$$

36. *Classroom Connection* While working on a problem, a fourth-grader asked whether 0 is an even number, an odd number, or neither. I asked what she thought, and she said that an even number is when you divide it into two equal pieces and an odd number is when you can't, so 0 is neither odd nor even. How might you respond to that child to help her realize that 0 is an even number?

37. *Classroom Connection* A third-grade teacher and her class were exploring numbers. One of her students argued that 6 could be both odd and even because it was made of three 2s. What might the child have been thinking? Are there more numbers that could be even and odd, according to this child's reasoning?

SECTION 4.2 PRIME AND COMPOSITE NUMBERS

WHAT DO YOU THINK?

- Why is 1 neither a prime nor a composite number?
- How are prime numbers the building blocks for the set of natural numbers?

4.5

Over the centuries, many famous mathematicians have been fascinated by prime numbers. The Greeks are believed to have discovered prime numbers and were also fascinated by them. The prime numbers are the building blocks of all natural numbers greater than 1. Children are often intrigued by prime numbers. Over the years, I have found that my strongest students are the ones who not only know facts and procedures but also have good number sense. One of the ways to develop this number sense in elementary children is through investigations that deal with prime numbers. Recall the statement

■ *Outside the Classroom* ■

Invasions of various insects have been noted for thousands of years. What is mathematically interesting is when these occur. It has been noted that various species of cicadas appear every 7, 13, or 17 years. A 17-year cycle has been noticed in eastern Tennessee since 1634, and the last occurrence was 1991. What is mathematically interesting about the cicada numbers is that they are prime numbers. Why might a species benefit by having a cycle that is a prime number rather than a nonprime number? Biologists have hypothesized that having a life cycle that is a prime number means that the cicada will encounter its main predators and parasites less often than if the cicada's life cycle were a composite number. For example, let's say there is a parasite that harms the cicada and the life cycle of that parasite is 2 years. With a life cycle of 17 years, it will be 34 years before the cicada encounters that parasite again.

from Chapter 3 that "(p)robably the major conceptual achievement of the early school years is the interpretation of numbers in terms of part and whole relationships." As in the previous section, we will again be focusing on multiplicative compositions and decompositions of natural numbers.

Let us begin our examination of prime and composite numbers by first revisiting Investigation 4.1. Are there any dates in the year 2097 that fit that pattern? Work on this and then read on. . . .

It turns out that there are no such dates, because 97 is a prime number. That is, it has only two factors: 1 and itself.

A natural number is a **prime** number iff it has exactly two factors: 1 and itself.

A natural number that has more than two factors is called a **composite** number. Note the relationship between the words *composite* and *composition*, a term we have encountered several times already in this course.

The number 1 doesn't fit into either of the sets above, and so we say that 1 is neither a prime number nor a composite number. Do you see why?

Determining Whether a Number Is Prime or Composite

A question that has fascinated both ancient and modern mathematicians concerns being able to determine whether a large number is a prime number or a composite number. We don't need a large number for this question to be difficult. For example, is the number 103 prime or composite? Work on this question for a minute or so and then read on. . . .

Immediately, we can see that no even number will divide 103. However, without sophisticated techniques, in order to determine if 103 is prime, we must check to see whether it is divisible by each odd number up to 51. Thus we would have to check 3, 5, 7, 9, 11, 13, 15, 17, 19, 21, 23, 25, 27, 29, 31, 33, 35, 37, 39, 41, 43, 45, 47, 49, and 51. Do you see why we can stop at 51?

There is a very elegant shortcut that saves us from having to test all these numbers. This shortcut was discovered over 2000 years ago and emerges from the following investigation.

INVESTIGATION 4.8

The Sieve of Eratosthenes

The sieve of Eratosthenes was developed by the Greek mathematician Eratosthenes, who lived about 230 B.C., as a tool for determining all prime numbers less than a given number. As usual, you will get far more from the following activity if you do it with pencil and paper rather than just reading it.

TABLE 4.2

1	2	3	4	5	6	7	8	9	10
11	12	13	14	15	16	17	18	19	20
21	22	23	24	25	26	27	28	29	30
31	32	33	34	35	36	37	38	39	40
41	42	43	44	45	46	47	48	49	50
51	52	53	54	55	56	57	58	59	60
61	62	63	64	65	66	67	68	69	70
71	72	73	74	75	76	77	78	79	80
81	82	83	84	85	86	87	88	89	90
91	92	93	94	95	96	97	98	99	100
101	102	103	104	105	106	107	108	109	110
111	112	113	114	115	116	117	118	119	120
121	122	123	124	125	126	127	128	129	130
131	132	133	134	135	136	137	138	139	140
141	142	143	144	145	146	147	148	149	150

In Table 4.2, cross out 1, which is neither composite nor prime.

Circle 2, which is the first prime number. Now cross out all multiples of 2.

Circle the next unmarked number (3), which must be prime, and cross out all multiples of 3. Did you make use of any patterns to help you cross out multiples of 3?

Circle the next unmarked number (5), and then cross out all multiples of 5.

Circle the next unmarked number (7) and stop. On the basis of the crossing out that you have already done, what do you predict will be the first multiple of 7 that you will have to cross out? Think and then cross out all multiples of 7, and then read on. . . .

You found that the first multiple of 7 that wasn't already crossed out was 49—that is, $7 \cdot 7$. The other multiples of 7 that you had to cross out were $7 \cdot 11$, $7 \cdot 13$, and $7 \cdot 17$.

Now circle the next prime number (11), and stop. Before you cross out all multiples of 11, what do you predict will be the first multiple of 11 that you will have to cross out? Think and then cross out all multiples of 11, and then read on. . . .

You found that the first multiple of 11 that wasn't already crossed out was 121—that is, $11 \cdot 11$. Now circle all the numbers in the table up to 121 that have not yet been crossed out. What do you think these numbers have in common? Think and then read on. . . .

DISCUSSION

If your intuition said that these are all prime numbers, you are right. Can you explain *why* all these numbers *must* be prime numbers? For example, can you explain why 149, for example, cannot be a multiple of 7, 11, 13, 17, 19, or any other prime number less than 149? Work on this question and then read on. . . .

One way of demonstrating why all the remaining numbers must be prime involves **indirect reasoning**, which is a method of logic in which we prove that something is false by assuming that it is true and then showing that this assumption leads to a contradiction.

Let us assume that one of the circled numbers in the sieve, for example 149, is not prime.

- 149 is clearly not a multiple of 2, because it's an odd number.
- 149 cannot be a multiple of 3, because it would have been crossed out with the multiples of 3.
- 149 cannot be a multiple of 5, because it would have been crossed out with the multiples of 5.
- 149 cannot be a multiple of 7, because it would have been crossed out with the multiples of 7.
- 149 cannot be a multiple of 11, because it would have been crossed out with the multiples of 11.
- 149 cannot be a multiple of 13, because we know from our investigation with the sieve that the first possible multiple of 13 that hasn't already been crossed out would be $13 \cdot 13$, which is 169.

In similar fashion, we can show that 149 cannot be a multiple of any other number. This disproves the initial assumption that 149 was not prime. Therefore, it is impossible that 149 is a composite number, so it must be a prime number.

> ### ■ *History* ■
>
> Many famous mathematicians have looked for functions that would produce only prime numbers. For example, consider all numbers of the form $M = 2^p - 1$, where p is a prime number. Numbers of this form that are prime are called Mersenne primes, after the monk Marin Mersenne (1588–1648). Can you find three Mersenne primes?
>
> Now it turns out that sometimes this function generates prime numbers and sometimes not.
>
> Pierre Fermat (1601–1665), arguably the king of number theory, played with the following function: $F(n) = 2^{(2^n)} + 1$. He was hopeful that this function would generate only prime numbers. For example, if $n = 2$, then $2^{(2^n)} + 1 = 2^{(4)} + 1 = 17$.
>
> This conjecture looked good until Leonard Euler (1707–1783) showed in 1732 that $F(5) = 4{,}294{,}967{,}297$ can be factored into $641 \cdot 6{,}700{,}417$. This was before any kind of calculating machine existed! No one has yet created a function that produces only prime numbers.

Actually, once we know that 13 is not a factor of 149, we are finished. Do you see why?

If we begin to cross out multiples of 13, we find that all the multiples of 13 from $1 \cdot 13$ to $12 \cdot 13$ have already been crossed out. Therefore, the first multiple of 13 that we needed to cross out is $13 \cdot 13$. But $13 \cdot 13$ is 169, which is greater than 149, and so we know that 13 cannot be a factor of 149. Similarly, all numbers greater than 13 (which must have squares that are greater than 169) cannot be factors of 149. We can turn this observation into a generalization: If the square of a number is greater than the number we are testing for, we can stop.

This very powerful generalization, in turn, can be stated as a rule:

Test for determining whether a number n is prime:
List all the prime numbers p that satisfy the equation $p \leq \sqrt{n}$ (or $p^2 \leq n$).
If none of those prime numbers divides n, then n is a prime number.

Prime Factorization

Recall the question about possible dates in 2096 in which the product of the day and month will be 96. We mentioned that another strategy would be to find all the factors of 96. One way to find all the factors of a number is to determine the **prime factorization** of that number—that is, to represent that number as the product of numbers, each of which is prime.

The prime factorization of 96 is $2 \cdot 2 \cdot 2 \cdot 2 \cdot 2 \cdot 3$. Do you see how this representation can help us to determine all the factors of 96? Think and then read on. . . .

We simply check to see whether our number can be constructed from the prime factorization. Clearly, 2 and 3 are factors of 96.

- 4 is a factor because $4 = 2 \cdot 2$.
- 5 is not a factor of 96 because 5 is not in this set.
- 6 is a factor because $6 = 2 \cdot 3$.
- 7 is not a factor of 96 because 7 is not in this set.
- 8 is a factor because $8 = 2 \cdot 2 \cdot 2$.
- 9 is not a factor of 96 because $3 \cdot 3$ is not in this set.
- 10 is not a factor of 96 because $2 \cdot 5$ is not in this set.
- 11 is not a factor of 96 because 11 is not in this set.
- 12 is a factor because $12 = 2 \cdot 2 \cdot 3$.

Because $8 \cdot 12 = 96$, we don't have to look any further. Do you see why?

Recall the first investigation, in which we saw that many dates came in pairs. In this case, each factor up to 12 is part of a pair: $1 \cdot 96$, $2 \cdot 48$, $3 \cdot 32$, $4 \cdot 24$, $6 \cdot 16$, and $8 \cdot 12$.

Thus the factors of 96 are 1, 2, 3, 4, 6, 8, 12, 16, 24, 32, 48, and 96.

Now it is simply a matter of determining which possibilities are valid dates; for example, 1/96/96 is not a valid date, nor is 2/48/96.

Determining the prime factorization of a number is not an easy task for many students, and there are many different methods. Take a few minutes to determine the prime factorization of 36 and 84. Then read the following discussion.

History

The number 1 was a puzzle to the early Greeks. They did not consider 1 to be a number at all because it was the "indivisible unit from which all other numbers arose."[3] However, it was not just the early Greeks who found 1 to be a "different kind of number." The German Kobel wrote in 1537 that "Wherefrom thou understandest that 1 is no number, but it is a generatrix, beginning, and foundation for all other numbers."[4]

---
[3]David Wells, *The Penguin Dictionary of Curious and Interesting Numbers* (New York: Penguin Books, 1986), p. 30.
[4]Ibid., p. 30.

Tree diagrams and prime factorization   One technique for determining the prime factorization of a number that appeals to many students is to use a tree diagram. The top of the tree is the original number. Each number is equal to the product of the two numbers immediately below it. Below are two tree diagrams that produce the prime factorization for 36—that is, $2 \cdot 2 \cdot 3 \cdot 3$.

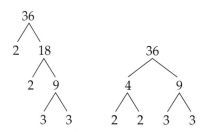

Division and prime factorization   Other students prefer a technique that looks like long division that goes up instead of down. The division is done mentally. Some students like to do it "in order"; that is, they divide by 2 until they can do it no more and then go to bigger and bigger prime numbers. When they do this, their final answer is already in order from smallest to largest.

$$\begin{array}{r} 7 \\ 3\overline{)21} \\ 2\overline{)42} \\ 2\overline{)84} \end{array} \quad 84 = 2 \cdot 2 \cdot 3 \cdot 7$$

The Fundamental Theorem of Arithmetic

If you compare the two tree diagrams above, you can see that we got the same prime factorization for 36 regardless of how we started. It turns out that this is true in all cases. This seemingly obvious statement actually has quite a formal name, the **Fundamental Theorem of Arithmetic**. It states that every natural number greater than 1 has one unique way of being represented as a product of its prime factors.

A Largest Prime Number?

"And the boy in Winnetka, Illinois, who wanted to know: Is there a train so long you can't count the cars? Is there a blackboard so long it will hold all the numbers?" (Carl Sandburg).

This quote brings to mind a related question: Is there a largest prime number? Do you think there is one or not? What arguments could be made either way?

The Fundamental Theorem of Arithmetic enables us to answer this question. It turns out that there is no largest prime number; there are infinitely many. The first proof of this conjecture occurs in Euclid's *Elements*. The proof uses indirect reasoning, in which we begin by assuming that there is a largest prime number, which we will call P.

Let us now create a number X, in the following way:

$$X = (2 \cdot 3 \cdot 4 \cdots P) + 1$$

We have assumed that P is the largest *prime* number, so X must be a *composite* number; that is, it must have at least one factor other than 1 and itself.

> ### Outside the Classroom
>
> The invention of radio early in the twentieth century had many ramifications for people's lives. During World War II, the ability to communicate by radio enabled commanders to be more effective both before and during battles. However, the other side was able to receive the messages too, as long as they knew the radio frequency on which the messages were transmitted. Therefore, important messages were sent in code. Functions were created to encode messages. One very simple coding function is to replace each letter with the next letter in the alphabet. Using this code, what is the following message: I B Q Q Z C J S U I E B Z? The answer is at the end of the next paragraph. As time went on, each side became better at cracking the other side's codes. A team of British analysts developed a machine to crack German codes, which became known as the German Enigma Cipher Machine.
>
> More recently, several M.I.T. scientists have developed what is called a safe coding system. The process involves selecting two very large prime numbers. The coding key is the product of these two numbers. Because it is very difficult to factor extremely large numbers, it is virtually impossible to crack such codes. Answer to coded message: happy birthday!

Let F represent the smallest *prime factor* of X. Because F is a prime number and P is the largest prime number, F must be a prime number less than P; that is, it must be one of the numbers 2, 3, 4, ... up to P.

However, it cannot be one of these numbers. Do you see why? Think before reading on. . . .

We constructed our number X in such a way that it is not divisible by 2, 3, 4, or any number up to P. That is equivalent to saying that 2 is not a factor of X, 3 is not a factor, and so forth. Thus the smallest possible prime factor of X is greater than P, which is supposed to be the largest prime number. This contradiction shows that the assumption that there is a largest prime number cannot be true.

INVESTIGATION 4.9

Numbers with Personalities: Perfect and Other Numbers

Some of the names that have been given to certain kinds of numbers convey human characteristics: for example, *perfect*, *amicable*, *sociable*, *weird*, *narcissistic*, and *irrational*. We will examine some of these numbers, because they provide rich ground for developing mathematical thinking; they are also good "rainy-day activities" when you teach.

A. A number n is a **perfect** number iff the sum of its *proper divisors* is equal to n. A **proper divisor** of a number is a divisor that is less than the number. Do you see why 6 is the first perfect number?

DISCUSSION

Six is the first perfect number because 1, 2, and 3 are the proper divisors of 6 and their sum is 6.

238 CHAPTER 4 / Number Theory

> B. Perfect numbers lead to two other kinds of numbers, which are called abundant and deficient. Before looking at the definitions of these terms, try to define them yourself. . . .

A number n is an **abundant** number iff the sum of its proper divisors is greater than n.

A number n is a **deficient** number iff the sum of its proper divisors is less than n.

Make a table listing the first 30 natural numbers. Put an A next to ones you think will be abundant, a D next to ones you think will be deficient, and a ? next to ones you feel could go either way. Try to articulate the reasoning behind your predictions. Then read on. . . .

DISCUSSION

All prime numbers are deficient by definition. Do you see why? Because the only proper factor of a prime number is 1, the sum of the proper factors of a prime number is 1. Of the nonprime numbers less than 30, the deficient numbers are 4, 8, 9, 10, 14, 15, 16, 21, 22, 25, 26, 27.

Does this list shed light on any other relationships? What does it show?

Notice that the subset of the square numbers in the above set {4, 9, 16, 25} has only deficient numbers. Also, except for 6, any doubles of members of the set of prime numbers less than 30 {2, 3, 5, 7, 11, 13, 17, 19, 23, 29} will also be deficient. Why is this?

The following numbers are abundant: 12, 18, 20, 24, and 30. What do you notice about this set?

One pattern that students often observe that leads to a hypothesis is that all of these numbers are even. Is this just coincidence, or is it true that all abundant numbers are even numbers? What do you think? This question will be left unanswered for you to solve on your own.

The active reader will have noticed that we overlooked 28 in our lists of deficient and abundant numbers. Why?

28 is the second perfect number!

Summary

In this section, we have examined another way of classifying natural numbers: prime and composite (and, of course, 1, which is neither prime nor composite). The Sieve of Eratosthenes not only gave us a tool to determine whether a number is prime but also helped us to develop and refine other problem-solving tools. We have also found that each number has a unique prime factorization, which is another kind of decomposition (funny how this idea keeps coming up!). It is critically important to note that the decomposition of composite numbers into their prime factors is a multiplicative decomposition; this is one of many places where seeing a number in additive or multiplicative terms is not a matter of preference but a matter of structure.

We have explored several strategies that enable us to determine the prime factorization of a natural number. When you teach children, you will find that the Fundamental Theorem of Arithmetic, like the commutative property, is not immediately obvious to children, and in both cases, an understanding of these ideas makes work with mathematics much easier. Determining the prime factorization is like looking through a microscope to see the atomic structure of a molecule. In one sense, prime numbers are the building blocks for whole numbers. We have also seen that the concept of prime and composite numbers has applications outside the classroom.

The notions of prime and composite enable us to understand better Explorations 4.1 and 4.2. Think back to the ones that you did. How does this section add to your understanding of those explorations? Then read on. . . .

In Exploration 4.1 (Taxman), we realize that the first move is to select the largest prime number. We realize that numbers with many factors (like 24) are better left for the end of the game. Do you see why? In Exploration 4.2 (Factors), we found that whereas all prime numbers have only two factors, the set of composite numbers is not so simple. Although most composite numbers have an even number of factors, not all do, and some composite numbers have many factors. Some have so many factors that we call them abundant numbers!

EXERCISES 4.2

1. How many factors does a prime number have?

2. Write the first 20 prime numbers.

3. Find the prime factorization of the numbers below:
   a. 48   b. 75   c. 92
   d. 144   e. 196   f. 504
   g. 756   h. 14,586

4. In each case below, determine whether the number is prime or composite. You will probably want to use a combination of divisibility tests, common sense, and concepts from our work with prime numbers.
   a. 61   b. 65   c. 71
   d. 89   e. 223   f. 347
   g. 437   h. 961   i. 3,437

5. Which of the numbers between 511 and 525 are prime? Briefly explain your reason for denoting each number as either composite or prime.

6. Consider the numbers representing the years in the second half of the twenty-first century. Which of those numbers are prime?

7. To determine whether 617 is a prime number, it suffices to check that no prime number less than ___ is a factor of it. Explain your answer.

8. A **superprime** is defined in the following way: It is a prime number such that each time a digit is removed (one at a time), starting at the right side, the remaining number is still prime. For example, 2391 is a superprime because 239 is a prime and 23 is a prime and 2 is a prime. Find three three-digit superprime numbers. Describe strategies that helped in your search.

9. There is an all-star team of abundant numbers called **superabundant** numbers. Here are the membership criteria for inclusion in this exclusive club:

   Take the sum of all of a number's divisors, including itself. Divide by this sum by the number itself.

   If this quotient is greater than the quotient for any natural number less than this number, then the number is called superabundant.

   a. Design and run a spreadsheet to determine which of the first 50 natural numbers are superabundant, and then write directions for your program so that others may use it.
   b. Describe any new patterns or observations about characteristics of superabundant numbers that you found in making and analyzing this table.
   c. Make a line graph of the first 50 natural numbers in which the horizontal axis consists of successive natural numbers and the vertical axis represents the value of the quotient used to determine whether the number is superabundant. What new observations come from the graph?
   d. Predict the next superabundant number and explain the reasoning behind your prediction.

10. Construct a number that has exactly five divisors.

11. a. Find the smallest natural number that is divisible by the first ten counting numbers.
    b. How else could this question have been worded?

240 CHAPTER 4 / Number Theory

12. a. What is the smallest four-digit number that has exactly three factors? Justify your answer and explain how you arrived at your solution.
   b. What is the largest three-digit number that has exactly four factors? Justify your answer and explain how you arrived at your solution.

13. What is the smallest composite number that is divisible by none of the primes 2, 3, 5, and 7? Briefly explain your answer.

14. Here are partial directions for creating five consecutive composite numbers. First, begin with $2 \cdot 3 \cdot 4 \cdot 5 \cdot 6$. Then use the divisibility theorem from the previous section; that is, if $a|b$ and $a|c$, then $a|(b + c)$. Can you use these hints to create five consecutive composite numbers?

15. An organization with 100 people will be divided into smaller committees.
   a. What are the possible committee sizes if every committee must be the same size?
   b. What are the possible committee sizes if the organization wants to have some large committees (each the same size) and some smaller committees (each the same size), and the larger committees must be at least twice the size of the smaller ones?
   c. Create at least one more restriction on part (b) that would lead to fewer possible answers.

16. a. If 21 divides m, what else must divide m?
   b. If 36 divides m, what else must divide m?

17. If you examine all the prime numbers less than 100, some of them are "next-door neighbors"—for example, 11 and 13, 17 and 19, 71 and 73. Mathematicians call these twin primes or prime twins. What about prime triplets? Only one set of prime triplets exists: 3, 5, 7. Prove that there cannot be another set of prime triplets.

18. This is a famous problem in mathematics education. Suppose you lined up 1000 dolls in a row, all facing up. Then suppose you walked down the line and turned over every other doll. Now, one-half of the dolls would be facing up (those representing odd numbers) and one-half of the dolls would be facing down. Now suppose you walked down the line again, and this time turned over every third doll. That is, if it was facing up, you would turn it so that it was facing down; if it was facing down, you would turn it so that it was facing up. Continue in this manner 1000 times. At this point, which dolls will be facing up and which dolls will be facing down?

The following suggestions may help you with the problem: Some students find it helpful to make the problem smaller—for example, 50 dolls or 25 dolls. Some students find it helpful to model the problem. What other objects could you use to represent the dolls? Most students find it helpful to make a table to represent their findings.

19. a. Describe the patterns you find in the following figure.
   b. Explain how to make the figure as though you were talking on the phone to someone.
   c. Find a shortcut for finding the sum of the numbers in any hexagon.
   d. Find a shortcut for finding the sum of any row.

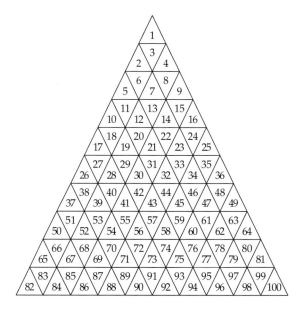

20. The following questions have come from students over the years. Explore these questions with a partner. After your exploration, summarize and justify your findings.
   a. Of the first 100 numbers, 25 are prime. How many of the second hundred numbers will be prime? Will the frequency (density) of prime numbers increase, decrease, or stay at about 25 percent as the numbers get larger and larger?
   b. Can you predict when twin primes will occur? How many are there in the first 100 numbers? Will there be the same number of twin primes in the next 100 numbers?
   c. The largest difference between two prime numbers less than 100 is 8; it is between the 89 and 97. What is the largest difference you can find between any two prime numbers?
   d. Are there more primes with a 1, a 3, a 7, or a 9 in the ones place?

21. These are some famous conjectures, some of which have yet to be proved or disproved! Explore these conjectures with a partner. Then explain why you think the conjecture is true.
   a. Any prime greater than 3 is within 1 unit of a multiple of 6.
   b. Every even number can be expressed as the difference of two primes.
   c. This conjecture comes from Fermat: Any prime number of the form $p = 4n + 1$, $n \in N$, can be expressed as the sum of two squares in only one way.

d. This conjecture is called Goldbach's conjecture after Christian Goldbach (1690–1764): Any even number greater than 2 can be expressed as the sum of two primes. To this day, no one has either found a counterexample or proved this conjecture.

e. Another conjecture made by Goldbach: Any odd number greater than 5 can be expressed as the sum of three prime numbers.

f. Every triangular number can be represented as the sum of a square number and two triangular numbers.

22. The largest known prime number has 9,152,052 digits. How many pages of paper would it take just to write this number out?

SECTION 4.3 GREATEST COMMON FACTOR AND LEAST COMMON MULTIPLE

WHAT DO YOU THINK?

- What does it mean to say that the GCF and the LCM let us decompose composite numbers?
- How are the GCF and the LCM related to each other?

One of the many decisions I had to make when writing this book was where to place the section on the greatest common factor (GCF) and least common multiple (LCM). In elementary school, these concepts are used primarily when adding and simplifying fractions, and so this section could be placed in Chapter 5. However, these concepts have applications beyond fractions problems, and one of the goals of this book is to get students to see that mathematics is more than just procedures and that many mathematical ideas are connected to other mathematical ideas. Therefore, we will first examine these concepts in this chapter and then apply them to fraction ideas in Chapter 5.

INVESTIGATION 4.10

4.4

Cutting Squares Using Number Theory Concepts

Let's say that a teacher has a rectangular sheet of cardboard 420 centimeters long and 378 centimeters wide and that he wants to cut that sheet into many squares, all of the same size. What are the dimensions of the largest possible square (whose length is a whole number) that will create no waste? Think about this problem and then read on. . . .

DISCUSSION

STRATEGY 1: *Use guess–check–revise*

Looking at the numbers, we can see that they are both divisible by 2, so we could make 2 by 2 squares. Using the divisibility rule for 3, we can see that they are also divisible by 3, so we can make 3 by 3 squares. We can proceed in this manner, checking for divisibility by 4, 5, 6, and so on, until we find that we can go no further.

STRATEGY 2: *Use prime factorization*

What if we found the prime factorization of both numbers? How might that help us? Think about this before reading on. . . .

$$378 = 2 \cdot 3 \cdot 3 \cdot 3 \cdot 7$$
$$420 = 2 \cdot 2 \cdot 3 \cdot 5 \cdot 7$$

Because we are looking for a number that divides *both* 378 and 420, an equivalent statement is that we are looking for a number that is a factor of both. In fact, we are looking for the largest such common factor. By looking at the prime factorization of both numbers, we can see that they have in common a 2, a 3, and a 7. How does this help? If we multiply the three numbers, the product of these three numbers (42) will also be a factor of both 378 and 420.

In fact, 42 is the *greatest* number that divides both 378 and 420. If we express this conclusion using the term *factor* instead of *divides*, we can say that 42 is the greatest factor that 378 and 420 have in common. A more concise way to express this is to say that 42 is the *greatest common factor* of 378 and 420.

The Greatest Common Factor

Whenever we examine two natural numbers, we can create a set of numbers called their common factors. In some cases, the only common factor is 1; for example, examine 7 and 10. However, in many cases, the numbers have several common factors. The greatest of these common factors is called the **greatest common factor** (GCF). We use the notation **GCF(a, b)** to express the GCF of two natural numbers a and b.

Note: In a similar fashion, we can speak of the GCF of three or more numbers. This will be explored in the exercises.

Exploring the greatest common factor concept Before we examine techniques for finding the GCF of two or more numbers, let us take some time to explore the concept. This is analogous to the work we did in Chapter 3 when we explored the concepts of addition and subtraction to understand the algorithms better, and then explored the concepts of multiplication and division to understand those algorithms better. What can we say about the size of the GCF of *any* two numbers a and b? For example, will it always be smaller than the two numbers? Can it be in between two numbers? What do you think? One student compared this question to thinking of the qualifications for a job. For example, what qualities would you expect in an excellent elementary teacher: likes children, has eyes in the back of the head, doesn't require much sleep? Take a few moments to think about this before reading on. . . .

The GCF of two numbers is generally smaller than either of them, because it has to be a factor of each. However, there are some cases in which the GCF of two numbers is *not* smaller than either of them. Can you think of such a case? Think and then read on. . . .

Although the GCF can never be larger than either of the numbers, it can be equal to the smaller of the two numbers. For example, the GCF of 4 and 8 is 4.

In a manner similar to a detective story, we can make predictions about the GCF if we know just a little bit about the two numbers. For example, if you know that two numbers are prime, what can you say about their GCF? Think and then read on. . . .

Prime numbers have only two factors, and one of them is 1. Because the other factor will be the number itself, we can conclude that if the two numbers are prime, their GCF is 1.

What can you say about the GCF of two even numbers? Think and then read on. . . .

> ■ *Mathematics* ■
>
> Most work in mathematics classes involves quantitative thinking, that is, work with quantities. When working with quantities, it is easy for the focus on ideas to take second place to procedures. Thus, it is important also to develop qualitative thinking, as we are doing here; that is, asking questions where we are focusing on quantities (of the ideas) and not just procedures.

Their GCF will be at least 2. Why is this?
What about the GCF of two odd numbers? Think and then read on....

This clue alone is not enough for us to say much, except that the GCF will be an odd number also. Why is this?

INVESTIGATION 4.11

Methods for Finding the GCF

Let us now investigate how we might determine the GCF of two numbers. Rather than give an efficient procedure right away, we will take some time to build the foundation of this procedure, much like taking care while constructing the foundation of a house.

Using only the definition of GCF, how would you determine GCF(45, 60)? Work on this any way you want to. My only recommendation is that you think about the meaning of whatever you do, as opposed to random guess and test. Think and then read on....

DISCUSSION

STRATEGY 1: *Use factorization*

We could, as we just saw, determine all the factors of each number and then find the largest of the common factors:

Factors of 45 = {1, 3, 5, 9, 15, 45}

Factors of 60 = {1, 2, 3, 4, 5, 6, 10, 12, 15, 20, 30, 60}

Common factors = {1, 3, 5, 15}

We see from this list that 15 is the GCF of 45 and 60.

STRATEGY 2: *Use intuition or number sense*

A student who is highly intuitive and has good number sense might just know that 15 divides both these numbers. The fact that 15 divides both numbers simply means that 15 is a common factor. How might you reason that, in fact, 15 is the GCF? Think before reading on....

Let us represent the results of dividing each number by 5 (which we know is not the GCF) and by 15. What do you notice?

$45 = 5 \cdot 9$ $45 = 15 \cdot 3$
$60 = 5 \cdot 12$ $60 = 15 \cdot 4$

When we divide 45 and 60 by 5, we are left with 9 and 12. When we divide 45 and 60 by 15, we are left with 3 and 4. One difference between 9 and 12 and 3 and 4 is that 3 and 4 have no common factors.

When two numbers have no factors in common other than 1, they are said to be **relatively prime**. One of my students came up with an interesting paraphrase of this concept: If two numbers are relatively prime, they are prime to each other.

Because 3 and 4 are relatively prime, 15 is the GCF of 45 and 60. Do you see why?

STRATEGY 3: *Repeatedly divide by prime numbers*

What about people who did not have the intuition and number sense to see this? Is their only recourse the long way we saw above? Another procedure involves an adaptation of the long-division algorithm. The following problem illustrates a systematic application in that we begin with the smallest prime divisor and then move up. That is, we first divide both numbers by 3. At this point, we move up to 5 because 15 and 20 are both divisible by 5. The resulting quotients, 3 and 4, have no factors in common. The GCF of 45 and 60 is the product of their common factors: $3 \cdot 5 = 15$.

$$
\begin{array}{r}
3,\ 4 \\
5\overline{)15,\ 20} \\
3\overline{)45,\ 60}
\end{array}
$$

STRATEGY 4: *Use prime factorization*

This strategy uses the Fundamental Theorem of Arithmetic, which we explored in Section 4.2. We first determine the prime factorization of each number and then look for common factors. If we look at the prime factorizations of 45 and 60 and circle the factors that the two numbers have in common, we have the following:

$$45 = 3 \cdot \boxed{3 \cdot 5}$$
$$60 = 2 \cdot 2 \cdot \boxed{3 \cdot 5}$$

We can further refine this procedure by using exponents:

$$45 = 3^2 \cdot 5^1$$
$$60 = 2^2 \cdot 3^1 \cdot 5^1$$

The GCF is determined by examining those factors that both numbers have in common and then taking the *smallest* exponent in each case. The common factors of 45 and 60 are 3 and 5. The smallest exponent of 3 is 1, and the smallest exponent of 5 is 1. Thus $3 \cdot 5$ is the GCF.

CLASSROOM CONNECTION

Using Cuisenaire rods, how many different ways can you make a 12 train and an 8 train using the same colors? (See Figure 4.10.)

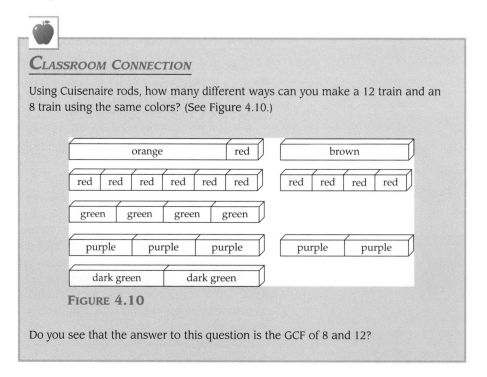

FIGURE 4.10

Do you see that the answer to this question is the GCF of 8 and 12?

Least Common Multiple

Let us begin our investigation of the least common multiple (LCM) by having you think about the following question before reading on: What do you think "least common multiple" means? Write a definition in informal language.

Let us examine this concept word by word, using a specific example. Suppose we wanted to find the least common multiple of 8 and 12. At the most basic level, we can generate the set of multiples of 8 and 12 and then find the smallest number that the two sets have in common.

$$\text{Multiples of } 8 = \{8, 16, 24, 32, 40, 48, 56, 64, 72, 80, 88, 96, 104, \ldots\}$$

$$\text{Multiples of } 12 = \{12, 24, 36, 48, 60, 72, 84, 96, 108, 120, 132, \ldots\}$$

What multiples do the two numbers have in common?

These two numbers have many common multiples: $\{24, 48, 72, 96, \ldots\}$. Because the least of the common multiples is 24, we say that LCM(8, 12) = 24.

Whenever we examine two natural numbers, we can create a set of numbers called their common multiples. The least of these common multiples is called the **least common multiple** (LCM). We use the notation **LCM(a, b)** to express the LCM of two natural numbers, a and b.

Note: In a similar fashion, we can speak of the LCM of three or more numbers. This will be explored in the exercises.

Exploring the least common multiple concept Looking at the process we used to find the LCM, the active reader might think at this point, "Yes, this will work, but what if the numbers are larger? This approach could get tedious pretty fast!" This is true. However, as we did with the greatest common factor, let us engage in some qualitative thinking about LCM to elicit some of the general attributes of the LCM. What can we say about the size of the LCM of *any* two numbers a and b? For example, will it always be greater than the two numbers? Can it be in between the two numbers? What do you think? Take a few moments to think about this before reading on. . . .

The LCM of two numbers is generally greater than either of them, because it has to be a multiple of each. However, there are some cases in which the LCM of two numbers is *not* greater than either of them. Can you think of such a case? Think and then read on. . . .

When one number is a multiple of the other, the LCM will be equal to the larger number; for example, LCM(5, 10) = 10. Why is this?

Now let me give you two more numbers. Without doing any computation, what can you tell me about the LCM of 18 and 40? Think and read on. . . .

Here are several possible responses:

- It is greater than 40.
- It is less than or equal to $18 \cdot 40$, or 720.
- It is a multiple of both numbers.
- It ends in a zero, because all multiples of 40 end in 0.

Strategies for Finding the LCM

Now let us examine how we might find the actual LCM of 18 and 40. Again, there are many ways to determine the LCM of two numbers. We will examine several ways.

One way of determining the LCM of 18 and 40 is to construct the LCM by beginning with one of the numbers and applying our understanding of LCM. This process is illustrated in the following discussion:

Reasoning

The LCM *must* contain all the factors of 18.

Now, in order *also* to be a multiple of 40, the LCM will have to contain all the factors in the prime factorization of 40.

Looking now at the factors of 18, what factors of 40 are we missing? Think and then read on. . . .

The work

$18 = 2 \cdot 3 \cdot 3$,

LCM(18, 40) must contain: $2 \cdot 3 \cdot 3$

$40 = 2 \cdot 2 \cdot 2 \cdot 5$

Reasoning

We need to put two more 2s and one 5 into our prime factorization of LCM(18, 40).

Another way of illustrating this process is to note the prime factorizations of 18 and 40 and realize that the least common multiple must contain all the factors in either number with no redundancies. That is, the LCM needs to contain three 2s, two 3s, and one 5.

The work

LCM(18, 40) must contain: $\mathbf{2} \cdot \mathbf{2} \cdot 2 \cdot 3 \cdot 3 \cdot \mathbf{5}$

That is, LCM(18, 40) = 360.

$ 2 \cdot 3 \cdot 3$
$2 \cdot 2 \cdot 2 \cdot \cdot 5$

Using Prime Factorization and Exponents to Find the LCM

There is a more formal way to find the LCM, and this is connected to one of the ways in which we found the GCF. This method comes from representing the prime factorization of each number in exponential form:

$18 = 2 \cdot 3 \cdot 3 = 2^1 \cdot 3^2$
$40 = 2 \cdot 2 \cdot 2 \cdot 5 = 2^3 \cdot 5^1$

When finding the GCF, we took the smaller exponent of all common factors. What do you think we will do when finding the LCM? Think and read on. . . .

In order for a number to be the LCM, it must contain *all* the factors in either number. For example, because the prime factorization of 18 contains a 2, the LCM must contain a 2. However, because the prime factorization of 40 contains three 2s, the LCM must contain three 2s. Thus, when we examine the prime factorization of each number, whenever there is a common factor, we must take the greater exponent.

Using this method, we find that LCM(18, 40) = $2^3 \cdot 3^2 \cdot 5^1$. Do you see why?

The only factor that 18 and 40 have in common is 2, and the greatest exponent above 2 is 3 (meaning that 2 · 2 · 2 is a factor of 40). Therefore, the prime factorization of the LCM must contain 2^3. The noncommon factors are 3 and 5, so 3^2 and 5^1 are also placed in the prime factorization of the LCM.

Note: This procedure with exponents makes sense to some students and not to others. The point is not that everyone should use this procedure because it is the *best* procedure. Rather, the point is that this is a very efficient procedure *if* you feel comfortable with it. If you don't feel comfortable with this procedure, then you can work with it until it makes more sense, or you can use another procedure.

In the remainder of this section, we will develop and explore relationships between GCF and LCM.

CLASSROOM CONNECTION

Below are two different ways in which the concept of LCM can emerge in elementary school. Teachers may give their students a 100 chart and ask them to cross out multiples of 8 with one color and cross out multiples of 12 with another color, or to draw a circle around multiples of 8 and a square around multiples of 12, as in Table 4.4. When students discover that some numbers have both a circle and a square around them, the teacher can ask, "How would you describe those numbers?" In this manner, the students have constructed the concept of LCM in their heads. Then, when the more formal definition is presented, they already have experience that connects to this idea.

TABLE 4.4

1	2	3	4	5	6	7	⑧	9	10
11	☐12	13	14	15	⑯	17	18	19	20
21	22	23	⓸24	25	26	27	28	29	30
31	㉜	33	34	35	☐36	37	38	39	㊵
41	42	43	44	45	46	47	㊽	49	50

Cuisenaire rods can also be used to introduce this concept to children. The teacher might ask the students to select a 4 rod and a 6 rod and then ask, "How many different ways can you make a 4 train and a 6 train that have the same length?" (See Figure 4.11.) Do you see how the answer to this question connects to the LCM concept?

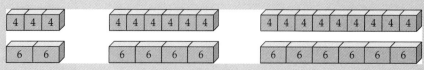

FIGURE 4.11

INVESTIGATION 4.12

Relationships Between the GCF and the LCM

It is important to note that although the GCF and LCM are *different* concepts, they are also closely related. One of the goals of the next few investigations is for you to deepen your understanding of each of these concepts and also to understand more deeply how they are related.

Complete Table 4.5 and note what patterns you observe and what hypotheses you make. Think and then read on. . . .

TABLE 4.5

a	b	GCF(a, b)	LCM(a, b)
4	6	2	12
6	10		
8	12		
9	12		
10	15		

DISCUSSION

If you take any two numbers, it just so happens that the product of those numbers is always identical to the product of their GCF and LCM; that is, for any natural numbers a and b,

$$a \cdot b = \text{GFC}(a, b) \cdot \text{LCM}(a, b)$$

Using Venn diagrams The proof of this is beyond the scope of this book, but Venn diagrams can help us to understand *why* this relationship is true. However, using these diagrams here requires a slightly unorthodox use of the concepts of set, intersection, and union. This connection among GCF, LCM, and sets came from a tutorial session in my office with several students; one student was trying to explain the concepts to another student and suddenly "saw" the connection among GCF, LCM, and intersection and union. One of the outcomes of our discourse was a deeper understanding of the connection between the two concepts.

Let us examine this connection now. Consider the prime factorization of 18 and 40 as sets; that is,

Prime factorization of 18 = {2, 3, 3}

Prime factorization of 40 = {2, 2, 2, 5}

If we place these elements in a Venn diagram, we have Figure 4.12. How does this Venn diagram connect to the GCF of 18 and 40? How does it connect to the LCM of 18 and 40?

From this unorthodox representation, the GCF of 18 and 40 can be seen as the **intersection** of the two sets; that is, GCF(18, 40) = 2. The LCM of 18 and 40 can be seen as the **union** of the two sets; that is, LCM(18, 40) = 3 · 3 · 2 · 2 · 2 ·5 = 360.

In this case, we see that the factors fall within three distinct regions of the Venn diagram. The first region contains those numbers that are factors of 18 but not of 40 (3 · 3), the second region contains common factors (2), and the third region contains those numbers that are factors of 40 but not of 18 (2 · 2 · 5).

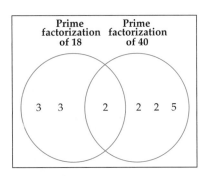

FIGURE 4.12

CLASSROOM CONNECTION

For some mathematicians, the unorthodox use of Venn diagrams is heresy. However, for many mathematics educators, there is much value in having students invent (sometimes unorthodox) procedures and algorithms. This is alluded to in several places in the NCTM standards. *Principles and Standards for School Mathematics* encourages having students invent procedures, representations, and notation. The professional standards speak of "invented and conventional terms and symbols" among Tools for Enhancing Discourse (*Professional Standards*, p. 52).

Now examine the equation connecting the two concepts in this light: $18 \cdot 40 = \text{GCF}(18, 40) \cdot \text{LCM}(18, 40)$. We can see that 18 and 40 both consist of numbers from two regions, the GCF consists of numbers from the common region, and the LCM consists of numbers from all three regions.

Yet Another Way to Find the LCM

This relationship between the GCF and the LCM has a very practical application. It yields another way to find the LCM of two numbers. Do you see how? Think and then read on. . . .

Consider, for example, finding LCM(40, 72). The equation in Investigation 4.12 is an algebraic equation, and thus we can use the rules of algebra to solve the equation for LCM(a, b). Doing so, we have

$$\text{LCM}(a, b) = \frac{a \cdot b}{\text{GCF}(a, b)}$$

As you may have already discovered, finding the GCF of two numbers is generally much easier than finding the LCM. Thus, to find LCM(40, 72), we have only to find GCF(40, 72) and then solve the equation. Because GCF(40, 72) = 8, we have

$$\text{LCM}(40, 72) = \frac{40 \cdot 72}{8}$$

If you pause for a few seconds, you should be able to see how you can determine this computation in your head. Do so and then read on. . . .

If you first divide the 72 by 8, you then have the relatively simple multiplication problem $40 \cdot 9$, which is 360.

INVESTIGATION 4.13 | Going Deeper into the GCF and the LCM

The following problem provides an opportunity to develop your problem-solving toolbox while applying your understanding of GCF, LCM, and their relationship.

If the GCF of 45 and x is 9, and the LCM of 45 and x is 135, find x.

Take a few minutes to work on this problem on your own and then read on. . . .

DISCUSSION

There are several different ways to solve this problem.

STRATEGY 1: *Think of the characteristics of GCF and LCM*

I want to take this time to emphasize the value of thinking *before* jumping into a problem. Think about the characteristics of x. What do we know about x from the given information? First, let's look at the other two numbers.

$$45 = 3 \cdot 3 \cdot 5$$
$$135 = 3 \cdot 3 \cdot 3 \cdot 5$$

If 9 is the GCF of 45 and x, that means x is a multiple of 9. Do you see why? If the LCM of 45 and x is 135, that means that the prime factorization of x has to have at least three 3s. Do you see why?

Using this line of reasoning, we find that our first candidate for the answer is a multiple of 9 that has three 3s: $3 \cdot 3 \cdot 3 = 27$

It also turns out, not surprisingly, that 27 is the correct answer. The value of doing a bit of qualitative thinking at the beginning of the problem recalls the old proverb "A stitch in time saves nine!"

STRATEGY 2: *Build x from the ground up*

We could also proceed by building this number x.

Since we know that the GCF of 45 and x is 9, we know that x must be at least 9, that is 3×3.

Since we know that the LCM of 45 and x is 135, we know that:

$$45 = 3 \times 3 \times 5$$
$$x = 3 \times 3 \times ?$$
$$135 = 3 \times 3 \times 3 \times 5$$

The meaning of LCM is that it has all the factors of either number. Since 135 has $3 \times 3 \times 3$, x must be at least $3 \times 3 \times 3 = 27$. The only other possibility now for x would be $3 \times 3 \times 3 \times 5$, but the GCF of 45 and 135 could be 45, so x cannot be 135.

Relationships of Operations on the Set of Natural Numbers

One of the important "big ideas" of arithmetic is the part–whole aspect of numbers—what some people refer to as understanding how numbers are decomposed and composed. In the lower elementary grades, the primary focus is on additive decompositions. For example, a student who understands the various ways in which 10 can be decomposed—$9 + 1, 8 + 2, 7 + 3, 6 + 4, 5 + 5$, and so on—knows more than just "number facts." This decomposition is a critical part of knowing how to add and subtract when regrouping and is a critical part of good mental arithmetic and estimating. In the upper elementary grades, understanding multiplicative decompositions is developed. For example, 12 can be decomposed multiplicatively in several ways—$12 \cdot 1, 6 \cdot 2$, and $4 \cdot 3$, as well as its prime decomposition, $2 \cdot 2 \cdot 3$.

Additive and multiplicative decomposition are "big ideas" that occur repeatedly throughout mathematics. Understanding additive and multiplicative relationships enables us to compare amounts in two different ways.

For example, if an item costs $10 in one store and $15 in another store, we can compare the two prices in additive terms—it costs $5 more in the second store—or we can compare the two prices in multiplicative terms—it costs $1\frac{1}{2}$ times as much or it costs 50 percent more in the second store. These comparisons have to do with how we choose to decompose the amounts: 15 as 10 + 5 or 15 as 3 · 5 and 10 as 2 · 5. These two different ways of comparing amounts multiplicatively will be developed in Chapter 5 and Chapter 6.

Because people learn and perceive ideas in different ways, I also offer the diagram in Figure 4.13, which illustrates the relationships among these three ideas. The figure shows that when we examine the way two numbers are related, we can decompose them additively or multiplicatively. In the former case, relating two numbers often leads to statements that use the words *more than* or *less than*. In the latter case, relating two numbers often leads to statements involving concepts and terms such as *divisibility*, *ratio*, *percent more*, and *percent increase*.

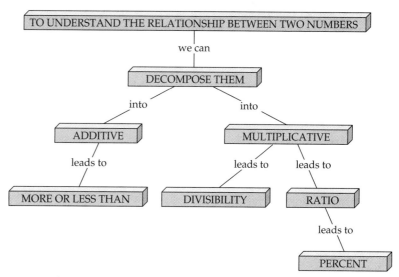

FIGURE 4.13

Summary

In this section, we have investigated the concepts of the greatest common factor and the least common multiple. In developing more efficient strategies for determining the GCF or the LCM of two or more numbers, we used concepts from the first two sections of the chapter (divisibility, prime, and prime factorization), and we again used the idea of decomposition. In this case, the additive decompositions were not relevant; rather, the multiplicative decompositions enabled us to determine the GCF and LCM and to understand better how these two ideas are related.

These ideas of GCF, LCM, and relatively prime shed new light on our understanding of African sand drawings in Exploration 4.4. The number of paths needed to trace a sona was the GCF of the dimensions of the rectangular array for that sona.

EXERCISES 4.3

1. Find the following:
   a. GCF(30, 75)
   b. GCF(45, 54)
   c. GCF(12, 35)
   d. GCF(27, 189)
   e. GCF(75, 144)
   f. GCF(105, 132)
   g. GCF(156, 910)
   h. GCF(630, 1848)

2. Find the following:
   a. GCF(12, 30, 75)
   b. GCF(12, 333, 8415)

3. Find the following:
   a. LCM(12, 20)
   b. LCM(30, 75)
   c. LCM(44, 66)
   d. LCM(462, 630)

4. Find the following:
   a. LCM(12, 30, 75)
   b. LCM(20, 30, 40)

5. a. Find two numbers such that their GCF = 2.
   b. Find two numbers, both greater than 100, such that their GCF = 2.
   c. Find two numbers such that their GCF = 6.
   d. Find two numbers, both greater than 100, such that their GCF = 6.

6. Find the GCF and the LCM of the following pairs of numbers:
   a. 25, 40
   b. 72, 108
   c. 252, 420
   d. 3780, 3960

7. The GCF of two numbers is 18. Their LCM is 630. What are the two numbers?

8. The GCF of 66 and x is 11; the LCM of 66 and x is 858. Find x.

9. The GCF of two numbers m and n is 12, their LCM is 600, and both m and n are less than 500. Find m and n.

10. A child has a large supply of dominoes, each of which measures 32 millimeters by 56 millimeters. She wants to lay them out to form a solid square, and she wants them all to be laid out horizontally. What will be the dimensions of the smallest square that she can form, and how many dominoes will it require?

11. George vacuums the rugs every 18 days, mows the grass every 12 days, and pays the bills every 15 days. Today he did all three. How long will it be before he has another day like today?

12. In each of the following, explain your answer.
    a. What is the relationship between a and b if GCF(a, b) = a?
    b. GCF(a, a) = ?
    c. If GCF(a, b) = b, then LCM(a, b) = ?

13. Classify each of the following as true or false, assuming that a, b, and c are integers. If a statement is true, justify it. If it is false, give a counterexample, and then revise the statement to make it a true statement.
    a. If a and b are different and prime, then GCF(a, b) = 1.
    b. If a and b are relatively prime, then LCM(a, b) = ab.
    c. If a and b are different and are both even, then GCF(a, b) = 2.
    d. If GCF(a, b) = 2, then a and b are both even.
    e. LCM(a, b) | ab.

14. Ann and Bob are cycling on a track. Ann completes one lap every 12 seconds, and Bob completes one lap every 15 seconds. When will Ann lap Bob, assuming that they started together?

15. A manufacturer sells widgets, each of which is packaged in a box whose dimensions are 4 centimeters by 6 centimeters by 10 centimeters. Design the dimensions of a box to ship widgets. Your box must hold at least 1000 widgets.

16. I am thinking of a two-digit number in which the sum of the digits is 9 and the number is even. These clues alone do not create one answer. Add one more clue so that there is only one answer.

17. a. What is the largest square that can be used to fill a 6 by 10 rectangle? For example, you would need sixty 1 by 1 squares to make a 6 by 10 rectangle. However, if you started with a 3 by 3 square, you would not be able to fill the rectangle using only 3 by 3 squares. The most concise way to ask this question is "What is the largest square that can be used to tile a 6 by 10 rectangle?" We will study tiling in Chapter 9.
    b. What is the largest square that can be used to fill a 12 by 18 rectangle?
    c. Now comes the more general question: How would you go about finding the largest square that can be used to fill an x by y rectangle?

18. *Classroom Connection* Obtain a set of Cuisenaire rods from your instructor or from the library on your campus.
    a. What is the shortest train that can be measured by both the purple rod and the dark green rod? *Note:* A longer train can be measured by a shorter train if you can make the longer train by using only the shorter train. Thus a 9 rod cannot be measured by a 2 rod, but it can be measured by a 3 rod.
    b. What is the shortest train that can be measured by both the dark green rod and the brown rod?
    c. Now comes the more general question: How would you go about finding the shortest train that can be "measured" by both a rod that is x units long and a rod that is y units long?

19. a. *Classroom Connection* What is the longest train that will measure the brown rod and a train that is 12 units long?
    b. What is the longest train that will measure a train that is 15 units long and a train that is 20 units long?
    c. Now comes the more general question: How would you go about finding the longest train that will measure a train that is *x* units long and a train that is *y* units long?

20. We have talked about the GCF of two or more numbers and the LCM of two or more numbers. Why don't we talk about the LCF (least common factor) or the GCM (greatest common multiple) of two or more numbers?

Chapter Summary

1. Prime numbers are the building blocks for the set of natural numbers.
2. Divisibility, GCF, and LCM enable us to decompose composite numbers. If elementary students understand these concepts, they can more easily compose and decompose mathematical expressions.
3. These decompositions are multiplicative rather than additive.
4. These mathematical concepts and a good problem-solving toolbox are essential to being able to see patterns that lead to hypotheses, which are then tested and either confirmed or revised.
5. Number theory connects with other mathematical concepts and fields.
6. Exploring numbers and number patterns can be fascinating and enjoyable.

BASIC CONCEPTS

Section 4.1 Divisibility and Related Concepts

divides *213*
factor *213*
multiple *213*
divisible *213*
divisor *213*
odd and even numbers *218*

Divisibility relationships:
If $a|b$ and $a|c$, then $a|(b+c)$. *221*
If $a|b$, then $a|bc$. *225*

Divisibility Rules:
Divisibility by 2 *223*
Divisibility by 5 *223*
Divisibility by 10 *223*
Divisibility by 3 *223*

Section 4.2 Prime and Composite Numbers

prime *232*
composite *232*
sieve of Eratosthenes *233*
indirect reasoning *234*
test for determining whether a number is prime *235*
prime factorization *235*
the Fundamental Theorem of Arithmetic *236*
perfect numbers, proper divisor, abundant numbers, deficient numbers *237–238*

Section 4.3 Greatest Common Factor and Least Common Multiple

greatest common factor (GCF) *242*

Methods for determining the GCF
listing all factors of each number, using prime factorization and reasoning, using factor trees, using intuition, repeated division by prime numbers, using prime factorization and exponential notation

least common multiple (LCM) *245*

Methods for finding the LCM
listing a sufficient number of multiples of each number, using factor trees, using prime factorization and reasoning, using prime factorization and exponential notation

$$\text{LCM}(a, b) = \frac{a \cdot b}{\text{GCF}(a, b)}$$ *249*

relatively prime *243*

Chapter 4 Review Exercises

1. Use divisibility rules to answer the following. Briefly explain how you arrived at your conclusion. Simply dividing the large number by the smaller is not acceptable.
   a. Is the number 222,666 divisible by 3? Why or why not?
   b. Is the number 222,666 divisible by 6? Why or why not?
   c. Is the number 222,666 divisible by 4? Why or why not?

2. The text described several equivalent ways of describing the situation when one number divides another. Write two statements that are equivalent to the statement that 4 divides 12.

3. Find a number between 4300 and 4325 that is divisible by 3, 4, and 5.

4. True or false? If a divides c and b divides c, then ab divides c. Why?

5. a. Find a number that has a remainder of 2 when divided by 3 and a remainder of 3 when divided by 4.
   b. Find a three-digit number that has a remainder of 2 when divided by 3 and a remainder of 3 when divided by 4.

6. Find a number between 90 and 100 that has exactly four factors.

7. What can you say about two numbers if their sum is odd and their product is even? Justify your answer.

8. What is the units digit of 7^{100}?

9. Represent with a Venn diagram the relationship among the set of numbers divisible by 2, the set of numbers divisible by 3, and the set of numbers divisible by 6.

10. Find the prime factorization of each of the following numbers.
    a. 72
    b. 252

11. Determine whether 347 is prime or composite.

12. What is the smallest composite number that is not divisible by 2, 3, 4, 5, or 6?

13. If 15 divides m, what else must divide m?

14. Find the following.
    a. GCF(24, 40)
    b. GCF(72, 200)
    c. GCF(24, 100, 144)

15. Find the following.
    a. LCM(18, 30)
    b. LCM(8, 12, 20)

16. Find two numbers a and b, both greater than 50, such that GCF(a, b) = 6.

17. The GCF of 45 and x is 9; the LCM of 45 and x is 270. Find x.

18. Indicate under what conditions the following statement is true: LCM(a, b) $|$ ab.

19. Your organization is having a fund-raiser and will be selling hot dogs. You anticipate selling about 200 hot dogs. You go to the store and find that the hot dogs come in packages of 12, but the hot dog buns come in packages of 8. What number, around 200, would give you the same number of hot dogs and hot dog buns?

CHAPTER 5

Extending the Number System

5.1 Integers

5.2 Fractions and Rational Numbers

5.3 Understanding Operations with Fractions

5.4 Beyond Integers and Fractions: Decimals, Exponents, and Real Numbers

For many thousands of years, whole numbers were adequate for most people's needs. However, the limitations of whole numbers became more and more problematic as time went on. In this chapter, we will examine four extensions of the set of whole numbers: negative numbers, fractions, decimals, and irrational numbers.[1] Most people do not realize how recent the invention of these kinds of numbers is and that these sets of numbers represent significant and important leaps in the development of mathematics. In this chapter, it is important that you come to appreciate the significance of these sets of numbers and that you understand how they are related to one another. In Chapter 3, we examined the various meanings of the four basic operations and how different algorithms enable us to compute quickly. In this chapter, we will examine what each of these new sets of numbers means. Knowing what they mean and understanding the four operations will then enable you to make sense of the computation algorithms that we use for integers, fractions, and decimals. As before, knowing why as well as how increases one's mathematical power:

> Teachers [of all grades] should be able to extend the number systems from the whole numbers to fractions and integers, then rationals and real numbers, including a discussion of the extension of the operations, properties, and ordering. Notions of fractions, decimals, percents, ratio, and proportion should be developed through problems with an applied flavor.
>
> (*Professional Standards*, p. 136)

---

[1] Although there are still numbers beyond irrational numbers, we will limit ourselves to the sets of numbers mentioned here because these are the sets that we focus on in K–8 mathematics.

SECTION 5.1 INTEGERS

WHAT DO YOU THINK?

- Why is it that the sum of two negative numbers is a negative number, but the product of two negative numbers is a positive number?
- How do operations with integers connect with whole-number operations? How do they differ?
- Why is ⁻8 less than ⁻7?

Our first extension of the set of whole numbers is the set of **integers**, which is simply the union of the set of positive integers, the set of **negative integers**, and zero. Although most of people's everyday use of mathematics involves positive numbers, we encounter negative numbers in various ways. One of the major goals of this section is for you to understand how the procedures for computing with positive numbers are related to computing when one or more of the numbers are negative.

Integer Connections

Before we examine operations with integers, let us take a little time to see how integers connect to the set of whole numbers we have been working with up to now.

- We began our study of numbers with the set of natural numbers, N.

    $N: \{1, 2, 3, 4, \ldots\}$

- With the invention of zero, we have the set of whole numbers, W.

    $W: \{0, 1, 2, 3, 4, \ldots\}$

- With the invention of negative numbers, we have the set of integers, I.

    $I: \{\ldots {}^{-}4, {}^{-}3, {}^{-}2, {}^{-}1, 0, 1, 2, 3, 4, \ldots\}$

Figure 5.1 shows a Venn diagram illustrating the notion of extending our set of numbers as our ancestors invented new kinds of numbers. Each set of numbers contains the previous set. We will extend this diagram as we discuss fractions, decimals, and irrational numbers.

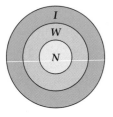

FIGURE 5.1

CLASSROOM CONNECTION

The term *integer* often causes some confusion for many students because the terms *integer* and *whole number* do not have identical meanings in mathematics and in everyday English. In mathematics, *integers* refers to the following set of numbers: $\{\ldots, {}^{-}4, {}^{-}3, {}^{-}2, {}^{-}1, 0, 1, 2, 3, 4, \ldots\}$. In everyday English, the description of this set is not identical to the description in the first sentence of this section. That is, in everyday English, one might say that the integers can be broken down into three sets—negative whole numbers, zero, and positive whole numbers. However, *whole numbers*, in mathematics, refers to the set $\{0, 1, 2, 3, \ldots\}$. Thus saying "negative whole numbers" is like saying "he goed to the store yesterday." My resolution to this dilemma is to use the terms *positive integers* and *negative integers* but not the terms *positive whole numbers* and *negative whole numbers*.

History

Like our base 10 numeration system, the set of integers was developed relatively recently. Negative numbers were either ignored or dismissed as absurd by most of the ancients. We find the earliest mention of negative numbers and how we might compute with them in the works of Brahmagupta (A.D. 628) in India. Other early work with the concept of negative numbers is found in the writings of al-Khwarizmi (A.D. 825) in Persia and Chu Shi-Ku (A.D. 1300) in China. The first modern mathematician to treat negative numbers seriously was Cardan in 1545. He accepted negative numbers as solutions of equations and, like the three authors above, stated the rules for computing with them. However, he referred to them as "false" numbers. Descartes, who developed the coordinate system, still referred to negative numbers as false numbers.

Integers in our world When do we use negative numbers? Stop to reflect on where you have encountered negative numbers before reading on....

People use negative numbers both on and off the job. For instance:

- Businesses use negative numbers to indicate a business deficit, or "negative profit."
- We often use negative numbers when describing change. For example, graphs often have negative numbers.
- We use negative numbers to indicate temperatures below zero.
- Locations below sea level are often represented with negative numbers.
- A golfer uses negative numbers to indicate a score below par.
- Physicists use negative numbers to indicate negatively charged particles.
- We use a coordinate system to indicate the position of an object in space. For example, the location of point P in Figure 5.2 is $(^-3, ^-2)$.

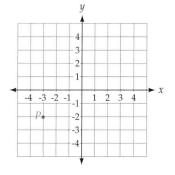

FIGURE 5.2

Representing Integers

There are many models for representing integers. We will focus on the number line model because it is the model to which most real-life applications connect.

Number lines can be represented horizontally or vertically, as shown in Figure 5.3.

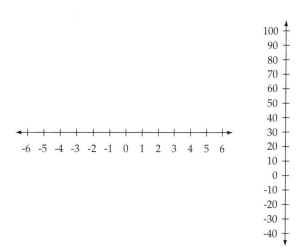

FIGURE 5.3

Some other models commonly used to represent integers and their applications include black and red chips, positively and negatively charged particles, and money (assets and liabilities). If you did Exploration 5.1, you used black and white dots to represent integers. The discrete model is a representation of integers that generally makes more sense to young children than the number line model.

Operations with Integers

In this section, we will examine the four operations with negative numbers. As you recall from Chapter 3, we developed models for each of the operations:

- addition: combine, increase
- subtraction: take away, comparison, missing addend
- multiplication: repeated addition, rectangular array, Cartesian product
- division: partitioning, repeated subtraction, missing factor

As we strive to develop ways to operate *meaningfully* with negative numbers, we will also examine which of these models for the operations work and which do not.

Many textbooks present the students with various algorithms (that is, rules) for adding, subtracting, multiplying, and dividing with negative numbers. The main problem with this approach is that students remember the rules only as long as they use them. I call this rented or Teflon knowledge; after a few months, the rules are gone. If the students do not have a strong understanding of where the rules came from, then they have nothing to show for all that work. Rather than present you with rules, we will investigate how we obtain sums, differences, products, and quotients with negative numbers. If you first solve, or at least attempt to solve, each problem *on your own*, you will find that not only will you be able to compute confidently with negative numbers, but also your problem-solving toolbox will be fuller. Furthermore, even if you forget the rules a year from now, you will be able to use your knowledge to re-create them.

Understanding Addition with Integers

5.1

We will use a problem-solving tool from higher mathematics to develop algorithms for adding integers: We will examine all possible combinations (cases) and then look for patterns. The four cases are represented below:

	a	b	Example
Case 1: Both numbers are positive.	+	+	3 + 4
Case 2: One number is positive, one number is negative, and the magnitude of the negative number is greater than the magnitude of the positive number.	+	−	3 + ⁻4
Case 3: One number is positive, one number is negative, and the magnitude of the positive number is greater than the magnitude of the negative number.	−	+	⁻3 + 4
Case 4: Both numbers are negative.	−	−	⁻3 + ⁻4

Language You may have noticed that sometimes the + and − signs have been raised (printed as superscripts) and sometimes not. This convention allows us to avoid meaningless sentences. For example, we will read the problem ⁻6 − ⁻8 as "*negative 6 minus negative 8*" instead of "minus 6 minus minus 8."

The words *plus* and *minus* will be used to refer to the *operations* of addition and subtraction. The words *positive* and *negative* will be used to refer to the *value* of the number.

For practice, translate each of the equations below into English. Check your translations with the sentences at the right.

$5 - {}^-4 = 9$ 5 minus negative 4 is equal to positive 9.
$32 + {}^-48 = {}^-16$ 32 plus negative 48 is equal to negative 16.

Let us now examine the first two cases.

Case 1   Both numbers are positive. We explored this case in Chapter 3.

Case 2   One number is positive, one number is negative, and the magnitude of the negative number is greater than the magnitude of the positive number.

Do several other examples and then try to express a rule that would apply to any addition problem in which the magnitude of the negative number is greater than the magnitude of the positive number. Then read on. . . .

We will examine a specific problem in detail and then look at generalizations. How would you represent the problem $3 + {}^-4$?

Number line model   From our work with number lines in Chapter 3, we learned that the problem $3 + 4$ can be represented as shown at the left in Figure 5.4. That is, we begin at zero and move 3 units to the right, then we move 4 more units to the right; thus we find that $3 + 4 = 7$. If a positive number is represented by an arrow pointing to the right (the positive direction), then a negative number is represented by an arrow pointing to the left (the negative direction). Thus the problem $3 + {}^-4$ can be represented as shown at the right in Figure 5.4. That is, we begin at zero and move 3 units to the right and then move 4 units to the left; thus we find that $3 + {}^-4 = {}^-1$. As you may have already found, sometimes the sum of a positive number and a negative number is positive (for example, $12 + {}^-5 = {}^+7$) and sometimes it is negative, as in Figure 5.4.

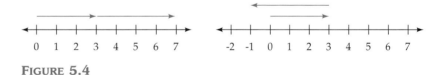

Figure 5.4

If you do a number of problems, you realize that the procedure for adding a positive number and a negative number feels a lot like subtraction. In fact, one description of a general rule is: Disregard the signs and subtract the *smaller* number from the *larger* number; the sign of the sum is the same as the sign of the *larger* number.

Absolute Value

There is a mathematical way to say, "Disregard the sign in front of the number," and that is to use the term *absolute value*. Let us first define this concept and then

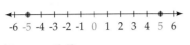

FIGURE 5.5

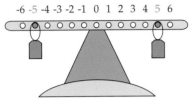

FIGURE 5.6

use it to state more precisely the rule for adding a positive number and a negative number.

One way to illustrate the concept of absolute value is to consider how far the number is from zero (the origin). The **absolute value** tells us the distance of the number from zero. For example, the numbers $^+5$ and $^-5$ are both the same distance from zero, so they both have the same absolute value, which is 5 (see Figure 5.5).

With notation, we say, $|^-5| = 5$. Similarly, $|^+5| = 5$.

In English, we say that the absolute value of negative 5 is 5, and the absolute value of positive 5 is also 5.

We can also use a balance scale (a common manipulative used in elementary schools) to illustrate the concept of absolute value. If we place a weight under $^-5$ and an equal weight under $^+5$, the scale will balance (see Figure 5.6).

There is also language for referring to pairs of numbers whose absolute values are equal: We say they are opposites or negatives of each other. Thus the **opposite** of $^+6$ is $^-6$ and the opposite of $^-6$ is $^+6$. Using another meaning of **negative**, we say that the negative of $^+6$ is $^-6$ and the negative of $^-6$ is $^+6$.

As you can readily see, when we add any integer and its opposite, the result is zero; that is,

$$a + {}^-a = 0$$

Thus we can say that every integer has an additive inverse.

In this text, we will use the term **additive inverse** instead of *negative* for two reasons. First, the term *negative* often creates a false impression. For example, if $x = {}^-6$, then $^-x = {}^+6$. In other words, when we are working with variables, the value of ^-x is often positive. Second, when working with fractions, we will develop a similar concept with a similar term: the *multiplicative inverse*.

The concept of absolute value lets us now state more precisely the rule for adding a positive number and a negative number: We first find the difference of the absolute values of the two numbers; the sign of the sum is the sign of the number with the larger absolute value.

The Other Two Cases of Integer Addition

Case 3 One number is positive, one number is negative, and the magnitude of the positive number is greater than the magnitude of the negative number (an example is $4 + {}^-3$).

When we look carefully at all integer addition problems that fall in this category, we find that the generalization stated in the preceding paragraph applies to this case too. In other words, the similarity between Case 2 problems ($3 + {}^-4$, $2 + {}^-8$, $^-7 + 3$, $^-9 + 4$) and Case 3 problems ($4 + {}^-3$, $7 + {}^-2$, $^-6 + 9$, $^-1 + 5$) is that one number is positive and one number is negative. Regardless of which number has the larger magnitude, we can use the same procedure to determine the answer.

This discussion connects to NCTM Standard 7 (Reasoning and Proof), to sets language and to NCTM Standard 9 (Connections). When we first examine all the possibilities when adding integers, we have four distinct subsets. However, when we look closely at two of these subsets, we find that their similarity (one positive, one negative) outweighs their difference (which number has the greater absolute value).

Case 4 In Case 4, both numbers are negative. Let us examine a specific problem in detail and then look for generalizations. Figure 5.7 shows a representation of $^-3 + {}^-4$ on a number line. Do several other examples and then try to

---

■ *Mathematics* ■

With one exception, all the properties of addition, subtraction, multiplication, and division that we developed in Chapter 3 hold *for all numbers*. That exception is the closure property. Whether or not a set is closed under an operation depends on the nature of the set. This concept will be pursued in an exercise.

express a rule that would apply to any addition problem in which both numbers are negative. Many students find that they can do this more easily if they have a context. For example, Jeremy borrowed 3 dollars from Annie and then borrowed 4 more dollars from her. How much does he owe her? Then read on. . . .

FIGURE 5.7

With the concept of absolute value, we can state the rule for adding two negative numbers precisely: To find the sum of two negative numbers, we first find the sum of the absolute values of the two numbers and then place a negative sign in front of this sum.

Understanding Subtraction with Integers

In Exploration 5.2, you examined integer subtraction with the take-away model and the comparison model. In the text, we focus on connecting whole-number subtraction to integer subtraction.

INVESTIGATION

5.1

Subtraction with Integers

Do the following subtraction problems yourself before reading on. As you work, check to make sure that you are using your understanding of integers rather than just guessing.

1. $14 - {}^-25 =$ 2. ${}^-5 - 17 =$ 3. ${}^-6 - {}^-8 =$ 4. ${}^-12 - 5 =$

DISCUSSION

If you found that you did better with adding negative numbers than with subtracting negative numbers, I have good news: We can define subtraction of negative numbers in terms of addition, which most people understand more easily.

To make the connection stronger, let us examine how we might use what we know about subtraction to determine the answer to $4 - 6$. In one sense, we cannot "take away" 6. However, if we think in terms of a checking account, if we take away 6 from 4, we will have a deficit of 2; that is, we have ${}^-2$. We could apply our work with subtraction and number lines from Chapter 3: When we subtract one number from another, we move to the left. When we begin at the point 4 and move 6 units to the left, we end up at the point negative 2, as shown in Figure 5.8.

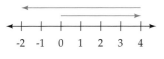

FIGURE 5.8

Thus we can conclude that $4 - 6 = {}^-2$.

However, this is similar to the addition problem $4 + {}^-6 = {}^-2$.

We use this knowledge to define subtraction of integers formally in terms of addition:

$$a - b = a + {}^-b$$

That is, subtracting is equivalent to adding the additive inverse.

Connecting Whole-Number Subtraction to Integer Subtraction

In Chapter 3, we examined different models for the four operations and different algorithms. In order for your knowledge of mathematics to be as connected as possible, it is important that you see how we apply these models to each new set of numbers.

Go back to Chapter 3 and examine how we defined subtraction there. In what ways are the two definitions of subtraction the same (that is, equivalent)? In what ways are they different? Why did we define subtraction differently there from the way we define it here? Think and then read on. . . .

Let us compare the two definitions:

Chapter 3: $a - b = c$ iff there is a number c such that $c + b = a$
Chapter 5: $a - b = a + {}^-b$

If you examine the actual problems involved in subtracting a positive number from a positive number, you find that we are not adding the opposite as much as we are taking away a positive amount or comparing the size of two sets (each with positive values). Therefore, although the definition given in this chapter seems simpler and more practical (to most of my students), it doesn't connect as well to subtraction with two positive numbers. Therefore, I chose to defer the definition until this chapter so that the work in Chapter 3 would be more focused on the context in which subtraction occurs when all three of the numbers (minuend, subtrahend, and difference) are positive numbers.

Connecting Subtraction Contexts to Algorithms

We have now developed efficient procedures for integer addition and subtraction. However, as you have discovered, in many real-life settings, translating words into a mathematical sentence is not always simple. Therefore, we will examine a few such settings to apply our knowledge.

Problem 1   Denine realized that she had overdrawn her checking account by $60, and she was fined $15 for a returned check. What is her present balance? Work on this problem and then read on. . . .

If we translate this problem into mathematical language, we find that we need to take away $15 from negative $60. Thus the problem is ${}^-60 - 15$. Using our understanding of subtraction, we translate this subtraction problem into the following addition problem: ${}^-60 + {}^-15$. Applying our understanding of integer addition, we have an answer of ${}^-75$; that is, her present balance is negative $75 (she is $75 in the red).

Problem 2   On one day the high temperature in Nome, Alaska, was ${}^-6$ degrees. On that same day, the high temperature at the North Pole was ${}^-64$ degrees. How much warmer was Nome than the North Pole? Work on this problem and then read on. . . .

If we translate this problem to mathematical language, we find that we are using the comparison model of subtraction. Thus the problem is ${}^-6 - {}^-64$, which we can translate as $-6 + {}^+64$, and the answer is 58; that is, it was 58 degrees warmer in Nome.

▪ History ▪

Our present notation, putting a minus sign in front of a negative number, was a long time in the making. The early Hindus designated negative numbers by circling them—for example, ⑤—whereas the later Hindus and Arabs represented them by placing a dot over the number—for example, 5̇—and the Chinese used red to represent positive and black to represent negative (just the opposite of what we use today). Cardan used the symbol m to represent a negative number—for example, m5.

We could also have interpreted this as a missing-addend problem: $^-64 + x = {^-6}$.
Using algebra, $x = {^-6} - {^-64} = 58$.

Understanding Multiplication with Integers

There are very few simple real-world problems in which we multiply and divide integers. Most cases of integer multiplication and division occur in solving equations. Because the ability to work confidently with algebraic language is critical in high school and college, understanding the procedures is important. "(S)tudents should see and expect that mathematics makes sense" (PSSM, p. 56). If you did Exploration 5.3, you realized that although the rules for integer multiplication are simple, justifying them is more difficult.

INVESTIGATION 5.2 — The Product of a Positive and a Negative Number

Consider the following problem: $5 \cdot {^-3}$. Can we apply our understanding of multiplication with positive numbers to determine the product? Think and read on. . . .

DISCUSSION

Applying the repeated-addition model of multiplication, $5 \cdot {^-3}$ literally means to add $^-3$ five times, that is, $^-3 + {^-3} + {^-3} + {^-3} + {^-3}$. We know from integer addition that the sum must be $^-15$. That is, we can deduce that $5 \cdot {^-3} = {^-15}$.

What about $^-3 \cdot 5$? Think and read on. . . .

Trying to apply repeated addition here is problematical; we must add 5 *negative 3 times*. However, the commutative property makes the task easier: $^-3 \cdot 5 = 5 \cdot {^-3} = {^-15}$.

From this discussion, we can recall the rule that you may have memorized in school: The product of a positive number and a negative number is a negative number.

Two negative numbers What about $^-3 \cdot {^-5}$? None of the models for positive whole-number multiplication adapt nicely to this problem, and the commutative property does us no good here. However, we can use our knowledge that a positive times a negative is a negative and make use of patterns:

We know that	$4 \cdot {^-3} = {^-12}$
Thus	$3 \cdot {^-3} = {^-9}$
Similarly,	$2 \cdot {^-3} = {^-6}$
And so	$1 \cdot {^-3} = {^-3}$
	$0 \cdot {^-3} = 0$ Recall the zero property of multiplication from Chapter 3.
	$^-1 \cdot {^-3} = ?$

What does $^-1 \cdot {^-3}$ *have to* equal if the pattern is to continue? Think and then read on. . . .

■ *Mathematics* ■

Leonard Euler (1770) produced one of the earliest proofs that $^-1 \cdot {^-1} = 1$. However, many great mathematicians thought that the idea of a negative number times a negative number was absurd. Many eighteenth-century algebra textbooks rejected multiplication of negative numbers because this was not a problem that they saw as having any real-life relevance.

As the number line in Figure 5.9 illustrates, in this case, each product is 3 more than the previous product, and 3 more than 0 is ⁺3.

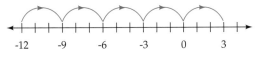

FIGURE 5.9

From this discussion, we can recall the rule that you may have memorized in school: The product of two negative numbers is a positive number.

Understanding Division with Integers

Just as we interpreted integer subtraction by applying our knowledge of the relationship between subtraction and addition, we can now understand integer division by applying our knowledge of the relationship between division and multiplication. From an intuitive perspective, you may sense that the multiplication rules translate quite directly into division:

- The quotient of a positive and a negative number is a negative number.
- The quotient of two negative numbers is a positive number.

We can apply the missing-factor model of division ($y \div n = x$ iff $x \cdot n = y$) to verify these rules.

For example, consider the problem ⁻12 ÷ 4. There is little doubt that the quotient is either ⁻3 or ⁺3. Many students simply guess, but we can apply this model to help us find the correct answer. Applying our definition of division, we can say that

$$^{-}12 \div 4 = x \quad \text{iff} \quad x \cdot 4 = {^{-}12}$$

What number times 4 is equal to negative 12? We know from multiplication that this number must be ⁻3. Therefore, ⁻12 ÷ 4 must equal ⁻3.

Similarly, consider ⁻15 ÷ ⁻3. Applying our definition of division, we can say that

$$^{-}15 \div {^{-}3} = x \quad \text{iff} \quad x \cdot {^{-}3} = {^{-}15}$$

We know from multiplication that this number must be 5. Therefore, ⁻15 ÷ ⁻3 must equal 5.

Summary

In this section, we have learned that the set of integers includes both the set of positive and negative numbers and zero. We have discovered that there are many ways to represent integers, and we have explored two models for representing integers: number lines in the text and dots in Explorations 5.1–5.3. We have adapted our understanding of the four basic operations, learned in Chapter 3, so that the algorithms for adding, subtracting, multiplying, and dividing integers make sense. In the course of exploring these operations, it became important to understand several new terms: *absolute value, opposite, negative,* and *additive inverse.*

In seeking to make sense of integer operations, we find that we cannot rely solely on common sense and manipulatives, as we did for operations with the set of natural numbers. For example, ⁻20 ÷ ⁻4 does not readily lend itself to a concrete, real-life problem or to a division-as-partitioning representation. At some point in the study of mathematics, the students must metaphorically "leap from the nest" and be able to use more analytic methods of reasoning to construct new understandings on the foundations of old ones.

Although operations with integers are not generally introduced until middle school, this section is important for several reasons. First, teachers need to have a sense of what lies beyond the concepts their students are learning at the present time. Even if you teach younger students, it is important that you know how the more sophisticated sets of numbers connect to the set of natural numbers and that the four basic operations, which students first learn in the context of natural numbers, will need to make sense with these more sophisticated kinds of numbers. Second, many students have trouble with algebra because there are too many rules to memorize. However, as you have seen in this section, we can apply our understanding of the meaning of the four operations and the properties of the four operations to understand the procedures for operating with integers. Third, you are likely to encounter gifted students. I happened to be visiting a second-grade teacher one day when one of her students invented the procedure for integer subtraction on the spot. The teacher was explaining how to subtract via regrouping with the problem 72 − 58. She said, "You can't take 8 from 2." He replied, "Yes, you can; you get minus 6; then you take 50 from 70, which is 20, and 20 minus 6 is 14!"

EXERCISES 5.1

1. Perform the computations:
   a. ⁻356 − ⁻138
   b. $\dfrac{^-36 + 48}{^-6}$
   c. 16 − (⁻3)
   d. ⁻217 + 139
   e. −2(3 − 10)
   f. ⁻3 · 5 + (⁻6 ÷ 2)
   g. 16 − (⁻3)²
   h. ⁻6 + (24 ÷ 3)
   i. ⁻124 − ⁻345
   j. (47 + ⁻11)/(⁻6)

2. Find the missing number:
   a. ⁻65 + □ = 173
   b. 36 − □ = 83
   c. 331 + □ = ⁻86
   d. ⁻812 + □ = ⁻223
   e. ⁻342 + □ = ⁻129
   f. ⁻12,348 − □ = ⁻348

3. Show two different ways to determine the following sum:
   ⁻19 + ⁻6 + ⁻22 + 8 + ⁻4 + 7 + 1

4. The origin of our symbols for + and − can be traced back to the practice of merchants in the Middle Ages, who used the signs p for *piu* (more) and m for *meno* (less) to indicate how much above or below a standard weight each sack was. Thus, if the standard weight was 100 pounds, a sack weighing 96 pounds would have m4 written on it. How much overweight or underweight is the following shipment? Describe at least two different ways you could have solved this problem.

5. The formula for converting from Fahrenheit to Celsius is given by $C = \dfrac{5(F-32)}{9}$. If the temperature outside is 23° Fahrenheit, what would the temperature be in Celsius?

6. At what temperature would the Fahrenheit and Celsius temperatures be equal?

7. Here are the low temperatures (Fahrenheit) for one week in Minneapolis, Minnesota: ⁻19, ⁻6, ⁻22, 8, ⁻4, 7, 1. What was the mean (average) low temperature for the week?

8. The Crabby Apple restaurant lost $2500 in January. If its net worth at the end of the month was ⁻$400, what was its net worth at the beginning of the month?

9. Jacob opens a savings account on January 1 with a deposit of $250. He has "direct deposit," in which $25 is deposited every other week. The bank also charges a $3 monthly processing fee. How much money will he have at the end of the year?

10. Have you ever been on an airplane and heard the pilot say that the plane would be a little late because it would be flying into a strong headwind or that even though the plane was taking off a bit late, you would be making up time because you would be flying with a tailwind? This problem asks you to analyze such a situation. You have the following data: A plane flying at its maximum speed can go 200 miles per hour with a tailwind or 160 miles per hour into a headwind.
    a. What is the wind speed?
    b. What would be the maximum speed of the plane if there were no wind?

11. If you are flying to a city 800 miles away, what will be the difference in flight time between flying into a 40-mph headwind and flying with a 40-mph tailwind if the plane's maximum speed with no wind is 160 mph?

12. Let's say the countdown for a space shuttle launch has begun. At "T minus 27 hours" (that is, 27 hours before launch), a problem occurs. If the technicians have not fixed the problem by T minus 8 hours, the launch will have to be scratched. How much time do the technicians have to correct the problem?

13. John had $123 in the bank but wrote a check for $56 and another check for $86. What is his current balance?

14. Many Americans do not realize that our calendar is not universal. For example, the copyright date for this textbook is 2007. However, this year would be reported as 5767 in the Jewish calendar. According to the standard calendar, when was the world created?

15. In Chapter 1, we investigated magic squares in which all rows, columns, and diagonals had the same sum. The magic square below is a subtraction magic square. Why is this?

15	7	8
5	4	1
10	3	7

    a. Make another subtraction magic square.
    b. Make a subtraction magic square in which all the numbers are negative numbers.
    c. Make a subtraction magic square in which four of the numbers are positive, four of the numbers are negative, and one number is zero.
    d. Write instructions explaining to someone how to make a subtraction magic square that works for any number.

16. Below are the low and high temperatures on a winter day in several cities. Make up and answer two questions from the chart.

City	Low	High
Anchorage	⁻28	⁻7
Minneapolis	⁻12	5
Honolulu	68	73
Miami Beach	68	87
Portland	26	46

17. Examine some daily newspapers or weekly magazines to find articles that require some knowledge of negative numbers. Either make up a story problem that might come out of the article or explain how a knowledge of negative numbers is needed in order to understand the article.

18. Consider the positive and negative values of the integers from 1 to 9:

    $$\pm 1 \quad \pm 2 \quad \pm 3 \quad \pm 4 \quad \pm 5 \quad \pm 6 \quad \pm 7 \quad \pm 8 \quad \pm 9$$

    The sum of the nine positive integers is ⁺45, and the sum of the nine negative integers is ⁻45. How many numbers between ⁻45 and ⁺45 can be made by adding and using each number or its opposite exactly once? For example, here is one combination whose sum is ⁻3.

    $$1 + 2 + 3 + 4 + 5 + 6 + {}^-7 + {}^-8 + {}^-9 = {}^-3$$

19. Let x and y represent any positive integers, $x \neq y$. For each of the operations below, tell whether the result will be always positive, will be always negative, or might be one or the other. Explain your reasoning.
    a. $|x - y|$
    b. $x^2 - y^2$
    c. $x^2 - xy + y^2$
    d. $x^2 + 2xy - y^2$

20. Let x and y represent *any* integers, $x \neq y$. For each of the operations below, tell whether the result will be always positive, will be always negative, or might be one or the other. Explain your reasoning.
    a. $|x - y|$
    b. $x^2 - y^2$
    c. $x^2 - xy + y^2$
    d. $x^2 + 2xy - y^2$

21. Sam was told by the doctor that he had to lose about 40 pounds. Six weeks ago he weighed 185 pounds. Below is a weekly record of his progress. How much does he weigh now? Do this problem at least two different ways.

Week	1	2	3	4	5	6
Change	⁻3	⁻2	⁺1	⁻6	⁺3	⁻2

22. The set of integers is closed under which of the four fundamental operations?

23. The term *wind chill factor* is an important concept for people living in or visiting a cold climate in winter. Most body heat is lost through the skin, and the wind can accelerate this cooling process. For example, if the air temperature is 20°F and there is a 30 mph wind, then the wind chill temperature is ⁻2°F. That is, you are losing body heat at the same rate as though it were 2 degrees below zero on a calm day! Examine the wind chill table below.

   a. If the air temperature is ⁻20°F and there is a 15-mph wind, what is the wind chill temperature?
   b. Describe two observations from the chart.

Wind Chill Temperature (°F)

Air temperature (°F)	Wind speed (mph)								
	0	5	10	15	20	25	30	35	40
32	34	32	27	24	21	17	14	12	10
28	30	28	23	19	15	12	9	6	4
24	26	24	19	14	10	7	3	0	−3
20	23	20	14	9	5	1	−2	−6	−9
16	19	16	10	5	0	−4	−8	−12	−17
12	15	12	6	0	−5	−10	−15	−19	−24
8	11	8	1	−5	−11	−16	−21	−26	−32
4	7	4	−3	−10	−16	−22	−28	−34	−40
0	3	0	−7	−15	−22	−28	−35	−42	−49
−4	−1	−4	−12	−20	−28	−35	−42	−50	−58
−8	−4	−8	−16	−25	−33	−41	−50	−59	−67
−12	−8	−12	−21	−30	−39	−48	−58	−68	
−16	−12	−16	−26	−36	−45	−55	−66		
−20	−16	−20	−30	−41	−52	−63			
−24	−20	−24	−35	−47	−58				
−28	−24	−28	−39	−52	−65				
−32	−27	−32	−44	−58					
−36	−31	−36	−49	−64					
−40	−35	−40	−54	−69					

Source: R. G. Steadman, "Indices of Windchill of Clothed Persons." *Journal of Applied Meteorology,* vol. 10, August 1971. Reprinted with permission from the American Meteorological Society.

SECTION 5.2 FRACTIONS AND RATIONAL NUMBERS

WHAT DO YOU THINK?

- Can you give a definition of *fraction*?
- Why is 1/5 less than 1/3?
- What is the difference between the whole and the unit?

The concept of rational numbers is one of the "big ideas" of elementary mathematics, and it is also one of the more abstract and difficult. As you will soon discover, fractions are more than just numbers, and the numerator and denominator are more than just the top and bottom numbers.

In this chapter, we will build on the part–whole relationships developed in Chapter 2, and we will continue to use composition and decomposition (first seen with expanded form in Chapter 2) to probe more deeply into the whys of rational numbers.

Operations with fractions are a relatively recent part of the history of mathematics. In Section 5.1, we discovered that the set of integers simply involves an extension of the number system. Fractions, on the other hand, are different from counting numbers and integers in a very significant way: *Two* numbers are needed to represent one amount! From another perspective, when we work with rational numbers, we change the question from *how many?* to *how much?* I hope that by the end of this section, you will see what a difference this makes!

268 CHAPTER 5 / Extending the Number System

CLASSROOM CONNECTION

Children encounter the notion of rational numbers and fractions even before they enter school through the notion of sharing between two people. That is, when one whole is divided into two equal parts, each person has one-half.

Two Different Sets

Before we begin, it is necessary to make a distinction between *fraction* and *rational number*.

A **rational number** is a number whose value can be expressed as the quotient or ratio of two *integers a* and *b*, represented as $\frac{a}{b}$, where $b \neq 0$.

A **fraction** is a number whose value can be expressed as the quotient or ratio of *any two numbers a* and *b*, represented as $\frac{a}{b}$, where $b \neq 0$. For example, $\sqrt{2}/3$ is a fraction but not a rational number.

In either case, *a* is called the **numerator** and *b* is called the **denominator**.

Technically, the set of rational numbers is a subset of the set of fractions, which can include amounts like $\sqrt{2}/2$ and $\pi/6$. In elementary school, children work with fractions that are rational numbers, and these will be our primary focus in this chapter. Therefore, we will generally use the term *fraction*.

Fractions in History

5.5

The notion of 3/4 may make sense to you now, but it is not easy for many children, and it is such an abstraction that it is relatively recent in human history.

The Egyptians expressed all fractions (with the exception of 2/3) as unit fractions—that is, fractions whose numerator is 1. They used the symbol ⌒, which they placed above a numeral to indicate a fraction. Thus 1/12 was written as ⌒/nII. The Egyptians' decision to represent fractional amounts using only unit fractions was a consequence of their difficulty with using two numbers to represent a single amount.

As we saw earlier, the idea of representing *all* amounts with whole numbers was very appealing to the ancient Greeks, and so they did not even consider the idea of creating numbers that were not whole numbers. Rather, they worked with ratios. For example, instead of saying that 2/5 of the students at a college are male, they would say that the ratio of males to females is 2 to 3. (We will examine ratios more closely in Chapter 6.)

The Romans also avoided fractions. We live with the effects of one of their ways of avoiding fractions: Rather than dealing with parts of a unit, they created smaller units. Their word for twelfth was *unica*, which is where our words *ounce* and *inch* come from.

Our present method of writing fractions (for example, $\frac{2}{3}$) was probably invented by the Hindus. Brahmagupta (A.D. 628) wrote $\begin{smallmatrix}2\\3\end{smallmatrix}$. The bar seems to have been introduced by the Arabs, but just as the base 10 system took many years to be accepted, it was some time before most writers used the bar when

> **■ Language ■**
>
> The origin of the word *fraction* is also interesting; it is derived from the Latin word *fractio*, which comes from the Latin word *frangere*, meaning "to break." In early American arithmetic books, the term *broken numbers* was often used instead of the word *fraction*. The first known mention of the actual word *fraction* was by Chaucer in 1321.[2]

---

[2]Louis Karpinski, *The History of Arithmetic* (New York: Russell & Russell, 1965), p. 127.

Section 5.2 / Fractions and Rational Numbers   269

expressing fractions. Part of the problem seems to have been caused by printing difficulties, and even today few word processors are able to express all fractions vertically; instead they write 2/3.

INVESTIGATION 5.3

5.8

Rational Number Contexts: What Does 3/4 Mean?

This investigation uncovers the different meanings of fractions. Although all of the following situations involve the fraction 3/4, they represent 3/4 in different contexts. One of the goals of our work with rational numbers is for you to understand the similarities and differences among these contexts. Think carefully as you work.

Represent each of the following situations with a diagram. Then read on. . . .

A. Joey grew 3/4 of an inch last month.

B. Four children want to share 3 pies equally. How much pie does each person get?

C. At a recent meeting, 3/4 of the participants were women. If there were 12 people at the meeting, how many were women?

D. At a college, 3/4 of the students are women.

DISCUSSION

Do you see how the pictures in Figure 5.10 connect to the verbal descriptions?

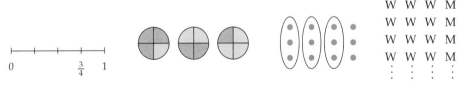

FIGURE 5.10

Each of the situations embodies one of the four major contexts in which we use rational numbers.

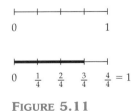

FIGURE 5.11

A. The first situation illustrates **rational number as a measure**. To measure the appropriate location of 3/4, we must divide (partition) the unit length (1 inch) into 4 equal lengths. The length of 3 of those equal lengths shows how much Joey grew (Figure 5.11).

When we are dealing with the fraction a/b in the context of measuring, we have some amount or object that has been divided into b equal amounts, and we are considering a of those amounts.

B. The second situation illustrates **rational number as a quotient**. The initial problem situation is 3 (pies) divided among 4 (people). The answer of 3/4 pie per person represents the value of each person's share; that is, the

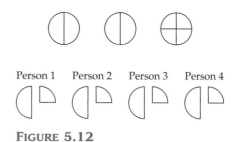

FIGURE 5.12

FIGURE 5.13

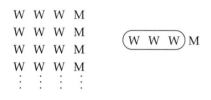

FIGURE 5.14

denominator once again is in relation to the unit (1 pie). If we divide each pie into 4 pieces, each person's share is 3 of those pieces (Figure 5.12).

When we are dealing with a fraction a/b in the context of a quotient, an amount a needs to be shared or divided equally into b groups.

C. The third situation embodies **rational number as an operator**. In this case, 3/4 acts as a function machine (recall "What's my rule?" from Chapter 2) by operating on the whole (in this case, 12 people). In this example, the input is 12 people, and 3/4 is the rule that tells us what to do—that is, take 3 of every 4 people. The output is thus 9 people (Figure 5.13).

The operator context can be thought of as a stretching or shrinking process; a/b is a function machine that tells us the extent to which the given object or amount is to be stretched or shrunk.

D. The fourth situation illustrates **rational number as a ratio**. That is, if 3/4 of the students are women, and if we count all the men as 1 group, then the number of women would be equal to 3 of those groups. In this case, we can represent this statement in fraction language—3/4 of the students are women—or we can represent the statement in ratio language—the ratio of women to men is 3 to 1—3:1 (Figure 5.14).

A ratio is a relationship between two quantities. Ratios can involve comparisons between parts and parts (for example, women:men = 3:1), comparisons between parts and wholes (women:total = 3:4), or comparisons between two different wholes (when we say that a car averages 30 miles per gallon, we are comparing the number of miles traveled to the number of gallons consumed).

We will examine the measure context in this section, the quotient and operator contexts in Section 5.3, and the ratio context in Chapter 6. However, we shall soon find that they are highly interconnected and that many problems can be interpreted in more than one context. See Exploration 5.13 (Meanings of Operations with Fractions).

What do these contexts have in common? There are certain important ideas that are in all four contexts:

- Each context can be interpreted in part–whole relationships. In each case a whole has been divided into 4 parts.
- Something is to be partitioned into parts of equal size (value).
  The something can have a value of 1, in which case the unit = the whole.
  The something can have a value ≠ 1, in which case the unit ≠ the whole.
- The numerator and denominator are like codes that tell us about the relative sizes of the parts and the unit, and the code is multiplicative in nature. For example, when we say 1/2, it is not the difference between the two numbers that contains the key to the value; rather, it is the fact that the value of the

denominator is twice the value of the numerator that contains the key. Thus 1/2 has the same value as 4/8, not 7/8.

It is also important to note that the diagrams in Figure 5.10 also illustrate three basic ways in which we can represent rational numbers: **length models** (e.g., number lines and Cuisenaire rods), **area models** (e.g., squares, circles, Geoboards, Pattern Blocks™), and **set models** (e.g., discrete circles and egg cartons). These are not the only representations (for example, we can speak of volume models), but these three have been found useful in helping children to understand fraction concepts and useful in terms of representing problems visually.

The unit and the whole are not always the same! Equating the unit with the whole is one of the most common misconceptions that people have about working with fractions. Let us revisit the four problems posed in connection with Figure 5.10. In each case, what is the whole? What is the unit? What do we mean by whole and by unit? Think and then read on. . . .

In the first case, the whole is 3/4 inch—that is how much Joey grew; the unit is 1 inch. In the second case, the whole is 3 pies; the unit is 1 pie. In the third case, the whole is 12 people; the unit is 1 person. In the fourth case, the whole is the number of students at the college; the unit is 1 person.

The whole is the given object or amount. The unit is that amount to which we give a value of 1—1 inch, 1 pie, 1 person. In some cases, the whole and the unit are the same. For example, if Lisa gets 1/2 of a pizza and Liam gets 1/3 of the pizza, the whole is 1 pizza and the unit is also 1 pizza. However, if Ramon buys $2\frac{1}{2}$ gallons of gas and his brother takes 1/2 gallon, the whole is $2\frac{1}{2}$ gallons and the unit is 1 gallon. The need to think about wholes and units comes up in Explorations 5.7, 5.8, 5.9, 5.13, 5.14, and 5.15 and throughout this chapter.

Let us now investigate the context of rational number as measure.

INVESTIGATION

5.4

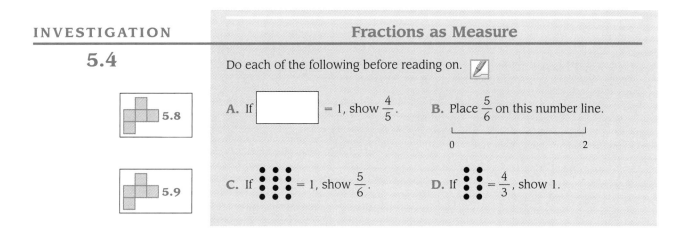

DISCUSSION

A. This question is relatively straightforward: Divide the rectangle into 5 equal regions, and shade in 4 of them [Figure 5.15(a)].

B. This question requires you to grapple with the difference between the unit and the whole.

In this case, the whole and the unit are not the same. This is important, because the meaning of the denominator *is in relation to the unit*. That is, to find the location of 5/6, we do not take the whole line and divide it into 6 equal lengths. Rather, we first must determine the unit length and then divide that length into 6 equal lengths. Recall the wire problem (Investigation 1.15). In that problem, the unit is 1 inch but the whole is 50 inches. This difference between the whole and the unit cannot be overemphasized [Figure 5.15(b)].

C. To make sense of this question, we must partition the dots into 6 equal-size groups. Recall the partitioning model of whole-number division. We then take 5 of those groups to show 5/6 [Figure 5.15(c)].

D. For many students, this question is the most difficult. There are different ways to make sense of this situation and answer the question. For example, we can focus on the numerator, which indicates that we have 4 equal parts. Thus each part—that is, 1/3—contains 2 dots. When we then focus on the denominator, we find that 3 of these equal parts represent the unit (that is, 3 of those parts have a value of 1).

On the other hand, we can interpret this statement from a ratio context. For example, if 8 dots have a value of 4/3, then 4 dots will have a value of 2/3, and so 2 dots will have a value of 1/3. Many people feel more confident once they get to a unit fraction; they reason that if 2 dots have a value of 1/3, then 6 dots will have a value of 3/3—that is, 1 [Figure 5.15(d)].

CLASSROOM CONNECTION

The need for fractions is introduced in *Inchworm and a Half* by Elinor Pinczes (Boston: Houghton Mifflin, 2001). An inchworm can measure fruits and vegetables by how many loops it takes her to go from one end to another. When she encounters fractions, smaller worms that are 1/2, 1/3, and 1/4 of her length help her out.

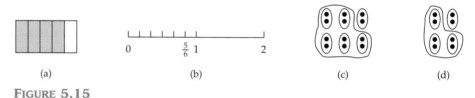

(a)   (b)   (c)   (d)

FIGURE 5.15

Let us work with another problem to develop these fraction ideas further.

INVESTIGATION 5.5

Rational Numbers and Multiple Representations

An organization has set $300,000 as their fundraising goal. A large sign with a big thermometer sits by the street in front of the building. After 24 days the thermometer at the left shows their progress. Describe the progress of the fundraiser in as many ways as possible. Then read on. . . .

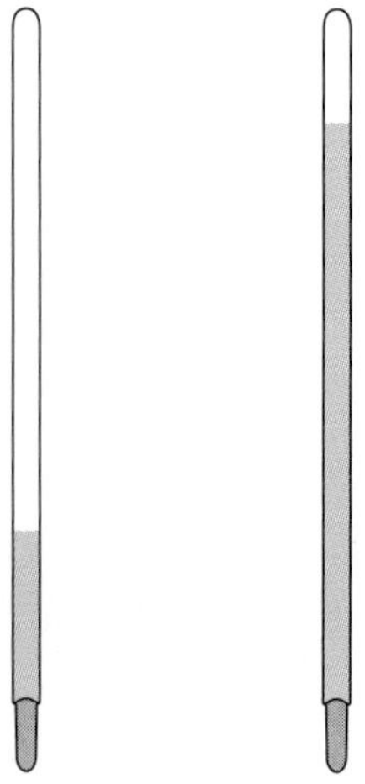

The notion of multiple representations and equivalence are in the foreground here and are essential for a deeper understanding of important mathematical ideas. Answer the following questions, each of which represents different interpretations and representations of the progress.

A. They are about 1/4 of the way. Describe how one might arrive at this answer.
B. What other descriptions are equivalent to 1/4?
C. At this rate, how long will it take them to reach their goal?

FIGURE 5.16

DISCUSSION

A. You could trace the picture on a piece of paper and fold. You could use the amount as a ruler and see how many of these lengths fill the thermometer. In this case, you are using this length as your unit.

B. We could also say they have made 25% of their goal, that their progress is 0.25 of the goal, or that they have raised about $75,000 so far. Conversely, we could say they have 3/4 of the way to go, 75% to go, 0.75 to go, or $225,000 more to raise.

C. If they have made 1/4 of their goal in 24 days, then they would make their goal in $4 \times 24 = 96$ days, or just over 3 months.

The thermometer at the right shows their progress after 49 days.

A. Determine a fraction to represent their progress at this point.
B. About how many dollars have they raised? Can you determine this answer mentally?
C. At this rate, how long will it take them to reach their goal?

DISCUSSION

A. We can use either of the methods described above. In this case, we would use the nonshaded length as our unit. Realizing that it is 1/5, we would conclude that the organization has reached 4/5 of their goal. We could also estimate thirds lightly, then fourths, then fifths, etc. until one division seems appropriate.

B. 4/5 of 300,000 is $240,000. Here is one way to determine the amount mentally. First, we can see that 1/5 of 300,000 is 60,000. Then we multiply 60,000 by 4 to get 240,000.

C. If 4/5 is equivalent to 49 days, then 1/5 is about 12 days, so 5/5 is about 60 days.

INVESTIGATION 5.6

Determining an Appropriate Representation

Jose paid $12 for a box of chocolates that weighed 3/4 pound. What is the price of 1 pound (at this rate)? Work on this and then read on. . . .

DISCUSSION

An area model is an appropriate representation, because boxes of chocolates are often rectangular in shape. If we look at the box in terms of weight, we have 3/4 of a pound. If we look at the box in terms of money, it costs $12. In one sense, we are saying that 3/4 of a pound is equivalent to $12. That is, there is a functional relationship between 3/4 (box) and 12 (dollars). There are many strategies for transforming this information into an answer.

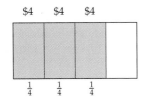

FIGURE 5.17

STRATEGY 1: "See" the multiplication in the question

Thinking of the problem as "we have 3 parts, which are equivalent to $12" invites the question "What number do we add three times to get 12?" In other words, "3 times what is 12?"

From this perspective, we realize that each part (that is, each quarter-pound) has a value of $4 (see Figure 5.17), and so one pound will cost $16.

STRATEGY 2: Apply the function concept

Input	Output	Reasoning
$\frac{3}{4}$ pound	$12	Essentially, we now have the rule (recall "What's my rule?").
$1\frac{1}{2}$ pounds	$24	We are using the idea of proportional reasoning, which we will explore in more detail in Chapter 6. If we double the amount, we must double the price.
3 pounds	$48	We double the amount again. Why do we do this?
1 pound	$16	Most people work more confidently with whole numbers and can now confidently reason that if 3 pounds cost $48, then 1 pound costs $16.

It is important to note that there is no one "right" way to use the function concept. We could also have reasoned that if 3/4 pound costs $12, then 1/4 pound must cost $4, and therefore 1 pound must cost $16.

Let us continue our investigation of the meaning of fractions with a story problem.

INVESTIGATION 5.7 — How Many Blue Balloons?

Josh goes through a large bag of balloons containing red and blue balloons. He finds that 1/8 of them are blue. He also notes that there are 120 more red balloons than blue balloons. How many blue balloons are in the bag? Work on this and then read on. . . .

DISCUSSION

Balloons are discrete amounts, and so we could represent this problem with a discrete model. However, this would be tedious; drawing 120 more red balloons would take some time! In this case, even though the actual problem involves discrete objects, a more appropriate model to represent the problem is an area model.

This idea of constructing appropriate representations for a problem situation is one that has been neglected in school mathematics and one that the NCTM is encouraging teachers to emphasize more. As mentioned in Chapter 1, Representation was not a separate standard in the 1989 NCTM Curriculum and Evaluation Standards. Partly for the reason described above, a decision was made to make Representation one of the basic standards in *Principles and*

Section 5.2 / Fractions and Rational Numbers   **275**

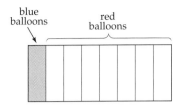

FIGURE 5.18

Standards for School Mathematics. Do you understand Figure 5.18? If you were unable to solve this problem, first try to solve the problem using Figure 5.18 and then read on. . . .

STRATEGY 1: *Use reasoning*

If we now try to connect the problem to Figure 5.18, *one* way (of many) to verbalize this would be to say that the blue balloons are represented by 1 box, and the red balloons are represented by 7 boxes. We unlock the secret to the solution when we connect the figure to the other piece of given information: There are 120 *more* red balloons than blue balloons. This means that 6 boxes must represent 120 balloons. (Write the amounts on Figure 5.18 as was done in Figure 5.17.) Consider why this is and then read on. . . .

From this, we can conclude that 1 box represents 20 balloons. Thus there are 20 blue balloons and $7 \cdot 20 = 140$ red balloons.

Check this solution with the original problem. Does this answer work?

If there are 20 blue balloons and 140 red balloons, then there are 160 balloons in all. Because 20 is 1/8 of 160, and 140 is 120 more than 20, our solution fits the given information.

STRATEGY 2: *Use algebra*

Those students who are fluent algebraically may have wondered why we didn't use algebra. If you are comfortable with algebra, it certainly is quicker in this case. However, there are several ways to do the algebra. Try to solve this problem algebraically and then read on. . . .

One algebraic solution path:

- We can let $x =$ *the number of blue balloons*.
- We can then represent the number of red balloons as $x + 120$.
- This means that $x + (x + 120)$—that is, $2x + 120$—represents the total number of balloons.
- Because 1/8 of them are blue, we can say that $x = \frac{1}{8}(2x + 120)$.
- It is important to note that we can connect language nicely to the equation; that is, the number of blue balloons is 1/8 the total number of balloons.
- When we solve this equation for x, we find that $x = 20$.

Another algebraic solution path:

- If 1/8 of the balloons are blue, that means 7/8 of them are red.
- If we let $x =$ *the number of balloons in all*, we have the following equation. That is, the difference between the number of red and blue balloons is 120.

$$\frac{7}{8}x - \frac{1}{8}x = 120$$

- Solving this equation for x, we get $x = 160$, and 1/8 of 160 is 20.

Not only are there different ways to solve this problem algebraically, but it is also important to note that x did not represent the same thing in the two cases we examined. This is another reason for taking the time on your paper to make sure that you (and the reader) know what the x refers to.

> **■ Mathematics ■**
>
> As a result of this course, when you see the word *equivalent*, you should think more than equivalent fractions. Stop and think where you have read the word *equivalent* before this page.
>
> We have consistently talked about equivalent representations from the very first Investigation in Chapter 1; conversions (e.g., two nickels is equivalent to one dime); the equivalence of everyday English to if-then statements; the equal sign and equivalence; the equivalence of 14 singles to 1 ten and 4 singles; the equivalence of 4 + 8 and 8 + 4; and the equivalence of 12 × 4 and 10 × 4 + 2 × 4, and more! This really is a big idea of mathematics!

Equivalent Fractions

Let us now examine the concept of equivalent fractions.

Two fractions are **equivalent fractions** if they have the same value.

When you made your own manipulatives in Exploration 5.6, you could *literally* see the equivalence of certain fractions as you developed strategies to cut the unit squares or circles into pieces. However, it is not always obvious that two fractions do indeed have the same value.

Almost all of the essential fraction ideas surface in the concept of equivalent fractions, and this concept permeates many of the fraction problems that we encounter. One of the biggest challenges in approaching this section is that you already know the procedures for finding equivalent fractions. For example, you know that 3/4 and 6/8 are equivalent fractions because you can multiply 3/4 by 2/2.

$$\frac{3}{4} = \frac{3}{4} \cdot \frac{2}{2} = \frac{6}{8}$$

As we have stressed repeatedly in this book, mathematical power and number sense come not just from knowing procedures but also from knowing why these procedures work and how they connect to other mathematical ideas. Therefore, we will do some work to make sense of the procedures for determining equivalent fractions.

Equivalence   Please address these two questions before reading on. . . .

- What does *equivalent* mean?
- Where else have you encountered the notion of equivalence in mathematics and in life outside school?

One way of looking at "equivalent" comes from taking apart the actual word: *equi-valent*, or equal value. We needed the notion of equivalence when we added and subtracted whole numbers with regrouping; for example, 1 ten and 2 ones is equivalent to 12 ones. We use equivalence with money every day; for example, one quarter is equivalent to 25 pennies.

Using models   Similarly, two fractions are equivalent if they have the same value. Using this notion, can you illustrate the equivalence of 3/4 and 6/8 using one or more of the fraction models we have discussed: area, length, or set? Do this before reading on. . . .

We can use the set model to illustrate equivalence. For example, consider a set of 8 dots [Figure 5.19(a)]. If we take 6 of them, we literally have 6/8 [Figure 5.19(b)]. However, we can also partition this set of 8 dots into 4 equal groups of 2 dots; if we then take 3 of these 4 equal groups, we have, by definition, taken 3/4 of the set [Figure 5.19(c)].

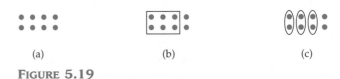

Figure 5.19

Classroom Connection

Many teachers use paper folding to introduce the idea of equivalent fractions. For example, take a piece of paper and fold it in half. Then fold it in half again. [*Note:* There are several ways to do this.] Shade 3 of the 4 regions to show 3/4. Now fold the paper in half again. Now we see that the paper has been folded into 8 regions and that 6 are shaded. Thus 3/4 = 6/8; that is, these two fractions represent the same *part* of the whole.

However, we could also have used an area model. For example, consider a rectangle representing 1 [Figure 5.20(a)]. We can partition the rectangle into 4 equal regions and shade in 3 of them to represent 3/4 [Figure 5.20(b)]. However, if we draw a horizontal line through the middle of the rectangle, the rectangle has now been partitioned into 8 equal regions. Because 6 of those regions have been shaded in, we have, by definition, 6/8 of the rectangle [Figure 5.20(c)].

(a)   (b)   (c)

FIGURE 5.20

Patterns and Order: Approaches to Procedures

Patterns in a sequence   As stated before in this book, mathematics has been referred to as "a science of pattern and order." Let us examine equivalent fractions from the perspective of pattern and order. Look at the sequence of equivalent fractions below. What "patterns" and "order" do you notice? What is the next fraction in the sequence? Why?

$$\frac{3}{4} = \frac{6}{8} = \frac{9}{12} = \frac{12}{16} \cdots$$

Some of the many aspects of pattern and order include the following:

1. As we move from one fraction to another, the numerator increases by 3 and the denominator increases by 4. In this sense, the 3/4 acts as an operator.

2. The numerator of each of the fractions is a multiple of 3, and the denominator of each of the fractions is the same multiple of 4.

3. All the denominators are even numbers, and every other numerator is an even number.

We can translate the first statement into notation:

Specific example: $\dfrac{3}{4} = \dfrac{3+3}{4+4} = \dfrac{6}{8}$

More general case: $\dfrac{a}{b} = \dfrac{\overbrace{a+a+\cdots+a}^{n \text{ times}}}{\underbrace{b+b+\cdots+b}_{n \text{ times}}} = \dfrac{an}{bn}$

Mathematics

When exploring equivalent fractions, children have to realize that there are additive and multiplicative relationships. For example, 6/8 is equivalent to 3/4 because we *add* 3 to the numerator and 4 to the denominator; 6/8 is equivalent to 3/4 because we have *multiplied* both numerator and denominator by 2. We find additive and multiplicative relationships throughout this book. For example, in Investigation 1.1, when we increase the number of pigs from 13 to 16, we can either see that this *adds* 6 feet or we can see that when we add 3 pigs, the increase in total feet is 3 *times* 2.

We can also translate the second statement into notation:

Specific example: $\dfrac{3}{4} = \dfrac{3 \cdot 2}{4 \cdot 2} = \dfrac{6}{8}$

More general case: $\dfrac{a}{b} = \dfrac{a \cdot n}{b \cdot n} = \dfrac{an}{bn}$

Another pattern Another aspect of equivalent fractions is used extensively in algebra. Look at the following pairs of equivalent fractions. What can you say about each pair?

$$\frac{3}{4} = \frac{6}{8} \qquad \frac{5}{6} = \frac{10}{12} \qquad \frac{3}{8} = \frac{9}{24}$$

In each case, the product of the first numerator and the second denominator is equal to the product of the second numerator and the first denominator. This relationship (frequently referred to as **cross multiplication**) is used (often incorrectly) as a shortcut in solving equations. It is expressed most abstractly as

$$\frac{a}{b} = \frac{c}{d} \quad \text{iff} \quad ad = bc$$

When you examine children's mistakes in the exercises for Section 5.3, you will see how this relationship is commonly misunderstood.

Addressing a Common Difficulty in Language

When trying to explain why 3/4 and 9/12 are equivalent, many students use the word *divide*—for example, "We divided the rectangle in half, and now there are 12 equal regions compared to 4 before." This choice of words is interesting and points to a problem that many children have in trying to understand equivalent fractions at a conceptual level. When we look at why fractions are equivalent, we commonly encounter this word *divide*. However, in the procedure for creating equivalent fractions, we *multiply* the top and bottom by the same number! If we physically divide, why do we mathematically multiply? Think about this and then read on. . . .

There is no simple answer to this question. It has to do with the reciprocal relationship between multiplication and division. Figure 5.21(a) shows 3/4. If we divide each of the regions by 2, we are also multiplying the total number of regions by 2 [Figure 5.21(b)]. We now have 6 out of 8 regions shaded—that is, 6/8. Starting with 3/4, we could have divided each region into 3 smaller pieces. By dividing each region into 3 smaller pieces, we are multiplying the number of pieces by 3, and we can name this shaded amount 9/12 [Figure 5.21(c)].

(a)   (b)   (c)

FIGURE 5.21

> ■ *Language* ■
>
> Just as we recommend not using the terms *carry* and *borrow* when regrouping during addition and subtraction of whole numbers, many mathematics educators recommend not using the term *reduce*, but rather using the term *simplify* for this process. Why might they make this recommendation?

Simplest Form

A fraction is in **simplest form** if the numerator and the denominator are relatively prime numbers. The notion of simplifying fractions is clearly connected to the concept of equivalent fractions, and we will discuss it here. One important connection is that when we are simplifying fractions, we are essentially finding an equivalent fraction in which the numerator and denominator are smaller (and thus simpler) numbers. For example,

$$\frac{15}{20} = \frac{15 \div 5}{20 \div 5} = \frac{3}{4}$$

As has been true for other procedures, there are many strategies for simplifying fractions. Before we discuss them, simplify the following fractions yourself: 24/40, 42/60, and 63/105. Did you use the same procedure in each case? Did you encounter any difficulties? How did you overcome those difficulties?

One strategy is to use guess–test–revise. For example, you might divide both numerator and denominator by 4, which produces 6/10. A quick glance reveals that this can be simplified further to 3/5.

$$\frac{24}{40} = \frac{24 \div 4}{40 \div 4} = \frac{6}{10} = \frac{6 \div 2}{10 \div 2} = \frac{3}{5}$$

Another strategy is to determine the prime factorization of each number and then cross out the common factors.

$$24 = 2 \cdot 2 \cdot 2 \cdot 3$$
$$40 = 2 \cdot 2 \cdot 2 \cdot 5$$

Let us examine more closely what happens when we are able to simplify the fraction in one step, as we did earlier to simplify 15/20. What is the relationship between the divisor and the original numerator and denominator in each of the three fractions given above (24/40, 42/60, and 63/105)?

$$\frac{24 \div 8}{40 \div 8} \qquad \frac{42 \div 6}{60 \div 6} \qquad \frac{63 \div 21}{105 \div 21}$$

In each case, to simplify the fraction in one step, we divide both the numerator and the denominator by their GCF. By doing so, we ensure that the resulting numerator and denominator will have no factors in common. In other words, a fraction is in simplest terms when the numerator and the denominator are relatively prime.

Developing fraction sense One goal of this work and Exploration 5.11 is to develop "fraction sense." Number sense simply cannot be directly taught; rather, it emerges from being aware of the connectedness of, and subtle relationships among, various concepts and procedures. We see this awareness all the time when we watch experts at work: the dancer who can "feel" the subtle rhythms in a new piece of music, the electrician who rewired my house and could tell so much about the wiring problems from so little evidence, the mechanic who can look at a nut and know that he needs a 5/16-inch wrench.

As you are coming to realize, the set of fractions is in many ways very different from the set of whole numbers. Another important difference between fractions and whole numbers is that the whole numbers are evenly spaced on the number line. This is not true of fractions.

INVESTIGATION

5.8

Ordering Rational Numbers

Arrange these fractions from smallest to largest without converting into equivalent fractions, converting into decimals, or drawing diagrams—that is, work by focusing on fraction concepts and reasoning tools. Think and then read on. . . .

$$\frac{3}{4} \qquad \frac{2}{5} \qquad \frac{5}{6}$$

DISCUSSION

One of the reasons for assigning Exploration 5.6 is to help you to develop a sense of the relative size of fractions through hands-on work with fractions with different denominators. There are a variety of ways in which we can validly answer the question. We will explore several of these ways with the intention of refining or expanding your "fraction sense" toolbox.

Some people start by picking two fractions whose order they know. In this example, let us start that way. We know that $\frac{2}{5} < \frac{3}{4}$. An area diagram would readily show this.

Since one goal of this course is to develop mathematical reasoning (Standard 7), let us examine a tool that goes beyond visual evidence (i.e. "it looks bigger"). This tool uses the fraction 1/2 as a reference point or bench mark.

We know that 3/4 is greater than 1/2 because 3 is more than 1/2 of 4; similarly, we know that 2/5 is less than 1/2 because 2 is less than 1/2 of 5. Thus we can conclude that $\frac{2}{5} < \frac{3}{4}$.

The next debate concerns 5/6 and 3/4. How would you explain which is larger? Think and then read on. . . .

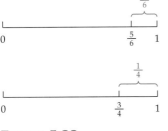

FIGURE 5.22

Looking at the two fractions, we see that they are both one piece away from 1. Sixths are smaller than fourths, so we can reason that 5/6 must be greater than 3/4 because the distance between 5/6 and 1 is 1/6, whereas the distance between 3/4 and 1 is 1/4 (Figure 5.22). We can combine $\frac{2}{5} < \frac{3}{4}$ and $\frac{3}{4} < \frac{5}{6}$ to conclude that $\frac{2}{5} < \frac{3}{4} < \frac{5}{6}$.

The Denseness of the Set of Rational Numbers

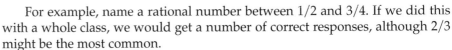 5.11

This kind of activity brings up a question that children sometimes ask: Can we name *any* point on the number line with a rational number? What do you think?. . .

For example, name a rational number between 1/2 and 3/4. If we did this with a whole class, we would get a number of correct responses, although 2/3 might be the most common.

Can you name another rational number between 1/2 and 3/4? How many can you name?

In fact, we can say that between any two rational numbers, there are an infinite number of rational numbers. Mathematicians refer to this property by saying that rational numbers are **dense**. Think of naming any point on a number line, knowing that no matter how close two rational numbers are, we can find an infinite number of rational numbers between those two rational numbers!

At the same time, there are points on the number line that cannot be named by rational numbers! Can you think of any such points? Think and then read on. . . .

Such points (for example, $\sqrt{2}$ and π) are called **irrational numbers**; these numbers cannot be expressed as a ratio between two integers. We will explore such numbers in Section 5.4.

Summary

This section and Explorations 5.5 to 5.11 lay the groundwork for understanding operations with fractions. In this section, we have seen that there are four major interpretations of fractions—as measures, as quotients, as operators, and as ratios. We have seen that we need to pay attention when we work with fractions because the fraction concept is not at all a simple one.

You have learned that a fraction is not simply a number; rather, a fraction expresses a relationship between two quantities. The numerator and denominator can be seen as a code that tells us the relative size of the fraction.

In order to work effectively with fractions as measures, we need a strong understanding of several basic ideas:

- We are dealing with part–whole relationships.
- Something is to be partitioned into parts.
- All of the parts must have the same value.
- We need to take care not to confuse the unit and the whole, which are the same in some situations but not in others.
- We can use various models to represent fractions and fraction situations. In this book, we emphasize number line, area, and discrete models.

We then applied these ideas in order to understand more fully the notions of equivalent fraction and simplifying fractions. A deeper understanding of rational numbers enables us to sense their relative size; this fraction sense is useful in estimating and problem-solving.

If you look back over the investigations in this section and Explorations 5.5 to 5.11, you will find that to the extent that you could apply these ideas, you were able to develop strategies that enabled you to solve the problems confidently. These ideas will continue to be important as we explore other fraction ideas and algorithms in the next section.

EXERCISES 5.2

1. In each of the questions below, justify your answer.

   a. If ▯▯ has a value of 4/5, draw and shade in the amount that would have a value of 1. Justify your answer.

   b. If ▦ $= 1\frac{1}{3}$, then shade in the amount equal to 1. Justify your answer.

   c. If ⁙ $= 3/4$, show 1.

   d. If ⁙ $= 5/4$, show 1.

   e. On the number line below, mark the approximate location of 1. Explain your reasoning.

   0      $\frac{2}{3}$

   f. On the number line below, approximate the value of x. Explain your reasoning.

   0   x 2

   g. Name a fraction between 2/3 and 3/4.
   h. Name a fraction between 1/4 and 1/5.
   i. Name a fraction closer to 1 than to 7/8.
   j. Name a fraction between 1/4 and 1/2 that has a denominator of 11.

2. Let's say you are making rulers for your students to help them learn fractions. You have decided to have a ruler marking every 1/6 inch and every 1/8 inch. How many different marks will there be per inch?

3. Draw three diagrams to represent each of the following fractions. Justify your diagrams.

	Length	Area	Discrete
$\frac{3}{5}$	├─┼─┼─┼─┤ 0 1		
$1\frac{2}{3}$			
$\frac{9}{4}$			

4. Let's say that Amy made 15 out of 21 free throws and Jane made 8 out of 12. Which student had a better rate of success? Explain your answer.

5. Two elementary school principals are comparing attendance figures for their schools.

School	Total enrollment	Absent
Wheelock	264	32
Fuller	402	58

Which school had the larger fraction of students absent? Justify your answer.

6. Write a fraction to represent the shaded portion. Justify your response.

7. Jamila was solving the problem of dividing a rectangular cake into eighths. She drew the diagram at the right. Although this is an unorthodox answer, do you agree with her that it is a mathematically correct solution? If so, explain why. If not, explain what you think is wrong about this.

8. What is the value of the fraction whose manipulative is midway between $\frac{1}{2}$ and $\frac{1}{4}$?

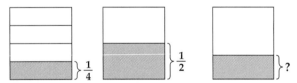

9. Look at the sequence below and look at the multiplication table in Chapter 3. Do you see a connection? Can you explain the connection, or is it just coincidence?

$$\frac{2}{3} = \frac{4}{6} = \frac{6}{9} = \frac{8}{12} = \frac{10}{15} = \cdots$$

10. Justify the cross-multiplication equation on page 278.

11. Simplify the following fractions.
    a. $\frac{25}{40}$ b. $\frac{32}{48}$ c. $\frac{54}{60}$ d. $\frac{26}{65}$
    e. $\frac{168}{216}$ f. $\frac{84}{132}$ g. $\frac{493}{510}$ h. $\frac{101010}{505050}$

12. Arrange the following fractions in order from least in value to greatest in value without converting to decimals or finding the LCM. Explain your reasoning.
    a. $\frac{3}{4} \quad \frac{4}{5} \quad \frac{5}{15}$ b. $\frac{6}{7} \quad \frac{5}{11} \quad \frac{2}{3}$
    c. $\frac{1}{3} \quad \frac{2}{5} \quad \frac{5}{8} \quad \frac{3}{4} \quad \frac{3}{50}$ d. $\frac{2}{5} \quad \frac{5}{6} \quad \frac{4}{7} \quad \frac{7}{8} \quad \frac{3}{10}$

13. Is 10/13 closer to 1/2 or to 1? Justify your choice.

14. Which of the following fractions is the farthest from 1? Explain why.

$$\frac{2}{3} \quad \frac{3}{4} \quad \frac{5}{6} \quad \frac{6}{7} \quad \frac{7}{8}$$

15. In the following exercise, $0 < a < b < c$. For the following pairs, tell which is larger and explain why, or explain why you cannot be sure which is larger. Try to answer the questions using only reasoning rather than plugging in numbers. You may then substitute actual numbers to check your reasoning.
    a. $\frac{a}{b}$ or $\frac{a}{c}$ b. $\frac{a}{b}$ or $\frac{b}{c}$ c. $\frac{a}{c}$ or $\frac{b}{c}$

16. If the numerator and the denominator of a fraction are increased by the same amount, is the new fraction greater than, equal to, or less than the original fraction? Justify your answer. Read this problem carefully; it is not a trick problem, but you need to make sure you interpret the wording correctly.

17. If $\frac{a}{b} = \frac{3}{4}$, will the value of $\frac{a+x}{b+x}$ be less than, equal to, or greater than $\frac{3}{4}$? Justify your answer by a means other than plugging in values for x.

18. The fraction of Fortune 1000 companies that employed undercover security agents in 1974 was 1/10. In 1985 it was 1/2. Compare the two numbers.

19. Let's say you decide to give blood at a blood drive. The body of an average adult contains about 5 quarts of blood. If you donate 1 pint of blood, what fraction of your blood do you give?

20. You are driving to Memphis, 115 miles away. You have 24 miles to go. Approximately what fraction of the trip is left? Justify your choice of fraction.

21. You have from 10 to 11:30 p.m. to do a project.
    a. At 11, what fraction of the time remains?
    b. At 11:20, what fraction of the time remains?

22. In 1950 the annual per capita consumption of eggs was 395; that is, the average person ate 395 eggs per year. In 1990 the per capita consumption had dropped to 236. Fill in the blank with a fraction: In 1990 the average person eats ____ as many eggs as the average person did in 1950.

23. Below are the median prices of homes sold in 1991, by region of the country:

South	$183,500
Northeast	$249,300
Midwest	$173,300
West	$322,000

The median selling price of a home in the Northeast was approximately what fraction of the median selling price of a home in the West?

24. PhonesRUs did a survey to determine telephone usage at a college. On the day the survey was done, a total of 5243 calls were made.
    a. If 1371 calls were made to directory assistance, approximately what fraction of the total calls were made to directory assistance? Justify your choice.
    b. The three pizza restaurants that delivered pizza received 737 calls. Approximately what fraction of the total calls were for pizza? Justify your choice.

25. In 2003, the average salary for beginning public school teachers was $31,704, and the average salary for all public school teachers was $46,597. The average salary for beginning public school teachers was approximately what fraction of the average salary for all public school teachers?

26. The data below show the area (in square miles) of several states and some of the Great Lakes:

    | Lake Superior | 31,800 |
    | Lake Huron | 23,010 |
    | All five Great Lakes | 94,710 |
    | New Hampshire | 9,279 |
    | California | 158,706 |
    | Tennessee | 42,104 |
    | Rhode Island | 1,212 |

    For each of the questions below, using mental arithmetic and estimation, find a fraction with a small, "friendly" denominator (2, 3, 4, 5, 6, 8, 10, or 12) that most closely approximates the actual fraction. Justify your reasoning.
    a. Lake Huron is approximately what fraction of the size of Lake Superior?
    b. New Hampshire is approximately what fraction of the size of Lake Superior?
    c. Lake Superior is approximately what fraction of the size of Tennessee?
    d. The area of all five Great Lakes is approximately what fraction of the size of California?

27. In 1996 there were approximately 4,800,000,000 people in the world. Using mental math and estimation, find a fraction to represent the fraction of the world's population that belonged to each of the following religions. Justify your reasoning.

    | Christianity | 1,548,500,000 |
    | Islam | 817,000,000 |
    | Hinduism | 645,500,000 |
    | Buddhism | 295,600,000 |
    | Judaism | 17,800,000 |
    | Total | 3,324,400,000 |

28. Each year the Weather Channel displays a map to indicate those portions of the country that will be celebrating a "white Christmas." If the shaded areas represent those regions that will have snow on the ground on Christmas Day, approximately what fraction of the contiguous United States will have a white Christmas? Explain your answer.

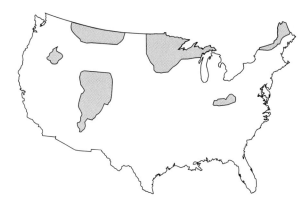

29. Give approximate values of this function for the given values of x:
    $$f(x) = \frac{2x - 1}{3x + 7}$$
    a. $x = 3$
    b. $x = 10$
    c. $x = 100$
    d. What do you observe about the change in the value of the function as x gets larger and larger?

30. The music teacher would like to have the same number of girls and of boys in the chorus. She finds that 5/8 of the chorus are girls but that if she can get 12 more boys, the chorus will have the same number of boys as of girls.
    a. How many students are in the chorus?
    b. If there are 216 children in the school, which unit fraction would you select to represent the fraction of students in the present chorus?

31. Darcy loves tomatoes, but she lives in Minnesota, where the growing season is shorter than in most of the country. Therefore, she begins her tomato plants inside. However, she tells her friend that there were some casualties this spring. One-third of the plants were destroyed by her cat. Furthermore, one-fourth of the remaining plants were destroyed by a disease. What fraction survived?

32. A few years ago, my son announced that he was 1/4 of my age. In how many years will he be 1/3 of my age? 1/2 of my age?
    a. Find one solution.
    b. How many solutions are there?
    c. Add one piece of information so that there is only one answer.

33. *Classroom Connection* I recall a teacher asking for fractions that were equivalent to 3/4 and one student replying $1\frac{1}{2}/2$. How would you respond to this answer? Justify your response.

34. The thermometers on the next page represent the fraction of the total goal raised in fundraising drives.
    a. What fraction is represented in the thermometer at the left?

b. What fraction is represented in the second thermometer from the left?

c. If the total is $158,000 and $37,000 has been raised, which of the three thermometers at the right most accurately represents the progress of the fundraising effort?

35. Below are two problems made by students to show fractional parts of a unit. Determine what fraction of the whole region is shaded, In (c), make up your own problem.

Source: From *Standards-Based School Mathematics Curricula: What Are They? What Do Students Learn?*, edited by Sharon L. Senk, Denise R. Thompson, Erlbaum, 2003, p. 111.

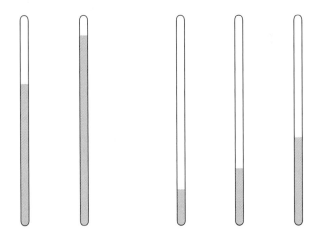

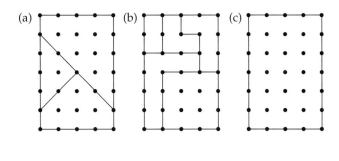

SECTION 5.3 UNDERSTANDING OPERATIONS WITH FRACTIONS

WHAT DO YOU THINK?

- Why *do* we need a common denominator to add (or subtract) two fractions?
- Why *don't* we need a common denominator to multiply (or divide) two fractions?
- If whole-number multiplication "makes bigger," why does fraction multiplication sometimes "make smaller"?
- Why isn't $3\frac{1}{4} \cdot 2\frac{1}{3} = 6\frac{1}{12}$

One of the sentences in the NCTM standards that has burned in my head is that students should "believe that mathematics makes sense." My earliest memories of teaching (tutoring while in college) are of surprise that this subject, which was so much fun for me, made almost no sense to many other people. There are many algorithms that enable us to manipulate fractions quickly: algorithms for adding, subtracting, multiplying, and dividing, and algorithms for translating improper fractions into mixed numbers and vice versa. However, as I have stressed in this book so far, it is simply not enough for an elementary teacher to know how to compute. It is crucial that the teacher also know the *whys* behind the *hows*. I have seen many elementary teachers have wonderful discussions when students asked why and the teacher knew what to do with that question!

Addition of Rational Numbers

Addition of fractions is one of those situations that mathematics is famous for—the procedure goes against what most people's common sense tells them to do. When we add fractions, we do not add the numerators and the denominators; on the other hand, when we multiply fractions, we *do* multiply the numerators and the denominators.

5.12

$$\frac{1}{2} + \frac{1}{3} \neq \frac{2}{5} \quad \text{but} \quad \frac{2}{3} \times \frac{4}{5} = \frac{8}{15}$$

Can you explain *why* we need to find a common denominator before we can add fractions with different denominators? Write (a first draft) in your notebook and then read on. . . .

INVESTIGATION 5.9

Using Fraction Models to Understand Addition of Fractions

Using the area model, the length model, or the set model, determine the sum of $\frac{1}{2} + \frac{1}{3}$ and then try to explain *why* we need to find a common denominator in order to add fractions. This is your "second draft" for justifying the need for a common denominator. Then read on. . . .

DISCUSSION

This is an excellent place to demonstrate the role of manipulatives.

Using pattern blocks In Exploration 5.6, you worked with two kinds of area manipulatives: squares and circles. Let us now consider pattern blocks (see Figure 5.23). How might you use pattern blocks to add 1/2 and 1/3?

FIGURE 5.23

If we let the hexagon represent 1, then the trapezoid is 1/2, the parallelogram is 1/3, and the triangle is 1/6. Using the notion of addition as combining, we can combine the two and we have 1/2 + 1/3 =   .

We could be humorous and say that parallelogram + trapezoid = baby carriage! The key question, though, is to determine the numerical value of this amount. The solution lies in realizing that to name an amount with a fraction, we must have equal-size parts. If you are familiar with pattern blocks, you realize that an equivalent representation is to cover this amount with 5 triangles, as in Figure 5.24.

FIGURE 5.24

Because the value of each triangle is 1/6, we can now say that $\frac{1}{2} + \frac{1}{3} = \frac{5}{6}$.

Using Cuisenaire rods This is a more different question when using Cuisenaire rods, because there is no "natural" choice for a unit. Thus you have to think of a color for which you can represent 1/2 of that color *and* 1/3 of that color. If students have worked with Cuisenaire rods in previous contexts (recall the discussion of GCF and LCM in Chapter 4), they will be familiar with the relationships among the various rods. One of many different solutions is to choose the dark green rod to have a value of 1. The red rod now has a value of 1/3, and the green rod has a value of 1/2. See Figure 5.25(a). As with pattern blocks, when we combine these two parts, in order to name the amount we have to find a way to represent this length with equal-size pieces, and 5 white rods have the same length. Because 6 white rods have the same length as the dark green rod, each of the white rods has a value of 1/6, and therefore 5 of them have a value of 5/6. See Figure 5.25(b).

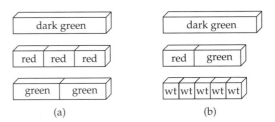

FIGURE 5.25

Looking back An important outcome of working with manipulatives is for the students to see that, for example, the yellow pattern block is not 6 but, rather, can be assigned any value, and the value of the other pieces is determined by their relationship to this block. So too with the Cuisenaire rods. Recall your work with the different models in Exploration 5.9.

Connecting Concepts to a Procedure for Adding Fractions

Let us now connect this work to the general procedure for adding fractions. Look at the addition problem below. Try to explain this procedure meaningfully—that is, to *justify* each step. Do this before reading on. . . .

$$\frac{2}{3} + \frac{1}{4} = \frac{8}{12} + \frac{3}{12} = \frac{11}{12}$$

In order to add two fractions that have different denominators, we determine the least common multiple of those two numbers, in this case 12. When working with fractions, we often refer to the LCM as the **lowest common denominator**. We now determine an equivalent fraction (with denominator 12) for each of the addends. We do this because all of the parts have to be the same size. We know that 2/3 is equivalent to 8/12 and that 1/4 is equivalent to 3/12. Now that both fractions have the same denominator (that is, all pieces are the same size), we can combine them. We find that we have $8 + 3 = 11$ pieces, each of which has a value of 1/12. Thus, the answer is 11/12.

You may remember and use the algebraic shortcut that is presented below without proof:

$$\frac{a}{b} + \frac{c}{d} = \frac{ad + bc}{bd}$$

If this doesn't make sense, try it for the numbers above; that is, substitute 2 for a, 3 for b, and so forth.

Which method (finding the LCM or using the algebraic shortcut) will be faster when the least common multiple is not the product of the two denominators? For example, $\frac{7}{8} + \frac{5}{6} = ?$ You might try it both ways to see which method you prefer.

Mixed Numbers and Improper Fractions

Up to this point, all of the addition problems have involved only **proper fractions**—that is, fractions whose value is between 0 and 1. Why do you think these fractions are called proper fractions? In many real-life situations, we encounter mixed numbers (for example, $5\frac{1}{4}$) and improper fractions (for example,

21/4). Take a couple of minutes to write down your own definitions of these two terms and then read on. . . .

An **improper fraction** is one in which the numerator is at least as large as the denominator.

A **mixed number** is a number that has a whole-number component and a fraction component.

The language we use to describe mixed numbers is important. How do you say $5\frac{1}{4}$ in English? Just as we found the value of expanded form helpful with whole numbers, there is an expanded form for mixed numbers. How do you think we would represent $5\frac{1}{4}$ in expanded form?

A recent National Assessment of Educational Progress contained the following item:

$5\frac{1}{4}$ is the same as:

(a) $5 + 1/4$ (b) $5 - 1/4$ (c) $5 \times 1/4$ (d) $5 \div 1/4$ (e) I don't know

Only 47 percent of the seventh-graders chose the correct response, (a). Still more startling, an even smaller percentage of eleventh-graders chose the correct response—only 44 percent. This lack of connectedness between concepts and procedures is one of the reasons why students do so much worse on algebraic equations with fractions than on algebraic equations with whole numbers. In this case, the problem is not the algebra so much as it is the arithmetic!

INVESTIGATION 5.10

Connecting Improper Fractions and Mixed Numbers

Most of you remember the procedure for converting a mixed number into an improper fraction.

For example, to convert $3\frac{1}{4}$ into an improper fraction, we do the following:

$$3\frac{1}{4} = \frac{3 \cdot 4 + 1}{4}$$

However, did you ever stop to think about what this might feel like to a youngster? On the surface, this procedure is as far from common sense as almost any in mathematics: We multiply the *whole number* by the *denominator* and then we add the *numerator*, and then we put this number on top of the *original denominator*. This sounds a lot like "gazinta" and "bring down" from division.

How would you explain the why of this procedure to, let's say, a fifth-grade student? Write your first attempt and then read on. . . .

DISCUSSION

As is often the case, concrete representations are helpful to most students. Figure 5.26 shows two concrete representations of $3\frac{1}{4}$.

FIGURE 5.26

In each of the representations, we see that the process is similar to the kinds of regrouping we did with whole numbers. In converting from the mixed number to the improper fraction, we need to convert all the units to fourths because in order for the improper fraction, 13/4, to have meaning, all the pieces must be the same size—that is, they must be fourths. We can convert each 1 to four fourths and then add all the fourths to get 13/4. However, because one context for multiplication is repeated addition, we see that we are adding three 4's; hence 3 · 4. When we add these 12 fourths to the 1 fourth we already had, we have a total of 13 fourths.

$$3\frac{1}{4} = 1 + 1 + 1 + \frac{1}{4}$$

$$= \frac{4}{4} + \frac{4}{4} + \frac{4}{4} + \frac{1}{4}$$

$$= \frac{3 \cdot 4}{4} + \frac{1}{4}$$

$$= \frac{3 \cdot 4 + 1}{4}$$

$$= \frac{13}{4}$$

Classroom Connection

Many children and adults do the problem below in the following way:

$$23\frac{2}{5} = \frac{117}{5}$$
$$-17\frac{4}{5} = \frac{89}{5}$$
$$\frac{28}{5} = 5\frac{3}{5}$$

Some middle school teachers teach this procedure. Their rationale is that so many students struggle with fraction subtraction that they gave up and taught the students this "foolproof" way, even though it is cumbersome in many cases. The teachers just gave up and focused on the *how*. I hope that even though many students will say, "Just show me how," you will be one of those teachers who keeps awake or reawakens the curiosity that young children naturally possess but most people lose as they grow up.

Understanding Subtraction of Rational Numbers

For the most part, subtraction of fractions involves essentially the same processes as does addition. However, many students encounter difficulty in situations that involve regrouping. Therefore, it is important to examine this process also. Consider the following problem: $23\frac{2}{5} - 17\frac{4}{5}$. First do it on your own, writing down your justification of each step. Then read on. . . .

$$23\frac{2}{5} = 22\frac{7}{5}$$
$$-17\frac{4}{5} = 17\frac{4}{5}$$
$$\phantom{-17\frac{4}{5} =\ } 5\frac{3}{5}$$

Let us discuss the transformation from $23\frac{2}{5}$ to $22\frac{7}{5}$, because this is difficult for many children. How would you explain this to someone who does not understand it?

We must do some renaming because we cannot subtract 4/5 from 2/5. With fractions, instead of converting 1 ten into 10 ones (as we did with whole numbers), we convert 1 into five fifths (or six sixths, or ten tenths, depending on the common denominator). Once again, expanded form will help to illustrate the process and justify the equivalence of $23\frac{2}{5}$ and $22\frac{7}{5}$.

$$23\frac{2}{5} = 23 + \frac{2}{5} = (22 + 1) + \frac{2}{5} = 22 + \left(\frac{5}{5} + \frac{2}{5}\right) = 22\frac{7}{5}$$

Section 5.3 / Understanding Operations with Fractions   289

INVESTIGATION 5.11

Estimation and Mental Arithmetic: Sums and Differences of Fractions

Think of circumstances in which you or other people might need to find the sum of fractions or the difference between fractions. Try to come up with realistic problem situations.

When I do this exercise with students and teachers, we find that more often than not, an estimate rather than an exact answer is needed. Thus it is important to examine strategies for estimating. As we have seen before with whole numbers, this also involves examining mental arithmetic. Because there are many strategies, we will examine them using a process similar to the one we used when developing estimation and mental math strategies with whole numbers.

Part 1: Mental addition and subtraction with fractions

In each of the following problems, try to find the sum or the difference entirely in your head.

Briefly describe your strategies. Then read the discussion below.

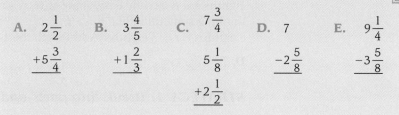

A. $2\frac{1}{2}$
    $+5\frac{3}{4}$

B. $3\frac{4}{5}$
    $+1\frac{2}{3}$

C. $7\frac{3}{4}$
    $5\frac{1}{8}$
    $+2\frac{1}{2}$

D. 7
    $-2\frac{5}{8}$

E. $9\frac{1}{4}$
    $-3\frac{5}{8}$

DISCUSSION

Keep the following points in mind for each problem:

- Deciding which strategy is better is often not as helpful as knowing which strategy works best for you.
- Not all possible useful strategies have been included here, because too many strategies (and 1500-page books) can overwhelm students!

A. $2\frac{1}{2} + 5\frac{3}{4}$

STRATEGY 1: Represent the sum with expanded form and use the commutative and associative properties

$$2\frac{1}{2} + 5\frac{3}{4} = \left(2 + \frac{1}{2}\right) + \left(5 + \frac{3}{4}\right) = (2 + 5) + \left(\frac{1}{2} + \frac{3}{4}\right) = 7 + \frac{1}{2} + \frac{3}{4}$$

That is, we quickly transform the problem into $7 + \frac{1}{2} + \frac{3}{4}$.

Decomposing $\frac{3}{4}$ into $\frac{1}{2} + \frac{1}{4}$, we then have $7 + \frac{1}{2} + \left(\frac{1}{2} + \frac{1}{4}\right) = 8\frac{1}{4}$.

STRATEGY 2: Add up and break into parts

$$\left(2\frac{1}{2} + 5\right) + \frac{3}{4} = 7\frac{1}{2} + \frac{3}{4} = 7\frac{1}{2} + \left(\frac{1}{2} + \frac{1}{4}\right) = 8 + \frac{1}{4} = 8\frac{1}{4}$$

B. $3\frac{4}{5} + 1\frac{2}{3}$

Use the shortcut

$$\frac{4}{5} + \frac{2}{3} = \frac{(4 \cdot 3) + (5 \cdot 2)}{15} = \frac{22}{15} = 1\frac{7}{15}$$

Add this to the sum obtained from the whole numbers and we have $4 + 1\frac{7}{15} = 5\frac{7}{15}$.

C. $7\frac{3}{4} + 5\frac{1}{8} + 2\frac{1}{2}$

STRATEGY 1: Look for compatible numbers

$\frac{3}{4}$ and $\frac{1}{2}$ can be added mentally to get $1\frac{1}{4}$; then add $1\frac{1}{4} + \frac{1}{8} = 1\frac{2}{8} + \frac{1}{8} = 1\frac{3}{8}$. Add this to the 14, and we have $15\frac{3}{8}$. *Note:* I find it easier to add $\frac{3}{4}$ and $\frac{1}{2}$ first rather than adding $\frac{3}{4}$ and $\frac{1}{8}$ and then $\frac{1}{2}$. Do you see why? Which would be easier for you?

STRATEGY 2: Convert to common denominator

Because the common denominator is relatively small and the conversions to eighths are relatively easy, some students find it easier to add all three at once, thinking something like this: "6 + 1 + 4 = 11, that means 11 eighths, which is 8 eighths (1) + 3 eighths. Add this to the 14 and we have $15\frac{3}{8}$ as the answer."

D. $7 - 2\frac{5}{8}$

STRATEGY 1: Break into parts and subtract

First, $7 - 2 = 5$; now take $\frac{5}{8}$ from 5 to get $4\frac{3}{8}$.

STRATEGY 2: Add up

$2\frac{5}{8} + 4 = 6\frac{5}{8}$. How much more do we need to get to 7? We need $\frac{3}{8}$ more. The answer is $4\frac{3}{8}$.

E. $9\frac{1}{4} - 3\frac{5}{8}$

Add up:

$3\frac{5}{8}$ plus 5 equals $8\frac{5}{8}$ plus $\frac{3}{8}$ equals 9 plus $\frac{2}{8} = 9\frac{1}{4}$. What you have to remember to add now is $5 + \frac{3}{8} + \frac{2}{8} = 5\frac{5}{8}$.

A number line helps some students to understand the process better (see Figure 5.27).

FIGURE 5.27

This is one place where many students find that the explorations pay rich dividends. Making the manipulatives (Exploration 5.6), grappling with relative sizes of fractions (Exploration 5.10), and developing fraction sense (Exploration 5.11) all focus on understanding the relationships involved in fractions. Having a good understanding of these relationships enables the student to decompose and then recompose the various fractions more confidently.

We can now apply these strategies to problems with a story.

Part 2: Estimating Sums and Differences with Fractions

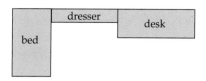

FIGURE 5.28

A. In their retirement, the parents of a close friend of mine have created a dollhouse to represent their dream house. They are shopping, and they see some miniature furniture that might fit in one of the bedrooms. The bed is $\frac{7}{8}$ inch wide; the dresser is $1\frac{1}{2}$ inches long, and the desk is $1\frac{3}{4}$ inches long. If the three articles are placed as shown in Figure 5.28, will they fit into the dollhouse bedroom, which is 4 inches wide? Try to add the three lengths in your head, and then read on. . . .

DISCUSSION

STRATEGY 1: *Make a quick estimate*

A quick estimation reveals that it will be close:

$$\frac{7}{8} + 1\frac{1}{2} + 1\frac{3}{4} \approx 1 + 1\frac{1}{2} + 1\frac{1}{2} = 4$$

Thus we need a strategy that will be more precise.

STRATEGY 2: *Add the two easier numbers*

$$1\frac{1}{2} + 1\frac{3}{4} = 1\frac{1}{2} + \left(1\frac{1}{2} + \frac{1}{4}\right) = 3\frac{1}{4}$$

We can use number sense here; that is, we know that $3\frac{1}{4} + \frac{3}{4} = 4$, and because $\frac{7}{8} > \frac{3}{4}$, we can conclude that the furniture is too big.

B. It rained almost every day this past week. Here are the amounts per day. How much rain fell during the week? First, make a rough estimate. Then make a refined estimate. Then read on. . . .

Monday	$1\frac{3}{4}$ inches	Friday	0 inches
Tuesday	$\frac{1}{2}$ inch	Saturday	$3\frac{3}{4}$ inches
Wednesday	$\frac{5}{8}$ inch	Sunday	$1\frac{5}{16}$ inches
Thursday	$\frac{7}{8}$ inch		

DISCUSSION

STRATEGY 1: *Make a rough estimate*

One way to get a rough estimate is to round each fraction to the nearest $\frac{1}{2}$ inch and keep a cumulative total:

$$2 + \tfrac{1}{2} = \mathbf{2\tfrac{1}{2}};\ 2\tfrac{1}{2} \text{ plus } \tfrac{1}{2} = \mathbf{3};\ 3 \text{ plus } 1 = \mathbf{4};\ 4 \text{ plus } 4 = \mathbf{8};\ 8 \text{ plus } 1\tfrac{1}{2} = \mathbf{9\tfrac{1}{2}}$$

STRATEGY 2: *Look for compatible numbers*

One way to get a closer estimate is to look for compatible numbers—numbers whose sum is close to 1 or numbers that you can quickly add mentally.

$$\underbrace{\left(1\tfrac{3}{4} + 3\tfrac{3}{4}\right)}_{5\tfrac{1}{2}} + \tfrac{1}{2} + \underbrace{\left(\tfrac{5}{8} + 1\tfrac{5}{16}\right)}_{\text{about 2}} + \underbrace{\tfrac{7}{8}}_{\text{about 1}} = \text{about 9}$$

Before we move on, take a few moments to reflect on the mental math and estimation strategies that you used and that were discussed here. What strategies have been adapted from ones we developed with whole numbers, such as compatible numbers? What strategies are new, such as the algebra shortcut?

INVESTIGATION 5.12 — Understanding Multiplication of Rational Numbers

The following multiplication problems have several purposes. First, they will enable you to connect the meaning of multiplication, which we developed with whole numbers, to fraction situations. Second, they will enable you to connect the meaning of fractions to fraction multiplication. Third, your ability to apply this knowledge to nonroutine and multistep problems should increase. Finally, you will be able to understand and explain things that teachers need to know—for example, why $3\tfrac{1}{2} \cdot 2\tfrac{1}{4} \neq 6\tfrac{1}{8}$.

For each of the following problems, represent the situation with a diagram and determine the answer from the diagram. As you are doing this, think back to the models we developed for whole-number multiplication. Which problems connect well to repeated addition? to a rectangular array? Which problems do not connect well? Why don't they?

A. Julio runs four times a week. His route is $2\tfrac{1}{4}$ miles long. How many miles does he run each week?

DISCUSSION

This connects nicely to multiplication as repeated addition: 4 times $2\tfrac{1}{4}$.

We can represent this problem on a number line, as in Figure 5.29.

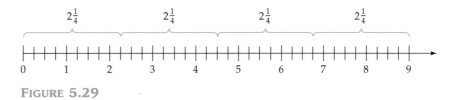

FIGURE 5.29

> **B.** A group of investors purchased a rectangular parcel of land that is 3/4 of a mile long and 2/3 of a mile wide. How many square miles did they buy?

DISCUSSION

This problem is definitely not repeated addition, but it does connect to the area model of multiplication. Here we can say that we want 2/3 of 3/4 (of 1 square mile).

However, getting the answer by applying our knowledge of multiplication and fractions is not terribly easy. If the large square in Figure 5.30 represents 1 mile by 1 mile, the shaded region represents the area of the land that the investors bought: 3/4 of a mile by 2/3 of a mile. But how do we give a name to this amount? Try to do this yourself and then read on. . . .

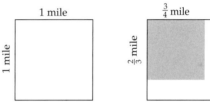

FIGURE 5.30

Just as we decomposed and recomposed numbers in working with whole numbers (when regrouping, for example), we need to decompose this (shaded) amount in such a way that we can recompose the amount as a set of equal-size pieces.

We can create equal-size pieces (needed to name a fraction) by dividing the square first into fourths (with vertical lines) and then into thirds (with horizontal lines). If we put the two diagrams together and shade in the area enclosed by 3/4 times 2/3, we have the figure on the right in Figure 5.31. Can you determine the answer now from the diagram? Can you justify that answer? Think and then read on. . . .

  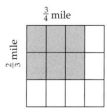

FIGURE 5.31

CLASSROOM CONNECTION

There is another difference between multiplying whole numbers and multiplying fractions that bears noting, for it often baffles children when they encounter fraction computations for the first time. One notion of multiplication that many people (often unconsciously) get from working with whole numbers is that "multiplication makes bigger and division makes smaller," but just the opposite is true when we multiply proper fractions!

Because our unit (1 square mile) has been divided into 12 equal-size regions, each region has a value of 1/12 square mile. The plot of land covers 6 of these rectangles, so its value is 6/12 or 1/2 square mile.

> **C.** If 2/3 of the class went on a field trip and there are 24 students in the class, how many students went on the field trip?

DISCUSSION

Problems like this one illustrate the third fraction context that we mentioned in Section 5.2: fraction as operator. The operator context can be thought of as a stretching or a shrinking process. In this case, we have a whole (24 students), and we are shrinking that whole—that is, we are considering a part (2/3) of that whole.

Let us now see how we can apply our knowledge of fractions and operations to solve this problem. Figure 5.32(a) represents the problem using a set model, with each dot representing one person. Because we have 2/3 of the people, the 2/3 tells us that we have to partition this amount into 3 equal groups and then take 2 of those groups. Figure 5.32(b) represents the problem using an area model. The whole rectangle represents the whole class. We have shaded in 2/3 of the rectangle. If the whole class represents 24 students, then when we divide the class into 3 equal regions, each region must represent 8 students (that is, $8 + 8 + 8 = 24$ or $3 \cdot 8 = 24$). Both models quickly produce the correct answer of 16 students.

> ■ *Language* ■
>
> When students memorize "*of* means multiply," they are memorizing a procedure but not understanding this interpretation of fractions.

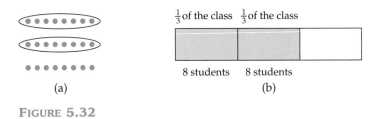

FIGURE 5.32

Connecting Multiplication of Fractions to the Multiplication Algorithm

As we did with multiplication of whole numbers, we will develop the algorithm with the rectangular model. Let us do so with the problem

$$\frac{3}{4} \cdot 2\frac{1}{2}$$

Let us first represent the problem with a diagram, determine the answer from that diagram, and then look to connect this solution to the algorithm. First, check to make sure that you understand that Figure 5.33 represents $\frac{3}{4} \cdot 2\frac{1}{2}$. Then determine the answer from the diagram before reading on. . . .

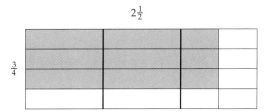

FIGURE 5.33

As we know, all the parts must be the same size in order for us to name the amount with a fraction. We can readily see (from Figure 5.34) that the numerator of our product must be 15 (that is, there are 15 parts). But what is the denominator? (Please resist the temptation to answer this question by doing the algorithm. In my own class, this question is actually a method I use to assess how deeply my students understand the meaning of denominator.) Then read on. . . .

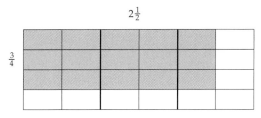

FIGURE 5.34

I usually get three different responses for the denominator: 24 because there are 24 parts in the "whole" diagram, 15 because there are 5 times 3 parts in the $2\frac{1}{2}$, and 8 because it takes 8 parts to make a 1 by 1 square (the unit).

The correct answer is 8. Why is this? Recall that the value of the denominator is determined by its relationship to the unit. That is, the denominator shows how many pieces it takes to have a value of 1, rather than how many pieces are in the whole.

Therefore, we have determined that

$$\frac{3}{4} \cdot 2\frac{1}{2} = \frac{15}{8} = 1\frac{7}{8}$$

Now let us examine the algorithm and look for connections between the algorithm and the concepts. We must first convert $2\frac{1}{2}$ to an improper fraction, and then we simply find the product of the numerators and the product of the denominators:

$$\frac{3}{4} \cdot 2\frac{1}{2} = \frac{3}{4} \cdot \frac{5}{2} = \frac{3 \cdot 5}{4 \cdot 2} = \frac{15}{8}$$

Can you justify this procedure? Can you explain the *whys* behind these *whats*? Think. Look back to Figure 5.34. Then read on. . . .

If we look at the shaded region in Figure 5.34, we have 3 rows, each containing 5 regions; that is, we have 3 times 5 regions. Just as we found that the whole-number multiplication algorithm automatically regrouped for us, so too the fraction multiplication algorithm automatically creates equal-size pieces. Similarly, we have 4 · 2 pieces in each unit.

Justifying the Equality of Equivalent Fractions

In Section 5.2, we examined equivalent fractions. We can now justify the procedure we used earlier. The process is shown at the left for a specific case and at the right for the general case.

Specific case	Justification	General case
$\frac{3}{4} = \frac{3}{4} \cdot 1$	Multiplicative identity	$\frac{a}{b} = \frac{a}{b} \cdot 1$
$= \frac{3}{4} \cdot \frac{2}{2}$	Substitution	$= \frac{a}{b} \cdot \frac{c}{c}$
$= \frac{3 \cdot 2}{4 \cdot 2}$	Definition of multiplication of fractions	$= \frac{a \cdot c}{b \cdot c}$
$= \frac{6}{8}$		

The general case also emphasizes that the numerator and denominator of the new fraction are simply multiples of the numerator and denominator of the original fraction.

INVESTIGATION 5.13

Division of Rational Numbers

As with most concepts being considered in this course, you probably remember the procedure. This recalls a famous saying of unknown origin: "Ours is not to reason why, just invert and multiply."

$$\frac{3}{4} \div \frac{5}{6} = \frac{3}{4} \times \frac{6}{5} = \frac{18}{20} = \frac{9}{10}$$

Now, though, ours *is* to reason why. Exploration 5.14 examined an alternative algorithm first expressed by another Indian mathematician, Brahmagupta, more than a thousand years ago. Let us begin our investigation of the fraction division algorithm by first examining fraction division problems in context.

■ History ■

The Hindu mathematician Mahavira (A.D. 850) first stated a rule for dividing fractions: "After making the denominator of the divisor its numerator and vice versa, the operations to be conducted then are as in the multiplication [of fractions]." It was not until the sixteenth century that this procedure was "discovered" in Europe.[3]

Recall the two contexts for division with whole numbers: partitioning and repeated subtraction.

Make up a story for each of the two division problems below.

First, solve the problem using what you know about division and fractions.

Then respond to the following questions: Which of your stories were from the partitioning model, and which were from the repeated-subtraction model? If you made up a story connected to one model, could you make up another story for the same problem using the other model?

A. The divisor is a whole number: $3 \div 4$
B. The divisor is a proper fraction: $3 \div 2/5$

[3]Louis Karpinski, *The History of Arithmetic* (New York: Russell & Russell, 1965), p. 128.

FIGURE 5.35

Mathematics

As mentioned in Section 5.2, fraction as division is another of the rational-number contexts. That is, the division problem $a \div b$ and the fraction a/b are equivalent. This relationship between fractions and division is often problematic for children, and many do not realize that this is not a commutative relationship. Many middle school children cannot accurately translate $4\overline{)3}$ into 3/4, and vice versa.

5.15

DISCUSSION

A. Most people make up a story for $3 \div 4$ using the partitioning model, and most of the stories involve sharing.

Let us consider one such story and examine a solution: Basil has 3 pints of ice cream that he wants to share equally with 3 friends—that is, among 4 people. How much is each person's share?

Figure 5.35 shows a partitioning (dealing) solution. That is, if we divide each pint into 4 equal parts, each person gets 1 part per pint. Each person's share thus consists of 3 parts, each of which is 1/4 pint. Therefore, each person's share is 3/4 of a pint.

Some people literally interpret this problem as a multiplication problem. If we are dividing the ice cream among 4 people, then each person gets 1/4 of the ice cream. This connection between the two operations produces the following equality: $3 \div 4 = 3 \cdot \frac{1}{4}$, which we just mentioned. We will return to this relationship between multiplication and division shortly. However, first let us examine a story for the other problem.

B. One story for $3 \div 2/5$: Jake is stranded in the middle of the desert. He has 3 quarts of water, and he figures that he will drink 2/5 of a quart each day. How many days' supply does he have?

Solve the problem using only your knowledge of fractions and your knowledge of division. Then read on. . . .

This is a repeated-subtraction problem because we have an amount (3 quarts) and we are specifying the size of the group (2/5 of a quart) to be repeatedly subtracted. The answer will be the number of groups we have.

I have chosen to represent this problem with a number line, although an area or set model could also have been used. Each day Jake uses 2/5 of a quart. We see from Figure 5.36 that he is able to drink 2/5 of a quart each day for 7 days. However, we have a small problem: We have repeatedly subtracted 2/5 of a quart, but we have 1/5 of a quart left over. Does this answer contradict the answer that comes from using the algorithm, which tells us that $3 \div \frac{2}{5} = 7\frac{1}{2}$? How do we reconcile the difference between the answer from the diagram and that from the algorithm? Think and then read on. . . .

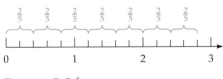

FIGURE 5.36

It turns out that both answers are correct—they represent different ways of describing the remainder; recall the remainder discussion in Chapter 3 and Exploration 5.15. The 1/5 tells us that the remainder is 1/5 of a quart; thus we can say that at the end of 7 days, he will have 1/5 of a quart left. The $7\frac{1}{2}$ tells us that the remainder is equivalent to 1/2 day's water. This underscores the point made in Investigation 1.15 (How Many Pieces of Wire?) that the key to a problem is often knowing what the computation means.

Virtually all stories for this problem involve repeated subtraction. In the first edition of this textbook, I stated, "In fact, I cannot make up a realistic story for this problem using the partitioning model of division." Then I read a book called *Knowing and Teaching Elementary Mathematics* by Liping Ma. In her book,

Ma asked a number of U.S. and Chinese elementary school teachers to make up a story problem for $1\frac{3}{4} \div \frac{1}{2}$. I was stunned to read that most of the Chinese teachers' stories involved the partitioning model. Although my conception of this model is not exactly the same as Ma's—who describes it as "finding a number such that 1/2 of it is $1\frac{3}{4}$" (p. 74), I found her example and her book to be very educational. I recommend the book to you.

Making Sense of the Standard Division Algorithm

5.14

Although using the invert-and-multiply algorithm is rather straightforward, understanding why it works is another matter. Unlike most of the other algorithms we have examined, it does not lend itself to a diagrammatic representation. Therefore, we will begin to build our understanding of it by examining the similarities between how division and multiplication are connected and how subtraction and addition are connected.

We need to introduce a new concept to do so. Just as we discovered in the first section of this chapter that every integer has an additive inverse, every nonzero rational number has a **multiplicative inverse**. These parallel concepts are shown side by side below:

^-a is the additive inverse of a	$\frac{1}{a}$ is the *multiplicative inverse* of a, $a \neq 0$
$a + {}^-a = 0$ for any integer a	$a \cdot \frac{1}{a} = 1$ for any rational number, $a \neq 0$

Let us now recall the way in which we defined integer subtraction in terms of addition, and then we will define fraction division in terms of fraction multiplication.

$a - b = a + {}^-b$	$a \div b = a \cdot \frac{1}{b}$
That is, to subtract b, we add the (additive) inverse of b.	That is, to divide by b, we multiply by the (multiplicative) inverse of b.
In other words, one way to interpret *subtraction* is to *add* the *additive inverse* of the second number.	In other words, one way to interpret *division* is to *multiply* by the *multiplicative inverse* of the second number.

Thus,

$$a \div \frac{x}{y} = a \cdot \frac{y}{x}$$

Revisiting properties Let us now examine how the properties that we developed in Chapter 3 extend to operations with fractions. The commutative and associative properties hold for addition and multiplication but not for subtraction and division. The distributive properties hold for fractions. Both identity properties hold: 0 is the identity for addition and 1 for multiplication. The zero property of multiplication holds. We still have additive inverses, and we have now acquired the multiplicative inverse for all numbers except 0.

INVESTIGATION 5.14

Estimating Products and Quotients

Let us now examine some estimation strategies for multiplication and division.

In each of the problems below, do the following: First obtain a rough estimate (5 to 10 seconds), and then try to get as close as you can to the actual answer—either a refined estimate or, in some cases, an exact answer (computed mentally).

A. Anastasia walks at a pace of $3\frac{1}{2}$ miles per hour for 2 hours and 15 minutes. How far does she walk during this time?

DISCUSSION

Note: We first need to convert the time to fractional form; that is, we are estimating $3\frac{1}{2} \cdot 2\frac{1}{4}$.

STRATEGY 1: Bound the answer

One kind of estimating involves bounding the answer. For example, the answer will be at least 6. We get this by focusing only on the whole numbers. We call 6 a lower bound.

An upper bound can be obtained by rounding both numbers to the next higher whole number. The upper bound is thus $4 \cdot 3 = 12$.

STRATEGY 2: Get rid of one of the proper fractions (that is, rounding)

Another method of making a rough estimate involves getting rid of one of the proper fractions. We can more easily determine $3\frac{1}{2} \cdot 2 = 7$ or $3 \cdot 2\frac{1}{4} = 6\frac{3}{4}$.

STRATEGY 3: Use partial products

We can estimate and round the four partial products:

$$3\frac{1}{2} \times 2\frac{1}{4} = \left(3 + \frac{1}{2}\right) \times \left(2 + \frac{1}{4}\right) = (3 \times 2) + \left(3 \times \frac{1}{4}\right) + \left(2 \times \frac{1}{2}\right) + \left(\frac{1}{2} \times \frac{1}{4}\right)$$
$$\approx \quad 6 \quad + \quad 1 \quad + \quad 1 \quad + \quad 0 \quad \approx 8$$

B. The Jones family finds that they spend about 2/3 of their income on food and rent. If their monthly income is $1635, how much is their monthly food and rent bill?

DISCUSSION

Rough estimate (using compatible numbers):

$$\frac{2}{3} \times 1500 = 1000$$

More refined estimate (using compatible numbers and expanded form):

$$\frac{2}{3} \times 1650 = \frac{2}{3}(1500 + 150) = 1000 + 100 = 1100$$

C. Marvin has 23 yards of cloth with which to make costumes for the play. Each costume requires $3\frac{1}{4}$ yards of material. How many costumes can he make?

DISCUSSION

Rough estimate (using bounding):

$$23 \div 3 = 7\frac{2}{3}$$

$$23 \div 4 = 5\frac{3}{4}$$

The estimate is about 6 or 7 costumes.
Refined estimate (using repeated addition):

$$3\frac{1}{4} \times 2 = 6\frac{1}{2}, \text{ so}$$

$$3\frac{1}{4} \times 4 = 13, \text{ so}$$

$$3\frac{1}{4} \times 8 = 26$$

Because $26 - 3 = 23$, this estimate yields 7 costumes.

D. Shelly has 25 pounds of dog food, and each day she feeds her dog 3/4 pound. How many days' worth of food does she have?

DISCUSSION

Rough estimate (using bounding):

$$25 \div 1 = 25 \quad \text{and} \quad 25 \div \frac{1}{2} = 50$$

Thus she has enough for more than 25 days and less than 50 days.
Refined estimate (using the algorithm and the commutative and associative properties):

$$25 \div \frac{3}{4} = 25 \times \frac{4}{3} = \frac{100}{3} = 33\frac{1}{3}$$

which is the actual quotient. Thus, she has 33 days' worth of dog food.

This problem can also be estimated using multiplication as adding up. Do you understand why this solution path yields an estimate of about 32 days?

3/4 × 2 = 1 1/2
3/4 × 4 = 3
3/4 × 8 = 6
3/4 × 16 = 12
3/4 × 32 = 24

Applying Fraction Understandings to Nonroutine Problems

We have carefully examined the different meanings that fractions can have, and we have examined the four operations in terms of fractions. In the following nonroutine problems, you will need to determine which computation(s) to do and how to interpret the computation(s).

INVESTIGATION 5.15

When Did He Run Out of Gas?

Jeremy started on a trip from Mobile, Alabama, to New Orleans, approximately 120 miles away. His car ran out of gas after he had gone one-third of the second half of the trip. How far is Jeremy from New Orleans? Work on this problem yourself and then read on. . . .

DISCUSSION

This is a case in which a good diagram practically solves the problem by itself (see Figure 5.37).

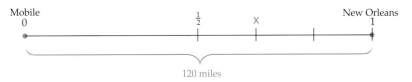

FIGURE 5.37

If half of the trip is equal to 60 miles, then each of the "thirds of the second half" is equal to 20 miles. X marks the spot of "one-third of the second half of the trip." So Jeremy is still 40 miles from New Orleans.

INVESTIGATION 5.16

They've Lost Their Faculty!

Suppose you are reading the newspaper and you see that American State College has had to reduce its faculty by one-sixth because of an economic crisis in the state. The article says that American State College now has 350 faculty members. A curious reader might ask, "How many did they have before the cut?" Think about this problem before reading on. . . .

DISCUSSION

STRATEGY 1: *Represent the situation with a diagram*

The large rectangle in Figure 5.38 represents the original faculty.

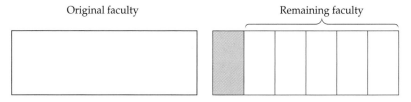

FIGURE 5.38

If the college lost one-sixth of its faculty, then we can divide that rectangle into 6 equal pieces and take away 1 piece, which represents one-sixth of the faculty. Thus the 5 remaining boxes represent the remaining 5/6 of the faculty.

Because we know that there are 350 faculty members now, the value of those 5 boxes is 350. But if the value of 5 boxes is 350, then the value of 1 box is 70. Now we can answer the problem. Do you see why?

Because the value of the original faculty was represented by 6 boxes, there were $6 \times 70 = 420$ faculty.

STRATEGY 2: *Use algebra*

Can you find an algebraic solution? Work on it and then read on. . . .

One algebraic solution comes from letting x = the number of the faculty before the cut. The remaining faculty represents 5/6 of the original, so 350 is 5/6 of x; that is, $\frac{5}{6}x = 350$.

We solve the equation by multiplying both sides by 6/5:

$$\left(\frac{6}{5}\right)\frac{5}{6}x = \left(\frac{6}{5}\right)350 \quad \text{and} \quad x = 420$$

Summary

Let us take a few moments to look back on this section. What have you learned? What fraction ideas make more sense now than they did before? What fraction ideas still feel difficult?

In this section, we have focused on connecting conceptual knowledge to procedural knowledge—on understanding the whys behind the how of the algorithms for operations with fractions.

We have found that the contexts in which we first studied these four operations in Chapter 3 do not all apply when we are working with fractions; for example, the partitioning model of division does not readily apply when the divisor is not a whole number. In making sense of algorithms, we need to pay attention to the very fraction ideas that we developed in Section 5.2: that the parts in a fraction have the same value, that working with fractions involves parts and wholes, and that we have to pay attention to units and wholes (because they are not always the same).

When we explored mental arithmetic and estimation with fractions, we found that many of the strategies developed in Chapter 3 applied directly to computing with fractions—for example, the commutative and associative properties, compatible numbers, adding up, partial products, and expanded form.

When we applied our knowledge of operations to nonroutine and multistep problems, we found that we needed to think about how we might represent the problem. We also had to make sure that we kept basic fraction ideas in mind—for example, the distinction between the unit and the whole.

Now that we have a deeper understanding of fractions, we are ready for our next extension of the number system—to decimals and real numbers.

EXERCISES 5.3

1. Perform the following computations.

   a. $23\frac{7}{8} + 56\frac{5}{6}$

   b. $-3\frac{5}{12} + 1\frac{9}{40}$

   c. $7\frac{3}{4} + 8\frac{3}{5} + 7\frac{3}{10}$

   d. $213 - 21\frac{5}{16}$

   e. $9\frac{1}{4} - 12\frac{5}{6}$

   f. $12\frac{1}{3} \times 5\frac{3}{8}$

   g. $6\frac{3}{4} \times 14\frac{2}{3}$

   h. $36 \times 3\frac{3}{4}$

   i. $\frac{3}{10} \times 45$

   j. $\frac{3}{4} \div 6$

   k. $2\frac{3}{8} \div 6\frac{1}{3}$

2. Determine each of the following mentally, and briefly explain your solution.

   a. $5\frac{3}{4} + 7\frac{5}{8}$

   b. $8\frac{4}{5} + 6\frac{2}{3}$

   c. $16\frac{3}{4} + 15\frac{1}{2} + 7\frac{1}{4}$

   d. $26 - 6\frac{3}{5}$

   e. $7\frac{1}{3} - 3\frac{1}{2}$

   f. $8\frac{2}{3} + 7\frac{5}{6} + 4\frac{2}{3} + 2\frac{2}{3}$

   g. $3\frac{2}{3} \cdot 12$

   h. $2\frac{1}{2} \cdot 3\frac{1}{4}$

3. For each of the following, determine which of the choices represents the better estimate. Then explain your choice.

   a. $5\frac{5}{8} + 4\frac{3}{42}$ Greater than 10 or less than 10?

   b. $\frac{7}{8} + \frac{3}{4} + \frac{1}{16}$ Greater than 2 or less than 2?

   c. $\frac{1}{4} + 1\frac{9}{10}$ Greater than 2 or less than 2?

   d. $8\frac{1}{2} - 2\frac{2}{3}$ Between which two numbers: 5 and $5\frac{1}{2}$ or $5\frac{1}{2}$ and 6?

   e. $8\frac{3}{4} \times 2\frac{7}{8}$ Greater or less than 20?

   f. $4\frac{7}{8} \div 8\frac{7}{8}$ Greater or less than 1/2?

4. Without actually multiplying, can you tell which of the following two is greater? Explain your reasoning.

    $26\frac{5}{7} \times 9$ or $26 \times 9\frac{5}{7}$

5. Without doing any computing, order the following from least to greatest. Explain your reasoning.

    2/3 + 3/4   2/3 − 3/4   2/3 × 3/4   2/3 ÷ 3/4

6. Use a number line model or a set model to find $\frac{2}{3} + \frac{1}{4}$. Justify each step in the process.

7. Represent and solve each of the following problems with diagrams. Explain your solution path.

    a. 2/3 + 5/6   b. 2 1/3 + 3/4   c. 3/8 − 3/4
    d. 2/3 × 1/2   e. 3/4 × 6   f. 3/4 × 5 1/2
    g. 3/4 ÷ 6   h. 7/8 ÷ 3/8   i. 5 1/2 ÷ 3/4

8. The shapes below refer to Pattern Blocks.
    a. If hexagon + rhombus = 1, what is the value of triangle?
    b. If rhombus + rhombus = 1, what is the value of hexagon?
    c. If hexagon + triangle = 2, what is the value of trapezoid?
    d. If hexagon − triangle = 1, what is the value of trapezoid?

For Exercises 9 to 17, first estimate the answer, next explain your estimate, and then determine the "exact" answer. Finally, explain whether the "exact" answer is an exact number or still an estimate.

9. Sunflower seeds are sold in packages that weigh $3\frac{1}{4}$ ounces. If there is a supply of 66 ounces of sunflower seeds, how many packages of seed can be made? How many ounces of seeds will be left over?

10. What is the weight (in ounces) of a $4\frac{1}{2}$ inch by $7\frac{3}{4}$ inch rectangle of sheet metal if one square inch weighs 1/8 ounce?

11. Betty is stacking boxes at a factory. Each box is $13\frac{3}{4}$ inches high. The ceiling is 16 feet. How many boxes can she stack in one pile?

12. Javier is 2/3 as tall as his dad. Javier is 49 inches tall. How tall is his dad?

13. A recent survey concluded that 3/4 of the teachers in a certain school district had at least 10 years of teaching experience. If there are 152 teachers in the district, approximately how many have at least 10 years of teaching experience?

14. We find from the *Unofficial U.S. Census* that more than 136,800,000 Americans admit to doodling. Of the doodlers, slightly less than 2/3 said they doodle while talking on the telephone. Approximately how many Americans doodle while talking on the telephone?

15. Below is a picture of the gas gauge in Jackie's car. If Jackie figures that her car averages 34 miles per gallon and her car has a 12-gallon tank, about how many miles does she have left on this tank before she runs out of gas?

16. John is making a bookcase. The bookcase is $72\frac{3}{8}$ inches tall and he wants 5 shelves, equally spaced. Tell him where to drill the holes on the side. Each shelf is 3/8 inch thick. Your solution needs to contain a description that is clear enough so that he can understand it.

17. Julianna has $22\frac{1}{2}$ yards of material. If it takes 3/4 yard of material to make a pair of shorts, can she make enough shorts for 30 children?

18. An analysis of first-year students at a college revealed that 1/4 of the first-year women were from homes where both parents were professionals. Of these, 3/5 were interested in the same profession as one or both of their parents. If this latter group is made up of 18 students, how many first-year women are there?

19. A small office purchases bottled drinking water. The water comes in containers that hold 5 gallons. Two-thirds of the office container was drained last week to put out a fire in the wastebasket.
    a. How much water (in ounces) is left in the container?
    b. How many 12-ounce glasses can be filled with the water that remains in the container?

20. Let's say that a certain antibiotic kills approximately 1/2 of the germs on the first day, 1/2 of the remaining germs on the second day, and so on.
    a. After 10 days, what fraction of the germs remain?
    b. If you had 20 million germs in your body at the beginning, how many of them will be alive at the end of the tenth day?

21. A fruit juice drink is 5/7 water and 1/6 fruit juice (by weight). Other additives (primarily sugar) make up what fraction of its weight?

22. In a certain town, 2/3 of the women are married and only 1/2 of the men are married. What fraction of the community is single? What assumptions do you make in order to solve the problem?

23. A manufacturing company finds that approximately 3 out of every 1000 spark plugs it produces are defective. Last year the company produced 186,000 spark plugs. Approximately how many were defective?

24. In 1991, 207,610 rapes were reported to police. Estimates of the fraction of rape victims who report the rape range from 1/2 to 1/6. If 1/2 of all rape victims reported the rape to police, approximately how many women were raped in 1991? If only 1/6 of all rape victims reported the rape to police, approximately how many women were raped in 1991?

25. This is the kind of problem that really does happen—some data are lost, but other data can be used to retrieve them. Here are the employment data for a state:
    - 5/8 of the population is urban, 3/8 is rural.
    - 1/10 of the urban population is unemployed, and 1/6 of the rural population is unemployed.

    What fraction of the work force is unemployed? Illustrate this problem and solution with a diagram.

26. The Petersons have an apple-pressing machine that they use to make apple juice. The first pressing gets about 1/3 of the total juice in the apples. Each succeeding pressing extracts about 1/3 of the remaining juice in the apples. How many pressings will it take to get at least 3/4 of the apples' juice? How many pressings will it take to get at least 9/10 of the juice?

27. Irene has discovered a rule that works for some addition problems. For example, consider 1/5 + 1/2. The rule is this: The numerator of the answer is equal to the sum of the denominators of the two fractions, and the denominator of the answer is equal to the product of the denominators. What do you think of her rule? Describe the set of fractions for which this rule works. Can you modify her rule so that it works for a larger set of problems?

28. Find the difference: $\dfrac{bc}{a} - \dfrac{bd}{ac}$

29. Make up a realistic story problem for the following multiplication-of-fractions contexts:
    a. A proper fraction times a proper fraction
    b. A proper fraction times a whole number
    c. A proper fraction times a mixed number
    d. A mixed number times a mixed number

30. Demonstrate how to obtain the following product two different ways, neither of which can be the standard algorithm: $3\frac{1}{2} \times 2\frac{1}{3}$. Then explain why $6\frac{1}{6}$ is not the correct answer, as though you were talking to a student for whom $6\frac{1}{6}$ is the obvious answer.

31. Solve the following problems using only the area model.
    a. $\dfrac{2}{3} \times 2\dfrac{3}{4}$ b. $2\dfrac{2}{3} \times 3\dfrac{1}{2}$

32. Pete has just made a batch of Rice Krispies marshmallow treats, one of my favorites! The banking pan measures 13" by 9". He has decided that each treat will be 1 1/2" by 1".
    a. First, *estimate* how many treats he can get. Explain your estimate.
    b. Now determine exactly how many treats can he get from one pan.
    c. Why is the correct answer not (13 × 9) ÷ 1 1/2?
    d. Change the dimensions of the pan or the size of the treats so that you can get the answer by dividing the area of the pan by the area of one treat.

33. Ginger has $8\frac{1}{3}$, yards of silk cloth from India. She has decided to divide the cloth equally into 3 parts, one for each of her siblings. How much cloth does each person get? *Do this problem without using an algorithm*—that is, using reasoning. You may want to use a diagram, but you are not required to do so.

34. When we convert 13/5 to $2\frac{3}{5}$, which model of division do we use? Explain your answer.

35. An early division algorithm in India is actually simpler than the current algorithm. It involved converting the fractions to fractions with a common denominator, as we do with addition, and then simply dividing the numerators. For example,

    $$\dfrac{3}{4} \div \dfrac{5}{6} = \dfrac{9}{12} \div \dfrac{10}{12} = \dfrac{9}{10}$$

    Make up some other problems to see whether this algorithm works only with proper fractions or with all fraction problems. Why does it work?

36. Make up a realistic story problem for $2\frac{1}{2} \div 3/4$ and solve the problem using the meaning of fractions and division (that is, without the algorithm).

37. a. How would you figure 26 × 11/12 on a calculator? Show two ways.
    b. How would you figure 26 ÷ 11/12 on a calculator?

38. Selecting among the numbers 1 through 9 and repeating none of them, fill in the boxes below to make the sum as close as possible to 1, but not equal to 1.

    $\dfrac{\Box}{\Box} + \dfrac{\Box}{\Box}$

39. Use each of these numbers exactly once to make the following equation true: 1, 2, 3, 5, 11, 15.

    $\dfrac{\Box}{\Box} + \dfrac{\Box}{\Box} = \dfrac{\Box}{\Box}$

40. Selecting among the numbers 1 through 9 and repeating none of them, fill in the boxes below to make the smallest possible difference greater than zero.

    $\dfrac{\Box}{\Box} - \dfrac{\Box}{\Box}$

41. Selecting among the numbers 1 through 9 and repeating none of them, make the following equation true.

     $= \dfrac{1}{2}$

42. Selecting among the numbers 1 through 9 and repeating none of them, make the largest possible quotient.

    $\dfrac{\Box}{\Box} \div \dfrac{\Box}{\Box}$

43. Selecting among the numbers 1 through 9 and repeating none of them, make the quotient closest to one-half.

$$\frac{\square}{\square} \div \frac{\square}{\square}$$

44. *Classroom Connection* In each case, describe the error that the student made, then describe the mathematical idea(s) the student did not correctly understand.

a. $\frac{2}{3} + \frac{3}{8} = \frac{5}{24}$

b. $\frac{3}{5} + \frac{2}{5} = \frac{5}{10}$

c. $\frac{1}{5} + \frac{2}{3} = \frac{3}{8} + \frac{10}{8} = \frac{13}{8} = 1\frac{3}{8}$

d. $7\frac{2}{5}$
   $-3\frac{4}{5}$
   $\overline{\ 4\frac{2}{5}}$

e. $8\frac{5}{8} - 2\frac{2}{3} = 6\frac{3}{5}$

f. $7\frac{1}{8} = 6\frac{11}{8}$
   $-3\frac{5}{8} = 3\frac{5}{8}$
   $\overline{\ 3\frac{6}{8} = 3\frac{3}{4}}$

g. $\frac{2}{3} \times \frac{3}{4} = 89$ and $\frac{5}{6} \times \frac{3}{8} = 220$

h. $\frac{2}{3} \times 5 = \frac{10}{15}$

i. $\frac{2}{3} \cdot \frac{3}{8} = \frac{9}{16}$

j. $\frac{8}{15} \div \frac{2}{5} = \frac{4}{3}$

k. $\frac{8}{15} \div \frac{2}{5} = \frac{15}{8} \times \frac{2}{5} = \frac{30}{40} = \frac{3}{4}$

l. $8\frac{1}{8} \div 2\frac{1}{4} = 4\frac{1}{2}$

45. *Classroom Connection* In the December 2000 issue of *Teaching Children Mathematics*, the authors examined third- and fourth-graders' responses to the question "What is the value of $2\frac{1}{2} + \frac{3}{4}$?"

a. This is the diagram that accompanied one child's explanation. Describe what might have been the child's reasoning behind this diagram. Then describe the mathematical ideas that the child used—properties of addition, expanded form (taking apart and putting numbers back together), meanings of fraction, etc.

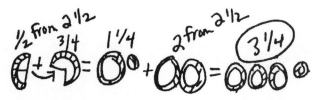

Source: Reprinted with permission from *Teaching Children Mathematics,* copyright © 2000 by the National Council of Teachers of Mathematics.

b. Below is another child's solution path: "I took the two off of the two and one half. Then I added the three quarters, and that's two and three quarters. From the one half I took a half away from that. So one quarter plus two and three quarters is three. Then you add the other half of the one half and it's three and one fourth." In this case, write the numbers and/or diagrams that might accompany this verbal description. Then describe the mathematical ideas that the child used—properties of addition, expanded form (taking apart and putting numbers back together), meanings of fraction, etc.

46. *Classroom Connection* In the November 1999 issue of *Mathematics Teaching in the Middle School*, the invention of a procedure for subtracting fractions is described. This procedure resembles an invented method for subtraction of whole numbers described in Chapter 3. Here is how Krystal's method works:

Find LCM	Subtract the whole numbers, and subtract the fractions	Simplify the result
$7\frac{1}{4}$ $-2\frac{1}{3}$	$7\frac{3}{12}$ $-2\frac{4}{12}$	$7\frac{3}{12}$ $-2\frac{4}{12}$ $\overline{5\frac{-1}{12}}$	$5 - \frac{1}{12} = 4\frac{11}{12}$

a. Make up and solve more problems until you are comfortable with this algorithm.

b. Does it generalize? That is, will it work for all fraction subtraction problems? Justify your response.

c. Justify the algorithm. That is, explain why it works.

d. Compare this algorithm to the one you are used to.

Source: Reprinted with permission from *Mathematics Teaching in the Middle School,* copyright © 1999 by the National Council of Teachers of Mathematics. All rights reserved.

47. *Classroom Connection* In the February 1998 issue of *Mathematics Teaching in the Middle School*, the authors showed two diagrams made by students to illustrate how they determine the answer to 1/2 of 3/4. This is before they learned the algorithm. Provide the answer and justification that come from each diagram.

a.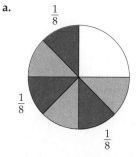

b.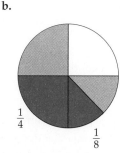

Source: Reprinted with permission from *Mathematics Teaching in the Middle School,* copyright © 1997 by the National Council of Teachers of Mathematics.

48. *Classroom Connection* The following problem appeared in "The Thinking of Students" in the April 2002 issue of *Mathematics Teaching in the Middle School*: "Gordon's Sporting Goods has sold all its wind-breaker jackets except one. It was so ugly that no one wanted to buy it. So Gordon's put the jacket on sale at one-fourth off its original price. The jacket did not sell, so Gordon's marked the sales price down one-third off. Still no one wanted to buy such an ugly jacket, so Gordon's marked the second sales price down one-half off. Finally, customer Ernie Harrison bought the jacket for his cat Wilson to sleep on. Ernie paid only $12 for the world's ugliest jacket. What was the original price of the jacket? What did the jacket look like?"

   a. Solve the problem yourself.
   b. Read the article and present one of the students' solutions, both the how and the why (that is, the justification of each step).

   Source: Reprinted with permission from *Mathematics Teaching in Middle School*, copyright © 2002 by the National Council of Teachers of Mathematics.

49. *Classroom Connection* In the 1999 NCTM Yearbook, *Developing Mathematical Reasoning in Grades K–12*, Deborah Schifter discusses several solution paths to the following problem that was given to a sixth-grade class: "You are giving a party for your birthday. From the Ice Cream Factory, you order 6 pints of ice cream that they make. If you serve 3/4 of a pint of ice cream to each guest, how many guests can be served?"

   a. First, do the problem yourself.

   Below are summaries of different students' solutions. In each case, describe where the number sentence comes from. In other words, supply the why behind the what.

   b. $24 \div 3 = 8$.
   c. $8 \times \frac{3}{4} = 6$, so the answer is 8.
   d. $\frac{3}{4} + \frac{3}{4} + \frac{3}{4} + \frac{3}{4} + \frac{3}{4} + \frac{3}{4} + \frac{3}{4} + \frac{3}{4} = 6$, so the answer is 8.
   e. $6 - \frac{3}{4} - \frac{3}{4} - \frac{3}{4} - \frac{3}{4} - \frac{3}{4} - \frac{3}{4} - \frac{3}{4} - \frac{3}{4} = 0$, so the answer is 8.

   Source: Reprinted with permission from *Developing Mathematical Reasoning in Grades K–12*, copyright © 1999 by the National Council of Teachers of Mathematics.

50. *Classroom Connection* A student says that she has found a quick way to divide fractions. She demonstrates her method as follows:

   $9\frac{1}{4} \div 3\frac{3}{4} = 3\frac{1}{3}$, because $9 \div 3 = 3$ and $\frac{1}{4} \div \frac{3}{4} = \frac{1}{3}$

   $20\frac{3}{5} \div 4\frac{3}{5} = 5\frac{1}{1}$, because $20 \div 4 = 5$ and $\frac{3}{5} \div \frac{3}{5} = 1$

   Unfortunately, her idea is not correct. Where is the conceptual error?

51. *Classroom Connection* In *Young Mathematicians at Work: Constructing Fractions, Decimals and Percents*, the authors describe a fifth-grade class where the children are exploring multiplication with fractions. The current problem is $15\frac{1}{2} \times 4\frac{1}{2}$. The children do not yet know the standard algorithm and are using their understanding of multiplication, of whole numbers, and of fractions to find the answer. Below are the beginning steps of three solution paths that enabled the children to determine $4\frac{1}{2} \times 15$. In each case, explain how to use this problem to determine $4\frac{1}{2} \times 15$, and then supply additional steps to find the answer to $15\frac{1}{2} \times 4\frac{1}{2}$.

   a. One group began with $4\frac{1}{2} \times 60 = 270$.
   b. Another group began with $9 \times 30 = 270$.
   c. A third group began with $15 \times 36 = 540$.

   Source: Reprinted from *Young Mathematicians at Work: Constructing Fractions, Decimals, and Percents* by Catherine Twomey Fosnot and Maarten Dolk. Copyright © 2002 by Catherine Twomey Fosnot and Maarten Dolk. Published by Heinemann. Used with permission.

52. *Classroom Connection* When faced with the problem $1/2 \times 1/4 = 1/8$, a child asks, "If multiplication is repeated addition, then what is this?" How might you respond?

53. Do the following problems in base 5, as opposed to converting to base 10 and solving in base 10.

   a. The Marble Palace has just received a shipment of 4400_5 marbles in four colors—blue, red, clear, and yellow. One-half of the marbles are blue, and there are half as many red marbles as blue marbles. If $1/10_5$ of the marbles are clear, how many of the marbles are yellow?

   b. The Marble Palace has just received a shipment of 440_5 marbles in four colors—blue, red, clear, and yellow. One-third of the marbles are blue, and three-quarters of the *remaining* marbles are red. If 20_5 of the marbles are clear, how many of the marbles are yellow?

54. Do the following division problem and represent the answer as a mixed number in reduced form. Explain how you obtained your answer without resorting to base 10.

   $13_5 \overline{)2130_5}$

SECTION 5.4 BEYOND INTEGERS AND FRACTIONS: DECIMALS, EXPONENTS, AND REAL NUMBERS

WHAT DO YOU THINK?

- What would Alphabitian decimals look like?
- Why is there no "oneths" place?

In this section, we will extend the system of numbers yet again—to decimals and to irrational numbers (that is, numbers that cannot be expressed as the quotient of two integers, such as $\sqrt{2}$ and π). There is a name for the set of numbers that includes both rational and irrational numbers: the set of real numbers (an interesting choice for this set!). There are more sets of numbers beyond the real numbers, but they are not considered until high school.

Decimals

Technically, the set of decimals is not an extension of the set of fractions. Rather, our use of decimal numbers comes out of an alternative way of representing fractions, using the advantages of our base 10 system. In fact, the word *decimal* comes from the Latin word *decem*, which means "ten." Thus decimal numbers are any numbers that are written in terms of our base 10 place value. Conventionally, we refer to decimals as fractions whose denominator has been converted into a power of 10.

The history of decimals   If you stop to think about it, decimals could not have been invented before our base 10 system. Credit for bringing decimals into everyday life goes to Simon Stevin. His book *The Art of Tenths or Decimal Arithmetic* was first published in 1585.

> "We will speak freely of the great utility of this invention . . . for the astronomer knows the difficult multiplication and divisions which proceed from the progression with degrees, minutes, seconds, and thirds. . . . This discovery . . . teaches (to tell much in one word) to compute easily, without fractions, all computations which are encountered in the affairs of human beings, in such a way that the four principles of arithmetic which are called addition, subtraction, multiplication, and division . . . are able to achieve this end." He went on to say that this work treats of "something so simple that it hardly merits the name of invention."[4]

The use of decimals was spurred primarily by a desire to make computation easier. For example, which computation would be easier to do: $37\frac{3}{4} \times 6\frac{4}{5}$ or 37.75×6.8? You may want to do both. How much faster was the decimal computation?

Even today there is not one universal notation. For example:

United States	England	Continental Europe
54.23	54·23	54,23

---

[4]Louis Karpinski, *The History of Arithmetic* (New York: Russell & Russell, 1965), p. 131.

No decimals? Imagine our world without decimals! As you discovered earlier in this course, the base 10 numeration system has been in widespread use for less than 500 years. The widespread use of decimals is even more recent.

What would life in the United States be like if no one had invented decimals—that is, if we expressed amounts using only whole numbers and fractions? Would things just look different (for example, 43 dollars and 26 cents instead of $43.26), or would some things be impossible?

You might find it helpful to look at cases in which we use decimals and to explore alternatives, as shown above for $43.26. Spend some time writing your answers to these questions before reading on. . . .

As you may have realized as you thought about these questions, much of our use of decimals involves measurement:

- We measure mass: for example, the package weighs 2.4 ounces.
- We measure distance: for example, Joan lives 3.7 miles from school.
- We measure time: for example, the world record for the 100-meter dash is 9.78 seconds.

Decimals are used in many aspects of daily life to communicate. For example,

- The U.S. birth rate is 14.1 per 1000 population.
- The average life expectancy in the United States is 76.9 years.
- Interest rates are expressed with decimals, such as 7.5 percent.
- The new city budget is $3.48 million.

Developing understanding of decimals The next 14 investigations look at decimals from several perspectives: examining different characteristics of the decimal system, why the algorithms work, rounding, scientific notation, and using decimals when solving problems. Using your increasing problem-solving, communicating, and reasoning abilities and your increasingly better connected network of mathematical concepts, you will be able to strengthen your own understanding of these concepts.

Connecting decimals to integers and fractions Many students who plan to teach elementary school do not feel that investigations with decimals are relevant because they believe that they will never teach decimals. In fact, although formal work with decimals is generally taught in middle school, informal work with decimals is an important part of the elementary school curriculum. Thus, in order to prepare elementary students for the decimal instruction they will receive in middle school, the elementary teacher needs to understand the many relationships between decimals and whole numbers and between decimals and fractions. The following investigations will help you to understand both similarities and differences between decimals and these other two sets.

Connecting Decimals and Integers

There are many important similarities and differences between the set of decimals and the set of integers. Before reading on, stop to think about the two sets of numbers. What similarities and differences can you see?

From one perspective, the set of decimals can be seen as an extension of base 10 counting numbers.

...	thousands	hundreds	tens	ones	tenths	hundredths	thousandths	...
	10^3	10^2	10^1	10^0	10^{-1}	10^{-2}	10^{-3}	
	1000	100	10	1	0.1	0.01	0.001	

That is, the system of decimals and the system of integers are both base 10 place value systems. The value of each digit depends on its place, and the value of each place to the right is one-tenth the value of the previous place. Investigation 5.17 helps to develop your decimal sense and give you a better "feel" for the relative size of tenths, hundredths, thousandths, and so on. Rather than state the differences, we will let them emerge through investigation.

INVESTIGATION

5.17

Base 10 Blocks and Decimals

One of the themes of this book has been the importance of multiple representations for understanding mathematical concepts. In order to develop your understanding of decimals, we will investigate two representations: base 10 blocks (see Exploration 5.16 for more work on this model) and number line models (after Investigation 5.20). Each representation highlights different aspects of decimals and their relationship to other systems.

Let's say that a sample of gold weighs 3.6 ounces. How would you use base 10 blocks (shown in Figure 5.39) to represent this amount? Note that the "small cube" was formerly called a "unit." Work on this and then read on. . . .

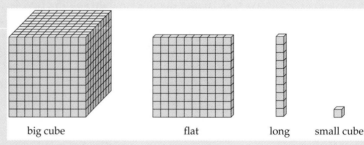

FIGURE 5.39

DISCUSSION

Any solution requires us to designate one of the pieces as the unit. For example, if we designate the big cube as having a value of 1, then we would represent 3.6 ounces as shown in Figure 5.40. Do you see why?

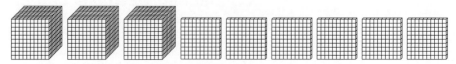

FIGURE 5.40

However, we could also have selected the flat to be the unit. If we had, then we would represent 3.6 ounces as shown in Figure 5.41.

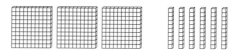

FIGURE 5.41

Zero and Decimals

We will examine the role of zero in decimals in the following two investigations.

INVESTIGATION 5.18 — When Two Decimals Are Equal

What if we added one or two zeros at the end of 0.2? Would that change its value? That is, do 0.2, 0.20, and 0.200 have the same value? Many sixth- and seventh-graders (and a surprising number of adults) believe that 0.2, 0.20, and 0.200 not only look different but also have different values because of the added zeros. How would you convince such a person that the value is not changed? Work on this and then read on. . . .

DISCUSSION

There are several ways to justify the equality. We will consider two below.

> ■ *Mathematics* ■
>
> The idea of multiple representations connects to our work with whole numbers. For example, just as
>
> $24 = 20 + 4$
>
> can be represented by 2 tens and 4 ones or by 24 ones, so too
>
> $0.24 = 0.2 + 0.04$
>
> can be represented by 2 tenths and 4 hundredths or by 24 hundredths.

STRATEGY 1: *Connect decimals to fractions*

- 0.2 is equivalent to the fraction 2/10.
- 0.20 is equivalent to the fraction 20/100, which can be shown to be equivalent to 2/10.
- 0.200 is equivalent to the fraction 200/1000, which can be shown to be equivalent to 2/10.

This strategy uses an important mathematical property. The **transitive property** is stated below for any three numbers, a, b, and c.

Transitive property of equality:

If $a = b$ and $b = c$, then $a = c$.

Transitive property of inequality:

If $a < b$ and $b < c$, then $a < c$.

Applying the transitive property of equality, we say that because 0.2, 0.20, and 0.200 are all equal to 2/10, they are equal to each other.

STRATEGY 2: Use manipulatives

In this case, let a square represent a value of 1. Such manipulatives are sold commercially as Decimal Squares©. We can divide this square into 10 equal pieces, 100 equal pieces, or 1000 equal pieces, as in Figure 5.42.

Shade in 0.2 of the first square, 0.20 of the second square, and 0.200 of the third square.

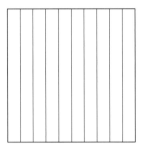
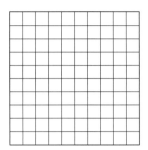
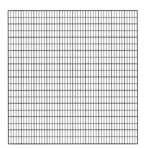

FIGURE 5.42

We find that in each case, the same area is shaded. Connecting this result to the area model of fractions, we conclude that 0.2, 0.20, and 0.200 are all equivalent decimals.

INVESTIGATION 5.19

When Is the Zero Necessary and When Is It Optional?

An important difference between decimals and whole numbers has to do with our old friend zero. Examine each number below. In each case, explain whether you think the zero in the number is necessary, optional, or incorrect. If you think it is necessary, explain why. If you think it is optional, explain why the zero doesn't matter. If you think the use of the zero is incorrect, explain why the zero should not be there.

A. 2.08   B. 0.56   C. .507   D. 20.6   E. 3.60

■ Outside the Classroom ■

We see examples of pseudo-precision every day. For example, a poll reported that 83.4 percent of people in the Northeast described the winter 1993–1994 as "the worst I have ever experienced." Given that polls generally have an error margin of several percent, this is an example of pseudo-precision. It would have been more valid to report that about 83 percent of the persons polled agreed with that statement.

DISCUSSION

A. With 2.08, the zero is necessary; 2 wholes, 0 tenths, and 8 hundredths.

B. With 0.56, the use of the zero is optional; it is a convention, like shaking hands with the right hand instead of the left.

C. With .507, the zero is necessary: 5 tenths, 0 hundredths, and 7 thousandths.

D. With 20.6, the zero is necessary: 2 tens, 0 ones, and 6 tenths.

E. In one sense, the zero here is optional. Mathematically, 3.6 = 3.60; you can verify this using the strategies in Investigation 5.18. In another sense, however, it depends on how the number is being used. For example, if we ask how long the room is, what is the difference between a response of 3.6 meters and a response of 3.60 meters?

A response of 3.6 meters implies that the room has been measured to the nearest tenth of a meter; that is, the length of the room is closer to 3.6 meters than to 3.5 or 3.7 meters. If the person doing the measuring has used an instrument that makes possible only this amount of precision but nevertheless reports the length as 3.60, I call this an example of pseudo-precision.

Some differences between decimals and integers This investigation illustrates some important differences between decimals and integers. With decimals, when we add a zero at the right end of the number, the value is unchanged; for example, 48.6 = 48.60. As you know, this is not true with integers; for example, 48 ≠ 480. Two other differences between the two systems are worth noting. One has to do with language: With whole numbers, the suffix for each place is *-s*; with decimals, it is *-ths*—for example, the hundreds place versus the hundredths place. Another difference is that there is a ones place but not a oneths place; this lack of symmetry has been noted by more than one youngster. How would you explain it?

INVESTIGATION 5.20

Connecting Decimals and Fractions

As stated before, each decimal can be translated into a fraction. You may be familiar with common translations between decimals and fractions:
0.5 = 1/2, 0.25 = 1/4, and so on.
  Many problem-solving situations require us to convert a fraction into decimal form or vice versa.

A. How many of the following common fractions can you convert to decimals immediately? How can you determine the decimal equivalent of others without using a calculator?

$$\frac{1}{2} \quad \frac{1}{3} \quad \frac{1}{4} \quad \frac{1}{5} \quad \frac{1}{6} \quad \frac{1}{8} \quad \frac{1}{10} \quad \frac{1}{16} \quad \frac{1}{50}$$

$$\frac{1}{100} \quad \frac{1}{1000}$$

DISCUSSION

Either by hand or with a calculator, all but 1/3 and 1/6 can be readily converted into decimal form. When we convert 1/6 into a decimal (by dividing 1 by 6), we find that it repeats; that is, the value is 0.1666666. . . . There are two ways to write repeating decimals. We can write 1/6 in decimal form as 0.166... or as $0.1\overline{6}$. The bar shows the part that repeats. For example, we would write 348/999 as 0.348348348... or as $0.\overline{348}$.

B. Now let us examine translating decimals into fractions. Translate the decimals 0.1, 0.007, and $0.\overline{18}$ into fractions before reading on. . . .

■ **Outside the Classroom** ■

Many mathematical concepts are misused in our society. One classic example of misuse of decimals occurs in baseball. Statistics for pitchers include the total number of innings pitched. There are 3 outs per inning, so if the pitcher pitched 7 innings and got 1 out in the eighth inning before being relieved, we would record his total innings pitched as $7\frac{1}{3}$ innings. If he had gotten two outs before being relieved, we would have recorded his total innings pitched as $7\frac{2}{3}$ innings. However, what the newspapers print is 7.1 and 7.2. Explain why this is mathematically incorrect. Why do you think they do this?

DISCUSSION

Most people have no trouble with the first two: 1/10 and 7/1000. The last decimal is equivalent to 2/11. Someone (unknown to history) discovered the process that enables us to conclude that there is a unique fraction associated with every repeating decimal. The process works like this:

Let $x = 0.18181818\ldots$ (that is, $0.\overline{18}$).

Now multiply both sides by 100, so that we have

$100x = 18.\overline{18}$

Do you see how these two equations enable us to determine the exact value of $0.\overline{18}$? Think and then read on. . . .

The key is to subtract the first equation from the second. The repeating part drops out, and we have $99x = 18$. When we solve the equation for x and simplify the fraction, we find that $0.181818\ldots$ is simply 2/11 in disguise! You can check this by doing the division $2 \div 11$.

Decimals, Fractions, and Precision

We have used number lines as a model to represent whole numbers and fractions. The number line can also be used to represent decimals, using the analogy of a zoom lens. Suppose you had never heard of decimals and had to determine the length of the wire in Figure 5.43 as accurately as you could. Do this and then read on. . . .

FIGURE 5.43

> **■ Mathematics ■**
>
> Do you see how the number line model connects to the density principle for fractions that we discussed in Section 5.2?

A visual inspection of Figure 5.43 shows that the wire appears to be closer to $4\frac{1}{3}$ inches than to $4\frac{1}{4}$ inches. With decimals, we do not look at the most appropriate unit fraction; instead, we look at the fractional length in terms of tenths, hundredths, thousandths, and so forth. Now imagine using a zoom lens (see Figure 5.44). To the nearest tenth of an inch, how long is the wire?

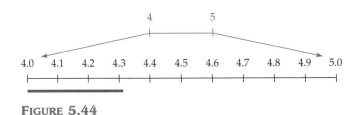

FIGURE 5.44

Because the length is closer to 4.3 inches than to 4.4 inches, we say that the length is 4.3 inches. If we want more precision, we can keep zooming in, in which case we move more decimal places to the right. Theoretically, we can continue this magnification process indefinitely. Realistically, electron microscopes are able to measure distances of about one ten-millionth of an inch, or 0.0000001 inch.

INVESTIGATION 5.21

Ordering Decimals

Just as young children need time to develop proper ordering relationships with whole numbers (for example, that 30, not 31, comes after 29), older children need time to develop ordering relationships with decimals. (Exploration 5.16 provides more decimals to order.)

Order the following decimals, from smallest to largest: 0.39, 0.046, and 0.4. Justify your solution. There are several different strategies. If you answer the question using one strategy, go back and try to use a different strategy. Do this before reading on. . . .

DISCUSSION

STRATEGY 1: *Use equivalent fractions*

$$0.39 = \frac{39}{100} = \frac{390}{1000}$$

$$0.046 = \frac{46}{1000}$$

$$0.4 = \frac{4}{10} = \frac{400}{1000}$$

Thus the order is 0.046, 0.39, 0.4.

STRATEGY 2: *Line up the decimal points and add zeros*

$$0.39 = 0.390$$
$$0.046 = 0.046$$
$$0.4 = 0.400$$

STRATEGY 3: *Represent them on a number line*

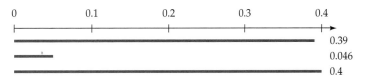

INVESTIGATION 5.22

Rounding with Decimals

The process of rounding with decimals is very similar to the process of rounding with whole numbers. In fact, if you *really* understand place value, then rounding with decimals is very straightforward. As you may have found with the Right Bucket game (Exploration 5.19), being able to round decimals is important when estimating.

Round each of the numbers to the nearest hundredth and to the nearest tenth:

A. 3.623 B. 76.199 C. 36.215 D. 2.0368

E. As you work on these, take mental stock of your confidence level. If you do not feel 100 percent confident, what models (discussed earlier) might you use to increase your confidence?

■ Beyond the Classroom ■

If you stop and think about it, rounding of decimals occurs frequently. For example, the rate I am charged for electricity is 11.146 cents per kilowatt-hour. If someone asked me what the electric rate is, however, I would say about 11 cents per kilowatt-hour. When doing tax forms, the IRS allows people to round some amounts to the nearest dollar. Thus I could record postage expenses as $75 instead of $74.68.

DISCUSSION

A. 3.623 to the nearest hundredth is 3.62, and to the nearest tenth is 3.6.

B. 76.199 to the nearest hundredth is 76.20, and to the nearest tenth is 76.2.

C. 36.215 to the nearest hundredth is 36.22, and to the nearest tenth is 36.2.

D. 2.0368 to the nearest hundredth is 2.04, and to the nearest tenth is 2.0.

E. Some students find the number line or physical models useful in determining how to round. One other common strategy goes like this: Look at 76.199. If we look at only two decimal places, we have 76.19. Therefore, the question can be seen as: Is 76.199 closer to 76.19 or to 76.20?

If we insert zeros so that all three decimals are given in thousandths, we have

76.200
76.199
76.190

Do you see how this strategy helps?

The number 36.215 requires us to deal with a 5 in the relevant place. To the nearest hundredth, this question becomes a decision between 36.21 and 36.22. Remember that with whole numbers, the decision was not to round up whenever we encounter a 5, but rather to round so that the last nonzero digit is an even number. We will apply that rule here. Thus we round 36.215 up to 36.22, but we round 36.245 down to 36.24.

Many students round confidently except when they encounter a zero. For example, many students feel comfortable stating that 2.0368 to the nearest hundredth rounds to 2.04 but do not feel comfortable stating that 2.0368 to the nearest tenth rounds to 2.0. If you feel this discomfort, I suggest using a number line, the base 10 blocks, or the strategy mentioned at the beginning of this discussion so that your understanding of rounding becomes more owned than rented!

INVESTIGATION 5.23 — Decimals and Language

I saw the following line in a newspaper: "The School Board has proposed a budget of $24.06 million." Express the School Board's proposed budget as a whole number. Do this before reading on. . . .

DISCUSSION

I have received many different answers from students in my own class, including the following:

24 million 600 thousand dollars

24 million 60 thousand dollars

24 million 6 hundred dollars

24 million and 6 dollars

24 million dollars and 6 cents

Let us look at several strategies that students have used to determine the correct answer.

STRATEGY 1: Connect decimals to fractions

Some students who have successfully used this strategy did not know how to represent 24.06 million at first, and so they began with decimal/fraction connections that they did know.

24.5 million	= 24,500,000	because 0.5 = 1/2
24.25 million	= 24,250,000	because 0.25 = 1/4
24.1 million	= 24,100,000	because 0.1 = 1/10 [This is often a hard place for many students to start, but it makes sense now, in light of the first two steps. Does it make sense to you?]
24.06 million	= 24,060,000	because it must be less than 24,100,000

STRATEGY 2: Connect decimals to expanded form

Some students who have successfully used this strategy did not know how to represent 24.06 million at first and found the solution by analyzing the meaning of 0.06 million.

24.06 million means 24 whole millions and six hundredths of another million.

Now we have to find the value of six hundredths of a million.

Using a calculator or computing by hand yields

$$0.06 \times 1,000,000 = 60,000$$

Thus, applying our knowledge of expanded form, we get

$$24.06 \text{ million} = 24,000,000 + 60,000 = 24,060,000$$

Once again we come back to the importance of being able to compose and decompose numbers: whole numbers, integers, fractions, and now decimals.

■ *Language* ■

The language we use when we write checks is connected to place value. How would you write this amount: $206.45?

It would be incorrect to say "two hundred and six dollars and 45 cents." Can you see why? To reduce confusion, the convention is to use *and* only once—just before giving the cents. Thus the conventional way to write $206.45 is "two hundred six dollars and 45 cents."

STRATEGY 3: Connect decimals to place value

If your understanding of place value is powerful, you know that the *places* of the digits will not change. Thus, when we represent 24.06 million as a whole number, we simply "add zeros" and we have 24,060,000. The *place* of the 6 does not change!

Operations with Decimals: Addition and Subtraction

With the widespread use of calculators, most people no longer do most decimal computations longhand. However, it is crucial to understand why the basic algorithms work because it is often necessary to interpret what the calculator displays; recall Investigation 1.15; the "answer" on the calculator display didn't indicate how much wire was wasted.

Few people make mistakes when adding or subtracting decimals if all the numbers have the same number of places, such as 3.24 + 5.56. However, a majority of middle school students have difficulty with a problem like the following: 7.8 + 0.46. Can you explain how to determine the sum and justify each step? You may recall being taught to "line up the decimal places," but why does that rule work?

We can justify the lining up of decimal places in terms of both whole-number computation and fraction computation. When adding whole numbers, we begin with the ones place and move to the left, regrouping as necessary. Adapting this procedure to decimals, we begin with the smallest place—in this case, the hundredths place. We have 6 hundredths in 0.46 and 0 hundredths in 7.8. Moving to the tenths place, we have 12 tenths (regrouped to 1 and 2 tenths). Moving to the ones place, we have 7 plus the regrouped 1, giving us a sum of 8 ones, 2 tenths, and 6 hundredths—that is, 8.26.

$$\begin{aligned}&7.80\\+\,&0.46\end{aligned}$$

From the perspective of fractions, we need common denominators. Thus, representing the decimals in expanded form and applying the commutative and associative properties, we have

$$\begin{aligned}7.8 + 0.46 &= \left(7 + \frac{8}{10}\right) + \left(\frac{4}{10} + \frac{6}{100}\right)\\ &= 7 + \left(\frac{8}{10} + \frac{4}{10}\right) + \frac{6}{100}\\ &= 7 + \frac{12}{10} + \frac{6}{100}\\ &= 7 + 1 + \frac{2}{10} + \frac{6}{100}\\ &= 8 + \frac{20}{100} + \frac{6}{100}\\ &= 8 + \frac{26}{100}\\ &= 8.26\end{aligned}$$

Multiplication with Decimals

Children often find multiplication with decimals confusing. Let us examine two problems that arise:

- How do we know where to put the decimal point?

- How can you multiply tenths by tenths and get hundredths—for example, 3.2 × 2.6 = 8.32?

Note: There are three ways to represent multiplication: 3.2 × 2.6, 3.2 · 2.6, and (3.2)(2.6). The dot, which we use to represent whole-number multiplication, can be confusing when we are showing decimal multiplication, and so we shall use the × or parentheses.

Let us examine the problem 3.2 × 2.6. Stop and look at the precision in this problem. We multiply two numbers that are accurate to the tenths place, but our answer includes a number in the hundredths place. For example, let's say the dimensions of a room are 3.2 meters by 2.6 meters, but its area is 8.32 square meters. If both the multiplier and the multiplicand are in tenths, why isn't the product in tenths also?

We look to a diagram for the explanation. We know from Chapter 3 that any multiplication problem can be represented by a rectangle; that is, the product of the two numbers is equal to the enclosed area. In this case, let us make a rectangle that is 3.2 meters long and 2.6 meters wide [Figure 5.45(a)].

Before reading on, please work on the following questions:

- How can you determine the area of the rectangle directly from the diagram?

- What are the connections between the diagram and the standard algorithm for multiplying decimals?

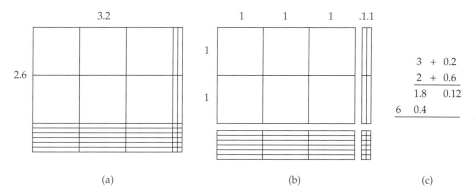

FIGURE 5.45

In this diagram, we have four distinct regions that correspond to the four partial products [see Figures 5.45(b) and (c)]. The top left region consists of 6 squares (each of which measures 1 meter by 1 meter). Thus the top left region has a value of 6 square meters.

The top right region consists of 4 longs, each of which measures 1 meter by 0.1 meter. Thus each long has a value of 0.1 square meter. The value of the top right region, then, is 0.4 square meter.

The bottom left region consists of 3 × 6 = 18 longs and has a value of 1.8 square meters.

The bottom right region consists of 12 little squares, each of which is 0.1 meter by 0.1 meter. Thus the value of each square is 0.1 × 0.1, or 0.01 square meter. Therefore, the value of this region is 0.12 square meter, because we have 2 × 6 = 12 of these little squares.

Adding up the value of the four regions, we have 6 + 0.4 + 1.8 + 0.12 = 8.32 square meters.

The calculation at the left below shows the sums of the four regions, and the calculation at the right shows the standard multiplication procedure. Do you see the connections?

```
    3.2          3.2
   ×2.6         ×2.6
    1.8          192
    .12           64
    6           8.32
    .4
   8.32
```

We are still applying the distributive property. Going back to the meaning of whole-number multiplication, the multiplication algorithm simply tells us the sums of the four regions. The beauty of the procedure is more evident when we realize that $3.2 \times 6 = 1.92$ and $3.2 \times 2 = 6.4$ don't "line up." However, when we move 64 over one place, the two partial sums line up just right. We are not generally aware of this lining up; we just remember how many decimal places to move over in the answer column.

Division with Decimals

Most of you know the procedure for dividing with decimals. For example, let us say that you travel 247.9 miles on 7.4 gallons of gas. To determine the miles-per-gallon, you divide 247.9 by 7.4. If you don't have a calculator, you know to move the decimal points over one place; you also know that the answer is 33.5 because you move the decimal point in the dividend straight up. This procedure is illustrated below. Can you provide a mathematical justification for moving the decimal point? Think and then read on. . . .

■ **Mathematics** ■

This is yet another example of the ideas of multiple representations and equivalence helping us to understand mathematical ideas more deeply!

```
                          33.5
   7.4)247.9   ⟶    74)2479.0
                          222
                          259
                          222
                          370
                          370
```

One key is the idea of multiple representations, which has been presented many times. We can also represent this division problem in the following form: 247.9/7.4. This form connects to the idea of equivalent fractions. If we multiply this decimal fraction by 10/10, we have

$$\frac{247.9}{7.4} \times \frac{10}{10} = \frac{2479}{74}$$

In other words, 247.9/7.4 has the same value as 2479/74. By moving the decimal point over, we create an equivalent computation in which the divisor (denominator) is a whole number.

INVESTIGATION 5.24

1.95
1.87
89¢ a pound
1.69
1.29

Decimal Sense: Grocery Store Estimates

In this and the next investigation, we will examine some estimation problems. We will pay attention to adapting estimation strategies developed with whole numbers and fractions and to using properties developed in Chapter 3 when appropriate.

In doing math outside the classroom, we generally find that we are more likely to solve decimal problems by estimating than by doing pencil-and-paper mathematics. Let's say that it is Friday afternoon. You drop by the supermarket to get a few items. Suddenly you realize that you didn't go to the bank, and you find that you have only $7.45. The milk is $1.95. The ice cream is $1.87. Grapes are 89¢ a pound, and you have just over a pound. The bag of potato chips is $1.69. The loaf of bread is $1.29. Do you have enough money? How would you estimate this amount in your head? Try it (see the column of numbers at the left) and then read on. . . .

DISCUSSION

STRATEGY 1: *Round up and use compatible numbers*

$1.95 is almost **$2**

$1.70 + 1.30 = **$3**

$1.87 + $.89 ≈ $1.90 + $.90 = **$2.80**

Thus we need almost $7.80, and we know that the actual sum is lower than this estimate. In situations like this, we intentionally overestimate because we don't want to find out at the checkout counter that we don't have enough money.

This next investigation offers several ways to apply ideas from different concepts we have studied.

STRATEGY 2: *Use leading digit and compatible numbers*

Looking at the "dollars column" (that is, the ones column), we have 4. Next, we look at the "dimes" (tenths) column.

This is what that strategy sounds like:

- 8 (dimes) + 2 (dimes) is one dollar.
- 9 (dimes) + 6 (dimes) is a dollar fifty, so that's two fifty.
- 8 (dimes) more makes three thirty, plus the 4 (dollars). That's seven thirty plus the cents. You don't have enough money!

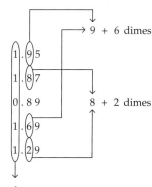

Just a reminder that there are many ways to estimate. The focus should not be on "getting" the "right" way, but rather on developing ways that make sense to you and that are reasonably sophisticated.

INVESTIGATION 5.25

Decimal Sense: How Much Will the Project Cost?

Let's say a contractor needs 308 sheets of plywood for a project. He can get them for $7.55 each at one store. He could also get them for $7.39 at another store, but he doesn't particularly like the owners, and more paperwork would be required. How much money would he save if he went to the second store? Estimate the answer as closely as you can without using a calculator or pencil and paper. Then read on. . . .

DISCUSSION

We could try to estimate 308×7.55 and 308×7.39, but this would be quite tedious. With a bit of reflection, we realize that what we need to know is not the approximate total cost but the difference. Because the difference is 16¢, or $.16, per sheet, we find that we really have to estimate $308 \times \$.16$ or $308 \times 16¢$.

STRATEGY 1: Converting to fractions

$$0.16 \times 308 \approx \frac{1}{6} \times 300 = \$50$$

STRATEGY 2: Use substitution and the associative property

$$0.16 \times 308 \approx 0.16 \times (100 \times 3)$$
$$= (0.16 \times 100) \times 3$$
$$= 16 \times 3$$
$$= \$48$$

INVESTIGATION 5.26

How Much Did They Pay for Their Home?

The Hill family is moving. The value of their home is two and a half times what they paid for it, and it is now worth $123,250. Approximately how much did they pay for it?

DISCUSSION

Before we can estimate, we have to translate the problem into a form that enables us to estimate.

We can translate this problem as

Buying price $\times\ 2.5 = \$123{,}250$

This is a division (missing-factor) problem, and the answer will be 123,250/2.5.

Section 5.4 / Beyond Integers and Fractions: Decimals, Exponents, and Real Numbers 323

STRATEGY 1: Missing-factor and guess–check–revise

What number times 2.5 is equal to $123,250?

First guess: 40,000; check: 40,000 × 2.5 = 100,000; revise.

Second guess: 50,000; check: 50,000 × 2.5 = 125,000.

The Hills paid about $50,000 for their house.

STRATEGY 2: Division and compatible numbers

$$\frac{120{,}000}{2} = 60{,}000 \qquad \frac{120{,}000}{3} = 40{,}000, \text{ so } \frac{120{,}000}{2.5} \approx 50{,}000$$

CLASSROOM CONNECTION

Consider the following examples from the National Assessment of Educational Progress. Students were given the following problem and asked to circle the correct product among five choices.

Select the correct answer to 3.04 × 5.3.

Many of you might say, "Well, 72 percent of all eleventh graders is not that bad" (see Table 5.1). But now look at a very similar problem, which actually should have been easier because the students were asked only to *estimate* the product.

TABLE 5.1

	Percent Correct	
	Grade 7	Grade 11
	57	72

Which of the following is the best estimate of 3.04 × 5.3?

As Table 5.2 shows, barely one out of five seventh-graders was able to see that 16 is the best estimate of 3.04 × 5.3! In fact, more seventh-graders chose 1.6, probably because it was the only answer that had a decimal point. Barely one out of three eleventh-graders recognized that 16 is the best estimate of 3.04 × 5.3!

TABLE 5.2

	Percent Responding	
	Grade 7	Grade 11
1.6	28	21
16	21	37
160	18	17
1600	22	11
I don't know	9	12

INVESTIGATION 5.27

How Long Will She Run?

A runner is running at 7.6 meters per second. At this rate, about how long will it take her to run 100 meters?

DISCUSSION

This problem is a classic example of how decimals can obscure simple relationships. The problem-solving strategy of using simpler numbers is powerful here. For example, suppose the runner was running at 5 meters per second. How long would it take her to run 100 meters? With these simpler numbers, it is almost immediately apparent that it will take 20 seconds; that is, divide 100 by 5. We can now apply this understanding to this problem and see that it will take her 100/7.6, or about 100/8, seconds, which is about 12 or 13 seconds. Over my teaching career, I have seen many students be very intimidated by "ugly" numbers.

INVESTIGATION 5.28 — Exponents and Bacteria

With the development of negative numbers and decimals, we are now poised to understand better the operation of exponentiation, which is an operation that we have already referred to (for example, in describing place value relationships) and that we will use in future investigations.

If a kind of bacteria doubles every hour, how many bacteria will there be after 24 hours if we begin with 1 bacterium? Think and then read on. . . .

History

In the 1500s, Francois Viete and Michael Stifel introduced symbols for unknowns and powers, for example, x and x^2. Before this time, writing algebraic expressions was very tedious. For example, the expression we write as $x^2 + 2x = 8$ would have been written as Zp 2Rm 8 in the 1500s.[5] To decipher the sixteenth-century code, you need to know that they let R represent the unknown amount; however, to represent the square of the unknown amount, they chose a different letter, Z (for zensus or census); then they used p for plus and m for minus. With this code, we can now translate Zp 2Rm 8 as "some unknown to the second power plus 2 of those unknowns minus 8." Can you imagine solving equations or factoring in the sixteenth century? With this primitive notation, it was very cumbersome.

DISCUSSION

There are several ways to address this problem. For example, we can make a table to represent the growth. The third column simply represents the number of bacteria, using exponents.

Does this help you? Using exponents, how would you represent the number of bacteria after 24 hours?

Number of hours	Number of bacteria	Number of bacteria
0	1	
1	2	
2	4	2^2
3	8	2^3
4	16	2^4
5	32	2^5
.	.	.
.	.	.

The number of bacteria after 24 hours will be 2 to the 24th power. Do you see why?

---

[5]H.A. Freebury, *A History of Mathematics* (New York: Macmillan, 1961), p. 92.

Section 5.4 / Beyond Integers and Fractions: Decimals, Exponents, and Real Numbers   **325**

Exponents and multiplication Just as multiplication can be seen as repeated addition, exponentiation can be seen as repeated multiplication, as in Figure 5.46.

$$b^x = \overbrace{b \cdot b \cdot b \cdot b \cdots b \cdot b \cdot b}^{b \text{ occurs } x \text{ times}}$$

FIGURE 5.46

In mathematical language, b is called the **base** and x is called the **exponent**. We say

$b^2 = b$ **squared**
$b^3 = b$ **cubed**
$b^4 = b$ to the fourth power
$b^x = b$ to the xth power

Computing exponents This then brings us to another problem: How do we determine the value of 2^{24}? What do you think? If you remember anything about exponents, explore alternatives for a while before reading on. If you don't recall anything about exponents, read the following discussion first and then try. . . .

In one sense, we want to know how to make the following computation less tedious:

$2 \cdot 2$

One (of several) possibilities is to translate this computation into an equivalent computation, using the associative property of multiplication:

$(2 \cdot 2 \cdot 2 \cdot 2) \cdot (2 \cdot 2 \cdot 2 \cdot 2) \cdot (2 \cdot 2 \cdot 2 \cdot 2) \cdot (2 \cdot 2 \cdot 2 \cdot 2) \cdot (2 \cdot 2 \cdot 2 \cdot 2) \cdot (2 \cdot 2 \cdot 2 \cdot 2)$

That is, it is much quicker to press $16 \cdot 16 \cdot 16 \cdot 16 \cdot 16 \cdot 16$ on a calculator than to press 2 twenty-four times.

If you have a scientific calculator, pressing the following entries yields the answer:

2 $\boxed{y^x}$ 24 = 16777216, that is, 16,777,216.

Negative exponents When the exponent is 2 or greater, we can interpret it as repeated multiplication. However, what if the exponent is less than 2? For example, what are the meanings and value of the following amounts: $2^1, 2^0, 2^{-1}, 2^{-2}$?

One way to understand the meaning and value of such exponents is to look at the connections among all exponents. For example, what is happening to the value of the expression each time we decrease the exponent? Try to verbalize this and then read on. . . .

$2^4 = 16$
$2^3 = 8$
$2^2 = 4$
$2^1 = ?$
$2^0 = ?$
$2^{-1} = ?$

One way to verbalize the pattern is to say that each time the exponent decreases, the value of the expression is divided by 2. It is reasonable to assume that this pattern will continue, and thus it is reasonable to conclude that the value of 2^1 is $4 \div 2 = 2$. This line of reasoning enables us to continue the progression. In this manner, we deduce that the value of 2^0 is $2 \div 2 = 1$.

Continuing this pattern, $2^{-1} = 1 \div 2 = 1/2$, and so $2^{-2} = 1/2 \div 2 = 1/4$.

Can you verbalize the pattern now so that you can give the value of 2^{-n}? Work on it and then read on. . . .

Once again, we rely on multiple representations. We can represent 2^{-2} as $1/4$, or we can represent 2^{-2} as $1/2^2$, because $4 = 2^2$. This now leads us to the generalization that

$$2^{-n} = \frac{1}{2^n}$$

INVESTIGATION 5.29 — Scientific Notation: How Far Is a Light-Year?

One of the ways in which we use exponents is to express very small and very large numbers. In many cases, the numbers that we use are so large or so small that computation becomes very cumbersome. For example, the U.S. federal debt in March 2006 was $8.2 trillion, which is 8,200,000,000,000. The wavelength of red light is 0.00000000000000586 meter. The following situation is a good introduction to the need for an alternative notation, which we call *scientific notation*.

The size of the universe is so huge that it is almost beyond the ability of the human mind to comprehend. One way in which astronomers address this hugeness is to measure astronomical distances not in miles or kilometers but rather in a unit called "light-years"—that is, the distance light travels in one year. The speed of light is 186,000 miles per second. Determine the length of a light-year. Work on this yourself before reading on. . . .

DISCUSSION

In this problem, dimensional analysis can greatly simplify the computations.

$$1 \text{ light-year} = \frac{186{,}000 \text{ miles}}{1 \text{ second}} \times \frac{60 \text{ seconds}}{1 \text{ minute}} \times \frac{60 \text{ minutes}}{1 \text{ hour}} \times \frac{24 \text{ hours}}{1 \text{ day}} \times \frac{365.25 \text{ days}}{1 \text{ year}}$$

Do the multiplication now on a calculator.

If you have an inexpensive calculator, you probably got a big E, representing "Error."

Many of you will have something like this:

5.8697 12   or   5.8697136 12

The more sophisticated calculators represent the computation in scientific notation, which enables us to represent very large or very small numbers without having to count all the zeros. The 5.8697 12 is the calculator's code for 5.8697×10^{12}.

Let's work with smaller and simpler numbers to ensure that you can understand how scientific notation works. Consider the number 34,000. One way to represent this amount that connects to place value comes from the realization that this number means 34 thousands. In symbols, we can thus say

$$34{,}000 = 34 \times 1000$$

We could also represent 34 as 3.4×10. Substituting 3.4×10 for 34, we have

$$3.4 \times 10 \times 1000$$

We can simplify this as

$$3.4 \times 10{,}000$$

Now we can substitute 10^4 for 10,000, and we have

$$3.4 \times 10^4$$

Formally, we say that a number is in **scientific notation** if it is in the form $a \times 10^b$, where a is a number between 1 and 10 and b is an integer.

The reasoning behind restricting a to a number between 1 and 10 is based on convention rather than mathematical structure. This convention makes it easier for us to compare amounts. For example, if we have 6.4×10^7 and 5.6×10^8, we immediately know that the second amount is larger. However, if we have 0.64×10^8 and 56×10^7, it is not as quickly evident that the second amount is larger.

Now let us return to our problem of representing the length of a light-year. Translate this amount into a whole number and then read on. . . .

One strategy is to write the decimal, and then move the decimal point 12 times:

5.869700000000

When we insert the commas, we have 5,869,700,000,000. How would we say this amount?

We would say 5.9 trillion miles.

Irrational Numbers

Thus far, our investigation of numbers has proceeded from counting numbers to whole numbers to integers to fractions and to decimals. There is one more set of numbers that we shall consider in this course. The discovery of this set of numbers is also one of the more dramatic stories in the history of mathematics. Many important contributions to mathematics came from a group of people who called themselves Pythagoreans; you first encountered them in Chapter 4. As you may recall, one of their core beliefs was that all the laws of the universe could be represented using only whole numbers and ratios of whole numbers. Now recall the Pythagorean theorem (for any right triangle with legs a and b and hypotenuse c, $a^2 + b^2 = c^2$) and our mention that the Greeks were the first people to prove this relationship. Ironically, this proof was part of the undoing of the Pythagoreans. Here is how it happened:

Consider a right triangle in which the length of each of the two sides is one inch, as shown in Figure 5.47.

FIGURE 5.47

If we call the hypotenuse x and apply the Pythagorean theorem, we have

$$1^2 + 1^2 = x^2$$
$$x^2 = 2$$

That is, x is the number that, when multiplied by itself, equals 2.

This problem was perplexing to the Pythagoreans, because try as they might, they could not find a rational number that, when multiplied by itself, came to *exactly* 2.

Using modern notation and base 10 arithmetic, we know that we can get *very* close:

$$1.4 \times 1.4 = 1.96$$
$$1.41421 \times 1.41421 = 1.9999899$$

However, there is no rational number (that is, a number that can be expressed as the quotient of two integers) that will produce a product of exactly 2. In modern language, we say $x = \sqrt{2}$.

Finally, one member of the sect proved that there is no rational number that will solve this problem. This discovery was like a child finding out that Santa Claus is not real or an adult finding out that there really are people from other planets. What it meant for the Pythagoreans was that one of their cornerstone beliefs—that all the laws of the universe could be represented as whole numbers and ratios of whole numbers—was not true.

What happened to the person who discovered the proof? Legend has it that he was set out to sea in a leaky rowboat! This is possibly the source of the saying "Don't kill the messenger who bears the bad news." It is also interesting to note that the word *irrational*, even in our present time, means "contrary to reason."

INVESTIGATION 5.30 — Square Roots

Students in elementary school may come upon situations in which they will encounter square roots that are not rational numbers, so it is important to investigate square roots briefly. Technically, any positive number has two square roots. For example, the square roots of 81 are $^+9$ and $^-9$. Because, in elementary school, square roots are explored only in the context of measurement, and because all lengths are positive, we will focus our explorations on the positive square roots of numbers, also called the **principal square root**.

Before the development of hand-held calculators, schoolchildren had to learn an algorithm for approximating the square root of a number. This algorithm was a standardized procedure that was taught and practiced in much the same way that long division was taught and practiced.

Without using the square root button on your calculator, determine the square root of 800 to the nearest hundredth. As you do so, use the tools that seem most appropriate. Then read on. . . .

DISCUSSION

To determine $\sqrt{800}$, a little estimating and mental math help. For example,

$$30 \times 30 = 900$$

Therefore, a reasonable starting point would be 29. However, if you apply the mental math skills developed in Chapter 3, you realize that 29^2 is still too high. Can you mentally compute 29×29?

We can represent 29×29 as $(30 - 1)(30 - 1)$ and apply the distributive property $900 - 30 - 30 + 1 = 841$. In any case, because we know that $30^2 = 900$ and $29^2 = 841$, a reasonable first guess is 28. If you want to follow the discussion below actively, please cover the table and think about *your* next guess and your reasoning before reading on. . . .

Guess	Result	Analysis
28	784	Too low. Because 800 is closer to 784 than to 841, the next guess should be closer to 28 than to 29.
28.4	806.56	Not bad! Let's try 28.3.
28.3	800.89	Clearly 28.2 will be too small. Next guess? 28.29? 28.28? Why?
28.29	800.3241	
28.28	799.7584	

Clearly $28.28 < \sqrt{800} < 28.29$. Without getting out your calculator or pencil and paper, which is closer? How did you figure it out?

The Real-Number System

With the set of irrational numbers, we can now extend our system of numbers to the set of **real numbers**, which is defined as the union of the sets of rational and irrational numbers. This is as far as we will go with number systems in this course. The next major expansion, which students encounter in high school, is imaginary and complex numbers—for example, the square root of $^-1$.

Let us look back at the numbers we have examined in this course with respect to how they are related to each other. We began our study of numbers with the set of natural numbers, which is the first set of numbers young children encounter. Our first extension of the number line was to add zero. The union of the set of natural numbers and zero is called the set of whole numbers. We then examined two kinds of numbers that children will encounter in elementary school: integers and rational numbers. Finally, we encountered numbers that cannot be represented as the ratio of two whole numbers—that is, irrational numbers. If you were to represent the relationships among these number systems visually, what kind of a diagram might you draw? Do this before reading on. . . .

Look at Figure 5.48. Could you explain this chart to someone who was familiar with the terms but not the relationships? Try to do so before reading on. . . .

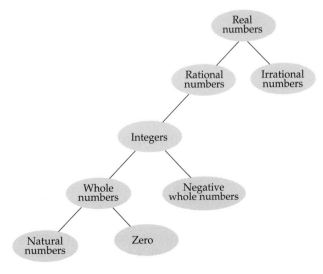

FIGURE 5.48

Properties of Real Numbers

As we have explored each of these sets of numbers, we have examined the properties of these number systems. These properties are not simply "technical stuff" that children should memorize because their teachers say so. When students understand and "own" these properties, their ability to work successfully with higher mathematics in general and algebraic concepts in particular greatly increases. We have used these properties to understand algorithms, and we have used these properties when estimating.

Now that we have finished our investigations of different sets of numbers, it is worthwhile to stop and reflect on what we have learned.

I hope that you will witness students discovering some of these properties (for example, "It doesn't matter whether you add 6 and 7 or 7 and 6, you get the same thing!") and that your students will be curious enough in your class to ask you questions like "Will you always get a number when you divide?" Which property is this student asking about?

This question connects to the closure property. The operations of addition, subtraction, and multiplication are closed for real numbers. That is, for any two real numbers, a and b,

$a + b$ is a real number, $a - b$ is a real number, and $a \cdot b$ is a real number.

Why do we not say that division is closed for real numbers?

If we were to say that division is closed, we would have to make one qualification. If we say that for any two real numbers a and b, a/b is a real number, we would simply have to note that b cannot equal zero.

Recall also that the closure properties for the four operations do not hold for all the number systems we have encountered. For example, the set of natural numbers is closed for addition and multiplication only. Do you remember why? The absence of closure with the natural numbers, whole numbers, and integers can be disturbing for youngsters. Do you see why the following two questions have to do with the closure property?

- "What if you subtract a big number from a little number?"
- "How can you divide 10 cookies among three people?"

Another property that is important for you to know about has to do with the density property of rational numbers, which represents a qualitative difference between the set of rational numbers and the three previous sets. However, as dense as the rational numbers were, we found that the number line was not yet complete, because there are points on the real number line that have no rational name. When we add the set of irrational numbers to the set of rational numbers, our number line is complete. This is known as the **completeness property**:

> There is a real-number name for every single point on the number line.

Summary

To understand decimals, we explored their relationship both to the set of integers and to the set of rational numbers. In doing so, we examined different ways to represent decimals, including base 10 blocks and number lines.

When examining connections with decimals, we found that zero presents special issues, just as zero did with integers. When we are representing an amount with a decimal, sometimes the placement of a zero is necessary and sometimes the placement is optional (a convention).

When examining connections with fractions, we found that some decimals terminate (1/2 = 0.5) but some do not (1/3 = 0.333...).

Understanding decimal computations relied on applying connections with whole numbers and with fractions. When adding and subtracting, we must add digits that represent the same place value, just as we did with whole numbers. To understand why we put the decimal point in a particular place when multiplying, we needed to examine decimal multiplication in the context of an area model, just as we did with multiplication of whole numbers. To justify the movement of the decimal point when we are dividing, however, we made use of the relationship between decimals and fractions.

The operation of exponentiation was introduced as repeated multiplication. Exponentiation enables us to understand a use of decimals that is important in many fields: scientific notation, which is used to represent very large and very small numbers.

Finally, we examined one more extension of the number system, irrational numbers, thus completing our journey through number systems in this course. The set of rational numbers is dense, but the set of real numbers is complete and sufficient for most everyday and business uses of elementary mathematics.

EXERCISES 5.4

1. Express each of the following as directed.
   a. 3/40 as a decimal
   b. 4/5 as a decimal
   c. 3/100 as a decimal
   d. 0.36 as a fraction in simplest form
   e. 0.005 as a fraction in simplest form
   f. 1.6 as a mixed number

2. Express each decimal in expanded form and as a fraction.
   a. 4.6
   b. 0.75
   c. 1.234
   d. 4.06

3. Write the following decimal numbers in words.
   a. 32.04
   b. 0.00004
   c. 508.604

4. Write the following using numbers:
   a. sixty-seven thousandths
   b. four thousand sixty and thirty-four thousandths.

5. Round the following decimals as directed:
   a. 2.36 to the nearest tenth
   b. 6.04 to the nearest tenth
   c. 2.398 to the nearest hundredth
   d. 5.0006 to the nearest hundredth
   e. 0.9876 to the nearest thousandth
   f. 6.1237 to the nearest thousandth

6. Circle the smaller of the pairs of numbers:
   a. 0.4   0.42
   b. 1.03   0.98899
   c. 0.05   0.058
   d. 0.302   0.087

7. Order the following decimals from smallest (at the left) to largest (at the right).
   a. 0.56, 0.058, 0.0084, 0.6
   b. 1.04, 0.065, 0.0086, 0.9

8. a. Name a decimal between 0.999 and 1.
   b. Name a fraction between 3.44 and 3.45.

9. In each part, draw a number line and place the given numbers on it.
   a. 3.2, 3.8, 2.9, 3.04
   b. 2.43, 2.4325, 2.4396, 2.441
   c. 0.004, 0.00397, 0.00394

10. Using grid paper, represent each of the problems below and explain how to determine the product by interpreting the grid.
    a. 0.3×0.2
    b. 1.2×0.4
    c. 3.4×2.6

11. Without doing any computation, determine which is bigger, $(0.8)^3$ or $(0.8)^2$? Briefly explain how you arrived at your decision.

12. a. Find two decimals whose product is 32.76.
    b. Find two decimals whose product is 4.76 and one of the decimals has a value less than 1.
    c. Find two decimals whose quotient is 0.36.

13. Recall the question about the proposed school board budget in Investigation 5.23. What if the superintendent recommended adding $380,000 to the proposed budget of $24.06 million? How much would the new budget be, both in whole-number form and in decimal form?

14. Let's say a company receives medicine in 1-gallon (128-ounce) jars and then sells the medicine in vials that hold 1.25 ounces. If we were to ask how many vials can be made from 1 gallon, we would divide 128 by 1.25 and obtain 102.4. What does the 102 mean? What does the .4 mean?

15. People frequently mispunch the calculator when doing problems.
    a. Let's say you were multiplying 24 boards at $4.50 per board and got $1080. What is the right answer? What did you punch?
    b. Let's say you are adding $3.45 + $12.45 + $16.23 + $34.45 + $23.45. You have just entered the last number and hit the = sign, and you realize that you punched $3.45 for the last number. What could you do with the number that now appears on the calculator rather than starting all over?
    c. Let's say you are finding the cost of 12 boards at $3.45 per board. You have just multiplied 12 × $3.45 to get $41.40. Now you decide that you want to buy 14 boards. There are several alternatives to multiplying 14 × $3.45. Can you find them?

16. Solve each of the following problems. Then explain how you knew whether to multiply or divide and, in the case of division, which number was the divisor and which was the dividend.
    a. A box of sand has a volume of 0.8 cubic meter. What is the volume of 0.23 of the box?
    b. If a motorcycle gets 40 miles per gallon, how far will it travel on 0.75 gallon?
    c. Let's say you make 5 liters of punch, and each cup holds 0.2 liter. How many cups can you make?
    d. A box of cookies weighs 0.8 pound. How many boxes can you make with 7.2 pounds of cookies?
    e. With 75 roses you can make 5 equal bouquets. How many roses will be in each bouquet?
    f. The price of 1 yard of fabric is $15.00. How much does 0.65 yard cost?
    g. A group of 12 friends together bought 5 pounds of cookies. How much does each person get?
    h. A rocket travels at a speed of 16 miles per second. How far does it travel in 0.85 second?
    i. A piece of elastic can be stretched to 3.3 times its original length. When fully stretched, it is 13.9 meters long. What was its original length?

Problems 17–22 involve translating between decimal and nondecimal systems.

17. Recall the discussion of reporting of innings pitched in baseball on page 313. Why do you think this is reported in a way that is mathematically inaccurate?

18. Our system of money is not a pure decimal system. Why not? Actually, this question has two parts:
    a. What aspects of our money system are not aspects of a pure decimal system?
    b. Why don't we use a pure decimal system?

19. Convet 7 minutes, 21 seconds to decimal representation, that is 7._ _ minutes.

20. Sam worked 4 days, 2 hours, and 15 minutes last week. Convert this to a decimal, that is, 4._ _ days. In this problem, one day is equal to 8 hours.

21. For a time after the French revolution in 1789, the French had a 10-day week, and each day was divided into 10 hours, each consisting of 100 minutes, each minute of 100 seconds. Thus, noon would have been 5 o'clock and 6 A.M. would have been 2.5 or perhaps two-fifty, since it represents one quarter of a day.

    Convert the following times into the French day.
    a. nine A.M.   b. four P.M.
    Convert the following times from the French into our time:
    c. eight o'clock   d. three thirty-three

22. Sherry is a hard-working student who is suffering from fatigue. The doctor says that she is not getting enough sleep. Below are her bedtimes for one work week. What is her average bedtime?
    10:30 P.M.   11:15 P.M.   11:45 P.M.   1:15 A.M.   11:00 P.M.

23. I am a little late and 30 miles from home. The speed limit is 55 mph. How much time will I save if I drive 65 mph the whole way?

24. In the *Guinness Book of World Records*, the record for dominoes toppling was 281,581 in 12 minutes 57.3 seconds. How many dominoes fell per minute? per second?

25. In the *Guinness Book of World Records*, the greatest distance a baby carriage was pushed in 24 hours was 342.25 miles. What was the average speed of the baby carriage? What was the average time per mile to the nearest second?

26. Perform the following computations mentally and describe your strategy.
    a. 6.47 − 3.95   b. 2.3 + 6.9 + 4.7
    c. 65.38 ÷ 1000   d. 0.75 × 23 × 16
    e. 8(8.5)

27. Estimate each of the following; then briefly explain and justify your estimate.
    a. 19.4 + 136 + 4.825   b. 23.4 + 24.6 + 5.7 + 34.4
    c. 23.34 − 22.56   d. $10.00 − $4.34
    e. 12.5 × 0.034   f. 2.3 × 6.7
    g. 34.2 × 0.007   h. 7.8 ÷ 3.12

28. Represent the following numbers in scientific notation:
    a. 123,456,789   b. 3,000,000,000,000,000
    c. 0.00000000056   d. 0.000000302

29. Perform the following computations using scientific notation:
    a. 123,000,000,000 × 34,000,000
    b. 123,552,000,000 ÷ 23,400,000
    c. 0.0000000034 × 0.0000000045
    d. $(0.00043)^3$

30. The width of a certain cell is 3×10^{-6} cm. If we placed 250 of these cells side by side, how long would that line be? Express this length without using scientific notation.

31. a. Explain how you would determine the length of a light-year if you had a calculator that did not "know" scientific notation—that is, that just diplayed an E sign for answers that were too big. Justify your work. What properties that we developed in Chapter 3 did you use?
    b. Determine the following product with such a calculator:
    $$14,000 \times 3,356,000 \times 7,890,000$$

32. a. Suppose a particular bacterium can divide into two bacteria every 45 minutes. If this process continues for 48 hours, how many bacteria will there be? Express your answer in scientific notation.
    b. What if we had started with 100 bacteria instead of 1? Predict the answer without doing any pencil-and-paper or calculator computation. Justify your reasoning.
    c. What if we had started with 1 bacterium that doubled every 16 minutes. How many would there be after 48 hours?

33. A manufacturing plant used 4.2 centimeters of wire on each item. If 3549 cm of wire are used in a day, how many items were manufactured? If the cost of the wire is 3.4¢ per meter, what is the cost per day of the wire?

34. A group of students held a car wash to raise money. They charged $2.50 for regular cars and $4.00 for trucks. They washed 150 cars and 68 trucks. If they spent $15 on soap, sponges, and other supplies, how much profit did they make?

35. At one point the telephone rates between Los Angeles and Tokyo were $9.50 for the first 3 minutes and $1.85 for each additional minute. What would be the cost of a 20-minute phone call?

36. The average retail price of a pack of cigarettes was $4.31 in 2006.
    a. If a person smokes 1 pack a day, about how much will it cost for a year?
    b. At this rate, how much for 50 years?

37. One day on Sesame Street, a friendly rabbit brought Big Bird a tube of toothpaste that contained one gallon of toothpaste. If regular toothpaste costs $1.49 for a 2.7-ounce tube, what is the value of the Seasame Street tube, to the nearest cent? [There are 128 ounces in a gallon.]

38. The national debt at one point in 2006 was just over $8.16 trillion. How could we represent this amount in a way that people could relate to?

39. You have decided to open a checking account with Xanadu County Savings and Loan. They offer you two options: (a) pay a flat $5.00 per month fee or (b) pay a monthly fee of $2 and 10¢ per check.
    a. If you write about 15 checks per month, which option is cheaper?
    b. How many checks per month would give you the same fee in both cases?

40. Judy recently took a trip in her car. The trip took 11 hours, including a lunch break of 1/2 hour. She filled the gasoline tank (11.3 gallons) and noted that the odometer read 38329.8. At the end of the trip, she refilled the tank (12.4 gallons), and the odometer read 38735.4. She paid 2.46\frac{9}{10}$ per gallon.
    a. How many miles per gallon did the car get on the trip (to the nearest tenth of a mile per gallon)?
    b. What was the cost of the gas?
    c. What was her average speed?

41. Most gasoline stations price gasoline out to 9/10 of a cent—that is, 2.46\frac{9}{10}$ per gallon rather than just $2.46 per gallon. How much extra money per day might a gasoline station expect to make from this extra 9/10 cent per gallon?

42. Oil companies argue that using super unleaded rather than regular unleaded pays for itself because you get better mileage. If this were the only benefit from using super unleaded and you got 20 miles per gallon with regular unleaded that cost $2.09 per gallon, how many miles per gallon would you need to get with super unleaded at $3.29 per gallon to make buying super unleaded worthwhile?

43. Let's say the cost of gas just went up by 5¢ per gallon. Approximately how much will this affect the average citizen?

44. Let's say your electric rate is 9.23¢/kWh, and you have 5 different 100-watt lights in the house that run from 5 P.M. until 11 P.M. (1 in the kitchen, 2 in the living room, 1 in the dining room, and 1 in the bedroom). How much do you pay per day for running these lights?

45. A laser printer prints characters at a density of 300 dots per inch, both horizontally and vertically.
    a. How many dots can be printed across the 8.5-inch width of a standard piece of paper?
    b. If every dot on a standard 8.5×11 inch piece of paper was printed, how many dots would that be? Express your answer in scientific notation.

46. Your school has just purchased a new high-tech mimeograph machine called a risograph. The accountant has determined that the school pays 5¢ per copy using the copier machine but only 2¢ per copy on the risograph. The school has kept track of the usage of the risograph for a week and has found the average number of copies per day is 1375. How much money per year will the school save with the purchase of the new machine? What assumptions did you make in order to solve this problem?

47. I bought a cheap candle for 89¢. It burns at the rate of 4.2 cm per hours. I bought a more expensive candle for $1.59, which burns at the rate of 2.8 cm per hour. Which is the better buy?
    a. What additional information would you need in order to answer this question?

b. Using the four numbers supplied in this problem, make a problem that can be answered, and answer it.

48. Jackie is thinking of making extra money by typing papers on her word processor. If she wants to make at least $8 an hour, how much should she charge per page? Assume that she can type 120 words per minute.

49. Henry is currently working at a job that pays $9.45 per hour, and he regularly works 40 hours per week. Another company has offered to pay him an annual salary of $24,000. How does this offer compare to his present yearly income? What assumptions did you make in order to solve this problem?

50. The actual solar year is 365 days, 5 hours, 48 minutes, and 46 seconds. The length of the year in the Julian calendar (from Julius Caesar) was established in A.D. 325 at $365\frac{1}{4}$ days; thus every fourth year was a leap year. By 1582 the use of the Julian calendar had caused the calendar date of the equinoxes to diverge considerably from the actual dates. The modern calendar omits the leap day from years ending in hundreds, unless they are divisible by 400. Assuming the Julian calendar began "correct"—that is, the first day of spring was at the vernal equinox—how far off had we become by 1582 when it was adopted?

51. The length of the tropical year is 365.24220 days, as compared to the length of 365.2425 days used by the Gregorian calendar.

a. What is the difference (in seconds) between the tropical year and the Gregorian calendar year?

b. How many years will it take for this difference to amount to 1 day?

52. *Classroom Connection* Why is there no oneths place?

53. *Classroom Connection* In each case, describe the error that the student made, and then indicate what mathematical idea(s) the student did not correctly understand.

a. $\begin{array}{r} .7 \\ +.5 \\ \hline .12 \end{array}$

b. $\begin{array}{r} 3.42 \\ 8\overline{)27.4} \\ \underline{24} \\ 3\ 4 \\ \underline{3\ 2} \\ 2 \end{array}$

54. After teaching a lesson on decimals, a sixth grade teacher asked the students to write out twenty-three hundredths. One student wrote .023. When asked why, the student replied that he had put 23 in the hundredths column! What is the student not understanding?

55. Think back to the Alphabitian system that we explored in Chapters 2 and 3. Answer each of the following questions without translating the amounts into base 10. That is, answer them as though you were an Alphabitian and that was the only numeration system you knew.

a. Draw a picture to represent the value of 0.A.

b. Add A.B + C.D.

56. **a.** Draw a picture to represent 0.1 in base 5.

b. Determine the value of 1.2 + 3.4 in base 5.

c. Determine the value of 2.4 × 3.2 in base 5.

d. Determine the value of $4\overline{)3044}$ in base 5. Represent the remainder in decimal form.

Chapter Summary

1. The set of integers is simply an extension of the set of whole numbers.

2. Operations with positive whole numbers can be adapted to work with integers.

3. A fraction is not simply a number; rather, a fraction expresses a relationship between two quantities. The numerator and denominator can be seen as a code that tells us the relative size of the fraction.

4. A fraction can be interpreted in four ways: as measure, as quotient, as operator, and as ratio.

5. Certain important ideas apply in all the rational-number contexts:
   - Something is to be partitioned into parts of equal size (value).
   - The something can have a value of 1, in which case the unit = the whole.
   - The something can have a value $\neq 1$, in which case the unit $\neq$ the whole.

6. The set of decimals has important connections with the set of integers and with the set of rational numbers.

7. All the sets of numbers that children will study in elementary school are subsets of the set of real numbers.

BASIC CONCEPTS

Section 5.1 Integers

integers 256
absolute value 260
negative 260
$a - b = a + {}^-b$

negative integers 256
opposite 260
additive inverse 260

Section 5.2 Fractions and Rational Numbers

rational number 268
fraction 268
numerator 268
denominator 268
rational number as a measure, as a quotient, as an operator, as a ratio 269–270
models to represent rational numbers: length models, area models, set models 271
equivalent fractions 276
cross multiplication 278
simplest form 278
rational numbers are dense 280
dense 280
irrational numbers 280

Section 5.3 Understanding Operations with Fractions

lowest common denominator 286
proper fraction 286
improper fraction 287
mixed number 287
multiplicative inverse 295
estimation and mental arithmetic with fractions 289, 299

Section 5.4 Beyond Integers and Fractions: Decimals, Exponents, and Real Numbers

decimals 308
transitive property 311
base 325
exponent 325
squared 325
cubed 325
scientific notation 326
irrational numbers 327
principal square root 328
real numbers 329
properties of real numbers 330
completeness property 330

CHAPTER 5 REVIEW EXERCISES

1. a. What is the additive inverse of 4?
   b. What is the multiplicative inverse of 4?

2. Perform these computations:
   a. $^-36 + {}^-38$
   b. $16 - ({}^-3)$
   c. $\dfrac{({}^-26 + 50)}{{}^-4}$
   d. $^-3 \cdot 5 + ({}^-6/2)$

3. Find the missing number: $^-8(\square + 5) = {}^-24$

4. Express $6 \times {}^-3$ as a repeated addition problem.

5. It is conventional to use $-$ to represent "minus" and $^-$ to represent "negative." Why don't we just use one symbol, writing, for example, $-3 - 4 = -7$?

6. Does the shaded part of the following figure below validly represent 3/4? Why or why not?

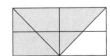

7. Name a fraction between 5/6 and 6/7 that is in simplest form. Justify your answer.

8. Name the fraction that tells the value of x on this number line. Explain your reasoning.

9. If this figure has a value of 1, show $\frac{1}{8}$.

10. If this figure has a value of $1\frac{1}{2}$, show 1.

11. If this figure has a value of $\frac{3}{4}$, show $\frac{1}{2}$.

12. In 2000 the average salary for beginning public school teachers was $28,986, and the average salary for all public school teachers was $43,250. In that year, the average salary for beginning public school teachers was approximately what fraction of the average salary for all public school teachers? Your answer needs to be a fraction whose denominator is less than or equal to 10. (Source: American Federation of Teachers, http://www.aft.org/research/survey01/tables.html)

13. For each of the pairs of fractions below, determine which of the fractions is larger without converting the fractions to decimals, finding the LCM, or drawing a diagram to see which is bigger. Briefly explain your reasoning.
    a. $\frac{9}{11}, \frac{13}{15}$
    b. $\frac{7}{12}, \frac{13}{28}$
14. When we express 1/2 as a decimal in base 10, we get 0.5. When we express 1/2 as a decimal in base 5, what do we get?
15. What does it mean to say that two fractions are equivalent?
16. Why do two fractions have to have a common denominator in order for us to add them?
17. An approach that children often take to find the product of, say, $3\frac{1}{4}$ and $2\frac{1}{2}$ is to multiply the two whole numbers and then multiply the two fractions; in our example this yields an answer of $6\frac{1}{8}$. Explain why this hypothesis is invalid.
18. Solve the problem 7 3/4 × 2 1/3 by a means other than the conventional algorithm of converting to improper fractions and multiplying the two numerators and the two denominators.
19. The points on the following number line are not drawn to scale. Without determining the actual product, determine the region in which $\frac{8}{15} \times \frac{5}{9}$ lies. Justify your reasoning.

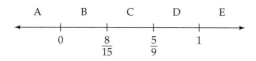

20. Pauline has 35 cups of flour. She makes cakes that require 2 1/4 cups each. If she makes as many such cakes as she has flour for, how much flour will be left over?
21. Without actually doing the problem $4\frac{3}{4} \div 8\frac{7}{8}$, would you estimate that the answer will be greater or less than $\frac{1}{2}$? Explain your reasoning.
22. Make up a realistic story problem for $2\frac{1}{2} \div \frac{3}{4}$ and solve the problem using the meaning of fractions and division (that is, without the algorithm). You need to explain which model of division you are using (partitioning or repeated subtraction), and you need to justify your solution.
23. In each case below, make as good an estimate as you can in your head. Briefly explain how you obtained your estimate.
    a. A campus newspaper reported that in a recent all-campus vote, 2/3 of all the students in a certain college favored the construction of a new recreational center. If 1754 students voted, approximately how many students voted for the new center?
    b. A silversmith has a sheet of silver that measures $9\frac{1}{2}$ inches by $6\frac{2}{3}$ inches. About how many square inches of silver does she have?
24. Express the following:
    a. 5/8 as a decimal
    b. 4/1000 as a decimal
    c. 0.24 as a fraction is simplest terms
25. Name a decimal between 0.4444 and 0.4445.
26. Order the following numbers from smallest to largest:
    5/11   0.045   0.454   ⁻0.54
27. Round 23.29 to the nearest tenth. Explain your reasoning.
28. A state budget is $4.306 billion. Express this amount as a whole number.
29. Perform the following computations, using scientific notation.
    a. 34,000,000,000 × 56,000,000
    b. 582,000 ÷ 23,000,000,000
30. Solve each of the following problems, and explain how you knew whether to multiply or divide.
    a. A tank of water has a volume of 0.7 cubic meters. What is the volume of 0.16 of the tank?
    b. Rosa is having a party and has bought 8 liters of punch. If each cup holds 0.15 liter, how many cups can she make?
    c. A box of candy weighs 0.6 pound. How many boxes can you make with 5.1 pounds of candy?
    d. The price of 0.8 ounce of a compound is $4.85. How much does 1 ounce cost?

CHAPTER 6

Proportional Reasoning

6.1 Ratio and Proportion
6.2 Percents

*I*n one respect, this chapter represents the culmination of the first five chapters. Over the course of the first five chapters, we have explored fundamental mathematical concepts and we have investigated applications of those concepts in real-life settings. In this chapter, we will focus explicitly on the idea of proportional reasoning. This is not a new idea—it is inherent in the concept of multiplication—but it has remained implicit up to now. The concept of proportional reasoning is a powerful one. In fact, some have said that proportional reasoning is both the capstone of [children's] elementary [school] arithmetic and the cornerstone of all that is to follow.[1] The NCTM asserts that "the ability to reason proportionally develops in students throughout grades 5–8. It is of such great importance that it merits whatever time and effort must be expended to [ensure] its careful development" (*Curriculum Standards*, p. 82). Let us examine why.

Additive Versus Multiplicative Comparisons

Much of our use of mathematics involves comparisons (How much more does this car cost than that one?) and change (How has the standard of living changed in the past 20 years?). To compare amounts and express change, we can use the mathematical tools called ratio, proportion, and percent. We shall see that these tools involve *multiplicative comparisons* between amounts.

In this chapter, we will examine two fundamentally different ways in which we can compare amounts and describe changes. For example, let's say we are comparing the cost of two cars, one of which costs $10,000 and the other $15,000. We can say that the second car costs $5000 *more than* the first car, that it costs $1\frac{1}{2}$ *times as much as* the first car, or that it costs 50% *more than* the first car.

The first and third descriptions both use the word *more*, but the first expresses the relationship in additive terms, and the second and third express the

---

[1] Richard Lesh, Thomas Post, and Merylyn Behr, "Proportional Reasoning," in *Number Concepts and Operations in the Middle Grades*, ed. James Hiebert and Merylyn Behr (Reston, VA: NCTM, 1988), p. 94.

relationship in multiplicative terms. The equations below illustrate why we use the terms *additive* and *multiplicative* to contrast these two ways to compare the amounts. To make the mathematics more visible, let $b = $ the cost of the second car and $a = $ the cost of the first car.

There are many ways to represent this relationship, using addition, subtraction, multiplication, division, and ratios (see the equations below). However, the first two are equivalent, and we call them both **additive comparisons**. The last three representations are equivalent, and we call them **multiplicative comparisons**.

$$b = a + 5000 \qquad b - a = 5000$$
$$b = 1\tfrac{1}{2}a \qquad \frac{b}{a} = \frac{3}{2} \qquad b:a = 3:2$$

We tend to use multiplicative comparisons more than additive ones. For example, we hear on television that a paper towel absorbs "50% more liquid than the leading brand."

In the first section of this chapter, we will develop the concepts of ratio and proportion and investigate real-life applications. Many real-life problems involving whole numbers or fractions actually have to do with ratios and proportions. In the second section, we will develop the concept of percent, the use of which pervades virtually everyone's daily life. The concept of percents rests firmly on the concept of ratio and proportion. Therefore, understanding how ratios and proportions operate is very important.

SECTION 6.1 RATIO AND PROPORTION

The Unit Concept Matures

WHAT DO YOU THINK?

- Are all ratios fractions? Are all fractions ratios? Why or why not?
- What do we mean by proportional reasoning?
- How do the concepts of ratio and proportion give us tools for making comparisons?

A noted educator and scientist once wrote that "it seems odd to refer to a relationship as a quantity."[2] For example, if we say that there are 25 students in a class, we can see or visualize them. Similarly, if we say that a "large" drink is 20 ounces, we can see or visualize that amount using our knowledge of measurement. We might use 1 ounce as our referent unit, or we might use a standard 12-ounce can as our referent unit; that is, "20 ounces is not quite 2 cans of soda." However, it is much harder to "see" rates—for example, this car gets 35 miles per gallon. That is, when we say "35 miles per gallon," the 35 actually expresses a *relationship* between two amounts (miles traveled and gallons consumed). As Schwartz notes, we "refer to [this] relationship as a quantity." However, 35 miles per gallon is much more abstract than 35 ounces or 35 people.

In the primary grades, children mostly work with whole numbers and with the operations of addition and subtraction. As their understanding of mathematical ideas grows, they move from counting physical objects to counting numbers themselves. In both cases, however, the unit is still a single whole entity (for example, 1 ounce or 1 person). With the introduction of multiplication and division, and then the introduction of rational numbers, both operations and numbers become more complex, as you saw in Chapters 3 and 5.

---

[2]Judah Schwartz, "Intensive Quantity and Referent Transforming Arithmetic Operations," in *Number Concepts and Operations in the Middle Grades*, ed. James Hiebert and Merylyn Behr (Reston, VA: NCTM, 1988), p. 43.

"Underneath all of the surface level changes is a fundamental change with far-reaching ramifications: *a change in the nature of the unit*. It is difficult to overestimate the significance of this basic shift. Many of the important differences in the subject matter between the primary and middle grades can be traced back to a change in the nature of the unit. Given the difficulty of mastering the concept of unit in whole-number situations, it is not surprising that changes in the nature of the unit in the middle grades bring new cognitive demands and renewed difficulties for students."[3] In the example above, we talked about 35 miles per gallon. In this case, our unit is 1 mile per gallon—a more complex unit than 1 ounce or 1 student. As noted in Chapter 3, one reason why multiplication and division problems are generally more difficult than addition and subtraction problems is that they involve a more complex unit. Thus it is crucial that we ask ourselves questions about the unit as we solve problems.

Ratios, Rates, and Proportions

A **ratio** is a relationship between two amounts or quantities. It can be expressed in the following equivalent ways: $a:b$, a/b, or $\frac{a}{b}$.

When the two amounts in a ratio represent different quantities, we often refer to such ratios as **rates**—for example, 30 miles/gallon or 55 miles/hour.

When we set two ratios equal to each other, we have a **proportion**. That is, $a:b = c:d$ iff $\frac{a}{b} = \frac{c}{d}$ and $b \neq 0, d \neq 0$.

Ratios, Rates, and Proportions in Mathematics and Real Life

Ratios and rates pervade mathematics. When two fractions are equivalent, we have a proportion. Ratio is inherent in the concept of place value: The ratio of the value of each place to the value of the place to its right is $10:1$. When we convert one unit to another, such as kilometers to miles, we use proportions. The concept of similarity involves equal ratios. When we make graphs, we use proportions. When we make scale drawings and scale models, we obey proportions. The essence of probability involves ratios. These are a few of the many aspects of ratio and proportion in mathematics.

Ratios, rates, and proportions show up in all kinds of real-world contexts.

- *Banking:* When I applied for a mortgage, the bank applied a ratio called the 28% rule: If the ratio of fixed monthly payments (mortgage, property tax, car payments, and so forth) to monthly income is more than $28:100$, the bank is not likely to make the loan.

- *Botany:* If we represent the number of complete turns made around the stem by t and represent the number of leaves between the two points as n, then the fraction t/n is called a divergency constant for that species (see Figure 6.1). Table 6.1 gives divergency constants for a number of different trees. The numerators and denominators are all Fibonacci numbers (see Exercise 33 in Section 1.4). This divergency constant, which can be expressed as a fraction, is actually a ratio. Do you see why?

■ Language ■

The term *ratio* comes from the Latin verb *ratus*, which means "to think or estimate." Many mathematicians in the sixteenth and seventeenth centuries used the word *proportion* for ratio. Even today you hear the two terms used interchangeably; for example, instructions for making a certain color might say, "Mix the two colors in the following proportion—3:2."

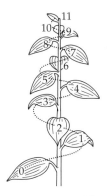

Source: Excerpted from *Symmetry: A Unifying Concept* © 1994 by Istvan and Magdolna Hargittai. Reprinted by permission.

FIGURE 6.1

---

[3] James Hiebert and Merylyn Behr, "Introduction: Capturing the Major Themes," in *Number Concepts and Operations in the Middle Grades*, ed. James Hiebert and Merylyn Behr (Reston, VA: NCTM, 1988), p. 2.

Table 6.1

Tree	Divergency constant
Elm	1/2
Beech, hazel	1/3
Apple, oak, apricot, poplar	2/5
Pear, weeping willow	3/8
Willow, almond, pussy willow	5/13

- *Commerce:* Many companies use rates to determine how much to charge for their goods or services. For example, Allied Shipping Company charges $7 for every 100 pounds of goods shipped.
- *Cooking:* If you want to make more than the amount given in the recipe, all the ingredients must be increased in the correct proportion.
- *Education:* One of the criteria used to determine the quality of a college is the student:teacher ratio.
- *Shopping:* If you get 250 tablets of generic aspirin for $2.49 and 100 tablets of name-brand aspirin for $3.89, how much cheaper is the generic aspirin?
- *Sports:* Football writers and announcers talk about the "turnover ratio," determined by subtracting the number of times a football team has lost the ball because of a fumble or interception from the number of times it has recovered the ball because of a fumble or interception. If a team has caused a total of 30 fumbles and interceptions and lost the ball 25 times because of fumbles and interceptions, the team is said to have a +5 ratio. Unfortunately, this is not a valid use of the term *ratio*. Can you explain why not? Would you recommend that these writers and announcers convert the numbers to a ratio or use the same numbers but invent a different term?

In many instances, quantities "should be" proportional but are not. Try to think of a few before reading on. . . .

Airline rates are often not proportional. Longer flights are often proportionally cheaper than short flights, and flights between major cities tend to be proportionally cheaper than flights between small cities.

The affirmative action debate for the past 30 years has been based on a belief that many people hold: that the ratio of a minority group in society and the ratio of members of that minority group in the work force or the schools should be equal. For example, the ratio of women:men is approximately 1:1, but the ratio of female college presidents to male college presidents is not 1:1, and so we say that a disproportionate number of college presidents are men.

Delving Further into Ratios

Look at the following ratio statements. If the statement can be expressed as a fraction statement, do so. If it cannot, try to explain why. Then read on. . . .

1. The ratio of males to females at Mountain State College is 3:2.
2. In order to get a stain out of a shirt, Lisa made a mixture of bleach and water in the ratio 3 parts water to 1 part bleach.

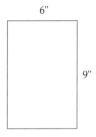

FIGURE 6.2

3. In the rectangle in Figure 6.2, the ratio of the width to the length is 2:3.
4. Fred's car gets 25 miles per gallon.

   Statement 1: We can say that 3/5 of the students are male and that 2/5 of the students are female.

   Statement 2: The stain mixture Lisa made was a 1/4 bleach solution. [*Note:* Many people would be more likely to call this a 25% bleach solution.]

   Statement 3: There are two ways in which we can translate this statement into fraction language. We can say that the width is 2/3 as long as the length. Alternatively, we can say that the length is $1\frac{1}{2}$ times as long as the width.

   Statement 4: There is no way to translate this statement into fraction language. Do you see why? In Statement 3, the two amounts (parts) had the same unit: length. In this statement, the two amounts are measured with different units: miles and gallons.

Table 6.2 summarizes some of these points.

TABLE 6.2

Statement	Diagram	Whole?	Can be expressed as fraction statement?
1	Males / Females	The student body	Yes
2	Bleach / Water / Water / Water	The mixture	Yes
3	6" × 9" rectangle	Either number can be seen as the whole	Yes
4	Miles: 25, 50, 75 / Gallons: 1, 2, 3	No whole	No

When Are Ratios Like and When Are They Unlike Fractions?

Before we move on to problem-solving situations, let us summarize the ways in which ratios are like fractions and the ways in which they are unlike fractions.

Ratios are like fractions in that they can be expressed as an ordered pair of numbers representing two sets or two amounts. Ratios are different from fractions in that fractions are restricted to part–whole relationships, whereas ratios are not.

Let us illustrate this difference by returning to the example of the water–bleach mixture.

The ratio of water to bleach is 3:1. However, there are actually several possible ratios:

3 water:1 bleach	1 bleach:3 water	3 water:4 total	1 bleach:4 total
part:part	part:part	part:whole	part:whole

Because a part–whole relationship exists, the water–bleach mixture can be expressed as a fraction.

In the case of the miles per gallon, we cannot make a fraction statement. However, the ratio is 25 miles:1 gallon, which is a whole–whole relationship even though the two amounts are measures of different units, namely miles and gallons.

Note also that some ratios represent discrete amounts, such as students, whereas others represent measured amounts, such as bleach, water, length of a side, miles, or gallons.

Let us now examine these various concepts and ideas that underlie what we call *proportional reasoning*. This first investigation represents a classical problem with a history to it.

INVESTIGATION 6.1

Unit Pricing—Is Bigger Always Cheaper?

Let's say you want to buy some laundry detergent. The small jug costs $2.99 for 36 fluid ounces, and the large jug costs $3.79 for 48 fluid ounces. Which is the better buy?

DISCUSSION

STRATEGY 1: Use fractions

In this case, the additive comparison is not the one that is useful. That is, we get 12 more ounces for 80 more cents. However, we can use this information to make a valid multiplicative comparison.

The larger size gives you 1/3 more detergent, so if it costs 1/3 more, then the value of both sizes will be the same. Does the larger jug cost 1/3 more? Because 1/3 of $2.99 is $1.00, the two sizes will be the same value if the large size costs $4. Note that here we are rounding $2.99 to $3.00. Because $3.79 is less than $4, we conclude that the larger size is a better buy. Why is this?

Perhaps you used this strategy in Exploration 6.3 to determine which box of pancake mix was the better buy. This strategy, used intuitively by many students,[4] is called the **factor-of-change method**. In this case, the factor of change was 1/3. That is, if we multiply the 36 ounces and the 299 cents by the same factor, we produce an equivalent ratio. The idea of equivalent fractions helps us to understand why this method works.

$$\frac{299 \text{ cents}}{36 \text{ ounces}} \times \frac{1\frac{1}{3}}{1\frac{1}{3}} = \frac{400 \text{ cents}}{48 \text{ ounces}}$$

---

[4]Kathleen Cramer, Thomas Post, and Sarah Currier, "Learning and Teaching Ratio and Proportion: Research Implications," in *Research Ideas for the Classroom: Middle Grades Mathematics*, ed. Douglas T. Owens (New York: Macmillan, 1993), pp. 159–178.

STRATEGY 2: *Use ratios and a calculator*

$$\frac{\$2.99}{36 \text{ ounces}} = 0.0830556$$

$$\frac{\$3.79}{48 \text{ ounces}} = 0.0789583$$

I have deliberately not placed labels on the two amounts. What do the two decimals mean? Think and read on. . . .

The meaning of 0.083 is $.083 per ounce, or 8.3 cents per ounce. From a meaningful interpretation of these ratios, we see that the larger size is a better buy because its unit price is less; that is, $.079 per ounce is less than $.083 per ounce.

This strategy, also intuitively used by many students, is known as the **unit-rate method**. That is, we determine the cost of 1 unit (in this case 1 ounce).

STRATEGY 3: *Another way to express the ratios?*

Maria says she used the previous strategy but "upside down." What do these ratios mean? What do you think of her idea? Is this a variation of the unit-rate method?

$$\frac{36 \text{ ounces}}{\$2.99} = 12.0 \qquad \frac{48 \text{ ounces}}{\$3.79} = 12.7$$

In this case, we have 12.0 ounces per dollar versus 12.7 ounces per dollar. This is a variation of the unit-rate method. In this case, the unit is 1 dollar instead of 1 ounce; that is, the ratios tell us how much detergent we get for a unit of money. Had we selected cents as our unit, we would have had the ratios 0.120 ounce per cent versus 0.127 ounce per cent.

STRATEGY 4: *Solve a proportion*

We can let x represent how much a 48-ounce jug of equivalent value would cost:

$$\frac{\$2.99}{36 \text{ ounces}} = \frac{x \text{ dollars}}{48 \text{ ounces}}$$

When we solve for x, we find that $x = \$3.99$; that is, if the 48-ounce jug cost $3.99, the two jugs would have the same value. Because the 48-ounce jug costs less than $3.99, it is a better buy.

> ■ **Mathematics** ■
>
> Many states now require grocery stores to show the unit price below each item. One of the reasons for this law is that many companies realized that people tended to believe that larger packages were cheaper. Even today with unit prices, you can find instances in most stores where the larger amount is not cheaper.

INVESTIGATION 6.2

How Much Money Will the Trip Cost?

Most real-life uses of ratios and rates involve proportions. For example, let's say you are planning a 2000-mile trip and want to estimate how much money you will spend on gas. How would you do that? Work on this problem on your own and then read on. . . .

DISCUSSION

As you may have realized, the cost of gas for the trip would depend on two variables: how many miles per gallon your car gets and the cost of gasoline. Let's say your car averages 25 miles per gallon and you estimate that gas will cost $2.39 per gallon on the trip. Now estimate the cost of gas for the trip and then read on. . . .

The rate at which your car uses gas is expressed by the ratio 25 miles : 1 gallon. The following proportion enables us to determine how many gallons of gas you will need at that rate of consumption:

$$\frac{25 \text{ miles}}{1 \text{ gallon}} = \frac{2000 \text{ miles}}{x \text{ gallons}}$$

In other words, what number do we need to divide 2000 by so that the ratio is still 25 to 1?

We find x by solving the proportion for x: $25x = 2000$; therefore $x = 80$ gallons.

We can now use another proportion to find the cost of 80 gallons.

$$\frac{\$2.39}{1 \text{ gallon}} = \frac{y}{80 \text{ gallons}}$$

We find y by solving the proportion for y: $y = (\$2.39)(80)$; therefore, $y = \$184$.

We could have solved the entire problem in one step using dimensional analysis. The units for the answer will be dollars. Therefore, we begin with the amount that is represented by dollars: the cost of gasoline. If we multiply the ratio representing the cost of gasoline by the ratio representing the consumption of gasoline, look what happens:

$$\frac{\$2.39}{\text{gallon}} \times \frac{1 \text{ gallon}}{25 \text{ miles}} \times 2000 \text{ miles} = \$184$$

Rates as Functions

You may recall working with proportional functions in Chapter 2. If you did Exploration 6.4, you explored several different proportional functions. Rates express a functional relationship between two variables. For example, if my car averages 25 miles per gallon, then there is a functional relationship between the miles I drive and the gallons of gas my car consumes (see Figure 6.3). In this case, we say that these variables are *directly proportional* to each other. Do you see why?

If we take any two points on the graph in Figure 6.3—for example, the points (1, 25) and (4, 100)—and represent them as ratios—25 miles/1 gallon and 100 miles/4 gallons—the ratios are equal; that is, 25/1 = 100/4. When two ratios are equal, by definition, we have a proportion.

Now that we have worked with ratios and proportions, let us revisit some investigations from previous chapters for which ratios and proportions are appropriate strategies for answering the questions.

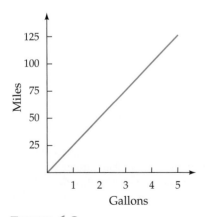

FIGURE 6.3

INVESTIGATION 6.3

Reinterpreting Old Problems

A. In Investigation 5.7, Josh finds that 1/8 of the balloons are blue and that there are 120 more red balloons than blue balloons. How many blue balloons are in the box? In Section 5.2, we explored several different ways to solve this problem. How could you solve this problem using ratios and proportions? Work on this and then read on. . . .

DISCUSSION

If 1/8 of the balloons are blue, then the ratio of blue balloons to red balloons is 1:7. Why is this?

We need to focus on the ratio of blue to red balloons; that is, blue:red = 1:7. If there are 120 more red balloons than blue balloons, we can express this relationship by letting x = the number of blue balloons and $x + 120$ = the number of red balloons. Because the ratio is still 1:7, we have the proportion

$$\frac{x}{x + 120} = \frac{1}{7}$$

Solving the proportion for x, we have

$$7x = x + 120$$

from which we find that $x = 20$.

Thus there are 20 blue balloons and 140 red balloons.

B. In Section 5.2, Investigation 5.6, Jose paid $12 for a 3/4-pound box of chocolates. What is the price of 1 pound (at this rate)?

DISCUSSION

Many students find this problem challenging as a fraction problem but can solve it more easily if they use decimals and proportions:

$$\frac{\$12}{0.75 \text{ pound}} = \frac{x \text{ dollars}}{1 \text{ pound}}$$

C. In Investigation 5.14, Marvin has 23 yards of cloth to make costumes for the play. Each costume requires $3\frac{1}{4}$ yards of material. How many costumes can he make?

DISCUSSION

In Section 5.3, we solved this problem by interpreting it as division (repeated subtraction). However, it can also be interpreted as a proportion. In this case, 3.25 yards can be seen as the unit—that is, the number of yards needed for 1 costume.

$$\frac{1 \text{ costume}}{3.25 \text{ yards}} = \frac{x \text{ costumes}}{23 \text{ yards}}$$

INVESTIGATION 6.4 — Using Estimation with Ratios

The following investigations consist of problem-solving situations that can be answered with an estimate (rather than an exact answer). Assume that you have no calculator or pencil and paper.

A. It took Rene 24 minutes to go 15 miles from Hingham to Marshfield. What was Rene's average speed for the trip? What is your first impression (5-second estimate)? What is your best estimate? Can you get the "exact" answer? Why is the "exact" answer probably not exact? Work on these questions and then read on. . . .

DISCUSSION

The problem can be represented as the following proportion: 15 miles is to 24 minutes as x miles is to 60 minutes (that is, 1 hour).

$$\frac{15 \text{ miles}}{24 \text{ minutes}} = \frac{x \text{ miles}}{60 \text{ minutes}}$$

STRATEGY 1: Use equivalent fractions

We can see that 15/24 is just under 16/24, which is equivalent to 2/3, and 2/3 of 60 = 40. Thus a quick estimate (if you connect this question to equivalent fractions) is "slightly under 40 miles per hour."

STRATEGY 2: Use the missing-factor model of division

We also can get a decent estimate by asking: 24 times what is equal to 60? This strategy is related to the compatible-numbers strategy we have used before; it would not have been so straightforward had the number of minutes been, say, 22.

Because 24 times $2\frac{1}{2}$ is equal to 60, we now have to determine 15 times $2\frac{1}{2}$, but that is relatively simple using multiplication as repeated addition: $15 + 15 + 7\frac{1}{2} = 37\frac{1}{2}$, which is the exact answer—well, sort of.

The exact answer of $37\frac{1}{2}$ miles per hour is itself an approximation. Why is this? The 24 minutes and 15 miles are rounded numbers. In this case, the "exact" numbers were 23 minutes and 45 seconds and 15.2 miles. Even if we now computed Rene's "exact" average speed, that is,

$$\frac{15.2}{23.75} = \frac{x}{60}$$

and solved for $x = 38.4$ miles, the term *average speed* is itself a mental construct, not something that exists in an observable sense. At times Rene was going

55 miles per hour; at times Rene was going 20 miles per hour. We will study the concept of average in more detail in Chapter 7.

> **B.** It is Christmas time, and virtually every store in town is selling wrapping paper. If Zoe doesn't care about looks and just wants the cheapest, which should she buy? Work on this and then read on. . . .

- Store A: 36 square feet for 88¢.
- Store B: 50 square feet for $1.79.
- Store C: 150 square feet for $2.99.

DISCUSSION

If we divide the cost (in pennies) by the square feet, the resulting ratio will tell us the cost per square foot. The ratio that is the smallest number will represent the cheapest paper. Try this on your own and then read on. . . .

In this case, if the prices are not terribly close, you may need only one run-through.

- Store A: 88/36 = more than 2—that is, more than 2¢ per square foot.
- Store B: 179/50 = more than 3—that is, more than 3¢ per square foot.
- Store C: 300/150 = 2¢ per square foot. We can use 300 instead of the exact 299¢ because 1¢ makes virtually no difference in this case. There is no need to do more arithmetic here, because store C clearly has the "best" ratio.

> **C.** Ellie bought a new car on August 16. On November 1 the odometer read 2650. Ellie's insurance company gives a discount if she puts less than 10,000 miles per year on the car. If she continues driving at this rate, will she qualify? Work on this and then read on. . . .

DISCUSSION

The time between August 16 and November 1 is very close to $2\frac{1}{2}$ months, so we can represent this problem by the following proportion:

$$\frac{2650 \text{ miles}}{2\frac{1}{2} \text{ months}} = \frac{x \text{ miles}}{12 \text{ months}}$$

We can once again obtain a quick estimate by using the idea of equivalent fractions: What multiple of $2\frac{1}{2}$ will be closest to 12? A bit of mental arithmetic yields the realization that $2\frac{1}{2} \times 5 = 12\frac{1}{2}$. Therefore, the estimated miles per year will be 2650×5. How can you calculate that amount mentally?

Because $2650 \times 10 = 26{,}500$, 2650×5 will be half that amount, or 13,250. At this rate, Ellie will travel about 13,250 miles in just over 12 months and won't qualify for the discount.

Another quick estimate can be obtained by connecting to compatible numbers: 2650 miles is just over 2500 miles, which is 1/4 of 10,000 miles. Thus, if she had traveled 2500 miles in 3 months, she would be on target to travel 10,000 miles in 12 months. Because she has traveled more than 2500 miles in less than 3 months, her ratio of miles:months is higher than 2500 miles:3 months.

Real-Life Applications

The next three investigations are not quite as simple and tidy as the previous ones and require a careful application of concepts we have studied thus far. I find that most real-life applications of proportion are more like the following investigations than like the previous ones. Just a reminder: Work on these actively—that is, get out a pencil and paper and a calculator and work on them yourself, as opposed to just reading what I have written. By doing so, you will not only deepen your understanding of proportional reasoning but also improve your computational ability, including estimation and mental math.

INVESTIGATION 6.5 — Is the School on Target?

An all-female college first began admitting men three years ago. Its goal for this, the third year, is to have a 2:1 ratio of females to males in the incoming freshman class. The admissions department has just reported a problem. When the college sent out letters of acceptance to this year's freshman class, the ratio of females to males was 2:1. However, the ratio of females to males in the 1000 students who indicated that they will be coming this fall is 7:3. If the college still wants to meet the 2:1 goal, how many more males will it need to accept to reduce the ratio of females to males to 2:1? Try to solve this problem on your own and then read on. . . .

DISCUSSION

The desired ratio of females to males is 2:1. If the ratio of females to males among the 1000 incoming freshmen is 7:3, that means that there are 700 women and 300 men. One way of representing the dilemma is to let x be the number of additional males the college needs to admit.

In other words, if the college admits x more males, the ratio of females to males will be $700:(300 + x)$, and this ratio needs to be equal to 2:1. Because the two ratios are equal, we have the following proportion:

$$\frac{700}{300 + x} = \frac{2}{1}$$

We find that $x = 50$.

This problem could also have been solved using common sense. Targeting a ratio of 2 females to 1 male can be interpreted as wanting the number of males to be one-half the number of females. Because 350 is half of 700, the college needs a total of 350 males—that is, 50 more males than it has presently accepted.

INVESTIGATION 6.6

Transferring Credits

I attended the University of Tennessee in my freshman year and then transferred to the University of Arizona. At that time Tennessee ran on a quarter system (the academic year consisted of three 10-week quarters and then summer school), and Arizona followed the more common fall and spring semesters, each being about 15 weeks long. At the University of Tennessee, I took 15 credits in the first quarter, 15 in the second, and 19 in the third, giving me a total of 49 credits for the year. When I transferred to Arizona, how many semester-hour credits did my 49 quarter-hour credits convert into?

DISCUSSION

STRATEGY 1: "See" the implied proportion

If we see that two ratios must be equal, we can immediately create a proportion: 49 quarter-hour credits is to x semester-hour credits as 3 quarters is to 2 semesters.

$$\frac{49}{x} = \frac{3}{2}$$

$$3x = 98$$

$x = 32\frac{2}{3}$ semester-hour credits

STRATEGY 2: Solve a simpler problem

We can solve a simpler problem and then generalize. If I had taken 16 quarter-hours for each of the three quarters, then that would have been equivalent to taking 16 semester-hours for each of two semesters. That is, 48 quarter-hours is equivalent to 32 semester-hours. This means that a conversion ratio is 48:32, which simplifies to 3:2. This can be stated in fractions by saying that we have to multiply the quarter-hour credits by 2/3 to get the semester-hour credits, because 2/3 of 48 is 32. In this case, 2/3 of 49 is $32\frac{2}{3}$.

INVESTIGATION 6.7

Stuck Behind a Truck

One day I was driving to see a student teacher who was teaching in a nearby town. I was late, and I got stuck behind a truck. The speed limit was 45 miles per hour, but I was stuck for 5 miles behind the truck, which went only 20 miles per hour. How much time did I lose?

DISCUSSION

As is true with many problems, we need to define the question. In this case, I am assuming that I would have averaged 45 miles per hour as opposed to 20 miles per hour for those 5 miles.

STRATEGY 1: Use ratios

At the rate of 20 miles per hour (that is, 20 miles in 60 minutes), I would go 10 miles in 30 minutes, or 5 miles in 15 minutes.

At the rate of 45 miles per hour (had I not been stuck behind the truck), I would go 5 miles in 1/9 of an hour. Why is this? If we round 1/9 to the nearest tenth of an hour, we have 0.1 hour, and 0.1 hour $\times$ 60 minutes per hour = 6 minutes.

Instead of taking 6 minutes, it took me 15 minutes, so I lost about 9 minutes.

STRATEGY 2: Set up proportions

$$\frac{45 \text{ miles}}{60 \text{ minutes}} = \frac{5 \text{ miles}}{x \text{ minutes}}$$

$$x = 6.7 \text{ minutes}$$

$$\frac{20 \text{ miles}}{60 \text{ minutes}} = \frac{5 \text{ miles}}{x \text{ minutes}}$$

$$x = 15 \text{ minutes}$$

Thus I lost $15 - 6.7 = 8.3$ minutes, or about $8\frac{1}{2}$ minutes.

INVESTIGATION 6.8

From Raw Numbers to Rates

We are bombarded with numbers and data about our world every day. Making sense of this information is not always easy. Sometimes the data are presented in raw numbers and sometimes in rates. Why do you think this is so? Let's examine some actual data to see how rates can help us to "see" data more clearly. In 2003, there were 599 homicides in Chicago and 596 in New York City. What can you conclude from these numbers?

DISCUSSION

Before we examine the numbers, it is important to note that the raw numbers are only one part of the making sense process. For example, I have friends in Chicago and New York City, and they tell me that they do not feel terribly unsafe, but they know when and where they can go in order to stay safe. Here, we will focus on the mathematics of the data. Let's say you were thinking of moving to either city. From a numerical perspective, what can you conclude? For exmple, would you say that since the two numbers are essentially equal, both cities are basically equally safe in this respect?

Most people will say no because New York City is much larger and therefore, proportionally, it is a safer city than Chicago with respect to homicides. How might we express this conclusion numerically?

The population of Chicago in 2003 was about 2.86 million and the population of New York City was 8.10 million. How can we use these numbers to compare the two cities?

We can find the fractions or ratio in both cities. What do the numbers mean?

$$\frac{599}{2.82} = 212 \qquad \frac{596}{8.10} = 74$$

There were 212 murders in Chicago per 1 million people and there were 74 murders in New York City per 1 million people. This calculation has changed the data from raw numbers, where we are essentially comparing apples to oranges, to rates, where we are comparing apples to apples.

We could also divide the Chicago rate by the New York City rate: 212/73.5 = 2.88. How can we put this number into a sentence?

> The homicide rate in Chicago is 2.88 times the rate for New York City. Rounding this to a whole number, we can say: The homicide rate in Chicago is almost 3 times the rate for New York City.

What about other cities? For example, El Paso, Texas, had 21 homicides in 2003 and a population of 586,000. When we convert raw numbers to rates, we determine an index number. In the case of crime rates, this number is generally 100,000 people. The rates for each city are determined by a proportion. For example, this is the computation for Chicago.

$$\frac{599}{2,820,000} = \frac{x}{100,000}. \text{ Solving for } x, \text{ we have } x = \frac{599 \times 100,000}{2,820,000} = 21.2$$

At this point, for many readers, this is just a computation. Let's investigate further what this means. I have used the term *equivalence* throughout this book and stated that it is one of the big ideas of elementary mathematics. The concept of equivalence is front and center again. Essentially, these computations involve transforming the number for each city into a rate (homicides per total population) for each city. This rate can be treated just like a fraction, and in order to compare the rates for different cities, we make the denominator the same for each city.

For example, for Chicago $\frac{599}{2,820,000}$ and $\frac{21.2}{100,000}$ are *equivalent* fractions.

Thus, we can present the following data. The murder rate, per 100,000 people was:

Chicago	21.2
New York City	7.4
El Paso	2.4

We can also divide each of these numbers by 2.4 if we wanted to compare them to El Paso.

Chicago	9
New York City	3
El Paso	1

In the exercises, you can examine various data about AIDS, pollution, prisons, and so on to see how rates can help us to interpret data about our lives.

INVESTIGATION 6.9

How Much Does That Extra Light Cost?

Let's say you are considering the installation of a 300-watt outdoor security light for your home. By how much will your electric bill increase?

DISCUSSION

In order to answer this question, what assumptions do we need to make? What data do we need?

Let us assume that the light will be on an average of 12 hours per day over the course of a year—less than 12 in the summer, more than 12 in the winter. We also need to know how much electricity costs. For this problem, use the rate of 12.325¢/kWh.[5] See whether you can do the problem on your own now, and then read on. . . .

STRATEGY 1: Multiply step by step

- 300 watts = 0.3 kilowatt, so 0.3 kilowatt per hour times 12.325 cents/kWh = 3.6975 cents per hour to run the light.
- 3.6975 cents per hour times 12 hours per day = 44.37 cents per day to run the light.
- 44.37 cents per day times 365 days per year = 16195.05 cents per year to run the light.

Note: This problem illustrates the need to make sure constantly that each step makes sense. Many of my students give me an answer of $16,195.05. However, the number 16195.05 refers to cents, not dollars. How do you convert cents into dollars?

STRATEGY 2: Use dimensional analysis

By aligning the rates correctly, we can solve the problem in one step:

$$\frac{0.3 \text{ kW}}{\text{hour}} \times \frac{12.325 \text{ cents}}{\text{kW}} \times \frac{12 \text{ hours}}{\text{day}} \times \frac{365 \text{ days}}{\text{year}} \times \frac{\$1}{100 \text{ cents}} = \frac{\$161.9505}{\text{year}}$$

Instead of doing each computation separately, what we do on the calculator is simply

.3 ⊗ 12.325 ⊗ 12 ⊗ 365 ⊘ 100 =

What appears on the calculator is 161.9505. Dimensional analysis tells us that the meaning of this number is dollars per year.

Dimensional Analysis Explained

Dimensional analysis was first introduced in Chapter 1. Can you now justify this method?

---

[5] A kilowatt is 1000 watts. For example, if you run a lamp with a 100-watt light bulb for 10 hours, you have used 1000 watt-hours or 1 kilowatt-hour.

The justification lies in the concept of equivalent fractions and rates. Usually, we think of finding equivalent fractions by multiplying the numerator and denominator by the same number. However, we can generalize this notion of equal number to equal amount.

When using dimensional analysis, we are treating the units as though they were numbers:

$$\frac{0.3 \text{ kW}}{\text{hour}} \times \frac{12.325 \text{ cents}}{\text{kW}}$$ Our initial problem.

$$= \frac{0.3}{1} \times \frac{\text{kW}}{\text{hour}} \times \frac{12.325}{1} \times \frac{\text{cents}}{\text{kW}}$$ Our work with multiplication of fractions helps us understand that this is an equivalent representation.

$$= \frac{0.3}{1} \times \frac{12.325}{1} \times \frac{\text{kW}}{\text{kW}} \times \frac{\text{cents}}{\text{hour}}$$ The commutative and associative properties tell us that we can change the order.

$$= \frac{3.6975}{1} \times \frac{\text{kW}}{\text{kW}} \times \frac{\text{cents}}{\text{hour}}$$ Our understanding of multiplication of fractions tells us that we have the amount 3.6975.

$$= \frac{3.6975}{1} \times 1 \times \frac{\text{cents}}{\text{hour}}$$ Any amount divided by itself is equal to 1.

$$= \frac{3.6975}{1} \times \frac{\text{cents}}{\text{hour}}$$ The identity property of multiplication tells us that when we multiply an amount by 1, the value is unchanged.

$$= \frac{3.6975 \text{ cents}}{\text{hour}}$$ Our knowledge of multiplication of fractions tells us that this is an equivalent representation.

Of course, when we use dimensional analysis, we simply compute and "cancel" units when appropriate.

Summary

In this section, we have explored the last interpretation of fractions that was introduced in Section 5.2: fraction as ratio. We have seen that not all ratios can be expressed as fractions—only those that are given in part–whole terms or can be translated into part–whole terms. We have seen that rates are a special kind of ratio, one that cannot be expressed in fractional terms. Rates also express a functional relationship between two variables.

In this section, we have made explicit the difference between additive and multiplicative comparisons. If you did Exploration 6.1, you probably realized how tempting it was to use additive comparisons to compare slopes. Like fractions, ratios and rates signify multiplicative relationships between two amounts. When two ratios are equal (or when two rates are equal), we have a proportion. Thus the concept of a proportion connects to the notion of equivalent fractions, which we studied in Chapter 5. In the next section, we will build the concept of percent upon the concept of proportion.

We have examined several real-life applications of ratios, rates, and proportions. In doing so, we found that it is critical to make sure that the computations make sense. We have given names to two methods that students often invent: the factor-of-change method and the unit-rate method. Dimensional analysis is one tool that helps us to solve proportion problems. Finally, we have seen that many estimation problems involve proportions.

EXERCISES 6.1

Exercises 1–16 present you with fairly straightforward ratio and proportion situations. In each problem, first obtain a rough estimate (5 to 10 seconds, entirely in your head); then obtain a refined estimate (30 or so seconds, with a minimum of paper and pencil) or the "exact" answer mentally. Then explain how you obtained the rough estimate and the refined estimate (or exact answer). Finally, explain whether the exact answer is, itself, actually exact or an approximation.

1. A quart (32 ounces) container of yogurt contains 920 calories. Approximately how many calories would there be in a 5-ounce serving?

2. a. If the Dow Jones Industrial Average in the stock market began the day at 12,000 and lost 800 points, how much would a proportional fall be if the stock market began the day at 1000?

   b. If the stock market was at 2700 and rose 75 points, how much would a proportional rise be if the stock market began the day at 10,000?

3. An advertisement says that 5 out of 8 dentists recommend the new zigzag toothbrush. If 264 dentists were interviewed, how many recommended the toothbrush?

4. An intravenous solution needs 2 liters (L) of glucose to be mixed with 7 units of blood. How much glucose is needed for 40 units of blood?

5. An employee making $24,000 was given a raise of $1000. All employees were given proportional raises.

   a. How much of a raise would an employee making $18,000 receive?

   b. How much of a raise would an employee making $30,000 receive?

   c. How much of a raise would an employee making $23,450 receive?

6. If $1\frac{3}{4}$ cups of flour are required to make 30 cookies, how many cups of flour (to the nearest 1/4 cup) are required for 96 cookies?

7. In the summer of 1991, Israel airlifted 14,000 Ethiopian Jews to Israel as immigrants. If the United States were to receive a proportional number of refugees at one time, how many would be received? Say the population of Israel is approximately 4.4 million and that of the United States is 270 million.

8. a. On one map, 1/3 inch represents 18 miles. If two cities are $2\frac{1}{2}$ inches apart on the map, what is the actual distance between them?

   b. On another map, 1 inch represents 65 miles. Los Angeles is about 1000 miles from Portland. How many inches apart would Portland and Los Angeles be on this map?

   c. On a map $\frac{2}{3}$ inch represents 84 miles. How far apart are two cities in reality if they are $4\frac{5}{8}$ inches apart on the map? You can solve this problem by any means except setting up a proportion and solving for x.

9. You can use proportions to estimate the height of a tree. John is 6 feet tall, and his shadow is $10\frac{1}{2}$ feet long. How high is a tree whose shadow is 90 feet long?

10. Jane finds that she can read 36 pages of a book in 40 minutes. At this rate, how long, to the nearest minute, will it take her to finish a book 473 pages long?

11. The other day Janet was using the rowing machine in the Fitness Center. Her goal was to row 2200 meters in 10 minutes. At exactly 7 minutes, she saw that she had rowed 1560 meters. Is she going to make her goal if she continues at this rate?

12. Sheila and Dora worked $3\frac{1}{2}$ hours and $4\frac{1}{2}$ hours, respectively, on a programming project. They were paid $176 for the project. How much did each earn?

13. A photograph is 3 inches high and 5 inches wide. If it is enlarged to be 7 inches high, how wide will it be (to the nearest 1/4 inch)?

14. The recommended dosage for a particular medication is 6.8 milligrams (mg) per pound of body weight per day, not to exceed 1000 mg per day. For a 125-pound patient, about what would you expect the doctor to recommend for a daily dosage?

15. A car and a train both begin at the same time and are moving at constant speeds, but the train is moving faster. For every 2 miles that the car travels, the train travels 3 miles.

    a. How far will the train have traveled when the car has traveled 30 miles?

    b. How far will the car have traveled when the train has traveled 40 miles?

16. What is the average speed in each of these situations?

    a. You travel 11 miles in 16 minutes.

    b. You travel 31 miles in 39 minutes.

    c. You travel 85 miles in 90 minutes.

    d. On a bicycle you travel $12\frac{1}{2}$ miles in 45 minutes.

    e. On a bicycle you travel $3\frac{3}{4}$ miles in $10\frac{1}{2}$ minutes.

17. Make up and solve a real-life problem that lends itself to needing a rough or refined estimate.

Exercises 18–37 are nonroutine, multistep problems. If it is possible to estimate the answer, do so and briefly explain how you obtained your estimate. If you feel that making an estimate is not practical, describe what measures you took to ensure that your answer is reasonable.

18. **a.** If you are traveling 65 miles per hour, how fast are you traveling in kilometers per hour, to the nearest whole number? (Fifty miles per hour is equivalent to 80 kilometers per hour.)

    b. If you are traveling 35 kilometers per hour, how fast are you traveling in miles per hour, to the nearest whole number?

19. It was reported that Ross Perot spent approximately $40 million of his own money in the 1992 presidential campaign. His total worth is reported to be approximately $4 billion. If your total worth were $50,000 and you spent the same fraction of your worth on an election as Perot did, how much money would you have spent?

20. In a healthy person, the ratio of red blood cells to other blood cells should be about 1 to 5000. Amy just got back a lab report that showed 300 red blood cells out of 230,000 blood cells. Is her red blood cell count low, high, or normal?

21. A car travels 60 miles per hour, and a plane travels 15 miles per minute. How far does the car travel while the plane travels 600 miles?

22. On a TV game show, a contestant makes $700 for every correct answer but loses $500 for every wrong answer. After answering 24 questions, Sarah broke even. How many questions did she answer correctly?

23. Five cups of a certain flour weigh 1 pound, and 1 cup of cornstarch weighs 1/4 pound. If the ratio of flour to cornstarch in a mixture is 2:1, how much would 1 cup of the mixture weigh?

24. This is a problem given to students in Baghdad over 2000 years ago. Two men sat down to eat, one with five loaves and the other with three, all the loaves having the same value. Just as they were about to begin, a third man came along and proposed to eat with them, promising to pay eight cents for his part of the meal. If they ate equally and consumed all the bread, how should the eight cents be divided?

25. In the women's downhill during the Olympics, the difference between first place and second place was 0.04 second. If the two skiers had been racing side by side, what would have been the distance between the two at the finish line? Assume that they were going 60 mph at the end of the race.

26. Convert each of the following figures to miles per hour.

    a. The world record for the men's 100 meters was 9.77 seconds, held by Asafa Powell. What was his average speed over the course of that race?

    b. The world record for the men's 10,000 meters was 26:17.53, held by Kenenisa Bekele of Ethiopia. What was his average speed throughout that race?

    c. The world record for the women's 100 meters freestyle swimming was 53.83 seconds, held by Inge de Bruijn of the Netherlands. What was her average speed over the course of that race?

27. **a.** In the *Guinness Book of World Records*, the record for handshaking is 16,615 in 7 hours and 25 minutes. How many is this per minute? The person shook one hand every ___ seconds.

    b. In the *Guinness Book of World Records*, the greatest distance pushing a baby carriage in 24 hours was 342.25 miles. What, to the nearest second, was the average time per mile?

28. At a certain college, the ratio of men to women is 9:4. If there are presently 360 women, how many additional women would it take to reduce the ratio of men to women to 2:1?

29. At a certain college, there are 7 men for every 5 women. If there are 420 more men than women, what is the total enrollment?

30. Yosha is working on a small project that is due in 2 days. She has spent $4\frac{1}{2}$ hours on it and figures that she is about 3/4 done. How many more hours will she need to spend?

31. A painting crew of 4 takes 5 days to do an apartment building. There are 40 apartment buildings in their current project.

    a. If the foreman wants to do the project in 20 days, how many people should he hire? What assumptions do you make in order to solve this problem?

    b. A crew of 3 could do one building in ___ days.

32. Ginger wants to fill her new swimming pool. She has two pumps; the large pump takes 40 minutes to fill the pool, and the small pump takes 60 minutes. How long will it take to fill the pool if both pumps are working?

33. The ratio of physicians to inhabitants of the United States is 1 to 549. The ratio of prison inmates to inhabitants of the United States is 1 to 497. Are there more physicians or prison inmates in the United States? Approximately how many are there of each?

34. Here are the directions from an oatmeal container:

Servings	1	2	3
Water or milk	$\frac{3}{4}$ cup	$1\frac{2}{3}$ cup	$2\frac{1}{3}$ cup
Cereal	$\frac{1}{2}$ cup	1 cup	$1\frac{1}{2}$ cup
Salt (optional)	Dash	$\frac{1}{8}$ tsp	$\frac{1}{8}$ tsp

   a. The proportions are not the same. Explain why not.

   b. What amount of each ingredient would you use to make 10 servings? Explain your work.

35. A recipe for chocolate chip cookies follows.

    $1\frac{1}{4}$ cups flour

    $\frac{1}{2}$ cup sugar

    $\frac{1}{2}$ teaspoon salt

$\frac{1}{2}$ cup butter

6 oz chocolate chips

1 teaspoon vanilla extract

1 egg

$\frac{1}{2}$ teaspoon baking powder

The recipe makes 4 dozen cookies.

   a. How much of each ingredient would you need if you wanted to make 10 dozen cookies?

   b. Unlike in the oatmeal problem, it is essential in this case that certain ingredients be in the "correct proportion." Explain what that phrase means.

36. An advertisement says that 5 out of 8 doctors recommend one brand of aspirin over another. In actuality, 325 doctors were interviewed, and 199 of the doctors recommended the first brand over the other. Is the advertisement accurate? Explain your response.

37. If the ratio of boys to girls in a class is 3 to 8, will the ratio of boys to girls stay the same, become greater, or become smaller if 2 boys and 2 girls are added to the class? Justify your response.

Exercises 38–43 require you to make some assumptions in order to determine an answer. Describe and justify the assumptions you make in determining your answer.

38. A worker estimates that she spends 75 minutes a day at the copying machine. How many hours would this be per year?

39. Two persons have the same yearly income, but one gets paid every other week, whereas the other gets paid twice a month. Which paycheck is larger?

40. One morning I was listening to National Public Radio, which was conducting its annual pledge drive. The announcer said, "Our goal this morning (from 6 A.M. to 9 A.M.) is to get 40 new pledges. We've received 15 pledges so far, so we're on track." I was puzzled because it was 7:30.

   a. Explain why the announcer was inaccurate from a mathematical perspective.

   b. Assuming that the announcer was aware that the math wasn't quite right, explain why he might have made that statement.

41. Sonja claims that she was 20 minutes late to work because she got stuck behind a truck the whole way. Is this plausible? If not, explain why. If so, make up some numbers to support your conclusion.

42. Determine the cost of flying from your nearest airport to two different cities. If we consider the fares as rates—that is, cost per mile—are the two rates proportional? If not, why do you think they are not proportional?

43. You are in charge of designing a 6-week summer school program at your college. Some classes will meet every day, and others will meet three times per week. How long will the classes that meet every day be? How long will the classes that meet three times a week be?

44. The following data give the numbers of people in prison in the United States for four different subpopulations.

   a. Just looking at these numbers alone, write a sentence or two comparing the different amounts.

   b. Now look at the second column, where the raw numbers have been turned into rates. Write a sentence or two comparing the different amounts.

   c. Describe how the rates can changes one's perception of people in prison.

	Number	Per 100,000 population
White male	454,300	465
Black male	586,300	3405
White female	39,100	38
Black female	35,000	185

45. On a scale drawing for the student union building, the height of the atrium is 8.2 centimeters. If the actual height is 16.4 meters, what is the scale of the drawing? In other words, how many meters are represented by 1 centimeter on the drawing?

46. When data are presented in either tables or graphs, sometimes we see raw numbers and sometimes the raw numbers have been translated into rates. Unfortunately, we have found that many people do not fully understand rates. Explain the reason why we have rates, as though in response to someone who asks, "Why not just use the raw numbers?"

47. The data below give the numbers of adults living with HIV in several countries.

   a. Just looking at these numbers alone, write a sentence or two comparing the different amounts.

   b. Now look at the second column, where the raw numbers have been turned into rates—in this case, percentages. Write a sentence or two comparing the different amounts.

   c. Describe how the rates can change one's perception of the AIDS epidemic in different parts of the world.

	Adults with HIV	Percent of adult population
South Africa	5,100,000	21.5
India	4,750,000	0.8
Nigeria	3,300,000	5.4
Zimbabwe	1,600,000	24.6
Tanzania	1,500,000	8.8
Ethiopia	1,400,000	4.4
Mozambique	1,200,000	12.2
Kenya	1,100,000	6.7
Democratic Repulic of the Congo	1,000,000	4.2
Zambia	830,000	16.5

48. Converting raw numbers to rates.
    a. There were 2,777,000 births in the United States in 1900 and 4,059,000 in 2000. Determine the birth rates, per 1000 population, for those two years. The U.S. population in 1900 was 76 million, and in 2000 it was 281 million.
    b. There were 1,344,520 reported violent crimes in the United States in 1980 and 1,425,486 in 2000. Compare the crime rate, per 100,000 people, for these two years. The population in 1980 was 227 million.
    c. There were 24,363 people in federal prisons in the United States in 1980 and 173,059 in 2003. Compare the rates, per 100,000 people, for these two years. The population in 2003 was 291 million.
    d. There were 54,633 deaths in motor vehicles in 1970 in the United States and 43,354 in 2000. Compare the rate, per 100,000 people, for these two years. The population in 1970 was 203 million.

49. Below are infant mortality rates for whites and blacks in the United States. Compare the improvement in the white rate to the improvement in the black rate.

	White	Black
1980	10.9	22.2
2003	5.7	14.0

SECTION 6.2 PERCENTS

WHAT DO YOU THINK?

- What does percent mean?
- To what other mathematical concepts is percent related?
- In what real-life contexts have you encountered percents?

You first learned percents in the sixth or seventh grade, and you see them every day in the newspaper. Yet the confident use of percents as a tool eludes most adult Americans. The concept of percents is not so much one of the big ideas of arithmetic as it is a tool that rests upon many big ideas of arithmetic. In this section, you will see how percents rest upon the notion of fractions (in terms of part–whole relationships) and proportions (in terms of equivalent ratios or equivalent fractions). You will learn to translate freely among percents, fractions, and decimals.

The Origin of Percent

The word **percent** literally means "per hundred" and comes directly from the Latin *per centum*.

There are direct and immediate connections among percent, equivalent fractions, and proportions. For example, let's say that a student correctly answered 12 out of 16 questions on a quiz. To determine the student's score as a percentage, we are asking what fraction with a denominator of 100 is equivalent to 12/16. This question can be stated as a proportion:

$$\frac{12}{16} = \frac{x}{100}$$

That is, what value of x makes $x/100$ and $12/16$ equivalent fractions? Solving for x, we find the student's grade is 75%.

■ *Mathematics* ■

We can also interpret this problem from the perspective of rates. That is, if the student were to continue to get problems correct *at this rate*, how many correct would she or he get out of 100 questions?

Uses of Percent

We use percents for a variety of purposes:

- To communicate—for example, we hear on the news that a fire is 30% contained.
- To make sense of situations—a manufacturer may say that the germination rate of a grass seed is 80%.

- To make decisions—should I refinance my house if the interest rate is down to 7.5%?
- To compare—in one high school of 345 there were 12 dropouts, and in another high school of 567 there were 17 dropouts; which school has the lower dropout rate? Converting these numbers to percents essentially treats them as rates. The first school has a dropout rate of 3.5% and the second school has a dropout rate of 3.0%.

> ### History
>
> The concept of percent is related to societies' need to compute interest, profit and loss, and taxes.
>
> When the Roman emperor Augustus levied a tax on all goods sold at auction . . . the rate was 1/100 . . . . In the Middle Ages, as larger denominations of money came to be used, 100 became a common base for computations. Italian manuscripts of the fifteenth century contained such expressions as "20p100," "x p cento," and "vi p c^o" to indicate 20 percent, 10 percent, and 6 percent. . . . The percent sign, %, has probably evolved from a symbol introduced in an anonymous Italian manuscript of 1425. Instead of "per 100," "P100," or "P cento," which were common at that time, this author used P^o_o. By about 1650 the ^o_o had become $\frac{o}{o}$ so per $\frac{o}{o}$ was often used. Finally the per was dropped, leaving $\frac{o}{o}$ or %.[6]

Let us use this basic understanding of percents to solve a few problems and then look at commonalities and differences among those problems.

INVESTIGATION 6.10 Who's the Better Free-Throw Shooter?

You are the coach of the girls' basketball team at a local middle school. It is the fifth game of the season, the game is tied, and there are only 5 seconds left on the clock. The referee has called a technical foul on the other team. You get to choose any one of your girls to take the shot. Basing your decision only on their free-throw shooting thus far this season, whom would you pick?

- Becky has made 8 of 12 free throws.
- Rachel has made 15 of 20 free throws.

DISCUSSION

STRATEGY 1: Use fractions

Represent the players' free-throw shooting as fractions.

$$8/12 = 2/3$$

At this rate, Becky would make 2/3 of 20 shots, which is $13\frac{1}{3}$. Thus Rachel is doing better.

---

[6]*Historical Topics for the Mathematics Classroom: 31st Yearbook* (Reston, VA: NCTM, 1969), p. 147.

CLASSROOM CONNECTION

When Investigation 6.10 is given to middle school students, many will pick Becky, saying that she has missed fewer shots. Do you see why? What is their misconception?

STRATEGY 2: *Make a proportion*

Some students feel more comfortable with proportions. How would you describe why this proportion will tell us the answer?

$$\frac{2}{3} = \frac{x}{20}$$

In this case, x represents how many baskets Rachel must have made in order to have the same ratio as Becky. When we solve for x, we get $x = 13\frac{1}{3}$. Because Rachel has made 15 out of 20, which is better than $13\frac{1}{3}$, she is doing better than Becky.

STRATEGY 3: *Convert fractions to decimals or percents*

$$\frac{8}{12} \approx 0.67 = 67\% \quad \text{whereas} \quad \frac{15}{20} = 0.75 = 75\%$$

In real life, what factors other than the players' free-throw percentages might a coach consider before making this decision?

INVESTIGATION 6.11

Understanding a Newspaper Article

A newspaper story reports that 8% of the 7968 students at Midvale College work full-time. How many students work full-time? First try to estimate the number of students and then determine the exact answer.

DISCUSSION

STRATEGY 1: *Use 10% as a benchmark*

A very rough estimate: 8% is close to 10%, which is 1/10, a fraction that we can do mental multiplication with rather simply. If we round 7968 to 8000, then 1/10 of 8000 is 800.

STRATEGY 2: *Use 1% as a benchmark*

We could mentally find 1% and use simpler numbers to build up to 8%:

1% of 7968 is about 80.
8% of 8000 is 80 × 8 = 640.

STRATEGY 3: *Find a close unit fraction*

We could use 1/12 for 8%. Do you see why?
In this case, we can use the compatible-numbers strategy from Chapter 3.

$$\frac{7968}{12} \approx \frac{8400}{12} = 700$$

When we compare this to 7200/12 = 600, we conclude that the actual number of students working full-time is between 600 and 700.

STRATEGY 4: Break the problem into parts

We could also adapt the decomposition strategy developed for whole numbers in Chapter 3.

10% of 8000 is 800.

5% of 8000 is 400.

Because 8% is between 5% and 10% (a little bit closer to 10%), we can conclude that 8% of 8000 is a little more than 600.

Converting among percents, decimals, and fractions When we estimate the answers to percent problems, it is helpful to know the basic conversions among percents, fractions, and decimals. Table 6.3 shows some of the more commonly used conversions.

TABLE 6.3

Fraction	Decimal	Percent
1/2	0.5	50%
1/3	0.333... or $0.\overline{3}$	$33\frac{1}{3}\%$
1/4	0.25	25%
1/5	0.2	20%
1/10	0.1	10%
1/100	0.01	1%

With this in mind, let us examine different ways to determine the "exact" number of students working full-time at Midvale College.

STRATEGY 1: Connect to the meaning of percent

We can represent the ratio of students working full-time to the total number of students as a fraction—that is,

$$\frac{\text{Number of students working full-time}}{\text{Total number of students}} = \frac{8}{100}$$

If we let x represent the number of students working full-time, we have

$$\frac{x}{7968} = \frac{8}{100}$$

Now we can solve for x and see that $x = 637.44$.

STRATEGY 2: Use an appropriate procedure

8% of 7968 = 0.08 × 7968 = 637.44

About 637 students work full-time.

STRATEGY 3: Use a diagram

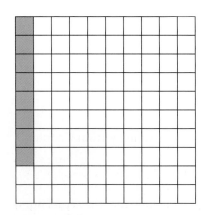

FIGURE 6.4

Figure 6.4 shows 8%. How can you use this diagram to solve the problem?

If the whole square represents the total student population, this means that 100 boxes represent 7968 students. The shaded area represents that fraction (percent) of the student body that works full-time.

To solve the problem using this diagram, we need to know the value of 8 boxes. If

$$100 \text{ boxes} = 7968$$

then

$$1 \text{ box} = 79.68$$

so

$$8 \text{ boxes} = 79.68 \times 8 = 637.44$$

An answer of 637.44 or even 637 is too precise; it is an example of what I call pseudo-precision. Why is this? Think before reading on. . . .

When the college reports that 8% of its students work full-time, if it is using mathematics appropriately, it is saying that the percent of the students who work full-time is closer to 8% than it is to 7% or 9%. That is, when the person in the college determined that 8% of the students work full-time, he or she took the number of students who work full-time and divided by the total number of students, and the number that appeared on the calculator was closer to 0.08 than it was to 0.07 or to 0.09.

In other words, the actual number of students who work full-time is between 7.5% of 7968 and 8.5% of 7968. That is, the actual number is between 598 and 677. Do you see why?

We have just found that any number of full-time students between 598 and 677 will round to 8% of 7968. Because we know only that 8% was reported, it would be more accurate to say not that 637 students work full-time but, rather, that the number of students who work full-time is between 598 and 677, or 637 ± 40.

INVESTIGATION 6.12

Buying a House

The Benanders are going to buy a house. Before giving a family a mortgage to buy a home, banks generally require that the total monthly payment (including property taxes) be no more than 28% of the family's gross monthly income. What must the Benanders' monthly income be in order for them to buy a home on which the monthly payment will be $800?

This is not an easy question. Please think about it, grapple with it, apply your understanding of percent and your problem-solving toolbox, and then read on. . . .

DISCUSSION

STRATEGY 1: Use guess–check–revise

One way of paraphrasing the problem is to say that if the bank computes 28% of the Benanders' monthly pay, this number must be at least $800 or the bank will turn them down. In other words, the bank takes their monthly income and multiplies it by 0.28. If the product is greater than $800, they qualify.

For many students, this line of reasoning leads to guess–check–revise, as shown in Table 6.4.

TABLE 6.4

Guess	Computation	Analysis
$3000	$3000 × 0.28 = $840	Too much. Guess less.
$2900	$2900 × 0.28 = $812	Too much. When the guess went down by $100, the payment went down by $28. If the next guess is $50 less, the payment will be $14 less, which will be too low (798). So make a guess of $40 less.
$2860	$2860 × 0.28 = $800.80	

It would take several more trials to get the exact answer, and that is one limitation of guess–check–revise: It is not practical when you need an exact answer. However, when determining eligibility, the bank is not interested in three decimal places. Banks have some flexibility; that is, if you are "close" to 28% and there are other factors in your favor (such as job stability, a promotion due, or an inheritance anticipated), then you are likely to qualify.

STRATEGY 2: *Rewrite the problem as an equation*

Another set of students paraphrases the problem something like this:

28% of their salary must be at least 800.

Changing the wording to be more "mathematical," we have

28% of what is $800?

The jump to an equation now is not as great a leap to make:

0.28 times x = 800, where x = their monthly salary

That is, if $0.28x = 800$, solve for x, and see that $x = 2857.14$. That is, their monthly income must be at least $2857.

STRATEGY 3: *Use a diagram*

We can also represent this problem with a diagram. Even if you "understand" how to get the answer, can you explain the "why" of Figure 6.5? Try to do so before reading on. . . .

In Figure 6.5, the whole box represents their monthly income, and the shaded area represents 28% of their income.

If we look at the problem from a part–whole perspective,

$$\frac{\text{Part}}{\text{Whole}} = \frac{28 \text{ boxes}}{100 \text{ boxes}} = \frac{\$800}{\text{total income}}$$

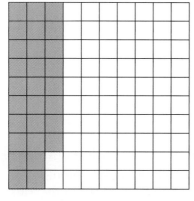

FIGURE 6.5

We can also interpret the figure in the following way:

If 28 boxes has a value of 800, then 100 boxes has a value of what?

If 28 boxes	represents a value of	$800, then
1 box	represents a value of	$\frac{800}{28} = \$28.571428$
100 boxes	represents a value of	$2857.14

Connections Between Percent and Other Mathematical Topics

How would you represent the three previous investigations in terms of part–whole relationships? How would you interpret them in terms of proportions?

One of the keys to seeing the interconnectedness of all three problems is to realize that in percent problems, there are two parts and two wholes.

For example, in Investigation 6.10, the 12 shots Becky attempted are the whole and 8 are the part she made. When we determine that this part–whole relationship is equivalent to 67%, we are saying that this is equivalent to 67 parts in a whole of 100 [Figure 6.6(a)].

To determine 8% of 7968 in Investigation 6.11, we need to find the part out of the whole of 7968 that is equivalent to 8 parts out of 100 [Figure 6.6(b)].

In Investigation 6.12, to determine the Benanders' monthly payment, we ask, 800 is 28% of what whole [Figure 6.6(c)]?

8 shots made

Part

Whole = 12 shots
(a)

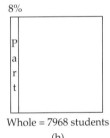
8%

Part

Whole = 7968 students
(b)

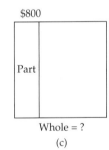
$800

Part

Whole = ?
(c)

FIGURE 6.6

If we now examine these problems in terms of proportions, in each case 100 represents the whole. From this perspective, we can see that the differences among the three problems have to do with what part or whole is missing.

$$\frac{\text{Part}}{\text{Whole}} \qquad \frac{8}{12} = \frac{x}{100} \qquad \frac{x}{7968} = \frac{8}{100} \qquad \frac{800}{x} = \frac{28}{100}$$

Now that we have investigated some of the basic aspects of percents, the following investigations can serve both as stretching problems and as a self-assessment of how well you "own" the concept (that is, how well you can apply your knowledge to nonroutine problems). If you did Exploration 6.5, you grappled with understanding percent problems and justifying different ways of solving them.

INVESTIGATION 6.13 Sale?

Jorge is excited. He just saw in the newspaper that Showroom Appliances is celebrating 20 years in business by offering 20% off on all merchandise. He just went there yesterday to buy a television set that was regularly priced at $260. What is the sale price of the television?

DISCUSSION

Probably the most common solution I see to this problem looks like this:

$$0.20 \times \$260 = \$52$$
$$\$260 - \$52 = \$208$$

That is, you determine the discount and then subtract the discount from the regular price. Jorge will pay $208.

However, Randi says that she solved the problem in one step:

$$0.80 \times \$260 = \$208$$

Do you understand why Randi did this? How would you explain it to someone else? Please reflect on these questions before reading on. . . .

One way to get beyond the "how" of this shortcut and into the "why" is to represent this problem with a diagram. In Figure 6.7, if the whole box represents $260 (the original price), what do the two shaded areas represent (that is, 20 boxes and 80 boxes)? Think before reading on. . . .

The 20-box shaded region represents the **discount**—that is, the amount by which the store will reduce the price, or how much Jorge will save.

The 80 boxes therefore represent the **sale price** of the television. Because it is the sale price that we are looking for, we do not need to find 20% of the price and subtract it from the original price; we can determine the sale price directly.

We can show that the two procedures are mathematically equivalent:

$$260 - 0.20(260) = 260(1 - 0.20) \quad \text{We are using the distributive property.}$$
$$= 260(0.80) \quad \text{Because } 1 - 0.20 = 0.80$$

Thus,

$$260 - 0.20(260) = 260(0.80)$$

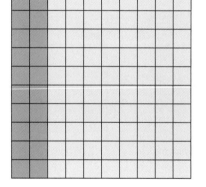

FIGURE 6.7

Percent Change

Many problems involving percents in real life are not simple problems like the ones we have investigated thus far but, rather, involve change. Such comparisons are called percent increase, percent decrease, percent change, percent error, percent faster or slower, or percent more or less.

We need to be able to understand change in order to interpret intelligently the changes in our society—in employment, the economy, AIDS, and other areas. *On the Shoulders of Giants: New Approaches to Numeracy*[7] is a book that was written by mathematicians to help people to see the "big ideas" of mathematics. The six chapters discuss pattern, dimension, quantity, uncertainty, shape, and *change*. Percent is one of many mathematical tools that can help us better understand and make sense of change.

I find that more students rent the concept of percent change than almost any other concept in the course. Therefore, it is even more imperative that you enter this section owning the knowledge in the previous discussions and that you engage in the investigations in this section and in the explorations. Explorations 6.7 and 6.8 require you to make sense of percent change and to examine ways of determining percent change.

INVESTIGATION 6.14

What Is a Fair Raise?

Let's say I am the president of a small company, Bassarear's Bagels, that has done very well in the last year. I have decided that I want to share my good fortune with my hard-working and devoted employees. I announce that I will be giving a $1-an-hour raise to everyone. Two days later I become aware of some grumbling. A number of employees are complaining that this is not fair. I am stunned. I thought that giving everyone the same raise was the epitome of fairness. Why are some employees grumbling? How would you explain this to me if you were one of my dissatisfied employees? Think and then read on. . . .

DISCUSSION

When I do this investigation with my own students, I receive many different reasons for the complaints:

- Workers with more seniority should receive "bigger" raises.
- Full-time workers should receive "bigger" raises.
- Hard-working workers should receive "bigger" raises.
- This raise is not fair to the people making more money.

For the concept of percent increase, I want to focus on the last reason. Why would a higher-paid employee feel that my raise was not fair to him or her? Imagine you are such an employee. How might you help me to see your point? Think and read on. . . .

Let's say the janitor is making $5 an hour and the manager of one of the stores is making $15 an hour. From a proportional perspective, the janitor is pretty happy because $1 an hour represents 1/5 of her salary. The manager, however, is not very happy, because $1 represents only 1/15 of her salary. From the proportional perspective, a "fair" raise would be one in which the ratio of raise to present salary would be equal for everyone; that is, all the raises would be proportional.

---

[7]Lynn Steen, ed., *On the Shoulders of Giants: New Approaches to Numeracy* (Washington, DC: National Academy Press, 1990).

Additive versus multiplicative increases The chart below shows the difference between *adding* the *same amount* to each person's wage (an additive increase) and *multiplying* each person's wage by the *same amount* (a proportional increase).

Original salary	Add the same amount	Multiply by the same amount
5	5 + 1 = 6	5(1.2) = 6
15	15 + 1 = 16	15(1.2) = 18

What if I were to be convinced that I should not add the same amount to each person's salary but, rather, should multiply each person's salary by the same amount? How would I announce that to the company in terms of percent? Think and read on. . . .

If we multiply the janitor's salary by 1.2, we are in effect increasing her salary by 1/5. That is,

$$5(1.2) = 5 + 0.2(5) = 5 + 1 = 6$$

If we multiply the manager's salary by 1.2, we are in effect increasing her salary by 1/5. That is,

$$15(1.2) = 15 + 0.2(15) = 15 + 3 = 18$$

Because 1/5 is equivalent to 20/100, we say that both employees have received a 20% raise; that is, their salaries are 20% greater than they were before.

In order to understand better the difference between additive and multiplicative changes, look at Figure 6.8, which shows the "before" and "after" salaries using both methods. What do you notice?

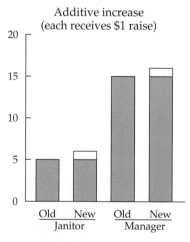
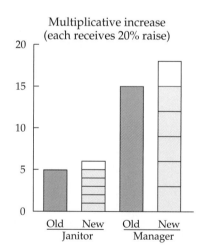

FIGURE 6.8

In the additive case (at the left in Figure 6.8), what is *equal* is the amount of the raise (represented by the white boxes). That is, both people received a raise of $1 per hour.

In the multiplicative case (at the right in Figure 6.8), what is *equal* is the ratio of the increase to the wage. In both cases, the amount of the raise (represented by the white boxes) is equal to 1/5 of the wage.

Section 6.2 / Percents   **367**

As I hope you are seeing, the idea of "equality" is as complex in mathematics as it is in history and political science. It is also important to know that not everyone would agree that the 20% raise is fair. Many people interpret this scenario as "the rich getting richer" because the difference between the janitor's and the manager's hourly pay was $10 and is now $12.

Now let us investigate a problem that is near and dear to the hearts, or at least the pocketbooks, of most students.

INVESTIGATION 6.15

How Much Did the Bookstore Pay for the Textbook?

When college bookstores purchase textbooks, they generally sell the books for 20% to 25% more than they paid for them. In other words, their markup is between 20% and 25%. Let's say that you paid $95 for a textbook and your bookstore marked up the price by 25%. How much did the bookstore pay for the textbook? Work on this problem before reading on. . . .

DISCUSSION

STRATEGY 1: *Act it out*

Let's say you are working in the bookstore. You would take the price of the book (which we don't know at this point) and find 25% of that number, and then you would add that to the price of the book. This sum would be $95. We can solve the problem by using guess–check–revise (see Table 6.5) or by forming an equation.

STRATEGY 2: *Use guess–check–revise*

TABLE 6.5

Guess	Work	Reflection
60	0.25(60) + 60 = 75	Too low.
70	0.25(70) + 70 = 87.50	Increasing the guess by 10 increases the sum by 12.50.
75	0.25(75) + 75 = 93.75	Close.
76	0.25(76) + 76 = 95	Done!

STRATEGY 3: *Form an equation*

If we let x = the price the bookstore paid for the book, many students can "see" the equation emerging by acting out the process. The bookstore employee finds 25% of x and then adds this amount to the amount the bookstore paid for the book; this sum must be $95.

That is,

$$0.25x + x = 95$$
$$1.25x = 95$$
$$x = 76$$

Percents Less than 1 or Greater than 100

In real life, we frequently encounter percents greater than 100% or less than 1%. Researchers and teachers know that when students do not thoroughly understand the concept of percents, their success rate with such problems goes down dramatically.

INVESTIGATION 6.16 — The Copying Machine

Clark Elementary School has decided to buy a new copying machine. One of the selling points of the new machine is that the manufacturer advertised that the school can expect about 2 paper jams per 1000 copies. After six months, the faculty wanted to see how the copier was actually doing. They had run 42,164 copies and had encountered 96 paper jams. Think about the following questions and then read on. . . .

A. Do you think the ratio of paper jams at Clark is consistent with the advertised paper jam rate? Is the school averaging about 2 paper jams per 1000 copies?

B. Represent the advertised paper jam rate as a ratio, a fraction, a decimal, and a percent.

C. Which one would you use if you were writing the ad for the copier?

D. How do you think the manufacturer determined the figure of 2 paper jams per 1000 copies?

DISCUSSION

A. One way (of several) to answer this question is to set up the following proportion:

$$\frac{96}{42{,}164} = \frac{x}{1000}$$

Could you explain this proportion to someone who didn't understand it? What does the x mean?

By setting the ratio 96/42,164 equal to x/1000, we are saying the rate of 96 jams per 42,164 copies is equivalent to how many jams per 1000 copies?

Solving for x, we find $x \approx 2.3$, which means that Clark is averaging about 2.3 paper jams per 1000 copies. Because the paper jam rate for Clark is closer to 2 jams per 1000 than to 3, we would conclude that the copier is acting as advertised.

B. All of the following are mathematically equivalent to a paper jam rate of 2 per 1000:

1 in every 500 copies will jam.

1/500 of the copies will jam.

0.2% of the copies will jam.

One-fifth of 1% of the copies will jam.

C. As an advertiser, I think I would say that the paper jam rate is one-fifth of 1%. This sounds smaller than 2 copies per 1000. However, I also know that most people will understand 2 copies per 1000 much better than one-fifth of 1%.

D. The manufacturer probably divided the total number of jams by the total number of copies. I would hope that this was done with many different copiers rather than just one copier, and I would hope that these copiers were not all brand new. The ratio that resulted from this calculation was closer to 2 per 1000 than to 1 per 1000 or 3 per 1000.

INVESTIGATION 6.17

132% Increase?

In 2005, I read that "the percent increase since 1975 in births by Cesarean section in the United States is 170." The article went on to say that 640,000 Cesarean sections had been done in 1975. I immediately wondered how many had been done in 2005, and that is the question I want you to answer now. Think and then read on. . . .

A. Before you solve the problem, try to make a rough or refined estimate.
B. Then solve the problem.

DISCUSSION

A. *A rough estimate:* Because a 100% increase is double the original number, the number of Cesarean sections more than doubled, so the answer will be more than 1,280,000 Cesarean births in 2005.
   A refined estimate:

   An increase of 170% = original amount + 100% + 70% (about 2/3)
   600,000 + 600,000 + 400,000 = 1,600,000 Cesarean births in 2005

B. <u>STRATEGY 1</u>: *Break the problem into parts*

   A 170% increase can be broken into a 100% increase and a 70% increase.
   A 100% increase is equivalent to doubling, which gives us 1,280,000.
   A 70% increase is 0.7(640,000) = 448,000.
   Therefore, a 170% increase brings us to 1,728,000 Cesarean births in 2005.

   <u>STRATEGY 2</u>: *Connect to a similar, simpler problem*

   For example, if the problem had said that there had been a 25% increase, many students would have found the answer by multiplying 640,000 by 1.25. Do you see why?
   If a 25% increase is determined by multiplying by 1.25, what would we do to determine a 170% increase? Think and then read on. . . .

   We would multiply 640,000 by 2.7.

Without an estimate, many students arrive at the wrong answer of 1,088,000 because they multiply 640,000 by 1.7 or they add 70% of 640,000 to 640,000. In this case, even a very rough estimate serves many students nicely: It tells them that the number of Cesarean births in 2005 must be at least 1.2 million. Students who do a very rough estimate and then obtain an actual answer of 1,088,000 see a contradiction between the estimate and the computed answer, and this contradiction causes them to go back and reexamine their work. This is one of the many useful aspects of estimating.

Interest

One of the ways in which almost everyone encounters percents is with interest—when you buy a car, when you buy a house, and when you don't pay off your entire credit card balance at the end of the month, you pay interest on the amount that you owe.

Most people have a very limited understanding of the consequences of interest. Let's say you buy a house for $225,000 and make a down payment of $25,000. You go to the bank and take out a 30-year mortgage for $200,000, and you are told that your monthly payment will be $1,151. Multiply this monthly payment by 360 (there are 360 months in 30 years) to determine how much you actually pay for the house (the $200,000 you borrowed plus all the interest). Then read on. . . .

No, you did not mispunch the calculator. Yes, you really will pay that much for the house! You actually paid $459,360 for the house! Thus, over 30 years, you pay $239,360 of interest!

Let us examine how interest is determined, first in a simple case and then in some instances that are more realistic. Several variables affect how much interest you receive (on an investment) or pay (on a loan): the original amount, called the principal (P), the annual interest rate (r), specified in percent, and the time of the investment/loan (t), specified in years.

Although simple interest—when the principal does not change over the course of the loan—is not common, it is a good starting point for understanding how interest is determined. For example, let's say that Jenna borrows $2000 from a relative and agrees to repay the loan after 1 year at the rate of 6%. At the end of the year, she pays $2000 + 6% of $2000; that is, she pays $2000 + 120 = $2120.

More commonly, interest is compounded; that is, the interest is determined at specified intervals and added to the principal at those times. For example, if Jenna's relative had specified that the loan be compounded semiannually, then to determine how much Jenna would owe at the end of the year, the interest would be determined every 6 months and added to the principal. Thus, after 6 months, she would owe $2000 + 0.03(2000) = $2060. Why is this?

The annual interest rate is 6%, so the semiannual rate is 1/2 of 6%, or 3%, because 6 months is 1/2 year. We determine how much she owes after the next 6 months as follows: $2060 + 0.03(2060) = 2060 + $61.80 = $2121.80. In this particular case, a 1-year loan compounded only semiannually, the difference between simple and compound interest is not huge. However, on most deposits and loans, the interest is compounded daily; that is, the annual interest rate is divided by 365.

In the following investigation, we will develop the formula for determining compound interest. In the second investigation, we will examine a very relevant situation.

INVESTIGATION 6.18

Saving for College

When Emily was born, her grandparents decided to contribute toward her college education by opening a savings account for $1000. If they add no other money, and if the account earns 6% compounded annually, how much money will there be in the account after 18 years?

DISCUSSION

This is a nonroutine, multistep problem. Therefore, let us go one step at a time.

After 1 year How much money would there be in the account after 1 year? Do this and then read on. . . .

There are several different methods that students will use.

We can find 6% of $1000 and then add that to 1000; that is, $1000 + 0.06($1000) = $1060. Or, as we saw earlier in this chapter, we can obtain $1060 in one step: $1000(1.06) = $1060. Do you see why?

This connection between procedures is important because in order to understand this problem, you need to be able to understand the second procedure; the first one is too cumbersome.

After 3 years Now determine how much money would be in the account after 3 years, and then read on. . . .

$1000(1.06) = $1060 after 1 year

$1060(1.06) = $1123.60 after 2 years

$1123.60(1.06) = $1191.02 after 3 years

Let us stop and analyze this strategy. In the first step, we multiplied 1000 by 1.06 to get 1060. In the second step, many students clear the calculator and then multiply 1060 by 1.06. Then, in the third step, they clear the calculator and then multiply 1123.60 by 1.06. If we examine this carefully, we can see a shortcut. See whether you can find it yourself before reading on. . . .

We could simply do this on the calculator:

1000 × 1.06 × 1.06 × 1.06

Do you see why? If you don't believe this, do it on the calculator to see that it does work. Then try to explain why it works before reading on. . . .

We can represent what we have learned as

$1191.02 = 1000(1.06)(1.06)(1.06)

We can now use our knowledge of exponents to say

$1191.02 = 1000(1.06)^3

The third, and most efficient, method for finding the amount after 3 years uses this knowledge. How would we enter the last expression in a scientific calculator? Try it before reading on. . . .

Simply enter 1000 × 1.06 y^x 3 to get the same answer.

Before we move on to solve the original question, this is a good time to understand the basic formula for use with interest. Look at how we determined the value (amount) in the bank after 3 years. Try to represent that procedure with a formula using the interest symbols introduced above, and then read on. . . .

$$\text{The money after 3 years} = (1000)(1.06)^3$$
$$= 1000(1 + 0.06)^3$$

That is,

$$A = P(1 + r)^t$$

After 18 years This discovery makes the original problem much less tedious to solve.

STRATEGY 1: *Use the y^x button*

The most straightforward solution is to use the y^x button on your calculator:

1000 ⊗ 1.06 ⓨˣ 18 = $2854.34

STRATEGY 2: *Use the memory key*

If your calculator lacks an exponentiation key, you could use the memory key M+ by doing the following:

First, activate the memory by pressing 1.06 and then M+.
Now press 1000 × MR = (you should see 1060, the amount after 1 year).

To get the amount for each succeeding year, you simply press × MR that many times.

$$\text{That is, you press } 1000 \times \overbrace{MR \times MR \times MR \cdots \times MR}^{18 \text{ times}} =$$

STRATEGY 3: *Use a spreadsheet*

A spreadsheet (on a computer) can also be used. When you open the spreadsheet, the columns are marked with letters—A, B, C, and so on—and the rows are marked with numbers. In column A, we enter the numbers 0 through 18 (see Figure 6.9). In column B, we enter 1000 into the first row, signifying the amount we have at the beginning.

In the B2 cell, we now enter "=B1*1.06"; that is, we tell the computer to multiply the amount in the B1 cell by 1.06. Now we will see 1060 in the B2 cell.

Now, we highlight the B column from row 2 through row 19 and select "Fill down." This command essentially tells the computer to repeat the computation—that is, to multiply the previous amount by 1.06. After we do this, the computer will display the amount at the end of each year!

One advantage of the spreadsheet is that we can play the "what if" game: what if they had started with $3000, what if the interest rate had been 7%, and so on.

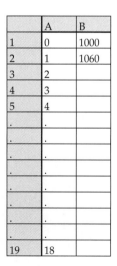

FIGURE 6.9

INVESTIGATION 6.19

How Much Does That Credit Card Cost You?

Most credit cards require a minimum monthly payment. If the customer is not able to pay the entire balance (for example, after a vacation or after Christmas), the customer pays at least this minimum amount; then a fee called a finance charge is determined on the basis of the interest rate the company is charging.

Let's say George found himself in just that situation. His VISA bill has come in, and the balance is $761.34. He decides he can pay $61.34. If the bank issuing the VISA card determines the finance charge at the annual rate of 18%, called the APR (annual percentage rate), what will his finance charge be? Think and then read on....

DISCUSSION

Because the balance is compounded monthly, the bank will determine George's finance charge by multiplying his unpaid balance ($700) by 0.015. Do you see where the 1.5% comes from?

To find the monthly rate, we divide the yearly rate by 12 (months): $18\% \div 12 = 1.5\%$, and $1.5\% = 0.015$.

Thus George's finance charge will be $700(0.015) = 10.50.

Summary

In this section, we began with the basic concept of percent, a part expressed as a hundredth, and then we connected this to the concepts of equivalent fractions and proportions. We found, however, that this simple notion of percent needs to be refined when we are dealing with percents greater than 100, with percent increase and decrease, and with nonroutine problems. This need for a richer concept of percent was also underscored in the explorations.

In examining percent problems, you saw applications of percents in several areas, including interest. You also saw the importance of estimating—sometimes because the question calls for an estimate instead of an exact answer, and sometimes because estimation helps you to check your solution or your reasoning.

EXERCISES 6.2

1. For each of the following questions, estimate first, and briefly explain how you determined your estimate. Then determine the actual answer. Compare your estimate to the actual answer. If your estimate was close, move on; if not, you might want to reread part of the section and/or consult with a friend or the instructor.

   a. What is 4% of 450?
   b. What is 23% of 85?
   c. What is 83% of $1460?
   d. What is 3.2% of 1700?
   e. What is 0.25% of 345?
   f. What is 120% of 200?
   g. 1200 is what percent of 1500?
   h. 30 is what percent of 35?
   i. 56 is what percent of 657?
   j. 1.2 is what percent of 4.6?
   k. 75 is what percent of 400?
   l. 175 is what percent of 120?

m. What is 3% more than 45?

n. What is 20% more than 400?

o. What is 16% more than 6.2?

p. What is 100% more than 35?

q. What is 10% less than 36?

r. What is 3% less than 45?

s. What is 1.2% less than 1200?

t. What is 80% less than 800?

In Exercises 2–15, first estimate the answer and briefly describe how you determined the estimate. Then determine the actual answer.

2. Missy Adams's gross pay is $1500 per paycheck. If her total payroll deductions are $340, her take-home pay is what percent of her gross pay?

3. All three second-grade teachers at Uphill Elementary School asked their students how many have pets at home. The response was that 23 out of 65 of the students have pets. What percent of the students have pets?

4. A victim won damages of $2.4 million from a company. If the lawyer's fee was 35%, how much did the lawyer get?

5. If 30% of the patients in a hospital have heart problems and there are 210 patients with heart problems, how many patients are there in the hospital?

6. Last year Mr. Rich paid $27,000 in income taxes because he was taxed at the rate of 38% of his gross income. What was his gross income?

7. **a.** Between 1990 and 2000, the price of an average house in a town increased from $42,000 to $97,000. What is the percent increase (rounded to the nearest whole number) in the average selling price?

   b. What is the percent increase for a house that sells for $60,000 and then later for $75,349?

   c. What is the percent increase for a house that sells for $71,450 and then later for $102,500?

   d. What is the percent increase for a house that sells for $62,000 and then later for $144,000?

8. Two dresses are on sale. The first was selling for $119 and is being marked down 40%. The second was selling for $79.99 and is being discounted by 20%. Which dress costs less now?

9. At birth a baby weighed 8 pounds 4 ounces. Two days later, the baby weighed 7 pounds 12 ounces. What percent of its weight had it lost? (This weight loss is normal—part of the adjustment to the living outside the mother's womb.) If a 160-pound adult lost the same proportion of its weight in 2 days, how much weight would the adult have lost?

10. *The Unofficial U.S. Census* reports that 34,177,500 of America's 139,500,000 cars are washed at least once each week and that 6,556,500 cars are never washed. Assuming that these numbers are relatively accurate:

    a. What percent of cars are washed each week?

    b. What percent of cars are never washed?

11. In 2003, about 290 million people lived in the United States, and the world population was about 6.32 billion. Approximately what percent of the world population lived in the United States?

12. In 2003, 17.6 percent of American children were living in poverty. If there were 73.3 million children, what number were living in poverty? Express the amount both as a decimal and as a whole number.

13. It has been estimated that 0.8% of Americans are homeless. Now 0.8% is a small percentage, but there are approximately 290 million people in the United States, and 290 million is a large number. Approximately how many Americans are homeless?

14. In 2000 only 3.4 million of the 11.9 million eligible voters aged 18 to 20 actually voted. What percent voted?

15. In an up-and-coming town, housing prices rose an average of 115% in five years. If a home cost $142,000 five years ago, how much would that home cost now?

Exercises 16–22 are nonroutine and/or multistep problems. If it is possible to make an estimate, do so and briefly explain how you obtained your estimate. If you feel that an estimate is not practical, describe what measures you took to ensure that your answer is reasonable.

16. A store reduced the price of a computer by 20% and sold it for $1760. How much did the computer originally sell for?

17. Janice Brady borrowed $4200 to buy a new car. If she pays the loan back in 1 year at 8.2% interest, how much will she actually pay for the car? (Assume simple interest.)

18. A store collected $53 in sales tax alone in one day. If the sales tax is 5.5%, how much did the store sell that day?

19. John paid $330 for a new mountain bicycle to sell in his shop. He wants to price it so that he can offer a 10% discount and still make 20% profit. At what price should he mark the bike?

20. A teacher was hired at a salary of $28,200 and is to receive a raise of 5.5% after her first year and a raise of 4% after her second year. What should her salary be after the 2-year period?

21. Your optimal exercise heart rate for cardiovascular benefits is calculated as follows: Subtract your age from 220 and then find 80% of the difference. Find the optimal heart rate for a 50-year-old and for a 20-year-old. The optimal rate for the 20-year-old is ___ % greater than the optimal rate for a 50-year-old.

22. A city budget is $21.43 million. The city council voted to increase the budget by 4.4%. What is the new budget, rounded to the nearest $1000?

23. In the 1980s, the company that built the Seabrook nuclear reactor in New Hampshire went bankrupt. When Northeast Utilities bought the company out, the agreement

was that Northeast could raise electricity rates at least 5.5% each year for the next 7 years.

a. If the Jones family paid an average of $83.00 per month for electricity last year and the rate rises 5.5% each year, what will they pay after 7 years?

b. If the Jones family's income is $40,000 this year and rises at a rate of 5.5% each year, what will be their income after 7 years?

24. The owners of High Hopes Apple Farm have a crop of approximately 1000 bushels of apples. They can have the apples picked by a regular labor crew for $2.50 per bushel and then sell the apples for $5.00 per bushel. Or they can let the public pick the apples at $4.00 per bushel. Which way will provide more income? One factor that they need to take into account is that when the public picks their own apples, the owners will lose about 20% of the crop from waste (eaten and dropped apples).

25. Let's say a map has a scale of 1:25,000. Two cities on the map are 3 inches apart. How many miles does this represent?

26. Arthur was not able to get his income taxes ready by April 15. The penalty for late returns is assessed at a 20% annual rate of interest. Arthur filed his income taxes on June 20 without having applied for an extension. He owed $2000 plus the penalty. How much did he owe?

27. Samantha Greene is on the negotiating team for the next teacher's contract. The school board is proposing a 6% increase for each of the next four years. The teachers' counterproposal is 12% this year and 4% for the next three years. Randi says "what's the difference; in either case, it's 24% over 4 years!" What would you say to Randi?

28. Several years ago, newspapers reported that the average unpaid balance on VISA cards was about $1500. Let's say that you are an average consumer, that your VISA card charges you at the rate of 18% per year, and that each month for 1 year, you have an unpaid balance of exactly $1500.

a. How much will your finance charges add up to for the year?

b. How much money would you save if you switched to a credit card that charged only an 8% annual interest rate?

29. Senior high school principals now average $82,225 a year. Their counterparts make $78,160 in junior high and $74,062 in elementary schools. The average U.S. teacher made $46,597 in 2004.

a. Compare the average salaries in percent-greater language; that is, compare all salaries to the $46,597 figure.

b. Compare the average salaries in percent-less language; that is, compare all salaries to the $82,225 figure.

30. Express 0.5% as a ratio in simplest terms. For example, 40% can be expressed as 40 out of 100, and in simplest terms as 2 out of 5.

31. Let's say that tests on a new vaccine show that less than 0.5% of all children have a serious reaction to the drug.

a. Gerry is not very mathematically literate. Help her to understand mathematically what this means.

b. What additional information would you like to have in order to help you decide whether or not to have your child take the vaccine?

32. The figures below are death rates for heart disease and cancer (per 100,000 people).

	1960	2000
Heart disease	559.0	257.5
Cancer	193.9	200.5

a. Describe the change in the death rates for heart disease and cancer since 1960, using percent language.

b. Compare the 2000 death rates for heart disease and cancer, using percent language.

33. Let's say you read that the rate of inflation in the United States last year was 5% and the rate of inflation in another country was 400%. Compare the two inflation rates in the way that will let the greatest number of readers relate to your statement.

34. A road sign says, "7% grade." What does that mean?

35. The U.S. Bureau of the Census reported that there were 35.9 million Americans living below the poverty level in 2004, 12.4% of the population. The U.S. Bureau of the Census also gives figures for persons living "below 125% of the poverty level." They reported that 49.7 million persons, or 17.1% of the population, were in this category.

a. If the poverty level for a nonfarm family of four in 2004 was $19,157, what would be the cutoff for a family of four "below 125% of the poverty level"?

b. Why do you think they came up with the reference point of "below 125% of the poverty level"?

36. First Bank Expressline made the following claim in an advertisement. The bank was offering 9.46% interest on the monthly unpaid balance. The chart below showed how much money a consumer with a $3500 balance would save by switching to First Bank's credit card. Is the ad valid?

	Current APR	Expressline savings
First Bank Expressline	9.46%	
Citibank	19.8%	$361.90
Discover	19.8%	$361.90
Sears	18.0%	$298.90
GM Card	16.4%	$242.90

Problems 37–42 require you to make some assumptions in order to determine an answer. Describe and justify the assumptions you make in determining your answer.

37. Let's say that you read in the newspaper that last year's rate of inflation was 7.2%.

    a. If your grocery bill averaged $325 per month last year, about how much would you expect your grocery bill to be this year?

    b. Let's say you received a $1200 raise, from $23,400 to $24,600. Did your raise keep you ahead of the game, or are you falling behind?

38. There was a proposal in New Hampshire in 1991 to reduce the definition of "drunk driving" from an alcohol blood content of 0.1 to 0.08. Explain why some might consider this a little drop and others might consider it a big drop. What do you think?

39. A recent news article stated that in 1986, women were paid only 64¢ for every dollar paid to men.

    a. Transform this statement to the following form: Men were paid ___% more than women.

    b. In 2004, the ratio had increased to 76¢ for every dollar paid to men. Describe the progress from 1986 numerically.

40. *Classroom Connection* Refer to Investigation 6.11. Jane still doesn't understand the problem. Roberto tries to help her make sense of the problem by saying that the 8% means that if we were to select 100 students at the college, 8 of them would be working full-time. What do you think?

41. Annie has just received a 5% raise from her current wage of $9.80 per hour.

    a. What is her new wage?

    b. What would this amount to over a year?

    c. What assumptions did you make in order to answer part (b)?

    d. What if the raise had been 5.4%?

42. Jack is building an office building. The local building code says that the window area in the building can be no more than 20% of the floor area.

    a. If the two-story building will contain 12 offices and have 3000 square feet of floor space, what is the maximum area of window area allowed?

    b. About how many windows would that be?

43. If Jonah puts $25,000 in the bank at 8% interest compounded quarterly (four times a year), how much will his investment be worth in 20 years?

44. If Liam puts $10,000 in the bank at 5% interest compounded quarterly (four times a year), how much will his investment be worth in 5 years?

45. How long will it take $1000 at 6% simple interest to double?

46. A company's sales increased from 2005 to 2006. It is possible to describe this increase either additively or multiplicatively. You will be asked to examine and then compare the two ways.

    a. We sold 34,234 more Bender Bobbers in 2006 than we did in 2005. What does that tell you?

    b. We sold 25% more Bender Bobbers in 2006 than we did in 2005. What does that tell you?

    c. If you were a stockholder in the company, which sentence would be more useful to you in the annual stockholder's report, the first sentence in part (a) or the first sentence in part (b)? Explain your choice.

47. Virtually all sunscreen lotions list the SPF (sun protection factor), which is an indication of how long you will be protected from sunburn when wearing the sunscreen. The amount of time you're protected is proportional to the SPF. If wearing SPF 8 sunscreen will protect your skin for 40 minutes, how long will SPF 30 sunscreen protect you?

48. One day a newspaper reported the following information, gathered from the National Restaurant Association, concerning the percentage of food budgets spent by the average person on eating out in different cities in the United States.

    a. Describe how the data might have been obtained.

    b. Explain why these data represent an example of pseudo-precision.

	Percentage of food dollars eating out
Miami	52.1%
Boston	49.5%
New York	48.9%
San Francisco	48%
Dallas/Ft. Worth	47.6%

49. *The Unofficial U.S. Census* determined the following data concerning the ages of cars on the road.

    a. First, estimate the percentage of cars that are 15 years old or older.

    b. Determine the "exact" percentage of such cars.

    c. How accurate do you think this number is? Justify your choice.

Age of car	Number on road
Less than 1 year	7,812,000
1–5 years	47,569,500
5–10 years	42,687,000
10–15 years	27,760,500
15 years or older	13,671,000

50. The July 20, 1996, issue of *America* had an article about the growing number of schools requiring school uniforms. The author describes a school where the amount of violence, especially between members of different gangs, was out of hand. A new principal tried various measures, "but the most effective change . . . was the one that cost the least, a school uniform policy that prescribes a white top—a white T-shirt, for instance—with black trousers or skirts. This innovation is said to have helped produce a 100 percent drop in violence at Farragut" (p. 20). Do you believe the number, or is there not enough information to determine whether the actual drop in violence is likely to be as reported? Why or why not?

51. The local gym, where I work out, has a ski-type exercise machine that saves wear and tear on my knees. The machine has different programs to choose from. The one I choose simulates going up and down hills. If I program the computer for 30 minutes, the screen looks something like the image that follows. Because there are 15 vertical bars, representing the elevation (and thus the difficulty) of the machine, I spend 2 minutes at each level. No matter how many minutes I select, there are 15 bars showing. Thus, if I select a shorter time period, the machine will move to the next level in a proportionately smaller time period.

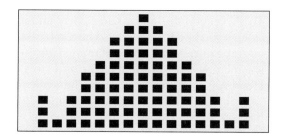

   a. One day I decide to exercise only for 20 minutes. How long do I spend at each height now?

   b. One day I decide to exercise for 45 minutes. How long do I spend at each height now?

52. During the Yankees' last weekend series, 46% of those attending Saturday's game were women, and 42% of those attending the double header on Sunday were women. Were there more women in the stadium on Saturday or on Sunday? Explain your answer.

53. Siena saw the following sign in an expensive clothing store: "Any dress in stock—take off 20% or $20 off." What would you advise Siena to do?

54. A manufacturing plant requires lengths of wire that are $3\frac{3}{4}$ inches long.

   a. If the wire comes in 48 inch coils, how many $3\frac{3}{4}$ inch segments can be made from one coil, and how much wire is wasted?

   b. What is the percent of waste?

   c. If the plant spent $40,000 each year on 48 inch coils of wire, how much of that money would be wasted?

   d. Determine a length of coil that would result in no waste.

   e. If your alternative lengths were 54 inches and 72 inches, would either of these result in less waste than purchasing 48 inch coils?

   f. Suppose we did this problem yesterday and Betty was absent. She solves the problem and says that 0.8 inch is wasted in each coil. Explain how you would convince her that 3 inches of wire is wasted.

55. Consider the following problem: x is what percent of y? One algorithm is

$$\frac{100x}{y}$$

Either explain the algorithm or use another perspective to provide a meaningful procedure for solving the problem.

CHAPTER SUMMARY

1. When we compare amounts, we can do so in additive or multiplicative ways. Both have value. In many cases, one is more appropriate.

2. Ratios, proportions, and percents involve multiplicative relationships between numbers.

3. Ratios are both like and unlike fractions. Ratios can be expressed in fraction notation, but many ratios involve part-to-part or whole-to-whole relationships. Not all such ratios can be translated into fraction language. Rates never can.

4. Seeing percent only as "what part of 100" is too limiting. For example, a 132% increase expresses a relationship between an original and a new amount.

BASIC CONCEPTS

Section 6.1 Ratio and Proportion

additive comparison 338
multiplicative comparison 338
ratio 339   rates 339
proportion 339   unit-rate method 343
factor-of-change method 342

Section 6.2 Percents
percent 357
converting among percents, decimals, and fractions 360
connections between percent and other mathematical topics 363
percent change 364
percents less than 1 or greater than 100 368
interest: principal, amount, rate, time 370

CHAPTER 6 REVIEW EXERCISES

1. In the fairy tale "Jack and the Beanstalk," the giant's tube of toothpaste holds one gallon. If the cost of Jack's toothpaste is $2.49 for 4.6 ounces, how much does the giant's tube cost if the prices are proportional?

2. An employee making $24,000 was given a raise of $1000. If all employees were given proportional raises, how much of a raise did an employee making $29,460 receive?

3. If the ratio of boys to girls in a class is 3 to 8, will the ratio of boys to girls stay the same, become greater, or become smaller if 2 boys and 2 girls are added to the class?

4. In a school with 560 students, the ratio of students to teachers is 20 to 1. How many new teachers need to be hired to reduce this ratio to 16 to 1?

5. On a map, $\frac{3}{4}$ inch represents 78 miles. How far apart are two cities in real life if they are $5\frac{3}{8}$ inches apart on the map? You can solve this problem by any means except setting up a proportion and solving for x.

6. The admissions department at a college figured that the ratio of females to males in the incoming freshman class is presently 7:3. If they have tentatively accepted 1000 students in the freshman class, how many more males would they need to accept to reduce the ratio of females to males to 2:1?

7. The following chart lists the U.S. birth rate (per 1000 total population) for the following years. However, there were more babies born in 2000 than in 1930. Explain how this is possible.

Year	Birth rate
1930	21.3
1940	19.4
1950	24.1
1960	23.7
1970	18.4
1980	15.9
1990	16.7
2000	14.1

8. When data are presented in either tables or graphs, sometimes we see raw numbers and sometimes the raw numbers have been translated into rates. Unfortunately, we have found that many people do not fully understand rates. Explain why we have rates, as though to someone who asks, "Why not just use the raw numbers?"

9. A United Way campaign set a goal of $1.6 million. If campaign workers have raised 85% of their goal, how much money do they still need to raise?

10. A mom makes a family favorite juice by mixing 6 cups of grape juice and 9 cups of orange juice. This weekend the family had a party, and mom decided to make a bigger batch. She combined 10 cups of grape juice and 13 cups of orange juice—that is, she increased the original recipe by 4 cups of each. However, it didn't taste quite the same. Explain why it didn't. How much orange juice *should* she have added? Explain your reasoning without using a proportion.

11. A city budget is $21.43 million. The city council voted to increase the budget by 4.4%. What is the new budget, rounded to the nearest $1000?

12. At birth a baby weighed 7 pounds 10 ounces. Three months later the baby weighed 12 pounds 12 ounces. What has been the percent gain in the baby's weight?

13. A pump can fill a pool in 4 hours. After 1 hour another pump is brought in. This pump, by itself, can fill the pool in 2 hours. How much longer will it take for both pumps, working together, to fill the pool?

14. In the *Guinness Book of World Records*, the greatest distance pushing a baby carriage in 24 hours was 342.25 miles. What, to the nearest second, was the average time per mile?

15. A victim won damages of $4.5 million from a company. If the lawyer's fee was 35%, how much did the lawyer get?

16. All the three second-grade teachers ask their students how many have pets. The result is that 43 out of 65 of the students have pets. What percent of the students have pets?

17. In 2002, 295.7 million of the world's 6.446 billion people lived in the United States. What percent of the world's people lived in the United States?

18. It is estimated that 2% of the students in a school have a hearing impairment. If there are 1500 students in the school, how many does this estimate suggest have a hearing impairment?

19. LaToya's gross pay is $1840 per paycheck. If total payroll deductions are $370, her take-home pay is what percent of her gross pay?

20. A certain item costing $60,000 appears in a $2.6 million budget. What percent of the total budget does this item represent?

21. The other night on my calculator I was doing one of the review problems, finding 5.5% of $156,250. My calculator said 859.375. I knew immediately that I had done something wrong. How did I know? What had I punched to get this answer?

22. An MP3 player was marked down 30% and sold for $87.95. What was its price before the markdown?

23. When is 20% off a better deal than $20 off?

24. Estimate the following and explain your reasoning. Estimate means that you get the answer, in your head, within 20–30 seconds.

   a. 16% of 450
   b. 123 is approximately what percent of 185?

25. In an up and coming town, housing prices rose an average of 115% in five years. If a home cost $142,000 five years ago, how much would that home cost now?

26. Fernando's salary in 1976 was $8400, and he bought a Honda Civic for $3200. His salary in 2006 was $59,000 and he bought a Honda Civic for $20,400. Has his salary kept pace with the price of cars?

27. The Addams family filled their oil tank just before a surge in oil prices. Their bill was $285 for 202 gallons. The Barnstables were not so lucky. They had to fill their oil tank in October, and their bill was $385 for 160 gallons. The price of oil went up by what percent between the time the Addams family bought oil and the time the Barnstables bought oil?

28. The pollution count in a bay was 72.0 parts per million. This was an increase of 9.2% over the previous year. What, to the nearest tenth of a part per million, was the pollution count in the previous year?

CHAPTER 7

Uncertainty: Data and Chance

7.1 The Process of Collecting and Analyzing Data
7.2 Going Beyond the Basics
7.3 Concepts Related to Chance
7.4 Counting and Chance

We have spent a considerable amount of time examining operations with numbers and different ways to represent amounts. We can add, subtract, multiply, and divide amounts; we can represent amounts and relationships between amounts with whole numbers, negative numbers, fractions, decimals, and percents. In this chapter, we will focus on another way in which we use numbers—to understand and deal with uncertainty.

In this world, nothing can be said to be certain except death and taxes.

BENJAMIN FRANKLIN

We expend a lot of energy trying to understand the uncertainty and chance in our lives. Let us consider a few examples of uncertainty and chance. What do the examples below have in common? Think and then read on. . . .

- Newspapers   Which candidate do you prefer?
  Are you for or against _____?
- Scientists   Are the levels of mercury in fish rising?
  How well are new AIDS drugs working?
- Educators   What is the best method for predicting success in college?
  How likely are our graduates to get a teaching job?
- Business   How much should we charge for this product?
  How many miles should we guarantee the tire for?
- Individuals   Which car should I buy?

As you can see from these examples, we collect data for a variety of reasons: to make predictions, to make decisions, to answer questions, and to better understand aspects of our life.

All of the examples above have two things in common: They all involve **uncertainty**, and they all involve collecting and analyzing data, which increases the chances that the decision or prediction will be a good one despite this uncertainty.

We often refer to that field of mathematics that focuses on collecting and interpreting data as *statistics* and to that branch of mathematics that focuses on determining the chance of something happening as *probability*.

380

However, as you may have discovered from the Chapter 7 Explorations, much of our interpretation of data involves chance, and much of our understanding of chance involves collecting and analyzing data. This is one of the reasons why I have chosen to have you examine these ideas within one chapter rather than in two separate chapters called "Statistics" and "Probability." Furthermore, just as you found that the words *carrying* and *borrowing* mask the essential connectedness of these two procedures, so, too, the terms *statistics* and *probability* tend to mask the essential connectedness of these two fields.

For example, when we say that the average annual snowfall in Buffalo, New York, is 92.2 inches,[1] that does not mean that you should expect 92.2 inches of snow in Buffalo this year (or 92 or 90 inches, for that matter). There are variations in any phenomenon; this number represents an average from data collected over 42 winters. We cannot say with certainty that we will get 92 inches of snow this year. We cannot even say with certainty that Buffalo will get between 82 and 102 inches this year. However, the average gives us an idea of what we might expect. In this sense, it tells us that when we look at the snowfall data for Buffalo, this amount is more common—it is in the middle of the data that have been collected.

Similarly, when we say that the chance of flipping a coin and getting tails is 50%, this does not mean that we will get 5 tails if we flip a coin 10 times. Even with this event, with less *variation* than snowfall in Buffalo, there is variation and a lack of certainty. I have flipped many coins (I was particularly fascinated by this and the dice probabilities as a youngster). I have occasionally gotten only 2 tails out of 10 throws, and it is statistically possible, though improbable, to flip a coin 10 times and have 0 tails.

In everyday life, *uncertain* has several connotations: not known or established, not determined, not having sure knowledge, and subject to change.[2] In one sense, our study of statistics and probability is meant to make our conclusions about phenomena that are not certain to occur more reliable and trustworthy, while at the same time we recognize the truth of Benjamin Franklin's statement.

Data Interpretation and Chance in Society

Understanding uncertainty is a crucial skill that all citizens need but few have. There is much evidence that many students understand the fundamental ideas of statistics and probability even less well than they understand other ideas in the elementary school classroom. Not surprisingly, misuse of statistics—by politicians, special interest groups, entrepreneurs, and others—is pervasive. Sometimes the misuse is deliberate—for example, to manipulate decisions and opinion. However, often the misinterpretation of data occurs either out of ignorance or because the problem is very complex.

At the same time, investigating situations involving uncertainty can be some of the most exciting and informative work that you (and the children you will teach) will do in mathematics classrooms. Toward this end, I find that students need to experience statistics and probability more directly, and so the

---

[1]*Information Please* (Boston: Houghton Mifflin, 1993), p. 392.
[2]*American Heritage Dictionary of the English Language*, 3d ed. (Boston: Houghton Mifflin, 1996), p. 1942.

investigations have been fashioned with that in mind. The NCTM also urges more hands-on work with all aspects of what we call statistical analysis:

> Teachers should have a variety of experiences in the collection, organization, representation, analysis, and interpretation of data. Key statistical concepts for all teachers include measures of central tendency, measures of variation (range, standard deviation, interquartile range, and outliers), and general distributions. Representations of data should include various types of graphs, including bar, line, circle, and pictographs as well as line plots, stem-and-leaf plots, box plots, histograms, and scatter plots.
> (From *NCTM Professional Standards*, p. 136, copyright 2006 by the National Council of Teachers of Mathematics.)

As you recall, the NCTM strongly emphasizes language in mathematics. Many students report being intimidated by the language of mathematics. I hope you are coming to realize in this book that mathematical language can actually make mathematical ideas easier to understand and grasp. For example, the terms *digit* and *place* make it easier to discuss numeration, computation algorithms, and estimation strategies. There are a number of words used in this chapter that are used in everyday life but that have a slightly different or more precise meaning in mathematics—for example, variation, distribution, *uncertainty*, *population*, *sampling*, *reliability*, *combination*, and *event*. To get too technical is to lose readers; to stay too informal is to cause confusion. Therefore, I have made an attempt to avoid both extremes.

SECTION 7.1 THE PROCESS OF COLLECTING AND ANALYZING DATA

WHAT DO YOU THINK?

- How can the way a set of data is represented affect the way it is interpreted?
- What do we mean by average?
- Why do mathematicians discourage reporting just the average when describing a population?
- Consider a graph that you have recently encountered. Was it accurate? How could you tell?

When we look at this thing we call statistics, we are talking about a process of moving from a question or a problem to gathering and analyzing data that will answer the question or solve the problem. There are four basic components in the process of using statistics to answer a question:

Four Basic Components

1. Formulating the Question
   The process begins with a question that can be answered by collecting data, for example, what are the chances that you will get a teaching job after graduating? Then we figure out how to change this question into a "statistical" question that is specific enough that we can collect useful data but yet does not so simplify the original question that the data aren't useful. We are creating data as much as we are collecting data.
2. Collecting the Data
   We need to decide what data we want to collect, and we need to design a plan for collecting appropriate data. That is, we want the data to be accurate and useful. There are many things we have to think about to ensure that we have accurate data. There are many methods used to collect data: observation, surveys, questionnaires, experiments, interviews, and simulations.
3. Representing and Analyzing the Data
   Raw data is simply a pile of numbers, like all the materials for a building project—nails, boards, shingles, etc. We need to organize the data and how

we organize the data depends on what we want to know. We can choose from many kinds of graphs and numerical methods, but there are no hard and fast rules for deciding which ones are most helpful with respect to our question. Each one is more useful for some situations and less useful, or even invalid, in other situations.

4. Interpreting and Presenting Our Results
   We interpret these graphs and numbers and turn them into conclusions. We determine what answers they gave us and what answers they didn't give us.

We will walk through the four components of this process with several different questions. Our first question involves categorical data, that is, data that are not numbers but categories.

INVESTIGATION 7.1

7.2

What Is Your Favorite Sport?

A common introductory activity with children is to have them formulate questions and then collect data with themselves as the *population*. Let's say this question has been presented: What is your favorite sport? What do we need to do to clarify this question so that when we ask everyone to answer the question, we will have useful data? Think and then read on. . . .

■ Language ■

The term **population** is used in statistics to refer to the complete set of people or things which we are studying. In Section 7.2, we will investigate how we can make conclusions about a population based on data from a *sample* of that population.

■ Outside the Classroom ■

For each investigation, at the side will be examples from outside the classroom. Some real-life questions with categorical data include:

- How do unemployment rates for people with less than a high school education, just high school, and college compare?
- How do infant mortality rates for different countries differ?

Refining the question Several questions emerge when we consider this question, for example, do you mean favorite sport to play or to watch? If we don't clarify this when we ask the question, people will not be answering the same question. Some will be saying their favorite sport to play and some will be saying their favorite sport to watch.

Let's say we decide to ask: What is your favorite sport to watch? This is still not 100% clear: Favorite to watch on TV or to watch in person? My favorite to watch in person is baseball, but my favorite to watch on TV is football. There are even more questions, what I call "yeah buts"; for example, "yeah, but what is a sport?" Is bowling a sport? Golf? Chess? Poker? In our clarification of the question, it is not so much what is right, but what do we want.

We also need to think about how to ask the question. For example, let's say we decide to ask about favorite sport to watch on TV. Do we want it open-ended, where people can give whatever sport they consider to be their favorite, or do we ask them to select among a list, for example, football, soccer, basketball, and baseball?

As you can see, we can easily get overwhelmed with the many complexities of a seemingly simple question. When this happens, it is helpful to go back to why you are asking the question. For example, if you are just curious as to what people might think of as sports, then you would make it as open-ended as possible. If you were asking this question just before the summer Olympics, you might ask: Which of the summer Olympic sports do you most like to watch? You might also consider having "none" as a legitimate response.

All this thinking in the first step of formulating the question!

Collecting the data With respect to collecting the data, we need to consider what additional data we want. For example, are we interested in investigating differences between boys and girls? If this were a high school

Outside the Classroom

An example of how important wording can be was a poll which concluded that 22% of American adults surveyed thought it was possible that "the Nazi extermination of Jews" never took place. These data were shocking to many people. However, when the pollsters were questioned, it was found that the question was worded this way: "Does it seem possible or does it seem impossible to you that the Nazi extermination of the Jews never happened?" The wording, with a double negative, certainly confused some people. Another poll was done where the question was asked: "Does it seem possible to you that the Nazi extermination of the Jews never happened, or do you feel certain that it happened?" In this poll, only 1% of Americans thought it was possible that the Holocaust never happened, while 8% were unsure.[3]

poll, would we want to see if there was a difference among freshmen, sophomores, juniors, and seniors?

We also need to think about the process. For example, do we write the questions down and have students write their answer? Do we want to ask them in person? As you might expect, asking in person might change the data. What if the asker is a popular person and a basketball player? What if you ask three kids at a table and the first two say football?

Representing the data   At some point, the question is written, the format determined, and the data collected. Let's say the question the college students asked was: What competitive, team sport do you most enjoy watching at your college?

A common and simple way to display the data is a **frequency table**. This is simply a table showing the number of times (called the *frequency*) that each category occurred.

Sport	Frequency
Football	5
Basketball	3
Soccer	4
Baseball	2
Rugby	1
Lacrosse	2

How might we represent these data? In this case, the two most common representations are a bar graph and a circle graph.

Bar graph   A **bar graph** is used to represent situations where the data are categories. The bar graph is pretty straightforward, though there are choices of order. We could arrange them randomly, alphabetically, or by popularity, depending on our preference. For example, we might put football, basketball, and baseball in order if we want to see how many of the class has one of the "big three" as their favorite sport.

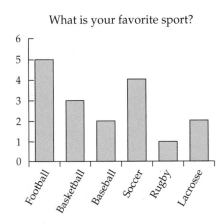

Figure 7.1

[3]*New York Times*, July 8, 1994.

> ### ■ Mathematics ■
>
> If you enter these data on a spreadsheet, and the football frequency (5) is in cell B2, then you could enter this equation in cell C2: = B2/17*100. That is, you are telling the computer to take the number in cell B2, divide it by 17 (to get the fraction) and then multiply it by 100, to get the percentage, which is 29. You can then click cell C2 and drag it down to have the computer automatically do the same for the rest of the data. If you have a calculator, you can divide 100 by 17 and then press M+. Then you don't have to do the entire computation each time. You can press 5 × MR = to get the percentage for football.

Circle graph

The **circle graph** is a bit more complex. We can turn the numbers into percents and then sketch a circle graph or we can turn them into degrees (fractions of a circle) and make the graph with a protractor. Can you convert the data into percentages and into parts of a circle (a whole circle is 360 degrees)? Try this before reading on.

	A	B Amount	C Percent	D Degrees
1				
2	Football	5	29	106
3	Basketball	3	18	64
4	Baseball	2	12	42
5	Soccer	4	23	85
6	Rugby	1	6	21
7	Lacrosse	2	12	42
			100%	360°

To convert to percentages, we first need to know the whole, which is 17 in this case. Thus, football = (5/17) × 100 = 29.41. Since we are sketching, we can easily round this to 29%.

To find the degrees, you multiply by 360. For example, the football slice is (5/17) × 360 = 106 degrees.

As with bar graphs, there are no hard and fast rules for ordering the slices. Again, it depends on what we want to know. If you are sketching the graphs by hand, there are many ways to do this. Let me walk you through one way. First, you can partition the circle into fourths. Thus each quarter circle is 25%. Then, depending on the nature of the data, you can divide each quarter into thirds (so each third is about 8%) or into fourths (so each fourth is about 6%).

We begin with the largest (29%), which is 25% plus 4%, so it makes sense to take the second quarter and break it into thirds (8%) and then divide the first of those thirds in half (4%) (see Figure 7.2a).

Basketball is next with 18%. We have the 4% slice left plus 8% plus 8% = 20%, so the basketball slice will be not quite the rest of this quarter circle. We can always check with our whole. That is, football plus basketball = 47%, so we need to be just under 1/2, and we are.

Baseball is 12%. In this case, rather than finding a 12% slice, I find it easier to find where 59% is, because that is how much of the circle I need after these three sports. Since 59% is 50% plus 9%, we can partition the third quarter circle into thirds (Figure 7.2.b). We have 50 plus 8, and so our line is just a "bit more" than this.

Next, we look at soccer and see that we will now have 82% of the circle: 75% plus 7. So it makes sense to partition that last 1/4 circle into fourths, 6, 6, 6, 6 (Figure 7.2.b). Soccer is thus about one slice past the 75% mark. We now have three 6% slices—6% for rugby and 12% for lacrosse.

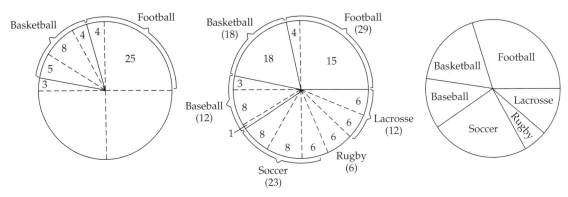

FIGURE 7.2

Stop for a moment and think about bar graphs and circle graphs. What are the advantages and disadvantages of each?

Most of my students find that bar graphs are easier to read than circle graphs. The bar graph tells you the number in each category (e.g., five people chose football as their favorite sport). However, you cannot tell immediately from a bar graph what percent each category represents, and this is one of the primary advantages of a circle graph. From the circle graph, we can quickly see that just over 1/4 of the people named football, and we can make statements about combinations of categories, for example, almost 1/2 of the people named football or basketball. One disadvantage of circle graphs is that they generally do not show the raw data, but rather the percentages, and so we lose the actual data. One cautionary note with circle graphs is that they are valid only when we have a whole for which it makes sense to find parts of, that is, percentages.

Analysis Now, we present our results. What we present depends on our intentions. We can simply present the frequency table and readers can see the number in each sport. The bar graph is a visual representation of the frequency table from which we can make various statements; for example, football and soccer were most popular. The circle graph enables us to make statements about what fraction or percentage of the whole each category or combinations of categories represent; for example, 59% chose the traditional "big 3" of football, basketball, and baseball. Or we could say almost 1/2 (41%) chose sports that haven't been around as long in the United States.

That's quite a bit of thinking just for one simple question! Let's do another question where the data are numbers.

INVESTIGATION 7.2

How Many Siblings Do You Have?

This is a commonly used question in elementary school classrooms to help children learn the process of statistics. In this case, the response is not a category but rather a number. How would you formulate this question? What are the aspects that need to be addressed so that everyone will be answering the same question; that is, what are the "yeah buts" for this question?

Section 7.1 / The Process of Collecting and Analyzing Data **387**

> ■ *Outside the Classroom* ■
>
> Some real-life examples like this investigation would be:
>
> - What is the size of the average American family? How is it different now from 50 years ago?
> - How many computers does the average family have?
> - What is the average family income?

Refining the question There are many considerations. For example: What is a sibling? Does adopted count? What about half-, step-, or foster siblings? What if someone had a sibling who died? Do we count that? There are no right answers to these questions. It depends on what you want to find out. If you want to know how many kids live full-time in the house, then you would ask how many siblings live with you all the time. If you wanted to know the number of biological siblings, you would ask for full and half-siblings. If you want to know how many people the child considers to be his or her siblings, you would ask: how many siblings do you have—people that you consider to be your brothers and/or sisters?

Collecting the data The data collection process for this question is fairly straightforward. However, we could expand the question to ask: How many siblings does your father have? Your mother? Of course, that could easily become complicated: what if the parents are divorced? What if the child feels closer to her step-father than to her biological father? But, as the saying goes, welcome to my world! *This* is the kind of complexity that is involved in virtually every statistical question that is investigated. Thus, one of the "big ideas" of this unit is for you to see statistics not as cut and dried, black and white, but as complex and inexact.

Representing the data Let's say we collected the data and here are our results.

0 1 4 0 1 2 3 1 7 3 1 3 1 2 1 0 12 1 2

A first step is to organize the data into a frequency table:

Number of siblings	Frequency (how many)
0	3
1	6
2	4
3	2
4	2
7	1
12	1

How might we represent these data, graphically and numerically? Make your own graphs and computations and interpretations before reading on. Really! You learn more by being more active readers.

> ■ *Language* ■
>
> Some books use the term **dot plot** because they use dots or small circles instead of X's.

DISCUSSION

Line plot One plot that is introduced in grade school is the **line plot**. A line plot for our data is shown below. What do you "see" from this graph? That is, what does it tell us with respect to our question that the raw data do not?

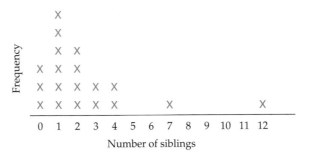

FIGURE 7.3

There are many ways to verbalize what we can interpret from the line plot. The **data** range from 0 to 12. We can see a **cluster** of data from 0 to 2. This could be verbalized as "most of the kids have 2 or fewer siblings." We see some **gaps** in the data, and we see some data that lie outside the rest. These are often called **outliers**. We could make quantitative statements about these data, for example, 68% of the students have between 0 and 2 siblings. We could represent this in fractional form also: almost 2/3 of the children have 0, 1, or 2 siblings.

We can make a **bar graph** from these data. Look at the bar graph below.

> ■ *Mathematics* ■
>
> What graph that we have studied most resembles the bar graph?
>
> Yes, the line plot. In one sense, the bar graph is simply a line plot where we enclose the area taken up by the individual *o*'s or *x*'s with a rectangle, that is, a bar.

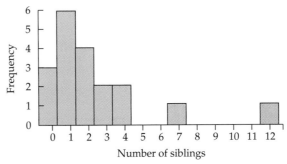

FIGURE 7.4

In the previous investigation, each bar represented the frequency of each category (that is, each sport). In this case, each bar represents the frequency of each number of siblings.

When some students make a bar graph from these data, their bar graph looks like the one below. Is this OK? Do you think both the first bar graph and the one below are valid, that is, it's a matter of preference which one we use? Or do you think only one is valid? If so, why?

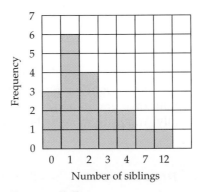

FIGURE 7.5

The first graph is valid and the second one is not. Do you see why? The second one is not considered valid because it hides or masks the *distribution* of the data. That is, there is a gap between 4 and 7 siblings and a gap between 7 and 12.

Another aspect of the bar graph that is confusing for many children and for some of my students is the *y*-axis which is labeled "Frequency." What does that mean?

It tells the frequency of each amount. That is, 3 students have 0 siblings, 6 students have 1 sibling, etc.

We could also make a **circle graph** from these data. Below are two circle graphs. One is in order of how many siblings. The other is in the order of greatest to least. Which do you prefer? Why?

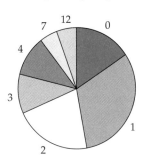
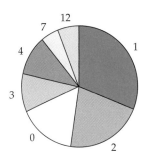

FIGURE 7.6

I prefer the former, because it enables me to quickly see other questions I might ask: How many children have 0 or 1 sibling? How many have more than 2 siblings? It is much easier to answer these kinds of questions if the slices of the circle are in numerical order.

Measures of Central Tendency

With numerical data (compared to categorical data in Investigation 7.1) we can examine *measures of central tendency* to see what they tell us about our question. You know the measures of central tendency as mean, median, and mode. We begin with how to find each:

Mean: Add each data value and divide by the number of data values.

Median: Arrange the data values in numerical order. The median is the middle data value. If there are an even number of data, then find the mean of the two closest to the middle. For example, if we have 3, 5, 6, 8, 12, and 15, there is no middle. Since 6 and 8 are closest to the middle, the median is 7.

Mode: The data value that occurs most often.

With our sibling data, we have

Mean = $(0 + 0 + 0 + 1 + 1 + 1 + 1 + 1 + 1 + 2 + 2 + 2 + 2 + 3 + 3 + 4 + 4 + 7 + 12)/19 = 2.47$
= $[(3 \times 0) + (6 \times 1) + (4 \times 2) + (2 \times 3) + (2 \times 4) + 7 + 12]/19 = 2.47$

Median = 2

Mode = 1

In this case, the mean is 2.47, the median is 2, and the mode is 1. So which one is correct?

Actually, they are all correct. The more useful question is "which one is better?" and the answer depends on what we are looking for. If we are looking to answer the question, "what is the most frequent number of siblings?" then the answer is the mode. If we are looking for the middle data value (half above and below), then our answer is the median. If we are looking for what is normally considered to be the "average," then the answer is the mean. As you will

CLASSROOM CONNECTION

Younger children, and even some adults, are bothered when they see the average not as a whole number. In the children's book, *The Phantom Tollbooth*, the average family has 2.58 children, and .58 of a child tells Milo that being a fraction of a person is actually nice: "Every average family has 2.58 children, so I always have someone to play with" (Norton Juster, *The Phantom Tollbooth* (New York: Random House, 1961), p.196).

discover as we consider more examples and in Exploration 7.2, each one of these numbers has advantages and disadvantages.

Let's stop for a moment to consider the question: why do we find the average in the first place?

If you are moving to a new town, you probably want to know the average price of a new home. When you are looking for a job, you might want to know the average starting salary of teachers in the state. In general, the average gives you a sense of the area where most of the data will lie, and it is generally close to the center of the data, which is why these three terms are called **measures of central tendency**. That is, if you look at the neighborhood in which the mean, median, or mode lie, that will generally be where a majority of the data values are found.

Conclusions So what have we discovered about the number of siblings in this group of people? Below are three of many conclusions we could draw from this set of data:

> The average number of siblings is either 1, 2, or 2.47, depending on which center we pick.
> More than half the class has 2 or fewer siblings.
> Two students come from much larger families than the rest of the class.

Deepening Our Understanding of Measures of Central Tendency

While most students have heard of mean, median, and mode before this course, "measures of central tendency" is a new concept. This concept is poorly understood and often misused in everyday life and is therefore worth more consideration. Think back on this investigation (and Explorations if you did them) and take a moment to note your responses to the following questions and then read on. . . .

> What does a measure of central tendency tell us about a set of data?
> Why do we determine one of these centers in the first place?
> What does it not tell us about a set of data?

Let me use an analogy to help answer these questions. If you saw a snapshot of my classroom, it would give you *some* information about my class. For example, you could determine the number of students, and you would see that the students are not sitting in rows. You might see me standing at the front of the room and you might see a computer projection on the screen. Similarly, the mean, median, or mode gives us a snapshot of a set of data.

Whichever one is selected is often interpreted as representative of the data or typical of the data. You hear this on television or read it in the newspaper: the typical teenager . . . ; the typical American family. . . . However, there is danger in this characterization. Do you see why?

Let us return to the picture of the classroom to help answer this question. There is much that the snapshot *does not tell* about my classroom, for example, what percent of the time I stand in front of the room or what tools I use (e.g., overhead projector, computer, etc.). Similarly, the mean, median, and mode do not tell us about the *distribution* of the data (the overall shape, clusters, gaps, outliers, the range).

Just as the snapshot of my classroom is an incomplete report of my class, so, too, the mean, median, and mode are an incomplete report of the data. Even worse is if someone generalizes from the picture, for example, concluding that I am one of those teachers who stands in front of the classroom all the time or that I use a computer most of the time. Similarly, as we will see as we examine

> **■ Mathematics ■**
>
> I want to underscore the term *measure* in talking about average. We tend to think of measuring in the sense of weight, distance, and volume. However, when we determine mean, median, and mode, we are essentially giving one measure of the set of data. There are many measures of a room: floor area, perimeter, volume, surface area. So, too, there are many measures of a set of data.

more cases, a perosn who is not statistically literate can make erroneous generalization from just seeing the mean, median, or mode.

To summarize: Measures of central tendency are simply one of many parts of an analysis of data. At best, they present an incomplete picture. At worst, they can lead to an erroneous sense of the set of data. Thus, a responsible report on a set of data will not give just the mean, median, or mode, but rather more information.

In some cases, the mean, median, and mode are almost identical, and in some cases not. We will examine the implications for that later. For now, let us consider situations where one is more useful than the others.

In some cases, we want to know the mean, which we find by taking the sum of the data and dividing by the number of data. If you took five tests in a course, your instructor would generally determine your average score by using the mean—adding up the scores and dividing by 5. The 2003 *Time Almanac*[4] reports that the mean number of people in U.S. households in 2000 was 2.62.

In some cases, we want to know the median, which we find by determining the numerical middle of the data. For example, if you determined the height of all the students in your class and ordered the numbers from smallest to greatest, the number in the middle would be the median. If the number of students in the class is an even number, then there is no middle number. In that case, you "split the difference" of the two numbers closest to the middle; technically, you take the mean of those two numbers. The *Statistical Abstract of the United States 2002*[5] reports that the median age in the United States in 1991 was 35.3 years.

In some cases, we want to know the mode, which is the datum that appears most often. The mode is often used when the characteristic we are studying is not a number. For example, if you were to collect data on what state the students in your class were born in, the mode would be the state that occurred most frequently. We often use the word *typical* with mode. For instance, the typical professional basketball player is over 6 feet. If you did Exploration 7.2, you defined *typical* for a specific population: you and your classmates in this course. When a newspaper reports that more women in the U.S. workforce were employed in administrative support than in any other category, it is saying that when we examine the population of working women and look at the characteristic called "type of job," "administrative support" is the mode.

One of the reasons for determining the average of a set of data is that one number or one phrase can give a quick summary of the data. The mean, median, and mode are all candidates to be considered as a representative of the data. In some cases, the mean, median, and mode are very close, but sometimes they are not. Let me connect this notation of representative of a population to a nonmathematical situation. When I was in the Peace Corps in Nepal, the country director told new volunteers that we needed to think about our behavior because we were representatives of the United States. In one sense, the director was referring to average. Because the people we worked with had had little or no previous contact with Americans, they would probably see whatever we did as typical. If we wore blue jeans or drank alcohol, many people would assume that all (or at least most) Americans wear blue jeans and drink alcohol. This is related directly to the notion of sampling—that is, making generalizations about a whole population on the basis of data from a sample of the population.

The concept of "mean" is frustrating for a college teacher, because so many students enter the course believing that *mean* and *average* are the same thing. This concept falls in the "rubber band family of learnings." Imagine the student as a rubber band. (Creativity and humor are important parts of teaching!) The

---

[4]*The Time Almanac 2003* (Boston: Information Please, 2002), p. 343.
[5]*A Statistical Abstract of the United States 2002* (Washington, DC: U.S. Department of Commerce, Bureau of the Census, 2001), p. 18.

professor teaching the new idea is stretching the student's understanding. However, researchers have found that in all too many cases, several months after the course the student's understanding is like the rubber band—it snaps back to its initial state. I discussed this earlier as the difference between rented and owned knowledge. The following investigation is one I first discovered in a methods textbook for elementary teachers; I have since seen variations of it in many places, including elementary school textbooks. My students enjoy it because it is fun and because they can quickly see that they can use it with their students also.

INVESTIGATION 7.3

Going Beyond a Computational Sense of Average

Almost all of you entered the course knowing how to determine mean, median, and mode. Many of you knew what median and mode meant. However, it is very likely that few of you could explain what the mean means. That is, you are doing a computation without knowing what it means, and that is not a great idea for future teachers! This investigation is designed to help you come to an understanding of the meaning of the mean.

Imagine that five elementary school children were asked how many movies they saw in the past year, and they responded: 7, 2, 9, 8, and 4. First, write down what you think the mean tells you about a set of data. . . .

Below is one physical representation of the data, using pennies to represent each movie seen. At the right is a bar graph where the bars are horizontal.

Do not compute the mean. Rather, draw a vertical line across the standard bar graph where, on the basis of your current sense of what the mean is, you "feel" the mean will be. Now, I want you to get some pennies and make the "graph" shown in Figure 7.7. If you don't have pennies, other coins will do. If you don't have coins, find a bunch of other objects (such as paper clips), or tear a sheet of paper into a number of small pieces of roughly the same size.

Now move the pennies so that all the bars are the same length. What did you just learn about the mean? Now read on. . . .

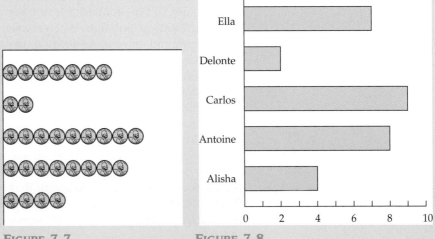

FIGURE 7.7   **FIGURE 7.8**

DISCUSSION

The mean can be viewed as the number you get when all the values are leveled off. In this case, if we "give" values from the larger amounts to the smaller amounts until all the amounts are the same, then the length of the bars and the number of pennies in each row are all the same. Before this course, most students' conception of the mean is purely procedural—you just do it. However, in order to have mathematical power, you also need conceptual understanding. If you have only procedural knowledge, you can do only problems "just like the ones in the book." The catch is that most real-life classrooms are never "just like the ones in the book." This one investigation does not give you a complete understanding of the mean. However, it can enhance your understanding of average and other data on population—an understanding that will enable you to interpret intelligently the graphs and data you find in newspapers and elsewhere and that will equip you to teach in such a way that your children will own what they learn.

Another Interpretation of Mean

The mean is also the center of gravity of a set of data; this can also be described as the *balance point* of the data. Think of a seesaw. If two people of the same weight sit the same distance from the center, it balances. If one person sits further from the center, the seesaw will not balance. If you imagine each of the children in this investigation sitting on a seesaw at the number corresponding to the number of movies they saw, the seesaw would balance at 6, the mean. That is, the two persons at 2 and 4 will balance the three persons at 7, 8, and 9 because the 2 and the 4 are further from the center than 7, 8, and 9.

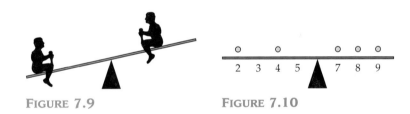

FIGURE 7.9   **FIGURE 7.10**

Let us now move on to another question and another set of data.

INVESTIGATION 7.4

How Many Peanuts Can You Hold in One Hand?

I have done this investigation with my students and with local elementary school children. You might do this also!

Refining the question   This question is pretty straightforward. However, the collection process is a bit messy!

Collecting the data   Most of the leaders in statistics education argue that one of the big ideas of statistics is **variation**. This has to do with the fact that most of the numbers we use when collecting and analyzing data are not the same. Sometimes they are not the same because of the **natural variation**

> ■ **Outside the Classroom** ■
>
> Similar, real-life questions include:
> - How many items (shoes, shirts, gloves) can a person make in one day?
> - How long will this battery last?

among individuals. Then there is variation that often happens when we take repeated measurements of an individual or an object. In this investigation, when a person grabs a handful of peanuts, they will not get the same number each time. When I first did this investigation, I did it five times and I got 32, 28, 22, 31, and 35 peanuts. This kind of variation is called **measurement variation**. We expect natural variation. However, when collecting data to answer a question, we want to minimize measurement variation. In this case, we have to think about the data collection process so that we minimize the measurement variation. Thus, we have to standardize the procedure for collecting the data.

How might you describe the procedure so that everyone is doing the same thing?

When we tried this, some people scooped the peanuts, that is, they reached in with palm up and open. Then they closed their hand slightly and slowly brought their hand up. Others reached in with their palms down and grabbed. I found that if I groped around, I could feel when I had about as many peanuts as possible. So we had to standardize the procedure. We decided on:

> Reach in with your palm down and grab as many peanuts as you can. You have to raise your hand from the bag within three seconds. Move your hand so that it is above the table. Whatever peanuts fall onto the table will be counted.

We also considered other "yeah buts" that could affect the reliability (consistency) of the data, for example, there were some empty shells and there were double shells and single shells. To reduce this variation, we could empty the peanuts on a table and then put into a bag only those peanuts that consisted of double shells.

Representing the data In this investigation, we will examine the data from my students. In Section 7.2, we will also examine the data from the third-graders in order to answer the question: How many more peanuts can an adult hold than a third-grader? Here are our data.

18, 18, 20, 22, 22, 22, 22, 22, 23, 25, 25, 25, 25, 25, 26, 26, 27, 27, 30, 30, 32, 32, 37

What do you see? What graphs might help us to understand the shape of the data—how the data values are distributed, clusters, gaps, outliers, range, etc.? What measures of central tendency are more useful?

As you are coming to see, the question is not which graphs and centers are right, but which are more useful. In the world beyond the classroom, people often "play with the data." That is, they try different representations to see what light they shed on the question being asked.

Let us consider a line plot as shown below. What does this representation tell us?

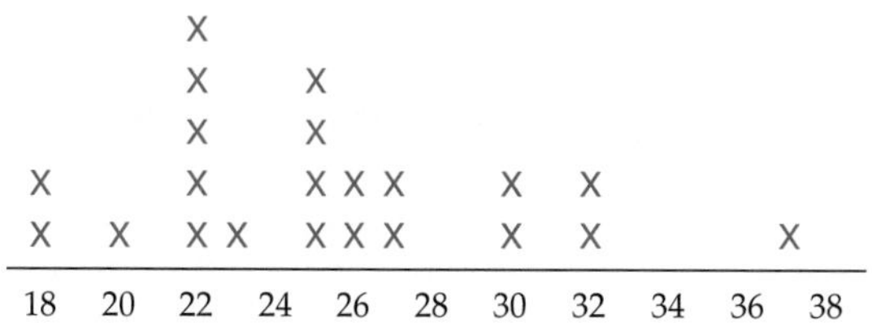

```
                  X
                  X    X
                  X    X
       X          X   X X X       X  X
       X   X  X X X X X X         X  X           X
       ─────────────────────────────────────────────
       18  20  22  24  26  28  30  32  34  36  38
```

FIGURE 7.11

From the line plot, we can make many statements:

The data range from 18 to 37 peanuts.

The data are pretty well spread out. That is, there is no primary cluster as there was in the previous set. About 1/2 of the students held between 22 and 26 peanuts.

The mean is 25.3. The median is 25. The mode is 22, well sort of. In this case, the mode is technically 22, but almost as many people held 25. So, 22 is not as "strong" a mode as 1 was for the number of siblings. We will discuss this idea more deeply in the next investigation.

The person that held 37 is a bit of an outlier in the sense that there is a fairly big gap between 37 and the next highest data value, 32.

Just as with the siblings data, we could make a bar graph. However, in this case, the data are more spread out. When the data are more spread out, in order to help us interpret the data better, it helps to put the data into intervals. In this case, we can make a grouped frequency table for the data. For example, we could do the following:

# of peanuts	# of people
15–19	2
20–24	7
25–29	8
30–34	4
35–39	1

■ Mathematics ■

Remember when we first discussed addition in Chapter 3 and we distinguished between numbers that represent discrete amounts (e.g., peanuts, trucks, etc.) and continuous amounts (e.g., inches, ounces, minutes)? In this case, the data (number of peanuts) are discrete amounts; that is, we can count them.

Histogram We can make a histogram from these data. A histogram is a bar graph where we have numerical data and where we eliminate the space between the bars. A histogram can be used to summarize numerical data that are on an interval scale, either discrete or continuous.

What do we gain from putting the data into intervals and then making a histogram?

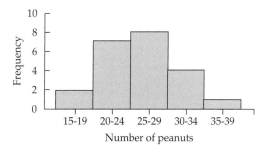

FIGURE 7.12

In this case, two conclusions we can make from the histogram are that the majority of the data are between 20 and 29 and that there are some below 20 and some above 29.

A question students often ask is: How did you know to put the data in these intervals (e.g., 15–19, 20–24, etc.). The answer is rather surprising to many students. There is no right answer. You have already seen throughout the first six chapters that mathematics is not as black and white as you thought. Statistics is the least black-and-white field of mathematics. When researching this book and talking with statisticians and mathematics educators in statistics, I encountered more cases of different opinions of what is "right" than in any other chapter.

In the case of intervals, the question is not "what is right?" but again "what is useful?" We generally group data using our base 10 system (e.g., 0–9, 10–19, 20–29, etc.). We could have done that with these data, but it wouldn't have been terribly useful:

# of peanuts	# of people (frequency)
10–19	2
20–29	15
30–39	6

Classroom Connection

In the *Explorations,* you have several opportunities to gather your own data. Most of my students find it more interesting to gather real data than to analyze some made-up set of data. So too with children. *Teaching Children Mathematics,* the NCTM monthly journal for elementary teachers, and other publications have had many articles written by elementary teachers about students gathering their own data. On one occasion, fourth-graders had complained that the playground was inadequate and unsafe. They gathered and analyzed data, presented it to the School Board, and experienced great satisfaction when a new playground was built!

That is, we didn't need a histogram to see that the vast majority of cases were in the 20s. In many cases, we want to view our data with a *finer grain.* In this case, the finer grain consisted of making the intervals 5 instead of 10. However, we could have picked smaller intervals (e.g., 18–20, 21–23, 24–26). In general, as the interval size increases, the graph gives us less information about the data. We will investigate intervals more deeply in the next investigation.

Generally, but not always, the intervals are equal. The reason for this is that just as the second graph for siblings in Investigation 7.2 could mislead people, unequal intervals have the potential for misleading people.

Conclusions

So what have we learned about how many peanuts can a person hold in one hand?

We have learned that there is quite a bit of variation: The greatest value is virtually double the smallest value.

The majority of the numbers are clustered between 22 and 26, and both mean and median are in the mid twenties.

In real-life, the process of gathering and analyzing data generally answers some questions and invites new questions. For example, is there a relationship between the size of one's hand and how many peanuts one can hold? (We will investigate this question in the next section.) How do we measure the size of a hand? Did some people get fewer peanuts because they didn't care? What if each person did this 10 times; would they get the same or close to the same number each time? If there is too much variation in our data, we are less confident about being able to make general statements about the data.

INVESTIGATION 7.5

How Long Does It Take Students to Finish the Final Exam?

Let us investigate one more set of data. In this case, this was a question for which I decided to collect data. I know that a fair exam is one in which there are questions about everything that was addressed during the semester. However, a fair exam would be very long. Therefore, a final exam has a certain inherent degree of unfairness. Thus, having more questions on the final exam increases the chance that the exam will be fair. However, if an exam is too long, students can get stressed. So I decided to gather data on this question. I told my students that they could take as much time on the final as they wished, and I recorded how long each student took to take the exam. The data below are the times, in minutes, that the students took on the exam:

62, 76, 87, 89, 93, 95, 98, 99, 101, 103, 105, 108, 111, 112, 115, 115, 116, 116, 124, 124, 126, 126, 130, 132, 132, 134, 137, 139, 139, 144, 146, 148, 148, 154, 154, 156, 160

What analyses might you do of these data to advise me? Take some time to explore the data, using knowledge that you already have. Summarize what you learned and state your conclusions, and then read on. . . .

■ Mathematics ■

What kind of graph does the stem-and-leaf plot remind you of? Rotate the stem-and-leaf plot a quarter turn counter clockwise. Can you see the relationship between this and a grouped frequency histogram?

TABLE 7.1

6	2
7	6
8	7 9
9	3 5 8 9
10	1 3 5 8
11	1 2 5 5 6 6
12	4 4 6 6
13	0 2 2 4 7 9 9
14	4 6 8 8
15	4 4 6
16	0

DISCUSSION

Examining the spread of the data A line plot (Figure 7.13) gives us a sense of the distribution without losing any data. In this case, the line plot doesn't tell us much beyond what we knew already. The range is so great that patterns in the data are not apparent.

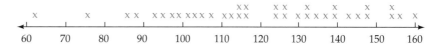

FIGURE 7.13

A **stem-and-leaf plot** helps us to organize the data (see Table 7.1). A stem-and-leaf plot (sometimes simply called a stem plot) display the values in rows. The numbers at the left are the stems and the numbers are the right are the leaves. Consider the two digits at the bottom of the stem-and-leaf plot: 6 and 2. The 6 is essentially a code that tells us that all the values on this row are in the 60s. Thus, the 2 next to the 6 represents a data value of 62.

As with the line plot, we don't lose any data (for example, we still know the minimum and maximum). In this case, the stem plot shows us that as we get closer to the middle of the data, the number of students is greater.

As we did with the peanuts data, by selecting an interval size, we can make a grouped frequency table for the data. The intervals are called classes. It is important to note again that there is no one "right" interval size. For example, we could choose an interval size of 10 minutes, in which case we have Table 7.2. This choice produces 11 classes. Alternatively, we could choose an interval size of 20 minutes, in which case we have Table 7.3, which gives us 6 classes.

TABLE 7.2	
Interval	Frequency
60–69	1
70–79	1
80–89	2
90–99	4
100–109	4
110–119	6
120–129	4
130–139	7
140–149	4
150–159	3
160–169	1

TABLE 7.3	
Interval	Frequency
60–79	2
80–99	6
100–119	10
120–139	11
140–159	7
160–179	1

From these data, we can make a histogram. Examine the two histograms in Figure 7.14. What do you see now that you didn't see before? What do they tell you about the data that is similar? What do they tell you about the data that is different? Then read on. . . .

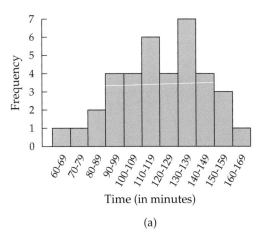

FIGURE 7.14

The histogram in Figure 7.14(a) indicates that the majority of the times are between 90 and 150 minutes, and it shows two peak intervals: 110–119 and 130–139. The histogram in Figure 7.14(b) indicates that a majority of the times lie between 100 and 139 minutes.

Note that with the second set of grouped frequencies, we could also make a circle graph for the data. This would more rapidly give a sense of what proportion of the class finished in each of the time intervals. Technically, we could do this with the first set, but a circle graph with 11 slices is a bit much.

Finding the center If you haven't already done so, estimate the median from the line plot or one of the histograms. Then read on. . . .

From the line plot (Figure 7.13), we can see that there are about as many data values above 120 minutes as below. From Figure 7.14(a), we can see that there are roughly as many data values above the 120–129 group as there are below. In fact, the median is 124 minutes. In this case, the mean is close, 120 minutes.

The strict interpretation of the mode is relatively meaningless in this case. There are several data values that occurred twice, but a frequency of only 2 in a set of 37 hardly makes a number a candidate for typical. Thus, when we make grouped frequency tables, we speak of a **modal class**—that is, the class that occurs most frequently. A look at Figure 7.14(a) reveals two intervals that stand out: the 110–119 interval and the 130–139 interval. Even though the 130–139 interval has one more data value than the other, both of them stand out from the other intervals, and so we can say that this set of data has two modal classes. In such a case, we say that the distribution is bimodal; that is, it has two modes.

We can draw several conclusions from the data and the graphs. When given more than two hours over half the class took the extra time. The "average" time for completion was about two hours, but it is ironic that although the mean and median are both very close to two hours, this is where the actual distribution dipped. I wondered whether that was just a coincidence or whether it was related to the traditional time limit. In any case, because over half the class took longer than two hours, I decided to make a shorter exam.

Measures of Center Revisited

It is crucial that you understand what centers tell us, what they don't tell us, and how they can be misleading. While centers can give us a snapshot of a population, they do not tell us anything about the variation in a set of data, about clusters, gaps, and the range.

While it is said that all people are created equal, it is not true that all three averages are equal. There is one property of the mean that bears explicit attention. The table below shows the mean, median, and mode for Investigations 7.2, 7.3, and 7.4.

7.2

	Siblings	Peanuts	Time
Mean	2.3	25.3	120
Median	1	25	124
Mode	1	22	120s

Mathematics

Here are two quotes about how the mean can be misleading!

Then there is the man who drowned crossing a stream with an average depth of six inches.
W.I.E. GATES

I abhor averages. I like the individual case. A man may have six meals one day and none the next, making an average of three meals per day, but that is not a good way to live.
LOUIS D. BRANDEIS

In the first case, the mean is significantly different from the median, and mode; in the other two cases, all three are similar. Exploration 7.2 addresses these issues in more depth. One significant way that the mean differs from the other two is that it is affected by extreme values. For example, if the child with 12 siblings had not been in the class, the mean would have been 1.8 instead of 2.3 (a 22% difference); however, the median and mode would still have been 1. Thus, when there is a population with extreme values, it can be misleading to give just the mean.

Before moving on, let us take some time to reflect on our understanding of measures of the center. For each of the three terms, describe the reasons for using it and its disadvantages. Then describe a real-life scenario in which that term would not be appropriate, and justify your choice. Then read on. . . .

Table 7.4 summarizes the main reasons for using each measure and some of the disadvantages of each.

TABLE 7.4

	Reasons for using	Disadvantage
Mean	It is often easier to compute with a large set of data. It is the center of gravity (or balance point) of the data.	It can give a distorted sense of the middle if there are extreme data (outliers). It might not make sense. For example, what does a mean shoe size of 6.3 mean?
Median	It is not affected by outliers. It is the midpoint of the data.	It is not always appropriate— for example, a baseball player's batting average.
Mode	It shows the most common datum. It is easy to determine from a graph.	It might be far from the center. There might be no modes or more than one mode.

Dispersion, Variation, and Distributions

Investigations 7.1 through 7.4 have focused on three tools that enable us to make statements about a set of data: graphs, measures of center (mean, median, and mode), and **measures of dispersion** (range, clusters, gaps, and outliers). Let us examine the ideas of variation and dispersion a little more deeply.

One of the themes of the course is the power of mathematics to help us make generalizations. When analyzing numerical data, the line plot or histogram will always have a shape, and there are several shapes that represent generalized cases of different patterns of variation in a set of data. Each of these general shapes has names and properties. We will explore them now.

There are many ways in which a set of data can be distributed. Another way of saying this is that graphs of populations can have many different kinds of shapes. In this course, we will focus on five **distributions**: uniform, skewed to the right, skewed to the left, bimodal, and normal. The graphs in Figure 7.15 represent idealized (smoothed) versions of these distributions. In real life the data are seldom so smooth. Can you think of a scenario for each of these graphs? Try to do so before reading on. . . .

Some students report having difficulty making sense of the preceding discussion. If this applies to you, the following may be helpful. The line graphs shown in Figure 7.15 can be thought of as evolving from bar graphs (with which most students report being more comfortable). For example, if we collected data on the number of siblings, we would have the bar graph shown at the left in Figure 7.16. If we made a line graph from those data, we would have the line graph shown at the right in Figure 7.16. If you look at the idealized graph, you can see that we can say these data are skewed to the right. That is, this graph has the characteristics of a "skewed to the right" distribution.

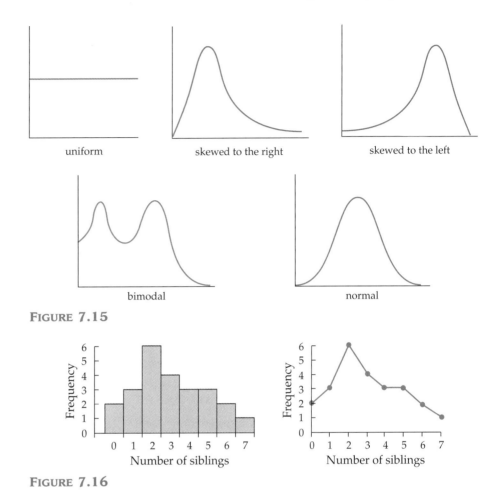

FIGURE 7.15

FIGURE 7.16

Table 7.5 gives one example for each of the five distributions.

TABLE 7.5

Distribution	x-axis (independent variable)	y-axis (dependent variable)
Uniform	Months of the year	Temperature in Hawaii
Skewed (right)	Salaries in a factory	Frequency
Skewed (left)	Class sizes in a high school	Frequency
Bimodal	Time to finish an exam	Frequency
Normal	Heights of people	Frequency

Take a minute or two to make sure you understand each of these examples. Otherwise, it's the classic "in one ear and out the other" story. I mention in the Instructor's Manual for this course that the NCTM process standards should permeate every class. Making sense of these distributions involves the standards Connections and Communication. In order to make sure that you understand these distributions, I am giving you examples to help you make connections. In order to make a better connection, you need to talk to yourself—to describe, in words of your own choosing, why each of these examples matches the distribution shown in the table. This is also what we call active learning, as opposed to passively reading the text and moving on.

For example, consider the row where "Salaries in a factory" is used to illustrate the distribution "Skewed (right)." Do you see that the graphical representation of this situation will be skewed to the right? Visualize a factory, who works there, what kinds of jobs are involved, and the like before reading on. . . .

If the shape of the graph of "Salaries in a factory" is skewed to the right, that means that the frequency of salaries will peak to the left of the middle, and the graph will slope more sharply to the left than to the right. In other words, there will be people much farther to the right of the center (making much higher salaries) than to the left of the center. From another perspective, the peak of this graph is not in the exact middle of the highest and lowest salaries but is closer to the lowest.

We can make some generalizations about using these terms to describe the center of a set of data:

- If the distribution of the data is skewed, the median will often be more representative than the mean. Do you see why?
- If the data are categories rather than numbers (for example, favorite TV show versus age), there is no mean or median, and so the mode is used to convey the center of the data—that is, the most commonly occurring datum. We might say, for example, that the typical American family eats hot dogs on the Fourth of July; this statement indicates that it has been determined that more families eat hot dogs on the Fourth of July than any other food.
- If the distribution is symmetric (for instance, normal) the mean, median, and mode will be close to one another.

Variation comes to play in another way with respect to distributions. Both of the graphs below represent data that are normally distributed. In the former case, the variation is small; in the latter case, the variation is large.

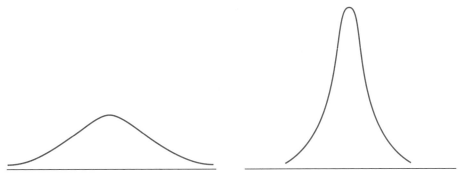

FIGURE 7.17

Summary

Now that we have examined the data collection and analysis cycle for five sets of data, let us reflect on what we have learned.

We have learned that we collect and analyze data for a variety of reasons: to help us make a decision, to help us make predictions, and to help us to understand situations.

We have learned that when we are collecting data, we often want a number that gives us a sense of that region where most of the data are likely to lie.

We have learned that there are different candidates for center, and each one has its pluses and minuses. Which one we select, or whether we report all three, depends on what we want to know about the population on which we are collecting data.

We have learned that any of the centers gives us only a partial sense of the population.

One chapter in a book called *Benchmarks for Scientific Literacy* focuses on the kind of mathematical knowledge especially useful in scientific inquiry. Below is an excerpt from the description of the students' need to be able to interpret data:

> One common misunderstanding is the belief that averages are always highly representative of a population; little or no attention is given to the range of variation around averages. . . . Because there is a persistent misconception, even in adults, that means are good representations of whole groups, it is especially important to draw students' attention to the additional questions, "what are the largest and smallest values?" and "how much do the data spread on both sides of the middle?"[6]

We have learned that variation exists everywhere. We have learned that there is naturally occurring variation—people prefer different sports, people have different size families, not everyone can pick the same number of peanuts, and not everyone finishes the exam in the same amount of time. There is also measurement variation. When asking a question, we want to minimize the measurement variation. If we have not done a good job of minimizing this kind of variation, then our results are suspect.

We have learned that there are different ways to represent data—tables, line plots, bar graphs, circle graphs, and histograms. We can also make graphs from grouping the results into intervals or classes.

We have learned that every population has a shape when we make a line plot or histogram. There are many features in interpreting the shape of a set of data:

All sets of data have a smallest and a largest value, a minimum and a maximum. We can subtract the minimum from the maximum and get the range.

Most sets of data have one or more clusters and one or more gaps.
Some sets have outliers.

We even have names for certain kinds of shapes: normal, skewed and so on.

Exploring Data with Larger Numbers and Different Settings

Up to this point, we have investigated data where the numbers are all relatively small. If you look at a newspaper, you will find data and graphs with large numbers. Now, we will examine how to make and interpret graphs when the numbers are large.

Lynn Steen edited a book called *On the Shoulders of Giants: New Approaches to Numeracy*, a book written by mathematicians to help mathematics educators and others realize the vision of the NCTM standards. As we have seen, a key aspect of this vision is to get people to realize that mathematics is more than just numbers, formulas, and computations. The six chapter titles are especially relevant to this section: Pattern, Dimension, Quantity, Uncertainty, Shape, and Change. Much of the reason why people collect and analyze data is to deal with *uncertainty* and to understand and predict *changes* in populations; we look for *patterns* and trends in the data; the *shape* of the data tells us much, and certain *quantities* (mean, standard deviation) also tell us much about the data. These ideas are what this chapter deals with: uncertainty and change, shapes, patterns, and quantities.

In each of the investigations that follow, you will initially be presented with a set of data or a graph. You will be asked to record your initial impressions and

---

History

The name of the book edited by Steen comes from a quotation from Isaac Newton, whom most people think of as the guy who discovered gravity when the apple fell on his head. However, he is also one of the creators of calculus, which he developed to explain his observations in physics. Newton remarked, "If I have seen farther than others, it is because I have stood on the shoulders of giants."

Source: Reprinted with permission from *On the Shoulders of Giants: New Approaches to Numeracy.* Copyright © 1990 by the National Academy of Sciences. Courtesy of the National Academy Press, Washington, DC.

---

[6]American Association for the Advancement of Science, *Benchmarks for Scientific Literacy* (New York: Oxford University Press, 1993), pp. 226–227.

conclusions. You will also be asked to note questions you have about the data or the graph. We will discuss two kinds of questions: questions about aspects of the data or graph that you don't understand and questions about the reliability and validity of the data—two issues that also come up repeatedly in this chapter's explorations. Questions about *reliability* ask whether two people collecting the data would get the same numbers. Questions about *validity* ask whether the methods used to collect the data are sound. Addressing all the questions that people have is beyond the scope of this book; therefore, the discussion of the data and graphs will focus on the questions that seem most important. Another reason for emphasizing questions is to encourage you to develop a critical mind when examining any data or graph.[7]

INVESTIGATION 7.6

Videocassette Recorders

Although videocassette recorders (VCRs) are almost everywhere now, they are a relatively recent invention. In this first investigation, we will examine two commonly used graphs and some of the problems involved in making them. Take a minute to examine the data in Table 7.6. First, describe the growth of VCRs, as mathematically as possible, as though you were talking to someone on the phone. Please be more precise than "Wow, they sure got popular fast." Second, describe any questions you have about the data. Then read on. . . .

TABLE 7.6

Year	U.S. households with VCRs ('000s)
1978	200
1979	400
1980	840
1981	1440
1982	2530
1983	4580
1984	8880
1985	17,600
1986	30,920
1987	42,560
1988	51,930
1989	58,400
1990	66,940

Source: The Universal Almanac 1992, copyright © 1992 by John W. Wright. Reprinted with permission of Andrews McMeel Universal. All rights reserved.

---

[7]Note that *critical* here does not refer to criticizing, but rather to being critical in the sense of "characterized by careful, exact evaluation and judgment" [*The American Heritage College Dictionary*, 3d ed. (Boston: Houghton Mifflin, 1996), p. 443]. Reproduced by permission. Copyright © 1996 by Houghton Mifflin Company. Reproduced by permission from *The American Heritage Dictionary of the English Language*, Third Edition.

DISCUSSION

At the most basic level, Table 7.6 shows that the number of U.S. households with VCRs increased every year. At a slightly more sophisticated level, using our knowledge of multiplication and estimation, we can say that the number of households with VCRs just about doubled every year until 1987.

Possible questions about this table include: What does ('000s) mean? What has happened since 1990? Because the table starts in 1978, does that mean that was the year in which VCRs were first sold? Who collected the data, and how did they get these numbers?

Many people wonder what ('000s) means. This is a convention that graph makers use when dealing with large numbers. There are two equivalent ways to "decode" this symbol: "Write three zeros after each number in the table to get the actual numbers" or "Each number is in the thousands; thus 200 means two hundred thousand." Recall our discussions about the choice of unit in Chapter 5. From this perspective, '000 means that the graph makers chose 1000 instead of 1 as the unit.

Graphing these data Now let us examine how a graph can help us to see the data better. What kind of graph do you think might best describe these data?[8] Make your own graph with graph paper first, before reading on. . . .

Look at the two graphs in Figure 7.18 and address the following questions: What do they tell us about the VCRs? Are both graphs "correct," or is one "better" than the other? Summarize the pros and cons of each graph. Then read on. . . .

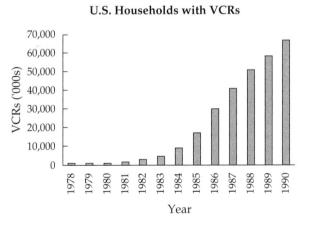 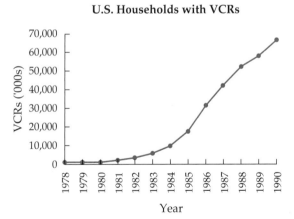

FIGURE 7.18

Both of these graphs are valid ways to represent these data. Many people prefer bar graphs to line graphs, finding the former easier to understand. The primary advantage of the line graph has to do with slope. You may recall from algebra that in a linear equation, the slope is constant, so a straight line indicates constant growth. However, when the slope keeps increasing, that indicates that

---

[8]The word *data* is the plural of the word *datum*. Therefore, it is correct to say, "Data *are* used" and "The data *support* the following conclusion."

the rate of growth is increasing, and we refer to such growth as exponential. The line graph in Figure 7.19 more clearly shows that the rate of increase started to slow down in 1987.

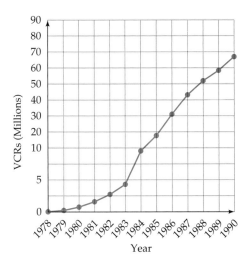

FIGURE 7.19

A common graphing mistake   A mistake that many students make when graphing is shown in Figure 7.19. Do you see why this graph is invalid? Think and then read on. . . .

The maker of the graph chose one unit for the bottom part of the graph and then chose another unit for the top part of the graph. That is, the student had two different vertical units on the same graph. For the first ten vertical intervals, the unit is 2.5 million; thereafter, the unit is 10 million. The rationale for the different scales is that the smaller numbers can be more accurately placed. However, it is not acceptable to change the unit (or scale) on a graph. Why do you think this is true?

Changing the unit conveys an invalid impression of the data. The graph above implies that the rate of increase slowed down after 1984, whereas it wasn't until 1987 that the rate of increase began to slow down.

On the other hand, there is no single "right" answer to what is the "best" unit. In general, the smaller the unit, the better we can see trends in the data. However, the smaller the unit, the bigger the graph.

Horizontal spacing   This issue of scale and unit applied to the horizontal spacing also. For example, one can choose to have more or less horizontal space between the years, as long as the spacing is constant—that is, each year is the same distance apart from the previous year. Although there are no rights and wrongs with respect to these decisions, it is important to note that different decisions about the choice of units cause the graph to look different. For example, the two graphs in Figure 7.20 represent the same data. In these two graphs, the vertical scales are identical. The difference is that in the graph at the right, the years are closer together. Do you see, though, why both graphs are considered correct?

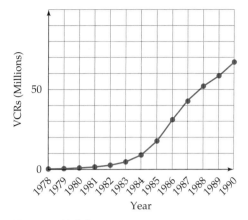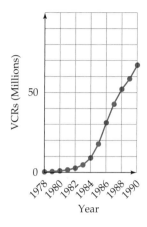

FIGURE 7.20

Although the two graphs appear very different, they are mathematically equivalent. For example, the number of VCRs in 1985 was double the number in 1984. In both graphs, the point representing 1985 is twice as high as the point representing 1984.

Missing years At this point a curious reader might be wondering what has happened since 1990. Since 1990, we don't have data for every year. We do have data for 1995 and 2000. However, before you look at the data, predict what you think they might be. This is called extrapolating—that is, predicting the future on the basis of current information and your interpretation of that information. . . .

The numbers for 1995 and 2000, respectively, are 79 and 86 million. If we include these data, what would the graph look like? Think about this before you read on. . . .

Figure 7.21 shows two commonly drawn graphs. What do you think about the graphs? Are they both valid or is only one valid?

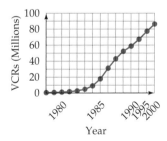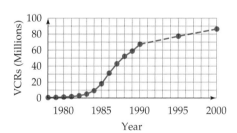

FIGURE 7.21

In some cases, as you have found already, two different methods can both be valid. However, in this case, the first graph is invalid. Do you see why?

It is invalid because the distance between 1990 and 1995 is the same as the distance between 1990 and 1989. However, 1990 and 1995 are 5 years apart, whereas 1989 and 1990 are only 1 year apart. The second graph shows visually what the numbers show—that the rate of increase in households with VCRs was slowing down. For example, from 1987 to 1988 the increase was about 9 million sets, then the next year there was an increase of about 6 million, and then there was an increase of about 9 million. Then, from 1990 to 1995 (5 years), the increase was only 12 million, or about 2.4 million per year. The second graph shows this slowing of the increase. I have put dotted lines between 1990 and 2000 because we do not know the numbers for each and every one of those years—there may have been spikes or valleys.

Finally, let us recall the discussion at the beginning of this chapter about uses of data. Other than satisfying curiosity, what use might these data and these graphs serve? Think and then read on. . . .

An individual deciding to invest in the stock market would probably be less interested in an area where growth was slowing down than in one where growth was picking up. Someone doing a report on the growth of the home entertainment industry might want to see how and when the VCR market grew.

Interpreting Graphs

In everyday life, we generally don't collect data and graph data as much as we interpret other people's graphs of data that they or yet *other* people have collected. The ability to interpret and to critique graphs is important. As we will find, graphs often distort the data. Sometimes this is intentional on the part of the people making the graph; other times it is unintentional, the result of carelessness or ignorance.

In Exploration 7.1, we critically examined data collected by other people and then constructed graphs. In the next three investigations, we will critically examine graphs made by other people. As you read each graph, I encourage you to think about four kinds of questions before you read the discussion. As always, if you write down your responses, you are likely to retain more from your work.

Conclusions

What conclusion(s) can I draw from the graph?
Do the conclusion(s) that I read seem reasonable?

Construction of the graph

Are the scales and the units clear or are they misleading?
Would another graph be more appropriate? Why or why not?

Reliability/validity

Do I have questions about how the data were obtained that could affect the accuracy of the data?

Further questions

Questions to help you better interpret or understand the data and graph.
Questions that this data set and graph provoke in you.
Let us begin with a graph that indicates hopeful news.

INVESTIGATION 7.7

Fatal Crashes

Examine the graph in Figure 7.22 and answer each of the four kinds of questions before reading on. . . .

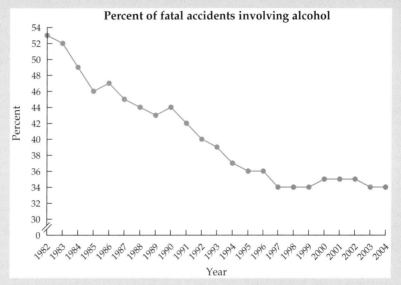

Source: NHTSA Fatality Analysis Reporting System (FARS), 2004.

FIGURE 7.22

Outside the Classroom

In the first edition of this textbook, I had data from a different source. The caption on the graph was "Percent of fatal accidents involving drunken drivers," and the data were from 1982 through 1990. When writing the second edition of the textbook, I could not find an updated set of these data, which had come from an article in *USA Today*. I searched the Web and, finally, I got the current data from the National Highway Traffic Safety Administration. However, the heading was "Percentage of traffic fatalities" and there were three columns: one in which no one involved in the crash had any alcohol in their bloodstream, one in which any person in the crash had a blood alcohol concentration between 0.01 and 0.07, and the third column (which I used) in which the highest blood alcohol concentration of any person in the crash was ≥ 0.08. In the discussion of these data, we are presuming that in virtually all of these cases, at least one of the people was drunk.

DISCUSSION

One student wrote the following: "The percent of fatal car crashes in which the driver was drunk fell dramatically between 1990 and 1997." What do you think of her summary? Think and then read on. . . .

Actually, there are several problems with the student's summary. First, the data don't seem to be restricted to car crashes. What kinds of "traffic fatalities" do you think count in these data? Think and then read on. . . .

Other kinds of traffic fatalities include crashes between two motor vehicles (trucks, cars, motorcycles) in which one or both drivers were drunk, single-motor-vehicle accidents in which the driver was drunk, and possibly accidents in which a motor vehicle hit a pedestrian or a person on a bicycle.

A second problem with the student's summary has to do with what "drunk" means. What do the graph makers mean by "drunk"? How did the people who recorded the data know that a driver was drunk? How were the data gathered? Think and then read on. . . .

It seems reasonable to expect that there are data for every motor vehicle accident in which there was a fatality. However, how did the people who recorded the data determine the number of such accidents in which at least one driver was drunk? Did a sobriety test or a blood test show that the person was drunk? Furthermore, the definition of *drunk* varies from state to state: In some states, a person with a blood alcohol level of 0.08% is considered drunk, whereas in other states the blood alcohol level has to be 0.10%.

Now let us examine the student's use of the word *dramatically* to describe the change in fatalities. Look back at the vertical axis of the graph—what does the jagged line just below 30% mean? Think and then read on. . . .

It means that this is a **truncated graph**—that is, the authors of this graph deleted the 0%–30% interval. Why might someone want to truncate a graph? Think and then read on. . . .

Sometimes graphs are truncated to save space. However, sometimes they are truncated to distort the data. Let us see how the graph would look if it had not been truncated (see Figure 7.23). How does the decline in the percentage of drunk drivers in fatal accidents look now? Think and then read on. . . .

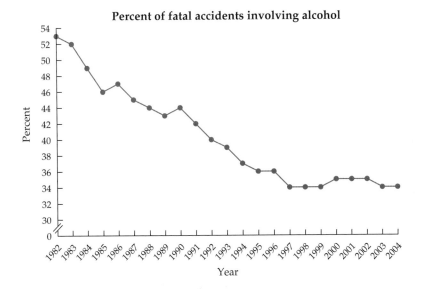

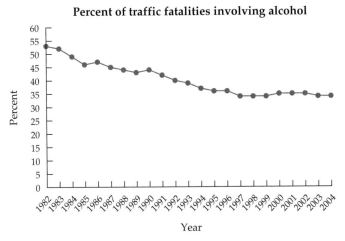

FIGURE 7.23

As you can see, the decline does not seem so great in the untruncated graph. The actual percent decrease from 53 to 34 is about a 36% decrease (that is, $\frac{53 - 43}{53} \approx 0.36 = 36\%$).

Which graph would you have picked if you were working for a beer company preparing an advertisement showing that drunk driving is on the decline? What if you were a member of SADD (Students Against Drunk Driving)?

What do the numbers mean? Let us examine two more aspects of this graph. In 2004, the percentage of traffic fatalities involving drunk drivers was about 36%. Write a sentence describing what that means, as though you were talking to someone who didn't understand. Then read on. . . .

Below are two of many possible responses.

- In 2004, 36% of all fatal motor vehicle accidents involved a drunk driver.
- In 2004, of every 100 fatal motor vehicle accidents, 36 involved a drunk driver.

Further questions Finally, let us examine additional questions that we might ask. What do you predict has happened since 2004? Do you think the percentage has leveled off or has continued to decline? Where could you go to find out? What other data would you like to see?

INVESTIGATION 7.8

Hitting the Books

Take a few moments to examine Figure 7.24. A critical aspect of this examination is asking yourself questions about conclusions, construction of the graph, and reliability/validity. Write your responses to these questions before reading on. . . .

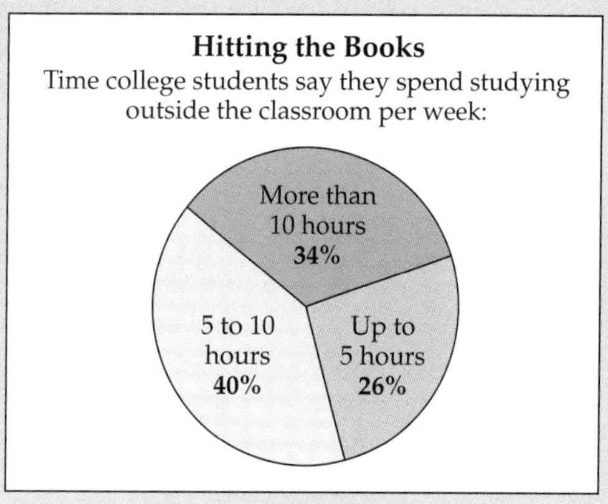

Hitting the Books
Time college students say they spend studying outside the classroom per week:

- More than 10 hours: 34%
- 5 to 10 hours: 40%
- Up to 5 hours: 26%

FIGURE 7.24

DISCUSSION

Describing the graph Critique the following two statements, which represent conclusions that people have made from this graph.

- Barely one-third of college students study more than 10 hours per week.
- Most college students average less than 2 hours a day on homework.

The first conclusion is simply taken straight from the graph: "Barely 1/3" is consistent with 34%. The second statement represents a valid interpretation of the graph, because less than half of the students spend more than 10 hours a week.

What do the data mean?   Now let us look beyond the first impressions to examine what the data mean. If you could meet with the people who collected these data, what questions would you ask? Think and then read on. . . .

Below are three of many possible questions that we might ask. As you read these questions and consider your own questions, ask yourself whether the answers to these questions would make a difference. Then read on. . . .

- How did you get the data (take a survey outside the dining commons, have questionnaires filled out in some classes)?
- What kind of students were included (full-time, a "representative" sample)?
- What did you ask? For example, did you ask, "About how many hours did you study last week?" or "Last week, did you spend less than 5 hours studying, between 5 and 10 hours, or more than 10 hours?"

As you may have seen in Exploration 7.2 (Typical Person), the way in which we gather data can make a big difference in how the graph appears. For example, if the questions were asked of students coming out of the dining commons, that means that off-campus students were not asked. And if off-campus students tend to study a lot more or less than on-campus students, this could change the data quite a bit. Similarly, if the data included both part-time and full-time students, the number of hours reported would naturally go down. (Why?) Finally, how we ask the question has an influence on our data. You might want to replicate this study on your own campus and compare the results of asking the question in two different ways.

Choice of graph   The makers of this graph chose a circle graph. What other choices would be appropriate, or is the circle graph the "best" choice? What do you think? . . .

Here it is a matter of personal preference. Circle graphs work well when we are looking at parts of wholes. In this case, the whole represents all students, and the makers of the graph have divided the whole into three subsets. However, using a bar graph would not be wrong.

Summary

Let us reflect on what we have learned from examining data gathered by other people.

We have learned that we collect and analyze data for a variety of reasons: to help us make a decision, to help us make predictions, and to help us to understand situations.

We have learned that in some cases, which graph is made is a matter of preference and that each graph has certain characteristics that we might want to use:

- Bar graphs enable us to compare the quantities of the categories.
- Circle graphs enable us to see what part each category is of the whole.

- Line plots enable us to see all data values and aspects of the shape of the data—distribution, clusters, gaps, and outliers.
- Histograms also enable us to see aspects of the shape of the data—distribution and clusters.
- Stem-and-leaf graphs can be used to organize the data and are like histograms turned on one side.
- Line graphs enable us to see change over time.

We have learned that we need to take care that our graph does not distort the data. For example, we need to be careful when choosing the units for our spacing, both horizontally and vertically, in line graphs and bar graphs.

We have also learned that when interpreting others' graphs, we need to think about what the data mean and how the data were gathered, and we need to realize that some graphs are distorted, ether intentionally or out of ignorance or carelessness. At the same time, we have learned that although a truncated graph does indeed distort the data, it sometimes offers advantages. We have learned that whether we are making graphs from data collected by other people or interpreting graphs made by other people, it is important to examine the data and the graphs with respect to fairness, appropriateness, reasonableness, reliability, and validity. "Students should begin to develop a critical attitude toward information presented in the media and learn to ask relevant questions before making judgments based on that information."[9]

Finally, we have learned that statistics is different from the other topics that we have studied in that in many cases, there is not an exact or precise answer, and there is often not a right or best graph to make. Thus, it is even more important that the student have an understanding of the ideas and the whys as opposed to simply knowing how to make the various graphs and how to determine the mean, median, or mode. The following quote nicely sums this:

Do not put your faith in what statistics say until you have carefully considered what they do not say.—William W. Watt

[9]Center for Statistical Education and the American Statistical Association, *Teaching Statistics: Guidelines for Elementary Through High School* (Palo Alto, CA: Dale Seymour Publications), p. 23.

EXERCISES 7.1

1. Here are the distances (in minutes) that students live from school:

5	30	60
5	30	70
5	30	70
5	30	85
10	40	90
10	45	120
20	50	125
20	60	140
25	60	200

   a. Determine the mean, median, and mode.
   b. What do the mean, median, and mode not tell us?
   c. Make a grouped frequency table with an interval of 30 minutes for these data.
   d. Make a histogram from these data.
   e. Summarize (2–3 sentences) what the histogram tells us.
   f. Make a grouped frequency table with an interval of 20 minutes for these data.
   g. Make a histogram from these data.
   h. Summarize (2–3 sentences) what the histogram tells us.
   i. How are the two histograms alike and different?

j. Make a boxplot for these data.

k. Summarize what the boxplot tells us about the data.

2. Children from two third grades collected data on the number and kind of pets that they had at home. There were 46 students in the two classes.
Here are the data:

Fish	64
Cats	32
Dogs	28
Birds	12
Hamsters	4
Gerbils	2
Guinea pigs	1
Horses	2
Ferrets	3
Snakes	1

   a. Make a bar graph for these data.

   b. Determine the mean, median, and mode for these data.

   c. Since there are a total of 149 pets and 46 students, could we say that the average student has 3 pets?

   d. What questions would you like to ask the students in order to better understand the data?

3. My students collected data on how many raisins were in a small box of raisins.

107	107	111	113
117	118	119	119
120	120	120	121
121	121	121	121
121	122	122	122
122	123	124	124
125	125	126	127
127	128	128	128
129	131	132	135
138			

   a. Determine the mean, median, and mode.

   b. What do the mean, median, and mode not tell us?

   c. Make a line plot from these data.

   d. Summarize (2–3 sentences) what the line plot tells us.

   e. What does it not tell us?

   f. Make a grouped frequency table with an interval of 10 for these data.

   g. Make a histogram from these data.

   h. Summarize (2–3 sentences) what the histogram tells us.

   i. Make a grouped frequency table with an interval of 5 for these data.

   j. Make a histogram from these data.

   k. Summarize (2–3 sentences) what the histogram tells us.

   l. How are the two histograms alike and different?

   m. If you were giving a report that described how many raisins are in a box, which graphs and which numbers would you include? Why?

   n. Make a boxplot for these data.

   o. Summarize what the boxplot tells us about the data.

   p. How could there be so much variation in the number of raisins?

4. The teachers at Gates Memorial High School got to talking about cars and how long they keep them. Someone wondered about the average age of the teachers' cars. Following is a graph representing the ages, rounded to the nearest year.

   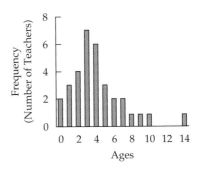

   a. Before computing, predict the mean and the median. Explain your reasoning.

   b. Determine the mean and the median.

5. When doing Exploration 7.2 (Typical Person), you found that you have to think about how to word a question. For example, if we simply ask, "How many brothers and sisters do you have?" there might be problems with half-siblings and step-siblings, adopted family members, and so on. Assume that you are in a group that has decided to ask a question about how much time people study. Describe the actual question you would ask and explain the thinking behind your phrasing. That is, convince the reader that the respondent will answer the question you think you are asking and that the data you get will be meaningful and analyzable (as opposed to problematic, as in the example cited above).

6. Another group in Exploration 7.4 asked, "How many drinks did you have in the past week?" Here are the data: 0, 0, 0, 0, 0, 0, 1, 2, 2, 3, 5, 5, 8, 12, 12, 15, 10–15. Imagine you are in that group and have been asked to present your conclusions to the class. In other words, what did you learn about the drinking habits? This includes making a graph—it can be a sketch, as opposed to a polished, precise graph. *Note*: You will

need to decide what to do with the 10–15 datum and to justify your decision.

7. In Exploration 7.2, a group in a previous class asked the following question: "What is your favorite sport?" Here are their data:

Basketball	5	Volleyball	4	Soccer	3
Tennis	3	Softball	3	None	2
Skiing	2	Rugby	2	Horseback	1
Gymnastics	1	Swimming	1	Cheerleading	1

   a. Select and make an appropriate graph.
   b. Justify your choice of graph.
   c. Critique the question. If you feel it is fine, explain why; if you feel it has problems, explain why and then suggest a rewording.
   d. Determine the mean, median, and mode. If any of these three is inappropriate, explain why.

8. Mr. Arnold asked his students how many brothers and sisters they had. The graph following summarizes what he found.
   a. Determine the mean and the median number of siblings.
   b. There is something wrong with the graph. Explain not only what is wrong with the graph but also why it is wrong.

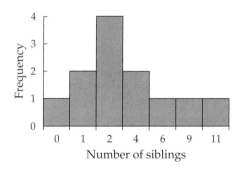

9. Willie has an average (mean) of 83.4 on the first five exams. What is the least he can score in order to have an average of 85 for all six exams?

10. Lucy took three tests. If her median score was 82, her mean score was 87, and the range was 17, what were her three test scores?

11. The mean of five numbers is 6. If one of the five numbers is removed, the mean becomes 7. What is the value of the number that was removed?

12. There is another way to figure the mean of a set of data. Here's how it works. Let's say we want to find the mean weight of these boxes of fruit: 26, 27, 38, 35, 34, 29. I estimate the average as 30 and then add or subtract the distance of each datum from my estimate. In this case, the arithmetic would be

$$-4 + -3 + 8 + 5 + 4 + -1 = 9$$

Next, $9/6 = 1.5$, and so the mean is 31.5.
   a. Explain why this procedure works.
   b. Make up another set of data and use this strategy to determine the mean.

13. Make up three questions such that in each case one of the centers (mean, median, or mode) would be most appropriate. In each example, all of the three must be able to be determined. ("What is the hair color of children at Riverside Elementary School?" is not acceptable because the mean and median cannot be determined.) Explain why the center you picked is, indeed, the most appropriate.

14. A student organization at Keene State College had an essay contest for Hunger and Homelessness Week one year. Students were asked to share their views on the following question: "The average age of a person experiencing homelessness is 9 years old. How does that make you feel and what would you do about it?"
   a. Which of the centers (mean, median, mode) do you think were used here?
   b. Explain how this statement could be true or argue that it is impossible.
   c. Research homelessness at the library or on the Web. Present the data that you find.

15. I recently read an article in which the following sentence appeared: "The average cost of a car is $19,000." How do you think that number was determined? In your answer, you need to state whether you think that figure represents the mean, median, or mode.

16. Why do you think many statisticians recommend that when describing a population, one should give at least three pieces of information: some sense of the "average," some sense of the spread, and some sense of how the data are distributed? For example, they would recommend not simply reporting that the average age of teachers in New Hampshire is 43 or that the average adult American goes to church 1.3 times a month (both figures are made up).

17. Nine people decide to share some candy in an unorthodox way. The most anyone has is 9 pieces and one person does have nine pieces. At least one person ends up with no candy (their choice). The average (mean) number of pieces is 4. Four is also the median. That means at least one person has 4 pieces. (Why?) More people have 2 pieces of candy than any other number.
   a. How many pieces of candy does each person have?
   b. Is there more than one solution? Explain why or why not.

18. This item is from a recent National Assessment of Educational Progress test (http://nces.ed.gov/nationsreportcard/pdf/demo_booklet/05SQ-grade12-part1.pdf). The table below shows the daily attendance at two movie theaters for 5 days and the mean (average) and the median attendance. Which statistic, the mean or the median,

would you use to describe the typical daily attendance for the 5 days at Theater A? Justify your choice.

	Theater A	Theater B
Day 1	100	72
Day 2	87	97
Day 3	90	70
Day 4	10	71
Day 5	91	100
Mean	75.6	82
Median	90	72

19. Imagine the exam scores of a class of 10 students. Make three very different distributions that all have a mean of 80.

20. Here are exams scores for a class of 20 students. 0, 40, 55, 65, 70, 70, 70, 75, 75, 75, 80, 80, 80, 80, 80, 85, 90, 90, 95, 100
    a. Determine the mean and median.
    b. Remove 2 quiz scores so the median remains the same.
    c. Remove 2 quiz scores so the median increases.
    d. Remove 2 quiz scores so the median decreases.
    e. Remove 2 quiz scores so the median stays the same and the mean decreases.

21. Create a set of 10 test scores for which the mean is higher than the median.
    a. Describe how the data need to be distributed in order for the mean to be higher than the median.
    b. Create a set of 10 test scores for which the mean is higher than the mode. Describe how the data need to be distributed in order for the mean to be higher than the mode.

22. Why is a bar graph less appropriate than a line graph for showing data that have changed over a period of time?

23. Describe how the valid changing of the vertical and/or horizontal scales of a graph can change the appearance of a graph.

24. When data are presented, either in tables or in graphs, sometimes we see raw numbers and sometimes the raw numbers have been translated into rates. Unfortunately, many people do not fully understand rates. Explain why we have rates, as though to someone who asks, "Why not just use the raw numbers?"

25. Find or make up a set of data for which:
    a. A bar graph would be an appropriate representation
    b. A circle graph would be an appropriate representation
    c. A line graph would be an appropriate representation

26. Thelma says that she assumes a circle graph is appropriate when the data are percentages. What do you think of Thelma's rule of thumb?

27. a. Make a line graph for the following data on U.S. cellular telephone subscribership.
    b. Describe what you learned or saw from the graph that you didn't see just from the data.
    c. Based on the graph predict the number for 2004.
    d. Write a 2–3 sentence summary that might appear in a newspaper article describing the growth of cell phone ownership.

Year	Number (000s)	Year	Number (000s)
1985	340	1995	33,786
1986	682	1996	44,043
1987	1231	1997	55,312
1988	2069	1998	69,209
1989	3509	1999	86,047
1990	5283	2000	109,478
1991	7557	2001	128,375
1992	11,033	2002	140,767
1993	16,009	2003	158,722

28. The table below shows data about daily use of cigarettes for U.S. eighth-, tenth-, and twelfth-graders.
    a. Make a line graph for these data.
    b. Describe what you learned or saw from the graph that you didn't see just from the data.
    c. Write a 2–3 sentence summary that might appear in a newspaper article describing changes in cigarette use among children.

	1997	1998	1999	2000	2001
8th	9.0	8.8	8.1	7.4	5.5
10th	18.0	15.8	15.9	14.0	12.2
12th	24.6	22.4	23.1	20.6	19.0

29. a. Make a line graph for the data shown in the table below.
    b. What did you learn or see from the graph that you didn't see just from the data?
    c. Write a 2–3 sentence summary for this graph that might appear in a newspaper article describing the changes in production of motor vehicles.
    d. Describe how the world production of motor vehicles changed during this time period, beyond saying simply that it has increased.

World Motor Vehicle Production, 1950–2001 (in thousands)					
Year	United States	W. Europe	Japan	World total	U.S. % of world total
1950	8,006	1,991	32	10,577	75.7
1960	7,905	6,837	482	16,488	47.9
1970	8,282	13,049	5,289	29,419	28.2
1980	8,010	15,496	11,043	38,565	20.8
1985	11,653	16,113	12,271	44,909	25.9
1990	9,783	18,886	13,487	48,554	20.1
1991	8,811	17,804	13,245	46,928	18.8
1992	9,729	17,628	12,499	48,088	20.2
1993	10,898	15,208	11,228	46,785	23.3
1994	12,263	16,195	10,554	49,500	24.8
1995	11,985	17,045	10,196	49,983	24.0
1996	11,799	17,550	10,346	51,332	23.0
1997	12,119	17,773	10,975	53,463	22.7
1998	12,047	16,332	10,050	53,841	22.4
1999	13,107	17,603	9,985	57,787	22.7
2000	12,832	17,678	10,145	59,704	21.5
2001	11,518	17,825	9,777	57,705	19.7
2002	12,328	17,419	10,240	59,587	20.7
2003	12,147	16,943	10,286	61,562	19.7
2004	12,021	16,982	10,512	64,388	18.7

Source: From *The World Almanac and Book of Facts,* 2006, p. 113. Copyright © World Almanac Education Group, Inc.

30. **a.** Make a line graph for the data shown in the table below.

b. What did you learn or see from the graph that you didn't see just from the data?

c. Write a 2–3 sentence summary for this graph that might appear in a newspaper article describing the increasing numbers of women in these four professions.

First Professional Degrees Earned in Selected Professions, 1970–2000												
Type of degree and sex of recipient	1970	1975	1980	1985	1990	1995	1997	1998	1999	2000	2001	2002
Medicine (M.D.):												
Degrees conferred, total	8,314	12,447	14,902	16,041	15,075	15,537	15,571	15,424	15,562	15,286	15,403	15,237
Percent to women	8.4	13.1	23.4	30.4	34.2	38.8	41.4	41.6	42.5	42.7	43.3	44.4
Dentistry (D.D.S. or D.M.D.):												
Degrees conferred, total	3,718	4,773	5,258	5,339	4,100	3,897	3,784	4,032	4,144	4,250	4,391	4,239
Percent to women	0.9	3.1	13.3	20.7	30.9	36.4	36.9	38.2	35.5	40.1	38.6	38.5
Law (LL.B. or J.D.):												
Degrees conferred, total	14,916	29,296	35,647	37,491	36,485	39,349	40,079	39,331	39,167	38,152	37,904	38,981
Percent to women	5.4	15.1	30.2	38.5	42.2	42.6	43.7	44.4	44.8	45.9	47.3	48.0
Theological (B.D., M.Div., M.H.L.):												
Degrees conferred, total	5,298	5,095	7,115	7,221	5,851	5,978	5,859	5,873	5,558	6,129	5,026	5,195
Percent to women	2.3	6.8	13.8	18.5	24.8	25.7	26.2	26.1	28.3	29.2	32.0	32.9

Source: Statistical Abstract of the U.S., 2006, p.189.

31. The table below shows the U.S. immigration rate by decade, from 1820 to 2000.

U.S. IMMIGRATION RATE BY DECADE, 1820–2000		
Period	Total number ('000s)	Rate per 1,000 U.S. pop.
1820–30	152	1.2
1831–40	599	3.9
1841–50	1,713	8.4
1851–60	2,598	9.3
1861–70	2,315	6.4
1871–80	2,812	6.2
1881–90	5,247	9.2
1891–1900	3,688	5.3
1901–10	8,795	10.4
1911–20	5,736	5.7
1921–30	4,107	3.5
1931–40	528	0.4
1941–50	1,035	0.7
1951–60	2,515	1.5
1961–70	3,322	1.7
1971–80	4,493	2.1
1981–90	7,338	2.9
1991–2000	9,095	3.2

Source: www.immigration.gov. and author's calculations.

a. Describe your first impressions from examining these data. Are there any surprises?

b. Make two separate line graphs, one using the total number and one using the rate per 1000 population. Describe the different impressions conveyed by the two graphs. Which graph do you think gives a "better" comparison of immigration to the United States? Why?

c. What did you learn about mathematics from this exercise?

d. What did you learn about immigration from this exercise?

32. The table following shows the number of people for whom the language cited is their first language. The first set of numbers were estimates as of 2000 from the 2002 World Almanac, and the second set of numbers are from the 2005 World Almanac.

a. Make a bar graph for the 2005 data including a title.

b. Describe what you learned or saw from the graph that you didn't see just from the data.

c. Would a circle graph be appropriate for these data? Explain why or why not.

d. What might have happened to make the two sets of numbers so different?

e. Approximately what percent of the world's population in 2005 do these languages represent?

	2000	2005
Chinese, Mandarin	874	873
Hindi	366	180
English	341	309
Spanish	322–358	322
Bengali	207	171
Portuguese	176	177
Russian	167	145
Japanese	125	122
German, Standard	100	95
Chinese, Wu	77	77

Source: From *The World Almanac and Book of Facts, 2006*, p. 731. Copyright © World Almanac Education Group, Inc.

INTERNATIONAL ARMS SALES, 1998–2001				
	1998		2001	
	Dollars	% of total	Dollars	% of total
United States	11,130,000	35.6	12,088,000	45.8
Russia	2,621,000	8.4	5,800,000	22.0
France	3,603,000	11.6	2,900,000	11.0
Germany	5,460,000	17.5	1,000,000	3.8
China	1,202,000	3.8	600,000	2.3
Great Britain	2,184,000	7.0	400,000	1.5
Others	5,023,000	16.1	3,600,000	13.6
Total	31,222,000	100	26,388,000	100

Source: Copyright © 2006 by The New York Times Co. Reprinted with permission.

33. Refer to the table on the previous page.
    a. Make two circle graphs from the % of total data.
    b. Summarize what you can conclude from the circle graphs.
    c. Make two bar graphs, either separate or side-by-side, from the Dollars data.
    d. Summarize what you can conclude from the circle graphs.
    e. Summarize the advantages and disadvantages of each.
    f. If you were to pick one graph, which would you pick? Why?

34. The following table records the highest education attainment levels for adults in the United States in 2000. Select and make an appropriate graph for the data below. Justify your choice.

Category	Percent
Not a high school graduate	15.8
High school graduate	33.1
Some college	17.6
Associate's degree	7.8
Bachelor's degree	17.0
Advanced degree	8.6

Source: Statistical Abstract of the United States 2002, (Washington, DC: U.S. Department of Commerce, Bureau of the Census, 2001), p. 140.

35. Select and make an appropriate graph for the data below. Justify your choice.

WORLD'S 8 MOST POPULOUS CITIES		
Rank	City and country	Population
1.	Bombay (Mumbai), India	11,914,398
2.	São Paulo, Brazil	10,406,166
3.	Seoul, South Korea	9,981,649
4.	Shanghai, China	9,220,000
5.	Jakarta, Indonesia	9,122,700
6.	Mexico City, Mexico	8,591,309
7.	Moscow, Russia	8,434,000
8.	Istanbul, Turkey	8,141,163

Source: Data from *The Time Almanac 2002* (Boston: Information Please, 2002), p. 711.

36. The table below shows the estimated number of HIV/AIDS cases in the world, as of 2004.
    a. Make a graph for these data.
    b. Describe what you learned or saw from the graph that you didn't see just from the data.
    c. Write a 2–3 sentence summary that might appear in a newspaper article about HIV/AIDS in the world.

Region	Current cases
Sub-Saharan Africa	25,400,000
South/Southeast Asia	7,100,000
Latin America/Caribbean	2,100,000
East Asia/Pacific	1,100,000
Eastern Europe/Central Asia	1,400,000
North America	1,000,000
Western/Central Europe	610,000
North Africa/Middle East	540,000

Source: Data from *World Almanac 2006* (New York: World Almanac Books), p. 857.

For problems 37–47, answer questions a through e and any additional questions that may be asked about that specific graph.

a. Describe in full sentences and in everyday English one major quantitative conclusion that you can make from the graph and data.

b. Critique the data from the perspective of reliability and validity. That is, state and explain at least one question you have that is about how the data were obtained—a question that could affect the accuracy of the data.

c. Critique the choice of the graph. If you think the choice is fine, explain why. If you think a different kind of graph would have been more appropriate, explain why.

d. Critique the construction of the graph—its scales, categories, and labels.

e. What other questions do you have about the data or the graph?

37.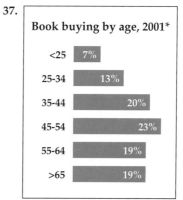

Newsweek. Copyright © 2003 Newsweek, Inc. All rights reserved. Reprinted with permission.

38.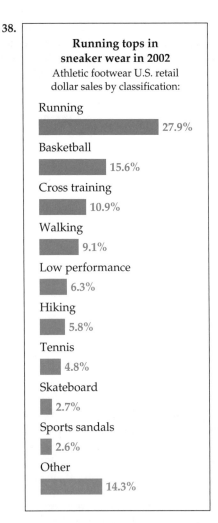

Source: USA Today, April 3, 2003. Reprinted with permission.

39.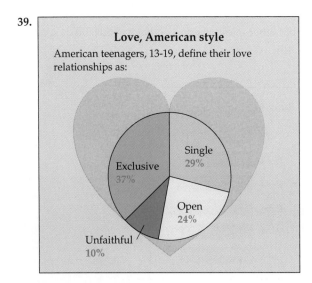

Source: Data from USA Today, April 1, 2003.

40.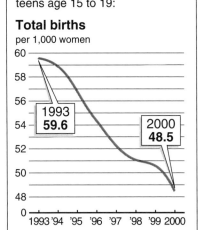

Source: AP Images.

41.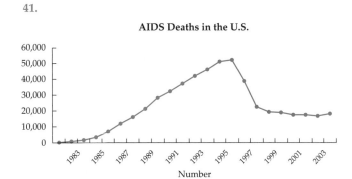

Source: 2006 New York Times Almanac. Copyright © The New York Times Company, 2006, p. 382.

42. The graph below shows who is getting AIDS and at what age.

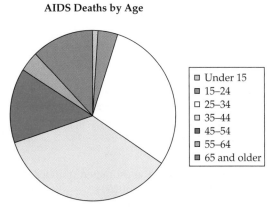

Source: Centers for Disease Control, 2004.

43. The graph below shows U.S. AIDS deaths by ethnicity. In addition to answering a through e, answer the following.

   f. The numbers of African Americans and Hispanics are disproportionate. Explain what this statement means.

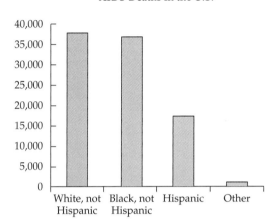

Source: Copyright © 2006 by the New York Times Co. Reprinted with permission.

44.

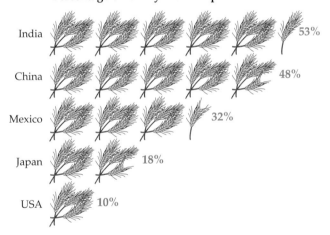

The graphs in Exercises 45 and 46 are from the TIMSS (Third International Mathematics and Science Study). TIMSS is the third international study of students' mathematics and science achievement. Students were assessed at fourth grade, at eighth grade, and at the end of high school. The study included about a half-million students from 41 countries. The National Center for Education Statistics (NCES) also sponsored a study of teaching methods in eighth-grade mathematics in Japan, Germany, and the United States. In this study, classes were observed and videotaped. The following two graphs are from that study.

45. Teachers were asked to tell the researchers what they wanted students to learn from the lessons that were videotaped. The researchers found that most of the answers fell into one of two categories: *skills*, where the answers focused on students being able to do something, and *thinking*, where the answers focused on students being able to understand something about a mathematical concept or idea.

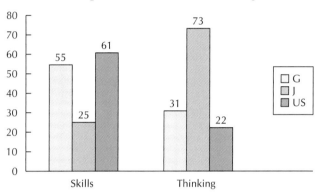

Source: http://nces.ed.gov/pubs99/timssvid/chap3.htm#12

46. In this case, the researchers examined the kind of work that students did during the lesson. They coded three types of student work: practicing routine procedures, applying concepts to new situations, and inventing new solution methods or thinking.

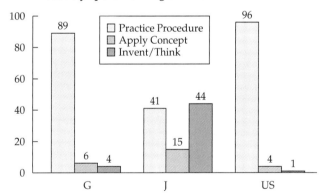

Source: http://nces.ed.gov/pubs99/timssvid/chap3.htm#65

47. a. Write a short paragraph that would accompany the following two graphs in an article describing changing patterns in immigration.

   b. Make a table for these graphs.

   c. Discuss the advantages and disadvantages of representing these data in a table as opposed to a graph.

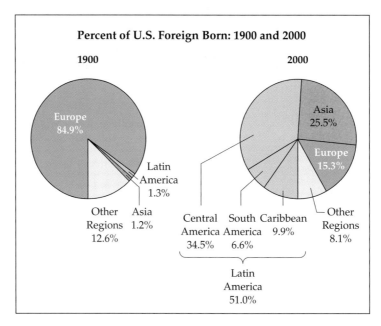

Source: *Time Almanac 2003*, p. 353. Pearson, Info Please. Reprinted with permission.

48. The graph shows the percentages of drivers of various ages who were involved in fatal motor vehicle accidents and had blood alcohol levels greater than 0.08. Write a short paragraph describing the differences in age groups.

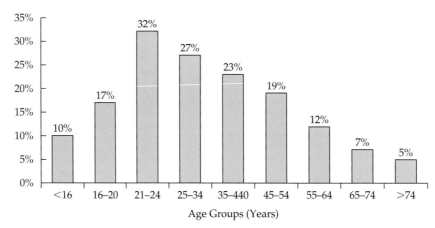

Source: U.S. Department of Transportation, National Highway Traffic Safety Administration, *Traffic Safety Facts, 2004* (Washington, DC: Author, 2000).

SECTION 7.2 GOING BEYOND THE BASICS

WHAT DO YOU THINK?

- How can we compare two sets of data?
- How can we measure how spread out a set of data is?
- How can we see how well two variables are related, like age and height?

In the first section, we examined basic concepts, graphs, and procedures to help us collect and analyze data and to interpret data that was collected by someone else. In this section, we will expand our inquiry.

Comparing Two Sets of Data

In many cases, we collect data because we are comparing two sets of data to one another. For example, we want to compare the test scores of two classes, or we want to compare how well a class performed before and after a presentation. There are two kinds of graphs (boxplots and scatterplots) that help us to understand questions of comparison.

INVESTIGATION 7.9

How Many More Peanuts Can Adults Hold Than Children?

When I had my students collect data about how many peanuts they could hold in one hand, we actually went to a nearby elementary school and collected these data with third graders. Below are the data for the children and also the data again for my students.

Children 11, 13, 14, 14, 16, 16, 16, 17, 18, 18, 19, 19, 20, 21, 22, 22, 23, 23, 24, 24, 24, 24, 25

College students 18, 18, 20, 22, 22, 22, 22, 22, 23, 25, 25, 25, 25, 25, 26, 26, 27, 27, 30, 30, 32, 32, 37

What kinds of statements can we make about the number of peanuts that children and adults can hold? Just from looking at the data, we can see that the college students could hold more peanuts—the minimum and maximum are greater. Since the numbers of students in each class are identical, we can see that at any point, from lowest to greatest, the college student data is greater.

We could make **back-to-back stem-and-leaf plots**. What additional information about the data can we see from the stem-and-leaf plots?

```
              4 4 3 1 | 1 |
      9 9 8 8 7 6 6 6 | 1 | 8 8
4 4 4 4 3 3 2 2 1 0 | 2 | 0 2 2 2 2 3
                      5 | 2 | 5 5 5 5 5 6 6 7 7
                          | 3 | 0 0 2 2
                          | 3 | 7
```

FIGURE 7.25

The stem-and-leaf plots show that the vast majority of the children are between 16 and 25 while the bulk of the adults are between 22 to 27. This corresponds to the notion that the average for the adults is about 5 to 7 greater than the average for the children.

Look at the line plots of the data. What else can we see or conclude from the line plots?

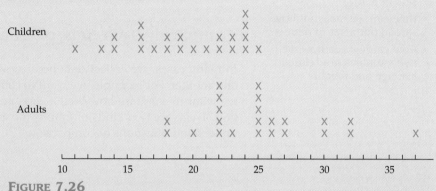

FIGURE 7.26

We already knew the minimum and maximum. What we get from the line plots is a visual sense of the variation or distribution of the data. We see that the kids are pretty steadily spread out from 11 to 24, no real huge clusters or big gaps or outliers. With the adults, we see a cluster around 22 and another cluster around 25 with some gaps and an 'outlier.' Some students have remarked that if you could pick up the children's data and move it over about 6 or so peanuts, the two sets would be very similar. As with the stem plots, we don't lose any data. That is, we can find the value of each child and adult.

Box-and-Whisker graphs   We can also make **box-and-whisker** graphs, also known as **boxplots**. Before getting into the details, let us examine the design of this graph, using the children's data. You already know what the median is, and so the idea of dividing a set of data into an upper half and a lower half makes sense. What if we divided each of these halves into half again? In this case, we would have divided the data into quarters, and we could say "this is where the bottom quarter of the data are, the next to the bottom quarter, the next to the top quarter, and the top quarter."

11, 13, 14, 14, 16, 1|6, 16, 17, 18, 18, 19,|19, 20, 21, 22, 22, 2|3, 23, 24, 24, 24, 25

While the "average" is a specific number (which can imply a false sense of exactness about the data), this division of the data into four quarters enables us to specify the middle region. In this case, this division shows us that half of the children held between 16 and 23 peanuts.

Now let's look at the actual procedure for making a boxplot.

1. Find the median of the data. In the case above, there are 22 pieces of data, so the median (19) is between the 11th and the 12th datum.
2. Find the median of the bottom half. In this case, there are 11 pieces of data in the bottom half, so the median of the bottom half (16) is the 6th datum from the bottom. This number is called the **lower quartile**.

3. Find the median of the top half. In this case, there are 11 pieces of data in the bottom half, so the median of the upper half (23) is the 6th datum from the top. This number is called the **upper quartile**.

   We now have five numbers: the minimum, lower quartile, median, upper quartile, and the maximum. In one sense, these five numbers give us a short and sweet summary of the data, more than just the average, but less than a line plot.

FIGURE 7.27

4. Draw two line segments (the whiskers) to show the bottom and top quarter of the data. Draw two rectangles (the boxes) to show the two middle quarters of the data.

FIGURE 7.28

Connecting this picture to the previous paragraph, you can see why some authors refer to the boxplot as the **five-number summary**. Now look at the boxplots of the children and college students, one above the other. What can you conclude just from looking at the boxplots?

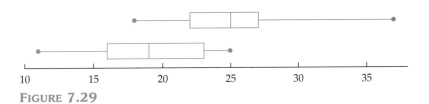
FIGURE 7.29

	Children	Adults
The range	11–25	18–37
The median	19	25
Bottom quarter	11–16	18–22
Middle half	16–23	22–27
Top quarter	23–25	27–37

> **CLASSROOM CONNECTION**
>
> The use of boxplots in the elementary classroom has increased dramatically in recent years. In use with children, this graph is commonly called the "five number summary."

Looking at the medians and/or looking at the various regions, it makes sense to say that the average adult can hold about 5 to 7 more peanuts. We can also make other statements. For example, the top children (23–25) are about where the average is for the adults.

We have used the term *outlier* informally thus far. To determine whether a datum is formally an outlier, we first find the difference between the upper and lower quartiles. We call this amount the **interquartile** range (IQR). In one sense, the IQR is like a mini-range—it is the range of the middle half of the data. To determine if a datum is an outlier, we first multiply the interquartile range by 1.5. If a datum is more than 1.5 IQRs lower than the lower quartile or 1.5 IQRs above the upper quartile, we call it an outlier. (It is important to note that not all statisticians use 1.5; even statisticians do not agree on everything!) Apply this procedure to the data on the college students to determine whether there are any outliers, and then read on. . . .

The difference between the upper quartile and the lower quartile is 5 (i.e., 27 − 22) and 5 × 1.5 = 7.5. Thus, to determine the outliers we subtract 22 − 7.5 (14.5) and add 27 + 7.5 (34.5). Thus, any number below 14.5 and greater than 34.5 is an outlier.

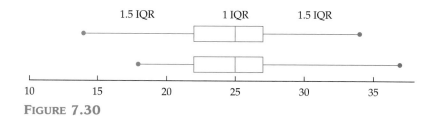

Figure 7.30

So the person who held 37 peanuts is officially an outlier!

Just as we have discussed what the average tells us and doesn't tell us about a set of data, let's look at what the boxplot doesn't tell us: it doesn't show us each individual. Some people would also say that with the boxplot we lose a sense of how the data are distributed, that is, clusters and gaps. Actually, you can get a sense of the distribution with the boxplot, and this has to do with the relative lengths of the boxes and whiskers.

Let's take a set of data where this will be more obvious. Look at the data on the size of classes at a small high school. Both the line plots and boxplots are given. From looking at the two graphs together, can you see what a short box or whisker means and what a long box or whisker means?

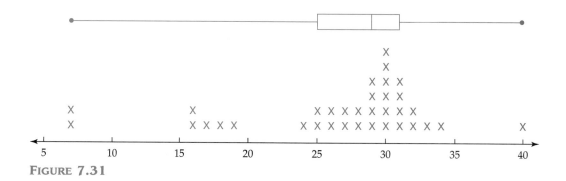

Figure 7.31

A short box or whisker tells us where clusters are, that is, the data are packed into a smaller region. A longer box or whisker tells us that the data are more spread out. Note that a long box or whisker does not necessarily mean there are outliers, because the data might or might not be spread out over this region.

While box and whisker graphs can be useful as a snapshot of one set of data, just like the average is a snapshot, the box and whisker graphs are most often used to give us a snapshot comparison of two sets of data.

INVESTIGATION 7.10

Scores on a Test

Let's say you have a friend, also a teacher, who has just finished grading his students' midterms. The scores are given below.

77, 96, 58, 100, 66, 76, 88, 73, 94, 75, 76, 84, 91, 74, 87, 92, 67

You ask how his students did, but the teacher does not know much mathematics. All he can say is that this is an unusual class. He can see that the scores range from 58 to 100. Take some time to analyze these data. What can you do with the data to help your friend better understand the overall picture of his students' performance? State your conclusions and then read on. . . .

DISCUSSION

Simply from examining the numbers as listed, we can quickly determine that the median score is 77. Either by observation or from a stem plot, we can also see that grades of C and A dominate the data. Thus grades of C and A are the modal classes for these data. (Recall that when we group the data, we speak of modal classes instead of modes. However, in both cases, the idea is the same: the value or interval that occurs most frequently.)

Look at the histogram and circle graph in Figure 7.32 What do they add to our understanding of the scores? Consider this, and then read on. . . .

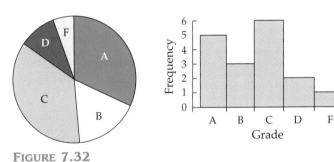

FIGURE 7.32

From the histogram, we can see that technically the modal class is a grade of C. However, because there is only one more C than A, and because these two intervals stand out, we call this distribution of scores bimodal. From the circle graph, we could frame our analysis optimistically (almost half the class got As and Bs) or pessimistically (more than half the class got C or lower).

What do the line plot and boxplot (Figure 7.33) add to our understanding of the distribution of the scores? Consider this, and then read on. . . .

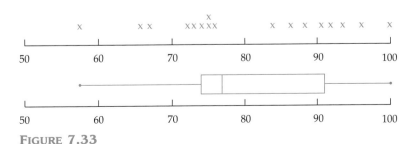

FIGURE 7.33

The line plot shows two major gaps, one from 58 to 66 and the other from 77 to 84. The line plot also shows a sense of a "top" group and a "bottom" group of the class. From your friend's perspective, this might help him understand why this is an unusual class. If he teaches to the top group, he is likely to lose the bottom group; on the other hand, if he teaches to the bottom group, he is likely to bore the top group.

In this case, because the number of data is small, the boxplot has no advantages over the line plot. From a teaching perspective, an examination of the boxplot is useful, so that you can use it and interpret it when working with larger sets of data. As stated earlier, the boxplot shows the range. Using the idea that the boxplot partitions the data into four roughly equal groups, we can focus on the two boxes and see that about 1/2 of the students scored between 75 and 90. We can also say that roughly 3/4 of the class scored above 75. Note that the left-hand box is narrower than the right-hand box. The narrowness of the left-hand box indicates a cluster of scores in that range—that is, the mid-70s.

You will be asked to compare these test scores to another set of test scores in the Exercises.

INVESTIGATION 7.11 Which Battery Do You Buy?

Let's say you were going to buy an expensive, long-life battery for your videocamera. Two companies claim that they have developed batteries that will last an average of 40 hours instead of lasting only a few hours. Both batteries cost the same. What other information would you want to know before making a decision?

Do you think it doesn't matter which you buy? Let's run some numbers. Say you had these data:

Battery A: 37, 42, 40, 38, 42, 40, 39, 38, 41, 40, 39, 41, 43, 39, 40, 40, 41

Battery B: 30, 48, 44, 36, 42, 36, 40, 34, 46, 38, 40, 40, 42, 44, 38, 40, 32, 50

What analyses of these data might you do to help you make your decision? Take some time to explore these two sets of data, using knowledge you already have. Summarize what you learned and state your conclusions, and then read on. . . .

DISCUSSION

If you did explore the data, you found that for each set, the mean, median, and mode were 40. In this case, the measures of the center tell us nothing about differences between the two sets of data. This occasionally happens with real-life data, and it illustrates a point that statisticians make: Measures of the center of a distribution give an incomplete picture of the data. Let us now discuss some graphs that help us to see how the data are spread out. In this case, because we are comparing two sets of data, we can make a back-to-back stem plot (Table 7.7).

TABLE 7.7

Battery A		Battery B
	5	0
3 2 2 1 1 1 0 0 0 0 0	4	0 0 0 0 2 2 4 4 6 8
9 9 9 8 8 7	3	0 2 4 6 6 8 8

The stem plot enables us to determine quickly the ranges of the two sets of data. We see that the data for Battery B are much more spread out (dispersed) than the data for Battery A:

- The range of the data for Battery A is 6 (37 to 43).
- The range of the data for Battery B is 20 (30 to 50).

Here are the line plots and box-and-whisker plots for the batteries. What can we conclude about the differences in the variation from these graphs?

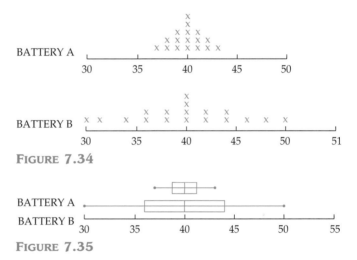

FIGURE 7.34

FIGURE 7.35

Both the line plots and the boxplots indicate that the data for Battery A are more clumped together, whereas the data for Battery B are more spread out.

We can make side-by-side histograms. These graphs show that the data for both cases are normally distributed but that the variation for Battery B is much greater.

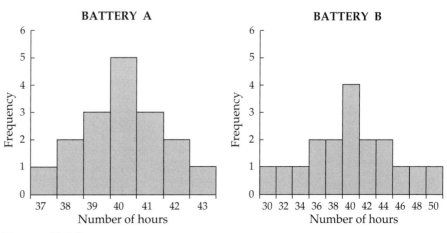

FIGURE 7.36

From this analysis of the data, most people would likely choose Battery A. Interestingly, their rationale would involve probability language (i.e., with Battery A the worst battery you are likely to get would last 36 hours while with Battery B the worst battery you are likely to get would last 36 hours).

Another Tool to Understand Variation

As mentioned before, a big part of statistical analysis involves looking at variation. How the data are varied or spread can affect what we can or cannot conclude, how accurately we can predict. Thus far we have several tools to describe the variation or spread in a set of data.

> Numerically, we have the range. The larger the range, the greater the variation.
> Visually, graphs can show clusters, gaps, and outliers.

There is another tool to help us describe variation. Just as average is a number that we can use to give the reader a sense of the typical or middle of a set of data, *standard deviation* is a number that can tell us how spread out the data are. Since standard deviations can be a rather imposing concept and procedure, I will present this concept in a way so that it makes sense and connects to the important ideas we have been discussing. In order to understand standard deviation, we first need to example more closely the normal distribution because standard deviation has most power with data that are normally distributed.

Normal Distribution

A line graph depicting a set of data that are normally distributed is often referred to as a bell-shaped curve; that is, the frequency values are highest in the middle, and the graph is symmetric. The discussion below is what is referred to as a "thought experiment"; we are not going to use data from an actual situation, but rather we are going to imagine data from an ideal situation. Suppose we have collected data on the heights of all students at a small college. As you have seen, we could make a histogram from these data. The graph at the left in Figure 7.37 shows how the actual histogram might look in this ideal case; that is, the graph is symmetric. If we trace a smooth line over the tops of the bars, we have the line graph at the right in Figure 7.37. When normally distributed data are visually depicted, the writers generally show a line graph. For example, in the case of the heights at the college, the figure on the right is a smoothed-out line graph for the heights. However, in actuality the heights have been measured in whole inches, and therefore the histogram on the left is a truer depiction of the data. Thus, whenever you see a "normal" curve, it will help if you realize that this curve is an idealization of a histogram. For example, we say that SAT scores and IQ scores are normally distributed.

> ■ *History* ■
>
> The beginning of modern statistics can be traced to the work of the Englishman John Graunt (1620–1674), who collected data on births and deaths. In his book *Natural and Political Observations of Mortality,* he presented such findings as the following: More boys are born than girls, more men than women die violent deaths, and the death rate in urban areas is greater than the death rate in rural areas.

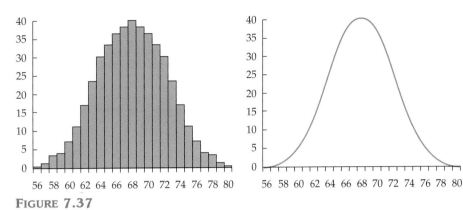

FIGURE 7.37

We are now ready to move forward.

Different Distributions

INVESTIGATION 7.12

Understanding Standard Deviation

As we saw in the previous investigation, we can see that the data for Battery B are more spread out than the data for Battery A. How else might we describe how spread out the data in these two sets are? Think and then read on. . . .

■ History ■

As you have discovered in this book, much of what we call elementary mathematics is relatively recent knowledge. So, too, is the idea of looking at distributions of data. In the 1830s, a Belgian mathematician, L.A.J. Quetelet, collected measurements of many kinds of data on people, including height, weight, length of arms, and intelligence. He did this for many people. What he found was that in most of these cases, the graphs were similar. What do you think all the graphs looked like? Think and read on. . . .

All of the distributions were similar to the normal curve.

DISCUSSION

One way to do so would be to note how far each datum is from the mean and then take the average (mean) of those numbers. This would give us an "average distance from the mean." Recall that in both cases, the mean life of the batteries was 40 hours. Determine the "average distance from the mean" for each of the two classes and then read on. . . .

Table 7.8 shows the computations. If we look at how far each number is from the mean and find the mean of *those* numbers, we can say that for Battery A, the average distance from the mean is 1.2 hours, whereas for Battery B, the average distance is 4.0 hours. Just as the mean, median, and mode are concepts that enable us to give a sense of the middle of the data with a single number, "average distance from the mean" enables us to compare the relative spreads of the data with a single number. Because 4.0 is more than 3 times 1.2, if someone were to tell you that the means from both sets of data were equal but that the average distance from the mean for Battery B was 4.0 hours compared with 1.2 hours for Battery A, this number would instantly tell you that the former set of data is more spread out.

Now the skeptical reader might be wondering, "Is it possible that real mathematicians do something this simple?" or "Why have I never heard the term *average distance from the mean* before?" The answer to the first question is yes, and we will defer the second question temporarily. What we call the

standard deviation is actually quite closely related to the average distance from the mean. For reasons that will be explained very shortly, the computation is slightly more complicated.

TABLE 7.8

BATTERY A		BATTERY B	
Number of hours	Distance from the mean	Number of hours	Distance from the mean
37	3	30	10
38	2	32	8
38	2	34	6
39	1	36	4
39	1	36	4
39	1	38	2
40	0	38	2
40	0	40	0
40	0	40	0
40	0	40	0
40	0	40	0
41	1	42	2
41	1	42	2
41	1	44	4
42	2	44	4
42	2	46	6
43	3	48	8
50	10		
Total distance from the mean	20	72	
Average distance	20/17 = 1.2 hours	72/18 = 4.0 hours	

Finding the standard deviation To determine the standard deviation, we can use much of the work we did for average distance from the mean. There are two more steps (see Table 7.9).

To determine average distance from the mean:	*To determine standard deviation:*
Find the distance from the mean of each datum.	Find the distance from the mean of each datum.
	Square each of these numbers.
Add this column of numbers.	Add this column of numbers.
Divide this sum by the number of data.	Divide this sum by the number of data.
	Take the square root of this number.

TABLE 7.9

		BATTERY A			BATTERY B	
	Score	Distance from the mean	Square of the distance	Score	Distance from the mean	Square of the distance
	37	3	9	30	10	100
	38	2	4	32	8	64
	38	2	4	34	6	36
	39	1	1	36	4	16
	39	1	1	36	4	16
	39	1	1	38	2	4
	40	0	0	38	2	4
	40	0	0	40	0	0
	40	0	0	40	0	0
	40	0	0	40	0	0
	40	0	0	40	0	0
	41	1	1	42	2	4
	41	1	1	42	2	4
	41	1	1	44	4	16
	42	2	4	44	4	16
	42	2	4	46	6	36
	43	3	9	48	8	64
				50	10	100
Sum		20	40		72	480
Sum/n		Average distance = 1.2	2.35		Average distance = 4.0	26.7
Square root			Standard deviation = 1.53			Standard deviation = 5.16

Now that we have examined the meaning of the standard deviation and developed the procedure for determining it, you are more likely to be able to make sense of the mathematical definition of the standard deviation.

If we have a set of data whose values are denoted $x_1, x_2, x_3, \ldots, x_n$, and the mean of these data is represented as $\bar{x}$, then the standard deviation is

$$\text{Standard deviation}^{10} = \sqrt{\frac{(x_1 - \bar{x})^2 + (x_2 - \bar{x})^2 + (x_3 - \bar{x})^2 + \cdots + (x_n - \bar{x})^2}{n}}$$

In this case, the standard deviation for the data for Battery A is 1.53, compared to the standard deviation of 5.16 for Battery B. Like the average distances from the mean, the two standard deviations tell us that the data for Battery B are much more spread out. Once again, the skeptical reader will ask, "If they both give us a sense of spread, why not use the average distance from the mean, because it is easier to understand and compute?"

[10] Technically, because we are finding the standard deviation of a sample, as opposed to the whole population, we should determine the standard deviation by dividing by $n - 1$ instead of n, as we would do in an ordinary average. A full explanation of why one should divide by $n - 1$ is beyond the scope of this book. Our focus is on understanding the concept, so I have chosen to divide by n, as have the authors of most elementary mathematics textbooks.

Using standard deviation The reason why we use the standard deviation has to do with some rather amazing generalizations that we can make about a set of data that is normally distributed. Let us look at those generalizations, make sure you understand them, and then see how we can apply these generalizations.

When a set of data is normally distributed, we can make the following conclusions (see Figure 7.38):

- 68% of the data lie within 1 standard deviation of the mean.
- 95% of the data lie within 2 standard deviations of the mean.
- 99.8% of the data lie within 3 standard deviations of the mean.

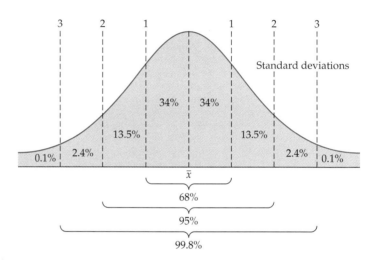

Figure 7.38

Let us apply these conclusions to our data for Battery A. Determine what percent of the data lie within 1 standard deviation of the mean. Then read on. . . .

In this case, the mean was 40 hours and the standard deviation was 1.5 hours. Thus we are looking for how many data points lie between $40 - 1.5$ and $40 + 1.5$—that is, between 38.5 and 41.5. We find that 11 of the 17 data points lie within this range, and 11/17 is equivalent to 65%. If we look at the data for Battery B and find how many data points lie within 1 standard deviation of the mean, we are looking for data points that lie between $40 - 5.2$ and $40 + 5.2$—that is, between 34.8 and 45.2. We find that 12 of the 18 data points lie within this range, and 12/18 is 67%.

What about other distributions? When data are not normally distributed, then these generalizations are not as useful. We can summarize our knowledge of how to use standard deviation by saying the following:

- When data are not normally distributed, the standard deviation is a tool that enables us to describe how spread out the data are or to compare the spreads of two sets of data.
- When data are normally distributed, the standard deviation is a unit for measuring dispersion; that is, approximately 68% of the data will lie within 1 standard deviation of the mean, and so on.

It is important to note that these numbers (68%, 95%, and 99.8%) are theoretical figures that are realized when the number of data is very large and when the data are "perfectly" normally distributed. However, even when the number of data is smaller and the distribution is not perfectly normal, we will generally find close to 68% of the data lying within 1 standard deviation of the mean.

Now that we have examined standard deviation, we can summarize the uses of measures of the center and of spread for different kinds of distributions. In general, the mean and standard deviation are used together to describe the center and spread of a set of data that is normally distributed. In general, the median and quartiles are used together to describe the center and spread of a set of data that is skewed.

Let us now consider some examples from real life that use the concept of standard deviation.

INVESTIGATION 7.13

Analyzing Standardized Test Scores

Let's say that a standardized test has a mean of 250 and a standard deviation of 50. Suppose 2000 students took the test, and their scores were normally distributed. Think about the following questions, and then read on. . . .

A. How many students would you expect to score over 350?
B. How many students would you expect to score at least 200?

DISCUSSION

A. Because a score of 350 is 2 standard deviations above the mean, we conclude that 2.5% of the scores will be over 350 [Figure 7.39(a)]. Then we simply need to compute 2.5% of 2000. We can use a calculator, or we can just as quickly do it mentally: 10% of 2000 is 200, so 5% is 100, and 2.5% is 50.

B. This question is slightly more complex. The answer is that about 1680 students will score at least 200. If you got this wrong, it is possible that you may understand the procedures but not the concepts related to standard deviation. Figure 7.39(b) illustrates one solution path. If we think of the results of the test in terms of a histogram, we can see that all the shaded bars represent scores over 200. However, when we superimpose this histogram on the graph showing the percent of scores with respect to standard deviations, we find that this means that 84% of the students scored at least 200.

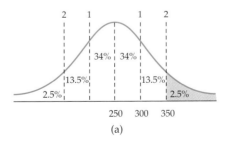
(a)

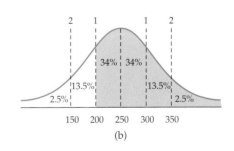
(b)

FIGURE 7.39

> ■ *Mathematics* ■
>
> When the SAT was first introduced, it was calibrated so that it had a mean of 500 and a standard deviation of 100. Over the past 20 or so years, the average score has declined considerably. In 1995, the Educational Testing Service "recentered" the test so that the mean would once again be 500. However, this provoked quite a controversy. Why do you think the ETS made the proposal? Why do you think some people objected? Think about these questions before reading on. . . .
>
> The rationale given by the ETS was that if 500 were the "average" score, then the scores would be more meaningful. That is, if a student's score was above 500, that student would know that the score was above average. Many people objected because such a move would make it appear that scores had not gone down.

The concept of standard deviation is used regularly by industries. Let us examine one such example.

INVESTIGATION 7.14

How Long Should the Tire Be Guaranteed?

A tire company has developed a new tire and has tested it extensively. The results of the tests showed that the "average" tire lasted 44,000 miles, that the distribution of wearing was normal, and that the standard deviation was 2500 miles. If the company guarantees that the tire will last 39,000 miles, what percent of the tires are likely to wear out before 39,000 miles and thus be subject to refund?

DISCUSSION

Again, a graph makes the solution to this problem much easier (see Figure 7.40). From the graph, we can conclude that 2.5% of the tires are likely to wear out before 39,000 miles.

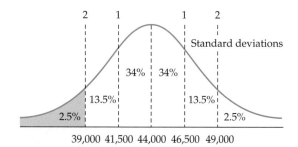

FIGURE 7.40

What if the company wanted to guarantee 40,000 miles? What percent of the tires are likely to wear out before 40,000 miles? In this case, you cannot get an "exact" answer. What could you do to increase the accuracy of your estimate? This question will be left as an exercise.

Scatter Plots

The last graphical representation of data we will examine is the **scatter plot**. Another theme is relationships. We have discussed relationships as one of the themes in mathematics throughout the textbook. On many occasions, when we examine a set of data or two variables, we expect a relationship. For example, we expect that older children are generally taller than younger children. On the other hand, sometimes we expect two variables to be related and they are not (e.g., air fares and distances between cities).

Let us begin with a situation where we expect and find a clear and strong relationship. The scatter plot in Figure 7.41 shows the ages and heights of a group of elementary school–aged children. How would you describe the shape of the graph?

> ■ *Technology* ■
>
> Given our discussion of truncation in the previous section, some students question why both axes are truncated (i.e., do not start at 0). Make your own scatter plot and begin both axes at 0. What do you notice?

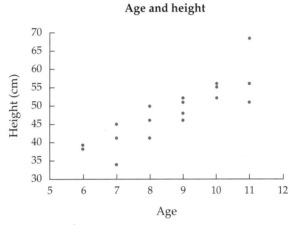

FIGURE 7.41

> ■ *Language* ■
>
> You can see that there are two outliers from this data set. How would you describe the outliers in everyday English?
>
> We would say that the outlier who is 7 years old is short for his or her age and the outlier who is 11 years old is tall for his or her age.

Some people say that as the age increases the height increases. Others will say that older children are generally, but not always, taller than younger children. Others will say that the data seem to move from the bottom left corner to the top right corner. This wording connects to a relationship you have explored before—linear relationships. The words *linear* and *line* clearly have the same root. Can you "see" a line that best fits the data, that is, a line where most of the points would either be on the line or very close? If you have a pencil, sketch such a line. When virtually all of the points lie on the line or very close, we say that there is a strong linear relationship between the variables. This is useful in science and profitable in business because it means we can predict the value of the dependent variable when we know the value of the independent variable.

Let us continue our investigation of scatter plots with the bungee jump, which interests kids and where the bungee jump operator has to be able to predict the value of the dependent variable (length of the cord) based on the value of the independent variable (weight of the jumper). The children set up a model of a bungee jump—a ruler, a rubber band, a paper clip, and weights that are attached to the paper clip. We know there is a relationship between the weight of a person and how much the bungee cord stretches. The owners of the bungee jump company had better get it right. If the cord stretches more than they predicted, the person dies! Here are the data and the scatter plot for these data (See Figure 7.42).

Number of weights	Stretch of the rubber band (cm)
1	0.9
2	1.2
3	1.5
4	1.8
5	2.0
6	2.4
7	3.0
8	3.3
9	3.7
10	4.1
11	4.5
12	5.0
13	5.5
14	5.9
15	6.5
16	6.9
17	7.4
18	8.1
19	8.6

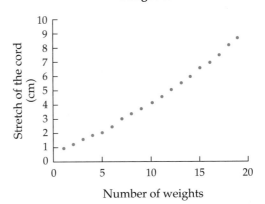

FIGURE 7.42

■ *Mathematics* ■

You might notice that the line that would "best fit" these data might not be a straight line but rather a curved line. We would need more data and a better model to be more certain. However, this observation is important because if this is true and the operator takes the average increase for the first 11 weights (about 0.35 cm for each increment in weight) and then uses that to predict how long the cord will be for the 19th weight, the operator would conclude that the length of the cord for that weight would be 7.3 cm when it is actually 8.6 cm!

How would you describe the relationship between weight and length?

There is a strong, positive relationship. While the length doesn't increase by the same exact amount each time, the increase is steady and similar. We can draw a straight line that would be a pretty good fit.

Let us consider another example where there is a relationship of a slightly different kind. The College Board has published average SAT scores for states for many years. There is quite a bit of variation and more than just "bragging rights" for states that do well. However, it is not quite so simple, because using those scores alone is like comparing apples and oranges. For example, let's say there are two similar high schools, but in one high school all seniors take the SAT while in the other school, only those going to college take the SAT. Obviously, we would expect the overall average to be higher for the second high school. The participation rates for states vary widely. Therefore, if we are going to compare scores, we had better check this out.

The scatter plot in Figure 7.43 shows the relationship between average SAT math scores for all 50 states in 2004 and the participation rate for the states. What do you see?

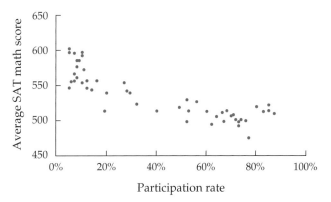

FIGURE 7.43

In this case, there is "sort of" a linear relationship, but not as strong as the bungee jump or even the ages vs. heights. Yet the data do generally move from the top right to the bottom left. That is, in general, states with lower participation rates have higher scores and states with higher participation rates have lower scores. When we have data like this, we describe the relationship as a negative relationship. The reason for this is simple and connects to basic algebra: A line that goes up as you go right has a positive slope and a line that goes down when you go right has a negative slope.

Let's go back to the peanuts data for our last example. It was noticed that there was variation in the number of peanuts that people could hold. Someone hypothesized that people with larger hands should hold more peanuts. We decided to investigate this hypothesis. First, we had to decide how to measure hand size. After some discussion, we decided to measure what is called the hand span: The person stretches their fingers as far apart as possible, and then we measure the distance from the top of the thumb to the top of the last finger. Here are our data and scatter plot (see Figure 7.44). What do you think? Do you see a correlation between hand size and number of peanuts held?

Hand size (cm)	Number of peanuts
15.5	25
17	22
17.5	25
18	18
18.5	23
19	22
19	30
19	26
19	32
19.5	29
19.5	22
20	26
20	22
20.5	20
20.5	25
21	32
21	29
21	18
22	22
23	31

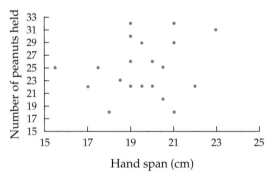

FIGURE 7.44

There really isn't much of a linear relationship here at all. You *could* try to draw a line from bottom left to upper right, but no matter what slope you chose, many points would not be very close to the line. In this case, statisticians say that there is either no relationship or a weak relationship.

Just as average is a number that can give us a "sense" of the center of a set of data and standard deviation is a number that can give us a "sense" of how spread out the data are, there is a number (between −1 and +1) called the **correlation coefficient** that can give us a sense of the extent to which there is a linear relationship. While the formula is beyond the scope of this course, we can say that the higher the absolute value, the stronger the linear relationship. A larger negative number implies an inverse relationship; a larger positive number implies a positive relationship.

Here are the correlation coefficients for the sets we have examined:

Age vs. height	0.86
Bungee cord weight vs. height	0.99
SAT scores vs. participation rate.	−0.85
Hand vs. peanuts	0.20

I know that standard deviation and correlation coefficients are two concepts that you will not teach in elementary school. In explaining why I have included them, let me paraphrase a quote from Chapter 1 of Explorations: The technical terms in mathematics are initially incomprehensible to those who have never seen them. The ideas themselves are not inordinately difficult and, in fact, the mathematical notation and formulas have been invented to make things easier for the people using them. Many of my students have told me that one of the more important learnings from this course is that they are coming to believe that mathematical ideas can make sense to "normal" people. Mathematicians have recognized the need to understand the ideas below and have found a way to quantify each of them. The result is an increased ability to understand and predict the world in which we live. In this chapter, you have seen the development of three numbers which we can use to measure a set of data.

The number	*What it tells us*
average	the region where most data are likely to occur
standard deviation	how spread out the data are
correlation coefficient	how closely we can predict the relationship (how close to linear)

If you have another area where you are competent (e.g., playing guitar, rock climbing, or skiing), you have experienced that sense of "wow, this is great" when your knowledge has developed to the extent that you can understand and use ideas or tools that just didn't make sense at first. So too with mathematics. It's really a lot more similar to other kinds of knowledge than most people think!

Inferential Statistics

In this section, we have begun to move from descriptive statistics, where we use data and graphs to summarize and describe a set of data, to inferential statistics where we take data from **sample** (part of the whole population of data), and interpret that data to make predictions. While inferential statistics is much more complex and abstract than descriptive statistics, we will take some time here to "uncover" some of the important ideas in inferential statistics.

Virtually all of your work with elementary school children will be with descriptive statistics. That is, you gather data for the entire population and you can describe what those data tell us about the population. One of your primary goals will be for students to understand the need for care in all four steps of the process.

The beginnings of inferential statistics are laid when the children begin to realize that all populations may not be the same. That is, the favorite sports of fourth-graders may not be the same as those of eighth-graders; the favorite sports of children in Los Angeles may not be the same as those of children in rural Georgia.

A next step is when they realize that we can't always gather data of the whole population.

One way not to get a representative sample is to get what we call a **convenience sample**. For example, let's say we wanted to determine how many

> ■ *History* ■
>
> A classic example of a sampling error occurred in the 1936 presidential election. The *Literary Digest* sent questionnaires to 10 million voters. Of the 2,266,566 responses, 1,293,669 were for Landon and 972,897 were for Roosevelt. That is, more than 57% were for Landon. In electoral politics, if a candidate receives more than 55% of the vote, this is considered to be a landslide. History shows that, in fact, Roosevelt received 62% of the vote, himself winning by a landslide! What happened?
>
> Analysis of the sampling process showed two flaws. First, the editors did not send the survey to a random sample of potential voters. They had gotten their sample (that is, the names and addresses) from telephone directories and lists of automobile owners. Second, less than one-fourth of the people responded to the survey. These two factors (the sample not being randomly determined and the response rate being too low) caused their error. Let us focus on the idea of obtaining a random sample. Why wasn't the *Literary Digest* sample a random sample? How could that have biased the data?
>
> In those days, a much smaller fraction of the population owned telephones and/or cars. Therefore, the sample was not representative but, rather, was biased toward upper-income voters.

hours a week college students study. It would be pretty convenient to sample students as they came out of a freshman dormitory or the library. But both of those samples would definitely not be representative. Can you explain why?

If your sample was taken outside the freshman dormitory, then you would be inferring that the overall distribution of freshmen is the same as for sophomores, juniors, and seniors. If your sample was taken outside the library, then you would be missing those students who never go to the library and underrepresenting those students who seldom go to the library.

Sometimes we need a **stratified random sample**, a sample in which there are distinct subsets of the population that need to be proportionally represented. For example, if we are surveying students on campus and 25% of the students are married, we might want to make sure that 25% of our sample are married students. In another situation, let's say that 50% of the voters in a district are white, 30% are black, and 10% are Hispanic. In this case, we will want to make sure that our sample contains these proportions also.

As we have said before, statistics does not have that same exactness that most other fields of mathematics have. Thus, the *very nature of sampling* means that our results do not have that same exactness as 2 + 2 = 4. When people taking polls give their results, they also give the **margin of error**. For example, if a poll states that 61% of the population agree with the president's position on a certain issue, they generally give the margin error. Let's say the margin of error is 3%. This means that 61% of the sample agreed with the president, and they are pretty certain that the actual percentage who agreed, if you asked everyone, would be within 3% of 61.

Another concept that needs to be addressed in sampling is the question of **sample size**. That is, how big a sample do you need in order to be confident that the statistics for this sample will be essentially the same as for the entire population? Even though I studied statistics, I must admit to still being surprised that a sample size of only a few thousand can be used to collect data that are then generalized for the entire population of the United States.

We also need to think about **response rate**. For example, let's say a large university decided to see what percentage of students who graduated in May got teaching jobs that September. Let's say they generated a sample of

> ■ *Mathematics* ■
>
> The margin of error is usually one standard deviation. Thus, if a poll says 61% plus or minus 3%, they are saying that there is a 95% probability that if we were to poll the entire population, the actual percentage would be between 58% and 64%.

100 students and sent out questionnaires and e-mails in September, and only 50% of the students responded. Would you feel comfortable making your inferences on those 50 students? How might the 50 nonrespondents be different from the 50 respondents?

Knowing human nature, it is very likely that some of the people didn't respond because they had not gotten jobs and were either depressed or embarrassed by that fact, and, in either case, did not want the university to know that they hadn't gotten a job.

INVESTIGATION 7.15 Comparing Students in Three Countries

Figure 7.45 is taken from a book entitled *Making the Grade in Mathematics: Elementary School Mathematics in the United States, Taiwan, and Japan*. Before reading on, look at the graph. Take some time to understand what information the authors are trying to present. What questions would you ask the authors in order to better understand the data? Then write down what you think they are saying. Then read on. . . .

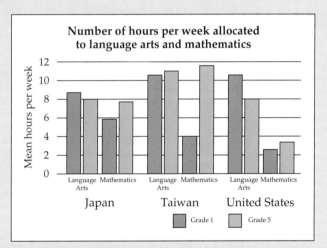

Source: Harold W. Stevenson, Max Lummis, Shinying Lee, and James W. Stigler, *Making the Grade in Mathematics: Elementary School Mathematics in the United States, Taiwan, and Japan* (Reston, VA: NCTM, 1990), p. 16. Reprinted with permission from *Making the Grade in Mathematics*, copyright 1990 by the National Council of Teachers of Mathematics. All rights reserved.

FIGURE 7.45

DISCUSSION

In their report, the authors state that three cities were selected: Minneapolis (U.S.), Taipei (Taiwan), and Sendai (Japan). They chose "representative samples of ten schools in each city. Within each school, we randomly selected two first- and two fifth-grade classrooms. . . . Thus, our report deals with data from 1440 children (240 first-graders and 240 fifth-graders in each of the three cities)."[11] The data were collected in the early 1980s.

[11] Harold W. Stevenson, Max Lummis, Shinying Lee, and James W. Stigler, *Making the Grade in Mathematics: Elementary School Mathematics in the United States, Taiwan, and Japan* (Reston, VA: NCTM, 1990), p. 2.

Note that there are several cases in which the authors needed to ensure that the samples were representative. All of these choices could affect the results: the choice of cities, the choice of schools, the choice of classrooms, and the choice of students.

For this investigation, we will focus on the differences among the countries in grade 5 mathematics. How would you summarize the differences? To give this question more context, assume that you are a writer for a national television network. The nightly newscast is going to have a feature story on the differences among the educational systems in these three countries. You are to write the story. At one point, this graph will appear in the background. What do you want the anchor to say? Think and then read on. . . .

First, let us make sure we are interpreting the information correctly. The numbers refer to the numbers of hours per week allocated to mathematics. The graph indicates that these are means (as opposed to medians or modes).

Now let us analyze the wording. Does "allocate" mean how much time is scheduled for mathematics, or does it mean how much time is actually spent on mathematics? I was not able to infer from the book exactly how the data for this graph were gathered. My interpretation (from reading the book and consulting the dictionary) is that the hours allocated to mathematics were determined by adding up the periods of time scheduled for mathematics in the teacher's plan book.

Now let us examine the hours allocated to mathematics each week in grade 5 and determine how we might convey that information to other people. Before we can make comparisons, we need to determine the actual times. Before reading on, determine the number of hours per week spent on math in grade 5 in each of the three countries.

Looking at the graph, we can conclude that in Japan, the number of hours allocated is about 8; in Taiwan, the number of hours is about 12; and in the United States, the number of hours is slightly over 3, or about $3\frac{1}{2}$ hours.

You have several options for how you write the news item:

1. *Simply report the numbers.* "In the ten Japanese schools, almost 8 hours per week was allocated to math in the fifth grade, compared to almost 12 hours per week in the ten Taiwanese schools and about $3\frac{1}{2}$ hours per week in the ten U.S. schools."
2. *Make additive comparisons.* "The Japanese schools allocate about $4\frac{1}{2}$ hours more per week than the U.S. schools, and the Taiwanese schools allocate about $8\frac{1}{2}$ hours more per week than the U.S. schools."
3. *Make multiplicative comparisons.* "The Japanese schools allocate more than twice as many hours per week as the U.S. schools, and the Taiwanese schools allocate more than three times as many hours per week as the U.S. schools."
4. *Make multiplicative comparisons using percent language.* "The Japanese schools allocate about 130% more time than the U.S. schools, and the Taiwanese schools allocate about 240% more time than the U.S. schools."

I hope you have seen that even when examining a reputable study, we need to examine critically and question the data and graphs that we read. Given their understanding of statistical analysis, the authors took pains to make the data as representative as possible. They are therefore confident that if they had had more money and had been able to sample more students in more places, those results would have been close to the results that they found.

Weighted Average

We will consider one more concept in this section: **weighted averages**. In our work thus far with the mean, we have encountered situations in which we computed the simple mean. However, there are many situations in which we use what we call weighted means. This concept is a bit more complex and requires a thorough understanding of the concept of *mean*. Let us investigate some examples.

INVESTIGATION 7.16 — Grade Point Average

Table 7.10 gives Ed's semester grades. First estimate his grade point average (GPA), and then compute his actual GPA. Recall that A = 4 points, B = 3 points, C = 2 points, and so on. How can you explain and justify your estimation and computations?

TABLE 7.10

Course	Grade	Credits
Mathematics 151	B	3
Philosophy 205	A	3
Computer Science 100	C	3
Biology 102	A	3

DISCUSSION

One way of estimating is to use the concept of balance; recall the seesaw analogy. Where would the balance point be in this case?

C	B	A
x	x	x
		x

Clearly it would be between the B and the A. If you place the wedge under the B, the seesaw will not balance there because the two As "weigh" more than the one C. Numerically, we can determine Ed's average by taking the mean of the four grades:

$$\frac{(3 + 4 + 2 + 4)}{4} = 3.25$$

What if the computer course had been only a 1-credit course? What would Ed's GPA be then? Think and read on. . . .

Intuitively, many students realize that his GPA will go up, since the low grade is now worth only 1 credit—that is, it doesn't "count as much" as the others. We can capitalize on this intuitive sense by connecting it to the idea of unit, which we have seen is one of the big ideas of mathematics. That is, we can count the B in mathematics (worth 3 points) three times, the A in philosophy 3 times, the C in computer science only once, and the A in biology three times: $(3 + 3 + 3) + (4 + 4 + 4) + 2 + (4 + 4 + 4) = 35$. We divide the sum by 10 (i.e., the number of credits Ed is taking), and so we find that his GPA is 3.5 (see Table 7.11).

CLASSROOM CONNECTION

There was a time when many (if not most) educators thought that grades should be normally distributed. Thus the grades were curved so that the "appropriate" proportion of students received grades of A, B, C, D, and F. In this course, I have found that the distribution of grades is often bimodal. That is, I often have a large proportion of students who do well (A), a large proportion of students who struggle to master the basic ideas (C), and a smaller proportion of students earning the other grades.

TABLE 7.11

Course	Grade	Numerical equivalent	Credits	Grade points
Mathematics 151	B	3	3	9
Philosophy 205	A	4	3	12
Computer Science 100	C	2	1	2
Biology 102	A	4	3	12
			10	35

We can understand the most efficient procedure for determining GPA by *connecting* our understanding that multiplication can be seen as repeated addition. We now multiply the numerical equivalent for each grade by the number of credits to get the grade points for each course. Then we add the grade points to get the total number of grade points for the whole semester and divide by the number of credits.

INVESTIGATION 7.17

What Does Amy Need to Bring Her GPA Up to 2.5?

Now let us use this knowledge to solve a problem many college students face: determining how their grades in the present semester will affect their overall GPA. At Keene State College, education majors need to have a 2.5 GPA in order to be able to student-teach. (You might want to check the policies at your college if you haven't already done so.)

Let's say Amy didn't do so well her first three semesters in college (Table 7.12). What GPA will she need this semester (she is taking 16 credits) in order to bring her overall GPA up to 2.5? Once again, estimate first. Then determine the exact answer.

TABLE 7.12

	Credits	GPA
Beginning of the semester	50	2.35
Spring semester	16	?
Total	66	2.50

DISCUSSION

We can use the balance idea again. The 50 credits Amy has already taken will count more than the 16 credits she is taking this semester. Because the ratio of 50 to 16 is approximately 3:1, we can represent the problem by placing three weights at 2.35 and the balance point at 2.5 (see Figure 7.46). How far away must this semester's GPA be in order to balance the 2.35?

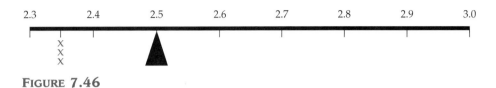

FIGURE 7.46

If we combine the idea of ratios and number lines with our "balance sense," we realize that the one weight (representing this semester) will have to be three times as far from 2.50 as is 2.35. Do you see why? If not, go out and play on a seesaw with some friends. Seriously! Therefore, we can estimate that she will need a GPA of about 2.95. Actually, because 16 is less than 1/3 of 50, it will have to be a bit farther to the right in order to balance the 50 credits at 2.35.

One way to determine the exact answer is to determine how many total grade points she has now (2.35×50) and how many total grade points she needs if she is to have a GPA of 2.5 at the end of the semester (2.5×66). The difference of these numbers will tell us how many grade points her 16 credits this semester have to provide.

That is, $2.35 \times 50 = 117.5$ and $2.5 \times 66 = 165$. We now divide the difference of 47.5 by 16 (the number of credits she is taking this semester), and we find that she must get a GPA of at least 2.97 to bring her GPA up to at least 2.5. As you can see, the estimate was very close.

Some students understand the solution better with a slightly different phrasing. At this point (50 credits), she has accumulated $50 \times 2.35 = 117.5$ grade points. At the end of the semester (66 credits), if she is to have a GPA of 2.5, she will need to have accumulated a total of $66 \times 2.50 = 165$ grade points. What does she need to accumulate this semester? She needs $165 - 117.5 = 47.5$ grade points. Now we compute: 47.5 grade points divided by 16 credits $= 2.97$.

Summary

In this section, we have examined how boxplots can help us to compare two sets of data and how scatter plots can help us to see how closely two variables or sets of data are related.

We have also examined different ways to quantify the dispersion (or spread) of a set of data, from simple ideas like range, clusters, gaps, and outliers, to more complex ideas like interquartile range and standard deviation. We have seen that the mean and standard deviation can be used together to describe the center and spread of a set of data that is normally distributed, and that the median and quartiles can be used together to describe the center and spread of a set that is skewed.

We have learned that the standard deviation is a measure of how spread out the data are. When a set of data are normally distributed, we can make powerful statements about subsets of the data.

We have seen how scatter plots can give us a visual sense of how related two variables are.

We began the transition from descriptive statistics to inferential statistics with the idea of sampling. The actual practice of gathering data on samples of population in order to make generalizations about the entire population is rather complex. However, we have explored some of the issues that are encountered in this practice. Taking measures to ensure that the sample is as representative of the total population as possible involves taking measures to ensure that the samples that are obtained are random samples. In cases where there are distinct subgroups within the population, this often involves stratified random samples.

We learned that statisticians need to think about the sample size when sampling. When conducting surveys, we also need to consider the response rate.

Finally, we examined the notion of weighed means. We connected the computation of weighted means and simple means, and we connected the concept of both kinds of means—that is, the notion of the balance point of the set of data.

EXERCISES 7.2

1. *Classroom Connection* On page 375 of the March 1999 issue of *Mathematics Teaching in the Middle School*, a class is exploring how data can be analyzed. They are given the following data from a hypothetical survey taken of 30 seventh-graders to find out how many hours of television these children watch in a week. Imagine you are asked to tell the principal what these data tell us about the television habits of seventh-graders at this school. Write that report. Then read the article to see what the children did!

1.5	21	12.5	0	2.5	15	23	19	4	14	8	16
13.5	16.5	6	4.5	9	18	5	10.5	8.5	6	3	9
11.5	3.5	19.5	13	10	9						

   Source: From the March 1999 issue of *Mathematics Teaching in the Middle School*, copyright 1999 by the National Council of Teachers of Mathematics.

2. Following are the heights of adults in a college class.

   a. Make a histogram for these data.

   b. Determine the mean and standard deviation.

   c. What percent of the values lie within 1 standard deviation of the mean?

Height (inches)	Frequency
59	1
60	1
61	3
62	8
63	11
64	13
65	17
66	14
67	11
68	7
69	4
70	2
71	1
72	1

3. Below are the ages of students in a class in mathematics for elementary teachers:

    18, 18, 18, 19, 19, 19, 19, 19, 19, 20, 20, 20, 21, 21, 21, 21, 21, 21, 23, 24, 24, 26, 27, 37, 42, 43, 46

   a. Use one or more graphs to show the data.
   b. What is the average age? Justify your answer.
   c. Describe the spread of the data in as many ways as you can.

4. Below are the annual salaries (in thousands of dollars) of workers in Bates Ball Bearing factory:

    12, 12, 13, 13, 13, 16, 16, 16, 16, 18, 18, 18, 21, 21, 21, 26, 26, 28, 31, 51, 51, 75

   a. Use one or more graphs to show the data.
   b. What is the average salary? Justify your answer.
   c. Describe the spread of the data in as many ways as you can.

5. Suppose the instructor in Investigation 7.11 gave the same test to another class whose scores are given below. Tell the instructor how the students did in each class and how the classes compare.

    Class 2: 96, 94, 93, 92, 89, 85, 85, 85, 83, 80, 79, 77, 77, 77, 75, 74, 74, 68, 67, 65, 65

6. A class of college students collected data about how many siblings they had and how many siblings their mother had.

   a. Determine the mean, median, and mode for the two sets.
   b. How would you answer this question: How has the average number of siblings changed from the two generations?
   c. Can we generalize from these data about how the number of siblings has changed from generation to generation? Why or why not?
   d. Make two line plots for these data. Write what you can conclude from analyzing those plots.
   e. Make two box-and-whisker plots for these data. Write what you can conclude from analyzing those plots.
   f. Determine the standard deviation for each set of data.
   g. What additional information would we need in order to make a scatter plot for the data?
   h. What would a scatter plot tell us?
   i. Write a short report that answers the question "What do we know about the number of siblings we have and our mothers have?" Include at least one graph.

	0	1	2	3	4	5	6	7	8	9	10	11
My students	3	6	4	2	1	1	0	0	0	1	0	0
Their parents	1	2	4	4	1	3	0	1	0	1	0	1

7. Below are the years of experience of teachers in two elementary schools. What do these data tell us about the experience of the teaching staff of the two schools?

   a. Determine the mean and median for the two schools.
   b. What would you say is the modal class for each school? Justify your answer.
   c. What do the mean and median tell us and not tell us about the data?
   d. Make a line plot for each school from these data.
   e. Summarize (2–3 sentences) what the line plots tells us.
   f. Make a grouped frequency table for these data.
   g. Make a histogram from these data.
   h. Summarize (2–3 sentences) what the histogram tells us.
   i. Make a boxplot for these data.
   j. Summarize what the boxplot tells us about the data.
   k. Compare the spread of the data for the two schools in as many ways as you can.

Years Teaching	0	1	2	3	5	6	7	8	10	14	20	25	31
School A	4	3	4	2	1	0	2	0	0	0	1	0	0
School B	1	2	1	2	0	2	1	2	1	1	1	0	1

8. The superintendent and school board have decided to collect data on the years of experience of teachers in the school district, for a variety of reasons—if you have too many veteran teachers, the payroll is higher; if you have too many teachers nearing retirement, it is hard to replace them. Here are the data for a school district concerning the number of years of teaching.

Years Teaching	0	1	2	3	4	5	6	7	8	9	10
Teachers	11	8	10	7	6	5	3	6	2	4	4

Years Teaching	11	12	15	16	17	18	19	20	21	22	23
Teachers	2	3	2	5	1	3	4	3	5	4	3

Years Teaching	24	25	27	28	29	30	32	33	34	35	39
Teachers	5	3	2	4	2	2	2	1	1	1	1

Source: http://exploringdata.cqu.edu.au. Reprinted by permission of Education Queensland.

   a. Make a grouped frequency bar graph from these data, and justify your choice of intervals.
   b. Make one other graph for these data. Explain why you thought that graph would be appropriate.
   c. Write a report (maximum of five sentences) that you would give to the school board. Select whatever

graph(s) and other information you would want to include.

d. Jacob says "Just give them the average. Why clutter their minds with graphs and all sorts of other information?" How would you respond to Jacob?

Washington Redskins	$117,962,286
Philadelphia Eagles	$104,977,331
Houston Texans	$ 97,473,626
Detroit Lions	$ 94,578,628
Seattle Seahawks	$ 94,006,282
Miami Dolphins	$ 93,937,308
New York Jets	$ 93,866,236
Minnesota Vikings	$ 92,407,989
Indianapolis Colts	$ 92,209,207
Chicago Bears	$ 87,826,859
Carolina Panthers	$ 87,807,573
Cleveland Browns	$ 87,728,284
Baltimore Ravens	$ 86,478,031
Kansas City Chiefs	$ 84,617,626
Atlanta Falcons	$ 82,711,268
Tampa Bay Buccaneers	$ 81,989,547
New York Giants	$ 81,657,826
Green Bay Packers	$ 80,383,682
Buffalo Bills	$ 80,170,229
Tennessee Titans	$ 79,003,839
Arizona Cardinals	$ 78,961,345
Pittsburgh Steelers	$ 77,955,021
Oakland Raiders	$ 77,369,122
New England Patriots	$ 76,999,180
St. Louis Rams	$ 76,389,455
San Diego Chargers	$ 76,253,021
New Orleans Saints	$ 73,324,022
Denver Broncos	$ 72,564,908
Jacksonville Jaguars	$ 72,113,009
Cincinnati Bengals	$ 68,811,884
Dallas Cowboys	$ 65,409,479
San Francisco 49ers	$ 63,033,817

9. a. Make a boxplot for the 2004 team payrolls of the National Football League and a boxplot for the 2005 team payrolls of Major League Baseball.

b. Identify any outliners.

c. What can you conclude from these data?

New York Yankees	$208,306,817
Boston Red Sox	$123,505,125
New York Mets	$101,305,821
Los Angeles Angels	$ 97,725,322
Philadelphia Phillies	$ 95,522,000
St. Louis Cardinals	$ 92,106,833
San Francisco Giants	$ 90,199,500
Seattle Mariners	$ 87,754,334
Chicago Cubs	$ 87,032,933
Atlanta Braves	$ 86,457,302
Los Angeles Dodgers	$ 83,039,000
Houston Astros	$ 76,779,000
Chicago White Sox	$ 75,178,000
Baltimore Orioles	$ 73,914,333
Detroit Tigers	$ 69,092,000
San Diego Padres	$ 63,290,833
Arizona Diamondbacks	$ 62,329,166
Cincinnati Reds	$ 61,892,583
Florida Marlins	$ 60,408,834
Minnesota Twins	$ 56,186,000
Texas Rangers	$ 55,849,000
Oakland Athletics	$ 55,425,762
Washington Nationals	$ 48,581,500
Colorado Rockies	$ 48,155,000
Toronto Blue Jays	$ 45,719,500
Cleveland Indians	$ 41,502,500
Milwaukee Brewers	$ 39,934,833
Pittsburgh Pirates	$ 38,133,000
Kansas City Royals	$ 36,881,000
Tampa Bay Devil Rays	$ 29,679,067

10. The table below shows the ages of the winners in the Best Actress category at the Academy Awards from 1928–2005. Has the age of the winners changed over time? Divide the data into 1928–1966 and 1967–2005.
   a. Make two stem-and-leaf plots.
   b. Summarize what they tell you about the question.
   c. Determine the mean, median, and standard deviation for the two sets.
   d. Summarize what these numbers add to our understanding of the response to the question.
   e. Make and justify at least one additional graph.
   f. How would you answer the question?

Year	Age	Year	Age	Year	Age	Year	Age
1928	22	1948	34	1968	61	1988	41
1929	37	1949	33	1969	26	1989	26
1930	30	1950	28	1970	35	1990	81
1931	62	1951	38	1971	34	1991	42
1932	32	1952	45	1972	34	1992	29
1933	26	1953	24	1973	26	1993	33
1934	31	1954	26	1974	37	1994	35
1935	27	1955	47	1975	42	1995	45
1936	26	1956	41	1976	41	1996	49
1937	27	1957	27	1977	35	1997	39
1938	30	1958	39	1978	31	1998	33
1939	26	1959	38	1979	41	1999	24
1940	29	1960	28	1980	33	2000	24
1941	24	1961	27	1981	30	2001	32
1942	39	1962	31	1982	74	2002	32
1943	24	1963	37	1983	33	2003	34
1944	29	1964	30	1984	49	2004	55
1945	37	1965	24	1985	38	2005	28
1946	30	1966	34	1986	61		
1947	34	1967	60	1987	21		

11. The American Federation of Teachers publishes the average salaries of beginning teachers and of all teachers by state. Here is the website where you can get the latest data: http://www.aft.org/salary/
   a. Make a boxplot of the data for beginning teachers and all teachers for the latest date available.
   b. Summarize what you can conclude from the boxplots.
   c. Do any additional exercises that your instructor assigns.

12. The Association of Community Organizations for Reform Now (ACORN) presented these to a Joint Congressional Hearing on discrimination in lending. The data concern the refusal rate for people applying for a mortgage. ACORN contends that these data show a pattern of discrimination.
   a. Make boxplots for the four sets of data.
   b. Summarize your conclusions.

	Refusal rates of applicants			
Bank	Minority	White	High-income minority	High-income white
Harris Trust	20.9	3.7	21.4	2.2
NCNB Texas	23.23	5.5	8.0	8.0
Crestar	23.1	6.7	11.3	3.6
Mercantile	30.4	9.0	17.3	5.5
First National Bank Commerce	42.7	13.9	38.0	7.6
Texas Commerce	62.2	20.6	33.3	10.3
Comerica	39.5	13.4	33.6	9.4
First of America	38.4	13.2	29.5	7.3
Boatman's National	26.2	9.3	21.7	7.4
First Commercial	55.9	21.0	39.1	15.8
Provident National	49.7	20.1	36.6	15.3
Worthen	44.6	19.1	28.6	10.1
Hibernia National	36.4	16.0	32.9	9.2
Sovron	32.0	16.0	21.0	13.0
Bell Federal	10.6	5.6	5.8	4.2
SEC PAC AZ	34.3	18.4	24.2	14.1
Core States	42.3	23.3	38.3	15.0
Citibank AZ	26.5	15.6	27.3	16.1
MF'ers Hanover	47.2	29.7	41.1	26.8

13. The graph below is a box-and-whisker graph for the lengths (in miles) of the 18 longest rivers in the world. What information does the graph give us?

14. Explain the similarities between a line plot and a box-plot. You are encouraged but not required to use data from a set of data such as that in Investigation 7.9 or 7.10.

15. Suppose the distribution of heights of a group of 400 children is normal, with a mean of 150 centimeters and a standard deviation of 12 centimeters. About how many of these children are taller than 162 centimeters?

16. The heights of 1500 boys at West High School were measured, and the mean was found to be 66 inches, with a standard deviation of 2.5 inches. If the heights are approximately normally distributed, 95% of the boys are between ___ and ___ inches tall. About how many of the boys are less than 5 feet tall? *Note:* You cannot get an exact answer here. Any answer will involve some estimating.

17. Let's say a student scored 45 on a math achievement test for which the mean was 36 and the standard deviation was 6. Judging on the basis of this performance, what score would you predict for her on a standardized test for which the mean is 100 and the standard deviation 15?

18. What happens to the mean and the standard deviation of a set of data when the value of each datum is increased by the same amount?

19. A tire company tested a particular model of super radial tire and found the tires to be normally distributed with respect to wear. The "average" (mean) tire wore out at 59,000 miles, and the standard deviation was 2500 miles.

   a. If 2000 tires are tested, about how many are likely to wear out before 54,000 miles?

   b. What if the company wanted to guarantee 55,000 miles? What percent of the tires are likely to wear out before 55,000 miles? In this case, you cannot get an "exact" answer. What could you do to increase the accuracy of your estimate?

20. A tire company tested another model of tire and found the tires to be normally distributed with respect to wear. The "average" (mean) tire wore out at 46,000 miles, and the standard deviation was 2400 miles. If 2000 tires are tested, about how many are likely to wear out before 45,000 miles? You cannot get an exact answer. I want your best estimate. Please use graph paper to help you estimate.

21. Jack and Jill go to different schools. Jack got an 82 on a test with a mean of 60 and a standard deviation of 15. Jill got a 78 on a test with a mean of 60 and a standard deviation of 9. If both tests are graded on a curve, who did better?

22. Wendy took three standardized aptitude tests: English, mathematics, and general information. The table below gives her score, the mean, and the standard deviation for each test. On which test did she do the "best"? Justify your answer.

Test	Wendy's score	Mean	Standard deviation
English	85	75	7.5
Math	63	55	6
General information	104	94	8

23. We found that in a normally distributed set of data, the mean, median, and mode are virtually identical. However, this is not necessarily true for other distributions. What about data whose distribution is skewed to the right? Can you predict the numerical order of the three averages? For example, will the mean be largest, followed by the median, followed by the mode? Or do you think "it depends"? Justify your answer.

24. Why do you think many statisticians recommend that when describing a population, one should give at least three pieces of information: some sense of the "average," some sense of the spread, and some sense of how the data are distributed? For example, they would recommend not simply reporting that the average age of teachers in New Hampshire is 43 or that the average adult American goes to church 1.3 times a month (both figures are made up).

25. Below are data about the calories and sodium (salt content) of beef and poultry hot dogs. Analyze the data to see how the distributions of the three sets of data are similar and different.

   a. Make two grouped frequency histograms for calories of beef and poultry hot dogs. Summarize what you can conclude from those graphs.

   b. Make two box-and-whisker graphs for calories of beef and poultry hot dogs. Summarize what you can conclude from those graphs.

   c. Which of the graphs (histogram or box and whisker) were more useful? Explain.

   d. Make a scatter plot to examine the relationship between calories and sodium content of the beef hot dogs. Describe the relationship you observe in words.

Beef		Poultry	
Calories	Sodium	Calories	Sodium
186	495	129	430
181	477	132	375
176	425	102	396
149	322	106	383
184	482	94	387
190	587	102	542
158	370	87	359
139	322	99	357
175	479	107	528
148	375	113	513
152	330	135	426
111	300	142	513
148	375	113	513
152	330	135	426
111	300	142	513
141	386	86	358
153	401	143	581
190	645	152	588
157	440	146	522
131	317	144	545
149	319		
135	298		
132	253		

26. Below are data on the lengths of feet (in inches) and heights of a sample of adults. Make a scatter plot for these data. Describe the degree of correlation, that is, positive or negative, and none, weak, or strong. Justify your conclusion.

Student	1	2	3	4	5	6	7	8	9	10	11	12	13
Foot	8.5	9.5	9.5	10	11	11	11	12	12	12.5	13	13.5	14
Height	60	65	66	70	71	72	73	71	72	71.5	70	72	74

27. **a.** Describe some situations where you would expect a scatter plot to show a strong positive correlation (e.g., age of husband and age of wife). Justify your answer.
   b. Describe some situations where you would expect a scatter plot to show a weak positive relationship. Justify your answer.
   c. Describe some situations where you would expect a scatter plot to show a strong negative relationship. Justify your answer.

28. The following scenarios have been taken from the *Instructor's Course Planner* for the textbook *Children*.[12] In each case, do you think the sample is a random sample or a biased sample? Justify your response.
   a. The researchers asked teenage boys about their driving records and habits by going to a movie drive-in, a local bar, the beach, and a baseball park.
   b. Parents at a PTA meeting were interviewed about the quality of the local public school system.
   c. A telephone survey assessed a community's attitudes toward welfare recipients.
   d. Children were asked their opinion of Santa Claus on December 28.

In Problems 29–33, you are given data from various surveys that I have read about. In each case:

a. State at least two questions you would ask the people who did the survey to satisfy yourself of the validity and reliability of the survey. Briefly explain the reasoning behind your questions.
b. Write the actual question that you would ask if you were to conduct a similar survey.
c. Make a graph for these data.
d. Justify your choice of graph.
e. What does the graph add to your understanding of the question? If you feel the graph adds nothing, explain why not.

29.
PERCENTAGE OF TELEVISION HOUSEHOLDS WITH:	
Number of TVs/VCRs	**Percent**
Two or more TVs	74
One TV	26
Two or more VCRs	26
One VCR	74

Source: Roper Organization survey.

[12]From *Instructor's Course Planner* to accompany *Children* 4e, by John Santrock (Dubuque, IA: Brown & Benchmark, 1995), p. 36. © 1995. Reprinted by permission of The McGraw-Hill Companies.

30.

HOW PARENTS WAKE UP THEIR CHILDREN	
Response	Percent
Call to them	43
Alarm clock	22
Kids wake on their own	16
Other	19

Source: Aunt Jemima survey poll of 400 parents of children under the age of 18.

31.

HOW OFTEN FAMILIES EAT DINNER TOGETHER	
Nights per week	Percent
0	5
1–2	17
3–4	26
5–6	30
7	22

Source: BKG youth survey for Kodak.

32.

HOW MANY FRIENDS PEOPLE SAY THEY HAVE	
Number of friends	Percent
0	1
1	2
2–5	36
6–10	25
11–20	18
More than 20	18

Source: MCI/Louis Harris survey.

33.

AVERAGE NUMBER OF COLDS PER YEAR	
Age group	Number
0–4	5
5–19	3
20–39	2
40 plus	2

Source: University of Michigan.

34. Several secretaries in an academic building on a college campus share a copy machine. They believe the copy machine is too slow, and they want the college to buy a newer, high-speed copy machine. Describe what kind of data they should collect and how they should present their data so as to best support their case.

35. In a special triathlon, Betty swam 1 mile in 25 minutes, biked 25 miles in 45 minutes, and then ran 5 miles in 40 minutes. What was her "average" speed for the race?

36. On a certain exam, Tony corrected 10 papers and found the mean for his group to be 70. Alice corrected the remaining 20 papers and found that the mean for her group was 80. What is the mean of the combined group of students?

37. Compute Ed's GPA, to two decimal places, for the semester (A = 4, B = 3, C = 2).

Course	Grade	Credits
Mathematics	B	3
Elementary methods	A	6
Computers in school	C	1
Biology	B	4

38. Perry has a 2.53 GPA after 112 credits. This is his last semester. He is taking 15 credits, and he has senioritis.
   a. What is the minimum GPA Perry can get and still graduate with at least a 2.5 GPA?
   b. He is taking five 3-credit courses. Give one scenario that will let him "squeak by."

   Grade
   Course 1
   Course 2
   Course 3
   Course 4
   Course 5

39. Joe has a 3.22 GPA after 46 credits. He says that he has a total of 148.12 grade points. Julie says that's impossible. What do you think?

SECTION 7.3 CONCEPTS RELATED TO CHANCE

WHAT DO YOU THINK?

- How can we figure out the probability that something will happen before it actually happens?
- How can we express that probability?

> The study of probability . . . should not focus on developing formulas or computing the likelihood of events pictured in texts. Students should actively explore situations by experimenting and simulating probability models. . . . Students should talk about their ideas and use the results of their experiments to model situations or predict events.
>
> (*Curriculum Standards*, p. 109)

If you did one or more of Explorations 7.12 through 7.17, you have conducted experiments and simulations and grappled with the fundamental probability ideas that we will now examine more formally.

Stop for a moment and think of situations in which probability enters into our lives. Then read on. . . .

We make and are affected by probability decisions every day. For example,

- If you decide to have two children, what is the probability that you will have a boy and a girl?
- What is the probability that you will get a teaching job after you graduate?
- If you are independently employed, should you buy health insurance? People's decisions are influenced by their estimate of the probability that they will have a catastrophic illness.
- If you drive over the speed limit or decide not to put money in the parking meter, what is the probability of your getting a ticket?
- What is the probability that everyone who buys an airline ticket will actually take the trip? Airlines regularly sell more tickets than they have seats because they know that the probability of this is very low.

These situations all have in common that their results are unknown. Much of our understanding of probability is based on collecting and analyzing data on occurrences of situations that, individually, are random. Analyzing the data helps us to see patterns in these occurrences and to make predictions. For example, because the probability that a 21-year-old male will have an automobile accident is much greater than the probability that a 50-year-old male will have an accident, car insurance rates for 21-year-old males are much higher than those for 50-year-old males.

Our work with data that are normally distributed connects to probability. For example, if a tire company's sample of tires lasted an average of 42,000 miles with a standard deviation of 2000, the company can say that there is an 84% probability that a specific tire will last at least 40,000 miles.

Random phenomena When mathematicians use the word *random*, we are referring to phenomena that are unpredictable individually but that have regular patterns when considered as a group or when considered over the long run. *Random* is not a synonym for *haphazard*; rather, it refers to phenomena for which we cannot make individual predictions. The following example nicely illustrates this point. If we drop a coin from a certain height, we can predict with a great deal of precision the time it will take to reach the ground. However, if we flip that coin, we can give only the probability that it will land heads or tails. We do not say that flipping a coin is haphazard but rather that it is random.[13] When we flip many coins, we find that there are many patterns, as you will see in this section.

In this section, we will investigate both well-defined questions (for example, the probability of tossing 4 heads in a row) and questions that are not as well defined (for example, insurance premiums).

Preliminary Terms and Concepts

We hear probability statements all the time. Let's say you turn on the television and hear the weather forecaster say, "There is a 25% chance of rain tomorrow." What does that statement mean? Write your thoughts before reading on. . . .

When we make a probability statement, we are giving a numerical value that represents the degree to which we believe that event will or will not happen.

Probabilities can be represented by percents, fractions, decimals, ratios, and odds. The following five statements are equivalent:

- There is a 25% chance of rain tomorrow.
- The probability of rain tomorrow is 1/4.
- The probability of rain tomorrow is 0.25.
- There is 1 chance in 4 of rain tomorrow.
- The odds against rain tomorrow are 3 to 1.

Outcomes and events There are two terms whose definition will make our discussion about probabilities much clearer: *outcome* and *event*. Answer the following two questions and then read on. . . .

- If you select a card from a regular deck of playing cards, what is the probability of drawing the queen of spades?
- If you select a card from a regular deck of playing cards, what is the probability of drawing any queen?

The probability of drawing the queen of spades is 1/52, and the probability of drawing any queen is 4/52, or 1/13.

We often use notation as a shorthand to express probabilities:

$$P(Q \text{ of spades}) = \frac{1}{52} \quad P(Q) = \frac{1}{13}$$

We refer to "queen of spades" as an outcome and "any queen" as an event. Let us examine the difference between the two. When we define a situation (in

---

[13]David Moore, "Uncertainty" in *On the Shoulders of Giants: New Approaches to Numeracy*, ed. Lynn Steen (Washington, DC: National Academy of Sciences, 1990), p. 98. Reprinted with permission from *On the Shoulders of Giants: New Approaches to Numeracy*. Copyright © 1990 by the National Academy of Sciences. Courtesy of the National Academy Press, Washington, DC.

this case, drawing a card from a deck), each possibility is called an **outcome**. The set of all possible outcomes is called the **sample space**. Thus, each outcome is an element of the sample space. Within the sample space, there are many possible subsets—for example the subset *E*:

E = {queen of diamonds, queen of spades, queen of hearts, and queen of clubs}

Any subset of a sample space is called an **event**.

When All Outcomes Are Equally Likely

In this case, we are examining situations in which each outcome is **equally likely**. When we are examining a situation in which all outcomes are equally likely, we can determine the probability of an event by computing the following ratio, expressed here both in probability language and in set language. If all outcomes in a sample space are equally likely, then the probability of event *E* is given by

$$P(E) = \frac{\text{number of favorable outcomes}}{\text{number of total outcomes}} \qquad P(E) = \frac{\text{number of elements in } E}{\text{number of elements in } S}$$

When All Outcomes Are Not Equally Likely

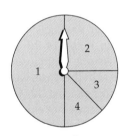

FIGURE 7.47

On many occasions, all outcomes are not equally likely. Consider the spinner in Figure 7.47. In this case, there are four outcomes: 1, 2, 3, 4. However, they are not equally likely. In this case, what is the probability of spinning an odd number?

$$P(\text{Odd}) = P(1) + P(3) = \frac{1}{2} + \frac{1}{8} = \frac{5}{8}$$

We can generalize from this example to state the probability of an event when all outcomes are not equally likely:

The probability of an event is equal to the sum of the probabilities of all outcomes in that event.

That is, if event *E* consists of *n* different outcomes $O_1, O_2, O_3, \ldots, O_n$, then $P(E) = P(O_1) + P(O_2) + P(O_3) + \cdots + P(O_n)$.

Note that this more general definition of probability works with both kinds of situations.

When we speak of the probability of an event happening, at one extreme we have **impossible events**—for example, the probability of a student having a GPA of 4.4. At the other extreme, we have **certain events**—for example, the probability that it will rain this year in Jacksonville, Florida. If an event is impossible, we say that the probability is 0. If it is certain, we say that the probability is 1. Thus, the probability of any event is between 0 and 1.

Theoretical and Experimental Probabilities

When working with probability situations, we distinguish between theoretical probabilities and experimental probabilities. In some situations involving

> ### ■ *History* ■
>
> There are some famous historical examples of empirical data concerning flipping a coin. The French naturalist Count Buffon (1707–1788) tossed a coin 4040 times; the results were heads 2048 and tails 1992, or 50.693 percent of the time. Statistician Karl Pearson tossed a coin 24,000 times; his results were 12,012 heads and 11,988 tails. While imprisoned during WWII, South African mathematician John Kerrich tossed a coin 10,000; his results were heads 5067 and tails 4933. The percentage of heads in the three cases were 50.69, 50.05, and 50.67.

probabilities, we can determine the theoretical probability—for example, the probability of a couple having 5 boys in a row or of rolling doubles 3 times in a row when playing Monopoly. In situations where we can determine the total number of outcomes and we can count the number of outcomes in a specific event, we refer to the probability as the **theoretical probability**. That is, the theoretical probability refers to our expectation, assuming that things turn out ideally. For example, we can say that when we flip a coin, theoretically the probability of its being heads is 1/2.

There are many real-life situations in which it is either impossible or very expensive to determine theoretical probabilities. In those situations, we determine the experimental probability by collecting and analyzing data. For example, manufacturers determine the probabilities that their products will last a certain time by testing some of the models. For example, if 90% of the sampled tires last at least 40,000 miles, the manufacturer will assume that close to 90% of all the tires of that model will last at least 40,000 miles. In this case, we would say that the experimental probability of a tire's lasting at least 40,000 miles is 90%. In situations in which the probability of an outcome or event has been determined by collecting data and determining the fraction of the time in which the outcome or event actually occurred, we refer to that fraction as the **experimental probability**. Some books and authors use the term **empirical probability** instead.

Let us now investigate some situations to deepen our understanding.

INVESTIGATION 7.18 Probability of Having 2 Boys and 2 Girls

7.13

Let's say a couple is planning to begin their family and would like to have 4 children. What is the probability that they will have 2 boys and 2 girls? What do you think? What problem-solving tools might help? Take some time to work on this question before reading on. . . .

DISCUSSION

Most initial hypotheses range from 20% to 50%. Let us investigate each of these.

Section 7.3 / Concepts Related to Chance **459**

CLASSROOM CONNECTION

In the May 1999 issue of *Mathematics Teaching in the Middle School*, Ann Lawrence uses two children's books, *The Giver* and *Twenty-One Balloons*, to engage her students in probability problems. In *Twenty-One Balloons*, a man invites 25 families to move to a remote island. He picks families that have exactly one girl and one boy. Ann began this unit by asking the children how often a family has exactly one girl and one boy.

7.17

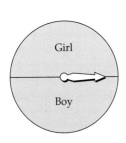

FIGURE 7.48

Those who answer 50% reason something like this: Each time, there is a 50% chance of getting a boy or a girl, so it makes sense that the "balanced" possibility (same number of boys and girls) will take place 50% of the time.

Those who answer 20% reason something like this: There are 5 possibilities (events)—4G0B (that is, 4 girls and 0 boys), 3G1B, 2G2B, 1G3B, 0G4B. Thus 2G2B is 1 out of 5, which is 20%.

Rather than present the answer just yet (presenting the answer too quickly means that more students are more likely to rent the solution rather than own it), let us do a **simulation**. The process of simulating a situation and then collecting and analyzing a large amount of data is called the Monte Carlo method. It is widely used in science and other fields where a problem is too difficult or complex to be addressed with other methods. The method has been used by mathematicians for some time but was given its current name by Stanislaw Ulam in 1946.

The simulation involves several steps.

Step 1: Make sure we clearly understand the problem.
Step 2: Select a method for modeling the problem.
Step 3: Gather the data.
Step 4: Analyze the data, which will result in an approximation of the actual probability.

The method is based on the **Law of Large Numbers**, which states that as we perform an experiment over and over, the experimental probability will converge on a fixed number. For example, if we toss a coin many times, the experimental probability of tails will converge on 0.5. If you have done any sampling, you know that, for example, if you toss a coin 10 times, you are likely to get tails 5 times, but this is not certain to happen. Even if you toss a coin 100 times, you are not certain to get tails 50 times. However, the law states that as you perform the experiment repeatedly, the experimental probability will converge—that is, it will work its way closer and closer to 0.5.

This is what scientists, economists, and mathematicians often do when they are beginning an investigation or when they have different hypotheses: They construct a model of the situation and then run a simulation. How might we run a simulation of this question? Please think for a minute yourself and then read on. . . .

There are many ways to simulate this scenario:

1. We could use a computer program to generate random digits. We could assign 0–4 to represent girl and 5–9 to represent boy.

2. We could roll dice. We could assign 1–3 to represent girl and 4–6 to represent boy.

3. We could write B and G on the same number of slips of paper, put them in a hat, and then draw pieces of paper out one by one. We would need to replace each piece of paper and mix the pieces after each drawing. Do you see why?

4. We could label a spinner as shown in Figure 7.48.

Your instructor may have you do this simulation in class. If not, do it yourself. As you may have learned from Explorations 7.12 through 7.17, we can use simulations not only to help us determine the experimental probability but also to understand better how to develop the theoretical probability.

With the knowledge gleaned from the simulation, let us now explore how the theoretical probability is determined.

STRATEGY 1: *Be systematic and look for patterns*

There are many ways to be systematic. Let us examine one way.

First, the couple could have all girls or all boys: GGGG and BBBB.

Note that we use letters to represent boy and girl rather than spelling them out. This saves time and makes patterns in the data easier to see.

Next, we could determine the different ways in which the couple could have 3 girls and 1 boy. If you did not do the problem this way, stop now and see whether you can determine the answer to this particular question. . . .

One way to do this systematically is to see that the boy could be in four different positions: oldest, second, third, youngest. Symbolically, this looks like

B G G G
G B G G
G G B G
G G G B

Do you see a pattern here? How would you describe it?

If you look only at the B, you can see it move diagonally. Do you see that this systematic approach also practically gives us the ways to have 3 boys and a girl?

If you stop and think for a minute, you might realize that because there are four different ways to have 3 girls, there must also be four different ways to have 3 boys. Do you see why? There are many symmetries in probability situations.

Finally, we need to determine how many different ways there are to have 2 boys and 2 girls. Do this first yourself before reading on. . . .

There are a variety of methods you might use to come up with the 6 different ways in which one could have 2 boys and 2 girls. (In probability language, we are saying that the event 2B, 2G consists of 6 outcomes.)

B G B G
G B G B
B G G B
G B B G
B B G G
G G B B

Because 6 of the 16 outcomes have 2 Bs and 2 Gs, the probability we sought is 6/16, or 37.5%.

STRATEGY 2: *Make a tree diagram*

Figure 7.49 is a tree diagram for this problem; it is a different way to represent and solve the problem. Can you interpret the tree diagram? Do you see why it is called a tree diagram?

CLASSROOM CONNECTION

When my daughter and son were small, they would come into our bed when they woke up. One morning, my daughter noted that there was a pattern. Our position was Emily, Dad, Mom, Josh. She said, "Girl, boy, girl, boy." She then jumped over me and said, "This is another pattern." I was expecting her to say, "Boys on the outside, girls on the inside." What she said was, "Oldest and youngest on the outside." I am older than my wife (by three months, which was amusing to our children when they were young: Sometimes I was "older" than my wife and sometimes we were the "same" age), and Emily is older than Josh.

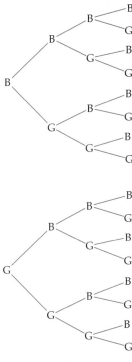

FIGURE 7.49

It is called a tree diagram because it branches out. We first begin with B and G to represent the possibilities with one child—a boy or a girl. Then a B and a G branch from the B and from the G. At this point, we have 4 outcomes: BB, BG, GB, and GG.

Now make a table of the 16 outcomes by using the tree diagram. (The table is not shown—you have to make it!) What patterns do you see in the table? Write the patterns you see and then read on. . . .

There is a pattern in each of the columns. The left column simply alternates between B and G. The next column alternates between 2 Bs and 2 Gs. The third column alternates between 4 Bs and 4 Gs. The last column consists of 8 Bs and 8 Gs.

Some students see the symmetric nature of the table. Pair each row in the table in the following way: row 1 and row 16, row 2 and row 15, and so on. Another way of seeing this is to fold the table in half and look at the rows that lie on top of each other. Each half is like a mirror image of the other: BBBB and GGGG, BBBG and GGGB, and so on.

Note that this table is not identical to the table that comes from Strategy 1, although they contain the same 16 outcomes. The other table exhibited different patterns. What is important here is that both tables represent a systematic approach to the problem. If a student were to generate the 16 outcomes in a more random fashion, there would probably be fewer patterns and hence a greater probability of missing one or more outcomes.

■ *Mathematics* ■

Blaise Pascal, whom we have encountered before, once said, "Chance favors the prepared mind." A powerful illustration of this occurred one year while my students were investigating this question. Some students noted that since the letters G and B are similar, it was hard to see patterns. One student replaced the Gs with 0 and the Bs with 1. The contrast between 0 and 1 is much greater than that between B and G. The student then listed all 16 outcomes on the board, noting that 0 represented B and 1 represented G. This notation made it easier to discuss patterns and also provoked another discovery. Does Table 7.13 remind you of something you have studied before?

TABLE 7.13

0 0 0 0
0 0 0 1
0 0 1 0
0 0 1 1
0 1 0 0
0 1 0 1
0 1 1 0
0 1 1 1
1 0 0 0
. . .
1 1 1 1

This table is simply the first 16 numbers in base 2!

Let us now examine two related questions for this couple, from which new probability concepts will emerge.

INVESTIGATION 7.19

Probability of Having 3 Boys and 2 Girls

What if the couple were planning to have 5 children? What is the probability of their having 3 boys and 2 girls? Work on this and then read on....

DISCUSSION

There are several different ways to answer this question. You can make a new table, extend the tree diagram, or extend the base 2 counting table and look for outcomes that show two 0s and three 1s or three 0s and two 1s. However, in any of these cases, obtaining the solution is relatively cumbersome. It becomes downright tedious if we extend the question further. For example, what is the probability of having 4 girls and 4 boys?

What is happening here is a classical example of a kind of situation that has contributed to the development of mathematics and science—a representation or a solution that worked well at some point becomes less useful at another point, and so a new approach or method is needed.

Table 7.14 represents another way of solving the problem: by looking at the number of outcomes in different events. I have filled in the first four rows of the table. Take a few minutes to make sense of the table.

TABLE 7.14

NUMBER OF GIRLS							Number of children
6	5	4	3	2	1	0	
					1	1	1
				1	2	1	2
			1	3	3	1	3
		1	4	6	4	1	4
	1					1	5

What this table shows is that if you have 4 children, there is 1 way to have 4 girls, there are 4 ways to have 3 girls (and 1 boy), there are 6 ways to have 2 girls (and 2 boys), there are 4 ways to have 1 girl (and 3 boys), and there is 1 way to have no girls (that is, all boys). Examine the numbers for 1 child, 2 children, and 3 children. If you didn't understand the table before, fill out the table now before reading on....

What patterns do you see? Describe them as though you were talking to someone on the phone....

There are small (local) patterns; for example,

- The bottom-left to top-right diagonal contains only 1s.
- The column representing 0 girls contains only 1s.
- The column representing 1 girl is the counting numbers: 1, 2, 3, 4, 5.

There are also large (global) patterns. For example, there is a connection between each row and the next row. Look at the third row and the fourth row. What do you see? Think and then read on. . . .

By now you may be smiling because you realize that you have seen this pattern before. Yes, this is simply Pascal's triangle, which you first saw in Investigation 1.5! Connecting this problem to Pascal's triangle makes our work much less tedious. If a couple has five children, what is the probability of 3 girls and 2 boys? Think and then read on. . . .

The cell representing 3 girls and 2 boys contains 10 outcomes, and there are 32 outcomes in all; therefore, the probability of 3 girls and 2 boys is 10/32, or about 31%. Similarly, the probability of having 2 girls and 3 boys is also 10/32. So now what is the probability that the couple will have 3 girls and 2 boys or 2 girls and 3 boys?

Mutually exclusive events We can simply add the two probabilities together. The justification for this goes back to set theory. The two events (3B, 2G) and (2B, 3G) represent disjoint sets. In probability language, we say that they are **mutually exclusive events**.

If events A and B are mutually exclusive, then $P(A \text{ or } B) = P(A) + P(B)$.

There is another global pattern that comes out of Table 7.14. It arises from adding a new column headed "Total number of outcomes." If you have not yet seen this pattern, add that new column and fill in the numbers. What do you see? Think and then read on. . . .

This column simply can be described most succinctly (in English) as "powers of 2."

INVESTIGATION 7.20

Probability of Having at Least 1 Girl

Let us examine another situation. Let's say a couple is planning to have 4 children. What is the probability that they will have *at least* 1 girl? Work on this and then read on. . . .

DISCUSSION

STRATEGY 1: *Refer to the table or tree diagram and count*

From the tables you created in Investigation 7.18, count those outcomes in which there is at least 1 girl and divide by 16.

STRATEGY 2: *Connect this problem to sets*

Let us explore this strategy because it has implications for other probability situations. What if you were asked, "What is the probability of having no girls?" Do you see how the answer to this question enables us to answer the question "What is the probability of having at least 1 girl?"

Do you see why the following formula is true: $P(\text{no girls}) + P(\text{at least 1 girl}) = 1$?

Consider the sample space for having 4 children, and put the outcomes into two subsets:

A = the set of outcomes containing no girls

B = the set of outcomes containing at least one girl

Complementary events In this case, all outcomes go into one subset or the other. Not only are the two subsets disjoint, their union is equal to the whole sample space. Since every outcome is in either set A or set B, that means that $A = \overline{B}$. In other words, these are complementary sets. In probability language, we say that these are **complementary events**. Not surprisingly, if we have two events such that event A and event B are complementary, then $P(A) + P(B) = 1$.

If we recall Investigation 7.19, we see that complementary events represent a special kind of mutually exclusive event.

In this case, the probability of A (the probability of having no girls) is much easier to determine than the probability of B (the probability of having at least 1 girl), and we can use basic algebra to transform the equation $P(A) + P(B) = 1$ into the equation $P(B) = 1 - P(A)$. We can now determine the probability of having at least 1 girl:

$$P(B) = 1 - P(A) = 1 - \frac{1}{16} = \frac{15}{16}$$

Let us now make use of our understanding of probability to investigate two related problems.

INVESTIGATION 7.21   50-50 Chance of Passing

PEANUTS reprinted by permission of United Feature Syndicate, Inc.

FIGURE 7.50

 7.12 Poor Peppermint Patty. Today there is a true-false quiz with 10 questions. She forgot to read the story on which the questions are based, and so she has to guess at every question (see Figure 7.50). Assuming that she has a 50% chance at each question, what is the probability that she will get 70% or better on the quiz? Think about this and then read on. . . .

DISCUSSION

STRATEGY 1: Make a tree diagram

In order to use a tree diagram for this problem, you would need a very big sheet of paper and a lot of time. However, the tree diagram points to another strategy. What is it?

STRATEGY 2: Be systematic and look for patterns

There is only one way she can get all 10 right. Why is this?

Now, how many ways can she get 9 out of 10 right? That is, how many different outcomes are there for getting 9 questions right? There are several ways to determine the number. Work on this question yourself before reading on. . . .

Some students reason like this: "She could get all but the last one (number 10) right, she could get all but number 9 right, she could get all but number 8 right, and so on." What pattern do you see here?

Some students see the systematic nature of this approach better by looking at, and filling out, the diagram below:

RRRRRRRRRW

RRRRRRRRWR

RRRRRRRWRR

etc.

Regardless of how we saw this problem, we find that there are 10 different ways in which she could get 9 out of 10 correct. More formally, we would say that there are 10 outcomes in the event "9 correct answers."

Now does the following row of Pascal's triangle make sense?

1   10   45   120   210   252   210   120   45   10   1

If not, analyze a 4-item test in which she guesses each time: There is 1 way to get all 4 correct, there are 4 ways to get 3 right, there are 6 ways to get 2 right, and so on. Can you use Pascal's triangle now to determine the probability of Patty's getting a score of 70% or greater? Think and then read on. . . .

If you have understood the application of Pascal's triangle to this problem, you have done the computation

$$\frac{(120 + 45 + 10 + 1)}{1024} = 0.17$$

Can you express the answer to the problem in a full sentence? Do so and then read on. . . .

There are many valid ways to express the answer. Here are two.

- "Peppermint Patty has a 17% chance of getting at least 70% right."
- "Peppermint Patty has about 1 chance in 6 of getting at least 70% right."

Does the answer surprise you? Is it higher or lower than you expected? Many teachers do not believe in true-false tests because they believe that a lucky student can appear to know more than that student really does. Does this investigation make you feel more or less disposed toward true-false tests? What if there were 20 questions and the student guessed at each one? Will the probability of getting at least 70% be greater or less than on a 10-item quiz?

466 CHAPTER 7 / Uncertainty: Data and Chance

INVESTIGATION 7.22

What Is the Probability of Rolling a 7?

If we roll one die, the probability of each outcome (that is, 1, 2, 3, 4, 5, or 6) is 1/6 (assuming a fair die). Suppose we roll 2 dice and find the sum of the numbers on the dice. What is the probability of rolling a 7? Work on this problem and then read on. . . .

DISCUSSION

There are several ways to be systematic, and there are several ways to represent this problem. Look at each of the following partially completed representations. How would you finish them?. . . Now, complete the tables.

Representation 1

Table 7.15 was created by thinking systematically.

TABLE 7.15

1, 1	2, 1	?, ?
1, 2	2, 2	?, ?
1, 3	?, ?	?, ?
1, 4	?, ?	?, ?
1, 5	?, ?	?, ?
1, 6	?, ?	?, ?

Representation 2

Table 7.16 was created by thinking systematically in a slightly different way.

TABLE 7.16

1, 1	1, 2	?, ?	. . .
2, 1	2, 2		
3, 1	3, ?		
4, 1	?, ?		
5, 1	?, ?		
6, 1	?, ?		

Representation 3

Figure 7.51 is yet another tree diagram!

History

The development of probability theory was spurred by gamblers' questions. The great Italian scientist Galileo (1564–1642), who is better known for getting in trouble with the Inquisition by insisting that the earth revolved around the sun, was asked why one is more likely to get a sum of 10 than a sum of 9 when rolling three dice and adding the numbers. The foundations for probability as a mathematical discipline were laid by the work of Blaise Pascal (1623–1662) and Pierre Fermat (1601–1665) as they responded to several questions asked by a professional gambler, Chevalier de Mere. One of those questions was: Why would one lose money by betting (even money) that you would get at least one double 6 when throwing 2 dice 24 times?

Many people object to students studying gambling theory in school, so we will de-emphasize the gambling connections between dice and probability. This is not hard because dice are also used in many board games.

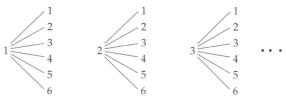

FIGURE 7.51

In each of these cases, we find that there are 36 distinct possible outcomes. This is a useful time to present formally the **multiplication principle**, which can be stated in outcomes or in probabilities:

Stated in number of outcomes If event A consists of a outcomes and event B consists of b outcomes, then the event "A then B" consists of $a \cdot b$ outcomes. In this case, event A is rolling a die and event B is rolling a die; both consist of 6 outcomes. Thus the event "roll a die and then roll it again" consists of $6 \cdot 6 = 36$ outcomes.

Stated in probabilities If the probability of event A is x and the probability of event B is y, then the probability of the event "A then B" is $x \cdot y$.

We can now count the number of outcomes that produce a 7 and answer the original question. The probability of rolling a 7 is 6/36, or 1/6.

What if we wanted to know the probability of *each* event—that is, of rolling a 2, a 3, a 4, and so on? How can you use your work on the previous question to answer this question? What patterns do you see in your previous work that would help you to make the table? Describe the patterns you see in your table and then read on. . . .

One way of describing what we find is to say that each time the sum increases, the number of outcomes increases by 1 until we get to (the event) 7. Then the number of outcomes decreases by 1 (for each event) until we get to 12 (see Table 7.17).

TABLE 7.17

Number (event)	Frequency
2	1
3	2
4	3
5	4
6	5
7	6
8	5
9	4
10	3
11	2
12	1

468 CHAPTER 7 / Uncertainty: Data and Chance

INVESTIGATION 7.23

What Is the Probability of Rolling a 13 with 3 Dice?

Let us extend the investigation now to three dice. Suppose we rolled 3 dice and added the 3 numbers; what is the probability of rolling a 13? Work on this question before reading on. . . .

DISCUSSION

First, we can use the multiplication principle to determine the total number of outcomes from rolling 3 dice—that is, the size of the sample space. There are $6 \cdot 6 \cdot 6 = 216$ possible outcomes!

How many of those outcomes produce a 13? Work on this and then read on. . . .

As you work, make note of what problem-solving strategies help you do the problem, help to keep you on track, and help you check your solution. Virtually all solutions to this problem require the solver to be systematic. However, there are different ways to be systematic, and you can use many different representations that will help answer this question. Below are the beginnings of two different strategies.

Representation 1

Table 7.18 is a table created by thinking systematically.

TABLE 7.18

661	652	643	553	?
616	625	?	?	
166	562			
	526			
	265			
	256			

Representation 2

Table 7.19 is another table created by thinking systematically.

TABLE 7.19

661	562	463	?
652	553	?	
643	544		
634	535		
625	?		
?			

Regardless of the representation used to solve the problem, we find that there are 21 outcomes that produce a sum of 13. Therefore, the probability of rolling a 13 is 21/216, or slightly under 1 in 10.

An alternative strategy is to construct the entire sample space. The completed tables are shown at the end of this chapter on pages 491–493. If you find all 216 outcomes in the sample space, then you can make use of patterns to count the number of outcomes in the event called "sum of the 3 dice equals 13." Making the table is an exercise in thinking about and looking for patterns that some people find fun and others find tedious. If you are in the latter group, you may look at the two different tables, each containing all 216 outcomes, at the end of the chapter. Look for patterns—my students and my children have found hundreds!

INVESTIGATION 7.24

"The Lady or the Tiger"

There is a famous story, written by Frank Stockton, called "The Lady or the Tiger," which I have modified slightly to make an interesting problem. It seems that the king and queen had arranged for their daughter to be married to a prince, but she fell in love with a peasant. When the king discovered this affair, he ordered that the peasant be thrown into a room full of tigers. However, in response to his daughter's pleas, he agreed to have the peasant walk through a maze to one of two rooms (see Figure 7.52). The princess would be waiting in one of the rooms, and the tigers would be in the other room. The princess asked if she could choose the room in which she would wait; the king agreed, for he believed that the chances were equal. If you were the princess, which room would you choose? Think about this and then read on. . . .

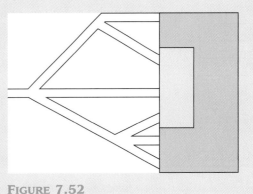

FIGURE 7.52

DISCUSSION

STRATEGY 1: *Solve this as a tree diagram*

Can you take your understanding of the tree diagrams used in other problems and apply it to this problem? Try to do so now; determine the probabilities of each event (the princess or the tigers), and then read on. . . .

Figure 7.53 shows 6 different doors (1 through 6) that open into two different rooms (A and B). The probability that the peasant will end up in room A = 1/6 + 1/3 + 1/9 = 11/18. Do you see why?

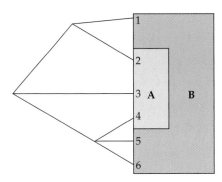

FIGURE 7.53

We can see why by breaking this problem into parts. We are assuming, of course, that at each point there is an equally likely chance that the peasant will take any path. The probability that, at the first junction, he will take the top path is 1/3; and the probability that, at the second junction, he will take the top path is 1/2. Therefore, using the multiplication principle, the probability that he will end up at door number 1 is (1/3)(1/2) = 1/6. We can determine the probabilities of arriving at the other five doors in a similar manner.

Because A and B are complementary events, we can conclude that the probability that he will end in room B is 1 − 11/18, or 7/18. Thus the princess should choose to wait in room A.

STRATEGY 2: Use an area model

At the beginning of the maze, there are three paths. Assuming that each is equally likely to be taken, we can represent them as having the same area (see Figure 7.54).

Figure 7.55 represents the outcomes of all of the paths with respect to the probabilities. Do you understand the diagram? How would you explain it?

The fraction of the area of the diagram that is allotted to the three As represents the probability that the peasant will end up at room A—that is, 1/6 + 1/3 + 1/9 = 11/18. Do you see why?

This next situation is somewhat different from the previous ones. Let us first explore the problem and then examine what is different about it.

Top path
Middle path
Lower path

FIGURE 7.54

A		B
A		
A	B	B

FIGURE 7.55

INVESTIGATION 7.25

 7.15

Gumballs

Josie asks her dad for money to buy 2 gumballs, 1 for her and 1 for her brother. There are only 4 white gumballs and 2 red gumballs left in the machine (see Figure 7.56), and Josie really likes the red ones because they are spicy. What is the probability that she will get at least 1 red gumball? Work on this question and then read on. . . .

FIGURE 7.56

DISCUSSION

STRATEGY 1: Do a simulation

We need to be careful, for there are ways to go wrong. However, if you write R on two slips of paper and W on four slips of paper and then simulate taking out two gumballs, after a number of trials your experimental probability will be more and more likely to be close to the theoretical probability. Some students find that they don't really understand the problem at first, but after several trials with the simulation, they suddenly understand the problem more clearly and can then go back and determine the theoretical probability.

STRATEGY 2: *Use reasoning*

One student claims that there are 4 possible outcomes—RR, RW, WR, WW—and so the probability of getting at least 1 red is 3/4. What do you think?

Plausible as this sounds, it is wrong. The mistake here is similar to the mistake that students often make in the boys-and-girls problem when they reason that there are five possible ways to have 4 children: 4G0B, 3G1B, 2G2B, 1G3B, and 0G4B. This line of reasoning is valid only if each outcome is just as likely to occur as any other.

If we think back to that problem (or to the dice problems) and look at ways to apply what we learned there to this problem, a tree diagram comes to mind. Explore a tree diagram before reading on. . . .

STRATEGY 3: *Make a tree diagram*

Figure 7.57 shows one way to represent this problem with a tree diagram. There are 6 trees because there are originally 6 gumballs. Each tree shows the outcomes if that gumball comes out first. That is, if a red gumball comes out first, then there are 5 gumballs left, 1 red and 4 white. However, if a white gumball comes out first, then there are 2 reds and 3 whites left.

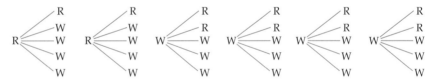

FIGURE 7.57

From Figure 7.57, we find that there are 30 outcomes and that the event "at least 1 red" occurs 18 times. Thus, the probability of getting at least 1 red gumball is 18/30, or 0.6.

STRATEGY 4: *Make a more sophisticated tree diagram*

We have seen that many developments in mathematics and the invention of many notations were spurred by a desire to find faster ways to determine answers and more efficient ways to communicate them. In this case, some people look at the tree diagram in Figure 7.57 and say, "There's a simpler way." This simpler tree diagram is shown in Figure 7.58. Study it and try to make sense of it. Can you see the solution contained within the diagram?

When we put the money in the machine, the probability of getting a red is 1/3, and the probability of getting a white is 2/3. Let's say we get a red the first time. In this case, the probability of getting a red is now 1/5, and the probability of getting a white is 4/5. Why is this? Similarly, let's say we get a white the first time. In this case, the probability of getting a red is now 2/5, and the probability of getting a white is 3/5.

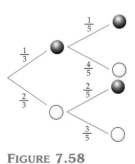

FIGURE 7.58

We can use the multiplication principle to determine the probabilities of each branch of the tree, and we can check our work, because the sum should be 1. Do you see why?

$$P(RR) = \frac{1}{3}\left(\frac{1}{5}\right) = \frac{1}{15}$$

$$P(RW) = \frac{1}{3}\left(\frac{4}{5}\right) = \frac{4}{15}$$

$$P(WR) = \frac{2}{3}\left(\frac{2}{5}\right) = \frac{4}{15}$$

$$P(WW) = \frac{2}{3}\left(\frac{3}{5}\right) = \frac{6}{15}$$

These events are mutually exclusive (why is this?), so we can conclude that the probability of getting at least 1 red is

$$\frac{1}{15} + \frac{4}{15} + \frac{4}{15} = \frac{9}{15} = 0.6$$

Dependent and independent events Now let us return to the question posed at the beginning of the investigation: How is this scenario different from the ones we have explored before? Think and then read on. . . .

With the gumballs, the probability of obtaining a red or white gumball the second time depended on what happened the first time. In the case of having boys and girls or throwing dice, the probability of the second stage of the activity does not depend on what happened first; that is, its probability is independent of the first result.

Two events are **dependent** if the probability of the second event is affected by the outcome of the first event.

Two events are **independent** if the probability of the second event is not affected at all by the outcome of the first event.

Fair Games

Another application of probability has to do with the idea of fair games and expected value, which has important real-life applications. Let us begin with a simple situation.

INVESTIGATION

7.26

Is This a Fair Game?

Consider the following game for two players. Each person spins the spinner (see Figure 7.59). If the spinner lands on the same animal twice, the first player wins. If it lands on a different animal each time, the second player wins. If you could choose, would you want to be the first or the second player, or doesn't it matter? Think about this and then read on. . . .

FIGURE 7.59

DISCUSSION

Can you see similarities between this problem and others we have investigated in this section? If not, look back and try to find connections. Then answer the question and read on. . . .

This problem is similar to the questions about the probability of having boys and girls. Do you see why?

Applying these similarities, we find that there are four equally likely outcomes in this game: (1) cat, cat, (2) dog, dog, (3) cat, dog, and (4), dog, cat. Thus each player has the same theoretical probability of winning.

Using probability language, we say that this is a *fair game*. How would you define *fair game* at this point? Write down your thoughts and then read on. . . .

INVESTIGATION 7.27

What About This Game?

A school is having a carnival to raise funds. In the following game, the player spins the spinner (Figure 7.60) and receives the dollar amount of the number on which the arrow lands. If it costs $2 to play the game, is this a fair game? What do you think?

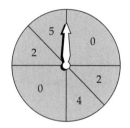

FIGURE 7.60

DISCUSSION

STRATEGY 1: Make a simulation

You could make a spinner like the one in Figure 7.60 and simulate 100 games. If you made approximately $200, the game is probably a fair game. If you didn't have a spinner, how else could you simulate this game?

First, we need to determine the probability of each outcome. That is, $P(0) = 1/2$, $P(2) = 1/4$, $P(4) = 1/8$, $P(5) = 1/8$. So you could make 8 pieces of paper and write 0 on four of them, 2 on two of them, 4 on one of them, and 5 on one of them, and then draw, making sure to put the piece of paper back each time.

STRATEGY 2: Determine the theoretical probabilities

We have spoken of theoretical and experimental probabilities earlier. Can we apply the concept of theoretical probability to theoretical winnings?

We can compute the theoretical winnings per turn by using the concept of weighted average (from Section 7.2):

$$0.125(5) + 0.125(4) + 0.25(2) + 0.5(0) = 0.625 + 0.5 + 0.5 = 1.625$$

That is, theoretically, 1/8 of the time the player will win $5; 1/8 of the time $4; 1/4 of the time $2; and 1/2 of the time nothing. Some students own this solution, whereas others rent it. If this solution falls into the category of "it sort of makes sense," what might you do to make it "really make sense"? Think and then read on. . . .

One strategy would be to act it out. For example, let's say the person played the game 80 times. Theoretically, how much money will the player win in 80 turns? (Do you see why I chose 80 turns instead of 50 or 100 turns?)

> **Outside the Classroom**
>
> Most states now have lotteries. In one lottery scenario, players pick four numbers between 1 and 50. The state then randomly selects four numbers from the sample space—for example, 2, 5, 23, and 47. In another lottery scenario, the players pick four one-digit numbers between 0 and 9. The state then randomly selects one digit at a time, replacing the selected number before drawing the second number. In this case, 1, 1, 2, 2 is a possible winning combination. The first situation is an example of dependent events, and the second is an example of independent events.

Theoretically, 1/8 of the time, or 10 different times (because 10 is 1/8 of 80), the player will win $5, and thus the player will win a total of $50 on those 10 turns. When we apply this strategy to the $4 and $2 outcomes, we find that theoretically the player's total winnings will be $130. Do you see that this is equivalent to the weighted average we obtained above?

If the player wins $130 after 80 turns, the winnings are an average of $1.625 per turn. Referring now to the original question, we find that theoretically, the player will lose money playing this game because it costs $2 to play the game. It is important to note that we are not saying that the player will lose. The player might be very lucky and play 10 games and win $50. However, playing a large number of games makes losing money more likely than winning it.

Can you revise your original definition of *fair game* so that it works for Investigations 7.26 and 7.27?

Expected Value

There is a term for the theoretical winnings per turn: **expected value**. The concept of expected value is an aspect of probability that has many real-life applications. For example, car insurance premiums are based on the insurance company's estimate of the probabilities of different groups of people having accidents. Our knowledge of expected value also arose from gambling; gamblers wanted to make sure that, over time, they would be more likely to make money than to lose money. If an experiment (activity) consists of n events, and each event has a specific probability and a specific payoff, we can determine the expected value of that experiment (activity) in the following manner:

Let us denote each event as $E_1, E_2, E_3, \ldots, E_n$.

Let us denote the probability of each event as $P_1, P_2, P_3, \ldots, P_n$.

Let us denote the payoff of each event as $X_1, X_2, X_3, \ldots, X_n$.

The expected value $= P_1 \cdot X_1 + P_2 \cdot X_2 + P_3 \cdot X_3 + \cdots + P_n \cdot X_n$

The concept of expected value is used to define a fair game formally. Before reading on, try to describe whether a game is fair or not using the concept of expected value. . . .

A game is "fair" if and only if the expected value of the game equals the cost of playing the game.

Let us now extend the concept of expected value to a real-life application.

INVESTIGATION 7.28 — Insurance Rates

Let's say that an insurance company is determining the rates for car insurance for the next year. The company has determined that for every 10,000 policyholders:

- 1 is likely to have an accident for which the company will have to pay a $200,000 claim.
- 10 are likely to have an accident for which the company will have to pay a $100,000 claim.
- 20 are likely to have an accident for which the company will have to pay a $50,000 claim.

- 100 are likely to have an accident for which the company will have to pay a $10,000 claim.
- 1000 are likely to have an accident for which the company will have to pay a $1000 claim.
- The remainder of the policyholders will have no claims for the year.

Do you understand how the insurance company will use the concept of expected value in determining its rates? If so, determine the expected value of the claims and then read on. . . .

If you don't see the connection between this situation and the others, please go back and reread the previous investigations and the definition of expected value. Do you see connections now?

DISCUSSION

This investigation has many connections to Investigation 7.27, in which each player asks, "Will the money I win (take in) be greater than the money that I put in?" Because the events did not all have the same probability, it was necessary to use the concept of weighted average, which was developed in Section 7.2.

In order to compute the expected value in the insurance situation, the insurance company wants to get an idea of the "average" payout per customer. This problem is a model of what insurance companies do, not an account of what they actually do. Given the existence of powerful computers, they can use more sophisticated techniques than we are using here. However, this simplified model is useful in understanding the concept of expected value and how it applies to the business world. When we say that 20 people are likely to have an accident for which the insurance company will have to pay $50,000, we are really saying that the company expects about 20 out of 10,000 policyholders to have accidents that will cost it in the neighborhood of $50,000. Thus, even before computing the expected value, there has already been some averaging and approximating.

In order to determine the expected value, the incidence of each event must be converted to probability language. For example, "1 in 10,000 claims will be in the neighborhood of $200,000" translates to "0.0001 probability of a $200,000 claim."

Thus,

$$\text{Expected payout} = 0.0001(200{,}000) + 0.001(100{,}000) + 0.002(50{,}000) + 0.01(10{,}000) + 0.1(1000) = 420$$

What does this 420 mean? Can you explain what it means in one or two sentences, as though you were talking to a fellow student who missed that class and doesn't quite own this concept? Write your rough draft and then read on. . . .

It means that if the accidents occur at the rate and amount that the insurance company is predicting, it will pay out an average of $420 per policy (per year). The actual rate the company charges will be determined by this expected value, the company's overhead (salaries, cost of operations, etc.), and the profit margin.

> ■ *Outside the Classroom* ■
>
> One of my sisters is an insurance agent. As I was consulting with her when writing this material, she told me something that surprised me. I had always thought that insurance companies would ideally like to have as large a percentage of policyholders as possible. That is, if a company had 35% of the market in a city or state, it would strive to keep increasing this percentage to get it higher and higher. However, this is not the case. Can you imagine why an insurance company would not want to have 100% of the policyholders in an area? Think and then read on. . . .
>
> The reason companies don't want to have all the clients has to do with natural disasters. Several major natural disasters have recently brought even some of the big companies to their knees: the Midwest floods in the summer of 1993, the California earthquake in 1994, and Hurricane Katrina in 2005. The total damage from Hurricane Katrina is approaching $200 billion. Had even the largest insurance company had most of the homeowners' policies in this region, it would have gone bankrupt. Therefore, insurance companies determine the maximum percentage of total policyholders in any particular area that they want to insure.

As before, some students may better understand the problem by determining the expected payoff for the 10,000 policies.

$$\$200{,}000(1) + \$100{,}000(10) + \$50{,}000(20) + \cdots$$

Do you see how this strategy and the previous strategy are connected?

Having performed several explorations and investigations on probability in this section, you are likely to have found that when we try to make predictions based on a small sample, we may or may not be close to the actual probability. For example, if you perform only 10 trials of having 4 children, you might get 2 boys and 2 girls 50% of the time, but that is not close to the theoretical probability of 3/8. In other words, you have found that if you want to be more confident of your answer, you need to perform more trials. However, this will not necessarily get you closer. For example, you could flip a coin 10 times and find 5 heads but flip the coin 100 times and find 60 heads. In this case, the experimental probability from the larger sample is not closer to the actual probability. Thus, we *cannot* say that the more trials we perform, the closer we will get to the actual probability. What we *can* say is that as we perform an experiment over and over, the experimental probability will converge on a fixed number. This generalization is called the Law of Large Numbers.

Summary

In this section, we have developed some basic language for exploring and describing probability situations. Probabilities can be expressed as rational numbers between 0 and 1 and can be expressed using percents, ratios, and odds. We can discuss probabilities using set language, in which case the set of all possibilities is called the sample space. We distinguish between outcomes, which are elements of the sample space, and events, which are subsets of the sample space.

We have learned other distinctions between different kinds of probability situations.

- In some cases, all outcomes are equally likely; in other cases, this is not true.
- Some events are independent and some are dependent.
- In all situations, we can determine the experimental probability of an event or an outcome by collecting data; in some cases, we can determine theoretical probabilities using different problem-solving tools.

We have examined several rules and formulas that have come out of making sense of situations and understanding connections between probability and other areas, such as sets, weighted average, and our understanding of operations.

- Complementary events: $P(A) + P(B) = 1$
- Mutually exclusive events: $P(A \text{ or } B) = P(A) + P(B)$
- The multiplication principle: If $P(A) = x$ and $P(B) = y$, then $P(A \text{ then } B) = x \cdot y$
- Expected value: $P_1 \cdot X_1 + P_2 \cdot X_2 + P_3 \cdot X_3 + \cdots + P_n \cdot X_n$

We have found that there are many different techniques we can use to solve probability problems: tree diagrams, formulas, area models, and simulations. These techniques work better when they are used with other problem-solving tools: making tables, looking for patterns, being systematic, and using reasoning.

Some real-life situations are similar to ones we have studied here—for example, the probabilities associated with genetics, as in Exploration 7.13. Some real-life situations are more complex than the ones we have discussed here, and additional tools are required to solve those problems.

EXERCISES 7.3

1. **a.** Which of the following would you choose if you were picking 1 ball and wanted black? Justify your choice.

   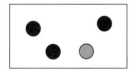 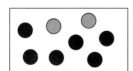

   b. What if you were picking 2 balls (without replacement) and wanted 2 black ones?

2. A drawer has 4 red socks and 4 blue socks.
   a. If 2 are drawn, what is the probability of a match?
   b. If 3 are drawn, what is the probability of a match?
   c. What is the probability of having all one color after 4 draws?

3. **a.** Drawing from a regular deck of playing cards, what is the probability of drawing a face card (jack, queen, or king) 3 times in a row? (Assume that jokers are taken out, and assume no replacement.)
   b. How would the probability change if you assumed replacement?
   c. Marla believes that the answer to the second question is 3/13 + 3/13 + 3/13. Describe how you might convince her that this answer is incorrect.

4. If it snows, there is a 33% chance that schools will close. The weather person says that there is a 25% chance of snow tomorrow. What is the probability that there will be no school tomorrow?

5. The American Lung Association states that approximately 1 out of every 17 deaths in the United States last year was from lung cancer. If there are 34 students in a class, does that mean that 2 people in this class will die from lung cancer?

6. Let's say there is a 15% chance of getting strep if you are exposed to a person with strep. Let's say there is a 50% chance that you were exposed.
   a. What is the probability that you will get strep?
   b. Let's say each person in your (4-person) family had a 50% chance of being exposed. What is the probability that you will all get strep?

7. Consider the following game: Each person rolls two standard dice and makes a proper fraction with the two numbers.
   - If the fraction is in simplest form, the first player wins.
   - If the fraction is not in lowest terms, the second player wins.
   - If the two numbers are identical, the second player wins.

   Is this a fair game?

8. Consider a pair of pyramid-shaped dice (that is, dice containing the numbers 1 through 4, rather than 1 through 6 as on regular dice. If you roll two dice at a time, what is the most likely sum and what is the probability of that sum?

9. What are the possible results from rolling two dice and subtracting the number showing on the face of one from that showing on the face of the other in such a way that a nonnegative number is obtained? What result is most likely, and what is its probability?

10. The following experiment consists of rolling a 12-sided die. The outcome set is {1, 2, 3, 4, 5, 6, 7, 8, 9, 10, 11, 12}. What probability would you assign to each of the following outcomes?
    a. An even number turns up.
    b. A prime number turns up.
    c. A divisor of 12 turns up.
    d. What is the most probable event if you roll two such dice and add the numbers? What is its probability?

11. Determine the probability of getting 8 heads in a row when flipping a coin 8 times. Give the answer both in decimal form and in ratio form. In decimal form, round the answer to two digits past the string of zeros, and write the decimal you obtained in words. By ratio form, I mean "the chance is 1 in ___."

12. If you flip a coin 10 times, what is the probability of getting at least 4 heads?

13. In the board game Monopoly, if you roll three doubles in a row, you go to jail. What is the probability of rolling three doubles in a row? Describe three ways in which you could represent this probability. Then select the one that you think would be most understandable to the general population. For example, if you were writing an article for the daily newspaper about math in society, which representation would you choose and why?

14. When my children were young, they played Junior Monopoly, which uses only one die. One day, Josh rolled a 2, then Emily rolled a 2, then I rolled a 2. Josh and Emily thought this was great fun. What is the probability of rolling three 2s in a row on three tosses of a die?

15. If we look at Table 7.23 (p. 492) and add the number of outcomes in each column, the numbers for the first six columns are 1, 3, 6, 10, 15, 21 which is like the Handshakes Problem in Exploration 1.1. However, the sum of the outcomes in the seventh column is 25 (not the 28 we got with handshakes). Can you explain why we get 25 instead of 28?

16. Make 2 six-sided dice so that only even sums are possible and each sum is equally likely. Your solution must be other than the trivial solution of "the same odd number on all 12 faces."

17. a. The following dates all have a similar pattern. Describe that pattern.

    6/12/72   3/25/75   5/19/95

    b. What is the probability of such a date in the year 1996?
    c. In which year of the century will such dates be most probable?

18. Recall the gumball investigation. What if Josie spent a third quarter? Do you think $P(WWR) = P(WRW) = P(RWW)$? Why or why not?

19. Let's say a four-year-old child sits down at the word processor and strikes four letters at random. What is the probability that the child will spell the word *math*? What assumptions did you make in answering this question? For example, is the probability of striking every key the same? That is, do you assume that they are all equally likely outcomes?

20. A popular Hanukkah game is called Dreidel. Each player begins with a number of pieces of gold (actually chocolate in a round, gold-colored wrapper in the shape of a coin). The game consists of spinning a top that has four sides.
    - If the top lands on the side labeled gimel, the player gets all the gold in the pot.
    - If the top lands on the side labeled shin, the player has to put one coin in the pot.
    - If the top lands on the side labeled hay, the player gets half of the gold in the pot.
    - If the top lands on the side labeled nun, the player gets nothing.

    a. What is the probability of getting no gimels after 4 turns?
    b. What is the probability of the top landing on each of the four sides on your first 4 turns?

21. Consider a multiple-choice quiz consisting of 5 questions with 3 possibilities per question. What is the total number of possible outcomes? What does "outcome" mean here?

22. Write a program to simulate flipping a coin 1000 times. Analyze the data, using what you learned about statistics in Section 7.2. For example, are the data normally distributed? Do 68% of the data lie within 1 standard deviation of the mean? Are the mean, median, and mode identical?

23. Let's say 1000 people play the carnival game in Investigation 7.27 over the course of the carnival. Theoretically, how much money will the school make from this game?

24. Let's say you wanted to make the carnival game in Investigation 7.27 a fair game. Describe one situation that would be a fair game. Explain why that game would be fair.

25. Write a computer program to simulate the carnival game in Investigation 7.27 and analyze the results.

26. There is a carnival game called chuck-a-luck. The game works this way:

    The player picks a number from 1 to 6. The operator then rolls 3 dice.
    If the player's number comes up all 3 times, the player wins $3.
    If the player's number comes up 2 times, the player wins $2.
    If the player's number comes up 1 time, the player wins $1.
    If the player's number doesn't come up, the player pays the operator $1.
    If the game costs $1, is this a fair game?

27. Let's say a state is considering introducing a new lottery game. The state will charge $1 for a ticket. Each week, one person's name will be selected at random. The amount of the person's prize will be determined in the following manner: Five Ping-Pong balls—four white and one black—will be placed in a hopper. The balls will be selected one at a time until all five balls are in a line.

    - If the black ball is first, the person wins $1,000,000.
    - If the black ball is second, the person wins $100,000.
    - If the black ball is third, the person wins $10,000.
    - If the black ball is fourth, the person wins $1000.
    - If the black ball is fifth, the person wins $500.

    a. Is this a fair game?
    b. The state expects to sell 250,000 tickets each week. Do you think this game will be a good money raiser? Why or why not?

28. Just as insurance companies use the concept of expected value in determining insurance rates, oil companies use expected value in determining whether or not to drill for oil. Let's say that an oil company is considering drilling for oil in a certain area. The company estimates that it will cost an average of $200,000 to drill each oil well in this area. It has determined that there are three kinds of outcomes: a dry well, a "medium" strike that would produce about $750,000, and a big strike that would produce about $2,500,000. The company has estimated that the probability of a medium strike is 1/5 and the probability of a big strike is 1/25. Should the company drill for oil? Determine the expected value, and then describe your conclusion as though you were talking to a group of speculators who don't understand expected value.

29. Describe a real-life scenario whose occurrence is a probability and whose probability you would be interested in determining.

    a. Define the scenario.
    b. Describe how you might go about determining the probability.
    c. Describe some aspects of the problem that would influence the actual probability.

    I have sketched one scenario below; this would be the beginning of a good response.

    a. The probability that I will get a job after I graduate.
    b. Interview professors who work with teacher certification and average their guesses; interview principals in this area and average their guesses; contact those students who got certified at KSC last summer and determine how many got jobs.
    c. Probabilities will differ depending on whether I will look only in this area, what my certification is (early childhood, elementary, middle, etc.), whether I am 22 or 32, and so on.

30. The inspiration for this problem came from one of my Middle School Mathematics Methods students. Let's say you have a bag that contains 10 colored beads. You can pull out one bead at a time and then put it back and then shake up the bag. Let's say you have done this five times and have pulled out a red bead three times, a blue bead, and a green bead. What are the facts and what inferences do you have about the contents of the bag? List and justify your facts and inferences.

31. *Classroom Connection* The inspiration for this problem came from a question asked of me by a parent. Let's say the local middle school has five sixth-grade mathematics teachers, two of whom are female, four seventh-grade mathematics teachers, one of whom is female, and four eighth-grade mathematics teachers, one of whom is female. Assuming that the students' placement is random, what is the probability that a student will have at least one female mathematics teacher during the student's three years at the school?

32. Here are some additional questions inspired by *The Giver* and *Twenty-One Balloons*. Parts (d) and (e) come directly from the article cited in Investigation 7.18.

    a. In a family of 6 children, what is the probability of having 3 girls and 3 boys?
    b. In a family of 8 children, what is the probability of having 4 girls and 4 boys?
    c. In a family of 10 children, what is the probability of having 5 girls and 5 boys?
    d. "As the number of children in a family increases, what happens to the probability of having the same number of boys and girls, and why does this happen?" (p. 508)
    e. From *The Giver*: "Each class of 50 children in the Community has 25 girls and 25 boys. Assuming that the probability of a woman giving birth to a boy or a girl is the same, how often do you think exactly 25 out of 50 babies will be girls?"[14]

[14]Lois Lowry, *The Giver* (Boston: Houghton Mifflin Company/Walter Lorraine Book, 1993).

SECTION 7.4 COUNTING AND CHANCE

WHAT DO YOU THINK?

- What does "counting" mean in the context of probability?
- What is the difference between a combination and a permutation?

 7.21   7.22

In many probability problems, simply determining the size of the sample space can be quite tedious. In this section, we will examine some algorithms that can make solving probability problems much easier. As has been the case throughout the book, the goal is not simply to "get" the algorithms but to understand why they work. There is a reasonable probability that some of you will never use the algorithms developed in this section with your students. However, there is a reasonable probability that all of you who go on to teach elementary mathematics will have your students engage in simpler counting investigations. What then can you expect to get out of this section?

These investigations will help you to apply problem-solving tools to probability concepts. They focus on representing a problem in useful ways and solving a problem by making a similar, simpler problem and then generalizing from the simpler problem to the actual problem. They call on your skills in recognizing patterns and then using those patterns to solve the problem. They build on your ability to communicate and reason mathematically. They encourage you to see connections among probability problems and connections to other problems we have already investigated.

INVESTIGATION 7.29 How Many Ways to Take the Picture?

Let's say your college has a chapter of Kappa Delta Pi, the national education society. Let's say this was a new society in your college, and in the first year there were 9 members. Let's also say that the members decided to have a group picture taken and wanted only one row—that is, they wanted everyone in the first row. In how many different ways can they line up for the picture? Work on this question yourself before reading on. . . .

DISCUSSION

This is one of the many problems in mathematics where many people find that if they see a solution, they see it right away, and if they don't, they're stuck. Even if you were systematic, you could become stuck trying to do this one:

1 2 3 4 5 6 7 8 9

1 2 3 4 5 6 7 9 8

etc.

Stuck, that is, unless you have problem-solving tools. If you are stuck, look at the steps for problem-solving on the inside front cover of the *Explorations* volume. What strategies might help? Then read on. . . .

STRATEGY 1: Make a simpler problem

There are several different strategies that can be successfully applied to this problem. The first one we will discuss is "make a similar, simpler problem and then work up." If you didn't make much progress on your own and you like this method, use it to finish the problem.

Let's see how it works. Let's start with a club with only 3 members:

1 2 3
1 3 2
2 1 3
2 3 1
3 1 2
3 2 1

This gives us a total of 6 different arrangements. Where did the 6 come from?

Most people see 3 groups of 2, and hence 3×2. Some people can jump to the solution of the original problem from here; others need more data from which to generalize.

If you don't yet feel confident that you can connect this to the original problem, try a club with 4 members. If you haven't done this, try it on your own before reading on. . . .

1 2 3 4
1 2 4 3
1 3 2 4
1 3 4 2
1 4 2 3
1 4 3 2

This gives us 6 possibilities in which person number 1 is on the left. Either by reasoning or by working out the other possibilities yourself, you can see that there will be 6 outcomes in which each of the other 3 members is on the left. This gives us a total of 24 outcomes. Where did the 24 come from?

As before, we have 4 groups, each of which consists of 2 groups of 3. If we put the numbers in order, we have either $4 \times 3 \times 2$ or $2 \times 3 \times 4$.

If you feel that you can jump from here to the solution to the problem, do so. If not, then you can either determine the number of possibilities for a group with 5 members or try another strategy.

STRATEGY 2: *Make a tree diagram*

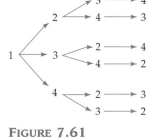

FIGURE 7.61

In this case, a tree diagram (Figure 7.61) is not terribly helpful for most students. In Section 7.3, we found that a tree diagram was helpful when we were looking at having 2 or 3 or 4 children. However, as the number of children increased, the tree diagram became less helpful and more tedious. I mention this here because many students who struggle in math often think that the better students and the teacher *always* know what to do and *always* pick the "right" strategy, and this is not true. I know many students who have done well in the course who admit that they frequently found that their first attempt at a challenging problem was not productive and that they needed to try another strategy.

STRATEGY 3: *Connect this problem to something familiar*

What about the multiplication principle? Many people understand this strategy better if we simultaneously use the "act it out" strategy.

Let's begin with the first person. We have nine possibilities. Once we have that person set, how many choices do we have for the second position? There are eight possibilities. Do you see why? Do you see how to finish the problem using this line of reasoning? Work on it before reading on. . . .

Using this line of reasoning, the number of possibilities will be 9 times 8 times 7 times 6 and so on, so that the total number of possibilities is

$$9 \times 8 \times 7 \times 6 \times 5 \times 4 \times 3 \times 2$$

Do you see how the first strategy hinted at this solution?

New terminology will make our discussion of this problem and other probability problems easier. How would you describe the computation above to someone—let's say a friend in class who didn't have this book? Do so and then read on. . . .

One way: "Start with 9, then 8, then 7, and keep reducing the number by one till you can't anymore. Now compute the product."

This kind of computation is common in probability and has a name: **factorial**. That is, $9 \times 8 \times 7 \times 6 \times 5 \times 4 \times 3 \times 2$ can be written as 9!.

Mathematicians insert a 1 at the end of this product, and define any number n factorial, $n!$, as

$$n! = n \times (n-1) \times (n-2) \times \cdots \times 3 \times 2 \times 1$$

> ■ *Language* ■
>
> Yes, a factorial contains an exclamation point; maybe the person who discovered this pattern was so surprised or excited that he or she decided to denote this algorithm with an exclamation point!

INVESTIGATION 7.30 | How Many Different Election Outcomes?

Let's say Kappa Delta Pi has its first election: for president and treasurer. How many possible ways can a president and treasurer be elected from a pool of 9 people? Work on this and then read on. . . .

DISCUSSION

STRATEGY 1: *Be systematic, make a table, and look for patterns*

TABLE 7.20

1, 2	2, 1	
1, 3	2, 3	
1, 4	2, 4	
1, 5	2, 5	etc.
1, 6	2, 6	
1, 7	2, 7	
1, 8	2, 8	
1, 9	2, 9	

Note that the numbers in Table 7.20 represent possible outcomes. For example, 1, 2 represents the outcome "person 1 as president and person 2 as treasurer." Why is {2, 1} not simply a duplicate of {1, 2}? Why did we skip {2, 2}? Can you finish from here?

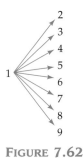

FIGURE 7.62

STRATEGY 2: Make a tree diagram

Some students find a tree diagram (Figure 7.62) very helpful. Can you finish from here?

STRATEGY 3: Connect to previous problems

We can use the multiplication principle again. Do you see how?

There are 9 different possibilities for president (each member), but only 8 possibilities for treasurer (assuming that no one will be both president and treasurer).

Without too much difficulty, most students can see this as 8 groups of 9, or 9×8.

This is like the beginning of factorial.

Electing 3 officers Now let's see how confident you are about this idea. What if the group decided to have a president, a treasurer, and a secretary? How many different outcomes are there for this scenario? Work on this and then read on. . . .

Each of the three strategies from before still applies (see Figure 7.63).

	8 groups						
7 groups	123	132	192	This process repeats 9 times; that is, there is another group of 8 × 7 outcomes that begins with 2, then another group of 8 × 7 outcomes that begins with 3, and so on.	213	231	291
	124	134	193		214	234	293
	125	135	194		215	235	294
	126	136···	195		216	236···	295
	127	137	196		217	237	296
	128	138	197		218	238	297
	129	139	198		219	239	298

FIGURE 7.63

The total number is $9 \cdot 8 \cdot 7$.

This question (about president, treasurer, secretary) is more complex than the question in Investigation 7.29 about the picture. Instead of looking at all the possible permutations of a set of specified size, which we did in the picture investigation, we are now asking how many different *permutations* of specified size we can make from a set of a given size. In this situation, the outcome "1 2 3" is considered to be different from the outcome "1 3 2." Do you see why?

Using symbols, we say that our election question seeks the number of different subsets of size r that we can make from a set of size n. We can also refer to this amount as the number of **permutations** of n things taken r at a time. The shorthand for this situation is ${}_nP_r$. (*Note:* Some books use the notation P^n_r.)

In the case of the different possibilities for president and treasurer, we were looking for ${}_9P_2$.

In the case of the different possibilities for president, treasurer, and secretary, we were looking for ${}_9P_3$.

Can you now generalize a formula for the number of permutations of a subset of size r from a set of size n? Try to do so and then read on. . . .

The generalization is

$${}_nP_r = n(n-1)\cdots(n-r+1)$$

If this makes sense, great. If not, substitute 9 for n and 2 (and then 3) for r so that it makes sense.

Some authors prefer a different formula, one that helps the learner to see how many terms to include and/or what the last term is. This formula rests on the following observation:

$$9 \cdot 8 \cdot 7 = \frac{9!}{6!}$$

(Active readers will make sure that they believe this equivalence!) In this situation, 6 is the number of people who will not be elected. This observation leads to an alternative formula:

$$_nP_r = \frac{n!}{(n-r)!}$$

INVESTIGATION 7.31

How Many Outcomes This Time?

Suppose Kappa Delta Pi decided to elect two members to send to the national convention. How many possible outcomes are there now? Think and then read on. . . .

DISCUSSION

Do you understand why this is not the same as the problem with the president and treasurer? If so, explain why. If not, think and then read on. . . .

Combinations Using the notation of the investigation with the president and treasurer, the outcomes {1, 2} and {2, 1} represent different outcomes. In the first outcome, student 1 is the president and student 2 is the treasurer, whereas in the second outcome, student 2 is the president and student 1 the treasurer. However, delegates {1, 2} and {2, 1} represent the same outcome. Mathematically, we distinguish between permutations and **combinations**. In probability situations in which order matters, we speak of permutations; when order doesn't matter, we speak of combinations. For example, if outcomes "*abc*" and "*bac*" are considered to be different, we are in a permutation situation; if they are considered to be the same, we are in a combination situation.

Using symbols, we can write the number of combinations of n things taken r at a time as $_nC_r$.

Now let us discuss the solution to how many outcomes there are for this question.

First, we will be systematic and look for patterns (see Table 7.21).

Can you finish the problem? Does this problem remind you of a problem or a pattern that we have encountered before? Think before reading on. . . .

If you thought about the handshakes problem (Problem 2 in Exploration 1.1) or the sum of the first 100 numbers (Investigation 1.4), you are right.

We find that there are 36 different outcomes—that is, 8 + 7 + 6 + 5 + 4 + 3 + 2 + 1. We would like to be able to generalize this procedure so that we can use it in a variety of situations; for example, what if we had a set of 20 elements (instead of 9) and we wanted all outcomes consisting of 3 elements at a time (instead of 2)? One way to determine the general procedure is to analyze the results of the simpler problems. Thus we will explore where the 36 comes from. Work on this before reading on. . . .

TABLE 7.21

1, 2	2, 3	3, 4	
1, 3	2, 4	3, 5	
1, 4	2, 5	3, 6	
1, 5	2, 6	3, 7	etc.
1, 6	2, 7	3, 8	
1, 7	2, 8	3, 9	
1, 8	2, 9		
1, 9			

From one perspective, we can see that 36 comes from $8 + 7 + 6 + 5 + 4 + 3 + 2 + 1$. Thus we could generalize that if there were 20 members and we wanted all outcomes of two at a time, there would be $19 + 18 + 17 + \cdots + 3 + 2 + 1$. What if we wanted all possible outcomes of three at a time? In order to develop a procedure that will work in more situations, we have to look more deeply. What other ways do you see to determine the 36? Think before reading on. . . .

Using the algorithm developed in Exploration 1.1, we can say that

$$_9C_2 = \frac{9(8)}{2} = 36$$

Now this should look similar to the permutation algorithm:

$$_9P_2 = 9(8) = 72$$

Why do we divide by 2? Think about this before reading on. . . .

Because each combination is connected to two permutations, we can see that in this case, the number of combinations will be half the number of permutations.

Do you feel able to make a hypothesis for the general combination algorithm, $_nC_r$? If so, make it and then read on. . . .

If not, work through the next problem and then try to make a hypothesis.

If 3 people go to the convention What if 3 people could go to the convention? How many different combinations are there? Complete Table 7.22 before reading on. What patterns do you see that could help you to complete the table or help you to determine the total?

TABLE 7.22

123	134	145	156	167	178	189
124	135	146	157	168	179	
125	136	147	158	169		
126	137	148	159			
127	138	149				
128	139					
129						
234	245	256	267	278	289	
235	246	257	268	279		
236	247	258	269			
237	248	259				
238	249					
239						
345	356	367	378	389		
346	357	368	379			
347	358	369				
348	359					
349						

In the first set of outcomes, where person number 1 is first, we have 28 total outcomes, and this number is the sum of the sequence $7 + 6 + 5 + 4 + 3 +$

2 + 1. In a similar fashion, the number of outcomes in the second set is equal to 6 + 5 + 4 + 3 + 2 + 1 = 21. From these patterns, it is reasonable to conclude that we can continue in this manner. Doing so, we find that there are 84 different outcomes for this situation—that is, 28 + 21 + 15 + 10 + 6 + 3 + 1. In everyday language, we say that there are 84 possible ways to send 3 people to the convention.

Determining a way of connecting this 84 to a formula that we can use is not obvious, but there are connections. What do you think? Can you generalize from the handshakes problem in Exploration 1.1? Work on this before reading on. . . .

There are many possible paths that will enable us to see the general procedure. Let us take one path.

First, let us summarize what we know:

We know that $_9P_2 = 9 \cdot 8 = 72$, that $_9C_2 = 9 \cdot 8/2$, and that there is a similarity between these two.
We know that $_9P_3 = 9 \cdot 8 \cdot 7 = 504$.
Therefore, we might suspect that $_9C_3$ might be something like $(9 \cdot 8 \cdot 7)/x$.
But we know that $_9C_3 = 84$, so we can make an equation and solve for x.
That is, $_9C_3 = 84 = (9 \cdot 8 \cdot 7)/x$.

Solving the equation, we find that $x = 6$. This still prompts the question "Why do we divide by 6?" What do you think? Work on this before reading on. . . .

In this case, each combination outcome is connected to 6 permutation outcomes.

$$\left.\begin{array}{l} 1\ 2\ 3 \\ 1\ 3\ 2 \\ 2\ 3\ 1 \\ 2\ 1\ 3 \\ 3\ 1\ 2 \\ 3\ 2\ 1 \end{array}\right\} \quad 1\ 2\ 3$$

Can you make the leap? Do you see a pattern?

$$_9C_2 = \frac{9 \cdot 8}{2}$$

$$_9C_3 = \frac{9 \cdot 8 \cdot 7}{6} = \frac{9 \cdot 8 \cdot 7}{1 \cdot 2 \cdot 3}$$

$$_nC_r = ? \qquad \text{What do you think?}$$

$$_nC_r = \frac{n \cdot (n-1) \cdots (n-r+1)}{r!} = \frac{_nP_r}{r!} = \frac{n!}{(n-r)!\,r!}$$

INVESTIGATION

7.32

Pick a Card, Any Card!

Consider a standard deck of playing cards. Pick two cards. What is the probability of getting two queens? Can we use any of the algorithms? Can you explain why one of the algorithms applies or why none of them applies? Work on this problem and then read on. . . .

DISCUSSION

This is a combination (not a permutation) problem. Do you see why?

$$_{52}C_2 = \frac{52 \cdot 51}{2} = 1326$$

Does this mean that the probability is 1/1326? What do you think?

The answer to this question is not 1/1326. The number 1326 means that there are 1326 different possible combinations. Because we want to know the probability of getting two queens, we also have to determine how many of those 1326 combinations involve two queens. What do you think? Do this now before reading on. . . .

One strategy is to list them systematically—for example, SD, SH, SC, HD, HC, DC. Another strategy is to use the appropriate combination algorithm appropriately. Do you see how $_4C_2$ applies to this part of the problem?

In either case, we find that there are 6 different combinations that involve 2 queens.

Thus we find that the probability of two queens is

$$\frac{6}{26 \cdot 51} = \frac{1}{13 \cdot 17} = \frac{1}{221}$$

INVESTIGATION 7.33

So You Think You're Going to Win the Lottery?

Since 1980, the number of states with lotteries has increased considerably. What factor(s) do you think might have led to the increase in states having lotteries?

During the 1980s, federal aid to states and cities shrank considerably. Many states "solved" this problem by establishing or increasing lotteries and designating the profits from the lottery for expenses such as education. The state lotteries have been called voluntary taxes. Can you see why?

Let's begin with a rather straightforward lottery in which players select four digits. The winning number is commonly selected in the following fashion: 10 Ping-Pong balls (each with a digit 0 through 9 written on it) are placed in a hopper. One Ping-Pong ball is selected. That digit now goes in the first place. That Ping-Pong ball is put back in the hopper, and the process is repeated three times.

What is the probability of picking a winning number? Work on this and then read on. . . .

■ *Outside the Classroom* ■

There are some facts about state lotteries that are troublesome.

In California, the poor spend 15 times more of their income, as a percentage, on lottery tickets than the rich do. . . . A Detroit study showed that in tough times, middle- and upper-class folk stopped playing the lotteries—but poor urban dwellers kept right on playing. . . . Other studies show that players are disproportionately minority people with less than average income and education.[15]

DISCUSSION

STRATEGY 1: Connect to the multiplication principle

Each time, there are 10 possibilities, so the sample space is $10 \times 10 \times 10 \times 10$. Because all outcomes are equally likely, the probability of any outcome is 1/10,000.

[15]*Utne Reader*, November/December 1993, p. 19.

STRATEGY 2: Act it out

What are some possible winning numbers?
Possibilities include 1234, 4612—in fact, any number from 0000 to 9999. How many combinations is that?
Can you extend this problem or think of similar problems?

Summary

In this section, we have examined a number of situations known as counting situations. That is, we have looked for patterns and order in those situations in order to find easier ways to count the number of outcomes in the sample space. Many children find such counting problems to be interesting. We learned the difference between combinations and permutations, and we examined how the formulas work.

Once again, it is important to emphasize that the algorithms themselves are useless if you do not understand what they mean. I strongly believe that there is too much blind application of algorithms in our schools. I see this regularly in schools that I visit, and I see many of my students doing this also, either out of math phobia or out of apathy. It is my hope that over time, more and more students and teachers will strive to make sense of situations in which mathematics applies.

EXERCISES 7.4

1. In how many ways can the following students' names be displayed on a poster: Amy, Betty, Carl, Ed, Frank, Gisela?

2. In how many ways can the names of 4 candidates be listed on a ballot for an election?

3. Find the number of different ways in which 4 flags can be displayed on a flagpole, one above the other, if 10 different flags are available.

4. Mrs. Olson has 7 brands of cat food and 3 cats, each of which receives 1 can per day. In how many different ways can she serve the cats on any day, assuming that no 2 cats get the same brand?

5. Those students who celebrate Christmas and come from large families may find this problem to be familiar. Let's say that a family of 5 decides that instead of each person buying a present for each other person in the family, all the family members will put their names in a hat. Each person then takes one name out of the hat and buys a present for that person. What is the probability of every family member getting his or her own name?

6. Several years ago I received as a gift a book of different animals. However, the picture of each animal's body had been cut into three pieces, and the pages of the book enabled me to make different combinations of animals. The name of the book was *Por-gua-can* because in the creature shown on the cover, the head was a porcupine's, the torso was an iguana's, and the feet were a pelican's. There were 9 animals. How many different animal combinations are possible?

7. The German club has 12 members. In how many different ways can a subset of 3 members be selected to go on a field trip?

8. A basketball team has 9 players, 5 of which are in the starting lineup.
   a. How many different starting lineups are possible?
   b. How many different starting lineups are possible if the star must be in the lineup?

9. Let's say you went out for ice cream. The store had 9 different flavors, and you wanted a triple-decker ice cream cone. How many different triple-decker ice cream cones are possible at this store?

10. Janine's boss has allowed her to have a flexible schedule. Her boss says she can pick whatever 5 days a week she will work.
    a. How many different work combinations can be made?
    b. How many choices give her consecutive days off?
    c. How many choices give her Wednesday off?

11. This unfortunate event happened to me when my two children were little. Each of them had a friend over for lunch. After lunch, we had popsicles for dessert. There were 9 popsicles in the freezer and 3 flavors: 3 grape, 3 cherry, and 3 orange. What is the probability that each child will get the popsicle that child wants?

12. Let's say a company has 10 members on its management team: 7 men and 3 women. There is a conference in Hawaii, and the company has decided to send 3 people. If the selection of the 3 people to go to Hawaii is determined by lottery, what is the probability that at least 1 woman will be selected?

13. You and a friend are at Paul and Elizabeth's Restaurant for a night of fine dining. You can choose among 2 appetizers, 4 main dishes, and 3 desserts.

    Appetizer: soup or salad
    Main dish: fish, chicken, beef, vegetarian
    Dessert: chocolate cake, apple pie, ice cream

    a. How many possible different dinners are there?
    b. If there are 3 choices of salad dressing and 4 choices of ice cream, how many possible dinner combinations are there?

14. If you pick a card from a regular deck of cards, what is the probability of getting a face card 3 times in a row? Assume that the face card is put back in after each draw, so that you are selecting from a full deck each time.

15. If you select 2 cards from a deck, what is the probability of getting 2 matching cards (2 kings, 2 fives, and so on)?

16. If you select 5 cards from a deck, what is the probability that you will get 4 of a kind? 3 of a kind? 2 of a kind?

17. Recall Investigation 7.33. Suppose the Ping-Pong ball was not replaced. Now what is the probability?

18. Explain why these two formulas are equivalent.

    $${}_nP_r = \frac{n!}{(n-r)!}$$ and $${}_nP_r = n(n-1)\cdots(n-r+1)$$

19. What do you think will be the value of ${}_nP_n$?

20. Little Caesar's pizza had a commercial on television that said that you could buy one pizza and get another one free, with up to 5 toppings per pizza. The dialogue on the commercial went something like this. The older person said, "5 toppings per pizza; that's 10 different pizzas to choose from." The little guy then said, "No, it's 1,048,576 combinations." What do you think?

21. Interview a nearby automobile dealership. Ask the people responsible for ordering how they determine which cars to order to have on hand. Compare the selection process for a small dealership that can stock only 5 Plymouth Voyager minivans with that for a large dealership that can stock 50 Plymouth Voyager minivans.

22. a. How many possible words can be spelled using the letters from the word *mathematics*?
    b. What word can you find with the highest ratio of actual (possible) words to the theoretical number of words?

23. *Classroom Connection* This question comes from "Figure This," a website for upper elementary and middle school children containing many problems that are particularly interesting and mathematically worthwhile. Here's the question: How many people would have to be in a school before at least two people had to have the same first and last initials?

    Source: Reprinted with permission from Figure This! (http://www.figurethis.org), copyright 2006 by the National Council of Teachers of Mathematics.

Chapter Summary

1. There are many ways to represent data. In some cases, what graph to use is a matter of preference; in other cases, some graphs are more appropriate than others.

2. Different graphs can give different impressions of a set of data. Graphs can be constructed to give the reader a distorted impression of the data!

3. When examining data that others have gathered, we need to examine the data carefully, thinking about the reliability and validity of the data, and about whether it is fair, appropriate, or distorted.

4. The terms *mean*, *median*, and *mode* all represent an "average." Each of the three terms has a conceptual base:
   - The mean is the center of gravity of a set of data.
   - The median is the numerical middle of the set of data.
   - The mode is the datum that occurs most often.

5. In some distributions, the mean, median, and mode are equal. In others, they may be very different, and some may not even be appropriate. In those cases, we need to examine which measure is the best representative for "average."

6. Averages are not always highly representative of a population. In most cases, when we are examining a population, we need to have more information about the population, depending on the questions we are asking. We may want to know the center of the data, or we may want to know about the spread of the data, in which case knowing the extreme values, the range, and the standard deviation is important.

7. When a set of data is normally distributed, we can make powerful generalizations and predictions about that set of data.

8. Random and haphazard are not the same thing. Many events that are random can be quantified.

9. When working with probability situations, sometimes we can determine both the theoretical and experimental probability of an event happening. Sometimes we can only determine the experimental probability.

10. Probability formulas and rules are easier to understand when we take the time to be clear about the terms being used, e.g. events vs. outcomes, dependent vs. independent events, complementary events, and permutations and combinations.

11. Care needs to be taken when conducting surveys so that the results are reliable.

BASIC CONCEPTS

Section 7.1: The Process of Collecting and Analyzing Data

uncertainty *380*
population *383*
frequency table *384*
bar graphs *384*
circle graph *385*
line plot *387*
dot plot *387*
cluster *388*
gaps *388*
outliers *388*
mean *389*
median *389*
mode *389*
measures of central tendency *389*
natural variation *394*
measurement variation *394*
grouped frequency table *395*
histogram *395*
modal class *397*
measures of dispersion *400*
different distributions: uniform, skewed to the right, skewed to the left, bimodal, normal *401*
line graph *405*
truncated graph *410*

Section 7.2: Going Beyond the Basics

box-and-whisker graph *424*
boxplot *424*
lower quartile *424*
upper quartile *425*
five-number summary *425*
interquartile range *426*
normal distribution *430*
standard deviation *432*
scatter plot *437*
correlation coefficient *441*
   sample *441*
convenience sample *441*
stratified random sample *442*
margin of error *442*
sample size *442*
response rate *442*
weighted average *445*
stem-and-leaf plot *423*

Section 7.3: Concepts Related to Chance

outcome *456*
sample space *456*
event *456*
equally likely and not equally likely *457*
impossible events *457*
certain events *457*
theoretical probabilities *458*
experimental probabilities *458*
empirical probability *458*
simulation *459*
mutually exclusive events *463*
complementary events *464*
multiplication principle *467*
dependent events *472*
independent events *472*
fair game *473*
expected value *474*
Law of Large Numbers *459*

Section 7.4: Counting and Chance

factorial *482*
combinations *469*
permutations *470*

COMPLETED PROBABILITY TABLES FOR INVESTIGATION 7.23, P. 468

TABLE 7.18

6 6 1	6 5 2	6 4 3	5 5 3	4 4 5
6 1 6	6 2 5	6 3 4	5 3 5	4 5 4
1 6 6	5 6 2	3 6 4	3 5 5	5 4 4
	5 2 6	3 4 6		
	2 5 6	4 6 3		
	2 6 5	4 3 6		

TABLE 7.19

6 6 1	5 6 2	4 6 3	3 6 4	2 6 5	1 6 6
6 5 2	5 5 3	4 5 4	3 5 5	2 5 6	
6 4 3	5 4 4	4 4 5	3 4 6		
6 3 4	5 3 5	4 3 6			
6 2 5	5 2 6				
6 1 6					

TWO DIFFERENT REPRESENTATIONS OF THE SAMPLE SPACE OF ROLLING 3 DICE

Note: The first column is simply to make communication easier; for example, we can talk about patterns in rows 19 to 24.

TABLE 7.23

1	1 1 1	2 1 1	3 1 1	4 1 1	5 1 1	6 1 1
2	1 1 2	2 1 2	3 1 2	4 1 2	5 1 2	6 1 2
3	1 1 3	2 1 3	3 1 3	4 1 3	5 1 3	6 1 3
4	1 1 4	2 1 4	3 1 4	4 1 4	5 1 4	6 1 4
5	1 1 5	2 1 5	3 1 5	4 1 5	5 1 5	6 1 5
6	1 1 6	2 1 6	3 1 6	4 1 6	5 1 6	6 1 6
7	1 2 1	2 2 1	3 2 1	4 2 1	5 2 1	6 2 1
8	1 2 2	2 2 2	3 2 2	4 2 2	5 2 2	6 2 2
9	1 2 3	2 2 3	3 2 3	4 2 3	5 2 3	6 2 3
10	1 2 4	2 2 4	3 2 4	4 2 4	5 2 4	6 2 4
11	1 2 5	2 2 5	3 2 5	4 2 5	5 2 5	6 2 5
12	1 2 6	2 2 6	3 2 6	4 2 6	5 2 6	6 2 6
13	1 3 1	2 3 1	3 3 1	4 3 1	5 3 1	6 3 1
14	1 3 2	2 3 2	3 3 2	4 3 2	5 3 2	6 3 2
15	1 3 3	2 3 3	3 3 3	4 3 3	5 3 3	6 3 3
16	1 3 4	2 3 4	3 3 4	4 3 4	5 3 4	6 3 4
17	1 3 5	2 3 5	3 3 5	4 3 5	5 3 5	6 3 5
18	1 3 6	2 3 6	3 3 6	4 3 6	5 3 6	6 3 6
19	1 4 1	2 4 1	3 4 1	4 4 1	5 4 1	6 4 1
20	1 4 2	2 4 2	3 4 2	4 4 2	5 4 2	6 4 2
21	1 4 3	2 4 3	3 4 3	4 4 3	5 4 3	6 4 3
22	1 4 4	2 4 4	3 4 4	4 4 4	5 4 4	6 4 4
23	1 4 5	2 4 5	3 4 5	4 4 5	5 4 5	6 4 5
24	1 4 6	2 4 6	3 4 6	4 4 6	5 4 6	6 4 6
25	1 5 1	2 5 1	3 5 1	4 5 1	5 5 1	6 5 1
26	1 5 2	2 5 2	3 5 2	4 5 2	5 5 2	6 5 2
27	1 5 3	2 5 3	3 5 3	4 5 3	5 5 3	6 5 3
28	1 5 4	2 5 4	3 5 4	4 5 4	5 5 4	6 5 4
29	1 5 5	2 5 5	3 5 5	4 5 5	5 5 5	6 5 5
30	1 5 6	2 5 6	3 5 6	4 5 6	5 5 6	6 5 6
31	1 6 1	2 6 1	3 6 1	4 6 1	5 6 1	6 6 1
32	1 6 2	2 6 2	3 6 2	4 6 2	5 6 2	6 6 2
33	1 6 3	2 6 3	3 6 3	4 6 3	5 6 3	6 6 3
34	1 6 4	2 6 4	3 6 4	4 6 4	5 6 4	6 6 4
35	1 6 5	2 6 5	3 6 5	4 6 5	5 6 5	6 6 5
36	1 6 6	2 6 6	3 6 6	4 6 6	5 6 6	6 6 6

TABLE 7.24

	3	4	5	6	7	8	9	10	11	12	13	14	15	16	17	18
1	111	112	113	114	115	116	126	136	146	156	166	266	366	466	566	666
2		121	122	123	124	125	135	145	155	165	256	356	456	556	656	
3		211	131	132	133	134	144	154	164	246	265	365	465	565	665	
4			212	141	142	143	153	163	236	255	346	446	546	646		
5			221	213	151	152	162	226	245	264	355	455	555	655		
6			311	222	214	161	216	235	254	336	364	464	564	664		
7				231	223	215	225	244	263	345	436	536	636			
8				312	232	224	234	253	326	354	445	545	645			
9				321	241	233	243	262	335	363	454	554	654			
10				411	313	242	252	316	344	426	463	563	663			
11					322	251	261	325	353	435	526	626				
12					331	314	315	334	362	444	535	635				
13					412	323	324	343	416	453	544	644				
14					421	332	333	352	425	462	553	653				
15					511	341	342	361	434	516	562	662				
16						413	351	415	443	525	616					
17						422	414	424	452	534	625					
18						431	423	433	461	543	634					
19						512	432	442	515	552	643					
20						521	441	451	524	561	652					
21						611	513	514	533	615	661					
22							522	523	542	624						
23							531	532	551	633						
24							612	541	614	642						
25							621	613	623	651						
26							622	632								
27							631	641								

Chapter 7 Review Exercises

1. The table shows the number of CDs shipped between 1992 and 2001.
   a. Make a line graph for these data.
   b. Write a brief summary that would accompany the graph.

SALES OF RECORDED CDS, BY UNITS SHIPPED (in millions) AND VALUE, 1992–2001											
	1992	1993	1994	1995	1996	1997	1998	1999	2000	2001	% CHANGE 2000–2001
Units shipped	407.5	495.4	662.1	722.9	778.9	753.1	847.0	938.9	942.5	881.9	−6.4%

Source: Recording Industry Assn. of America, Washington, DC (in millions, net after returns)

2. The table below shows the prevalence of smoking among U.S. adults.
   a. Make a line graph for these data.
   b. Write a brief summary that would accompany the graph.

SMOKING PREVALENCE AMONG U.S. ADULTS, 1965–1999 (as a percent of population, 18 years of age and older)			
Year	Overall population	Males	Females
1965	41.9%	51.2%	33.7%
1974	37.0	42.8	32.2
1983	31.9	34.8	29.4
1985	29.9	32.2	27.9
1990	25.3	28.0	22.9
1992	26.3	28.1	24.6
1994	25.3	27.6	23.1
1995	24.6	26.5	22.7
1997	24.6	27.1	22.2
1998	24.0	25.9	22.1
1999	23.3	25.2	21.6

Source: Health, United States, 2001, Centers for Disease Control and Prevention.

3. The table shows the most widely represented nationalities of travelers to the United States in 2000.
   a. Make a bar graph for these data.
   b. Write a brief summary that would accompany the graph.

TOP NATIONALITIES OF TRAVELERS TO THE U.S.		
2000 rank	Country of residence	2000 total
1.	Canada	14,594,000
2.	Mexico	10,322,000
3.	Japan	5,061,377
4.	United Kingdom	4,703,008
5.	Germany	1,786,045
6.	France	1,087,087
7.	Brazil	737,245
8.	South Korea	661,844
9.	Italy	612,357
10.	Venezuela	576,663

Source: U.S. Dept. of Commerce, International Trade Administration, Web: www.tinet.ita.doc.gov/research/reports.

4. The table shows the ages of AIDS patients in the United States in 1985 and 2000.
   a. Make a circle graph for each set of data.

b. Write a brief summary that would accompany the graph.

AIDS Deaths and New AIDS Cases in the U.S., 1985 and 2000

Age	1985	2000
13–19 years	27	146
20–29 years	1,501	3,332
30–39 years	3,588	12,554
40–49 years	1,632	9,724
50–59 years	597	3,403
60 years and over	159	1,092

Do the following for Exercises 5–7; see page 399 for more details.

   a. Write a brief summary of the graph.

   b. Critique the data from the perspective of reliability and validity.

   c. Critique the choice of the graph (circle, bar, histogram, or line).

   d. Critique the construction of the graph (scales, categories, and labels).

5. Examine the graph below.

   e. The United States created 24.4% of carbon dioxide emissions but is home to only 5% of the world population. Mathematically, we say that the United States's share is out of proportion. Explain that statement, as though to someone who has no idea what it means.

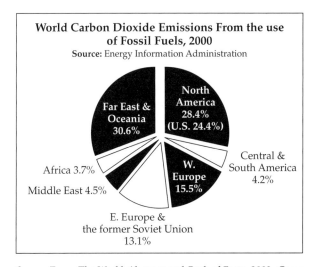

Source: From *The World Almanac and Book of Facts*, 2003. Copyright © World Almanac Education, Inc.

6. The graph below shows the number of Cesarean births per hundred births.

   e. Why do you think the makers of the graph didn't label the graph using the term *percentage*?

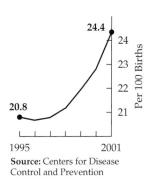

The Cesarean Spike

Source: Centers for Disease Control and Prevention

Source: Copyright © 2002 U.S. News & World Report, L.P. Reprinted with permission.

7. The following graph, showing quality of mathematical content in math lessons, is from the TIMSS videotape study of eighth-grade classrooms in Japan, Germany, and the United States, in which classes were observed and videotaped. (*Source:* http://nces.ed.gov/pubs2000/2000094.pdf)

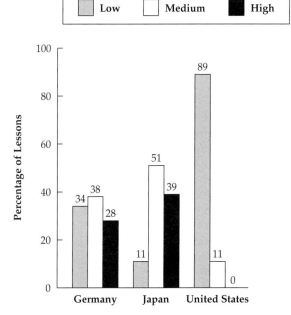

Note: Percentages may not sum to 100 due to rounding.

8. An elementary teacher had his students record the number of drops of water that they could drop onto a penny from an eyedropper until it spilled: 30, 29, 28, 27, 27, 25, 25, 25, 24, 24, 22, 21, 18, 16, 15.
   a. Use one or more graphs to show these data.
   b. What is the average? Justify your answer.
   c. Describe the spread of the data.

9. A company is testing a new dog food for overweight dogs. They found the following weight loss in a group of 10 dogs after one month. The data are pounds, to the nearest half pound: 0, 1, 2, 2, 3, 3.5, 4, 4, 4.5, 4.5, 5, 6, 6.5, 12.
   a. What is the mean weight loss?
   b. What is the standard deviation?
   c. What questions would you like to ask the company about the study in order to feel more confident about the validity and reliability of the data? Explain the reasoning behind your question(s).

10. Here are the grades on the midterm exam for two of my classes:
    Class 1: 76, 66, 73, 55, 68, 77, 75, 74, 98, 65, 54, 92, 71, 57, 90, 63, 69, 51, 70, 76, 66, 75, 68, 81, 82, 63, 66, 98
    Class 2: 90, 81, 64, 68, 84, 89, 95, 82, 72, 81, 83, 85, 88, 78, 100, 85, 88, 82, 72, 70, 65, 87, 70
    a. Make two box-and-whisker graphs. Write a brief summary that would accompany these graphs in an article.
    b. Make two line plots. Write a brief summary that would accompany these graphs in an article.
    c. Summarize the pros and cons of representing these data with box-and-whisker or line plots.
    d. Compare the average scores in the two classes. Justify your answer.
    e. Describe the spread of the data.
    f. Determine the standard deviations.
    g. What percent of the scores in each class lie within 1 standard deviation of the mean?

11. Willie has an average (mean) of 91.2 on the first 6 exams. What is the least he can score in order to have an average of 90 for all seven exams?

12. Suppose the distribution of heights of a group of 300 children is normal with mean of 160 centimeters and a standard deviation of 15 cm. About how many of these children are less than 145 cm tall?

13. The following graph shows the numbers of siblings reported by a class of students.
    a. How many students are there in the class?
    b. What is the mean?
    c. What is the median?

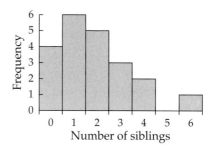

14. a. Describe a situation in which the median would be a more appropriate description of the center of a data set than the mean would.
    b. Describe a situation in which the mean would be a more appropriate description of the center of a data set than the median would.

15. If you randomly draw a circle from each box, what is the probability of your drawing two white circles?

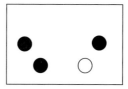 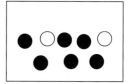

16. A drawer contains 6 red socks and 3 blue socks.
    a. If 2 are drawn, what is probability of a match?
    b. If 3 are drawn, what is probability of a match?
    c. What is probability of all one color after 4 draws?

17. Consider a pair of eight-sided dice—that is, dice that contain the numbers 1–8 as opposed to regular dice, which contain the numbers 1–6. If you roll two dice at a time, what is the most likely sum and what is the probability of getting that sum?

18. What is the probability of getting two odd numbers in a row if you throw two regular dice and add the two numbers?

19. If the sample space consists of the numbers 10–99, what is the probability of a number drawn randomly containing a zero? a five?

20. Consider the following game: Two dice are rolled, and then the numbers multiplied. If the product is odd, player A gets 1 point. If the product is even, player B gets 1 point. Is this a fair game? Why or why not? If it is not a fair game, change the rules to make it fair.

21. In how many ways can the following sports be displayed on a poster: soccer, basketball, baseball, lacrosse, and field hockey?

22. If you select two cards from a deck, what is the probability of your selecting two matching suits—for example, two clubs?

23. Janine's boss has allowed her to work a flexible schedule. Her boss says she can pick whatever 4 days a week she will work, excepting Sunday.

   a. How many different work combinations can be made?
   b. How many combinations will give Janine consecutive days off?
   c. How many combinations will give her Saturday off?

24. Let's say a state decided to have a license plate with the following format: one letter followed by three digits followed by one letter—for example, A123B. How many different license plates could be made with this format?

25. Below are measurements in centimeters of neck, waist, and wrist from a sample of adult females.

   a. Make boxplots for each set of data.
   b. Make two scatter plots: neck:waist and neck:wrist. Describe the degree of correlation, that is, positive or negative, and none, weak, or strong. Justify your conclusion.

neck	waist	wrist
36.2	82.2	17.1
38.5	83	18.2
34	87.9	16.6
37.4	86.4	18.2
34.4	100	17.7
39	94.4	18.8
36.4	90.7	17.7
37.8	88.5	18.8
38.1	82.8	18.2
42.1	88.6	19.2
38.5	83.6	18.5
39.4	90.9	19
38.4	91.6	17.7
39.4	101.8	18.8
40.5	96.4	18.2
36.4	92.8	16.9
38.9	96.4	17.3
42.1	92.8	19.3
38	96.4	18.5
40	97.5	18.2

CHAPTER 8

Geometry as Shape

8.1 Basic Ideas and Building Blocks

8.2 Two-Dimensional Figures

8.3 Three-Dimensional Figures

SECTION 8.1 BASIC IDEAS AND BUILDING BLOCKS

From a basic perspective, geometry is the study of shapes, their relationships, and their properties. What is the value of geometry? Who needs to know geometry and who has needed to know geometry? How has understanding of geometry shaped our lives? These are some of the questions that will be addressed in this and the next two chapters.

Here are a few examples of cases where people have needed geometry and this need has produced greater understanding:

- People in ancient Egypt developed methods to determine the boundaries of their land. In fact, the word *geometry* literally means "to measure the earth."
- Explorers needed maps to show where they had been.
- House builders have used many shapes, from Native American teepees, which look like cones but are more like many-sided pyramids, to African houses resembling cylinders, to the more familiar house structures with triangles and quadrilaterals, to geodesic domes with equilateral triangles.
- Inventors and engineers wanted to make machines and objects that would perform certain functions.
- Surveyors and planners needed to ensure that tunnels or railroad tracks built from both ends would actually meet in the middle.

- Artists wanted to portray convincingly what they saw with their eyes or visualized in their minds.
- Architects and builders needed buildings that would be both beautiful and strong.

*I*n the next three chapters, we will focus on three aspects of geometry: shapes, transformations, and measurement. Look at the pictures in Figure 8.1. What do you see?

FIGURE 8.1 Three natural spirals: (a) leaves of the sago palm, (b) horns of a mountain sheep, (c) the chambered nautilus.

Here, a shape that is called a spiral occurs in different parts of the world: in a plant and on two different animals. How is this shape formed? Why is this shape formed in these situations—that is, what purpose or function does this shape serve? Mathematicians, biologists, physicists, and others have asked and answered questions like these. Essential to answering such questions is a deep understanding of shapes, which also involves understanding their measures.

Let us begin our exploration of geometry by addressing the question "What is the value of geometry?" from three perspectives: fun, function (usefulness), and aesthetics. In each case, you will find that geometric structures (basic ideas) are important.

8.2 Why Geometry? Because It's Fun!

INVESTIGATION 8.1

Playing Tetris

Most students in my course have played or are familiar with the computer game called Tetris. Let's take a brief look at Tetris and see some of the mathematics that can be developed by playing and analyzing the game.

A shape comes down the screen, and you can move (transform) the shape in several ways: You can move it to the right or left, or you can rotate it 90 or 180 degrees. You get points when an entire row is filled. The game

500 CHAPTER 8 / Geometry as Shape

can be accessed on several websites; type "tetris" on a search engine and play some games if you like!

In the game that is in progress, there really is no good spot for the shape that is dropping down the screen, even if we turn it. If we could flip it, we would get this shape, but the Tetris game allows only turns and lateral movement, not flips.

Now, many people enjoy playing Tetris, but where is the mathematics in Tetris? Think of questions that could be asked and explored that would result in students learning more mathematics. Then read on. . . .

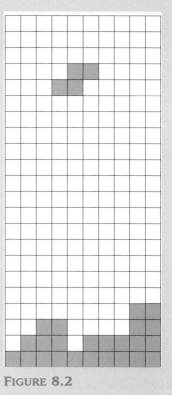

FIGURE 8.2

DISCUSSION

Below are ten questions that can be explored with Tetris; there are many more. Each of these questions can generate some important mathematics ideas and discussions. I have done many of these with fourth- and fifth-graders also. They love that the game can be played in math class, and they love the related explorations with tetrominoes, pentominoes, and hexominoes. Some of these questions appear in the *Explorations*.

Question		Content Area(s)
1. What is a tetromino (the mathematical name for the Tetris pieces)? If you just say, "four squares put together," then someone not familiar with the game could conclude that either of the shapes at the right is a tetromino.	FIGURE 8.3	Defining terms
2. What are good strategies for playing Tetris? Strategies come both from spatial sense (being able to visualize what the shape will look like when we slide and/or turn it to get it to the proper position) and probability sense (deciding how long to wait for the "right" piece and when to give up on filling a hole).		Spatial sense Probability
3. Am I getting better? We can measure improvement by recording our scores and comparing early scores to later scores.		Statistics
4. Are these two pieces "the same"? On the one hand, they have the same shape and size. On the other hand, if the figure at the left starts to come down the screen, you cannot turn it into the figure on the right; so in that sense, they are not the same.	FIGURE 8.4	Congruence Communication- defining terms

(*continued*)

Question	Content Area(s)
5. How many Tetris pieces are there? There are only five different (noncongruent) tetrominoes, but there are quite a few different pentominoes and a lot of different hexominoes!	Problem solving
6. Can we fill in a 5 × 4 grid with all of the Tetris pieces? There are many such puzzle questions that we can ask; some are in *Explorations*.	Spatial sense
7. Can we make bigger versions that are the same shape? For example, using all the Tetris pieces, can we make a bigger version of the L Tetris piece that has the same shape as the L piece?	Similarity
8. Which tetrominoes will cover a surface with no gaps? Will all of them? Will certain combinations?	Tessellation
9. If we wanted to make a set of tetrominoes for each of 24 students, how many sheets of Geoboard Dot Paper (end of *Explorations*) would we need?	Measurement Spatial sense
10. How does the Game Boy or computer version work? That is, the computer doesn't really rotate and translate the figures, but rather shades in different pixels on the screen!	Geometry and algebra

As you can see, a lot of mathematics can be developed by looking more closely at a game. There are many popular games that involve geometric thinking: Brick by Brick, Set, Soma cubes, Rubik's Cube, and online games. One of my favorites is Super Collapse.

Why Geometry? Because It Has Practical Applications

There are many machines, tools, and materials whose shape was determined by geometric properties. Let us explore a few here. The reason why manhole covers are round is explored in Exploration 8.4.

INVESTIGATION 8.2 — Different Objects and Their Function

In each of the cases below, there is a reason for the shape of the object. Can you figure it out? Look closely at the pictures and think of the reason for the shapes in each case before reading on. . . .

Why are nuts hexagonal?
FIGURE 8.5

Why is the plug pentagonal?
FIGURE 8.6

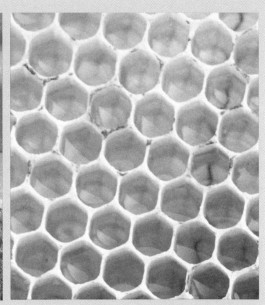
Why are the cells of a honeycomb hexagonal?
FIGURE 8.7

Why is a soccer ball made with hexagons and pentagons?
FIGURE 8.8

Why are there triangles in a bicycle?
FIGURE 8.9

DISCUSSION

Most nuts are hexagonal in shape for two reasons. First, opposite sides of hexagons are parallel, and the jaws of the crescent wrench are parallel. Thus you can open the wrench to the appropriate setting, slide it onto the nut, tighten, and then turn. But why not square nuts instead of hexagons? Stop and think before reading on. . . .

For a square nut, you would have to turn the nut 90 degrees (a quarter-turn) before you could slide the wrench off and then slide it back on. In tight places, often it is not possible to turn 90 degrees. With the hexagonal shape, you need only to be able to make a 60-degree rotation.

Why is the fire hydrant plug pentagonal? For the same reason that nuts are hexagonal! If you look at the pentagon, you will see that opposite sides are not parallel. Thus a large pipe wrench, widely available at hardware stores, will not work on a fire hydrant. A special wrench, not sold to the public, is needed.

Why do you see triangles in any bicycle? This question will be explored in Section 8.2.

Why is the soccer ball made of hexagons and pentagons? We will explore questions of covering surfaces with repeating shapes in Chapter 9.

Why is the honeycomb hexagonal? This will be explored in Chapter 10.

Why Geometry? Because We See It in Art All Over the World

In almost every case when archaeologists find artifacts from ancient cultures, we find shapes (squares, pentagons, hexagons, symmetric designs) on pottery, on cloth, and in religious artifacts. Were humans replicating shapes they saw in nature—honeycombs, snowflakes, spirals in many sea animals—or is this love of shapes and designs hard-wired into the human brain like language is? We don't know, but we do know that it is pretty universal.

Let us investigate just a few examples of the value of geometry in art. I remember taking an art history class in college. The professor showed us paintings made before artists fully grasped the geometric connections between the two-dimensional surface of the canvas and the three-dimensional scenes they were representing on that canvas. Figure 8.10 shows the consequences of not understanding the geometry.

FIGURE 8.10

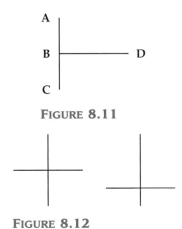

FIGURE 8.11

FIGURE 8.12

One of the most interesting aspects of art and geometry is seen in optical illusions. Which line in Figure 8.11 is longer, *AC* or *BD*? Look and write your answer before reading on. . . .

Most people perceive *DB* as longer than *AC*, but they are the same length. Why does *BD* appear longer? Many theories have been developed. The one that makes the most sense is that *DB* interrupts *AC*. To illustrate this, draw two lines of equal length that intersect each other (see Figure 8.12). In these cases, most people will see the two segments as having the same length.

Which inner circle in Figure 8.13 is larger? Look at Figure 8.13 and think before reading on. . . .

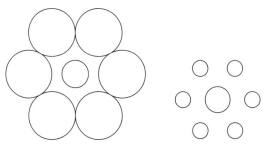

FIGURE 8.13

The inner circle on the right looks larger to most people because it is larger than its surrounding while the other circle is dwarfed by its surrounding. In fact, the inner circle at the left is slightly larger!

What do you see in the figures below? Think before reading on. . . .

Although Figure 8.14 consists only of lines and curves, most people instantly see "Hello." By making the colored shapes the top or the bottoms of the boxes in Figure 8.15, you can see either six or seven boxes.

FIGURE 8.14   **FIGURE 8.15**

Most local libraries have books in the children's section on optical illusions. You can also visit a large bookstore or go online to find more about the geometry of optical illusions and how they work. We will examine some of the work of M. C. Escher, one of the greatest mathematical artists of all time, in the next chapter.

In our limited space, we have examined only a bit of geometry from the perspective of art and aesthetics. In the next chapter, when we explore symmetry, we will examine geometry in fabrics and crafts that people make.

Children

Now that we have a few investigations under our belts, let's look at this notion of geometry from the child's perspective. You will be teaching children, so it is important to look at what we expect them to learn and what we know about how they learn as you are learning geometry in this course. Because this is a mathematics course (as opposed to a methods course, where you focus on learning how to teach geometry), we will focus primarily on the mathematical ideas. However, it is crucial that we keep the child's level of reasoning in mind. Toward that end, let us look at NCTM's Standards and what we know about the development of spatial thinking. Throughout the next two chapters, there will be sidebars about children, when relevant. More examples of children's thinking can be found in the exercises and on the web page.

 Instructional programs from prekindergarten through grade 12 should enable all students to:

- Analyze characteristics and properties of two- and three-dimensional geometric shapes and develop mathematical arguments about geometric relationships.
- Specify locations and describe spatial relationships using coordinate geometry and other representational systems.
- Apply transformations and use symmetry to analyze mathematical situations.
- Use visualization, spatial reasoning, and geometric modeling to solve problems (PSSM, p. 41; available online at **nctm.org**).

These are just the main ideas. When you read the standards for each grade area (pre-K–2, 3–5, 6–8, and 9–12), you will find a more detailed discussion and actual examples. Using the "Navigating Through Geometry" series of NCTM books is an excellent way for you to get an idea of the kinds of geometry explorations in school that are consistent with the NCTM standards.

Geometry and Spatial Thinking

Douglas Clements and Michael Battista have articulated this notion that we call "spatial sense." Just like "number sense," "spatial sense" is an important goal of school mathematics. However, it is generally left undefined and unarticulated.

One aspect of spatial thinking is **hand-eye coordination**. Most young children are still very much in the process of developing this ability. There are many activities that can help children to develop this ability, such as copying shapes that the teacher or other students have made with pattern blocks, Geoboards (such as those at the end of the *Explorations*), or other manipulatives.

Another important spatial thinking ability is called **figure-ground perception**, the ability to identify a figure against a complex background. One example that most of my students remember from elementary school is a picture for which the directions ask, "Can you find the 12 monkeys hidden in this picture?" The activity is far more than just a fun exercise. This ability to discriminate what you are looking for from extraneous lines or information is useful in everyday life and is especially important in many professions.

Another important ability is called **perceptual constancy**, the ability to recognize figures and objects when they are not in their familiar orientation. For example, many students see the triangle on the left in Figure 8.16 as an isosceles triangle but will say that the triangle on the right is not an isosceles triangle. Similarly, other students will see that the triangle on the right is an obtuse triangle but will fail to see that the triangle on the left is also an obtuse triangle. In fact, the two triangles are congruent.

FIGURE 8.16

Visual discrimination is the ability to find similarities and differences between or among objects. You have probably seen tasks designed to develop this ability in *Sesame Street*, commercial activity books, and schoolbooks. For example, five similar objects are shown, and the student is asked to pick the two that are identical.

Visual memory has to do with a person's ability to describe or draw accurately an object that is no longer in view. In Section 8.2, there will be an investigation that asks you to look at an object for only a few seconds and then try to draw that object from memory.

These are but a few of the many kinds of abilities that come under the larger heading of spatial thinking and that are important in everyday life and crucial in many occupations. Douglas Clements and Michael Battista state that "much of the thinking required in higher mathematics is spatial in nature,"[1] and they note that Einstein said that he thought not so much in words as in images. Ultimately, we want students to realize that virtually all fields of mathematics have spatial aspects and applications.

Now that you have seen that there are many uses of geometry and we have examined some aspects of what we want children to learn, let us begin at the beginning, so to speak, and examine the building blocks of geometry and geometric thinking. These ideas will be crucial in the following sections when we look at various shapes and their properties.

8.5 Back to the Greeks

Although the study of geometry is probably as old as the study of numbers, our modern understanding of geometry began with the Greeks over 2000 years ago when some people felt the need to go beyond merely knowing that certain facts or properties were true to being able to prove why they were true and that they were true everywhere. For example, they wanted to understand the Pythagorean theorem; why the diagonals of a square are the same length and perpendicular, whereas the diagonals of a nonsquare rectangle are the same length but not perpendicular, and so on. (See Figure 8.17.)

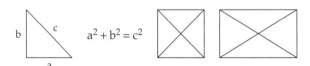

FIGURE 8.17

This focus on *why*, beyond knowing *that*, represented a huge leap in human history. Even so, modern geometry developed slowly over several centuries. Around 300 B.C. a man named Euclid decided to put together what was known

---

[1]Douglas H. Clements and Michael T. Battista, "Geometry and Spatial Reasoning," in *Handbook of Research on Mathematics Teaching and Learning*, ed. Douglas A. Grouws (New York: Macmillan, 1992), p. 442.

History

Actually, Euclid did try to define these basic terms! Here are his definitions of *point*, *straight line*, and *plane*:

  A point is that which has no part.

  A straight line is a line that lies evenly with the points on itself.

  A plane surface is a surface that lies evenly with the straight lines on itself.

at that time, but he didn't just record all the scattered information. He organized it as systematically as he could, and his book, *The Elements*, ranks as one of the most important and influential books of all time. It has sold almost as many copies as the Bible and has been translated into virtually every language. It was used as the geometry text in high school well into the nineteenth century. So let us take a look at what Euclid did. Along the way, we will also develop the building-block ideas for geometry, and we will look at how children's understanding of these ideas develops too.

As you know from high school, there are many definitions and theorems in geometry. Euclid realized that you can't define everything, and you can't prove everything. That is, you have to have some terms that are undefined, and you have to have some assertions that are simply accepted. If you are not convinced of this, try to define *round* without using the word *circle* or *circular*! The futility of such an approach is humorously treated in the B.C. cartoon in Figure 8.18. Thus certain terms, such as *point*, *line*, and *plane*, were left undefined, left as common-sense notions.

Source: By permission of John L. Hart FLP, and Creators Syndicate, Inc.
FIGURE 8.18

Similarly, we can't prove everything. Euclid and other Greek mathematicians thought, "What is the least that we can assume as self-evident, and then prove everything else on the basis of those assumptions?" These are known as the five postulates of Euclid:

1. To draw a straight line from any point to any other.
2. To produce a finite straight line continuously in a straight line.
3. To describe a circle with any center and any distance.
4. That all right angles are equal to each other.
5. That, if a straight line falling on two straight lines makes the interior angles on the same side less than two right angles, the straight lines if produced indefinitely, meets on that side on which are the angles less than the two right angles.

Most people find the fifth postulate difficult to understand, and many later mathematicians struggled over this one too. A common rewording of this postulate is attributed to an Englishman named John Playfair, who wrote, in 1795, "Given a line and a point not on the line, it is possible to draw exactly one line through the given point parallel to the line."

Whole new fields of geometry were born, some 2000 years after Euclid, when various people played the famous "what if" game with this famous fifth postulate: What if you could draw two lines through the given point parallel to the line? What if you couldn't draw any lines through the given point parallel to the line?

Basic Terms and Ideas

The following pages are admittedly not the most exciting pages of this text. Some readers have taken high school geometry and retained much of what they learned; others have forgotten much of that material; and still others did not take a geometry course in high school. Thus, how fast you read through this section will be related to how familiar these geometric ideas are to you.

It is important to note that points, lines, and planes are ideas—mental constructs only. That is, we cannot find a line in our physical world that extends infinitely in two directions. However, we can imagine them, and by imagining them we are able to develop powerful ideas and understandings that are very valuable in many ways. This notion of imagining is downplayed in much of schooling, which is a shame. One of the more famous quotations from Albert Einstein is "Imagination is more important than knowledge"!

Before we examine basic geometric definitions and concepts, let us briefly describe three fundamental undefined terms. In everyday life, we refer to the starting point of a race or a point on a map. Geometrically, a **point** has no dimension (length or width), but it does have a location. In everyday life, we refer to lines on a paper and lines of longitude. Geometrically, a **line** is straight and extends infinitely in two directions. We can also think of a line as an infinite collection of points that indicate a straight path. In everyday life, we refer to planes when we talk about floors and countertops. Geometrically, a **plane** is considered to be a flat surface that extends infinitely in all directions.

INVESTIGATION 8.3

Point, Line, and Plane

The purpose of this investigation is to have you come to appreciate the kind of thinking the Greeks started. You may find it helpful to use the point of your pencil or pen to represent a *point*, a ruler (or other object with a straight edge) to represent a *line*, and a piece of posterboard to represent a *plane*. By cutting slits in the posterboard, as shown in Figure 8.19, you can more concretely investigate the questions asked below.

Points and lines

A. Draw a point on a piece of paper. How many different lines can you draw that go through that point?
B. Draw two points on a piece of paper. How many different lines can you draw that go through both points?

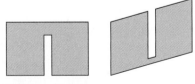

FIGURE 8.19

C. Draw three points on a piece of paper. How many different lines can you draw that go through all three points?

If you answer any of the questions, "It depends," can you be more specific? Write down your responses before reading on. . . .

FIGURE 8.20

FIGURE 8.21

FIGURE 8.22

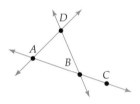

FIGURE 8.23

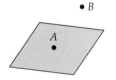

FIGURE 8.24

DISCUSSION

A. An infinite number of lines can go through any point (see Figure 8.20).

B. Only one line can go through any two points. That is, if we draw *any* two points, there will always be one and *only* one line that contains those two points (see Figure 8.21).

This powerful, but often only partially understood, conclusion is frequently stated as follows: Two points determine a line. Do you understand the equivalence of the following two statements? "Only one line will go through any two points" and "Two points determine a line."

C. How many lines we can draw through three points depends on the points. If the three points are on a line, then we can draw only one line (see Figure 8.22). However, if the three points are not on a line, then three points will determine three lines (see Figure 8.23).

This leads us to our first true definition: Collinear points are points that lie on the same line. As we have just seen, any two points are collinear.

On the other hand, if three (or more) points do not lie on a single line, then those points are said to be noncollinear. In Figure 8.23, points *A*, *B*, and *C* are collinear points, whereas points *A*, *B*, and *D* are noncollinear points.

There is a common misconception about collinear points that needs to be addressed. Draw a point on your paper and place your pencil somewhere above the desk. If the point on the paper is point *A* and the tip of the pencil is point *B* (see Figure 8.24), are points *A* and *B* collinear? Think before reading on. . . .

Yes; *any* two points are said to be collinear. Even when a line is not already drawn, if it is *possible* to draw such a line, then the points are collinear.

Planes

A. Now draw a point on a piece of paper. How many different planes can go through that point?

B. Draw two points on a piece of paper. How many different planes can go through both points?

C. Draw three points on a piece of paper. How many different planes can go through all three points?

If you answer any of the questions, "It depends," can you be more specific? Write down your responses before reading on. . . .

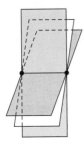

FIGURE 8.25

FIGURE 8.26

DISCUSSION

A. An infinite number of planes can go through any one point.

B. An infinite number of planes can go through any two points (see Figure 8.25).

C. The answer for three points depends on where you put the points. If we draw three noncollinear points, there is only one plane that will contain all three points. Do you see why? Think and then read on. . . .

One of the best ways to demonstrate this requires three people. First, draw one point on a sheet of paper. Next, the first person puts a pencil in the air (tip up). Then the second person puts another pencil in the air (tip up), making sure only that the three points are noncollinear. Now the third person takes a model of a plane (cardboard or a hard-cover book). How many different planes contain all three points? Just one!

When two or more points lie on the same plane, they are said to be **coplanar** points. As you have just seen, any two or three points are coplanar, because we can always find a plane that will contain them.

There is another way to see this concept concretely. Make a model of a three-legged stool like the one shown in Figure 8.26. You can be creative; for example, use three pencils, some glue, and a piece of cardboard. Put the model on a flat surface. Then make a model of a four-legged stool like the one shown and put it on a flat surface. What do you notice? Think before reading on. . . .

The three-legged stool will never rock. Many four-legged stools will. Why? How would you explain why to a friend who is not in this course?

Because the three points at the bottom of the stool are not collinear, there is only one plane that will contain them. However, the four points at the bottom of the four-legged stool may or may not all lie on the same plane. Thus we have a famous property: Three noncollinear points determine a plane.

Line Segments and Rays

In most everyday situations, we don't work with lines; instead, we work with line segments and rays.

We define a **line segment** to be a subset of a line that contains two points of the line (which we call the **endpoints**) and all points between those two points.

We define a **ray** as the subset of a line that contains a specific point (the endpoint) and all the points on the line that are on the same side of that point.

Table 8.1 shows the differences among a line, a line segment, and a ray.

TABLE 8.1

Term	Diagram	Notation
Line	$\xleftrightarrow{ABC}$	$\overleftrightarrow{AB}$ or $\overleftrightarrow{BA}$
Line segment	AB	$\overline{AB}$ or $\overline{BA}$
Ray	$AB\longrightarrow$	$\overrightarrow{AB}$ but not $\overrightarrow{BA}$

The notation to represent lines, line segments, and rays is a convention that has evolved. The conventional way to denote a line is by using two points on the line and a double arrow above the two letters. In Table 8.1, it would not be incorrect to refer to the line as $\overleftrightarrow{ABC}$, but it would be unconventional, like offering to shake someone's hand with your left hand.

The conventional way to denote a line segment is to use two letters, representing the endpoints, and a bar above the two letters.

Naming rays requires careful thought, for there is unconventional and then there is incorrect! For example, $\overrightarrow{AB}$ and $\overrightarrow{BA}$ are different rays. Why is this? The convention for naming a ray is to use two letters, the first of which is the endpoint of the ray. The second letter can be any other point on the ray and indicates the direction of the ray. For example, in Figure 8.27, we can speak of $\overrightarrow{DA}$, which is different from $\overrightarrow{AD}$. On the other hand, $\overrightarrow{DA}$ and $\overrightarrow{DM}$ are considered to be the same ray. Why is this? If you start at D and go through A, you produce the same ray as if you started at D and went through M. Thus we say $\overrightarrow{DA} \cong \overrightarrow{DM}$.

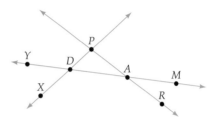

Figure 8.27

Relationships Between Lines

Figure 8.28

There are many possible ways in which two or more lines may be related to each other. For example, they may **intersect**, as in Figure 8.28. Two lines are said to intersect if they have exactly one point in common.

If two intersecting lines form right angles, they are said to be **perpendicular**. (See Figure 8.29.)

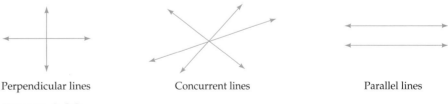

Perpendicular lines   Concurrent lines   Parallel lines

Figure 8.29

If three or more lines intersect at a single point (that is, have a point in common), they are said to be **concurrent** lines. (See Figure 8.29.)

As you know, not all lines intersect. If two lines lie in the same plane and never intersect, they are said to be **parallel** lines. (See Figure 8.29.)

There is a fourth possibility: two lines that do not intersect because they do not lie in the same plane. Such lines are called **skew** lines. For example, in Figure 8.30, $\overleftrightarrow{AD}$ and $\overleftrightarrow{EF}$ are skew lines.

Figure 8.30

Outside the Classroom

Some years ago, I was volunteering on a house-building project. It became necessary to cut a piece of lumber $5\frac{1}{2}$ inches wide into three strips of equal width. Easy, huh? So I divided $5\frac{1}{2}$ by 3 and got $1\frac{5}{6}$, but sixths are not on rulers. As I was trying to eyeball $1\frac{5}{6}$, a carpenter saw me and laughed. "Oh, you mathematicians know math theoretically, but not practically. Here's how carpenters do the problem." He laid the ruler on the lumber diagonally so that one end of the ruler was at the beginning and the other was at 6. He drew a mark at 2 and 4. Next he moved up the lumber a ways and repeated the process (Fig. 8.31). Then he connected the marks and the problem was solved!

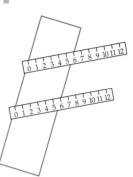

FIGURE 8.31

8.6 Angles

Although angles may seem pretty simple to an adult, it takes some time for children to understand the ideas about angles fully. In fact, there has been quite a bit of research on how children's understanding of angle develops. Let's begin with your responding to the question "What is an angle?" Write a definition of the term *angle* that you could give to a friend who never took geometry in high school. Think and then read on....

A common mathematical definition is that an **angle** consists of the union of two rays that have a common endpoint, which we call the **vertex** of the angle. Each of the rays is called a **side** of the angle (see Figure 8.32).

I will argue that many of the cases where we encounter angles do not consist of two rays or even line segments meeting at a common point. We see angles where two flat surfaces meet—for example, the wall and the ceiling; this is called a dihedral (three-dimensional) angle. From another perspective, the definition above, which is considered a static conception of angle, is not easily understandable to children, who more easily comprehend a more dynamic conception—that is, angle as movement. Think of situations where you use the term *angle* or *degrees* in the context of movement....

For example, instead of turn right, we say "turn 90 degrees" or "make a 90-degree turn." The saying "Our thinking was 180 degrees apart," mathematically means "in opposite directions." When turning a screw or a valve, we talk about a quarter-turn, which is 90 degrees. We talk about how much motion a damaged joint has "She can only raise her arm about 45 degrees."

While we will use the standard definition in this course, I will occasionally discuss the angles in a dynamic sense when that is appropriate and when we encounter situations where my own students have found the dynamic conception to make more sense. It is important to note that the two conceptions are not contradictory. In fact, mathematicians see both conceptions as coherent and consistent.

An angle partitions a plane into three disjoint sets: the angle itself, the **interior** of the angle, and the **exterior** of the angle (see Figure 8.33).

Now that we have defined angles, let us examine how they are named. Before reading on, though, make sure that you are actively thinking while

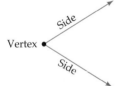

FIGURE 8.32

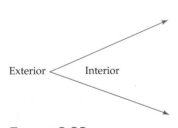

FIGURE 8.33

reading. What do you know about naming angles? Write down what you know before reading on. . . .

Naming Angles

We often refer to an angle by using its vertex. For example, we can call the angle at the left in Figure 8.34 angle *A*. We can also refer to angles by using a numbering system. In the diagram at the right in Figure 8.34, we can talk about angle 1, angle 2, angle 3, and angle 4.

Figure 8.34

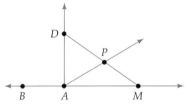

Figure 8.35

Both ways of referring to angles have limitations. Look at Figure 8.35. If someone asks you to look at angle *A*, what are the possibilities? How might we resolve this dilemma? Think before reading on. . . .

One way out of this dilemma is to use the symbol ∡ and three letters to name an angle. The order of the three letters is important. We start with a point on one side of the angle, then give the vertex, and then give a point on the other side of the angle (see Figure 8.36). These three points determine an angle in much the same way that three noncollinear points determine a plane. Thus you can see that ∡*DAM* and ∡*ADM* are different angles.

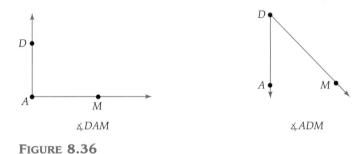

Figure 8.36

Measuring Angles

We measure angles by how "open" they are, and the tool we use is a **protractor**. The most common unit of measurement for angles is the **degree**.

Although protractors come in a variety of styles, the correct use of all protractors requires the following:

1. The vertex of the angle must lie at the center of the protractor, which is not always the center of the bottom edge of the protractor (as shown in Figure 8.37).

History

You may recall that there are 360 degrees in a circle. Do you know where the 360 comes from?

It comes from the Babylonians, who, for reasons unknown to us, decided (over 3000 years ago) to divide a circle into 360 parts. We know that 360 was an important number for them, perhaps because there are approximately 360 days in a year.

When the French converted measures of length, mass, and other quantities to the metric system, why didn't they convert measures of a circle? We know that the French attempt to metrify the week (10 days in a week and 10 hours in a day) failed. Perhaps the 24-hour day and 7-day week were just too deeply ingrained, and perhaps this was also true with angles. Actually, 360 has advantages over 100. Consider some common angles—for example, 60° and 45°. If there were 100 degrees in a circle, what would these angles be?

2. One ray of the angle must lie directly under the line that goes through a 0 point of the protractor (which is not always labeled).

3. The measure of the angle is read by looking at the number corresponding to where the other ray of the angle crosses the number line that goes around the protractor.

In the right-hand figure, the protractor has been moved in order to measure an angle neither of whose vertices is parallel to the bottom of the page. Measure the two angles and then check your measurements below or with a friend.

Outside the Classroom

As I was writing this edition of the text, I sprained my ankle badly and had to go for physical therapy. The PT used an instrument called a goniometer, which looked like a protractor, to measure the range of motion of my ankle to use as a baseline to measure my improvement. In the March–April 1997 issue of *Mathematics Teaching in the Middle School*, there is an article about goniometers, and an exercise appears at the end of this section.

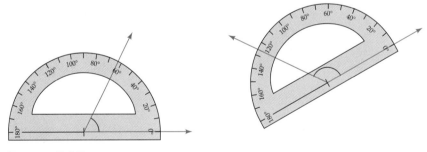

FIGURE 8.37

The measure of the angle at the left is 63 degrees, and the measure of the angle at the right is 126 degrees.

For the most part, the level of precision of measuring angles in this course will consist of rounding the measure to the nearest whole number. However, an angle can be divided into 60 minutes (denoted by ′), and a minute can be divided into 60 seconds (denoted by ″). Can you think of a situation in which this kind of precision would be necessary? For example, an angle of 0 degrees and 1 minute is 1/21,600 of a circle, too small to show on a piece of paper.

Section 8.1 / Basic Ideas and Building Blocks   515

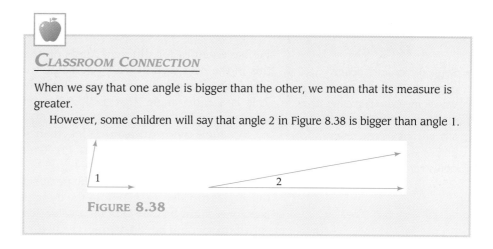

CLASSROOM CONNECTION

When we say that one angle is bigger than the other, we mean that its measure is greater.

However, some children will say that angle 2 in Figure 8.38 is bigger than angle 1.

FIGURE 8.38

INVESTIGATION 8.4

Measuring Angles

Estimate, and then measure with a protractor, the angles in the figures below.

Clock
FIGURE 8.39

Logo
FIGURE 8.40

Convex quadrilateral
FIGURE 8.41

Source: The Octagon logo is a Registered trademark of JPMorgan Chase & Co. and is used here with its expressed permission.

DISCUSSION

When we estimate, we often use benchmarks, just as we did with fractions in Chapter 5. Common benchmarks are 90°, a right angle, 45° (1/2 of a right angle), 30° (1/3 of a right angle), and 60° (2/3 of a right angle).

The angle of the clock at 4 o'clock should be 120 degrees. You can also calculate this, because the angle of the hands at 12 o'clock is 0 degrees and the angle at 6 o'clock is 180 degrees. Because 4 is 2/3 of 6, we need 2/3 of 180.

Each of the four trapezoids in the JPMorgan Chase & Co. logo have two 90-degree angles, one 45-degree angle, and one 135-degree angle. As you will see in the next section, we can further break these trapezoids into a rectangle and a right isosceles triangle.

The angles in the convex quadrilateral are 89°, 75°, 124°, and 72°.

Let us now examine the kinds of angles that we encounter in mathematics and in everyday life.

Classifying Angles

We have names for different kinds of angles. We will see the importance of this in the next section when we name triangles. We can classify angles with respect to an angle whose measure is one-fourth of a circle—that is, 90 degrees. This is a **right** angle. Just as 1/2 is a reference fraction, a right angle is a reference angle.

An angle whose measure is less than 90 degrees is called an **acute** angle.

An angle whose measure is greater than 90 degrees but less than 180 degrees is called an **obtuse** angle.

If angles are seen from a static perspective, these terms are often sufficient, but angles can also be seen from the dynamic perspective of *turns*. For example, someone might be told to open a valve "one-quarter of a turn," or 90 degrees. A half turn would be 180 degrees, a full turn would be 360 degrees, and one and a half turns would be 540 degrees. It sometimes makes sense to speak of angles with measures of 180 degrees or greater.

An angle whose measure is 180 degrees is called a **straight** angle.

An angle whose measure is greater than 180 degrees is called a **reflex** angle. Determine the measure of the reflex angle in Figure 8.42.

The measure of the angle is 225°. One way to determine the measure is to extend the horizontal ray to the right and measure the resulting acute angle and then add 180 to the measure of the acute angle. Another way is to measure the obtuse angle and then subtract it from 360. Do you understand both ways? Did you determine the measure in a third way?

Sometimes the relative location of two angles is important. For example, we speak of adjacent angles. What do you think adjacent angles are? Can you draw two adjacent angles? Can you write a definition? If you do, give it to someone who doesn't know the term and see whether that person understands the term in the way that you meant it. Then read on. . . .

Adjacent means "next to," but more specifically, two angles are adjacent if and only if they satisfy three conditions:

1. They have the same vertex.
2. They have a common side.
3. Their interiors are disjoint; that is, no point can be in the interior of both angles.

Do you see why the third condition is necessary? Can you find two angles in Figure 8.43 that have the same vertex and a common side but are not adjacent? . . . Think about it. Did you discover that ∡*DAM* and ∡*PAM* both share the side *AM* but are not considered adjacent?

We sometimes speak of **complementary** and **supplementary** angles.

If the sum of the measures of two angles is 90 degrees, we call them **complementary** angles.

If the sum of the measures of two angles is 180 degrees, we call them **supplementary** angles.

There is one more kind of angle to examine in this section: vertical angles. Whenever two lines intersect, four angles are formed. The angles that are opposite each other are called **vertical angles**.

That is, ∡1 and ∡7 in Figure 8.44 are vertical angles, and ∡2 and ∡8 are vertical angles. Most people can quickly see that each pair of vertical angles appear to be equal. Now this brings up one issue (language) and one question (proving they are equal).

> **■ Language ■**
>
> Some dictionaries define *obtuse angle* as an angle whose measure is greater than 90 degrees. In such cases, straight angles and reflex angles are special kinds of obtuse angles.

FIGURE 8.42

> **■ Outside the Classroom ■**
>
> Skateboarders and snowboarders use reflex angles to describe some of their moves. What do you think they mean when they talk about doing a frontside 540 or a backside 720?

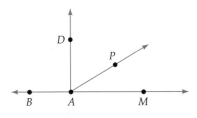

FIGURE 8.43

> ■ *Language* ■
>
> Many students argue convincingly that their lives in geometry would be much easier if such angles were called opposite angles.

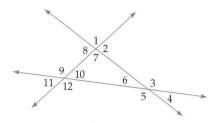

FIGURE 8.44

Technically, we say either that ∡1 and ∡7 are congruent or that their measures are equal.

The notation we use for congruence is ≅.

The notation we use for measures is $m\,∡$.

Saying ∡1 = ∡7 is like a child saying "I goed to the store yesterday" instead of "I went to the store." We understand the child, but we also correct the error. We will examine congruence in more detail in Section 8.2.

It certainly seems reasonable that vertical angles are always equal. Although this text is not emphasizing proofs, I believe a few proofs are helpful. How might we prove that the vertical angles *must* be equal? Think about this before reading on. . . .

A key to the proof is to realize that the members of each pair of adjacent angles in Figure 8.44 is supplementary. For example, ∡1 and ∡2 are supplementary. Do you see why? If you didn't see this, can you now see how we might show that ∡2 and ∡8 must be equal?

If you see the relationships, the proof is fairly straightforward:

Statement	Justification
$m\,∡1 + m\,∡2 = 180$	Together they form a straight angle.
$m\,∡7 + m\,∡2 = 180$	Together they form a straight angle.
$m\,∡1 + m\,∡2 = m\,∡7 + m\,∡2$	Transitive property; both sums are equal to 180.
$m\,∡1 = m\,∡7$	Algebra—we subtracted the same amount from both sides.

Again, it is this type of thinking that made what the Greeks did over 2000 years ago so different from what had been done before.

Summary

In this section, we have focused on the building-block language and concepts of Euclidean geometry: point, line, and plane. Line segments can intersect and join together to make plane figures, also called two-dimensional figures, which we shall explore in Section 8.2. We have examined different subsets of lines (line segments and rays) and different kinds of lines (parallel, perpendicular, concurrent, and skew). Plane figures can intersect to make space figures, also called three-dimensional figures, which we will explore in Section 8.3.

We have examined the different kinds of angles that can be formed by intersecting lines. We can classify angles by their measure: acute, right, obtuse, straight, and reflex. We can also classify angles by their relationship to other angles: complementary, supplementary, and vertical.

Exercises 8.1

1. **a.** What is a tetromino? Imagine defining this term to someone who has no idea what you mean.
   b. Write the definition of tetromino that could have resulted in someone thinking the two figures in Figure 8.3 were tetrominoes. Then fix the definition.

2. This exercise focuses on communication. For each of the following terms, suppose a friend in the class has come to you and told you that the term just doesn't make sense. Describe how you would help that person to make sense of the term. Your description must include your own definition of the term in your own words.
   a. Collinear points
   b. Concurrent lines
   c. Adjacent angle
   d. Skew lines
   e. Vertical angles

3. Optical illusions
   a. Which line is longer?

   b. Which circle is larger?

   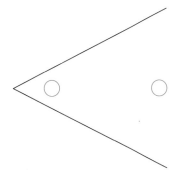

   c. Are the diagonal lines parallel?

   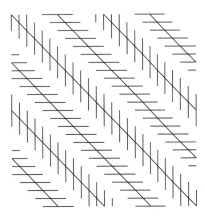

   d. Which segment on the top is a continuation of the segment on the bottom?

   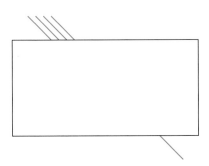

   e. What do you see?

   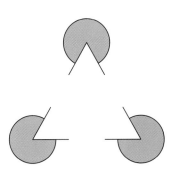

   f. Shade in the figure. What do you notice?

   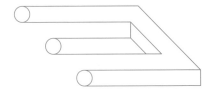

4. Name all the possible different rays that can be formed from the three points below.

5. How many different rays can be formed from four collinear points?

6. Sketch four lines such that three are concurrent with each other and two are parallel to each other.

7. True or false? If true, briefly explain why. If false, provide a counterexample.
   a. If two distinct lines do not intersect, then they are parallel.

b. If two lines are parallel, then they lie in the same plane.

c. If two lines intersect, then they lie in the same plane.

d. If a line is perpendicular to a plane, then it is perpendicular to all lines in that plane.

e. If three lines are concurrent, then they are also coplanar.

f. If two planes intersect, then the intersection is either a point or a line.

8. Refer to the cube pictured below, and use symbols such as $\overline{AB}$ to name the following:

a. Two parallel line segments

b. Two line segments that do not lie in the same plane

c. Two intersecting line segments

d. Three concurrent line segments that do not lie in a single plane

e. Two skew line segments

f. A pair of supplementary angles

g. A pair of perpendicular line segments

h. Are points A, B, and H coplanar points? Why or why not?

9. In the figure below, $\overleftrightarrow{AB}$ and $\overleftrightarrow{BC}$ are perpendicular lines.

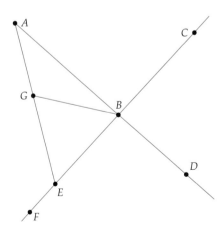

a. Name two complementary angles.

b. Name two supplementary angles.

c. Name two vertical angles.

d. Name two adjacent angles.

10. Estimate the measure of the following angles. Describe your reasoning process. Then measure the angles with a protractor and determine your percent error.

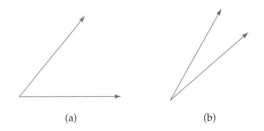

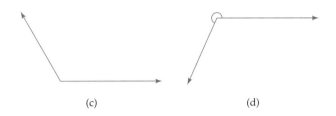

11. Clock problems.

a. How many times a day will the minute hand be directly on top of the hour hand?

b. What times could it be when the two hands make a 90-degree angle?

c. What angle do the hands make at 7 o'clock?

d. What angle do the hands make at 3:30?

e. What angle do the hands make at 2:06?

f. Make up and answer a problem like these. Then swap problems with a friend. Have the friend check your work and you check the friend's work.

12. Make at least five angles. Next estimate their measure, determine their measure, and determine your percent error. Repeat this process. If you note that the percent error seems to be decreasing, can you articulate any thinking processes that you are developing that are making your estimates more accurate?

13. Using only a straightedge and reasoning, try to make angles with the following measures. Then describe your reasoning process. After making each angle, check your work with a protractor and determine the percent error.

   a. 30°   b. 45°   c. 150°   d. 300°   e. 67°

14. a. Make your own goniometer (see p. 514). Briefly describe the thinking process behind your making of the goniometer.

   b. Compare goniometers with other members of your group or class. Which one(s) do you think will produce the most accurate results?

c. You will use one of the selected goniometers to measure the range of motion of an ankle. Each group separately will define "range of motion" and then determine a method for determining range of motion. Do that and then measure the range of motion of one of the member's ankles. Each person in the group will make one measurement, but do not say the number of your measurement until all members have measured.

d. Write on the board the three angles determined by your group. Observe the numbers of the other groups. Describe what you see.

e. Repeat the process after the entire class agrees on the same definition and the same measuring technique. Compare data. What is the average range of motion in your class?

f. What did you learn from this exercise?

15. Sketch a pair of angles whose intersection is
    a. exactly two points or explain why this is not possible.
    b. exactly three points or explain why this is not possible.
    c. exactly four points or explain why this is not possible.

16. a. Draw capital letters for the following letters: a, k, m, n, v, w, and z. Draw them large so you can measure angles.
    b. In the next step, you will measure your angles. Before you do, predict in which cases the angles will be the closest.
    c. Measure the following acute angles in each letter.

A K M
N V W Z

d. In the next step, you will compare your results with other members of your group or class. In which cases do you predict the standard deviation will be greatest? smallest?

e. Compare your measurements to those of other members of your group or class, as directed by your instructor. What did you find?

17. This exercise was taken from the March 2002 issue of *Discover* magazine. Below are the first five letters of the alphabet, designed to fit the vertices of a grid. Below them are the first four letters that were designed on other grids. Draw the e that goes with each type style, or font. As stated many times in this book, we want you to go beyond random trial and error and beyond the famous "just do it" commercial jingo. First, take some time and think. One strategy is to try to write the rules that the writer of the letter used. Another strategy is to try to articulate the common characteristics of the four letters.

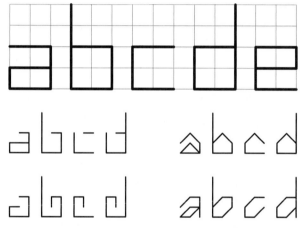

Source: Reprinted by permission of Scott Kim. Puzzle by Scott Kim, scottkim.com.

18. The terms *parallel* and *perpendicular* occur both in algebra (equations of lines) and in geometry (properties of many geometric figures). This exercise requires you to think about these concepts and some of the connections between algebra and geometry.
    a. On each of the Geoboards below, make a line parallel to the given line. Develop, describe, and justify a rule or procedure for making sure that the two lines are parallel. The rule or procedure should be more precise than "It looks parallel."

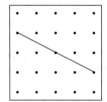 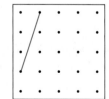 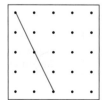 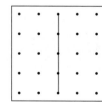 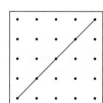 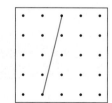

b. On each of the Geoboards below, make a line perpendicular to the given line. Develop, describe, and justify a rule or procedure for making sure that the two lines are perpendicular.

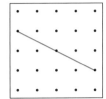 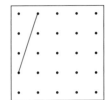 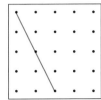 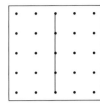 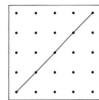 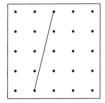

SECTION 8.2 TWO-DIMENSIONAL FIGURES

WHAT DO YOU THINK?

- In what ways are triangles and quadrilaterals different? In what ways are they alike?
- How are circles and polygons related?
- Can every polygon be broken down into triangles? Why or why not?
- Why do we use two words to name different triangles, but only one word to name different quadrilaterals?

Think of geometric figures that people generally find pleasing, such as those in Figure 8.45. What words would you use to explain why these objects are interesting or appealing? When you look at the various objects and pictures, what similarities do you see between certain objects and shapes—for example, triangles and hexagons?

When we discuss similarities and differences in my class, many geometric terms emerge in the discussion. Some students talk about similar shapes—for example, hexagons in honeycombs and snowflakes, squares in some pictures, and triangles in others. Some students observe that many of the shapes are symmetric. In explaining similarities, some students talk about the angles, the length of sides, or the fact that some figures look similar. Many students observe that even the more complex shapes can be seen as being constructed from simpler shapes, such as triangles and quadrilaterals.

(a)
Carpenter's Wheel quilt

(b)
lamp post

(c)
photograph of a snowflake

FIGURE 8.45

In a few moments, you will begin a systematic exploration of geometric shapes. Before you do, let us examine a very important framework for understanding the development of children's geometric thinking. This model was developed by Pierre and Dina van Hiele-Geldorf in the late 1950s and is widely used today. Essentially, the van Hieles found that there are levels, or stages, in the development of a person's understanding of geometry.[2]

Level 1: Reasoning by resemblance

At this level, the person's descriptions of, and reasoning about, shapes is guided by the overall appearance of a shape and by everyday, nonmathematical language. For example, "This is a square because it looks like one." Students at this level may be made aware of the various properties of geometric objects (for example, that a square has four equal sides), but such awareness can be overridden by other factors. For example, if we turn a square on its side, the student may insist that it is no longer a square but now is a diamond.

Level 2: Reasoning by attributes

At this level, the person can go beyond mere appearance and recognize and describe shapes by their attributes. A student at this level, seeing the figure above, can easily classify it as a quadrilateral because it has four sides. However, a student at this level does not regularly look at *relationships* between figures. A student who argues that a figure "is not a rectangle because it is a square" is reasoning at this level.

Level 3: Reasoning by properties

At this level, the student sees the many attributes of shapes and the relationships between and among shapes. A student at this level can see that the square and rhombus have many properties in common, such as opposite sides parallel, all four sides congruent, and diagonals that bisect each other and are perpendicular. This enables the student to understand that a square is simply a rhombus with one additional property—all the angles are right angles.

Level 4: Formal reasoning

Students at this level can understand and appreciate the need to be more systematic in their thinking. When solving a problem or justifying their reasoning, they are able to focus on mathematical structures.

The investigations in the text and the accompanying explorations have been designed to be consistent with this approach. Accordingly, as you are working in this chapter, reflect on your own thinking. Are you looking at the problem only on a vague, general level? Are you fixated on just one attribute? Are you seeing relationships among triangles, among quadrilaterals? As you move from *what* to *why*, are you able to move from solving a problem by random trial and error to being more systematic and careful in your approach?

With this model in mind, let us begin our exploration of shapes with an investigation my students have found to be both fun and powerful. Even more

---

[2]My writing of this section has been informed by Thomas Fox's "Implications of Research on Children's Understanding of Geometry" in the May 2000 issue of *Teaching Children Mathematics*, pp. 572–576.

important than knowing the names of all these different shapes will be knowing their properties and the relationships between and among the shapes. It is *this* knowledge that is used by artists, engineers, scientists, and all sorts of other people.

INVESTIGATION 8.5

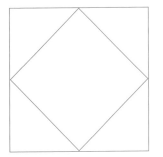

FIGURE 8.46

Recreating Shapes from Memory

For this investigation, you will want to have a pencil and an eraser.

A. Look at Figure 8.46 for about 1 second. Then close the book and draw Figure 8.46 from memory.
   Check the picture again for 1 second. If your drawing was incomplete or inaccurate, change your drawing so that it is accurate. Check the picture again for 1 second. Keep doing this until your drawing is complete and accurate.
   Now go back and try to describe your thinking processes as you tried to re-create the figure. From an information processing perspective, your eyes did not simply receive the image from the paper; your knowledge of geometry helped determine *how* you saw the picture. What did you hear yourself saying to help you remember the picture? Then read on. . . .

DISCUSSION

Some students see a diamond and 4 right triangles. Other students see a large square in which the midpoints of the sides have been connected to make a new square inside the first square. Yet other students see four right triangles that have been connected by "flipping" or rotating them.

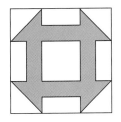

FIGURE 8.47

B. Now look at Figure 8.47 for several seconds. Then close the book and try to draw it from memory. As before, check the picture again for a few seconds. If your drawing was incomplete or inaccurate, fix it. Keep doing this until your drawing is complete and accurate. Then go back and describe the thinking processes you engaged in as you tried to re-create the figure. . . .

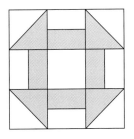

FIGURE 8.48

DISCUSSION

This figure was more complex. Some students see a whole design and try to remember it.
   Some decompose the design into four black triangles and four rectangles as shown in Figure 8.48.
   The figure can also be seen as being composed of 9 squares, which can also be seen in Figure 8.48. The four corner squares have been cut to make congruent

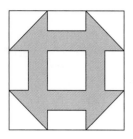

FIGURE 8.49

right triangles. Each of the other four squares on the border has been cut into two congruent rectangles.

Some students re-created this figure by seeing a whole square and then looking at what was cut out (see Figure 8.49). That is, they saw that they needed to cut out each corner, and they saw that they needed to cut out a rectangle on the middle of each side. Finally, they remembered to cut out a square in the center.

This is a famous quilting pattern called the Churn Dash. When people made their own butter, the cream was poured into a pail called a churn. Then the churner rolled a special pole back and forth with his or her hands. At the end of the pole was a wooden piece called a dash, which was shaped like the figure you saw. If you make many copies of this pattern and put them together, you can see what a Churn Dash quilt looks like.

There are several implications for teaching from this investigation. How a person re-creates the figure is related to the person's spatial-thinking preferences and abilities. Different people "see" different objects. That is, not everyone sees the figure in the same way. Although there are differences in how people re-create the figure, very few people can re-create the figure without doing some kind of decomposing—that is, without breaking the shape into smaller parts. Being able to do this depends partly on spatial skills and partly on being able to use various geometric ideas (congruent, triangle, square, rectangle) at least at an intuitive level.

Although some people manage to live happy, productive lives at the lowest van Hiele level, an understanding of basic geometric figures and the relationships among them is often helpful in everyday life (for example, in home repair projects and quilting) and in many occupations. Now that your interest in geometric figures has been piqued by this investigation and the pictures at the beginning of the section, let us examine the characteristics and properties of basic geometric shapes.

Before we examine specific kinds of polygons, beginning with triangles, the following investigation will serve to "open your thinking"—to get you to look at polygons not only through the zoom lens, which reveals specific properties and definitions, but also through the wide-angle lens, in which you see *all* the attributes. For example, a person is not just a person. She might be a mother, a sister, a daughter, a scientist, a Democrat, and so on. Should you become a friend of this person, you come to know her many facets. Similarly, if you become a "friend" of shapes, you come to know the many "sides" of the shapes with which you are working.

INVESTIGATION 8.6 — All the Attributes

Look at the polygons in Figure 8.50. Think of all the attributes, all the characteristics, anything about the polygons that might be important to state or measure. Write down your list before reading on. . . .

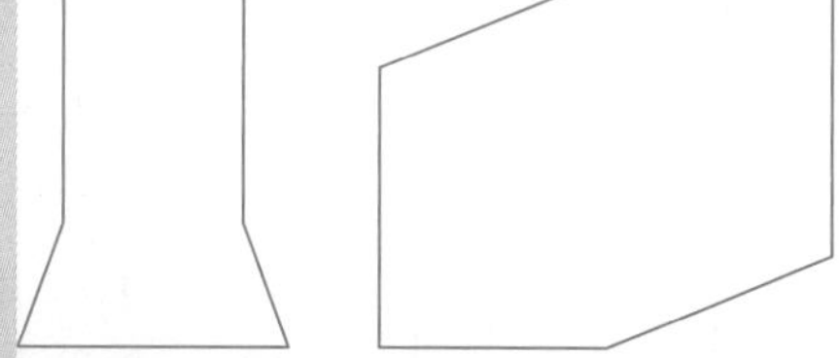

FIGURE 8.50

> ■ *Beyond the Classroom* ■
>
> This notion of objects having multiple attributes is not something applicable just to geometric shapes. Almost everything has multiple attributes. For example, every person has physical attributes (age, height, weight, etc.) and relational attributes (son, sister, mother, etc.). A garage has length, width, height, shape, surface area (if we want to paint it), volume (to consider storage), composition (wood or metal), number of windows, door (automatic or manual), and so on. A car—shape, interior room, size of trunk, size of engine, miles per galllon, repair record, and so on. When interacting with a person, or considering buying a car, or doing something with a garage, we focus on certain of those attributes. That is exactly what we do in geometry. Which attributes we focus on depend on the problem at hand.

DISCUSSION

As you compare the following lists to yours, read actively. If you made the same observation, did you use the same wording? If not, do you understand the wording here? If you missed one of these attributes, why? Do you understand it now? Are there other attributes that you noticed?

First figure	Second figure
6 sides	6 sides
Top and bottom sides parallel to each other	Top and bottom sides parallel to each other
	All three pairs of opposite sides parallel
Concave	Convex
2 sets of congruent sides	4 congruent sides; the other pair is also congruent
2 acute angles, 2 right angles, 2 reflex angles	4 obtuse angles, 2 right angles
3 pairs of congruent angles	Opposite angles congruent
1 line of reflection symmetry	No reflection symmetry
No rotation symmetry	1/2-turn rotation symmetry
Does not tessellate	Tessellates

A key idea here is to realize that there are lots of attributes and that knowing these attributes and combinations of attributes of a shape helps chemists and physicists to understand the behavior of a molecule or shape; helps builders to know which shapes work together better either in terms of structure and strength or in terms of appearance; and helps artists and designers to make designs that are the most appealing. As you read on, think of multiple attributes and of attributes that different objects have in common.

INVESTIGATION 8.7

Classifying Figures

Before we examine and classify important two-dimensional shapes, we first need to investigate the kinds of possible two-dimensional shapes. Although most of elementary students' exploration of two-dimensional figures will involve polygons and circles, it is important to know that these figures represent just a small subset of the kinds of figures that mathematicians study. Both circles and polygons are curves. A mathematical curve can be thought of as a set of points that you can trace without lifting your pen or pencil. If you watch young children making drawings, you discover that they make all sorts of curves!

As you might expect, if we look at any set of curves, there are many ways to classify them. As I have done throughout this book, rather than giving you the major classifications, I will engage you in some thinking before presenting them. Look at the 13 shapes in Figure 8.51 and classify them into two or more groups so that each group has a common characteristic. Do this in as many different ways as you can, and then read on. . . .

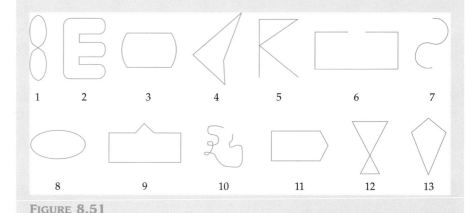

FIGURE 8.51

DISCUSSION

One way to sort the figures is shown below. How would you describe the figures in set A and the figures in set B? Do this before reading on. . . .

■ **Language** ■

What other words might you use to describe the intersecting and not intersecting subsets?

Some students use the phrase "trace over," and others talk about figures that "run over themselves" or "cross themselves." Other students talk about the set of figures that contain two smaller regions within each figure.

The figures in set A are said to be simple curves. We can describe simple curves in the following way: A figure is a **simple curve** in the plane if we can trace the figure in such a way that we never touch a point more than once. If you look at the figures in set A, you can see that they all have this characteristic; and all the figures in set B have at least one point where the pencil touches twice, no matter how you trace the curve.

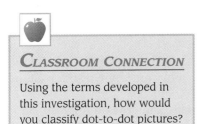

CLASSROOM CONNECTION

Using the terms developed in this investigation, how would you classify dot-to-dot pictures?

Now look at the curves in sets C and D. How would you describe the figures in set C and the figures in set D? Do this before reading on. . . .

The figures in set C are said to be closed curves. We can describe closed curves in the following way: A figure is a **closed curve** if we can trace the figure in such a way that our starting point and our ending point are the same. If you look at the figures in set C, you can see that they all have this characteristic; and no matter how you try, you cannot trace the figures in set D with the same starting and ending point.

Now look at the curves in sets E, F, and G. How would you describe the figures in set E, the figures in set F, and the figures in set G? Do this before reading on. . . .

These three sets are interesting for two reasons. First, these sets are likely to be generated in the classroom—both your classroom and the elementary classroom. Second, the language used to describe the three sets poses a challenge, for most people describe the figures in set E as consisting only of curvy lines, the figures in set F as consisting only of straight line segments, and the figure in set G as having both curvy and straight line segments. The challenge here is that when mathematicians use the word *curve*, this word encompasses both curvy and straight line segments—a curve is a set of points that you can trace without lifting your pen or pencil. There is nothing wrong with students' use of the terms *curvy* and *straight*. What is important is the realization that we are using the words *curve* and *curvy* in different ways. We do this all the time in everyday English. Recall the various uses of the word *hot* in Chapter 1: "It sure is a hot day." "I love Thai food because it is hot." "This movie is really hot!"

Most of our investigations of curves will focus on simple closed curves. Looking at the descriptions above, try to define the term *simple closed curve* before reading on. . . .

We will define a **simple closed curve** in the plane as a curve that we can trace without going over any point more than once while beginning and ending at the same point. The set of polygons is one small subset of the set of simple closed curves.

At this point, you might want to do the following activity with another student.

- Draw a simple closed curve.
- Draw a simple open curve.

- Draw a nonsimple closed curve.
- Draw a nonsimple open curve.

Exchange figures with another student. Do you both agree that each of the other's drawings matches the description? If so, move on. If not, take some time to discuss your differences.

There is an important mathematical theorem known as the *Jordan curve theorem*, after Camille Jordan: Any simple closed curve partitions the plane into three disjoint regions: the curve itself, the interior of the curve, and the exterior of the curve. See the examples in Figure 8.52.

> ■ **History** ■
>
> Generally, when a theorem is named after a person, it is named after the person who first proved the theorem. In this case, Jordan's proof was found to be incorrect, but the theorem is still named after him!

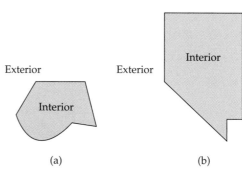

Figure 8.52

I know that many students' reaction to this theorem is "Why do we need to prove something that is so obvious?" As mentioned before, being critical ("to examine closely") is an attitude that I invite. In the examples in Figure 8.52, deciding whether a point is inside or outside is easy. However, look at Figure 8.53. Although this figure is a simple closed curve, it is a rather complicated figure, and such complicated shapes are encountered in some fields of science. Is point A inside or outside? How would you determine this? Think before reading on. . . .

Figure 8.53

Some wag once remarked that mathematicians are among the laziest people on earth because they are always looking for shortcuts and simpler ways to solve problems. Thus you may be wondering whether someone has

Classroom Connection

The active and curious reader here might realize that this procedure might have applications for solving mazes, and there is a branch of mathematics that does analyze mazes. Another active and curious reader might realize and/or remember that most children enjoy the challenge of solving mazes.

found an easier way to solve these problems. Look at Figure 8.53 to illustrate the method. Start at a point that is clearly outside the shape and draw a line segment connecting that point to the point you are looking at; it helps if you pick an outside point so that the line segment will cross the curve in as few points as possible. Each time you cross a point, it's like a gate—if you were outside, you are now inside; if you were inside, you are now outside. Thus it is a relatively simple matter to determine that point A is inside the curve.

Now go back to the first shape to see whether point A is inside or outside. As an active reader you will want to work on the figure to verify to your satisfaction that the method just described really does work.

Polygons

We are now ready to begin our exploration of **polygons**, which can be defined as simple closed curves in the plane composed only of line segments. Thus the simple closed curve in Figure 8.52(a) is not a polygon, whereas the simple closed curve in Figure 8.52(b) (which looks like the state of Nevada) is a polygon. On any polygon, the point at which two sides meet is called a **vertex**, the plural of which is **vertices**. The line segments that make up the polygon are called **sides**.

The word *polygon* has Greek origins: *poly-*, meaning "many," and *-gon*, meaning "sides." You are already familiar with many kinds of polygons. Just as we found in Chapter 2 that the names we give numbers have an interesting history, so do the names we give to polygons. The most basic naming classification involves the number of sides (see Table 8.2).

Now that we have a good general definition of the term *polygon*, we can spend time examining triangles, quadrilaterals, and a few specific kinds of polygons.

TABLE 8.2

Number of sides	Name
3 sides	Triangle
4 sides	Quadrilateral
5 sides	Pentagon
6 sides	Hexagon
7 sides	Heptagon
8 sides	Octagon
n sides	n-gon

Triangles

Triangles are found in every aspect of our lives—in buildings, in art, in science (see Figures 8.54 and 8.55). They are truly "building-block" shapes. Every bicycle I have seen has triangles. Bridges will always contain triangles. If you look at the skeleton of buildings, and the scaffolding around the building, you will always see triangles. Why? Rather than give you the answer, we will use the next investigation to think about this question.

FIGURE 8.54

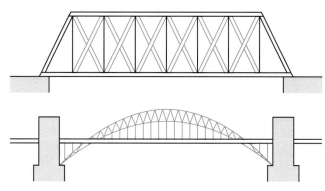

FIGURE 8.55

INVESTIGATION 8.8

Why Triangles Are So Important

Cut some strips of paper from a file folder or other stiff material. Punch a hole in the ends and use paper fasteners (improvise if you need to; for instance, you can use paper clips). Make one triangle and one quadrilateral, as shown in Figure 8.56. It need not be an equilateral triangle or a square. What do you see? . . .

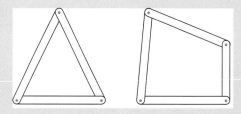

FIGURE 8.56

DISCUSSION

As you saw, the triangle won't move—we call it a rigid structure. However, the quadrilateral does move; it is not stable. Make another strip and connect two nonadjacent vertices of your quadrilateral. What happens now? It will remain in the shape. If it is a square, it will remain a square; if it is a parallelogram, it will remain a parallelogram. This is because the addition of that diagonal actually created two triangles, which, as you have found, are rigid structures. The next time you walk about campus and about town, look for triangles. You will suddenly see them everywhere!

As you already know, there are many kinds of triangles. A crucial goal of the next investigation is for your own understanding of triangles to become more powerful. Thus, as always, please *do* the investigation, rather than just reading through it quickly because "I already know this stuff."

INVESTIGATION 8.9

Classifying Triangles

You will find nine triangles in Figure 8.57. Copy them and cut them out, and then separate them into two or more subsets so that the members of each subset share a characteristic in common. Can you come up with a name for each subset? What names might children give to different subsets? Do this in as many different ways as you can before reading on. . . .

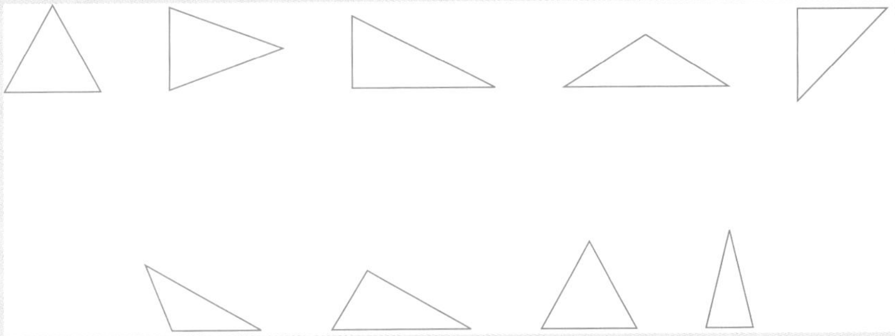

FIGURE 8.57

DISCUSSION

STRATEGY 1: Consider sides

One way to classify triangles is by the length of their sides: all three sides having equal length, two sides having equal length, no sides having equal length. There are special names for these three kinds of triangles.

- If all three sides have the same length, then we say that the triangle is equilateral.
- If at least two sides have the same length, then we say that the triangle is isosceles.
- If all three sides have different lengths—that is, no two sides have the same length—then we say that the triangle is scalene.

Which of the triangles in Figure 8.57 are scalene? Which are isosceles? Which are equilateral?

STRATEGY 2: Consider angles

We can also classify triangles by the relative size of the angles—that is, whether they are right, acute, or obtuse angles. This leads to three kinds of triangles: right triangles, obtuse triangles, and acute triangles.

- We define a right triangle as a triangle that has one right angle.
- We define an obtuse triangle as a triangle that has one obtuse angle.
- We define an acute triangle as a triangle that has three acute angles.

Many students see a pattern: A right triangle has one right angle, an obtuse triangle has one obtuse angle, yet an acute triangle has three acute angles. What was the pattern? Why doesn't it hold? Think before reading on. . . .

The key to this comes from looking at the triangles from a different perspective: Every right triangle has exactly two acute angles, and every obtuse triangle has exactly two acute angles; thus a triangle having more than two acute angles will be a different kind of triangle. This perspective is represented in Table 8.3. Does it help you to understand better the three definitions given above?

TABLE 8.3

First angle	Second angle	Third angle	Name of triangle	
Acute	Acute	Right	Right triangle	
Acute	Acute	Obtuse	Obtuse triangle	
Acute	Acute	Acute	Acute triangle	

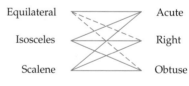

STRATEGY 3: Consider angles and sides

This naming of triangles goes even further. What name would you give to the triangle in Figure 8.58?

FIGURE 8.58

This triangle is both a right triangle and an isosceles triangle, and thus it is called a right isosceles triangle or an isosceles right triangle. How many possible combinations are there, using both classification systems? Work on this before reading on. . . .

There are many strategies for answering this question. First of all, we find that there are nine possible combinations (see Figure 8.59). We can use the idea of Cartesian product to determine all nine. That is, if set S represents triangles classified by side, $S = \{$Equilateral, Isosceles, Scalene$\}$, and set A represents triangles classified by angle, $A = \{$Acute, Right, Obtuse$\}$, then $S \times A$ represents the nine possible combinations.

FIGURE 8.59

> ■ **Mathematics** ■
>
> Did you notice the geometric balance in Figure 8.59, which represents the Cartesian product of the two sets of triangles?

However, not all nine combinations are possible. For example, any equilateral triangle must also be an acute triangle. (Why is this?) Therefore, "equilateral acute" is a redundant combination. However, it is possible to have scalene triangles that are acute, right, and obtuse. Similarly, we can have isosceles triangles that are acute, right, and obtuse.

Name the two triangles in Figure 8.60. Then read on. . . .

CLASSROOM CONNECTION

Some students have told me that this investigation is huge for them, because they had never thought about organization and classification with respect to triangles. Triangles just *were*!

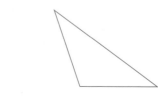

FIGURE 8.60

Both are obtuse, isosceles triangles. The orientation on the left is the standard orientation for isosceles triangles. Any time we position a triangle so that one side is horizontal, we call that side the **base**. As stated at the beginning of this chapter, students often see only one aspect of a triangle; for example, they see the triangle at the left as isosceles but not also obtuse, and they see the triangle at the right as obtuse but not also isosceles.

CLASSROOM CONNECTION

Children's development of triangles is fascinating. In several studies, children were given many shapes and asked to identify the shapes. Young children tend to identify the equilateral triangle in the "standard" position (base parallel to the bottom of the page) as a "true" triangle. They will often reject triangles such as the ones below because they are too pointy or turned upside down. Recall level 1 in the van Hiele model. One of my all-time favorite examples occurred when a first-grader was given the pattern shown in Figure 8.61 and was asked to continue the

FIGURE 8.61

pattern. After studying the pattern, she said, "Triangle, triangle, wrong triangle, triangle, triangle, wrong triangle, triangle. . . . The next shape is a right triangle!"[3] It is important to note that this child was doing wonderful thinking—she was seeing patterns, and she was looking at attributes, *and* she was at the beginning of her understanding of triangles.

As mentioned earlier, one of the new NCTM process standards is *representation*. As I have also mentioned earlier, one definition of understanding has to do with the quantity and quality of connections the learner can make within and between various ideas. In this case, a new representation of the relationship between and among triangles will help to deepen your understanding. More of these problems can be found in Exploration 8.13. These diagrams are also being used more and more in elementary schools because of their tremendous potential to help children see more relationships and connections.

INVESTIGATION

8.10

Triangles and Venn Diagrams

Recall our work with Venn diagrams in Chapter 2. Venn diagrams can help us to understand how concepts are and are not related. Let us take two kinds of triangles: right triangles and isosceles triangles. Draw several right triangles and draw several isosceles triangles. Which of these triangles could be placed in both categories? That is, which are right triangles and also

---

[3]From "Shape Up!" by Christine Oberdorf and Jennifer Taylor-Cox, published in the February 1999 issue of *Teaching Children Mathematics*, pp. 340–345.

isosceles triangles? Make a Venn diagram that illustrates the relationships between the right and isosceles triangles, and place each triangle in the appropriate region of the diagram.

Now draw several acute triangles and draw several equilateral triangles. Which of these triangles could be placed in both categories? How are these triangles related to each other? Make a Venn diagram that illustrates the relationships between the acute and equilateral triangles, and place each triangle in the appropriate region of the diagram.

DISCUSSION

When we have two groups of objects and look at how they are related, there are three possible relationships. They may be disjoint—each object is in one or the other group; there may be overlap—some objects are in both groups; or one group may be a subset of the other. Each of these relationships (remember the van Hiele levels) is represented with a different diagram, as shown in Figure 8.62.

Figure 8.62

Look at your Venn diagrams and think of what you just read. Do you want to change your diagrams? If you couldn't make the Venn diagrams because you just didn't understand the question, can you do so now? Do this before reading on. . . .

Because some triangles are both right and isosceles, those triangles can be placed in the center, showing that they belong to both sets (see Figure 8.63). Because equilateral triangles can contain only acute angles, all equilateral triangles are acute triangles. Hence the equilateral triangles are in a ring that is inside the ring that represents acute triangles.

> ■ *Mathematics* ■
>
> Interestingly, all right isosceles triangles have the same shape. That is, if you found the ratio of lengths of the sides of two right isosceles triangles, converted this to a percent, placed the larger one on a copy machine, and entered that percent in the reduction button, the smaller triangle would come out. We will explore this notion of similarity in Section 9.3.

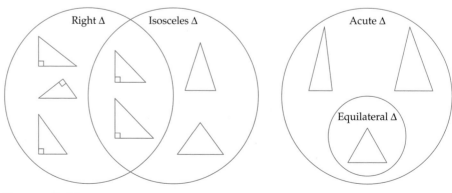

Figure 8.63

Triangle Properties

You probably remember that the sum of the measures of the angles in any triangle is 180 degrees. How could you convince someone who does not know that?

This is a case in which the van Hiele levels are instructive. A level 2 activity would be to have the students measure and add the angles in several triangles. If their measuring was relatively accurate, one or more students would see the pattern and offer a hypothesis that the sum is always 180.

An example of a level 3 activity, which I recommend if you have never done it, is to cut off the three corners of a triangle as in Figure 8.64 and then put the three corners together. What do you see?

An example of a level 4 activity would be a formal proof.

FIGURE 8.64

Special Line Segments in Triangles

There are four special line segments that have enjoyed tremendous influence in Euclidean geometry: angle bisector, median, altitude, and perpendicular bisector.

A **median** is a line segment that connects a vertex to the midpoint of the opposite side.

In triangle SAT (Figure 8.65), $\overline{TR}$ is a median. Hence $SR = RA$.

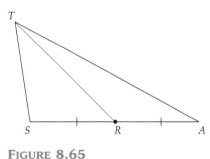

FIGURE 8.65

An **angle bisector** is a line segment that bisects an angle of a triangle.
In triangle ABC (Figure 8.66), $\overline{AD}$ is an angle bisector. Hence, $m\angle BAD = m\angle DAC$.

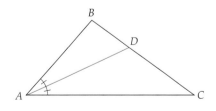

FIGURE 8.66

A **perpendicular bisector** is a line that goes through the midpoint of a side and is perpendicular to that side. In triangle PEN (Figure 8.67), $\overleftrightarrow{MX}$ is a perpendicular bisector of side $\overline{PN}$, because M is the midpoint of $\overline{PN}$ and $\overline{MX}$ is perpendicular to $\overline{PN}$.

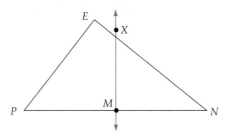

Figure 8.67

An **altitude** is a perpendicular line segment that connects a vertex to the side opposite that vertex. In some cases, as in $\triangle SAT$ in Figure 8.68, we need to extend the opposite side to construct the altitude.

In triangle ABC (Figure 8.68), $\overline{BF}$ is an altitude. Hence, $m\angle BFA = m\angle BFC = 90°$.

In triangle STA, $\overline{TP}$ is an altitude. Hence, $m\angle TPA = 90°$.

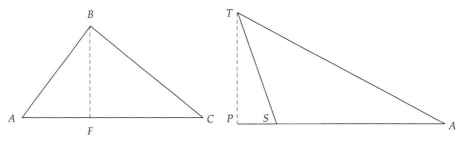

Figure 8.68

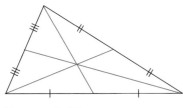

Figure 8.69

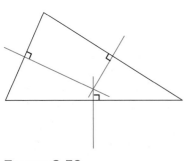

Figure 8.70

Many students have trouble with the idea of an altitude being outside the triangle. This happens when we have an obtuse triangle oriented with one side of the obtuse angle as the base of the triangle. If you are having trouble connecting the definition of *altitude* in these situations, I recommend the following: Trace the triangle and cut it out. Stand it up so that $\overline{SA}$ is on the plane of your desk and T is above that plane. Now draw a line from T that goes "straight down." What do you notice?

If triangle STA were large enough so that you could stand with your head at point T, the length of line segment $\overline{TP}$ would tell you how tall you were!

After reading the preceding discussion, a skeptical reader might ask, "So what? What is the value of these line segments, other than to mathematicians?" One response is that there is practical value, and another response is that there are some really cool relationships here. Let me explain.

The segment with the most practical value is the median. The point where all three medians meet is called the centroid and is the center of gravity of the triangle (see Figure 8.69). That is, if you found this point, made a copy of the triangle with wood, and placed the triangle on a nail at that point, the triangle would balance. Center of gravity is related to balance and is an important concept in the design of many objects—cars, furniture, and art, to mention just a few.

The point where the three perpendicular bisectors meet is called the circumcenter (see Figure 8.70). It turns out that this point is equidistant from each of the three *vertices* of the triangle. Thus you can draw a circle so that all three vertices of the triangle lie on the circle and the rest of the triangle is inside the circle. We say that the circle circumscribes the triangle.

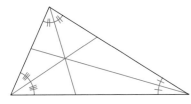

FIGURE 8.71

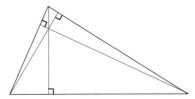

FIGURE 8.72

The point where the three angle bisectors meet is called the incenter (see Figure 8.71). It turns out that this point is equidistant from each of the three *sides* of the triangle. Thus you can draw a circle so that the circle touches (is tangent to) each of the three sides of the triangle; in this case, the circle is inside the triangle. We say that the circle is inscribed in the triangle.

The point where the three altitudes meet is called the orthocenter (see Figure 8.72). The orthocenter has connections to other geometric ideas. For example, there is a connection between the construction of a parabola and the orthocenter of a triangle.

There are two amazing things about these line segments. The first is that in any triangle there are three medians, three perpendicular bisectors, three angle bisectors, and three altitudes. In each case, the three line segments will *always* meet at a single point—they are concurrent. That is, the three medians will always meet at one point in any triangle, the three perpendicular bisectors will always meet at one point in any triangle, and so on. Now, except for the case of the equilateral triangle, these points are *not* the same point. In fact, in a scalene triangle, the four points are all different. However (and this is the other cool thing), there is a relationship among the centroid, circumcenter, and the orthocenter. In *any* triangle, they are collinear! Do you understand this? Figure 8.73 illustrates this fact, which Leonard Euler first discovered. We even call the line containing these three points an Euler line. If you are curious, you can type in these terms on a search engine. You will even find websites that have interactive features allowing you to see the situations where these four points are distinct but collinear and the situations where they converge into one point.

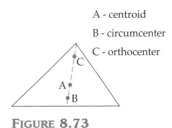

FIGURE 8.73

Congruence

As you have seen from your geometry explorations, questions sometimes arise about whether two figures are "the same" or not. Such observations and questions deal with the idea of congruence.

At an informal level, we can say that two figures are congruent if and only if they have the same shape and size. An informal test of congruence is to see whether you can superimpose one figure on top of the other. This is closely connected to how children initially encounter the concept and is related to the dictionary definition: "coinciding exactly when superimposed."[4] That is, if one figure can be superimposed over another so that it fits perfectly, then the two figures are congruent.

Formally, we say that two polygons are **congruent** if and only if all pairs of corresponding parts are congruent. In other words, in order for us to conclude that two polygons are congruent, two conditions have to be met: (1) Each corresponding pair of angles have the same measure and (2) each corresponding pair of sides have the same length. We use the symbol ≅ to denote congruence.

For example, in Figure 8.74, triangle *CAT* and triangle *DOG* are congruent iff $\angle C \cong \angle D$, $\angle A \cong \angle O$, $\angle T \cong \angle G$, $\overline{CA} \cong \overline{DO}$, $\overline{AT} \cong \overline{OG}$, and $\overline{TC} \cong \overline{GD}$.

[4]Copyright © 1996 by Houghton Mifflin Company. Reproduced by permission from *The American Heritage Dictionary of the English Language, Third Edition.*

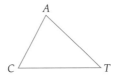
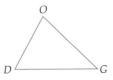

FIGURE 8.74

The notions of congruent and equal are related concepts. We use the term *congruence* when referring to having the same shape, and we use the term *equal* when referring to having the same numerical value. Thus we do not say that two triangles are equal; we say that they are congruent. Similarly, when we look at line segments and angles of polygons, we speak of congruent line segments and congruent angles. However, when we look at the numerical value of the line segments and angles, we say that the lengths of two line segments are equal and that the measures of two angles are equal.

Congruence is a big idea, both in geometry and beyond the walls of the classroom. Many important properties and relationships come from exploring congruence. The following investigations (and the related explorations) will help move your understanding of congruence to higher van Hiele levels, from the "can fit on top," geometric reasoning by resemblance, to understanding that all corresponding parts are congruent, to geometric reasoning by attributes, to geometric reasoning by properties, which we shall examine now.

■ Outside the Classroom ■

When do we need congruence in everyday life or in work situations? Take a few minutes to think about this before reading on. . . .

Congruence is important in manufacturing; for example, the success of assembly-line production depends on being able to produce parts that are congruent. Henry Ford changed our world by conceiving of making cars not one at a time but as many sets of congruent parts. For example, the left front fender of a 2003 Dodge Caravan is congruent to the left front fender of any other 2003 Dodge Caravan. One of the differences between a decent quilt and an excellent one is being able to ensure that all the squares are congruent. This is quite difficult when using complex designs. Most of the manipulatives teachers use with schoolchildren (Pattern Blocks, unifix cubes, Cuisenaire rods, and fraction bars) have congruent sets of pieces.

INVESTIGATION 8.11

Congruence with Triangles

For this you need a protractor and a compass, although you can improvise without them. Do both questions on a blank sheet of paper. First, see how many different triangles you can make that have the following attributes: The base is 50 millimeters (mm). The angle coming from the left side of the base is 30°, and the side coming from the left side of the base is 30 mm. Second, see how many different triangles you can make that have the following attributes: The base is 30 mm. The angle coming from the left side of the base is 30°, and the side coming from the right side of the base is 25 mm. Do this before reading on. . . .

DISCUSSION

There is only one triangle that can be drawn in the first case. However, in the second case, there are two possible triangles. Thus there is not enough information about the triangle to specify exactly one triangle (see Figure 8.75).

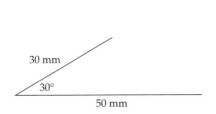 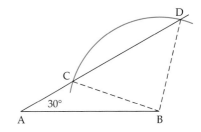

FIGURE 8.75

In Section 8.1, we said that two points *determine* a line; here, we are looking at what determines a triangle. This notion of *determining* or *specifying* is important in mathematics, both for congruence (When are two figures congruent?) and for definitions (How much do we need to specify to determine a shape?). For example, we can define a rectangle as a quadrilateral with four equal angles. Now a rectangle has many more properties than four congruent angles. However, mathematicians have discovered that this information—quadrilateral, four congruent angles—is sufficient so that only rectangles can be drawn that meet that criteria. Thus the notion of "determines" is an important one in mathematics. In elementary school we do not get terribly technical, but that is not the same as saying we just have fun and play around. When we ask children to explore well-focused questions, their understanding of shapes *and* relationships between and among shapes *and* their ability to see and apply properties can grow tremendously. When this happens, high school mathematics makes much more sense!

Quadrilaterals

We found that we could describe different kinds of triangles by looking at their angles or by looking at relationships among their sides. With quadrilaterals, which have one more side, new possibilities for categorization emerge: parallel sides, adjacent vs. opposite sides, relationships between diagonals, and the notion of concave and convex. Thus, how we go about naming and classifying quadrilaterals is not the same as how we name and classify triangles.

In this book, we will define the following kinds of quadrilaterals:

- A **trapezoid** (Figure 8.76) is defined as a quadrilateral with at least one pair of parallel sides.

FIGURE 8.76

- A **parallelogram** (Figure 8.77) is defined as a quadrilateral in which both pairs of opposite sides are parallel.

FIGURE 8.77

- A **kite** (Figure 8.78) is defined as a quadrilateral in which two pairs of adjacent sides are congruent.

FIGURE 8.78

> ■ *Language* ■
>
> Some books define a trapezoid as having exactly one pair of parallel sides. I have spoken with many mathematicians about this particular definition—some prefer the definition used here and some prefer the other one. There is no *right* answer, but rather reasons behind each one. There are two reasons why I use the definition given here. First, I would find it odd to define an isosceles triangle as having *at least* two congruent sides but then a trapezoid as having exactly one set of parallel sides. If we are going to say a trapezoid has exactly one pair of parallel sides, then an isosceles triangle should have exactly two congruent sides. There are reasons why *isosceles* is defined the way it is, reasons beyond the scope of this book, but I am persuaded by parallel usage here. Second, defining *trapezoid* this way enables us to represent relationships among quadrilaterals in a more elegant way.

> **Outside the Classroom**
>
> How many of each kind of quadrilateral can you find in everyday life?

- A **rhombus** (Figure 8.79) is defined as a quadrilateral in which all sides are congruent.

- A **rectangle** (Figure 8.80) is defined as a quadrilateral in which all angles are congruent.

- A **square** (Figure 8.81) is defined as a quadrilateral in which all four sides are congruent and all four angles are congruent.

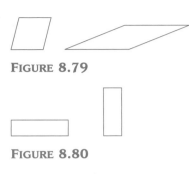

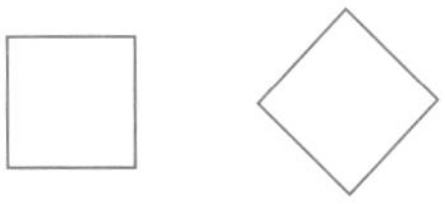

FIGURE 8.79

FIGURE 8.80

FIGURE 8.81

An active reader may have noted that there are many other possible categories of quadrilaterals. For example, there are many different quadrilaterals that have at least one right angle or exactly three congruent sides. Just as we called the triangle with no sides congruent a scalene triangle, we could call a quadrilateral with no sides congruent a scalene quadrilateral. If we have names for quadrilaterals with two pairs of adjacent sides congruent and quadrilaterals with two pair of opposite sides congruent, why stop there? The primary reason why we have names for the quadrilaterals described above is that these are the sets that mathematicians have found interesting, and specifying other groups didn't lead to anything beyond those groups; it just stopped, like a road that went nowhere. For example, we could define a rightquad as a quadrilateral with at least one right angle, and there are many different-looking quadrilaterals that would be rightquads. However, there is no name for this set of quadrilaterals because examining them as a class wasn't found to be useful. That is, it didn't lead to practical applications or to other discoveries.

Diagonals One characteristic of all polygons with more than three sides is that they have diagonals. The more sides in the polygon, the more diagonals. This term is probably familiar to most readers. However, before reading the definition of *diagonal* below, stop and try to define the term yourself so that it works for all polygons, not just squares and other quadrilaterals. Then read on. . . .

A **diagonal** is a line segment that joins two nonadjacent vertices in a polygon.

Figure 8.82 shows two different diagonals. One of the exercises will ask you to find patterns to determine the number of diagonals in any polygon.

> **Language**
>
> A few years ago, some students were stumbling over the question of how many diagonals a hexagon had. They contended that $\overline{FC}$ was not a diagonal of the pentagon at the right. Their argument was that it was a horizontal line and that diagonal meant going at a slant. At first, I was stunned. But then I realized that in everyday language, when we use the word diagonal, we do mean going at a slant. Brain research tell us that new information is almost always connected to something we know. The first diagonals we consider are those of a quadrilateral, and those diagonals are almost always shown in a 'diagonal' direction. The experience was humbling because it made me wonder how many students in my previous classes (almost 15 years' worth) had either not really understood that, in mathematics, a diagonal is *any* line segment connecting two vertices that are not adjacent.

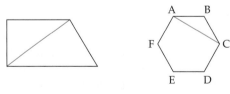

FIGURE 8.82

Angles in quadrilaterals We know that the sum of the measures of the angles of any triangle is 180 degrees. Will the sum of the measures of the four angles of *any* quadrilateral also be equal to one number, or will there be different numbers for different quadrilaterals? What do you think? What could you do to check your hypothesis? Once you believe that your hypothesis is true, how could you prove it? Think before reading on. . . .

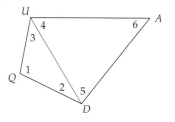

FIGURE 8.83

It turns out that for any quadrilateral, the sum of the measures of the four angles is 360 degrees. The following discussion is an informal presentation of one proof. If we draw a generic quadrilateral *QUAD* and one diagonal, as in Figure 8.83, what do you notice that might be related to this proof?

If you see two triangles and *connect* this to your knowledge that the sum of the measures of the angles of a triangle is 180 degrees, you have the key to the proof. That is, you can conclude that the sum of all six angles must be 360 degrees. However, these six angles are equivalent to the four angles of the quadrilateral!

INVESTIGATION 8.12

Quadrilaterals and Attributes

Look at the three quadrilaterals below. Think about their various attributes. Recall Investigation 8.6. Now answer the following question: Which two of the quadrilaterals in Figure 8.84 are most alike and why? Do your thinking and write your response before reading on. . . .

FIGURE 8.84

DISCUSSION

It is questions like this that help children, and adults, to move up the van Hiele levels. A good case can be made for different answers. Let us begin by simply listing various attributes.

1 pair of ≅ sides	2 pairs of ≅ sides	2 pairs ≅ of sides
1 pair parallel sides	0 parallel sides	0 parallel sides
2 right angles	1 right angle	0 right angles
1 obtuse angle	2 obtuse angles	0 obtuse angles
1 acute angle	1 acute angle	3 acute angles
0 reflex angles	0 reflex angles	1 reflex angle
convex	convex	concave

This list reflects various attributes on which we focused: congruence, parallel, angles, and shape (convex and concave, which will be formally defined soon). On the one hand, the two figures at the right are both kites and therefore "belong" together under that name. On the other hand, the two figures on the left are both convex and both have right angles, so there is much in common between the two of them also.

One of the most puzzling aspects of how mathematics has generally been taught is how much of it is simply learning and reciting facts and theorems that other people have learned. We ask our students to study mathematics, but we rarely let them *do* mathematics. If we were to teach art this way, students would learn techniques and be tested on how well they understood those techniques, but they would never get to do art. Most of the various groups advocating change in how mathematics is learned want to present students with problems where solving the problem means not simply applying an algorithm they have learned but, rather, involves what we call mathematical thinking. I have sought to give the activities in *Explorations* this flavor. It is more difficult to incorporate this "doing mathematics" flavor into the textbook, but the discussion of most investigations illustrates different solution paths to counter the widely held misconception that there is "one best way." To a greater degree than many of the investigations in the text, the investigation that follows has this flavor of *doing* mathematics.

INVESTIGATION 8.13

Challenges

This investigation brings together the notion of attributes and the notion of determinism (e.g., two points determine a line), which are two of the big ideas of geometric thinking. How many different kinds of quadrilaterals can you make that have exactly two adjacent right angles? Play around with this for a while on a piece of paper. Sketch different quadrilaterals that have exactly two adjacent right angles. What do you see? Can you make any conjectures? Can you prove them? Use whatever tools are available. As you do, try to push yourself beyond random trial and error to being more systematic, to thinking "What would happen if," to looking at your solutions to see what they have in common. . . .

DISCUSSION

Figure 8.85 shows three of many possibilities.

FIGURE 8.85

What do they all have in common besides two adjacent right angles? Think before reading on. . . .

They all have two parallel sides. That means they are trapezoids. A curious reader might now be asking whether it is possible not to get a trapezoid. What do you think? How might you proceed, rather than just using random trial and error? Think before reading on. . . .

If you took high school geometry, you might be remembering a theorem that said something like this: If two lines form supplementary interior angles on the same side of a transversal, then the lines are parallel. When we limit our investigation to two adjacent right angles, we make two interior angles that are also supplementary. Thus we know that the opposite sides of this quadrilateral must be parallel. Hence the condition of adjacent right angles *determines* a trapezoid. Although we can vary the height of the figure and the lengths of the opposite sides, we can get only trapezoids. Figure 8.86 illustrates this.

FIGURE 8.86

What if you make a quadrilateral where the two right angles are not adjacent to each other, but opposite each other? Is it possible to have a concave quadrilateral with two right angles? These questions will be left as exercises.

A critical reader might be thinking this is just an exercise for students. Mathematicians don't do this kind of stuff. But we do! A great deal of mathematics has been developed by mathematicians asking and exploring questions like "I wonder whether this is possible?" and "I wonder what would happen if. . . ." In Chapter 9, I will talk more about a housewife who became intrigued by the question of under what conditions a pentagon would tessellate. Her work was noticed by a mathematician who encouraged her, and the work she did made a significant contribution to the field of tessellations. My belief is that if you experience the *doing* of mathematics in this course, you will do this with your children, who will then retain the love of numbers and shapes that they almost universally bring to kindergarten and first grade!

We used Venn diagrams to deepen our understanding of triangles in Investigation 8.10. We will do so again with quadrilaterals.

INVESTIGATION 8.14

CLASSROOM CONNECTION

The act of classifying is one of the big ideas of mathematics and is found in every branch and at every level of instruction. It begins with attribute blocks and materials with preschool children. Materials such as buttons can be sorted, using more than one attribute, in many different ways (the shape of the button, the number of holes, the color, and so on). Plastic rings, large enough to hold many objects, are used to create Venn diagrams.

Relationships Among Quadrilaterals

Consider the Venn diagram and set of quadrilaterals shown in Figure 8.87. What attributes do all of the quadrilaterals in the left ring have? What attributes do all of the quadrilaterals in the right ring have? By the nature of Venn diagrams, the quadrilaterals in the middle section have attributes of both right and left.

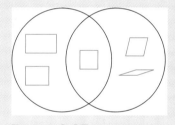

FIGURE 8.87

DISCUSSION

There are a number of ways to answer the question. Let us begin at the most descriptive level.

In the left ring, all of the shapes have four right angles, they all have opposite sides congruent, and they all have opposite sides parallel. Some students state it slightly differently, saying that the shapes all have two pairs of congruent sides and two pairs of parallel sides.

In the right ring, all the shapes also have opposite sides congruent and opposite sides parallel. Another attribute they possess is that all four sides are congruent. The shape that is in the center, by definition, must have the attributes of both rings—four right angles and four congruent sides.

There is a name for figures that are in both rings—squares.
There is a name for figures that are in the left ring—rectangles.
There is a name for figures that are in the right ring—rhombi.

Using set language from Chapter 2, we say that squares are the intersection of rectangles and rhombi. Some students will have noticed that all of these shapes are parallelograms. Thus we could actually add another ring encircling all the shapes in Figure 8.87. That is, all rectangles are parallelograms, all rhombi are parallelograms, and all rhombi are parallelograms.

Relationships Among Quadrilaterals

It turns out that we can view the set of quadrilaterals in much the same way we view a family tree showing the various ways in which individuals are related to others. Figure 8.88 shows one of many ways to represent this family tree for quadrilaterals. Take a few moments to think about this diagram and to connect it to what you know about these different kinds of quadrilaterals. Write a brief description. Does it make sense? Does it prompt new discoveries in your mind?

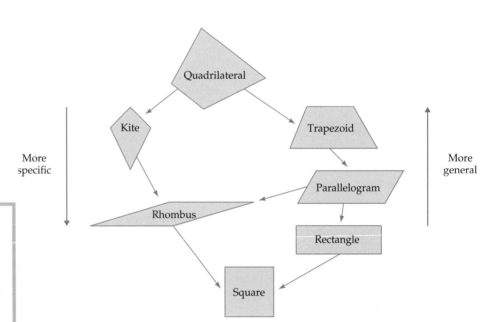

FIGURE 8.88

> **Mathematics**
>
> People not familiar with mathematics often express irritation with the mathematical statement that a square is a rectangle. However, it is really not that weird. Consider this question: Are humans primates? Are humans mammals? If you have had any biology, the answer is, "Of course!" Same thing. We can go from general (polygon) to a subset (quadrilateral) to a yet smaller subset (rectangle) to an even smaller subset (square). Similarly, we can go from general (animal) to a subset (mammal) to a smaller subset (primate) to a yet smaller subset (human).

One way to interpret this diagram is to say that any figure contains all of the properties and characteristics of the ones above it. The quadrilateral at the top represents those quadrilaterals that have no equal sides, no equal angles, and no parallel sides; this is analogous to the scalene triangle. The kite and the trapezoid represent two constraints that we can make: two pairs of congruent, adjacent sides or one pair of sides parallel. If we take a kite and require all four sides to be congruent, we have a rhombus. If we take a trapezoid and require *both* pairs of opposite sides to be parallel, we have a parallelogram. If we require the angles in a parallelogram to be right angles, we have a rectangle. If we require all four sides of a parallelogram to be congruent, we have a rhombus. Both the rhombus and the rectangle can be transformed into squares with one modification—requiring the rhombus to have right angles or the rectangle to have congruent sides. A key point is to begin to see connections and relationships among figures. Many students find, in this course, that their picture of

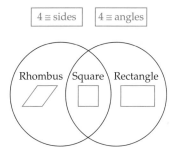

FIGURE 8.89

> **■ Language ■**
>
> Your instructor might have had you explore the definition of *convex* in Exploration 8.9. You may be surprised to know that there are multiple ways to view this idea. A common alternative to the definition given at the right is that a figure is convex if and only if it has no reflex angles. First make sure you understand the definition. Do you see the equivalence of the two definitions? Do you like one better than the other? Once again, mathematics is not as cut and dried as many people believe!

FIGURE 8.91

geometry changes from looking like a list of definitions and properties to looking more like a network with connections among the various figures. This quadrilateral family tree can also help students to realize why mathematics teachers say that a square is a rectangle *and* it is a rhombus: It has all the properties of each! In everyday language, we say that a square is a special kind of rhombus and a special kind of rectangle. In mathematical language, we say that the set of squares is a subset of the set of rhombuses and a subset of the set of rectangles. The Venn diagram in Figure 8.89 illustrates this relationship.

Convex Polygons

Another concept that emerges with polygons having four or more sides is the idea of convex. Before reading on, look at the two sets of polygons in Figure 8.90, convex and concave (not convex). Try to write a definition for *convex*. Then read on. . . .

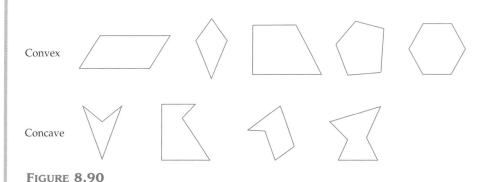

FIGURE 8.90

For many students, defining *convex* is much like defining balls and strikes in baseball. As one umpire once said, "I knows it when I sees it." Many students focus on the word *concave* and say that the figure has at least one part that is caved in. Such a description is acceptable at level 1 on the van Hiele model, but we need a definition that is not as vague as "caved in."

Examine the following definition to see whether it makes sense. Think before reading on. . . .

> A polygon is **convex** if and only if the line segment connecting *any* two points in the polygonal region lies entirely within the region.
> If a polygon is not convex, it is called **concave** or nonconvex.

Looking at diagonals is an easy way to test for concave and convex. If *any* diagonal lies outside the region, then the polygon is concave. In polygon *ABCDE* in Figure 8.91, the diagonal *AD* lies outside the region.

Other Polygons

Although most of the polygons we encounter in everyday life are triangles and quadrilaterals, there are many kinds of polygons with more than four sides. Stop for a moment and think of examples, both natural and human-made objects. Then read on. . . .

All of the figures in Figure 8.92 are polygons.

- The stop sign is an octagon—an eight-sided polygon.
- The common nut has a hexagonal shape—a six-sided polygon.
- The Pentagon in Washington has five sides.

Let us examine a few important aspects of polygons with more than four sides.

First, we distinguish between regular and nonregular polygons. What do you think a regular pentagon or a regular hexagon is? How might we define it? Think about this and write down your thoughts before reading on. . . .

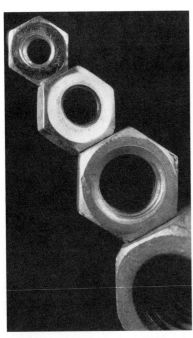
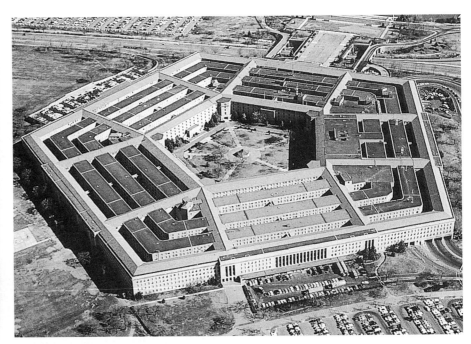

FIGURE 8.92

A **regular polygon** is one in which all sides have the same length and all interior angles have the same measure.

What do we call a regular quadrilateral? What about a regular triangle?
Is it possible for a regular polygon to be concave?
Think about these questions before reading on. . . .

A regular quadrilateral is called a square. A regular triangle is called an equilateral triangle. A regular polygon cannot be concave.

A critical reader might be wondering whether you can have a polygon—let's say a pentagon—where all the sides have the same length but not all the angles have the same measure. And what about the converse: Can you have a pentagon where all the angles have the same measure but not all the sides have the same length? What do you think? This will be left as an exercise.

INVESTIGATION 8.15

Sum of the Interior Angles of a Polygon

We found that the sum of the degrees of the interior angles of any triangle is 180, and the sum for any quadrilateral is 360. What do you think is the sum of the interior angles of a pentagon? Can you explain your reasoning? Can you find a pattern in this progression that will enable you to predict the sum of the interior angles of any polygon—for example, one with 10 sides or with 100 sides? Work on this before reading on. . . .

DISCUSSION

From Table 8.4, many students can see that the sum increases by 180 each time but cannot come up with the general case. The solution to this question comes from connecting the problem-solving tool of making a table to the seeing of patterns to seeing "increases by 180" as equivalent to "these are multiples of 180." We can represent this equivalent representation in a fourth column that contains 180, 2 · 180, 3 · 180, and so on. What do you see now? Then read on. . . .

TABLE 8.4

Sides	Sum	Reasoning
3	180	
4	360	Increases by 180 each time
5	540	
6	720	

The number we multiply 180 by is 2 less than the number of sides in the polygon. Therefore, the sum of the angles of a polygon having n sides will be equal to $(n - 2)180$.

The next step is to ask why. That is, this investigation takes an unproved hypothesis that the sum will increase by 180 degrees each time we add a side. If that hypothesis is true, then it follows that the sum for any polygon will be $180(n - 2)$, but *why* is this true? What do you think? That question will be left as an exercise.

There are two kinds of angles that we will add to our consideration. We have spoken of interior angles; we can also speak of exterior angles. Draw what you think would be the exterior angles of the polygons in Figure 8.93. Then read on. . . .

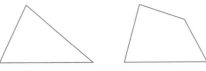

FIGURE 8.93

This is one of those cases where mathematics is counterintuitive. An **exterior angle** is an angle formed by a side of the polygon and the extension of the side adjacent to that side. Now go back and see whether you can draw the exterior angles for the two polygons. How many exterior angles does a polygon have? Then, and only then, look below. . . .

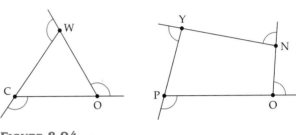

FIGURE 8.94

The diagrams in Figure 8.94 illustrate the exterior angles. As you may have guessed, there is an exterior angle associated with each interior angle.

The sum of the interior angles of a polygon is $180(n-2)$. What do you think is the sum of the exterior angles of a polygon? . . .

Amazingly, the sum of the exterior angles of any polygon is 360 degrees. You can see the proof for yourself by placing a pencil on vertex C of the triangle above. Now "walk" the pencil around the triangle. That is, move it to vertex O, now rotate the pencil so you are pointing toward W; note that the amount of turn is exactly the exterior angle at O. Now walk the pencil to W. Turn the pencil at W so you are pointing toward C; again, the amount of turn is equal to the exterior angle at W. Finally walk to C and turn to face O. You find that the pencil has made one complete turn. Do the same with quadrilateral PONY. Once again, you find you make one complete turn.

All polygons have interior and exterior angles. When we focus on regular polygons, there is an additional kind of angle. A **central angle** of a regular polygon consists of a vertex at the center of the polygon and the two sides connecting the center to adjacent vertices of the polygon (see Figure 8.95). The size of the central angle is not difficult to determine if you recall that one complete turn around a point is equal to 360 degrees. The measure of the central angle of a regular n-gon (a polygon with n sides) is equal to $360/n$ degrees. This and other theorems developed in this section are used to advance our understanding of the structure of geometry, which is then used by people in various fields. For example, graphic artists use their sense of geometric structure to make very beautiful and intricate designs, both on everyday objects like fabric and posters and as pieces of art.

> ■ *Mathematics* ■
>
> If you thought all mathematics proofs were like the two-column proofs in high school geometry, you were mistaken. Many, many mathematical proofs are fun and based on common sense that is accessible to all people!

FIGURE 8.95

Section 8.2 / Two-Dimensional Figures   549

Curved Figures

There is one more class of two-dimensional geometric figures that we need to discuss: those figures that are composed of curves that are not line segments.

How many words do you know that describe such shapes? Think and then read on. . . .

There are many such geometric figures—for example, circle, semicircle, spiral, parabola, ellipse, hyperbola, and crescent (see Figure 8.96).

FIGURE 8.96

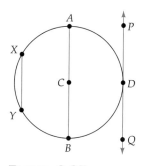

FIGURE 8.97

In this course, we will focus on the simplest of all curved geometric figures: the circle. Stop for a moment to think about circles. How would you define a circle? Try to do so before reading on. . . .

A **circle** is the set of points in a plane that are all the same distance from a given point, the center.

You are probably familiar with the following basic vocabulary for circles. In Figure 8.97, C is the *center* of the circle.

The line segment $\overline{CA}$ is called a *radius*, the plural of which is *radii*.
The line segment $\overline{AB}$ is called a *diameter*.
The line segment $\overline{XY}$ is called a *chord*.
The line $\overleftrightarrow{PQ}$, which intersects the circle only at point D, is called a *tangent*.

An *arc* is any part of a circle. We use two letters to denote an arc if the arc is less than half of the circle. However, for larger arcs, we use three letters. Do you see why?

We do this to distinguish between the arc at the left ($\overarc{AD}$) in Figure 8.98 and the arc at the right ($\overarc{ABD}$), because both of them have points A and D as endpoints.

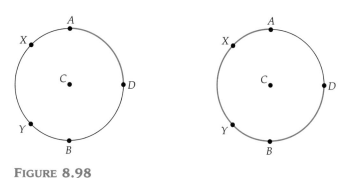

FIGURE 8.98

On the basis of these descriptions and your previous experience with circles, try to write a definition for each of these terms. Then compare your definitions with the ones below. . . .

A **radius** of a circle is any line segment that connects a point on the circle to the center.

A **diameter** of a circle is any line segment that connects two points of the circle and also goes through the center of the circle.

A **chord** is any line segment that connects two points of the circle. Thus, a diameter is also a chord.

A **tangent** line intersects a circle at exactly one point.

An **arc** is a subset of a circle—that is, a part of a circle.

Coordinate Geometry

■ *History* ■

For many hundreds of years, algebra and geometry were developed for the most part separately. It was in the 1600s that René Descartes began the mathematical work of connecting geometry to algebra.

We have explored some of the territory of two-dimensional geometry. There is so much more! Most of our focus has been on polygons, especially triangles and quadrilaterals. We hope your understanding has moved to higher levels in the van Hiele model. One of the themes of this book is multiplicity—multiple solution paths to most problems, multiple connections between and among ideas, and multiple ways of representing many ideas. One of the most powerful ways to illustrate the value of multiplicity is through coordinate geometry. Up to now, all of our exploration has been without coordinates—that is, we have simply used sketches of geometric figures. Let us look at what is added when we place the figures in a coordinate plane.

A brief review of the Cartesian coordinate system

Any point on a plane can be represented by an ordered pair. The first number represents the point's horizontal distance from the center of the coordinate plane, which is called the **origin** and is denoted by the ordered pair (0, 0). The second number represents the point's vertical distance from the origin. At some time during the development of mathematics, mathematicians adopted the convention that right is positive, left is negative, up is positive, and down is negative.

As we noted at the beginning of this chapter, a powerful contribution of the Greeks was to move us from the *how* to the *why*. In this chapter, we have examined some geometric proofs, and you have explored proofs in *Explorations*. It turns out that some geometric proofs that are very difficult in "normal" representation are actually quite simple with coordinate geometry. Before we get to some proofs, let us do one investigation to review your skills with the coordinate system.

INVESTIGATION 8.16   What Are My Coordinates?

First, let's play a game. I'm thinking of a rectangle. Three of its coordinates are (3, 5), (7, 5), and (3, 10). What is the fourth coordinate? Do this yourself before reading on. . . .

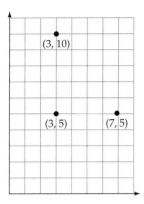

FIGURE 8.99

DISCUSSION

Using their basic understanding of the properties of rectangles and their intuitive, visual understanding of the coordinate plane (see Figure 8.99), most people will deduce that the fourth coordinate is (7, 10).

Now, let us look at how to make this pattern even more obvious. Let us move the parallelogram so that the bottom left vertex is at the origin (see Figure 8.100). That is, we will move the whole rectangle 3 units to the left and 5 units down. In Chapter 9, we will call this a $(-3, -5)$ translation. Now what are the coordinates of the rectangle? Find this before reading on. . . .

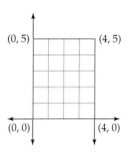

FIGURE 8.100

With this strategic placement of the coordinates, the assertion that either the x values or the y values are the same is even easier to see.

We need to develop one more idea before we jump more deeply into coordinate geometry, and that is to learn how to find the distance between two points on the coordinate plane. If you took geometry in high school, you probably memorized this formula. However, if you read this carefully, you will find that it doesn't have to be memorized. Although it is not simple, it is not hard to understand.

INVESTIGATION 8.17

Understanding the Distance Formula

Let us consider two random points on the plane: $Y = (x_1, y_1)$ and $S = (x_2, y_2)$. If we consider the line segment connecting these two points to be the hypotenuse, we can draw the two sides of the triangle (see Figure 8.101). What must be the coordinates of the third vertex, E? Remember the discussions in this chapter about determinism—in this case, both the x and y values of the third vertex are determined. Think before reading on. . . .

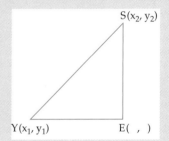

FIGURE 8.101

DISCUSSION

The coordinates of that third vertex are (x_2, y_1). Do you see why? If not, recall that this vertex is the same horizontal distance from the origin as the top vertex; thus it has the same x value. Similarly, it is the same height above the origin as the left vertex; thus it has the same y value.

Now we can apply the Pythagorean formula: $c = \sqrt{a^2 + b^2}$. This formula simply says that in any right triangle, the length of the hypotenuse is equal to the square root of the sum of the squares of the two sides.

If this seems a bit overwhelming, look at Figure 8.102 and connect it to Figure 8.101. That is, in order to find the distance from Y to S, we have to find the distance from Y to E and the distance from E to S. Then we will square those distances, add them, and take the square root.

But the distance from Y to E is easy because it's a horizontal line. It is just the difference between the x values—that is, $x_2 - x_1$. Similarly, the distance from E to S is the difference between the y values—that is, $y_2 - y_1$.

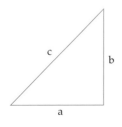

FIGURE 8.102

Thus the distance from Y to S, substituting our distances into the Pythagorean theorem, must be $\sqrt{(x_2 - x_1)^2 + (y_2 - y_1)^2}$.

Phrased in English, if we square the x distance and the y distance, add them, and take the square root, we have the distance between any two points.

Now we are ready for one of the proofs. Recall that a parallelogram is defined as a quadrilateral in which opposite sides are parallel. Using only this knowledge, we can *prove* that the opposite sides of a parallelogram must be congruent.

INVESTIGATION 8.18

The Opposite Sides of a Parallelogram Are Congruent

Let us begin by sketching a parallelogram on the coordinate plane (see Figure 8.103). Since the opposite sides are parallel and since, from an algebraic perspective, parallel lines have the same slope, we know that the slopes of the opposite sides are equal. Thus, if we place one vertex at the origin and the next vertex at $(c, 0)$, we have one side lying on the x axis. Because we know that PL and NA must be parallel, we can draw NA parallel to the x axis. Thus, if we let N be represented as the point (a, b), we know that the y coordinate of A must also be b. Do you see why?

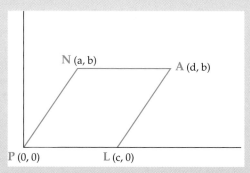

FIGURE 8.103

DISCUSSION

If not, recall that slope is the ratio of rise to run. That is, $\dfrac{y_2 - y_1}{x_2 - x_1}$.

We know that the slope of PL is 0. Thus the slope of NA must also be zero. If both N and A have the same y coordinate, b, then the slope is $(b - b)/(d - a) = 0$.

Because PLAN is a parallelogram, the other two sides must also be parallel—that is, they must have the same slope. Let us first determine the slope of AL and PN.

The slope of PN is $\dfrac{b - 0}{a - 0} = \dfrac{b}{a}$.

The slope of AL is $\dfrac{b - 0}{d - c} = \dfrac{b}{d - c}$.

What does this tell us about the relationship among a, d, and c?

It means that since the two lines have to have the same slope and thus be parallel, $d - c$ must be equal to a. That is, $a = d - c$, which is equivalent to $d = a + c$.

We can now substitute $(a + c)$ for d and have the new representation of the coordinates of the points of the parallelogram (see Figure 8.104).

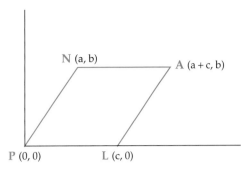

FIGURE 8.104

Next we must prove that the lengths of opposite sides are equal. Let's begin with the easier case, the horizontal sides. The distance from P to L is $c - 0$; that is, it is c. The distance from N to A is $(a + c) - a = c$.

What about the distance from P to N and from L to A? Using the distance formula from Investigation 8.17, we can find the length of PN—that is, the distance between P and N.

$$PN = \sqrt{(a - 0)^2 + (b - 0)^2} = \sqrt{a^2 + b^2}$$

Now, we find the distance from L to A.

$$LA = \sqrt{((a + c) - c)^2 + (b - 0)^2} = \sqrt{a^2 + b^2}$$

Because the two distances are the same, the lengths of the two line segments are congruent. Therefore, we have proved that the opposite sides of a parallelogram are the same length.

INVESTIGATION 8.19

Midpoints of Any Quadrilateral

Now let us investigate one of my favorite theorems in geometry. I want you to see it first; then we will prove it. On a blank piece of paper, draw a quadrilateral, any quadrilateral. Find the midpoints of each side. I recommend using the metric side of your ruler. If you don't have a ruler, you can still find the midpoints. Do you see how? . . . Yes, you can fold the paper to find the midpoints of each side. Now make a new quadrilateral by connecting the four midpoints consecutively. What do you see?

Do another one. This time, make a scalene quadrilateral—no congruent sides, no parallel sides. What you just observed will always happen. That is, in all cases, you will wind up with a parallelogram. Now, let's prove that.

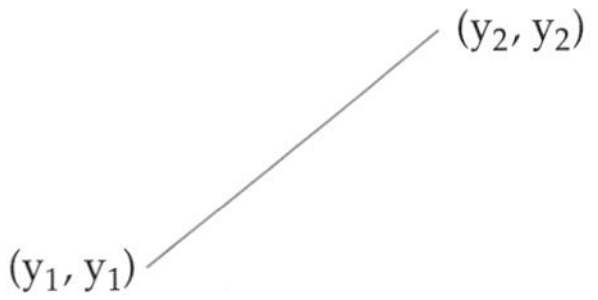

FIGURE 8.105

DISCUSSION

Before we can do the proof, we simply need to know how to find the midpoint of a line segment. In this case, the theorem makes intuitive sense, and I will simply present it. The midpoint of the line segment connecting points (x_1, y_1) and (x_2, y_2), shown in Figure 8.105, is $\left(\dfrac{x_1 + x_2}{2}, \dfrac{y_1 + y_2}{2}\right)$. That is, it is the average (mean) of the x values and of the y values.

We will prove the theorem for all quadrilaterals. As we saw from the quadrilateral hierarchy, once we prove a theorem for one quadrilateral, it is true for all quadrilaterals below that quadrilateral. Because we are dealing with any quadrilateral, we cannot assume any properties—congruence or parallel. Thus we will set one vertex at the origin (0, 0) and one vertex on the *x* axis (*a*, 0). Then the other vertices are at arbitrary points (*b*, *c*) and (*d*, *e*). See Figure 8.106.

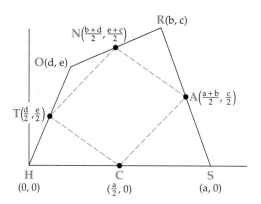

FIGURE 8.106

We will label the original quadrilateral *HORS* and the quadrilateral formed by the midpoints *CANT*.

Using the midpoint formulas, we determine the coordinates of the midpoints and then connect them. We now need to show that the opposite sides must be parallel—that is, that they have the same slope.

Let us begin with *CA* and *TN*.

$$\text{Slope of } CA = \frac{\frac{c}{2} - 0}{\frac{a+b}{2} - \frac{a}{2}}$$

A little algebra makes this next step much easier. If we multiply the top and bottom of this expression by 2, we don't change the value.

Doing this, we have

$$\text{Slope of } CA = \frac{c - 0}{(a+b) - a} = \frac{c}{b}$$

Now for *TN*.

$$\text{Slope of } TN = \frac{\frac{e+c}{2} - \frac{e}{2}}{\frac{b+d}{2} - \frac{d}{2}}$$

If we multiply the top and bottom of this expression by 2, we have

$$\text{Slope of } TN = \frac{(e+c) - e}{(d+b) - d} = \frac{c}{b}$$

Thus we have shown that the slope of $CA = c/b$ and that the slope of $TN = c/b$. That is, these two line segments are parallel.

By a similar means, we can show that the slope of *CT* and the slope of *AN* are equal.

Thus we have proved that if you take the midpoints of *any* quadrilateral and connect them in turn, you will always get a parallelogram. To prove that by other means is much more tedious and difficult.

There is much more to coordinate geometry! We will revisit coordinate geometry in Chapter 9, and you will find coordinate geometry in the upper grades of elementary school!

Exercises 8.2

1. Write down all the attributes of each of the figures.

   a.   b.   c.

2. In each case below, which two figures are most alike? Explain your reasoning.

   a.

   b.

   c.

3. *Classroom Connection* This question appeared on the Seventh National Assessment of Educational Progress. In what ways are the two figures below alike? In what ways are they different? You can see the "correct" answers at the back this book.

   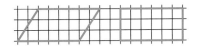

4. Fill in the table below.

Shape	Number of diagonals
Quadrilateral	2
Pentagon	
Hexagon	
Heptagon	
Octagon	
n-gon	

5. Describe all the geometric shapes you see in the quilt designs below:

   a. 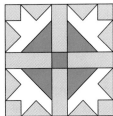 b.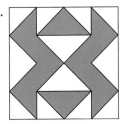

   Fool's Puzzle

   c.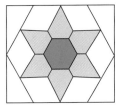

   Diamonds in the Sky

   Source: Jinny Beyes, *The Quilter's Album of Blocks & Borders* (Delaplane, VA: EPM Publications, Inc., 1986), p. 185. Reprinted with permission.

6. *Classroom Connection* This question was posted in *Teaching Children Mathematics*. The children's responses can be found in the April 2004 issue on pp. 403–406: "Tiffany is creating different shapes on her geoboard. . . . [S]he shares her shape with her friend Leticia [who] notices that she can see triangles and rectangles in the design."

   a. How many different triangles can you find in the design?
   b. How many different rectangles can you find?
   c. What other polygons do you see in the design?
   d. Create your own design. What shapes can you see? What other questions would you ask a friend about your design?

   Source: Reprinted with permission from *Teaching Children Mathematics*, April 2004, pp. 403–406, copyright 2004 by the National Council of Teachers of Mathematics.

   e. (My extension) What is the polygon with the most number of sides that you can find in this design?

   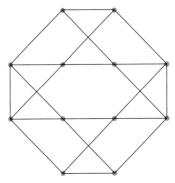

7. Name at least six different polygons that you can see in this shape. Trace and number each shape.

   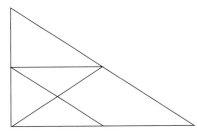

8. Try to make each figure on Geoboard Dot Paper *and* on Isometric Dot Paper. If you can make the figure, do so and explain why your figure is an example of the speci-

fied triangle or quadrilateral. If you cannot make it, explain why you think it is impossible to make the figure on that type of grid. Your reasoning needs to be based on properties and attributes (see the van Hiele discussion), as opposed to, for example, "It's a right triangle because it looks like a right triangle."

a. acute scalene triangle
b. right isosceles triangle
c. obtuse isosceles triangle
d. equilateral triangle
e. trapezoid
f. kite
g. parallelogram
h. rectangle
i. rhombus
j. square
k. square with no sides parallel to the sides of the paper.

9. For each figure below, write "polygon" or "not a polygon." If it is a polygon, also write "convex" or "concave."

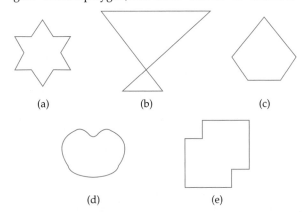

10. The definition of a regular polygon states that all sides have the same length and all interior angles have the same measure. Why is the second part of the definition necessary? That is, why can't we just say that a polygon is a regular polygon if all the sides are the same length?

11. How many different quadrilaterals can you make that have at least one pair of adjacent congruent sides? Sketch and label your figures. For example, you can make many different trapezoids, but they are all trapezoids. See the figures below.

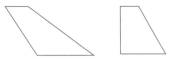

12. a. Is this figure a kite? Why or why not?

b. Is this figure a rectangle? Why or why not?

c. Is this figure an isosceles triangle? Why or why not?

13. Write directions for making the following figures. Following your directions, the reader should be able to make the same figure.

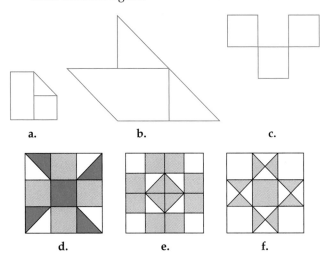

14. In each case below, determine whether the two figures are congruent only by using your mind. That is, you cannot trace one figure and see whether it can be superimposed on the other figure. Describe your reasoning—that is, how you arrived at your conclusion.

a. Below are two parallelograms made on a Geoboard.

b. Below are two figures made with tangram pieces.

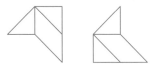

c. Below are two pairs of hexominoes.

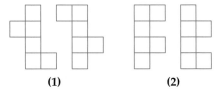

15. Describe all quadrilaterals that have these characteristics. If there is more than one, say so.

a. A quadrilateral with opposite sides parallel
b. A quadrilateral with 4 right angles
c. A quadrilateral with all sides equal
d. A quadrilateral in which the diagonals bisect each other
e. A quadrilateral in which the diagonals are congruent
f. A quadrilateral in which adjacent angles are congruent
g. A quadrilateral in which opposite angles are equal
h. A quadrilateral in which no sides are parallel

i. A quadrilateral with 4 congruent sides and 2 pairs of congruent angles

j. A quadrilateral with 4 congruent angles and 2 pairs of congruent sides

16. Draw each of the following or briefly explain why such a figure is impossible.
    a. An isosceles trapezoid
    b. A concave quadrilateral
    c. A curve that is simple and closed but not convex
    d. A nonsimple closed curve
    e. A concave equilateral hexagon
    f. A concave pentagon having three collinear vertices
    g. A pentagon that has 3 right angles and 1 acute angle

17. In each case, explain and justify your answer—that is, indicate why you think there are none, just one, or many.
    a. How many different hexagons can you draw that have all sides equal but not all angles equal?
    b. How many different hexagons can you draw that have exactly 2 right angles?
    c. How many different hexagons can you draw that have exactly 3 right angles?
    d. How many different hexagons can you draw that have exactly 4 right angles?
    e. How many different hexagons can you draw that have exactly 5 right angles?
    f. Can you make a trapezoid with no obtuse angles?
    g. How many different kinds of quadrilaterals can you make that have exactly two opposite right angles?
    h. Can you make a concave quadrilateral with exactly 2 right angles?
    i. Can you make a concave pentagon with exactly 2 right angles?

18. Use a Venn diagram to represent the relationship between:
    a. scalene and obtuse triangles.
    b. equilateral and isosceles triangles.
    c. parallelograms and rectangles.
    d. rectangles, rhombi, and squares.

19. Write in the labels for each set in the problems below. Justify your choice. Add at least one new figure to one of the regions.
    a.

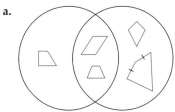

b.

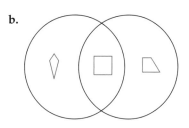

20. Show that the family tree for quadrilaterals also holds for the characteristics of the diagonals.

21. In this section, we found that we could name some quadrilaterals in terms of their "ancestors." For example, we could say that a rhombus is a parallelogram with four equal sides. What other quadrilaterals could we describe in terms of their ancestors?

22. Draw a triangle. Find the midpoints of each side. Connect those points. What is the relationship between the new triangle and the original triangle? Prove it.

23. Take a blank piece of paper. Fold it in half and then fold it in half again. Draw a scalene triangle and then cut it out. You should now have four congruent triangles. For each of the following questions, explain your reasoning.
    a. Put these triangles together to make a large triangle.
    b. Is there only one way or are there different ways?
    c. How could you prove that the new triangle is actually a triangle, as opposed to a figure that is almost a triangle?
    d. Does the new large triangle have anything in common with the original triangle?

24. What is the relationship between the number of sides in a polygon and the total number of diagonals in that polygon?

25. Prove that the sum of the interior angles of any polygon is equal to $180(n - 2)$, where n represents the number of sides in the polygon.

26. Describe all possible combinations of angles in a quadrilateral (for example, acute, acute, acute, obtuse). Briefly summarize methods you used other than random trial and error.

27. Trace this circle following #28 onto a blank sheet of paper. Describe as many ways as you can for finding the center of the circle. In each case, explain why the method works.

28. Trace the circle onto a blank sheet of paper. After cutting out the circle and finding the center, fold the paper so that the top point of the circle just touches the center of the circle. Fold the circle again so that another point on the circle just touches the center *and* the two folds meet at a point. The two folds will be congruent. Fold the circle one more time so that your three folds will all be congruent. What kind of triangle has been created inside the circle? Prove that you will get this kind of triangle by making these three folds.

29. A 3RIT is a figure made from 3 right isosceles triangles. On the left are two examples of RITs, and on the right are two figures that are not RITs.

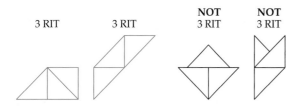

a. Write a definition of 3RIT.
b. Find as many different 3RITs as you can. Sketch (or tape, or glue) them on a separate piece of paper.
c. Are the two 3RITs at the right different or not? Explain why you believe they are (or are not) different.
d. Find as many 4RITs as you can.

30. Make as many different hexiamonds as you can. A hexiamond is made by joining six equilateral triangles—when a side is connected, a whole side connects to a whole side. Briefly (in two or three sentences) describe your method(s) for generating as many shapes as possible. Describe any method(s) other than random trial and error; there are many different ways to be systematic. Cut your hexiamonds out and tape them to a piece of paper. (There are more than 9 and fewer than 16.)

31. Find the following points on a sheet of graph paper:
    A (3, 5)   B (5, 5)   C (5, −5)   D (−2, −5)
    E (−2, 5) F (0, 5)   G (0, −3)   H (3, −3)
    Now connect the points in order. What do you see?

32. Find the distance between the pairs of points:
    a. (2, 3) and (5, 7)   b. (4, 7) and (10, 10)

33. Find the midpoints of the line segments connecting these pairs of points:
    a. (0, 4) and (4, 10)   b. (3, 4) and (7, 12)

34. a. Three vertices of a kite are (6, 8), (9, 11) and (12, 8). What are the coordinates of the fourth vertex?
    b. The two vertices that form the base of an isosceles triangle are (−5, 3) and (2, 3). What are the coordinates of the other vertex?
    c. The coordinates of the endpoints of the hypotenuse of a right triangle are (7, 5) and (3, 1). Find the other vertex. There are two possible solutions.
    d. Three vertices of a parallelogram are (0, 0), (4, 0), and (0, 6). Find the fourth vertex. There are three possible solutions.
    e. A rectangle is oriented so that its sides form vertical and horizontal lines. Two coordinates of a rectangle are (−3, −2) and (7, 6). Do you have enough information to determine the other two coordinates? If so, find them. If not, explain why not and describe what information you would need (for example, a third coordinate, the length of one side, the relative location of one of the points).
    f. Make up a problem similar to the ones above and solve it.

35. We can play a variation of a child's game called "What am I?" that I will call "Where am I?"
    a. I am a square. The intersection of my two diagonals lies at the point (3, 3), and the length of each of my sides is 6. My sides form horizontal and vertical lines. Where am I?
    b. I am an isosceles triangle. The midpoint of my base is the point (7, −2), my base forms a horizontal line, and my vertex is at the point (7, −9). Oh, I almost forgot to tell you. I am upside down. Where am I?
    c. I am a right isosceles triangle. I have an area of 50 square units. The coordinates of the vertex at which the two sides meet is (0, 0) and my sides form horizontal and vertical lines. Do you have enough information to determine the other two coordinates? If so, find them. If not, explain why not and describe what information you would need.
    d. Make up a problem similar to the ones above and solve it.

36. Determine the coordinates of the vertices of tangram pieces if the bottom left-hand corner is at the origin and if the length of each side of the square is 8.

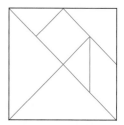

37. Determine the coordinates of the vertices of the following quilt pattern.

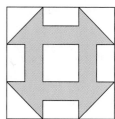

38. If we set the base of a triangle on the *x* axis, we can specify the coordinates of any triangle as follows: (0, 0), (*r*, *m*) and (*e*, 0). However, if we have an isosceles triangle, there is a relationship between some of these coordinates that enables us to be more specific, just as we were in Investigation 8.18 with a parallelogram. If $\overline{BA}$ and $\overline{BC}$ are the congruent sides, there is a relationship between *e* and *r* that enables us to specify *e* in terms of *r*. What is the relationship?

39. a. Prove that if we connect the midpoints of the two congruent sides of an isosceles triangle, the length of the line segment connecting those two points is one-half the length of the base.

    b. Is this true just for isosceles triangles or for all triangles? Support your conclusion.

40. Draw a right triangle, and find the midpoint of the hypotenuse; then connect the midpoint and the other vertex of the triangle. What do you think is the relationship between the length of that segment you just constructed and the length of the hypotenuse. Use coordinate geometry to find the answer.

41. a. Show that the diagonals of a square have the same length.

    b. You may recall from high school algebra that two lines are perpendicular if the product of their slopes is equal to -1. Use this knowledge to show that the diagonals of a square are perpendicular.

SECTION 8.3 THREE-DIMENSIONAL FIGURES

WHAT DO YOU THINK?

- What do pyramids and cones have in common?
- How are properties of two-dimensional objects and three-dimensional objects related?
- How can you represent three-dimensional objects in two-dimensional space?

This is the dimension in which we live. Almost everything we interact with is three-dimensional: people and pets, buildings, our rooms, our cars, what we see when we look around. Look at the six pictures on the next page. What do you see? Think, jot some notes, and then read on. . . . Four of the pictures are of objects made by humans, and two are natural—one organic and one inorganic. The snail (in its genetic makeup) and the crystal (in its molecular makeup) have the "blueprints" for the ultimate shape of the shell and of the crystal. Both the pyramid and the Parthenon are over 2000 years old, and yet the geometry that was used to make them is relevant to building today. The soccer ball and the design on the Shrine of Shah Nimatuollāhi represent the solution to two different questions about connections between two dimensions and three dimensions. In the former case, someone discovered that a specific combination of two polygons (a regular pentagon and a regular hexagon) will produce a nearly perfect ball (sphere). In the latter case, the designers had to figure out how to take a two-dimensional tessellation design and "fold" it around the roof of the shrine so that it would "work" in three dimensions.

Just as we discovered patterns and relationships among many two-dimensional objects, there are many patterns and relationships among three-dimensional figures, also called space figures. With respect to practical matters, understanding the geometry of human-made objects helps us to make them work better and, in the case of objects such as bridges, overpasses, and airplanes, to make them work more safely. Geometry also helps us to understand natural phenomena better—for example, why certain animals have the shapes they have. An understanding of shapes has many applications in science. For example, many carcinogens are virtually identical in size and shape to other compounds, and thus they fool the body into thinking they are not harmful. The silicon chip has the same structure as the diamond, except that there are silicon atoms instead of carbon atoms at these positions.[5] With respect to aesthetics, geometry helps us to understand why some shapes are so appealing to people and to understand patterns within those shapes (see Figure 8.107).

[5]Marjorie Senechal, *On the Shoulders of Giants*, p. 173. Reprinted with permission from *On the Shoulders of Giants: New Approaches to Numeracy*. Copyright © 1990 by the National Academy of Sciences. Courtesy of the National Academy Press, Washington, D.C.

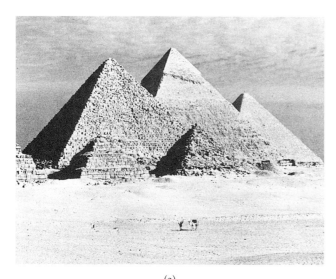
(a)
Pyramids at Gizeh

(b)
the Parthenon

(c)
nautilus shell

(d)
pyrite crystals

(e)
soccer ball

(f)
Shrine of Shah Nimatuollāhi

Figure 8.107

Section 8.3 / Three-Dimensional Figures

> ■ *History* ■
>
> The massive pyramid at Gizeh is one of the seven wonders of the ancient world. Even today we are not certain how it was constructed. Many of the blocks weighed more than 10 tons! The pyramid was originally covered with white marble and must have dazzled like a mirror in the desert. It was so well built that some of the edges fit together so well that a razor cannot be inserted into the space between two blocks.

In this section, we will begin simple and build up. Let's do a "What do you see?" investigation here, as we did in Section 8.2. As your ability to "see" three-dimensional objects improves—that is, your ability to see all the various attributes of a solid object, relationships between those attributes, and/or relationships between that object and other similarly shaped objects—your appreciation of the three-dimensional property grows too.

INVESTIGATION 8.20

What Do You See?

Examine a cube carefully (see Figure 8.108). What do you see? Write down all the attributes you can think of before reading on. Next, look at the "box" (see Figure 8.109). Which of the attributes of a cube does the box possess? What different attributes does it have? . . .

DISCUSSION

Cube

Box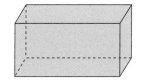

FIGURE 8.108

FIGURE 8.109

A cube has 6 faces, though you might have called them sides.	A box also has 6 faces.
All of the faces are squares, which also means right angles, parallel sides, and so on.	Some of the faces are squares and some are rectangles. As you discovered in the last section, our definitions enable us to say that all of the faces are rectangles.
All of the faces are congruent.	Not all of the faces are congruent, but some are.
We have names for opposite sides: front-back, side-side, top-bottom.	The opposite sides are congruent.
There are 12 edges on the cube.	The box, like the cube, has 12 edges.
There are 8 vertices.	The box, like the cube, has 8 vertices.

This investigation serves as what is called an advance organizer of this section. That is, it got you to grapple with many of the important ideas that we will examine in more detail. Our first order of business toward that end is to learn some of the language we will use with three-dimensional objects. Rather than just presenting you with the new terms, we will do another investigation in which you will be asked to think about what language and concepts from our work with polygons will make sense with our work with polyhedra, and where we need new terms or where the addition of one dimension "changes" things. For example, the addition of one side resulted in a way of naming quadrilaterals different from the way we named triangles.

INVESTIGATION 8.21

Connecting Polygons to Polyhedra

Just as we examined families (subsets) of triangles and quadrilaterals, we will now investigate families of three-dimensional geometric figures (see Figure 8.110). If you did Exploration 8.14, you grappled with describing and classifying three-dimensional figures.

Let us explore the connection between polygons and *polyhedra*, which will be loosely defined (for now) as three-dimensional figures made up of polygons.

The second column of Table 8.5 describes several attributes of polygons, which we investigated in Section 8.2. Which of these attributes do you think hold for polyhedra or can be modified to describe different kinds of polyhedra? Fill in as much of the third column as you can. The questions below are given to help you focus on the connections between how we see and define two- and three-dimensional figures. After you have completed as much of the third column as possible, compare your hypotheses with those of another student. Then read on. . . .

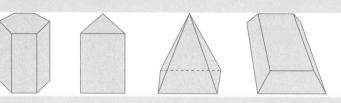

FIGURE 8.110

CLASSROOM CONNECTION

In the beginning, children view solids (three-dimensional shapes—3D) as entities instead of seeing the parts of the solid as a collection of related shapes.

- If a polygon is defined as a simple closed curve, is there an analogous definition for a polyhedron?
- All polygons have vertices, line segments, and angles. Do these terms work for describing component parts of polyhedra? Do we also need new terms?
- Can we classify polyhedra by the number of sides?
- If there are regular polyhedra, how might they be defined?
- If there are convex polyhedra, how might they be defined?

TABLE 8.5

	Polygons (two-dimensional)	Polyhedra (three-dimensional)
What they are	Simple closed curves	
Definition	Union of line segments	
Component parts	Vertices	
	Line segments	
	Angles	
	Other?	
Classification	By number of sides	
	Regular vs. not regular	
	Convex vs. concave	

DISCUSSION

In Section 8.2, we began with simple closed curves that partitioned a plane (two dimensions) into three disjoint sets: the curve, inside, and outside.

Though we will not rigorously define simple closed surfaces, we can say that they partition space (three dimensions) into three disjoint sets: the surface itself, inside, and outside (see Figure 8.111).

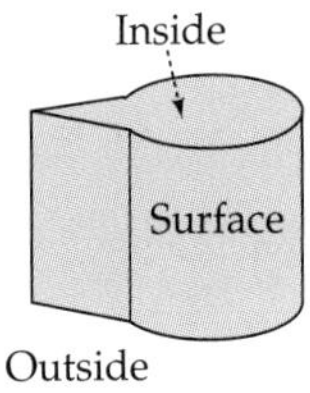

We will use the term *space figure* to describe any three-dimensional object.

We will use the term *polyhedron* (the plural is *polyhedra*) to describe those simple closed surfaces that are composed of polygonal regions.

We will use the term *solid* to describe the union of any space figure and its interior.

FIGURE 8.111

Component parts Just as the component parts of polygons have special names, so do those of polyhedra.

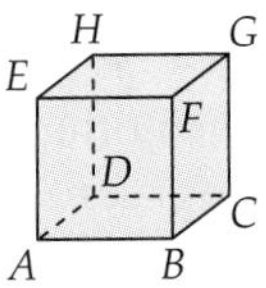

Each of the separate polygonal regions of a polyhedron is called a *face*; for example, square $ABFE$ is a face of the cube in Figure 8.112.

The sides of each of the faces are called *edges*; for example, $\overline{AB}$ is an edge of the cube in Figure 8.112.

The *vertices* of the polyhedron are simply the vertices of the polygonal regions that form the polyhedron; for example, E and F are vertices of the cube in Figure 8.112.

FIGURE 8.112

Convex and concave Just as polygons can be convex or concave, so can polyhedra. Before reading the definition of a convex polyhedron, think back to the definition of a convex polygon and see whether you can modify that definition for three-dimensional objects. Then read on. . . .

> **▪ Language ▪**
>
> Some mathematics dictionaries and textbooks define *edge* as the side of a polyhedron. Some focus on the term *line segment*. Yet others define *edge* as the intersection of two faces. On one mid-term exam, I asked my students to define *edge*. Interestingly, all three of these perspectives and interpretations appeared in the students' definitions!

A polyhedron is convex if and only if any line segment connecting two points of the polyhedron is either on the surface or in the interior of the polyhedron (see Figure 8.113).

 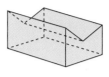

Convex   Concave

FIGURE 8.113

INVESTIGATION 8.22 — Features of Three-Dimensional Objects

Look at the picture of a box and a ramp (see Figure 8.114). In what ways are they "the same"? That is, what characteristics do they have in common that not all three-dimensional objects have? In what ways are they different? Do this before reading on. . . .

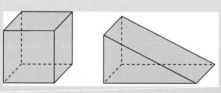

FIGURE 8.114

DISCUSSION

Some of the things they have in common:

    All the faces (sides) are polygons.
    In both cases, at least some of the sides are quadrilaterals.
    At least one pair of sides are congruent and parallel to each other. In the ramp, the two triangles on the side are parallel and congruent.

Some of the differences between them:

    In the box, there are an even number of faces, and opposite faces are congruent. In the ramp, only the triangle faces are opposite. The other three faces are noncongruent rectangles.
    The numbers of faces, edges, and vertices are different.
        The box: 6, 12, 8
        The ramp: 5, 9, 6
    However, the relationship between the numbers of faces, edges, and vertices is the same.
    See Exploration 8.15 (Relationships Among Polyhedra) for more on this.

Families of Polyhedra

Now let us investigate some of the families of polyhedra. Take a few minutes to examine the figures in Figure 8.115. How are these figures alike? How are they different? Write your thoughts in your notebook before reading on. . . .

FIGURE 8.115

All of these figures have at least two sides that are parallel; some students would say that the top and bottom sides are parallel. And the faces are all polygons. All of these figures are called prisms. We use the word **prism** to describe all polyhedra that have two parallel **bases** that are congruent polygons. It is a convention to call the other faces of prisms **lateral faces**.

What one shape can be used to describe the lateral faces of *all* prisms? In other words, all lateral faces of all prisms are _____. Think and read on. . . .

In all prisms, the lateral faces are parallelograms. In some cases, all of the lateral faces are rectangles. How would you describe the differences between those prisms whose lateral faces are nonrectangular parallelograms and those whose lateral faces are rectangles?

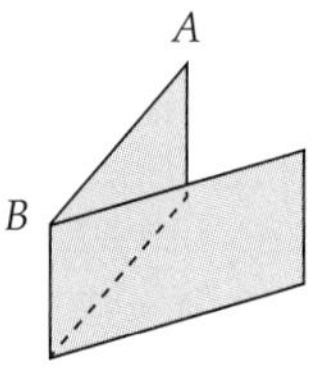

FIGURE 8.116

In the latter case, the plane of the base and the plane of the lateral faces are perpendicular (see Figure 8.116). We could also say that the *dihedral angle* formed by either base and any lateral face is a right angle. (A **dihedral angle** is simply a three-dimensional angle—that is, an angle whose vertex is a line and whose sides are planes.)

Thus we can define a **right prism** as a prism in which the lateral faces are rectangles. Alternatively, we could define a right prism as a prism in which the angle formed by either base and any lateral face is a right dihedral angle.

A prism that is not a right prism is an **oblique prism** (see Figure 8.117).

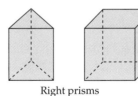

Right prisms   Oblique prism

FIGURE 8.117

■ *Language* ■

As you can see from this presentation of prisms, the box and ramp in Investigation 8.22 are both prisms. That is, they both have one pair of congruent bases that are parallel to each other.

Long before they study formal geometry, many children know the names for two special kinds of prisms.

Although this is not a term mathematicians use, what we call a **box** is actually a prism in which all six faces are rectangles. If all six faces are squares, we call the figure a **cube** (see Figure 8.118).

Cube   Rectangular prism

FIGURE 8.118

Pyramids

Let us consider now another family of polyhedra. You may recognize the polyhedra in Figure 8.119 as pyramids. How might we define that term? Make your own definition and then read on. . . .

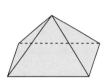
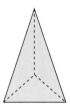
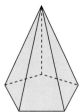

FIGURE 8.119

We use the word **pyramid** to describe those polyhedra whose base is a polygon and whose faces are triangles that have a common vertex. That common vertex is called the **apex** of the pyramid.

An alternative way to think of a pyramid is to start with any polygon and a point above the plane of the polygon. Now connect that point to each vertex of the polygon.

Most of the pyramids you have seen in pictures (or in person, if you are lucky) have square bases. However, the base can be any polygon. A pyramid is named according to its base: triangular pyramid, square pyramid, and so on.

As you have seen from our work with two-dimensional objects, the question of examining how objects are alike and how they differ is an important part of the learning process. Remember that it begins in preschool, where the teacher might give the children an assortment of buttons and have the children put them in piles so that each button belongs in one pile. The next investigation involves looking at similarities and differences.

INVESTIGATION 8.23 Prisms and Pyramids

Look at the set of prisms in Figure 8.115 and the set of pyramids in Figure 8.119. Note that these are just some examples of prisms and pyramids. What attributes do all prisms and pyramids have in common? Write your thoughts before reading on. . . .

DISCUSSION

In all prisms and all pyramids:

> There are bases, although prisms have two and pyramids have one.
> There are faces, edges, and vertices.
> The bases and faces are polygons.

Regular Polyhedra

In Section 8.2, we discussed regular polygons. How might we define a regular polyhedron? Try to do so before reading on. . . .

One of the ways we classified polygons was by the number of sides: triangles, quadrilaterals, pentagons, hexagons, and so on. We could speak of a regular hexagon and a nonregular hexagon. However, that doesn't work with polyhedra. Do you see why?

We define a **regular polyhedron** as a convex polyhedron in which the faces are congruent regular polygons and in which the numbers of edges that meet at each vertex are the same.

Stop! Does that definition make sense? The concept of a regular polyhedron is one of the more abstract in the book. What does it mean to say that "the numbers of edges that meet at each vertex are the same"? If you are still not sure, read on, but then go back and check to see whether this definition jibes with that of regular polyhedra.

Which of the prisms and pyramids we have discussed so far do you think might be regular polyhedra? Think before reading on. . . .

A cube is a regular polyhedron. A triangular pyramid composed of equilateral triangles is a regular polyhedron and has a special name, **tetrahedron**. The origin of the name is Greek: *tetra* ("four") and *hedron* ("face").

A fact that surprises many people is that there are not a large number of regular polyhedra. In fact, there are only five regular polyhedra: the tetrahedron, the cube, the **octahedron** (with 8 triangular faces), the **dodecahedron** (with 12 pentagonal faces), and the **icosahedron** (with 20 triangular faces) (see Figure 8.120). The solids made from the regular polyhedra are called Platonic solids after the Greek philosopher Plato.

■ Mathematics ■

The Five Regular Polyhedra

Tetrahedron
3 edges meet at each vertex

Octahedron
4 edges meet at each vertex

Cube
3 edges meet at each vertex

Icosahedron
5 edges meet at each vertex

Dodecahedron
3 edges meet at each vertex

■ History ■

Over 2000 years ago (long before we knew about atoms), many Greek philosopher-scientists believed that there were four basic elements out of which all things arose: earth, air, fire, and water. Some of the Greeks believed that the smallest particle of earth had the form of a cube, the smallest particle of air had the form of an octahedron, the smallest particle of fire had the form of a tetrahedron, and the smallest particle of water had the form of an icosahedron. The dodecahedron was associated with the universe, probably because it was the last solid discovered, although it has been speculated that it is associated with the universe because it has 12 faces and there are 12 signs in the zodiac.[6]

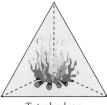
Tetrahedron

Octahedron

Cube

Icosahedron

Dodecahedron

Source: The five regular solids drawn by Johannes Kepler in *Harmonices Mundi, Book II,* 1619.

FIGURE 8.120

---

[6]H.A. Freebury, *A History of Mathematics* (New York: Macmillan, 1961), p. 36; H.A. Eves, *An Introduction to the History of Mathematics* (New York: Holt, Rinehart and Winston, 1969), p. 68.

▪ Outside the Classroom ▪

These regular polyhedra occur in nature:

- Crystals of salt and of pyrite are formed in the shape of a cube.
- Crystals of chrome alum are formed in the shape of a tetrahedron.
- Crystals of pyrite have been found in the shape of an octahedron.
- Skeletons of microscopic sea animals have been found in the shape of a dodecahedron and in the shape of an icosahedron (see Figure 8.121). I didn't make this up![7]

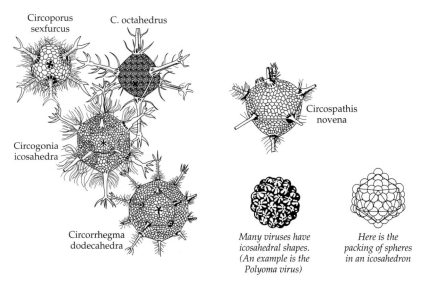

Source: Radiolarians—*Kunstformen der Natur*, Vols. 1–10, Ernst Häckel des Bibliographischen Instituts, Leipzig, Germany, 1899–1904; Polyoma virus–Drawn after K. W. Adolph, D. L. D. Caspar, C. J. Hollingshed, E. E. Lattman, W. C. Phillips, and W. T. Murakami, *Science*, Vol. 203, p. 1117, 1979; Packing of spheres—Drawn after A. L. Mackey, *Acta Crystallographica*, Vol. 15, p. 916, 1962 as seen in *Symmetry: A Unifying Concept.* © 1994 by Istvan and Magdolna Hargittai. Reprinted by permission.

FIGURE 8.121

Relationships Among Polyhedra

We examined relationships among quadrilaterals and among polygons in Section 8.2, and we examined relationships among polyhedra in this section, for example, the close connection between prisms and cylinders. One, of many more, relationships we will consider here was discovered by Leonard Euler and bears his name. If you did Exploration 8.15, you discovered this on your own. If not, you will see it here. It is included both because it is a famous formula and because it is one that children can discover, with some guidance, and it is another example of the rich interconnectedness that permeates mathematics which is not well appreciated by most people.

Any polyhedron has a certain number of faces, vertices, and edges. For example, count the number of faces, vertices, and edges of the cube on page (565), the square pyramid on page (562), and the truncated pyramid (the polyhedron to the right of the square pyramid) on page (562). Then read on. . . .

---

[7]*Historical Topics for the Mathematics Classroom: Thirty-first Yearbook* (Reston, VA: NCTM, 1969), p. 220.

The cube has 6 faces, 8 vertices, and 12 edges.
The pyramid has 5 faces, 5 vertices, and 8 edges.
The truncated pyramid has 6 faces, 8 vertices, and 12 edges.

In Chapter 2, you found that in each growing pattern, there is a relationship that enables us to construct a formula to predict the number of dots or squares in the nth figure in a growing pattern. Similarly, there is a relationship among the number of faces (F), vertices (V), and edges (E) in any polyhedron, and knowing this relationship enables us to construct a formula that connects the number of faces, vertices, and edges. Can you guess it from these three examples?

What Euler discovered is that the sum of the number of vertices and faces is always two more than the number of edges, and this is true for all polyhedra. In symbols, we write:

$$V + F = E + 2$$

Connecting Two-Dimensional Representations to Three-Dimensional Objects

Although we live in a three-dimensional world, much of our interaction with this world is on the two dimensions of books, magazines, and newspapers and on the two dimensions of computer and television screens.

There are many ways in which the two-dimensional and three-dimensional worlds connect. All buildings, from small sheds to large skyscrapers, are designed before they are built. For major projects, scale models are built. To enable the architects and the engineers to communicate, blueprints are designed and studied. So that the electricians, plumbers, and other members of the building team will know where to place the appropriate wires and fixtures, other kinds of drawings are used. Each of these drawings requires someone to think about the object in three dimensions and then represent that information two dimensionally, although computer simulation is changing the nature of these representations.

And this is only one example. Archaeologists work with the three dimensions of the excavation site and the two-dimensional representations of the "dig." Painters need a thorough understanding of geometry so that their paintings (on a two-dimensional surface) will look like the three-dimensional objects or landscapes that they are representing. In this section, we will examine several ways in which the two-dimensional and three-dimensional worlds connect: **cross sections**, **nets**, and simple (**isometric**) drawings.

First, we will focus on simple buildings, the kind that can be made with cubes. Most elementary classrooms have blocks, and many powerful geometric ideas can be developed by playing with blocks. One of those is for children to build block buildings and then give directions for making the buildings.

INVESTIGATION

8.24

Different Views of a Building

Look at the building at the right. Following are the profile views of the building from the front, from the right, from the back, from the left, and from the top (imagine flying over the building as you approach it from the front). Look at those views. Can you see how those views have been made? For example, can you see why the front view consists of three cubes stacked on one another and then a stack of two cubes to the right?

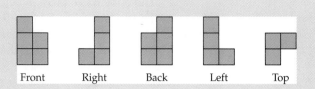

Front   Right   Back   Left   Top

Look again at those views. What do you see? Recall that we found in high school geometry that in order for two triangles to be congruent, all three pairs of corresponding sides were congruent and all three pairs of corresponding angles were congruent. When we examined triangles more closely, we didn't need to know all six pieces of information in order to say that two triangles were congruent. For example, if all three corresponding sides were congruent, that was sufficient for us to conclude that the two triangles were congruent. The analogous question here: Do we need all five views in order to make the figure? Why or why not? A slightly simpler, but related question: Are some of the views related to each other? Think about these questions before reading on. . . .

DISCUSSION

As you may have noticed, in this case, the right and left views are mirror images of each other. Similarly, the front and back views are mirror images of each other. Do you think this is true just in this case, in some cases, or all cases?. . .

It turns out that it will be true in all cases. Thus we can cut out two pieces of information. For the sake of convention, we will denote the front and right-side views. What about the top? Is that really necessary? For example, if you were given only the front and right-side views of the building above, could you make the building? Think and read on. . . .

There is another building that has the same front and right-side views as the one pictured above. It is shown below. However, its top view is different. Both the building and its top view are shown below. Thus the front, right, and top views are all necessary. A curious reader might be wondering whether the top, front, and right views will be sufficient in all cases. That is a great question and will be left as an exercise.

> ■ *Outside the Classroom* ■
>
> After finishing this section in my course, a student showed me a card from the game Mindtrap, in which the players are given two views of an object and have to sketch the object. We both found the question difficult, and we disagreed with the answer!

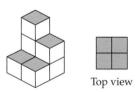

Top view

Another way to represent our three-dimensional block buildings would be to sketch them as was done above. Before proceeding further, cover the rest of this page and look at the block building shown above. Sketch it on a blank sheet of paper. Then read on. . . .

Having looked at representing a building with three views, let us look at how we can draw the building. Some of you remember being told to make a cube by first drawing two overlapping squares. If you try to use Geoboard Dot Paper, you won't draw very good models. But it turns out that Isometric Dot Paper will enable us to draw pictures of buildings quite nicely.

INVESTIGATION 8.25

Isometric Drawings

As we noted in Section 8.1, paintings became much more realistic when artists learned perspective during the Renaissance. I'm not sure when isometric drawings were developed, but they offer us a tool to sketch simple polyhedra. First, look at the Isometric Dot Paper in Figure 8.122. What do you see?

The dots are not in a rectangular array. Rather, the dots in each row are staggered so that when you connect dots, you form equilateral triangles. Believe it or not, this equilateral triangular array is an efficient way to draw objects whose angles are right angles. Now go back and try to sketch the block building from the previous investigation on the isometric grid below. Then read on. . . .

FIGURE 8.122

DISCUSSION

Some people find this easy to do. I wasn't one of them! Now that I have done it hundreds of times, it is easy for me, but I can remember just not being able to figure it out and then finally "getting it"—only to discover some months later, when I was faced with the task again, I had forgotten it. Therefore, I understand completely if you are one of the readers who feels baffled at this point. Let us begin at the beginning. Figure 8.123(a) below shows 1 cube. Figure 8.123(b) shows a column of 2 cubes, and Figure 8.123(c), show a column of 2 cubes next to a column of 3 cubes. Finally, the earlier block figure illustrates how to put columns of cubes side by side, especially when there is a blank spot in the building.

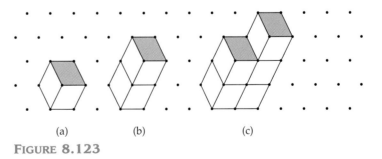

(a)   (b)   (c)

FIGURE 8.123

Now that we have examined buildings from different views and isometric drawings, let us examine another connection between two- and three-dimensional objects. A cross section of a solid is what the exposed face would look like if we sliced through the solid. Because there are many ways that we might slice through a solid, the shape of the cross section will depend on the nature of the slice. Let's examine a few.

INVESTIGATION 8.26

Cross Sections

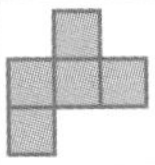

If we sliced the cube as shown in the Figure 8.124, what would the cross section look like? If we sliced the cube as shown in the Figure 8.125, what would the cross section look like? If we sliced the cube as shown in the Figure 8.126, what would the cross section look like? Could we get other shapes from slicing through a cube?

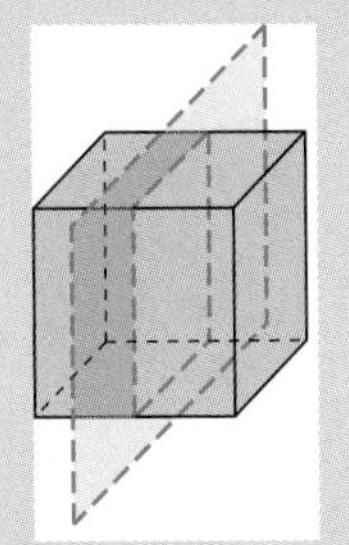

FIGURE 8.124 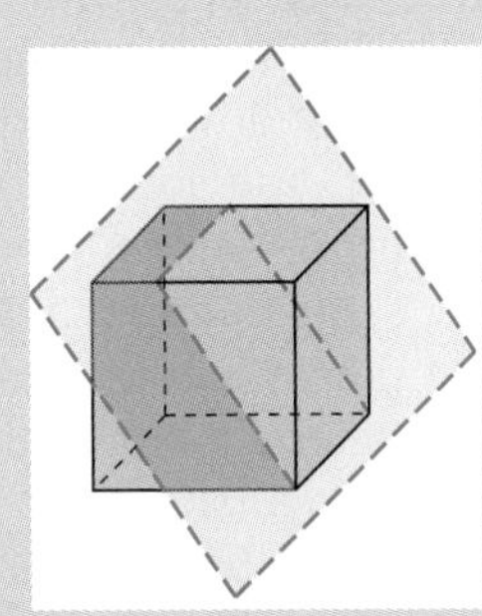   **FIGURE 8.125** 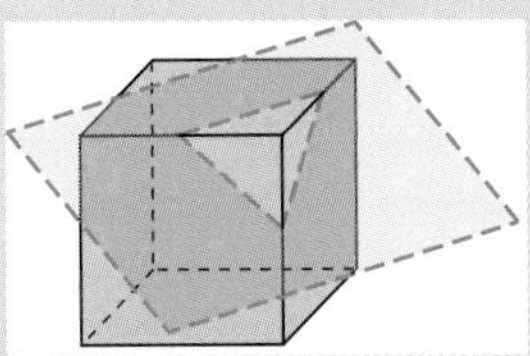   **FIGURE 8.126**

CLASSROOM CONNECTION

Many textbooks introduce the notion of cross sections to children through silhouettes and shadows.

DISCUSSION

In the first case, the cross section is a square. In the second case, the cross section is a rectangle. In the third case, the cross section is a triangle. There are different ways of slicing that will result in different rectangles and triangles.

One last connection between two- and three-dimensional figures that we will explore here is nets. A net is simply a two-dimensional representation of a three-dimensional object, in which:

1. Every face of the object is represented.
2. If you cut out the net and fold along the edges, it will fold up into an actual object.

The figure at the left in Figure 8.127 is a net of a cube, whereas the figure at the right is not. If you fold the first figure up, you will get a cube. Do you see that? If not, try to fold it in your mind. One way is to make use of the properties of a cube. I have labeled the faces: Bo, T, F, Ba, S, and S for bottom, top, front, back, and sides. Does that help? In the second case, if you cut out the figure, it won't fold up. Two faces will overlap.

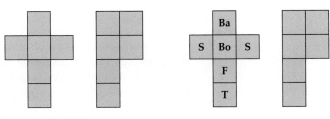

FIGURE 8.127

INVESTIGATION 8.27

Nets

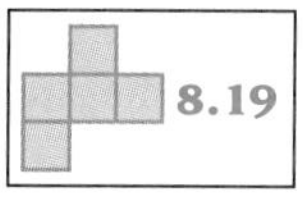

One of the nets that people experience regularly (especially if they recycle) is a flattened cereal box. One net for a standard cereal box is shown in Figure 8.128. What do you notice about this net? This includes, but is not restricted to, the question "What attributes and characteristics do you see?"

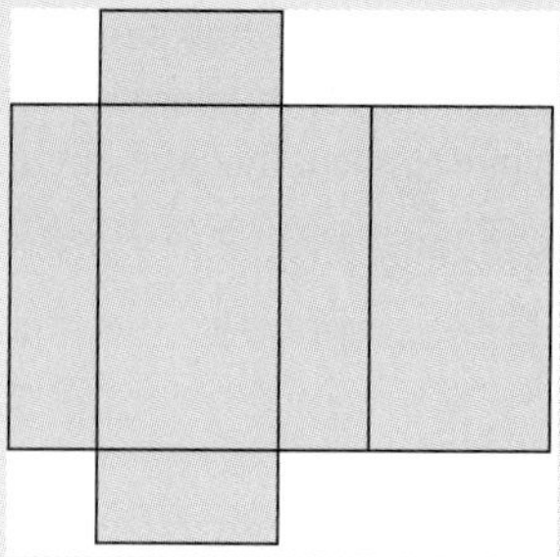

FIGURE 8.128

DISCUSSION

One of the key aspects of "really" understanding nets is to see certain attributes of nets and then to connect that information. One important observation is that the net has six faces. If you recall, cubes have six faces. The cereal box is a rectangular prism, and thus it has many of the attributes of a cube. Another observation is there are three pairs of congruent faces.

If we think of a cereal box, we think of front, back, sides, top, and bottom. If we label those faces on our net (see Figure 8.129), this leads to another observation: Two congruent faces are never side-by-side. Do you see why?

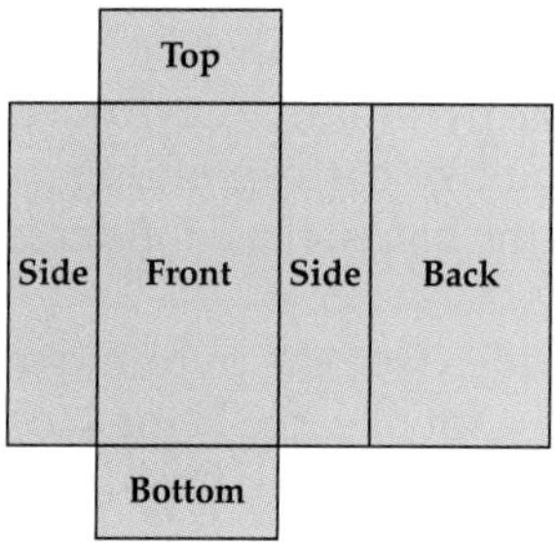

A good way to deepen your understanding of the connection between the three-dimensional and two-dimensional worlds is to sketch several other nets for the cereal box. Try this yourself before reading on. . . .

If your understanding of nets is not well connected, this is a very difficult task. If it is connected, the job is much easier. One thing that makes this task easier is to realize that each face of the box is connected to three other faces. Thus we can take our original net and slide the bottom underneath the back, as shown in Figure 8.130(a)—it still folds up. In the original net, the top and bottom were connected to the front. However, on the actual box, they are also connected to the sides. Figure 8.130(b) represents that connection. Finally, we can move the back so that it is connected to the bottom, as shown in Figure 8.130(c). My students have worked on this problem and have found many, many nets!

FIGURE 8.129

■ *Mathematics* ■

If you did Exploration 8.19, you found that one category of hexominoes that fold up to a cube consists of those with four cubes in a column and one cube on either side of the column. Because cubes and boxes are both rectangular prisms, some of the properties of a cube also hold true for the box. In this case, that categorization of hexominoes holds. Look at the three nets for the cereal box above. Do you see "four faces in a row and then one face on either side?"

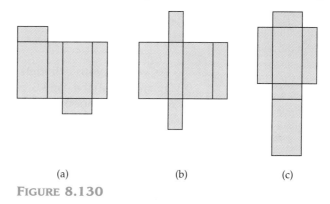

(a) (b) (c)

FIGURE 8.130

Cylinders, Cones, Spheres

The polyhedra we have defined thus far have all been simple, closed surfaces in which all the faces are polygons. There are three other kinds of three-dimensional

> **■ Mathematics ■**
>
> Technically, the base of a cylinder can be any simple, closed curve. In everyday life, cylinders usually have circular bases.

objects that are commonly found and that elementary schoolchildren study. *Cylinders*, *cones*, and *spheres* are related to polyhedra we have studied. Before we examine these three, stop for a moment and consider which polyhedra are related to cylinders, which to cones, and which to spheres. Then read on. . . .

Think of a prism with more and more sides (see the prism at the left in Figure 8.131). At some point, a prism with a lot of sides begins to look more like a cylinder than like a prism. From one perspective, we can think of a cylinder as a prism in which the bases are circles. Technically, this is not true, because the bases of prisms are polygons, and a circle is not a polygon.

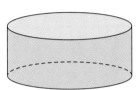

FIGURE 8.131

> **■ Mathematics ■**
>
> We can also define a cylinder by using set language (remember Chapter 2?) and say that a cylinder is constructed by starting with a simple, closed surface and a line segment that has one endpoint on the surface and the other endpoint not on the surface. A cylinder consists of the union of: the surface, the union of all line segments congruent to and parallel to the given line segment, and the surface formed by connecting the other endpoints of each of the line segments.

Thus we will describe a **cylinder** more formally as a simple, closed surface that is bounded by two congruent circles that lie in parallel planes.

Earlier, we talked about right prisms and right pyramids. A cylinder is a **right cylinder** if and only if the line segments joining two corresponding points on the two bases are perpendicular to the planes of the bases. If a cylinder is not a right cylinder, it is called an **oblique cylinder**.

Now think of a pyramid with more and more sides (see the pyramid at the left in Figure 8.132). At some point, a pyramid with a lot of sides begins to look more like a cone than like a pyramid. From one perspective, we can think of a cone as a pyramid in which the base is a circle. Technically, this is not true, because the base of a pyramid is a polygon, and a circle is not a polygon.

CLASSROOM CONNECTION

When children are asked to describe a cone, they say things like "A triangle with a flat bottom," "A round triangle," "A large circle with smaller and smaller circles on top until it reaches a point," "A cylinder, triangle, and a circle in one." From *Examining Features of Shape: Casebook* by Deborah Schifter, Virginia Bastable, and Susan Jo Russell, with Danielle Harrington and Marion Reynolds (Parsippany, NJ: Dale Seymour, 2002), p. 26.

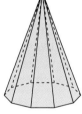

FIGURE 8.132

As we did with cylinders, we can define the term *cone* by using set language and say that a cone is constructed by starting with a simple, closed surface and a point not on the circle. A cone consists of the union of: the surface and the union of all line segments connecting that point and the surface.

Thus we will describe a **cone** more formally as a simple, closed surface whose base is a simple, closed surface and whose lateral surface slopes up to a vertex that we call the **apex**.

If the apex of the cone lies directly above the center of the base, then we call it a **right cone**. If a cone is not a right cone, it is called an **oblique cone**. In everyday life, we generally experience only right cones and right cylinders. Therefore, in this book, we will use the terms *cones* and *cylinders* unless referring to oblique cones or cylinders.

Finally, think of a polyhedron that has more and more and more faces. You might want to look at the five regular polyhedra in Figure 8.120 and imagine

> **Mathematics**
>
> Technically, the base of a cone can be any simple, closed curve. In everyday life, cones usually have circular bases.

starting with a dodecahedron made of clay and then slicing the various faces at an angle to make more and more faces. Eventually, the figure would begin to resemble a sphere more than a polyhedron.

A sphere is also conceptually related to a circle. Can you apply the earlier definition of a circle to define the term *sphere*? Try to do so before reading on. . . .

A **sphere** is the set of points in space equidistant from a given point, which is called the **center**.

How would you define the radius and diameter of a sphere? Try to do so before reading on. . . .

Any line segment joining the center of the sphere to a point on the surface is called a *radius*.

Any line segment whose endpoints lie on the surface of the sphere and that contains the center is called a *diameter*.

EXERCISES 8.3

1. Given the tetrahedron at the right, name the following:
   a. A face
   b. A vertex
   c. An edge

   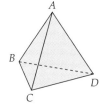

2. Identify the numbers of vertices, edges, and faces of the following figures.

   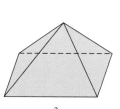
   a.

   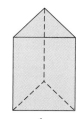
   b.

   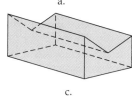
   c.

3. Consider a prism whose base is a regular *n*-gon—that is, a regular polygon with *n* sides. How many vertices would such a prism have? How many faces? How many edges? You may want to start with a triangular prism, square prism, pentagonal prism, and so on, and look for patterns.

4. Consider a pyramid whose base is a regular *n*-gon—that is, a regular polygon with *n* sides. How many vertices would such a pyramid have? How many faces? How many edges?

   a. Describe the relationship between the numbers of vertices of an *n*-gon prism and of an *n*-gon pyramid.

   b. Describe the relationship between the numbers of faces of an *n*-gon prism and of an *n*-gon pyramid.

   c. Describe the relationship between the numbers of edges of an *n*-gon prism and of an *n*-gon pyramid.

5. a. What attributes do all cylinders and all prisms have in common that not all polyhedra have?

   b. What attributes do all prisms have that only prisms have?

   c. What attributes do all cylinders and cones have that not all three-dimensional figures have?

6. Name the figures below.

   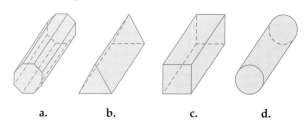
   a.   b.   c.   d.

7. Which of the polyhedra below are convex?

   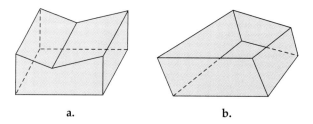
   a.   b.

8. Draw a nonconvex rectangular or pentagonal prism.

9. We defined a regular polyhedron as a convex polyhedron in which the faces are congruent regular polygons and in which the numbers of edges that meet at each vertex are the same. Carlos says that instead of saying that

the numbers of edges that meet at each vertex are the same, we could have said that the numbers of faces that meet at each vertex are the same. What do you think? Support your choice.

10. Can you make a pyramid in which the triangular faces are not all congruent?

11. Write a definition of *diagonal* for polyhedra.

12. There is a relationship between the number of diagonals and the base of a prism. That is, triangular prisms have a certain number of diagonals, square prisms have a certain number of diagonals, and so on. Determine this relationship so that you can answer the following question: How many diagonals does a prism have whose base is a regular polygon with *n* sides? [Your instructor may or may not give you hints for this problem. If not, I suggest looking at the 4 Steps for Problem Solving on the inside front cover of the *Explorations* volume.]

13. At the center of every tissue of toilet paper is a cardboard cylinder. Find and examine one of these cylinders. You can see a curved line running along the face of the cylinder.

    a. If you cut the cylinder along this line, what would the unfolded shape look like? Predict the shape and explain your reasoning.

    b. Why do you think these cylinders are manufactured this way instead of having a vertical cut?

14. Is there one geometric shape that describes *all* the sides of (right) pyramids? If there is, name it and justify your answer. If there is more than one shape, describe the shapes and justify your response.

15. a. Write directions for making each of the following block buildings any way you want.

    b. Write directions for the same block buildings, using a different method.

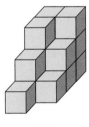
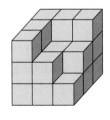

(1)　　　　　　　　(2)

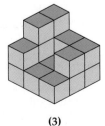

(3)

16. Sketch the figures below on Isometric Dot Paper.

a.　　　　b.

17. Sketch the front, side, and top views of the buildings below.

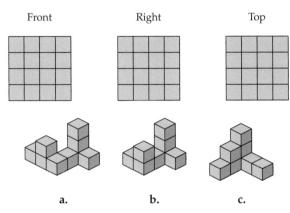

a.　　　　b.　　　　c.

18. Below are three nets for a cereal box. They count as one family because all it takes is a simple transformation—in this case a translation (slide)—to change one into another. Draw three more nets for a cereal box that are all in different families. Each net needs to have 6 whole faces; that is, do not cut one face into two or more pieces—otherwise, we have an almost infinite number of possibilities.

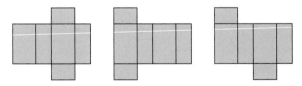

19. Using Polyomino Grid Paper, make as many nets as you can for the triangular prism shown below. There are 5 faces in the figure—the 2 triangular bases and 3 sides. Do not cut any of the faces—this would create many, many possible nets. That is, each net will consist of 5 distinct faces, joined together. The three lateral faces are all 1 × 3 rectangles, and the two bases are equilateral triangles. There are fewer than 10 possible nets.

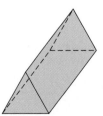

20. How many different heptominoes (made from 7 squares) are there that have 5 squares in a column and 1 square attached to opposite sides of the column? One is sketched below. Templates are given to make it easier to

draw others—you will not necessarily need all of them. Describe your method, other than random trial and error. How confident are you that you have drawn all of them? Why?

21. **a.** Look at the triangular prism shown below. If you were to make a net for this polyhedron, what would be the dimensions of the ramp?

    b. Look at the figure at the right below. If you were to make a net for this polyhedron, what would be the exact dimensions of the "roof"?

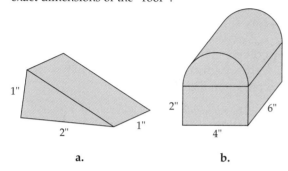

a.   b.

22. Below are drawings of two polyhedra. Draw the top, front, and side views for each polyhedron.

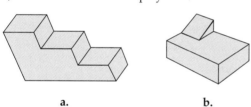

a.   b.

23. Which of the nets below is a possible net for a cereal box?

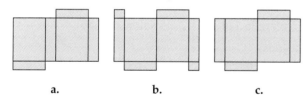

a.   b.   c.

24. Which of the nets below is a possible net for an oatmeal box (which has the shape of a cylinder)?

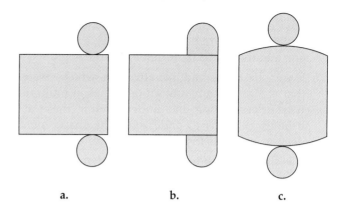

a.   b.   c.

25. How would you make the octahedral die shown below?

26. Most doors are rectangles. Occasionally, however, you see a door whose shape is a nonrectangular polygon. In *Lord of the Rings*, the door to Frodo's house was round. Explain why we don't see round doors very often!

Chapter Summary

1. Important spatial thinking abilities include: eye-motor coordination, figure-ground perception, perceptual constancy, visual discrimination, and visual memory.

2. The van Hiele levels of geometric thinking help us to see that the level of our geometric thinking determines what we see (for example, the churn dash investigation) and how powerfully geometric ideas can be used.

3. Our current knowledge of geometry took thousands of years to develop and is the result of intuitive thinking, inductive thinking, and deductive thinking. The Greeks were the first people to develop mathematical systems, that is, a coherent system of mathematical ideas.

4. Our understanding of geometry has been built carefully on a foundation of axioms, undefined terms, definitions, and theorems. The building blocks of this knowledge are the ideas about points, lines, planes, and space.

5. Shapes are found everywhere, and virtually all shapes have functional and/or aesthetic value.

6. Each geometric shape represents the common characteristics of a set of objects. For example, the set of objects we call prisms can look very different to a novice, but they are all prisms because they have two bases that are polygons, parallel, and congruent, and they have faces that are parallelograms.

7. Every shape has multiple attributes which means that we can classify any set of shapes in multiple ways. Recognizing and understanding these attributes helps us to understand the shape more deeply and leads to practical applications.

8. Classifying leads us to deeper understanding of mathematical structure which leads to greater mathematical power, for example, that the sum of the angles of any polygon is equal to $180(n - 2)$.

9. Looking for and recognizing relationships within and between shapes also leads to understanding of mathematical structure; for example, the many relationships among quadrilaterals (which we will further develop in Chapter 9), and seeing the relationships between prisms and cylinders.

10. Coordinate geometry is a useful tool for understanding attributes of shapes and relationships among shapes.

11. There are rich connections between two-dimensional and three-dimensional shapes. This knowledge has many practical uses and applications—drawings, representations of buildings, and nets.

BASIC CONCEPTS

Section 8.1 Basic Concepts of Geometry

hand-eye coordination **505**
figure-ground perception **505**
perceptual constancy **505**
visual discrimination **506**
visual memory **506**
Point, *line*, *plane*, *on*, and *between* are undefined terms. **508**
Sets of points may be collinear or noncollinear. **509**
Sets of points may be coplanar or not; lines may be coplanar or not. **510**

line **510** line segment **510**

ray **510** endpoint **510**
intersect **511**
Relationships between lines: perpendicular, parallel, concurrent, skew **498**
Angles have vertices and sides. **512**
Angles have an interior and an exterior. **512**

Naming angles
with one letter, one number, three letters **513**

Measuring angles
protractor **513** degree **513**

Kinds of angles
right **516** acute **516**
obtuse **516** straight **516**
reflex **516** adjacent **516**
complementary **516** supplementary **516**
vertical **516**

Section 8.2 Two-Dimensional Figures

Kinds of lines and curves
closed curves **527** simple curves **526**
Any simple closed curve partitions the plane into 3 disjoint regions. **528**
polygon **529** sides **529**
vertex (vertices) **529** base **529**
congruent **529**

Classifying triangles
By length of sides: equilateral, isosceles, scalene **531**
By size of angle: right triangle, obtuse triangle, acute triangle **531**

Special line segments in triangles
angle bisector **535** median **535**
altitude **536** perpendicular bisector **535**

Congruence
congruent **537**

Quadrilaterals
trapezoid **539** parallelogram **539**
kite **539** rhombus **540**
rectangle **540** square **540**

Polygons
convex or concave **545** regular polygon **546**
diagonal **540** interior angle **547**
exterior angle **548** central angle **548**

Properties
The sum of the measures of the angles of a convex polygon = $180(n - 2)$. **547**

Circles
circle **549** radius (radii) **550**
diameter **550** chord **550**
tangent **550** arc **550**

Section 8.3 Three-Dimensional Figures

space figure **563** polyhedron **563**
polyhedra **563** solids **563**

Parts
face **563** edge **563**
vertex **563**

Classifying
  convex 564
  concave 564

Polyhedra
  prism 565
  base 565
  lateral face 565
  right prism 565
  box 565
  cube 565
  dihedral 565
  oblique prism 565

Pyramid
  pyramid 566
  apex 566

Regular polyhedra
  tetrahedron 567
  octahedron 567
  icosahedron 567
  dodecahedron 567

Relationships among polyhedron
  Euler's formula 569

Connecting two-dimensional representations to three-dimensional objects
  different views 569
  cross sections 569
  isometric drawings 569
  nets 569

Cylinders, Cones, Spheres
  cylinder 574
  oblique cylinder 574
  right cone 574
  sphere 575
  apex 574
  right cylinder 574
  oblique cone 574
  center 575

CHAPTER 8 REVIEW EXERCISES

1. Without using a protractor, draw an angle that is approximately 30 degrees. Explain your reasoning. Now measure the angle. Repeat the process for a 120-degree angle.

2. In the figure at the right, $\overleftrightarrow{AB}$ and $\overleftrightarrow{BC}$ are perpendicular lines.
   a. Name two complementary angles.
   b. Name two supplementary angles.
   c. Name two vertical angles.
   d. Name two adjacent angles.

3. True or false? If true, briefly explain why. If false, provide a counterexample.
   a. If three distinct lines intersect, then they are coplanar.
   b. If two lines do not intersect, then they are parallel.

4. Why is it necessary to start with undefined terms in geometry?

5. A teacher defined *triangle* as a shape made by three line segments. By that definition, the shape at the right is a triangle. Fix the definition so that it works.

6. How many different quadrilaterals can you make that will fit on a 3 × 3 Geoboard?

7. How could you convince someone that the sum of the angles of any quadrilateral is 360 degrees?

8. What attributes do all rectangles have in common?

9. For each of the following, draw the figure or explain why it is impossible.
   a. A triangle that is isosceles and obtuse.

b. A quadrilateral that has exactly one set of parallel sides and two right angles.

c. A nonregular hexagon that has all sides congruent.

d. A pentagon with three right angles and two sets of parallel sides.

e. A concave pentagon.

10. How many right angles can a hexagon have?

11. Write down all the attributes of each of the following figures. Then write down the attributes they have in common.

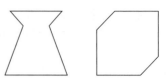

12. Name six different polygons that you can see in this figure in such a way that the reader can easily find the polygons that you have found.

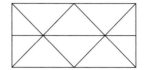

13. The four figures at the left below are Kiwis. The four figures in the middle group are not Kiwis.

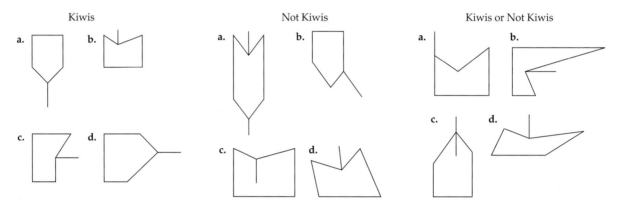

a. Judging on the basis of the four Kiwis and the four not-Kiwis, list the attributes that all Kiwis possess.

b. Which of the four figures in the group at the right are Kiwis? Justify your answers.

14. Write directions for making the figure below. Following your directions, the reader should be able to make the same figure.

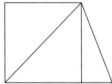

15. Is the Venn diagram below a valid representation of the relationship between parallelograms and rectangles? If yes, explain why. If not, explain why not.

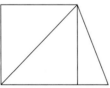

16. Draw a Venn diagram to represent the relationship between regular polygons and convex polygons.

17. Find the distance between the following pairs of points: (2, −4) and (4, 1).

18. Find the midpoint of the line segments connecting these pairs of points: (0, 3) and (5, 12).

19. Three vertices of a square are (5, 0), (5, 6) and (8, 3). What are the coordinates of the fourth vertex?

20. I am a parallelogram. Two of my sides are parallel to the bottom of the paper. The vertices of my bases are at (2, 3) and (8, 3). The slope of the line segments that form my sides is 0.75, and the length of each side is 5 units. What are my other two vertices?

21. Identify the number of vertices, edges, and faces of the accompanying figure.

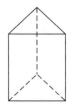

22. What attributes do all cones and pyramids have in common that not all three-dimensional figures have?

23. a. Write directions for making the block building at the right any way you want.

    b. Write directions using a different method.

24. Sketch the figure at the right on Isometric Dot Paper.

25. Sketch the front, side, and top view of the building above.

26. Following are the top view, the front view, and the right-side view of a building created only by cubes.

    a. Without making the building, predict how many cubes it will have. Explain your prediction.

    b. Predict the left-hand view. Explain your prediction.

    c. Sketch the building on Isometric Dot Paper.

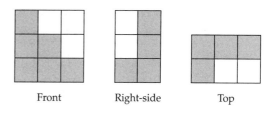

Front   Right-side   Top

27. a. Draw a net for a prism whose base is a right isosceles triangle.

    b. Draw a different net for the same prism.

28. Write a definition of *diagonal* for polyhedra.

29. Describe the polygon formed by the following cross sections.

    a. You cut a cross section that is perpendicular to the base of a square pyramid but does not pass through the apex of the pyramid.

    b. You cut a cross section that is perpendicular to the base of a circular cylinder.

30. Below is a drawing of a polyhedron. Draw the top, front, and side views for this polyhedron.

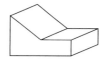

CHAPTER 9

Geometry as Transforming Shapes

9.1 Congruence Transformations

9.2 Symmetry and Tessellations

9.3 Similarity

9.4 Topology

*I*n Chapter 8, we focused on understanding shapes—attributes of shapes, different kinds of shapes, and relationships among shapes. In Chapter 9, we are going to look at what happens when we *transform* shapes; the knowledge we gain from these explorations will help us better understand the patterns that emerge when shapes are put together, as in quilts, floor patterns, art, and other situations. In both chapters, there are certain themes that run through our explorations: the notion of deconstructing—which leads to awareness of multiple attributes; classifying—which leads to understanding of structures; and looking for patterns and relationships—which leads to generalizations and, again, to understanding of structures.

Look at the pictures in Figures 9.1 to 9.3. What do *you* see? Note your ideas before reading on.

FIGURE 9.1

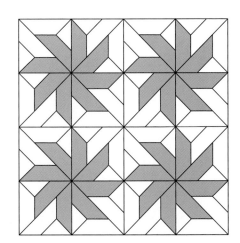

FIGURE 9.2

Source: M.C. Escher's ceiling at Philips International. © 2003 Cordon Art B.V.—Baarn—Holland. All Rights Reserved. Photo courtesy Philips International B.V.

FIGURE 9.3

Figure 9.1 shows an Islamic design composed entirely of two shapes: an 8-pointed star and a 15-sided polygon. Each star can be moved to the position of a nearby star by sliding it—horizontally, vertically, or diagonally. Each 15-gon can be moved to the position of a nearby 15-gon by turning it.

Figure 9.2 illustrates one of many parquet floor designs. Most people see pinwheels composed of 8 congruent isosceles trapezoids. Some people see these embedded in squares. Other people see congruent pinwheels of a different color. Others see white pinwheels composed of 4 white isosceles triangles. Each dark pinwheel can be moved to the position of a nearby light pinwheel by sliding it. Each trapezoid in a pinwheel can be moved to the position of an adjacent trapezoid by a turn.

In Figure 9.3, focus on the butterflies. Though the butterflies are in different positions and of different sizes, they all have the same basic shape. This is true for each of the animals pictured. The arrangement of the animals is not random but intentional. If you focus only on the butterflies, what do you notice? From one perspective, we can say that every butterfly has a twin, and each butterfly can be moved from its position to its twin's position by flipping, turning, or sliding or by a combination of these moves.

Transformations

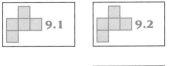

These three figures illustrate several mathematical transformations. However, before we get into the mathematical aspect of transformations, what do you think when you hear this word?

The interesting examples offered by my students include the transformation of a frog into a prince in fairy tales and the transformation of a working-class girl into a high-society lady, as seen in *My Fair Lady* with Audrey Hepburn and in *Pretty Woman* with Julia Roberts. Virtually all examples of transformation give a sense of movement and change. In mathematics, there are many kinds of transformations. Figure 9.4 illustrates several. Look at the various transformations of the letter P. How would you describe the transformation? Write your thoughts before reading on. . . .

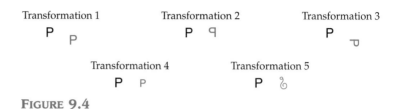

FIGURE 9.4

Most people find the first four transformations fairly straightforward but find the last one very difficult. Rather than give the answer at this point, let me ask a more refined question: What is the same and what is different about each figure and its image after the transformation? Think about it, and then read on. . . .

Table 9.1 shows the informal names for the transformations, the formal names, what is changed, and what is unchanged.

TABLE 9.1

Informal name	Formal name	What is changed?	What is unchanged?
1. Slide	Translation	Position	Size, shape
2. Flip	Reflection	Position	Size, shape
3. Turn	Rotation	Position	Size, shape
4. Shrink	Similarity transformation	Position and size	Shape
5. Distortion	Topological transformation	Position, size, and shape	Neighborhood

In a moment, we will zoom in on the first three transformations. In Section 9.3, we will focus on transformations in which a figure is reduced or enlarged so that the shape is similar but the size has changed. This set of transformations is called the set of similarity transformations. You may recall studying similar triangles in high school. You can explore topological transformations on the website. Ironically, young children's initial understanding of spatial ideas is more topological than Euclidean.

In elementary school, we want students to explore these important ideas of congruence and similarity in different ways, so that when they investigate these ideas more formally in high school, they will have had concrete *and* substantive interactions with them.

SECTION 9.1 CONGRUENCE TRANSFORMATIONS

WHAT DO YOU THINK?

- How are translations, reflections, and rotations related?
- How might you describe the translation, reflection, or rotation of a three-dimensional figure in space?
- How are the operations translation, reflection, and rotation like the operations addition, subtraction, multiplication, and division?

At this point, you have an informal understanding of slides, flips, and turns. In the next few pages, we will examine these transformations more precisely. The following investigations will all be driven by this question: How can we describe what we have to do to move an object from here to there on a plane?

Translations

When you see the word *translation*, you probably think of translating from one language to another: The translator replaces English words and phrases with Spanish words and phrases, for example, but the meaning of the phrase remains the same. In geometry, a translation involves moving an object along a straight line and not turning it.

Each of the designs in Figure 9.5 shows translations. Figure 9.5(a) is from the Native American Yuchi tribe and depicts storm clouds. Figure 9.5(b) shows a quilt border called Orange Peel. Figure 9.5(c) shows a papercutting, which is a common elementary school activity. As noted above, a translation is a special kind of congruence transformation. In each of the designs, we have congruent figures in the design, and we can imagine picking up part of the figure, moving it in a straight line, and then putting it back down on top of another congruent part of the figure.

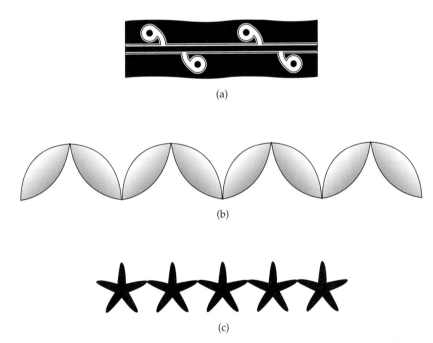

Source: (a) Le Roy H. Appleton, *American Indian Design and Decoration* (New York: Dover Publications, 1971), Plate 15 (in insert). (b) Jinny Beyer, *The Quilter's Album of Blocks & Borders* (Delaplane, VA: EPM Publications, Inc., 1986), p. 185. Reprinted with permission.

FIGURE 9.5

586 CHAPTER 9 / Geometry as Transforming Shapes

INVESTIGATION 9.1 | Understanding Translations

FIGURE 9.6

In order to make intricate designs, we often need to be precise in describing the translation. Let's say you were talking on the phone to a friend and you wanted to describe the translation of the triangle *CAT* in Figure 9.6 from its original location to its new location. The original triangle is *CAT*; its translation image is *C'A'T'*. How would you do this? You may assume that you and your friend have any resources you need: scissors, ruler, compass, graph paper, etc. Think and then read on. . . .

DISCUSSION

One common response is to say that we slide the triangle from one location to the other. The mathematical word for slide is *translation*, and we talk about a figure and its **image**. There are many different ways to describe a translation. Each of these descriptions represents different perspectives and preferences concerning how to communicate what has happened to triangle *CAT*. Some of the perspectives are easier to describe if we draw the two triangles on graph paper (see Figure 9.7). What they all have in common is the realization that when we translate a figure, every point on the figure moves the same distance and in the same direction.

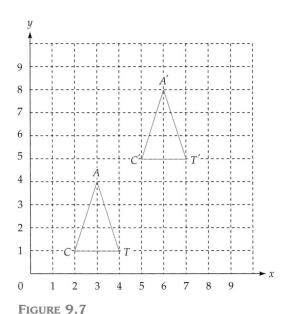

FIGURE 9.7

DESCRIPTION 1: Use "taxicab language"

Just as we could ask a cabbie to "go three blocks east and four blocks north," we could say that each point has been moved 3 units to the right and 4 units up. For example, copy triangle *CAT* onto another sheet of paper and cut it out. Then place it on the paper above and move the whole triangle, first 3 units to the right and then 4 units toward the top of the paper.

DESCRIPTION 2: Invent new notation

We can use notation to express this same idea more succinctly:

$$(x, y) \rightarrow (x + 3, y + 4)$$

This notation gives the directions "go 3 units to the right and 4 units toward the top of the page" very succinctly.

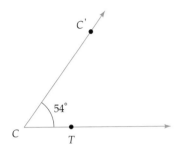

FIGURE 9.8

DESCRIPTION 3: Specify the distance and the angle

We could say that the figure has been moved a distance of 5 units at a 54-degree angle from the *x*-axis (see Figure 9.8). That is, C and C' are 5 units apart; the distance between any point on triangle CAT and the corresponding point on triangle C'A'T' is 5 units. Similarly, the angle formed by the rays CC' and CT is equal to approximately 54 degrees. I encourage you to check this for yourself with a ruler and a protractor.

DESCRIPTION 4: Use vectors

FIGURE 9.9

Another alternative is to use a vector. By definition, a **vector** has a length and a direction. For example, the instructions for the translation could be shown by drawing one vector, as in Figure 9.9.

We will formally define **translation** as a transformation on a plane determined by moving each point in the figure the same distance in the same direction.

Properties of Translations

Now let us determine some of the properties of translations. We know from Chapter 8 that two points determine a line. Therefore, if we connect each vertex in the triangle TAR to its image, T'A'R', we have three line segments: TT', AA', and RR' (see Figure 9.10). What relationship do you notice among these line segments?

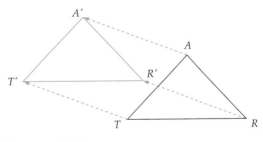

FIGURE 9.10

The three line segments are all congruent (same length) and all parallel (same direction). How might this make the task of drawing the translation image easier? Think and then read on. . . .

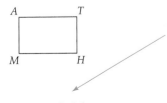

FIGURE 9.11

Translate the rectangle in Figure 9.11, using the translation arrow shown.

We can draw four lines, each going through a vertex, that are parallel to the translation arrow. If we copy the length of the arrow using a compass, we can quickly mark off the same lengths on the lines to determine the vertices of M'A'T'H' (see Figure 9.12).

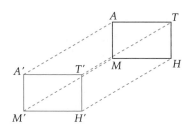

FIGURE 9.12

Reflections

Like translations, reflections can be accomplished in a variety of ways. (See Exploration 9.4 for more work with reflections.) One of the simplest is by folding paper. Trace the triangle *FOX* and the line below from Figure 9.13 onto a blank sheet of paper, and then fold the paper at the line.

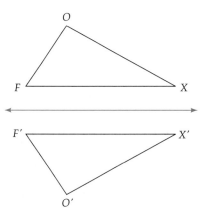

FIGURE 9.13

Now trace over the figure. This is like making a carbon copy of the figure. When you fold the paper on the line, each point on the triangle coincides with its image on the reflected triangle. Do you think we need to trace the whole figure? What do you think is the minimum tracing needed?

If you trace the points *F*, *O*, and *X*, you will have the points *F'*, *O'*, and *X'*, and we now need connect only these three points because we know that three noncollinear points determine a triangle.

INVESTIGATION

9.2

Understanding Reflections

Figure 9.14 shows a trapezoid that has been reflected (flipped) in three different ways. In each case, bold lines denote the original figure and dotted lines denote the flipped image.

A. Find the line of reflection in each case.

B. Can you discover a rule that would allow us to find the line of reflection in all cases? Stop, think, and write.

CLASSROOM CONNECTION

Although many elementary classrooms still use paper folding to help children understand reflections, many classrooms use other means—computer software, acetate sheets called Patty Paper, and Miras, which are made of red Plexiglas. The bottom edge of the Mira acts as the reflection line. When the Mira is placed on the paper, the image of the object can be seen and then traced on the other side of the Mira.

DISCUSSION

You can just read on and find the rule, or you can experience the kind of mathematical thinking that so many men and women and children have found very exciting.

A. You may find it helpful to start by tracing these figures on tracing paper and (by folding the paper) finding the line that makes the original figure coincide with its reflected image. Do you see what is happening? Try making figures of your own on paper; fold the paper and then trace the reflection image of the figure you drew. Do this several times. Play the "what-if" and "is it possible" games: What if the line of reflection went through the figure? Is it possible to have the line of reflection go right through the figure? Islamic artists, M. C. Escher, and many others have discovered so much by playing these two games.

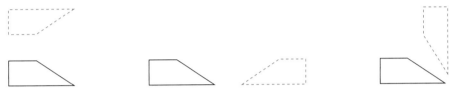

FIGURE 9.14

Paper folding is one of many ways to draw reflections. What other ways can you think of?

B. *Properties of reflections.* What is true for all reflections is that if we connect any point on the original figure with the corresponding point on the reflected figure, the line of reflection is the perpendicular bisector of that line segment. See Figure 9.15. That is, $A'X = XA$, and $\overline{AA'}$ and line l are perpendicular.

This realization leads to a more formal definition of reflection. A **reflection** is a transformation that maps a figure so that a line, called the **line of reflection**, is the perpendicular bisector of every line segment joining a point on the figure and the corresponding point on the reflected figure.

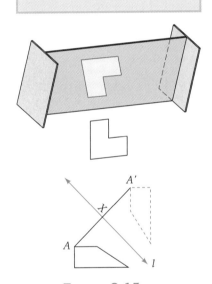

FIGURE 9.15

INVESTIGATION 9.3

Understanding Rotations

This investigation uncovers ways in which we can describe rotations. In each of the diagrams, the letter P has been rotated. However, as with translations and reflections, not all rotations are the same. In order to describe any rotation precisely, what do we need? You can play with this on your own by tracing the figures and playing around with them. . . .

> As with the concept of reflection, you might want to explore this yourself. You can do this by drawing the figure on a piece of paper and then tracing it on another piece of paper. Line up the two pieces of paper so that the top figure lies on the bottom figure, place the tip of your pencil or pen on the paper at some point, and then rotate the bottom sheet of paper. Draw the rotation image by tracing over the image you see from the bottom sheet. This is easier if the top sheet is a sheet of tracing paper, but it works with regular blank paper, too.

DISCUSSION

In order to rotate any figure (in a plane), we need to select a *center of rotation* (that is, the point that does not move), and we need to decide how much to rotate the figure. "How much" is determined by specifying an angle.

Look at the figures on the previous page and see whether you can guess the center of rotation and the degree of rotation.

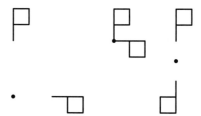

In the first figure, the center of rotation is the point that is approximately 17 mm below the bottom of the figure and 2 mm to the left, and the degree of rotation is 90 degrees clockwise. If you trace the figure, place your pencil on the point, and turn the paper 90 degrees, the figure will now be in the new position.

In the second figure, the center of rotation is the bottom of the figure, and the degree of rotation is 90 degrees clockwise.

In the third case, the location of the center of rotation is 5 mm below the figure, and the rotation is 180 degrees. Note that in the first and second figures above, the letter P was rotated 90 degrees clockwise. What both of these rotations have in common is that the orientation of the rotated figures is the same; in this case, both rotated images are lying on their sides. How, since the center of rotation was different, the distance and direction from the original figure are not the same.

Formally, we say that a **rotation** is a transformation on a plane determined by holding one point fixed and rotating the plane (in our case, the paper) about this point by a certain number of degrees in a certain direction. The fixed point is called the **center of rotation**.

These three transformations—translation, reflection, and rotation—are known as congruence transformations because the images are congruent to the original figure.

Before we go on, it is important to note one important attribute common to each of these three transformations. Recall that we found that when we translate a figure, each point on the figure is translated the same distance. Similarly, when we reflect a figure across a line, any point and its image are the same distance from the line of reflection. What do you think is equidistant with rotations?

When we rotate a figure, any point and its image are the same distance from the center of rotation. Thus the sense of "equidistance" is something all

> ■ **Mathematics** ■
>
> Technically, the center of rotation is always on the perpendicular bisector of the line joining two corresponding points. For example, if you connected the bottom points on the two figures and then draw a perpendicular line through the midpoint of that line, the center of rotation would be on that line.

congruence transformations share. As you may have seen in Exploration 9.9 (Tessellations), and as you will see in the explorations in the next section, many art designs involve one or more of these transformations.

Let us now examine some of the relationships among these three transformations.

INVESTIGATION

9.4

Understanding Translations, Reflections, and Rotations

Now that we have investigated translations, reflections, and rotations, let us stop to check for understanding. In each case below, describe the transformation that has moved the dark pentomino to the position of the light pentomino.

a. b. c.

d. e. f.

DISCUSSION

a. The pentomino was reflected across a horizontal line below the pentomino.

b. The pentomino was rotated 75° counterclockwise.

c. The pentomino was translated in a diagonal direction.

d. The pentomino was rotated 180 degrees.

e. The pentomino was reflected across a line making a 45-degree angle with a horizontal line.

f. There is no translation, reflection, or rotation that will accomplish this move. You could accomplish it in two moves by first translating the figure to the right and then reflecting across a horizontal line between the two figures. Or you could reverse the order: First reflect the pentomino across the horizontal line and then translate the image to the right.

The combination of translation and reflection described in part (f) is called a glide reflection and will be discussed shortly. The key to understanding this transformation is to realize that the translation and the reflection line must be parallel.

Combining Slides, Flips, and Turns

Now that we understand these three basic transformations of the plane, let us examine combinations of these transformations. There are many ways in which we can combine these transformations. For example, we could reflect a figure across a vertical line and then translate it 4 units to the right. Or we could reflect a figure across a vertical line and then reflect the image across another vertical line. We call any combination of transformations a **composite transformation**.

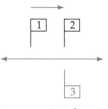

FIGURE 9.16

There is a special name for the composite transformation of a translation followed by a reflection with the condition that the translation vector and the line of reflection are parallel (that is, point in the same direction). We call such a composition a **glide reflection**. Recall the last problem in the previous investigation. I remember finding this transformation a bit elusive in terms of *really* understanding it, so let's take a minute to examine it. Figure 9.16 shows a flag being translated to position 2 and then being reflected across a horizontal line (which is parallel to the translation vector). The transformation of the flag from position 1 to position 3 is a glide reflection. Note that the translation vector does not need to be horizontal—it can be in any direction. But whatever the direction of the translation vector, the reflection line must be in the same direction.

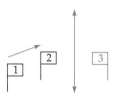

FIGURE 9.17

Let me give you an example of a translation followed by a reflection that is *not* a glide reflection. Figure 9.17 shows a flag being translated to position 2 and then being reflected across a vertical line that is *not* parallel to the translation vector. The composite transformation of the flag is *not* a glide reflection, because the translation vector and the reflection line are not parallel.

Probably the most famous example of glide reflection—and one that has helped many students get the sense of glide reflections—is a diagram of footprints on the sand (see Figure 9.18).

FIGURE 9.18

You can see the reflection (the left and right feet), and you know (from walking on a beach on the ocean or on a lake) the feeling of gliding along the beach. Also, you can connect your experience to the mathematical idea—that the line of reflection and the glide line are parallel (in the same direction), which is a *requirement* for all glide reflections.

Look at Figure 9.19, which is taken from a woodcut by M. C. Escher. Do you see the glide reflections? How would you explain them?

Any black bird can be mapped onto any white bird by translating the bird and then reflecting the translated image across a vertical line. Of course, to do it exactly, we would need to specify the translation vector (direction and distance) and the reflection line (direction and distance from the translated image of the black bird).

At this point, there are many possible "what-if" and "is it possible" questions about these transformations. Which ones come to your mind? Think before reading on. . . .

Such questions include the following:

1. Does it matter which one you do first when you do a composite transformation?

2. Is there a least number of steps that will enable you to map one figure onto a congruent image?
3. How many different congruence transformations are there?

Source: M.C. Escher's "Study of Regular Division of the Plane with Birds." © 2003 Cordon Art B.V.—Baarn—Holland. All Rights Reserved.

FIGURE 9.19

Rather than answer these questions directly, let us use the following investigation to examine them.

INVESTIGATION

9.5

Connecting Transformations

In each of the three pictures below, describe how you would get the flag in position 1 to position 2. Then read on. . . .

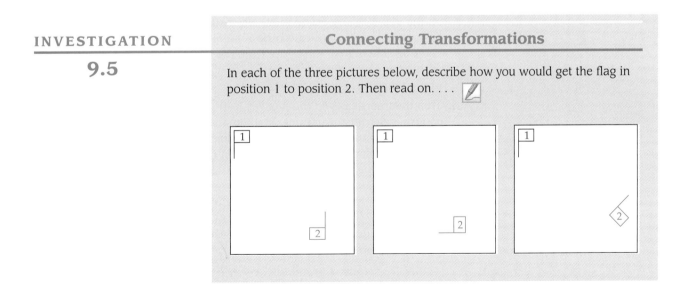

DISCUSSION

The curious reader may have realized that there is not one right answer for any of these! For example, in the first picture, you could get from the first to the second position by first reflecting the flag across a vertical line and then reflecting the image across a horizontal line. Or you could reverse the order, doing the horizontal reflection first and the vertical reflection second. Or you could get from the first to the second position by a single rotation. Can you find the center of rotation?

Let us use this example to answer the first question above. In this case, it did not matter which reflection (vertical or horizontal) we did first. This leads most people to conclude that the operation of reflection is commutative. This is connected to our discovery, in Chapter 3, of the commutative properties of addition and multiplication. However, if you conclude that there is a commutative property of reflection, you are wrong! Two reflections produce the same image *only* if the lines intersect at right angles. If the two reflection lines are parallel, we do not get the same image when we reflect across line m and then across line n as we do when we reflect across n and then m. Thus the operation of reflection is commutative only if the lines of reflection intersect at right angles. This is just one of many examples in mathematics where we must think carefully before generalizing!

In the second picture, you could translate the flag to the right, then rotate the image 90 degrees clockwise, and then flip the image across a horizontal line. Or you could rotate the figure 90 degrees clockwise and then do a glide reflection by translating the image so that it is directly above the flag in position 2 and then reflecting this image. Or you could move from the first to the second position by a single reflection. (Can you find the line of reflection?)

The third situation is more complicated. You could reflect the flag across a horizontal or vertical line, then rotate the image the appropriate amount, and then translate this image the appropriate amount. However, there *is* a glide reflection that enables us to move the flag from position 1 to position 2.

Let us now formally answer the three questions posed above.

1. Sometimes the order in which a composite transformation is done does matter. Recall the operations on whole numbers where the order does not matter for addition and multiplication but does for subtraction and division.

2. Imagine any figure on a plane. Now move that figure to any other position on the plane in any orientation. No matter how complicated the shape, and no matter how complicated the relationship between the two, we will always be able to map the figure from the first to the second position with one of the four congruence transformations in one step. This conclusion has been proved and this is called a theorem.

3. This theorem leads to the answer to the third question: These are *the* four congruence transformations. Mathematically, we use the term *sufficient*. That is, these four transformations are sufficient to describe the movement of any figure on a plane to any other place on the plane. Formally, we say that this set of congruence transformations is closed with respect to composition. That is, we can replace any combination of any two congruence transformations (that is, a composite transformation) with a single transformation. For example, a 47° clockwise rotation followed by a reflection across a specified line can be done in one step as one of the four congruence transformations. This idea is further explored in the exercises.

Operations on Shapes and Operations on Figures

It is beyond the scope of this book to go into more detail about other properties of operations on shapes, but I did want to give you a glimpse of the deeper mathematical structures here. At this point, let us summarize some important connections between operations on numbers and operations on shapes. In both cases, there are many important subsets. In both cases, there are many operations that we can perform. In both cases, we can make tables for those operations (Exploration 3.6, Exploration 9.8). In both cases, we can learn a lot by taking apart (decomposing) numbers and shapes in various ways. For example, we can take apart 45 as $40 + 5$ or as $9 \cdot 5$. In Chapter 8, you found that we can take apart any polygon into a number of triangles. Finally, just as we talked about properties of operations with numbers (such as closure, commutativity, associativity, identity, and inverse), we encounter the same properties when we do operations on shapes. Table 9.2 summarizes these connections between numbers and shapes, with examples for each.

TABLE 9.2

	Numbers	Shapes
Subsets	N, W, I, Q, R natural numbers, whole numbers, integers, rational numbers, real numbers	Triangles, quadrilaterals, regular polygons, concave, etc.
Operations	$+, -, \times, \div$, exponentiation, averaging	Translation, reflection, rotation, glide reflection, similarity transformation, topological transformation
Tables and patterns in tables	Addition and multiplication tables	Multiplication tables for symmetries of various shapes
Decompose/compose	Numbers can be decomposed additively and multiplicatively.	Any polygon can be decomposed into triangles.
Properties	Closure, commutativity, associativity, identity, inverse	Closure, commutativity, associativity, identity, inverse

Congruence

In Chapter 8 we discussed and worked with different concepts of congruence. At the most basic level, we can say that two figures are congruent if they coincide when we superimpose one on the other—in other words, when we move one figure (by translation, reflection, rotation, or glide reflection) so that its image coincides with the other figure. Now that we know that we can move any figure to any position through some combination of transformations (the theorem just cited), we can use this theorem to give a more general definition of congruence:

Two figures are **congruent** iff there is a translation, reflection, rotation, or glide reflection that maps one figure onto the other.

INVESTIGATION 9.6

Transformations and Art

Transformations have a lot to do with the patterns we see in tiles, quilts, and other art, especially Islamic art and the work of M. C. Escher. Look at the two patterns in Figure 9.20, which we have seen before. Do you see translations, reflections, and/or rotations in these figures? Write your thoughts and then read on. . . .

Parquet 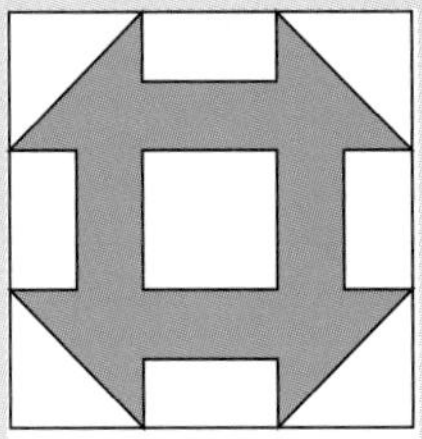   Churn Dash

FIGURE 9.20

DISCUSSION

In the parquet pattern, we can see that each light trapezoid can be mapped onto a dark trapezoid through a translation along a diagonal line. Each triangle can also be mapped onto another triangle through a 90-degree rotation. Furthermore, if we rotate the figure 90 degrees (clockwise or counterclockwise), the whole figure maps onto itself. If you don't see this, make a copy of the figure, superimpose the figure on the figure in the book, and rotate it.

From another perspective, you can see that the "pinwheel" part of the parquet pattern can be generated by taking only one-eighth of the pinwheel (that is, one trapezoid) and rotating it 45 degrees seven times (see Figure 9.21).

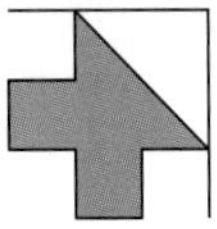

FIGURE 9.21

In the churn dash pattern, each of the white triangles can be mapped onto each of the other white triangles across a line of reflection. If we rotate the churn dash 90 degrees, with the center of rotation being the center of the figure, it maps onto itself. Furthermore, the churn dash can be generated by taking one-quarter of the figure and rotating it 90 degrees (clockwise or counterclockwise) three times (see Figure 9.22) or by reflecting it across a horizontal line and then reflecting the original figure and its image together across a vertical line.

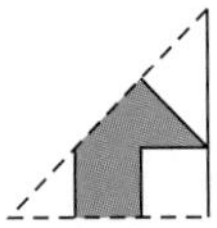

FIGURE 9.22

Actually, an even smaller piece of the figure (half of the piece of the churn dash in Figure 9.22) will generate the whole churn dash design (see Figure 9.23). Don't just take my word for it. Do you see how?. . .

FIGURE 9.23

If we reflect Figure 9.23 across the diagonal dotted line, we get Figure 9.22. One fun way to verify that Figure 9.23 will generate the churn dash is to tape two small mirrors together and place the two mirrors along the dotted lines of Figure 9.23. Like magic, you will see the whole churn dash! From one perspective, we can call this region the unit of the churn dash; from another perspective, mathematicians call this region a **fundamental domain** or **fundamental region**—that is, a region that under some combination of transformations will produce the whole pattern. This piece of the churn dash is the smallest part of the figure that will generate the whole figure. This looking for the smallest piece of a figure or a design that will generate the whole figure or design occurs in many parts of mathematics, not just with shapes; it is called finding the fundamental region of a pattern. I hope you can see that there might be some value in analyzing a pattern to find the fundamental region of the pattern (you might also think of this as the unit of the pattern). This notion of fundamental region is actually part of a bigger idea, called the "fundamental region of a group acting on a set," that occurs in several different branches of mathematics.

Summary

In this section, we have examined four fundamental ways in which we can transform a geometric figure: translation, reflection, rotation, and glide reflection. We have uncovered connections among these transformations. For example, performing two specific reflections is equivalent to performing a specific rotation. These connections have led to larger generalizations. We have discovered that there are many similarities between numbers and shapes: They have important subsets, we do operations on them, we can make tables for the operations, we can decompose numbers and shapes, and there are important properties that help us better understand our operations on numbers and shapes.

There is one more connection: All of these operations (addition, subtraction, multiplication, division, translation, reflection, and rotation) are functions. Do you see why? In each case, we can specify the operation by matching any input with a unique output.

These investigations of geometric transformations of two-dimensional figures in a plane can be extended. What if we examined transformations of geometric figures into different planes? What if we examined transformations of three-dimensional objects? These investigations occupy the attention of many mathematicians and have applications in other fields. Scientists in many fields are benefiting from a deeper mathematical analysis of three-dimensional tessellations. For example, the way in which atoms are packed helps to determine the properties of a compound.

EXERCISES 9.1

1. Below are a figure and a translation vector. Determine the translation. Explain at least two ways to describe the translation other than providing the vector.

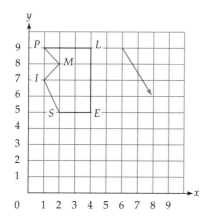

2. Below are a figure and a translation vector. Without using pencil or pen, determine whether the image overlaps the original figure or not. Explain your reasoning.

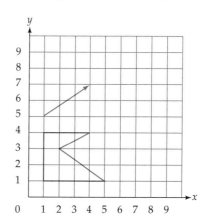

3. Find the image of the kite reflected across line *r*.

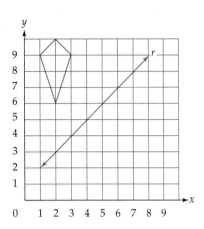

4. The figure below shows quadrilateral *ABCD* and a line of reflection. Determine the coordinates of *A'*, *B'*, *C'*, and *D'* using only reasoning (that is, without folding). Explain your reasoning.

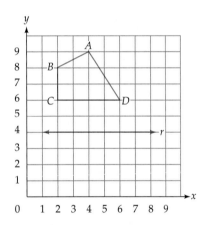

5. Below are a figure and its image. Without folding, sketch the approximate location of the line of reflection. Explain your reasoning.

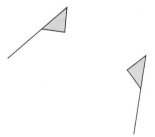

6. Without actually doing the reflection, predict the cases in which the following two images are reflections of each other across the line. Explain your reasoning.

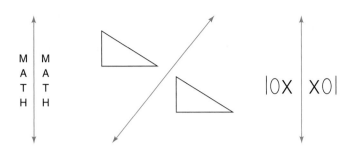

7. Determine the rotation image of triangle *MAT* if the triangle is rotated 90 degrees clockwise about point *A*.

8. Determine the rotation image of trapezoid *FARM* if the figure is rotated 90 degrees counterclockwise about point *F*.

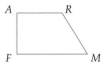

9. Determine the rotation image of the figure below if it is rotated 180 degrees clockwise about point *A*. Explain your reasoning.

10. The following two exercises appeared in *Geometry and Spatial Sense*, which was one of the books in the Grades K–6 Addenda Series published in the 1990s. Both were suggested for second-graders to help them explore the idea of reflection. The additional part for you in both cases is first to predict where the mirror line is and to explain why you predict that mirror line will produce the number of dots shown.

    (1) Place a mirror on or near the three circles shown below in such a way that you can see the numbers of circles shown:

      a. Six circles
      b. Five circles
      c. Four circles
      d. Three circles
      e. Two circles
      f. One circle

(2) For second-graders: Place a mirror on the Pattern Block design shown at the right to get the following figures.

a. b. c. d.

Source: Reprinted with permission from *Geometry and Spatial Sense,* copyright © 1993 by the National Council of Teachers of Mathematics.

11. *Navigating Through Geometry in Kindergarten–Grade 2* has some fascinating explorations for young children using either mirrors or Miras. One is called Monster Molly. The children explore how the placement of a mirror can produce different pictures. Molly is shown at the left. Where would you place the mirror to produce the image on the right?

Source: Reprinted with permission from *Navigating Through Geometry in Kindergarten–Grade 2,* copyright © by the National Council of Teachers of Mathematics.

12. **a.** Determine the coordinates of the vertices of the rotation image of the following trapezoid *ABCD* if it is rotated 90 degrees clockwise about vertex *A*. Explain your reasoning.

   b. Determine the coordinates of the vertices of the rotation image of the following trapezoid *ABCD* if it is rotated 90 degrees clockwise about point *X*. Explain your reasoning.

   c. Determine the coordinates of the vertices of the rotation image of the following trapezoid *ABCD* if it is rotated 90 degrees clockwise about point *Y*. Explain your reasoning.

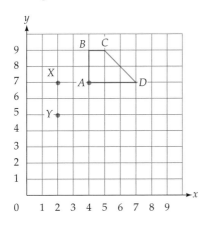

13. The diagram shows two triangles, *CAT* and *DOG*. It appears that they are translations of each other. Prove that they are congruent.

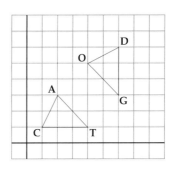

14. Who says geometry isn't practical? Describe where to aim the ball (white circle) to get into the hole (black circle) in a miniature golf course.

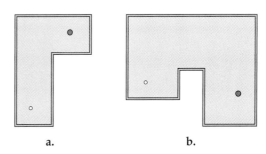

a. b.

15. Describe how to get the shape from position A to position B in as few moves as possible, using translations, rotations, reflections, and/or glide reflections.

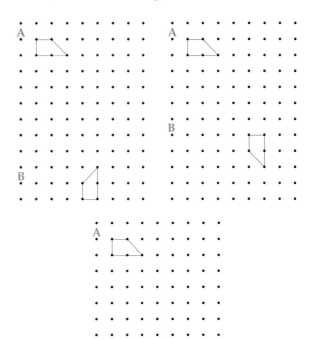

Exercises 16 and 17 involve various composite transformations. In each case:

a. *First predict the relative location and sketch the image of the figure after the transformations. Explain your reasoning.*

b. *Perform the two transformations. If your prediction was correct, great. If not, explain the error in your reasoning. Explain also your plans for "correcting" that error.*

c. *Predict whether the doubly transformed image will be the same if you do the transformations in the opposite order. Then do the two transformations. If your prediction was correct, great. If not, explain the error in your reasoning. Explain also your plans for "correcting" that error.*

16. First transformation: translation ⟶

Second transformation: reflect the figure across line *l*.

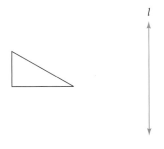

17.

First transformation: translation

Second transformation: reflection across line *l*

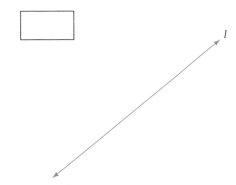

18. Under what circumstances will a glide reflection (translation then reflection) be commutative? That is, when will the translation followed by a reflection produce an image that coincides with the reflection followed by the translation? Explain your reasoning.

19. a. Determine the location of the image of the trapezoid after it is reflected across line *l* and its reflection image is then reflected across line *m*.

b. Predict whether the reflection of the trapezoid across line *m* and then across line *l* will give you the same image. Explain your reasoning.

c. Test your conjecture, and correct it if necessary.

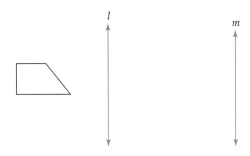

20. a. Determine the location of the image of the rectangle below after it is reflected across line *l* and its reflection image is then reflected across line *m*.

b. Predict whether the reflection of the rectangle across line *m* and then across line *l* will give you the same image. Explain your reasoning.

c. Test your conjecture, and correct it if necessary.

21. a. Determine the location of the image of the trapezoid after it is reflected across line *l* and its reflection image is then reflected across line *m*.

b. Predict whether the reflection of the trapezoid across line *m* and then across line *l* will give you the same image. Explain your reasoning.

c. Test your conjecture, and correct it if necessary.

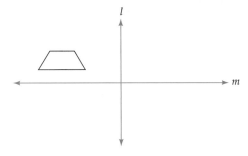

22. a. Rotate △*DIG* 90° clockwise about point *D*. Then reflect the rotated image across line *l*.

b. Is there another transformation or composition mapping that will produce the same effect?

23. Consider parallelogram STRU, whose diagonals intersect at M.
   a. Describe a transformation that would map $\overline{ST}$ onto $\overline{RU}$.
   b. Describe a transformation that would map ∡SRU onto ∡TSR.
   c. If we reflect $\overline{UM}$ across line $\overleftrightarrow{SR}$, will the image be $\overline{MT}$? If so, explain why. If not, explain why not.

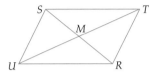

24. Consider isosceles triangle PRO, where M is the midpoint of $\overline{PO}$.
   a. Describe a transformation that would map $\overline{PR}$ onto $\overline{OR}$.
   b. What transformation would map △PMR onto △OMR?

25. The figure below shows triangle CAT, which has been reflected across line m.
   a. What relationships do you see between the coordinates of the vertices in △CAT and those of its reflection image △C'A'T'?
   b. What generalizations can you make from this exercise? You may want to gather more evidence on your own—that is, to make and reflect different figures.

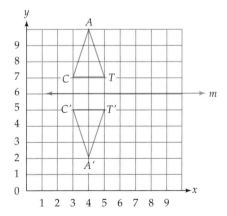

26. Look back to Figure 9.1. Describe any translations, reflections, rotations, and glide reflections that will move part of the figure on top of another part of the figure. For translations, supply a vector or other means of communicating the translation; for reflections, describe the line(s) of reflection; for rotations, describe the angle of rotation.

27. Follow the directions in Exercise 26 for the quilt patterns below.

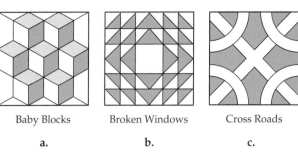

Baby Blocks   Broken Windows   Cross Roads
a.   b.   c.

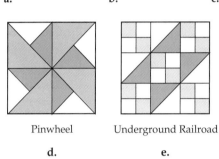

Pinwheel   Underground Railroad
d.   e.

28. Follow the directions in Exercise 26 for the Islamic designs below.

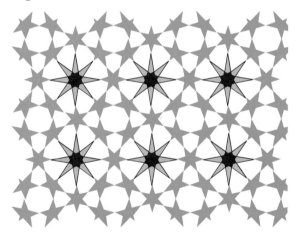

Source: From *Introduction to Tessellations* by Dale Seymour and Jill Britton. © 1989 by Dale Seymour Publications, an imprint of Pearson Learning. Used by permission.

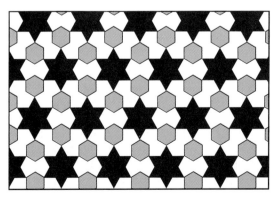

Source: From *Introduction to Tessellations* by Dale Seymour and Jill Britton. © 1989 by Dale Seymour Publications, an imprint of Pearson Learning. Used by permission.

29. Make the quilt blocks below using a computer software program. Describe how you used transformations to do so. In each case, you do not have to draw the whole block. That is, you can make a part of the block and then finish the block by translating, rotating, and/or reflecting part(s) of the pattern.

   a. Baby Blocks (see Exercise 27)

   b. Broken Windows (see Exercise 27)

   c. Pinwheel (see Exercise 27)

   d.
   Snail's Trail

   e.
   Does and Darts

   f.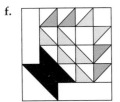
   May Basket

30. Earlier in the book, we discussed palindromes; for example, 1331 is a palindrome. There are word palindromes too. Some close friends of mine, math educators too, named their children HANNAH and AVIVA. We can make a new subset of palindromes, called reflection palindromes, that are legible when reflected. For example, if we reflect 1331 across a vertical line, it will read 1331, and if we reflect 1331 across a horizontal line, it will read 1331. But if we reflect AVIVA across a vertical line, we still have AVIVA. Explore different palindromes and reflections. Describe the characteristics of palindromes (both number and letter) that will be legible after reflections across horizontal lines, those that will be legible after reflections across vertical lines, and those that will be legible after reflections across either line.

31. a. If we reflect the angle formed by 3 o'clock through a vertical line through the center of the clock, what will be the time?

   b. If we rotate the angle formed by 3 o'clock 90 degrees clockwise, what will be the time?

   c. Make up and solve a similar problem.

32. The table is a base 10 multiplication table in which only the ones digit is given. For example, because 8 times 2 = 16, only the 6 is shown.

1	1	2	3	4	5	6	7	8	9	10
1	1	2	3	4	5	6	7	8	9	0
2	2	4	6	8	0	2	4	6	8	0
3	3	6	9	2	5	8	1	4	7	0
4	4	8	2	6	0	4	8	2	6	0
5	5	0	5	0	5	0	5	0	5	0
6	6	2	8	4	0	6	2	8	4	0
7	7	4	1	8	5	2	9	6	3	0
8	8	6	4	2	0	8	6	4	2	0
9	9	8	7	6	5	4	3	2	1	0
10	0	0	0	0	0	0	0	0	0	0

   a. Describe translations, reflections, and rotations within this table. For example, the first half of the 2s row (2, 4, 6, 8, 0) can be translated (5 boxes horizontally). Without transformation language, we would say the first half of the 2s row is the same as the second half of the 2s row. For the purposes of this exercise, consider only the location of the digit, not its spatial orientation.

   b. Make a multiplication table for base 12 in which only the ones digit of the product is given. Describe the transformations that can map one row or column onto another row or column. What similarities and differences do you see between the two tables?

33. The object of the game below is to move the arrow from the top-left corner to the bottom-right corner, using transformations. The only restriction is that you cannot go back to a square once occupied.

   a. Describe how to move the arrow from the top-left corner to the bottom-right corner in as few moves as possible.

   b. Describe how to move the arrow from the top-left corner to the bottom-right corner in as few moves as possible, with the additional restriction that each move has to be to a square that has at least one point in common with the square in which the figure resides.

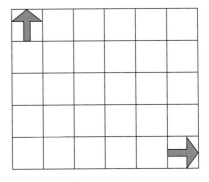

SECTION 9.2 SYMMETRY AND TESSELLATIONS

WHAT DO YOU THINK?

- Does a figure have to be closed to have reflection symmetry? Why or why not?
- Does symmetry have practical applications?

"Symmetry, as wide or as narrow as you may define its meaning, is one idea by which humanity through the ages has tried to comprehend and create order, beauty, and perfection."[1]

Our discussion of symmetry will evolve from our examination of symmetries of shapes into the examination of symmetries of patterns, which are more complicated but also more interesting to most people. Before presenting a definition of symmetry, let us explore this idea, which is not new to you.

Symmetry pervades our lives, and many people find symmetric figures appealing. We see symmetry in both natural objects and human-made objects—snowflakes, starfish, flags, logos, quilts, and many kinds of art, to name but a few. Look at Figure 9.24. Each of its elements is a well-known symbol or figure. For each of them, write what it is about that figure that is visually appealing or that attracts your attention. Describe the symmetry within each figure. Then read on. . . .

CLASSROOM CONNECTION

Snowflake Bentley is the story of William Bentley, born in 1865, who was fascinated by snowflakes and who worked for many years to develop a way to photograph actual snowflakes. In the process, he learned about the many variables that affect the way that snow crystal branches grow. His book *Snow Crystals*, which has over 2000 photographs of snowflakes and other snow crystals, is widely available and is fascinating to children and adults alike.

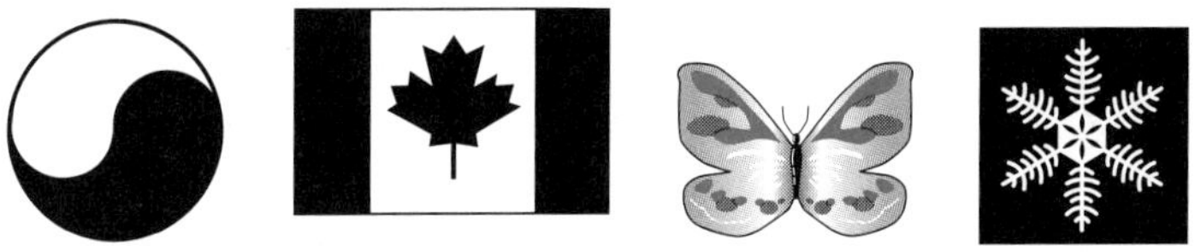

Chinese Yin - Yang "Unity of opposites"   Canadian flag   Butterfly   Snowflake

FIGURE 9.24

I think it is very possible that the first mathematical acts of human beings had to do not with numbers but with symmetry. As we saw in Chapter 2, many cultures survived even into the twentieth century without numbers or with very primitive number systems. However, the archaeological records of virtually every culture show symmetry in some of that culture's artifacts: its baskets, its pottery, the artifacts created for celebrations and rituals. I can think of few religious symbols that do not have symmetries within them.

Symmetry also abounds at the microscopic level. For example, many chemical molecules are symmetric, as are many viruses. In fact, the most common shape that viruses take is the icosahedron (remember Chapter 8?); it characterizes the viruses that cause herpes, chicken pox, and warts.

Symmetry abounds at the macroscopic level as well: Volcanoes are cone-shaped, stars are spherical, honeycombs are hexagonal, and galaxies tend to be spiral or elliptical; all of these shapes are distinguished by their symmetry. "Any understanding of nature must include an understanding of these patterns."[2] For example, the hexagonal shape of the honeycomb is the solution to the problem of how to get the maximum storage capacity from the minimum amount of building materials!

---

[1]Hermann Weyl, *Symmetry* (Princeton, NJ: Princeton University Press, 1952), p. 5.
[2]Ian Stewart, *Nature's Numbers* (New York: HarperCollins, 1995), p. 83. Excerpt from *Nature's Numbers* by Ian Stewart. Copyright © 1995 by Ian Stewart. Basic Books/Perseus Books Group.

Symmetry also adds to our understanding of patterns made by humans. For example, understanding symmetry can make it easier to make many very exotic-looking prints on skirts and quilts and can also increase our appreciation of their beauty (as we see the multiple symmetries and transformation relationships of various shapes within the design).

Let us then investigate this aspect of mathematics more deeply, to better understand what we mean by the term *symmetry* and what kinds of symmetries there are. We will investigate four kinds of symmetry: translation, reflection, rotation, and glide reflection.

Let us begin with a definition of **symmetry**: a transformation that places the object directly on top of itself.

Reflection and Rotation Symmetry

On the basis of our work with reflections in the previous section, what do you think reflection symmetry is? How would you explain reflection symmetry in your own words? Can you create a figure that you think will have reflection symmetry? Remember that when the learner is actively connecting what he or she already knows to the new ideas, the knowledge is much more likely to be owned rather than rented. Therefore, really try to answer these questions before reading on. . . .

A figure has *reflection symmetry* if there is a line, called the **line of symmetry**, that can be drawn through the figure such that when we fold the paper on the line, the part of the figure on one side will lie directly on top of the other part.

Now let us take a look at rotation symmetry. Thinking of the definition of reflection symmetry and what you know about a rotation transformation, how would you describe rotation symmetry? Can you create a figure that you think will have rotation symmetry? Try to do so before reading on. . . .

Imagine tracing a figure and putting the traced image directly on top of the figure. Informally, we say that the figure has *rotation symmetry* if we can pick up the tracing, turn it by some amount less than a full turn, and place it down so that it lies directly on the original figure.

INVESTIGATION 9.7

Reflection and Rotation Symmetry in Triangles

Many of the geometric shapes we investigated in the previous chapter have reflection and/or rotation symmetry. Describe the reflection and rotation symmetries you see in the equilateral triangle, the isosceles triangle, and the scalene triangles below. Then read on. . . .

DISCUSSION

We find that the equilateral triangle has three lines of symmetry, the isosceles triangle has one line, and the scalene triangle has none.

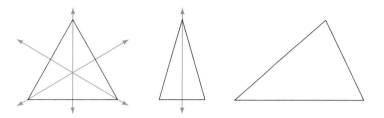

With respect to rotation symmetry, only the equilateral triangle has rotation symmetry. How would you describe that symmetry? Think before reading on. . . .

Finding the Amount of Rotation in Rotation Symmetry

■ *Language* ■

Some readers may be wondering why I didn't say that the equilateral triangle has $\frac{1}{3}$-turn rotation symmetry. Actually, I could have said so, but the word *rotation* is unnecessary.

Some people think of rotation symmetry in terms of how much we have to turn the figure in order for its image to lie directly on the figure. If we turn the equilateral triangle 1/3 of a whole turn (the center of rotation being the center of the triangle), the image will fit on top of the triangle. If we turn it another 1/3 turn, the image will fit on top of the triangle again. If we turn it another 1/3 turn, then we are back to its original position. Thus we can say that the equilateral triangle has $\frac{1}{3}$-turn symmetry.

Other people prefer to specify rotation symmetry by the number of degrees needed to make the rotated image fit on top of the figure. In this case, if we rotate the equilateral triangle 120 degrees, the image will fit on top of the figure. If we rotate the equilateral triangle another 120 degrees, the image will fit on top of the figure again. If we rotate the equilateral triangle another 120 degrees, the figure will be back to its original position.

Some of my students are very puzzled at this point and ask, "How did you figure out that it is 120 degrees?" Did you also wonder this? Could you explain why, as though to a ten-year-old (who will probably ask you the question someday)? If not, think before reading on. . . .

Other students find the following illustration useful. The triangle at the left in the figure below is in its original position, with the vertex angles labeled 1, 2, and 3, respectively. Imagine turning the triangle clockwise until the image is in the same position as the original triangle. The second figure represents the new position. Rotate the triangle again until the image is in the same position as the original triangle. The third figure represents the new position. Rotate the triangle again until the image is in the same position as the original triangle. Because it took three rotations to get back to the original position, each rotation has to equal 360/3 = 120 degrees.

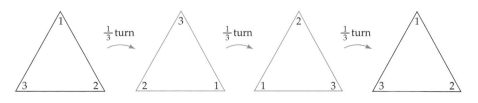

INVESTIGATION 9.8

Reflection and Rotation Symmetry in Quadrilaterals

What about quadrilaterals? Determine the reflection and rotation symmetry of the kite, parallelogram, rhombus, rectangle, and square. I encourage you to use whatever means you need in order to understand this investigation thoroughly. You might trace the figures on a blank sheet of paper, on tracing paper, or on an overhead transparency. Work on this before reading on. . . .

DISCUSSION

Kites have one line of symmetry, parallelograms have no lines of symmetry, rhombi have two, rectangles have two, and squares have four. The most common mistake my students make is to think that parallelograms have reflection symmetry. If you got this one wrong, make several parallelograms on a blank sheet of paper (tracing paper is even better), fold the parallelogram on either diagonal, and hold the paper up to the light. You will clearly see that the parallelogram does not fold onto itself in either case.

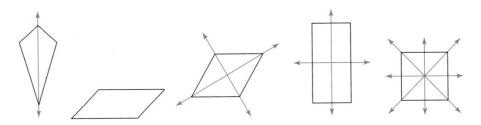

With respect to rotation symmetry, a kite has no rotation symmetry, a parallelogram has $\frac{1}{2}$-turn (or 180-degree rotation) symmetry, a rhombus has $\frac{1}{2}$-turn (or 180-degree rotation) symmetry, a rectangle has $\frac{1}{2}$-turn (or 180-degree rotation) symmetry, and a square has $\frac{1}{4}$-turn (or 90-degree) rotation symmetry.

As you may have already observed, many figures have 180-degree rotation symmetry but no other rotation symmetry. A figure has 180-degree rotation symmetry if it looks the same when turned halfway around. Because this particular kind of rotation symmetry is so common, it has its own special name: **point symmetry**. These three terms are synonymous: $\frac{1}{2}$-turn symmetry, 180-degree rotation symmetry, and point symmetry.

One last note on rotation symmetry. When I ask my students to tell me whether or not a figure has rotation symmetry, many of them will say that a figure has 360-degree rotation symmetry. What does it mean to say that a figure has 360-degree rotation symmetry? Think and read on. . . .

To say that a figure has 360-degree rotation symmetry means that if we turn it completely around, it will look the same. However, this is true for *any* figure.

Because any figure has 360-degree rotation symmetry, we do not use this term when describing the rotation symmetries of a figure.

INVESTIGATION 9.9

Reflection and Rotation Symmetry in Other Figures

The accompanying cartoon is amusing *and* presents us with a great problem. Assuming that the vultures are all congruent and that each vulture grabbed one piece of pizza right in the middle, describe the rotation and reflection symmetries in this idealized diagram. Then read on. . . .

■ Mathematics ■

It is important to note that when we say that a rectangle has two lines of symmetry, we mean that subset of rectangles that are not squares. Technically, squares are rectangles, so technically, some rectangles (those that are also squares) have four lines of symmetry.

In Section 8.2, we created a family tree for quadrilaterals. By definition, each descendant had the same properties as its immediate ancestor plus at least one new characteristic, thus creating a new class of shapes. When we looked at properties of the diagonals of quadrilaterals, we found that the same family tree applied. That is, every descendant had at least the same properties for its diagonals as its ancestor. As you can see from this exercise, every quadrilateral has at least as many reflection and rotation symmetries as any quadrilateral above it in the "family tree."

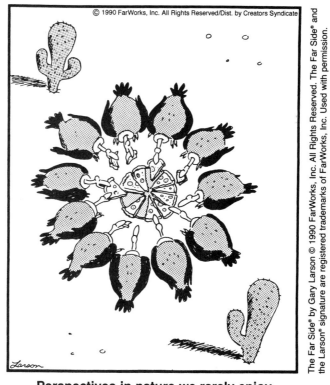

Perspectives in nature we rarely enjoy

DISCUSSION

This problem gives many students fits. Even when they count the number of vultures, the number 11 is just not a number that makes it easy to describe the symmetries. This is a good spot to try to sell you on a problem-solving strategy that is underutilized by students: making a simpler problem and using deductive or inductive reasoning to build up to the complexity of the given problem. Below are regular polygons with 3, 4, 5, 6, 7, and 8 sides. First draw the lines of symmetry for each figure. What do you see? . . .

Many students realize that the lines of symmetry in the figures with 3, 5, and 7 sides (that is, with an odd number of sides) are somehow "different" from those in the figures with an even number of sides. How would you describe this difference? . . . .

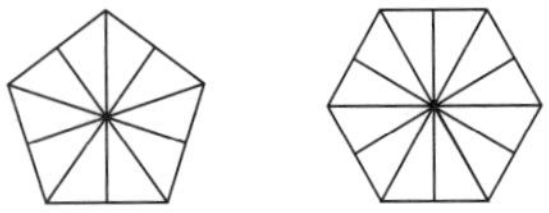

FIGURE 9.25

When the polygon has an odd number of sides, the lines of symmetry all connect a vertex of the polygon to the middle of the opposite side. When the polygon has an even number of sides, half of the lines of symmetry connect two opposite vertices and half of the lines of symmetry connect the midpoints of two opposite sides (see Figure 9.25). However, in all cases, the number of lines of symmetry is equal to the number of sides in the figure. Thus it seems reasonable to conclude that the original figure has 11 lines of symmetry, and you should be able to draw all 11 lines. Now, what about the rotation symmetry?

If you connect the discussions in the previous investigations to this problem, you should be able to deduce that because there are 11 vultures, this figure has 11-fold symmetry, or that the angle of rotation is 360/11.

INVESTIGATION 9.10 Letters of the Alphabet and Symmetry

Examine the letters of the alphabet. Which letters have rotation symmetry? What kind(s) of rotation symmetry? Which letters have reflection symmetry? What kind(s) of reflection symmetry? Work on this and then read on. . . .

A B C D E F G H I J K L M N O P Q R S T U V W X Y Z

DISCUSSION

When we examine the letters in terms of reflection symmetry, we find that there are three kinds of lines of reflection: vertical, horizontal, and diagonal.

The following letters have **vertical line symmetry**: A, H, I, M, O, T, U, V, W, X, Y. I think it is interesting that over half of the letters with vertical line symmetry occur at the end of the alphabet.

The following letters have **horizontal line symmetry**: B, C, D, E, H, I, O, X. Here we have the opposite phenomenon: four consecutive letters near the beginning of the alphabet.

Three letters have **diagonal line symmetry** (depending on how they are drawn): O, Q, and X.

The following letters have point symmetry: H, I, N, O, S, X, Z.

Note: The symmetry of some letters depends on how they are drawn. For example, B and K do not always have horizontal line symmetry and Q has diagonal line symmetry only if it is made as a circle (instead of oval-shaped) and if the line segment is on a diameter of the circle.

As we noted in Chapter 1, representation is a new process standard. Above, I represented the solution to the question in a list form. In Figure 9.26 are two Venn diagrams. The first shows just the letters that have line symmetry, and the second shows all the letters. Do these diagrams change or add to your understanding of which letters have what kind of symmetry?

Section 9.2 / Symmetry and Tessellations   **609**

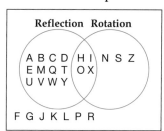

FIGURE 9.26

What Are Patterns?

We find patterns in our clothing, on quilts, on ceilings, on floors, on walls, and on the borders of many objects—in short, all over the place. To define the word *pattern* is very difficult, so let us take some time to develop this idea. . . .

INVESTIGATION 9.11 — Patterns

One of the reasons I became a mathematician is patterns, both numerical and visual. Patterns have been a recurring theme in this book. Looking for patterns is a tool for solving problems, and we have seen patterns in addition and multiplication tables, as well as other numerical patterns. When we talk about geometry, we almost inevitably encounter the word *pattern*. Let us explore this idea of pattern, partly because it's fun and most children absolutely love playing with patterns, but also because the exploration will deepen your understanding of geometry and its influence on so many aspects of our lives.

A. Below are three "brick" patterns. Look at them and think what they all have in common, and then try to come up with a definition of pattern. Then read on. . . .

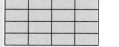      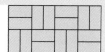

DISCUSSION

There are three attributes in all patterns. First, there is a unit, and, in some cases, multiple units. Second, there is a repetition of that unit, and the unit is repeated in some organized way.

In the first case, the unit is a rectangle, and it is repeated (translated) in a simple way. In the second case, the unit is still a rectangle; in this case, each row is staggered (translated) slightly—this pattern is used in brick walls and buildings because it is stronger. In the third case, we *can* see one brick as the unit;

however, most people will think of the unit as two bricks side by side. The principle of organization here is rotation. Each pair of bricks is laid side by side, but rotated 90 degrees.

B. Now look at the three patterns below. In each case, identify the unit and describe the way in which the unit is repeated to make the pattern. It may help to think of writing directions for someone who doesn't understand this idea of unit of a pattern. . . .

Sources: Top left—From *Geometry from Africa,* by Paul Gerdes. Copyright © 1999. Reprinted by permission of the Mathematical Association of America. Right—From *Introduction to Tessellations* by Dale Seymour and Jill Britton. © 1989 by Dale Seymour Publications, an imprint of Pearson Learning. Used by permission. Bottom left—a placemat in a restaurant.

DISCUSSION

The case at the top left is an example of an African (Kente) cloth pattern. The unit pattern is not so simple here. From one perspective, the unit pattern is the staircase shown at the right. If you put that staircase at the top left corner of the pattern, and then rotate it 90 degrees counterclockwise and move it to the right, you get the next colored staircase. If you rotate this staircase 90 degrees and move it to the bottom, you get the third staircase. Repeat this one more time and you get the last staircase. If you then draw two horizontal lines to bound the pattern, the white staircases are literally formed by the space. Alternatively, you can think of a pair of staircases (shown at the right) as the unit pattern. In this case, the whole pattern is still formed by rotating the pattern 90 degrees counterclockwise and moving it to the right, then down, and then to the left.

In the case at the right, the pattern consists of the "checkerboard" made by the four smaller parallelograms (two white and two dark) and the two longer parallelograms, one below and one to the right of the checkerboard. The organization is to slide this figure diagonally.

In the third case at the bottom left, the unit is the S-shaped design shown at the right. When the pattern goes around a corner, the S shape needs to be modified a bit. It will be left as an exercise to reproduce this pattern. Once you see the unit, then you can see that the pattern is simply made by sliding this shape.

Mathematics

The careful reader may have realized that we also speak of numerical patterns. Recall our work in Chapters 1 and 2. Do those numerical patterns have repetition? Is there a unit for those patterns? Is there some organization of those numbers? Or do we have to have two definitions—one for shape patterns and one for number patterns? What do you think?

This analysis of patterns works fine for certain kinds of patterns, such as checkerboard, herringbone, and quilt patterns. However, this definition always provokes a response in some of my students who consider the definition very carefully. They ask questions like this: "What about patterns you see on a sand dune, or patterns of light on the surface of a lake, or patterns you can see when you ride a canoe on a pond and there is no wind, or the patterns of ripples of sand along the seashore? Aren't these also patterns?" What a wonderful question!! Although these examples don't fit this crisp characterization we have described, there is some sense of repetition and organization in each case. The last thing I want to do is perpetuate the belief many students bring to mathematics class that mathematics is fundamentally "different" from the real world. Thus, even though we will focus here on that subset of patterns where we can find the unit and analyze the organization, I don't want to exclude this larger sense of patterns.

Patterns in Nature

Patterns appear everywhere in the natural world. These two examples are taken from a fascinating article by Marvin Harrell and Linda Fosnaugh in the May 1997 issue of *Mathematics Teaching in the Middle School* (pp. 380–389).

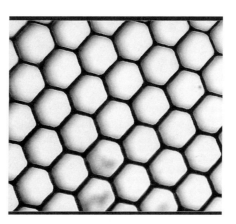
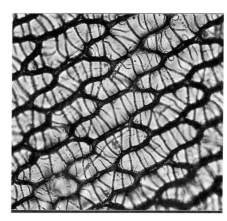

Bumblebee's eye at over 100 magnification   Sphagnum moss at over 200 magnification

Mathematics

Just as our crisp characterization of the concept of pattern left out a lot of interesting patterns and was restricted to "perfect" patterns, so too did traditional geometry "leave out" a lot of shapes that aren't perfect or nearly perfect—shapes like clouds, smoke, ocean currents, and weather patterns. A whole new field of geometry has been born since I graduated from college in the 1970s. This field, called fractal geometry, came into being in part because classical Euclidean geometry is very limited in its ability to help us to understand these nonperfect shapes.

We are almost ready to examine symmetries of patterns. However, before we do so, we need to revisit a concept that was presented in Chapter 8—infinity. Recall our discussion in Chapter 8 of the concept of a plane. That is, a mathematical plane is a flat surface that extends infinitely in all directions. Obviously, all examples of planes in a book must be finite, and thus any diagram of a plane is actually a piece (technically, a subset) of a plane. Similarly, mathematical patterns are theoretically infinite. Thus, when you see patterns below and are asked to translate, rotate, and reflect them, you need to realize that, theoretically, there are no edges to the pattern.

Symmetries and Patterns

We have now laid the groundwork for understanding and describing symmetries of patterns. Recall our definition of symmetry: a transformation that places

▪ Language ▪

Many people use the words *design* and *pattern* interchangeably. I have learned otherwise from friends who do design. A nice definition of *design* is "a set of shapes, generally contained in a larger shape where the shapes are arranged for a specific purpose—functional or aesthetic." Many designs have patterns in the design, but the design itself is generally not a pattern in the sense in which we use the term here.

the object directly on top of itself. Thus a pattern has **translation symmetry** if we can lift the pattern, translate it some distance and in some direction, and set it down on top of itself. A pattern has **reflection symmetry** if we can find a line such that if we reflect the pattern across that line, the pattern will fit directly on top of itself. A pattern has **rotation symmetry** if we can find a point such that if we turn the pattern around that point, the pattern will fit directly on top of itself. A pattern has **glide reflection symmetry** if we can find a translation and a line such that if we translate the pattern some distance and in some direction, and then reflect the pattern across a line that is parallel to the translation vector, the pattern will fit directly on top of itself.

Let us now focus on the subset of patterns whose characteristics can be clearly described. When an artist or a manufacturer wants to create a new design, the possibilities are virtually limitless. However, as soon as we replace *design* by *pattern*, the designers are limited by the mathematical laws that underlie the formation of patterns. Let us see what this means with one kind of pattern.

Now that we have the idea of translation symmetry, we can give a definition of **pattern**: a figure with translation symmetry. Do you see the connection between this definition and the earlier discussion of characteristics of patterns: having a unit that is repeated in an organized manner?

INVESTIGATION 9.12

Symmetries of Strip Patterns

We will begin by looking at the mathematically simplest kinds of patterns, which are often called border or strip patterns because the motif (unit) is repeated in only one direction. With your understanding of translations, reflections, rotations, and glide reflections, determine the translation, reflection, rotation, and glide reflection symmetry for each strip pattern below.

A. What kind(s) of symmetry does the following strip pattern have? Think before reading on. . . .

DISCUSSION

This pattern has translation symmetry because we can translate this pattern so that we can place it on top of itself. The translation vector is parallel to the bottom of the page, and the length of the vector is equal to the distance between any two shapes.

One way to verify this symmetry is to trace the flags on a blank sheet of paper or on an overhead transparency. You can then see that you can move the pattern over 1 unit and it fits onto itself. Remember that, just as the plane is considered to be infinite, extending in all directions, so too are patterns considered to be infinite. Thus, although my figure and your tracing will have a left-most and a right-most flag, the "pattern" is considered to have no ends.

B. What kind(s) of symmetry does the following strip pattern have? Think before reading on. . . .

DISCUSSION

This pattern has translation symmetry. This pattern has two vertical lines of symmetry. There are actually two "different" sets of vertical lines that will enable you to put the figure onto itself if you trace this pattern. One of the lines has the flags facing the line, whereas the other line has two flags pointing away from the lines. If you don't see the vertical lines, turn to page 644, where the symmetries for these strips are shown.

Although learning why there are two "different" sets of vertical lines of symmetry for this pattern is not one of the big ideas of this section, coming to understand and appreciate mathematical language is one of the big ideas of the book. Thus, let us take just a few moments to pursue why we say there are two different vertical lines of symmetry. Actually, mathematicians say not that the two lines of symmetry are different but rather that they are inequivalent. Two lines of symmetry are equivalent if a symmetry of the pattern can move one reflection line to the other. For example, if you draw one vertical reflection line on the pattern above and then trace the figure and draw another vertical reflection line on the top sheet, these two lines of symmetry are equivalent if you can move the figure so that when the one mirror line is placed on top of the other, the figure fits on top of itself.

C. What kind(s) of symmetry does each strip pattern have? Think before reading on. . . .

D.

E.

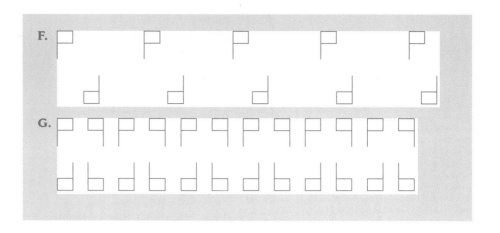

DISCUSSION

C. This pattern has translation symmetry. This pattern has horizontal line symmetry. If you trace this pattern and reflect the pattern across a horizontal line, it will fit on top of itself.

D. This pattern has translation symmetry. This pattern has both vertical and horizontal line symmetry. This pattern has 180-degree rotation symmetry.

E. This pattern has translation symmetry. This pattern has glide symmetry. If you reflect the pattern across a horizontal line and then move the pattern in the horizontal direction, it will fit on top of itself.

F. This pattern has translation symmetry. This pattern has $\frac{1}{2}$-turn symmetry or 180-degree rotation symmetry or point symmetry.

G. This pattern has translation symmetry. This pattern has vertical line symmetry and glide symmetry. If you reflect the pattern across a vertical line, it will fit on top of itself. If you reflect the pattern across a horizontal line and then move the pattern in the horizontal direction, it will fit on top of itself.

You can look on page 645 to see the actual lines of reflection, centers of rotation, and glide vectors.

The Seven Symmetries of Strip Patterns

I find it amazing that any strip or border pattern will fit into one of the seven symmetry types illustrated in A through G above. That is, just as the four congruence transformations are complete in that no more are needed to describe how to map any two congruent figures in a plane onto each other, these seven symmetry types are all that are needed to describe any border pattern.

Let us use a table and introduce some notation to increase the number of readers who understand this. First, you will notice that every one of the strips had translation symmetry. Therefore, we are technically looking at that subset of strip patterns that have translation symmetry. Actually, you will be hard pressed to find a quilt or building or other strip that does not have translational symmetry. As Ian Stewart put it, "Something in the human mind is attracted to symmetry."[3]

---

[3]*Nature's Numbers*, p. 73. Excerpt from *Nature's Numbers* by Ian Stewart. Copyright © 1995 by Ian Stewart. Basic Books/Perseus Books Group.

A few words about notation in the far right column of Table 9.3: We will use the letter m to indicate a line symmetry (m representing mirror), the letter g to represent a glide symmetry, the number 2 to indicate twofold turn symmetry, and the letter l to indicate a lack of line symmetry.

TABLE 9.3

Pattern	TYPE OF SYMMETRY				Notation to describe
	Vertical line	Horizontal line	$\frac{1}{2}$-turn	Glide	
A					ll
B	Yes				ml
C		Yes			lm
D	Yes	Yes	Yes		mm
E				Yes	lg
F			Yes		l2
G	Yes		Yes	Yes	mg

Some active readers look at this table and argue that there should be more than seven types of symmetry. For example, what about gm? What about m2? These are wonderful questions. It will be left as an exercise to explain why those kinds of symmetry are not found in mathematics textbooks.

Wallpaper Patterns

Let us examine another kind of planar pattern called wallpaper patterns, ones in which the motif (unit of repetition) can be repeated in two directions. It is important to note that the term *wallpaper pattern* refers to any pattern that can fill the plane if extended. Thus the three patterns that follow, which are often seen with bricks, are called wallpaper patterns by mathematicians.

Extending patterns in two dimensions leads to more complexity; consider, for example, all the wallpaper patterns that have been created—thousands, perhaps millions of different ones. However, when we classify them by the kinds of symmetry they possess, any wallpaper pattern falls into one of exactly 17 categories. I find that fascinating! It is a relatively recent discovery, though this issue puzzled mathematicians and scientists for centuries. Interestingly enough, the same 17 categories can be used to classify all crystals. Examining each of these 17 categories is beyond the scope of this book, but the interested reader can search for wallpaper patterns on the Web. You will find many descriptions.

INVESTIGATION

9.13

Analyzing Brick Patterns

A. What kinds of symmetries does this simple brick pattern have? Think before reading on. . . .

[Grid pattern figure]

DISCUSSION

It might not surprise you to find that this figure has lots of symmetry!

It has translation symmetry. In fact, the figure can be translated horizontally, vertically, diagonally down to the right, and diagonally up to the right. Thus this figure has four different lines of translation symmetry.

This figure has two vertical lines of symmetry. *Note:* Technically, you could say it has infinitely many vertical lines of symmetry. Recalling the discussion of inequivalence in the previous investigation, we say that this pattern has two inequivalent vertical lines of symmetry.

Similarly, the figure has two inequivalent lines of horizontal symmetry.

The pattern has $\frac{1}{2}$-turn rotation symmetry, and there are three inequivalent centers of rotation. That is, there are three different centers of rotation that cannot be placed on top of one another by any symmetry of the pattern. See page 646 to see the three centers of rotation. (The reader who finds the notion of inequivalent reflection lines and centers of rotation difficult to understand completely should not be discouraged: These are relatively advanced ideas.)

Finally, when looking for glide symmetry, we find that there are no glide lines that are not already mirror lines. Thus we say that this figure has no nontrivial glide symmetry.

B. What about this slightly more complex brick pattern? What symmetries does it have?

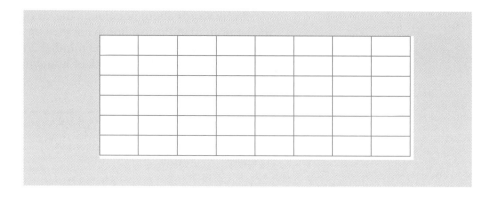

DISCUSSION

This pattern has translation symmetry. Like the previous pattern, this one can be translated horizontally, vertically, diagonally down to the right, and diagonally up to the right. Thus this figure has four different lines of translational symmetry. However, if you draw the translation vectors, you will find that they are not identical to the four in part A. You can check your thinking on this page.

This figure has one line of vertical symmetry.

The figure also has one line of horizontal symmetry.

The pattern has $\frac{1}{2}$-turn rotation symmetry, and there are three inequivalent centers of rotation. That is, there are three different kinds of points on which you can place your pen, rotate the figure 180 degrees, and have the pattern fit on top of itself.

Finally, this figure has one nontrivial glide symmetry. Did you find it? See page 646 to check your thinking.

C. Finally, what about this more complex brick pattern? What symmetries does it have?

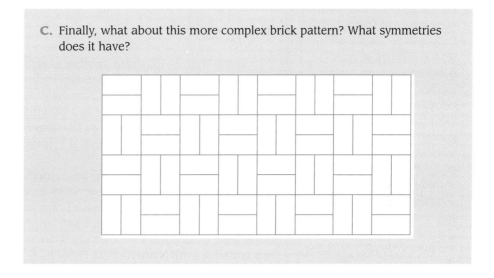

DISCUSSION

This pattern has translation symmetry. Like the two preceding patterns, this pattern can be translated diagonally down to the right and diagonally up to the right, although the four translation vectors are not identical to either of the four above.

This figure has one line of vertical symmetry.

Similarly, the figure has one line of horizontal symmetry.

The pattern has $\frac{1}{2}$-turn rotation symmetry, and there are three inequivalent centers of rotation.

Finally, this figure has two inequivalent glide lines. Can you specify the reflection and slide lines? See page 646 to check your thinking.

Symmetry Breaking

Our discussion of symmetry would be incomplete if we did not also talk about symmetry breaking. Although there is indeed "something in the human mind that is attracted to symmetry" and symmetry is one important factor in most people's sense of beauty, at the same time, perfect symmetry is repetitive and predictable. I will not attempt a rigorous definition of symmetry breaking but, rather, will give some examples and descriptions. From one perspective, symmetry breaking occurs when we take a pattern that has "lots" of symmetry and do something to reduce the symmetry. Consider Figure 9.27(a). This figure has four lines of symmetry and fourfold symmetry. If we add two colors [Figure 9.27(b)], we still have fourfold symmetry, but now we have no lines of symmetry. Thus we have broken the symmetry. One basic quilt design can be modified in many ways that break the symmetry. If we color in the center [Figure 9.27(c)], we now have only twofold symmetry and no lines of

■ Outside the Classroom ■

The notion that symmetry is a mathematical concept as well as an aesthetic one was brought home to me several years ago. I was asked to facilitate a four-day workshop for college professors who were invited to Dartmouth College to get a taste of several different courses designed to bridge mathematics and other disciplines—courses that would be more appealing as general education mathematics courses than the traditional offerings. In my workshop, there were 24 college professors: about half were Ph.D. mathematicians and half were art professors. My job was to keep them from killing each other. Seriously, part of my job was to help them recognize what they had in common. One of the many things they had in common was a love of patterns *and* a love of breaking patterns.

symmetry. Finally, we can redraw the lines in the middle to give a pinwheel effect. This design now has fourfold symmetry [Figure 9.27(d)].

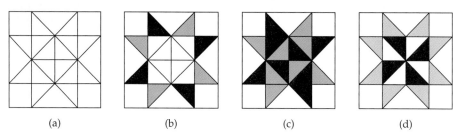

FIGURE 9.27

These figures illustrate another way of describing symmetry breaking: A symmetry is expected, but that expectation is not met. If you look closely at Oriental carpets, which often have very intricate designs and plenty of symmetry, you will see that there is a playfulness with symmetry that results in many intriguing patterns within the larger, more obvious patterns. "In art, too, it seems that the approximation of symmetry, rather than its precision, teases the mind as it pleases the eye."[4]

In nature, symmetry is always imperfect, just as you will find in the next chapter that measurements, by definition, are always approximations. However, just as a mathematical model of a problem will nicely represent the important or "big" ideas of a problem or phenomenon, while at the same time idealizing or simplifying other parts, so too do physicists, chemists, biologists, and other scientists often speak in terms of symmetries in their work as though the symmetries were perfect. Did you ever wonder why our faces are not perfectly symmetric but rather "almost" symmetric. At the left in Figure 9.28 is a photograph of Edgar Allen Poe. The middle figure shows what he would look like if you reflected his right side, and the figure at the right shows what he would look like if you reflected his left side.

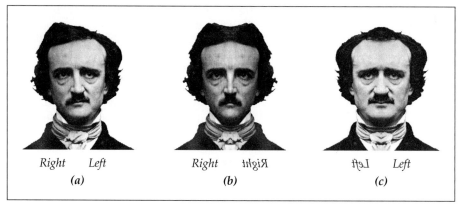

FIGURE 9.28

Symmetry for Three-Dimensional Objects

Thus far we have examined symmetry with two-dimensional figures. However, we see symmetry in many three-dimensional figures. Buildings often have symmetry; think of the Taj Mahal or the U.S. Capitol building. Everyday objects such as forks, nuts and bolts, and windows have symmetry. Many natural forms, including starfish, honeycombs, pyrite, and crystals, have symmetry.

---

[4]http://forum.swarthmore.edu/geometry/rugs/symmetry/breaking.html.

Let us briefly examine symmetry of three-dimensional objects. We will focus on reflection and rotation symmetry.

When we defined reflection symmetry with two-dimensional figures, we used a line. Therefore, it should make sense that when we define reflection symmetry with three-dimensional figures, we will use a plane. Because we have added one more dimension to the figure (two to three), we add one more dimension to the instrument of reflection (one to two).

If there is a plane that can be drawn through a three-dimensional figure so that one half of the figure is a mirror image of the other, then the figure has reflection symmetry. The plane is called the **plane of symmetry**.

Similarly, when we defined rotation symmetry with two-dimensional figures, we used a point (a one-dimensional object). We define rotation symmetry with three-dimensional figures using a line (a two-dimensional object).

If there is a line that can be drawn, around which a three-dimensional figure can be rotated so that it coincides with itself, then the figure has rotation symmetry. The line about which the figure is rotated is called the **axis of symmetry**.

Figure 9.29 shows two three-dimensional objects that have reflection and/or rotation symmetry. In each case, determine and try to describe the plane(s) of symmetry and the axis (or axes) of symmetry. Then read on. . . .

Right isosceles triangle prism   Nightstand

FIGURE 9.29

The prism has 180-degree rotation symmetry; the axis of symmetry is a line that is parallel to and midway between these bases and which intersects the vertex of the right dihedral angle. The prism has two planes of symmetry: a horizontal plane parallel to and midway between the bases and a vertical plane that contains the lines of symmetry of the two bases. The nightstand has one plane of symmetry, a vertical plane.

Tessellations

 9.9

For thousands of years, people have chosen to cover surfaces (for example, floors, walls, tables, beds, or roads) with patterns. We will use transformational geometry, including our knowledge of symmetries of patterns, to examine one kind of covering called tessellations. Look at the pictures in Figure 9.30, all of which are examples of tessellations. From these pictures, what do you think *tessellation* means? That is, a pattern that tessellates is one in which . . .

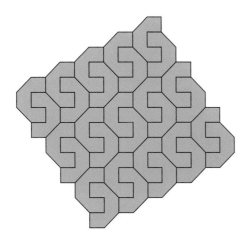

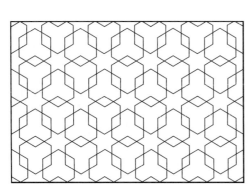

Figure 9.30

In preparing for this book, I read more than 10 authors' descriptions of tessellation. It was interesting to note that none of the definitions were identical, unlike the definitions of, say, isosceles triangle! However, all of the definitions had two elements in common: There can be no gaps, and there can be no overlap. We will say that a figure or a combination of figures **tessellates** the plane if a regular repetition of the figure or figures covers the plane so that there are no gaps and no overlapping of figures. Look at the four designs in Figure 9.30. Do you see that all of these designs qualify as tessellations? However, only certain geometric figures and combinations of figures will tessellate, and mathematics holds the clue to why.

One difficulty that people commonly encounter when first making sense of this definition has to do with the borders. When discussing tessellations, we will not be concerned with whether the borders are smooth or straight. For example, if we were covering a floor with square tiles, unless both the length and the width of the floor were equal to a multiple of the length of the tile, the

■ *Language* ■

The word *tessellate* comes from the Latin word *tessella*, which referred to small tiles that the Romans used for pavement and to decorate buildings.

Mathematics

Where else have you seen tessellations in this book? Figures 9.1 and 9.2 are tessellations, and the three brick designs in Investigation 9.13 are tessellations.

person laying the floor would have to cut the square to fit the edge of the room. Thus, when we say that a pattern tessellates, we are talking not about the edges of the design, but about the design itself.

We will investigate some basic ideas related to two-dimensional tessellations here. Exploration 9.9 carries these ideas further.

INVESTIGATION 9.14

Which Triangles Tessellate?

One starting point for investigating tessellations is to ask which figures will tessellate. We will begin with the triangle (see Figure 9.31). What triangles tessellate? Just a reminder that you will learn more if you really do think before reading on. . . .

Note your first thoughts as predictions. Then make copies of some triangles and test your ideas. One simple way to do so is to fold a piece of paper in half, then in half again, and then in half again. Now if you make a triangle and then cut the figure, you will have 8 copies of the triangle.

FIGURE 9.31

DISCUSSION

Many students are surprised to find that all triangles tessellate! Did you predict this? If not, make some copies of different triangles and confirm this fact for yourself. Then come back to the text. Now that you realize that all triangles tessellate, can you justify this—that is, explain why? Work on this question before reading on. . . .

Let us begin with a scalene triangle, shown in Figure 9.32(a). I have labeled the three angles of the triangle. If we rotate the triangle 180 degrees (or reflect it vertically and then horizontally), we can join the triangle and its image together [see Figure 9.32(b)]. The combining of the triangles creates a parallelogram. If we take this parallelogram and translate it, we now have four triangles [see Figure 9.32(c)]. Figure 9.33 shows that we can extend this pattern in all directions infinitely.

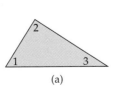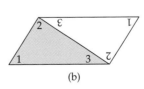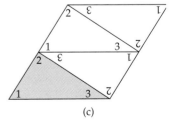

(a)   (b)   (c)

FIGURE 9.32

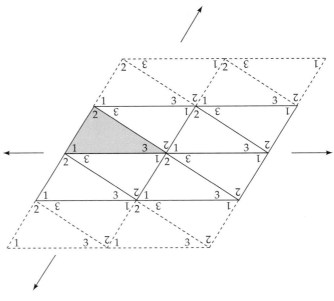

FIGURE 9.33

FIGURE 9.34

Our eyes tell us that it "looks as if" the triangle tessellates. We will now take some time to confirm what our eyes tell us. Again, this is not a formal geometry course, so we will give not a two-column proof but a more informal justification. Figure 9.34 repeats Figure 9.32(c), but with some vertices labeled to make communication easier. We can see that we have angles ∡1, ∡3, and ∡2 meeting at vertex B.

We know from Chapter 8 that the sum of the angles of any triangle is 180 degrees, and therefore ∡ABC = 180. However, this means that $\overleftrightarrow{AC}$ is a straight line. Looking back at Figure 9.33, we can see that there will be six angles at *any* vertex (∡1, ∡2, ∡3, ∡1, ∡2, and ∡3), and the sum of these six angles is 360. As you might be thinking, 360 is an important number; that is, one complete revolution about a point is 360 degrees. Thus we can repeat this triangle in such a way that there are no gaps or overlap, and so the triangle tessellates.

In a similar fashion, we can show that any quadrilateral also tessellates the plane. Does this surprise you? If so, make a number of different quadrilaterals to convince yourself that, indeed, *any* quadrilateral does tessellate. Justification of this will be left as an exercise.

CLASSROOM CONNECTION

Recall the discussion of the van Hiele levels in Chapter 8. Where is your level of understanding of the justification that all triangles tessellate?

INVESTIGATION

9.15

Which Regular Polygons Tessellate?

Another common starting point for understanding tessellations is to examine that subset of all polygons called regular polygons—that is, polygons in which all sides are the same length and all angles have the same measure. Which regular polygons will tessellate? Work on this before reading on. You might want to use Figure 9.35 as a template to make copies, or you might choose to apply your understanding of the properties of regular polygons. In either case, do some thinking on your own before reading on. . . .

FIGURE 9.35

DISCUSSION

It turns out that only three regular polygons tessellate: the equilateral triangle, the square, and the regular hexagon. (Note that an equilateral triangle is a regular triangle and that a square is a regular quadrilateral.) The reason has to do with the sum of the angles.

Whether or not you concluded that these three are the only regular polygons that tessellate, take a few minutes to see whether you can justify this conclusion—that is, explain why it is true. Then read on. . . .

If you recall from Chapter 8, the sum of the angles of any polygon = $(n - 2)180$, where $n =$ the number of sides in the polygon. From this formula, we can find the measure of each interior angle of regular polygons, shown in Table 9.4. Can you see how these numbers help us to understand why only three regular polygons tessellate? Think before reading on. . . .

TABLE 9.4

Figure	Sum of angles (degrees)	Measure of each interior angle (degrees)
Equilateral triangle	180	60
Square	360	90
Regular pentagon	540	108
Regular hexagon	720	120
Regular octagon	1080	135

Table 9.5 helps us to understand why the equilateral triangle, square, and regular hexagon tessellate, but no other regular polygons will. If we place three pentagons at a common vertex, the sum of those angles is 324. This is not quite 360; however, there is not enough room to place a fourth pentagon at that vertex (see Figure 9.36). Similarly, if we place two octagons at a common vertex, the sum of the two angles is 270, and there is not enough room to place a third octagon at that vertex.

TABLE 9.5

Figure	Number of angles at common vertex	Sum of those angles (degrees)
Equilateral triangle	6	360
Square	4	360
Regular pentagon	3	324
Regular pentagon	4	432
Regular hexagon	3	360
Regular octagon	2	270
Regular octagon	3	405

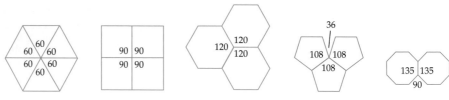

FIGURE 9.36

Figure 9.37 shows what these three tessellations look like when multiple copies are joined together. These will look familiar to you from common ceiling or floor patterns, and the hexagon tessellation is constructed by honeybees. We call these three tessellations *regular tessellations*, which are defined as tessellations composed of congruent regular polygons.

Classroom Connection

In *A Cloak for the Dreamer*, three sons make quilts for the Archduke. The first two sons use rectangles that tessellate. The third son makes circles that don't tessellate, so there are holes in the coat. They salvage his work by turning the circles into hexagons, which do tessellate. How could you cut a hexagon out of a circle?

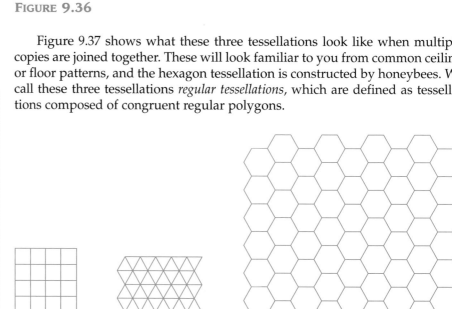

FIGURE 9.37

If you are playing the "what-if" game in your mind, you may have realized that $270 + 90 = 360$ and conjectured that a combination of octagons and squares *will* tessellate. This is true, as Figure 9.38 illustrates. There are many combinations of regular polygons that will tessellate. Mathematicians have names for certain subsets. One that we will consider here is the **semiregular tessellation**—a tessellation of two or more regular polygons that are arranged so that the same polygons appear in the same order around each vertex point. The tessellation at the left in Figure 9.38 is a semiregular tessellation because the two figures are a square and a regular octagon. The tessellation at the right in Figure 9.38 is not a semiregular tessellation because the two figures are a square and a nonregular octagon. As you can see, by varying the lengths of the sides of the octagons (that is, the ratio of the noncongruent sides), we can change the appearance of the tessellation.

■ Outside the Classroom ■

This semiregular tessellation is a common kitchen floor design.

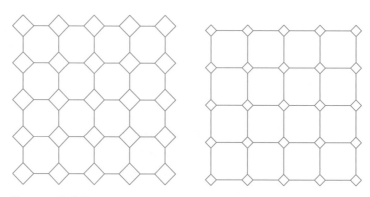

FIGURE 9.38

There are only eight possible semiregular tessellations. There is another whole class of tessellations made with regular polygons called demiregular tessellations. Here, however, we will move from making tessellations with regular polygons to making tessellations with nonregular polygons. This is where the artists and designers come in!

INVESTIGATION 9.16

Tessellating Trapezoids

In this investigation, you will deepen your knowledge about why certain figures tessellate. First, though, cover Figure 9.39 before reading on. This will better enable you to own the knowledge you will construct here. Now, recall the "brick" rectangles in Investigation 9.13 and our discussions about multiple attributes in Chapter 8. One reason why the brick rectangle has multiple tessellation patterns is that the length of the rectangle is exactly twice the width. Look at the trapezoid below and describe its attributes. Then read on. . . .

With respect to angles: the trapezoid has two right angles, one acute angle, and one obtuse angle.
The measures of the acute and obtuse angles are 45 degrees and 135 degrees, respectively.
The right angles are adjacent (as opposed to opposite).
With respect to sides: the trapezoid's left side and top base are congruent, and the bottom base is twice the length of the top base.

Now make about 12 or more copies of this trapezoid. You can use Geoboard Dot Paper, graph paper, or blank paper and a ruler. How many different ways can you get this trapezoid to tessellate? In other words, how many different tessellation patterns can you make?

DISCUSSION

The specific attributes of this trapezoid open the possibility of multiple tessellation patterns:

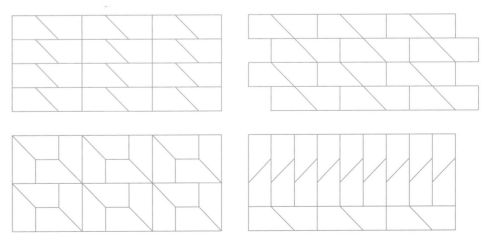

FIGURE 9.39

Once we move out of regular polygons, verifying that a pattern tessellates becomes more complicated. One way to make the task easier is to make sure that the sum of the angles at any point in the pattern is 360 degrees. To make *this* task easier, we use the term *vertex point*, which we will define as any point on the figure where polygons share a common vertex. Mathematicians have proved that all vertex points in regular and semiregular polygons are identical. In those cases, we need only verify that the sum of the angles of one vertex point is 360 degrees to prove that the figure tessellates. Look at the figures on page 625 and determine how many different vertex points there are in each figure. Determine the sum of the angles at those points. . . .

The figures below show the vertex points for two of the figures. In the figure at the left, you can see that A and B are equivalent vertex points. The sum of the angles at these vertices is 45 + 135 + 180 = 360 degrees. The sum of the angles at vertex C is easily seen to be 360 because we have four right angles. In the figure at the right, there are two different vertex points. The sum of the angles at point A is 45 + 135 + 45 + 135 = 360 degrees, and the sum of the angles at point B is 90 + 90 + 180 = 360 degrees.

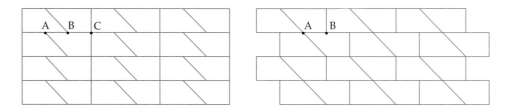

Now, we revisit this notion of attributes. *Any* trapezoid with two adjacent right angles can be tessellated in the fashion shown by the first two figures below. Do you see why? Since the sum of the angles of a trapezoid is 360 degrees and two of the angles are 90 degrees, the other two angles must be supplementary. This knowledge, coupled with a straight angle measuring 180 degrees, enables us to verify that the sum of the angles at any vertex point is 360 degrees.

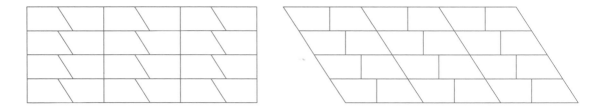

However, the other two designs made with the trapezoid at the beginning of this investigation will not work for any trapezoid, as the figures below illustrate.

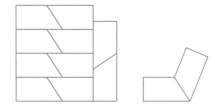

INVESTIGATION 9.17

More Tessellating Polygons

What about other shapes? Although the regular pentagon does not tessellate, many pentagons do. Similarly, many arrows tessellate. Can you figure out why the pentagon and the arrow below tessellate? What attributes do they have? Think before reading on. . . .

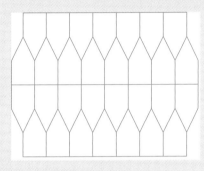
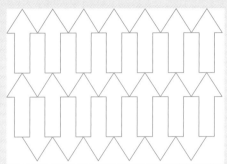

■ Mathematics ■

You may have already noticed this, but you can see hexagons in the pentagon tessellation because the pentagon has two sides that are parallel and congruent. This observation opens the way for looking at different nonregular hexagons that will tessellate. If you look back on the brick pattern in Figure 9.30, you can see that each group of four pentagons forms a hexagon. This is what artists and designers do—they get paid to play with possibilities! By viewing the problem with a wide angle lens, they see all sorts of things that many of us do not, even though they are right in front of our eyes!

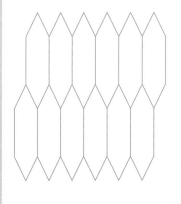

DISCUSSION

Some readers think that I should have told the angles, but in this case, other attributes of the figures—symmetry and perpendicularity—eliminate the need to know specific angles.

For the pentagons, there are two distinct vertex points. Make sure that you see them. If you are not sure, check another vertex point that you think might not be equivalent and examine its attributes—what angles and sides meet at that point.

The first vertex point is simply the meeting of four right angles, so it is easy to verify that the sum of the angles at that vertex point is 360 degrees. For the other vertex point, we can use our knowledge of geometry! First, we can break the pentagon into a square and a triangle (see Figure 9.40). Because the pentagon is symmetric, we know that the triangle is isosceles, which means the base angles are congruent. Now we connect algebra to geometry. Let the measure of these two base angles be a. What is the measure of the third angle? . . .

Since the sum of all three angles is 180, if we call that third angle b, we have the equation $a + a + b = 180$, which means $b = 180 - 2a$.

Thus the sum of the angles at this vertex point is

$$90 + 90 + a + (180 - 2a) + a$$
$$= 180 + 180 + 2a - 2a$$
$$= 360$$

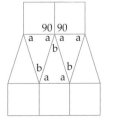

FIGURE 9.40

Because the arrows can also be broken apart into rectangles and isosceles triangles, the verification that it tessellates is similar and will be left as an exercise.

I need to sound a cautionary note about making tessellating figures. The shape below has some nice rotational symmetry and all its angles are 90 degrees and multiples of 90 degrees. At first glance (see the figure at the right below), it appears that it will tessellate.

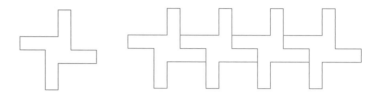

However, when we try to make a second row, to continue the tessellation, we find that there is simply no way to place copies of that figure—by translating, rotating, or reflecting—so that there will be no gaps, as the figure at the left below shows. However, if we continue with this idea, we do make a pattern that does tessellate; in this case, the unit of the tessellation is the original figure and a square. Recall the semiregular tessellation that was earlier discussed that consisted of a regular octagon and a square.

Thus far, we have restricted our investigation of tessellations to polygons. However, many nonpolygonal figures will tessellate. Recall our discussion of Escher and Figure 9.19 with the tessellating swans. M. C. Escher popularized tessellations that move from mathematics into art. How did he do it? Making interesting tessellation figures is not as hard as you think. One kind is shown here. Exploration 9.9 takes this idea further. Begin with a square (although you could begin with any parallelogram). Make a shape inside the square (it need not be a polygon) that has one side on the edge of the square. Translate this shape to the opposite side. You can do this more than once. The example below connects to jigsaw puzzles, in that I have cut out a circular shape from two sides and then translated that side.

> ■ *Language* ■
>
> Here is where the language of Chapter 8 makes communication easier. When I said, "you could begin with any parallelogram," this tells you, because you know that rectangles and rhombuses are also parallelograms, that I am including them too.

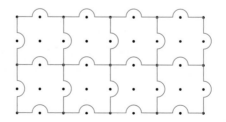

Exploration 9.9 offers many interesting ways to extend your understanding of tessellations.

INVESTIGATION 9.18

Generating Pictures Through Transformations

Let us apply our understanding of translations, reflections, rotations, and glide reflections to see how a tessellation figure is composed. Figure 9.41 shows one of my favorite tessellation figures, one that appears in many Islamic designs. Let's say this was a jigsaw puzzle and all the pieces were congruent. How would you put the puzzle together? I have numbered some of the pieces to make communication simpler. Make your own plan before reading on. . . .

■ Mathematics ■

Recall our discussion of unit in Chapter 5. We will also discuss unit in the next chapter when we discuss measurement. One of the goals of this course is for you to see that unit is not just "inch, foot, ounce, etc." What do you think *unit* means?

DISCUSSION

We have learned that if we have two congruent figures in the same plane, they can be mapped onto each other by some combination of translation, reflection, rotation, and glide reflection. In this pattern, each of the figures has line symmetry. As we look for patterns within this tessellation, we notice that each figure has many reflection neighbors and many rotation neighbors. For example, figures 1 and 7 are reflection images of each other; so are figures 2 and 8, figures 3 and 9, and figures 4 and 5, to name but a few. Similarly, the members of these pairs are also rotation images of each other. If we consider figures 2, 4, 5, and 8 as a unit, then we can generate the whole pattern by translating this unit in a diagonal direction. If you turn the paper 45 degrees, it is much easier to see that these four figures can easily generate the whole pattern!

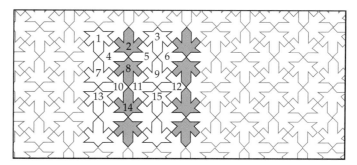

Source: From *Introduction to Tessellations* by Dale Seymour and Jill Britton. © 1989 by Dale Seymour Publications, an imprint of Pearson Learning. Used by permission.

FIGURE 9.41

Summary

We have found that each of the transformations from Section 9.1 leads to a type of symmetry—translation, reflection, rotation, or glide reflection. One of the more important goals is for you to see symmetry as a movement—as a verb, rather than a noun. When we see symmetry in an object, we do so by moving that object or parts of that object in our minds. Perhaps that is why symmetry is so appealing: It requires activeness on the part of our minds to appreciate it. When we find that an object has symmetry, we are using the notion of congruence; that is, we have found that we can move the figure (or part of the figure) and superimpose the figure on itself. We have seen that symmetry has many applications in the world of art (quilts, mosaics), in everyday life (wallpaper and tiles), and in science (symmetry at the atomic level). We have seen that symmetry helps us to understand better how and why tessellations work.

Finally, we have introduced the notion of three-dimensional symmetry, which has many important applications in everyday life and in science. I have talked with scientists about the usefulness of symmetry in their fields. A chemist told me that molecules of similar symmetry give similar spectra and that advances in our mathematical understanding of symmetry have helped organic chemists to understand the structure of organic compounds. At a molecular level, the atomic structure of various elements and compounds determines how they will be packed together. As Marjorie Senechal has written, "if all we learn about symmetry is to identify it, we miss the whole point. Symmetry is an effect, not a cause."[5]

[5]Marjorie Senechal, *On the Shoulders of Giants*, p. 153. Reprinted with permission from *On the Shoulders of Giants: New Approaches to Numeracy*. Copyright © 1990 by the National Academy of Sciences. Courtesy of the National Academy Press, Washington, DC.

EXERCISES 9.2

1. Describe the rotation and reflection symmetries in the figures below.

   a.
   b.
   c.
   d.

2. Describe the rotation and reflection symmetries in the figures below.

   a.

   Maple leaf
   Canada

   b.

   "Life"
   Ancient Egypt

   c.

   "Unity is Strength"
   Ghana

   d.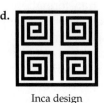

   Inca design

3. Describe the rotation and reflection symmetries in the flags below.

   a.
   Barbados

   b.
   Switzerland

   c.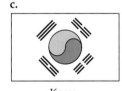
   Korea

4. Describe the rotation and reflection symmetries in the logos below.

   a.
   Chevrolet

   b.
   Volkswagen

   Sources: (a) Courtesy of Chevrolet Motor Division, General Motors Corporation. (b) Volkswagen logo courtesy of Volkswagen of America.

   c.

   d.

   Source: The Octagon logo is a Registered trademark of JPMorgan Chase & Co. and is used here with its expressed permission.

5. Describe the rotation and reflection symmetries in the quilt patterns below.

   a.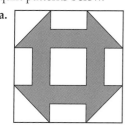
   Churn Dash

   b.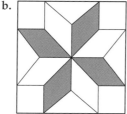
   Eight Point Star

   c.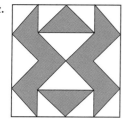
   Fool's Puzzle

   d.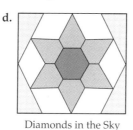
   Diamonds in the Sky

   e.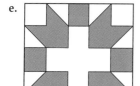
   Hearth and Home

6. a. Describe the rotation and reflection symmetries of the tangram pieces.
   b. Make a design that has symmetry with some or all of the tangram pieces. Explain the symmetry.

7. Classify the pentominoes by the kind(s) of symmetry that they have. Explain the different subsets.

8. Describe the rotation and reflection symmetries of the ten digits in base 10.

9. Recall the family tree for the properties of quadrilaterals in Chapter 8. When we look at the symmetries of the quadrilaterals in the family tree, will that same family tree still hold true? That is, does each descendant quadrilateral have the same symmetries as its ancestor plus at least one more symmetry? Justify your answer.

10. Tell whether the statements below are true or false. If a statement is true, explain why. If it is false, provide a counterexample.
    a. If a figure has rotation symmetry, it must also have reflection symmetry.
    b. If a figure has point symmetry, it must also have rotation symmetry.
    c. If a figure does not have point symmetry, it cannot have reflection symmetry.

11. Create a figure (not one copied from this book) that has the symmetry described below. In each case, describe your thought processes, including ideas that did not pan out.
    a. 60-degree rotation symmetry
    b. 120-degree rotation symmetry
    c. 90-degree rotation symmetry
    d. Two lines of symmetry
    e. Three lines of symmetry
    f. At least one line of symmetry and some kind of rotation symmetry
    g. Reflection symmetry but not rotation symmetry
    h. Rotation symmetry and exactly one line of reflection

12. a. Complete the figure following so that it has point symmetry but no other rotation symmetry.

    b. Complete the figure following so that it has two lines of symmetry.

    c. Complete the figure following so that it has two lines of symmetry.

    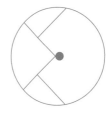

    d. Complete the figure following so that it has point symmetry and no reflection symmetry.

13. In each part, the square piece of paper shown at the right is folded along the dashed lines and then cut as shown at the left. Predict what the paper will look like when unfolded.

    a.

b.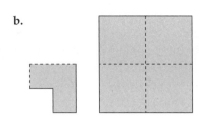

14. Sketch where you would cut the folded paper shown at the left in order to make the shape shown at the right when you unfold the paper. The dashed line represents the fold line.

a.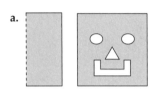

b.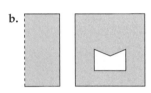

c.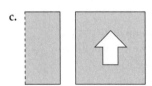

15. Sketch where you would cut the folded paper shown at the left in order to make the shape shown at the right when you unfold the paper. The dashed lines represent the two fold lines.

a.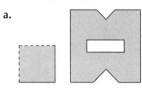

b.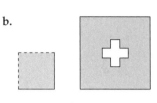

c.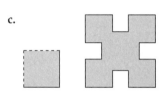

16. Describe the symmetries of the following strip patterns.

a.

b.

c.

d.

e.

f.

17. Describe the symmetries of the following pictures, which were taken at the Taj Mahal.

   a.

   b.

   c.

   d.

18. In each case, make your own original strip pattern with the symmetry indicated, and justify that it does indeed have this symmetry.

    a. ll **b.** lm **c.** ml **d.** mm **e.** l2 **f.** mg **g.** lg

19. Make the following strip pattern using grip paper or using a computer graphics program. Explain your solution path.

20. Describe the rotation and reflection symmetries in the Navajo designs as they are drawn below. Then, imagining them to extend to the right and left indefinitely, describe any translation symmetries.

    a.    b.

    c.

21. Make the designs shown in Exercise 20 using computer software. Describe how you made the designs.

22. a. Describe the symmetries of the quilt at the left below. Ignore the border.
  b. Describe the symmetries of the quilt at the right below. Ignore the border.

Source: Indian Pine Quilt, artist unidentified, Maine, 1880–1890. Cotton with cotton embroidery. 86″ × 82″. Collection American Folk Art Museum, New York. Gift of Cyril Irwin Nelson in memory of his grandparents, Guerdon Stearns and Elinor Irwin (Chase) Holden, and in honor of his parents, Cyril Arthur and Elise Macy Nelson. 1982.22.1

  c. Describe the symmetries of the quilt blocks selected by your instructor from the Sampler quilt at the right.

23. Look at the pattern below.
    a. Find a shape and an image of that shape after a translation. Describe the shape and specify the translation.
    b. Find a shape and an image of that shape after a reflection. Describe the shape and specify the reflection.
    c. Find a shape and an image of that shape after a rotation. Describe the shape and specify the rotation.
    d. Find a shape and an image of that shape after a glide reflection. Describe the shape and specify the glide reflection.

Source: From *Introduction to Tessellations* by Dale Seymour and Jill Britton. © 1989 by Dale Seymour Publications, an imprint of Pearson Learning. Used by permission.

24. Give instructions for making the patterns below as though you were talking to someone on the phone. Use symmetries when possible.

a.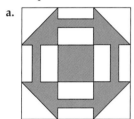
Shoo Fly

b.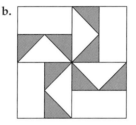
Amish Design

c.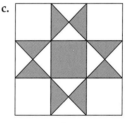
Star

25. Design a flag that has
    a. Rotation symmetry
    b. Reflection symmetry
    c. Rotation symmetry but no reflection symmetry
    d. Reflection symmetry but no rotation symmetry

26. Examine a deck of cards and separate them into subsets that have the same kinds of symmetry.

27. The word "HIDE" has one (horizontal) line of symmetry. This means that if we flip the word over, it will still spell HIDE. Find another word that has horizontal symmetry.

28. The word "WITH" has one (vertical) line of symmetry if we write it vertically. Find another word that has vertical symmetry.

    W
    I
    T
    H

29. Find a two-dimensional or three-dimensional figure on campus or from your home that has some symmetries. Sketch the figure and describe the symmetries.

30. Find examples of children's toys that have symmetries.

31. Describe the symmetries in a nut and explain why a nut might have these symmetries.

32. Copy the following shapes onto Geoboard Dot Paper and determine whether each shape does or does not tessellate.

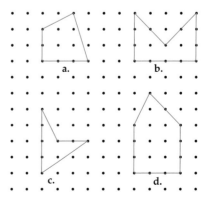

33. Copy the following shapes onto Isometric Dot Paper (dots arranged in a grid of equilateral triangles) and determine whether each shape does or does not tessellate.

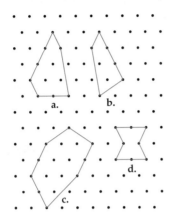

34. Verify that the arrow pattern in Investigation 9.17 does indeed tessellate.

35. Look at the tessellation patterns in Figure 9.30. Describe the symmetries of each of the patterns.
    a. The design based on the letter C
    b. The Japanese pavement design
    c. The Islamic design made from hexagons
    d. The Islamic dancers

36. Select one pair of figures from Investigation 9.18.
    a. Describe a translation that will map that pair onto another pair.
    b. Describe a reflection that will map that pair onto another pair.
    c. Describe a rotation that will map that pair onto another pair.
    d. Describe a glide reflection that will map that pair onto another pair.

37. Which of the 12 pentominoes will tessellate?

38. You have learned that a regular hexagon will tessellate. Below are two conjectures. There are three parts to what you turn in: your response: true or false; your justification of your response; your work—that is, the hexagons that you made up and tested to see whether they tessellated or not.
    a. If a hexagon has line or rotation symmetry, then it must tessellate.
    b. All convex hexagons will tessellate.

39. Look at the tessellation below, a hexagon that has been decomposed into six congruent hexagons called chevrons.
    a. Can you identify two chevrons in which the one is a translation of the other? Describe the translation.
    b. Can you identify two chevrons in which the one is a reflection of the other? Describe the reflection.
    c. Can you identify two chevrons in which the one is a rotation of the other? Describe the rotation.
    d. Can you identify two chevrons in which the one is a glide reflection of the other? Describe the glide reflection.

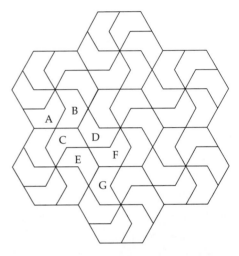

40. Which of the following figures tessellate? If you believe a figure tessellates, you need to show why. If you believe a figure does not tessellate, you need to explain why not.

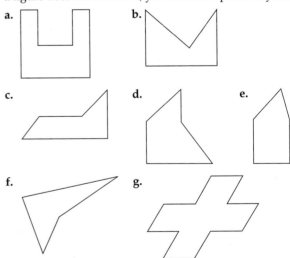

41. Describe each unit and the means by which it is repeated to form the patterns in Figure 9.30.
    a. The C shape
    b. The brick tile
    c. The pattern at the bottom left
    d. The pattern at the top right

42. Describe each unit and the means by which it is repeated to form the following patterns.

Source: Excerpted from *Symmetry: A Unifying Concept*. © 1994 by Istvan and Magdolna Hargittai. Reprinted by permission.

SECTION 9.3 SIMILARITY

WHAT DO YOU THINK?

- What does it mean to say that two figures or two objects are similar?
- Are an object and its shadow similar? Why or why not?

In Section 9.1, we explored several transformations that let us change the position of an object but preserve its size and shape. That is, all of the images were congruent to the original figure. In this section, we will investigate a transformation in which the size of the figure is changed but the shape is not; we will call these transformations **similarity transformations**.

Most people are comfortable with the idea of saying that two objects are similar if they have the same shape but different size. We encounter the notion of similarity frequently in daily life: When we use maps, when we use a photocopier to shrink or enlarge a picture or drawing, and in scale models. In each case, there is a scale: The map might say 1 inch = 50 miles. The photocopier

CLASSROOM CONNECTION

You have used the concept "similar" since you were young. "Similarity seems to be a very fundamental concept. Preschoolers understand that miniature animals, doll clothes, and play houses are all small versions of familiar things. The fact that even such young children know what these tiny objects are supposed to represent shows that they intuitively understand change of scale. Building and taking apart scale models of towers, bridges, houses, and shapes of any kind give the child—of any age—a firm grasp of this idea."[6]

will use percentages—for example, make a copy 75%. And scale models may include the information that they exhibit a 1:100 scale—that is, the actual object is 100 times the size of the scale model (see Figure 9.42).

FIGURE 9.42

We will investigate the mathematical idea of similarity, which simply makes the notion of "same shape, different size" more precise from three different perspectives.

First, let us use the ideas developed in Chapter 8 about the multiple attributes of shapes.

INVESTIGATION

9.19

Understanding Similarity

Figure 9.43 includes four pairs of shapes that are mathematically similar. Look at each pair. What is it that makes the larger polygon similar to the smaller one? Look at angles and sides. Measure them with a protractor and a ruler.

---

[6]Marjorie Senechal, *On the Shoulders of Giants*, pp. 141–142. Reprinted with permission from *On the Shoulders of Giants: New Approaches to Numeracy*. Copyright © 1990 by the National Academy of Sciences. Courtesy of the National Academy Press, Washington, DC.

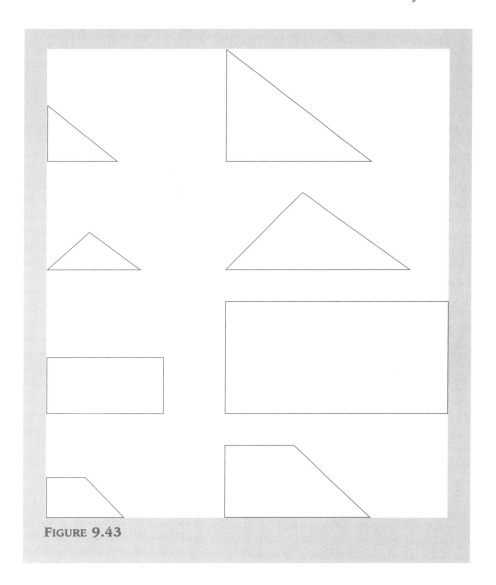

FIGURE 9.43

DISCUSSION

Angles In Chapter 8, we defined two polygons as being congruent iff all corresponding angles were congruent and all corresponding sides were congruent. What relationships do you notice between angles and sides of the similar figures that might enable us to define two polygons as being similar? Write down your present hypothesis before reading on. . . .

It can be easily determined that in each of the four cases in Figure 9.43, the corresponding angles of the similar figures are equal. If you worked with the Pattern Blocks in Exploration 9.11, you may have come to the same conclusion. You can confirm this relationship either with a protractor or by placing the smaller figure on top of the larger one and verifying by inspection.

Sides What about the sides? Obviously, the sides are not congruent. What relationship do they have? Think about this before reading on. . . .

One observation is that if both figures are aligned as shown, then corresponding sides are parallel. Another possible observation involves the lengths. Measure the lengths of corresponding sides. Do you notice anything? Do this before reading on. . . .

If we measure the sides of each figure and compare the lengths of corresponding sides, we find that the ratios are equal. For example, the lengths of the right triangle are 15 mm, 20 mm, and 25 mm. The lengths of the larger right triangle are 30 mm, 40 mm, and 50 mm. That is, each of the sides of the large triangle is twice the length of the corresponding side of the smaller triangle. Another way of expressing this is to say that the ratios of corresponding sides are the same.

This common ratio of the lengths of corresponding sides is called the **scale factor**.

From Chapter 6 we know that when two or more ratios are equal, we can use the term *proportional*. Thus, when two figures are similar, the corresponding sides are proportional. It is this language that is generally used to define similarity, because it takes fewer words.

Two polygons are **similar** iff corresponding angles are congruent and corresponding sides are proportional.

We use the symbol ~ to denote similarity. For example, in Figure 9.44, $\triangle ABC \sim \triangle XYZ$.

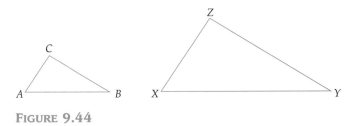

Figure 9.44

In the following investigation, we will use a technique developed many centuries ago by artists.

INVESTIGATION 9.20

Similarity Using an Artistic Perspective

We will develop the concept with a simple shape. The figure below shows how an artist might enlarge a figure. The small rectangle has been enlarged, using a scale factor of 3. Do you understand how this works? Could you use this method to enlarge another figure? Investigate this method before reading on. . . .

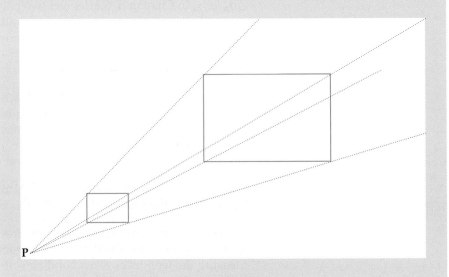

DISCUSSION

We draw the rectangle and a point P on a sheet of paper and draw lines that go through each of the four vertices of the rectangle. We measure the distance from the point P to the top left vertex of the rectangle, and then we mark a point on that line that is three times the distance from point P. If we do this for the other three vertices of the rectangle and connect the points, we have a new rectangle whose height is three times the height of the original rectangle.

Our last investigation adds the tool of coordinate geometry to the notion of perspective lines to bring out more relationships.

INVESTIGATION 9.21

Using Coordinate Geometry to Understand Similarity

Look at the two similar rectangles that have been drawn using perspective lines that have now been placed on the coordinate plane. What do you notice about the relationships among the coordinates? Think before reading on. . . .

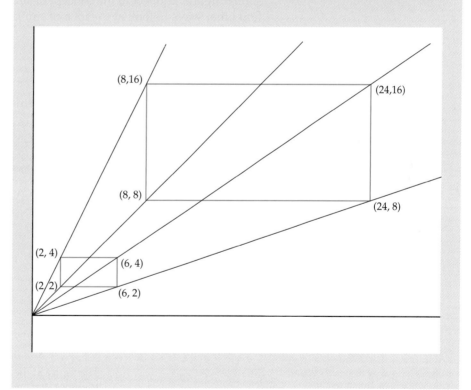

DISCUSSION

There are two ways to illustrate the connections. One is to notice that the relationship between the x- and y-coordinates of each pair is the same as that between the coordinates of any other pair. That is, the ratios $2:4$ and $8:16$ are the same—that is, $1:2$. By the same token, both $6:2$ and $24:8$ equal $3:1$.

Another way of showing the relationship is to multiply the coordinates of the first rectangle by 4:

$$4(2, 2) = (8, 8)$$
$$4(2, 4) = (8, 16)$$
$$4(6, 2) = (24, 8)$$
$$4(6, 4) = (24, 16)$$

This observation enables us to enlarge any object by any scale factor. For example, if we want to make it 2.5 times as tall, we place the original object on the coordinate plane, find the coordinates of each vertex, and then multiply them by 2.5.

Summary

In this section, we have investigated the idea of similarity so that your thinking about this concept could move up the van Hiele scale—that is, beyond "similarity means they look alike." We have also looked at how some ways to draw similar figures connect to the concept of similarity. In the explorations, you will find other ways to look at the idea of similarity.

EXERCISES 9.3

1. In each case below, the two polygons are similar. Find the length of the side labeled x.

   a.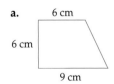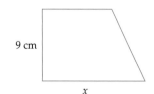

   b.

2. The two triangles at the right are similar, but they have gotten mixed up. Determine the corresponding parts and explain how you did so.

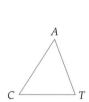 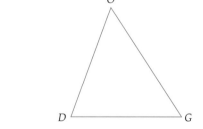

3. The figure at the right below has been created from four triangles that are congruent to the triangle at the left. Are the two triangles similar? Explain why or why not.

4. For each of the questions below, answer yes or no. In either case, justify your response.
   a. Are all isosceles triangles similar?
   b. Are all equilateral triangles similar?
   c. Are all squares similar?
   d. Are all rhombuses similar?
   e. Are congruent polygons similar?

5. Gerald has a photo that measures 8 × 11 inches and a frame that is 4 × 6 inches.
   a. Explain to him why we cannot reduce the photo so that it will fit perfectly in the frame.
   b. Explain to him his options for cutting so that the reduced photo will fit.

6. When we use an overhead projector in a classroom, is the image on the screen similar to the object on the overhead projector?

7. Below is a map of Utah. If the actual distance across the bottom of Utah is 275 miles, determine the scale of the map. That is, 1 inch = x miles.

8. The diagram below shows how we can find the distance across a lake using the principle of similar triangles.

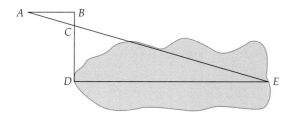

   a. Explain how points C, B, and A would be determined, as though you were talking to someone who understands the idea of similar triangles but has never heard of this method before.
   b. Explain why $\triangle ABC \sim \triangle EDC$.
   c. If $AB = 324$ feet, $BC = 103$ feet, and $CD = 212$ feet, what is the distance across the pond?

9. Explain how you could use shadows and similarity to find the height of a building.

10. A common standard setting for enlarging on a copy machine is 121 percent. What does this mean?

11. Find a quilt block that contains similar figures. Explain why the figures are similar.

12. Find a tessellation pattern that contains similar figures. Explain why the figures are similar.

13. Solomon Golumb invented reptiles, figures that both tessellate and also replicate themselves.
    Consider the trapezoid below.

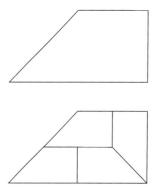

   This trapezoid tessellates. Now we will see how it "replicates itself." We can divide this trapezoid into four smaller congruent trapezoids, each of which is similar to the original—sort of like having children!
   a. Why is "reptile" an appropriate name for such figures?
   b. Prove that the four little trapezoids are similar to the larger trapezoid.
   c. Make another trapezoid that is a reptile.
   d. Describe the attributes that a trapezoid must have in order to be a reptile.
   e. How many different ways can you find in which this trapezoid will tile the plane?

14. The hexagon below is another reptile.

   a. Determine how to divide this hexagon into several little congruent hexagons, all of which are similar to the original.
   b. How many different ways can you find in which this hexagon will tile the plane?

Chapter Summary

1. Both congruence and similarity are kinds of transformations.
2. Congruence can be defined in Euclidean terms (the classical definition) and in terms of congruence mappings.
3. There are different ways to perform and describe different transformations. Which one(s) we use depends partly on preference and partly on purpose.
4. Translations, reflections, rotations, and glide reflections are congruence transformations—the figure is moved, but its shape and size are not changed.
5. The geometric transformations of translation, reflection, rotation, and glide reflection are operations, with many similarities to the operations of addition, subtraction, multiplication, and division.
6. Symmetry has to do with congruence within a figure.
7. The many symmetries in geometric figures include translation, reflection, rotation, and glide reflection symmetry. Many figures have more than one kind of symmetry.

BASIC CONCEPTS

Section 9.1 Congruence Transformations

Four transformations
   In everyday language: slide *584*   flip *584*
      turn *584*   shrinking *584*
   In mathematical language: similarity *584*
      translation *584*   reflection *584*   rotation *584*

Expressing translations:
   congruence transformation *585*
   similarity transformation *585*
   image *586*
   using taxicab language *586*
   specifying the distance and the angle *587*
   vector *587*   translation *587*
   line of reflection *589*   reflection *589*
   center of rotation *590*   rotation *590*
   composite transformation *592*
   glide reflection *592*
   congruent *595*
   Given two congruent figures on the same plane, one can be mapped onto the other by some combination of transformations *595*
   fundamental domain (region) *596*

Section 9.2 Symmetry and Tessellations

   symmetry *604*
   line of symmetry *604*
   vertical line symmetry *608*
   horizontal line symmetry *608*
   diagonal line symmetry *608*
   point symmetry *606*

Symmetry for two-dimensional figures
   translation symmetry *612*
   reflection symmetry *612*
   rotation symmetry *612*
   glide reflection symmetry *612*
   pattern *612*

Symmetry for three-dimensional objects
   plane of symmetry *619*   axis of symmetry *619*

Tessellations
   tessellation *620*   tessellate *620*
   semiregular tessellations *624*

Section 9.3 Similarity

   similar *640*   scale factor *640*
   similarity transformation *637*

SOLUTIONS FOR INVESTIGATION 9.12, PAGES 612–614

   v = translation vector
   m = mirror line
   · = center of rotation
   g = glide vector
   m_g = mirror line for glide reflection

A.

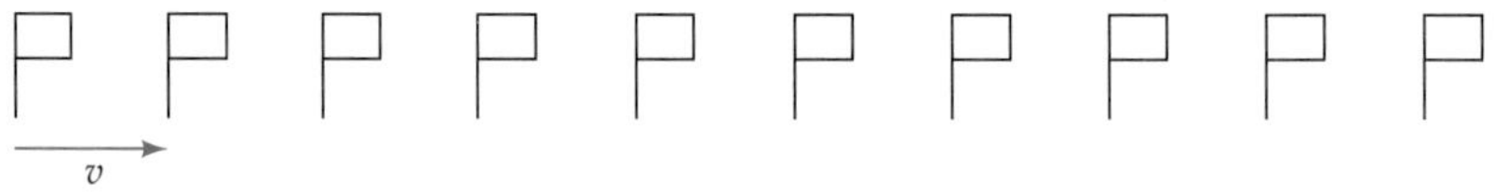

Chapter Summary

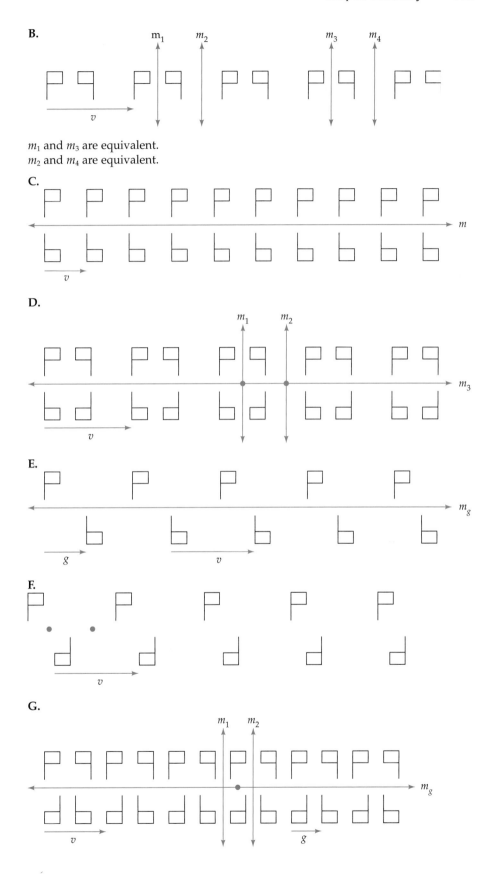

m_1 and m_3 are equivalent.
m_2 and m_4 are equivalent.

646 CHAPTER 9 / Geometry as Transforming Shapes

SOLUTIONS FOR INVESTIGATION 9.13, PAGES 615–617

m = mirror line
$\cdot$ = center of rotation
g = glide vector
m_g = mirror line for glide reflection

A.

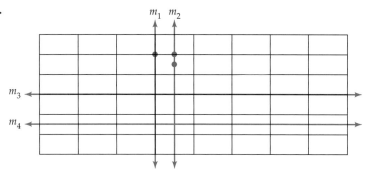

Translation vectors

B.

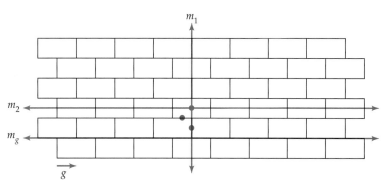

Translation vectors

C.

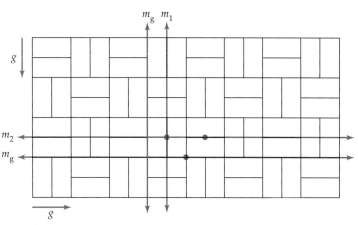

Translation vectors

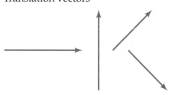

Chapter 9 Review Exercises

1. Perform the following transformation on the given figure, and give its new coordinates:
   a. Translation (4, 0)
   b. Rotate 90 degrees counterclockwise about the point (4, 3).
   c. Reflect the figure across the *x*-axis.

   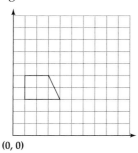
   (0, 0)

2. Without folding, determine the reflection image of the triangle at the right. Briefly justify your method.

3. Sketch the image by rotating the figure 180 degrees around point *P*.

4. a. Find the image of the figure across line *a* and then across line *b*.
   b. Find another way to get the same outcome.
   c. Find the image of the figure across line *c* and then across line *d*.
   d. Find another way to get the same outcome.

   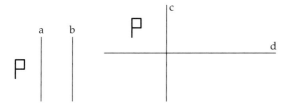

5. Describe how to get the shape from position A to position B in as few moves as possible, using translations, rotations, reflections, and/or glide reflections.

   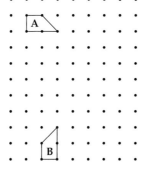

6. How are rotations and reflections alike? How are they different?

7. If a glide reflection is a composite transformation, why is it included as one of the basic four congruence transformations? That is, why not call translation, reflection, and rotation the basic three?

8. Describe the reflection and rotation symmetries of the following figures.

   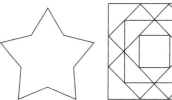

9. a. Complete the first figure so that it has reflection but not rotation symmetry.
   b. Complete the second figure so that it has reflection and rotation symmetry.
   c. Complete the third figure so that it has rotation but not reflection symmetry.

   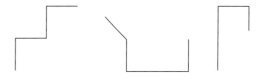

10. a. Make an original figure that has reflection but not rotation symmetry.
    b. Make an original figure that has reflection and rotation symmetry.
    c. Make an original figure that has rotation but not reflection symmetry.

11. a. What does it mean to say that a figure has rotation symmetry?
    b. What does it mean to say that a figure has reflection symmetry?

12. Describe the symmetries of the following strip patterns.
    a.
    b.
    c.

    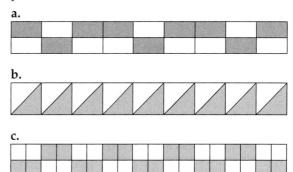

d.

13. In the tessellations below, describe the unit and the means by which it is repeated.

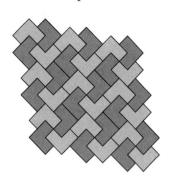

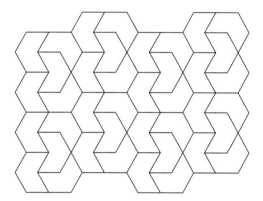

14. Determine whether each figure below tessellates or not. If it does tessellate, show the tessellation and explain why it tessellates.

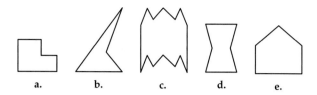

15. The following two pentagons are similar. Find the length of the side marked x.

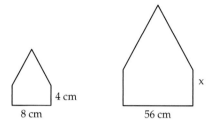

16. Are all right isosceles triangles similar? Explain why or why not.

17. Using the lines-of-perspective method, make a larger copy of the given figure so that each side of the copy is twice as long as each side of the given figure.

18. If the coordinates of a triangle are (3, 4), (8, 4) and (4, 7) and the projection point P is at the origin, predict the coordinates of the similar triangle whose sides will each be twice the length of the corresponding sides of that triangle. Explain your reasoning.

19. In base 5, there are only five numerals: 0, 1, 2, 3, and 4. Make an alternative set of base 5 numerals in which each numeral is topologically equivalent to the given numeral.

CHAPTER 10

Geometry as Measurement

10.1 Systems of Measurement

10.2 Perimeter and Area

10.3 Surface Area and Volume

*W*hat do you think of when you hear the term *measurement*? In what ways do you use measurement, and in what ways are you measured?

We measure and are measured all the time. When we ask how much or how many, we are often asking a question requiring a direct or indirect measurement:

- How much time will it take?
- How far is it from here to there?
- How much paint do we need to paint the house?

10.1

10.2

10.3

Long before they start school, children are fascinated by measurement, and some of their measurement questions leave parents confused, too:

The child asks, "How long until we get there?" The parent says, "One hour." The child asks, "Is that long?"

Or the parent divides the candy bar into two pieces (measures), and one child says, "His half is bigger than mine."

By the end of this chapter, you will come to understand that measurement is far more than just an examination of the quantities we use in everyday life, such as length, area, volume, and time. You will want to refer to the NCTM measurement standards before you have your children investigate this topic.

In this chapter, we will investigate several important aspects of measurement.

Big Idea	Example/Elaboration
1. Most objects have multiple attributes that are measureable.	Consider a garage that is not attached to a house. We might want to know its perimeter, for a fence; or the surface area of the sides, to paint it; or its volume, to know how much it will hold.
2. There is a process that we use to measure all attributes.	This three-step process is the same whether we are measuring distance or density.
3. The units that we use are arbitrary as opposed to absolute.	All units were invented by people.
4. Virtually all measurements are approximations.	There is no "exact" height of a building. How we report the height depends on the precision that is needed: 12 m, 12.2 m, 12.21 m, 12.214 m, etc.
5. Precision and accuracy are not the same thing.	A measurement can be very precise but still be inaccurate, or it can be accurate and yet not be very precise.
6. Different attributes of an object are generally not related in simple ways.	Doubling the area of a garden does not necessarily double the length of the fence around the garden.

SECTION 10.1 SYSTEMS OF MEASUREMENT

WHAT DO YOU THINK?

- What does it mean to measure?
- What are some difficulties we encounter when measuring?
- When do we need our measurements to be precise, and when it is acceptable to have just a "rough" measurement?
- How do we measure heights of mountains?

Visualize yourself measuring different things—for example, how far it is to your parents' home, how much paint you would need to buy to paint a house, how much a package weighs. We might want several measures of something. Consider a room. If we wanted to stencil the top of the wall around the room, we would want to know the room's perimeter. If we wanted to paint the room, we would want to know its surface area. If we wanted to store things in the room, we would want to know its volume. Think of the process of measuring in each of these situations. What is common to all of them? Think and then read on. . . .

In each case, we must

1. Identify what we want to measure—that is, identify the *attribute* we want to measure.
2. Select or determine the *unit* we will use to tell us "how much."
3. Determine the amount we have in terms of the units we have chosen.

Keep these three aspects in mind as you read the text and do the explorations.

Development of Measurement Systems

Measuring is probably one of human beings' earliest forms of mathematical activity. In the following pages, we will examine the development of the units and systems of measurement that people commonly use in everyday life.

Time One of the first quantities that people measured was time. Many thousands of years ago, people developed ways to determine the number of days in a year. Why do you think they wanted to know this? Think before reading on. . . .

There are many reasons, such as wanting to know when to celebrate certain rituals, when to hunt, or when to plant. Early divisions of the day were probably marked by three significant times: sunrise, midday, and sunset. Later, midnight became significant too. We say that 1 day = 24 hours, 1 hour = 60 minutes, and 1 minute = 60 seconds. However, this has not always been so. Where do you think 24 and 60 came from? Think and then read on. . . .

Over 3000 years ago, the Babylonians divided the day into 12 hours and the night into 12 hours. However, the length of an hour depended on the time of year; that is, in the winter, a day hour was shorter than a night hour. At some point, the Greeks decided to divide the entire day into 24 equal parts, but it was not until around 1330 that the hour was standardized—that is, that an hour in January and an hour in June were equal, and an hour in Stockholm and an hour in Venice were equal.

For centuries, people used the sundial to measure the passage of time. The earliest known sundial comes from Egypt and dates from about 1500 B.C. The sundial, of course, has several limitations: It is not useful on cloudy days or at night, and it is not useful for small amounts of time. An early attempt to address this limitation was the water clock. The first water clock came from Egypt and dates from about 1400 B.C., and the first evidence of an hourglass is from ancient Rome in about 300 B.C. Columbus used a sand clock to measure time on his voyage.[1] It was not until the 1700s that the mechanical watch (with springs) was both accurate enough and inexpensive enough to be used outside of scientific experiments. The present definition of a second is the duration of 9,192,631,770 periods of the cesium-133 atom![2] Believe it or not, there are machines that can count these periods. The ability to measure accurately time intervals of one-millionth of a second is crucial in fields like particle physics.

> ■ *Language* ■
>
> The origins of our words for the days of the week can be traced to the Babylonians and correspond to the seven "planets" they had identified: Sun, Moon, Mars, Mercury, Jupiter, Venus, and Saturn. Which ones do you recognize? What might have happened to the names of the other days?

Length The history of the various units that different peoples have selected to measure length (see Figure 10.1) is quite fascinating, even to those who are not "math people."

The earliest recorded linear unit of measurement was the cubit, the distance from the point of the elbow to the outstretched tip of the middle finger. The word *cubit* is derived from the Latin *cubitum*, meaning "elbow."

It should not be surprising that the foot was one of the units of measurement in ancient times. The Roman writer Plutarch stated that the foot was based on the actual length of Hercules' foot. It is likely that the French foot was originally the actual measurement of the length of King Charlemagne's foot.[3] In the tenth century, King Edgar I decreed that the yard would be the distance from the tip of *his* nose to the tip of the middle finger of *his* outstretched arm. The reason behind the decree was a desire to regulate trade in textiles, so that 1 yard of cloth would be approximately the same in all parts of England.

> ■ *Language* ■
>
> In the language of King Henry I (1496), the word for yard literally meant "stick." Thus our term *yardstick* is redundant. This is similar to speaking of the river that flows along the border between Texas and Mexico as the Rio Grande River. In Spanish, Rio Grande literally means "big river."

---

[1]Daniel Boorstin, *The Discoverers* (New York: Random House, 1983), p. 34.
[2]H. Arthur Klein, *The World of Measurements* (New York: Simon and Schuster, 1974), p. 163.
[3]Terry A. Richardson, *A Guide to Metrics* (Ann Arbor, MI: Prakken Publications, 1978), p. 2.

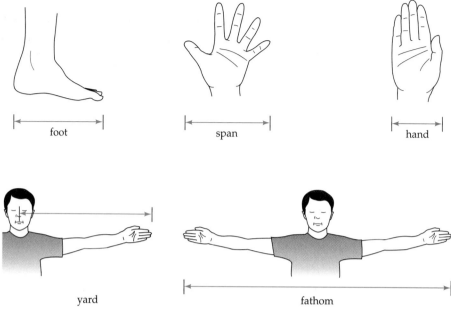

FIGURE 10.1

CLASSROOM CONNECTION

Virtually all mathematics textbooks now have young children first measure lengths with nonstandard units. For example, they might ask students to find how long the desk is in terms of paper clips. Why do you think teachers are being encouraged to start with nonstandard units?

There are several reasons for doing this. First, it helps to keep the focus of the students' thinking on the attribute that is being measured and learned. If you did Exploration 10.2, you may have seen this concept in action. For example, it is much simpler (conceptually) to line up and count paper clips than it is to line up and count units on a ruler. Second, the answers can be more understandable to the children—for example, the length of the soccer field in terms of footsteps or the number of body lengths of one student in the class.

CLASSROOM CONNECTION

Eunice Hendrix-Martin uses the children's book *Biggest, Strongest, Fastest* to spark interest in measurement. In order to get a sense of the size of large animals, the students determined the length of the animals in relation to the students' height. For example, a blue whale is about as long as 24 students. First the students had to determine the average height of a student in the class, and then they had to use proportional reasoning to determine the length of the whale in "student" units. The article describing this activity appears in the April 1997 issue of *Teaching Children Mathematics*, pp. 426–430.

The Romans divided 1 foot into 12 units (*unciae*, from which our word *inch* comes). The Roman pace consisted of 5 feet—that is, the distance of a left step and a right step—and their mile (*milia passuum*) was 1000 paces. Do you see a connection between their mile and our mile? When Alexander the Great conquered the known world, he had professional pacers to measure how far his army had traveled so that their maps would be accurate. These pacers were trained to make each step virtually the same length regardless of terrain: flat, uphill, or downhill. Table 10.1 lists common modern-day units of length.

TABLE 10.1

Units	Abbreviation	Relationship to other unit(s)
1 inch	in	
1 foot	ft	12 inches
1 yard	yd	3 feet = 36 inches
1 mile	mi	5280 feet

Weight Two ancient units for weight were the barleycorn and the seed of the carob plant (from which comes the word *carat* for the measure used to weigh gold and diamonds). Over the centuries, three different systems of weights evolved: the avoirdupois system for everyday use, the troy system to weigh precious metals and gems, and the apothecaries' system for very small amounts. However, conversion was confusing. For example, in the troy system, there were 12 ounces in a pound, whereas in the avoirdupois system, there were 16 ounces in a pound!

Volume and capacity Over time, two related systems of measure for what we call volume evolved, depending on whether the thing being measured was dry or wet. For dry volume, some basic units were the ounce, pint, quart, peck, and bushel. For liquid volume, there were a host of units. Thomas Jefferson almost certainly memorized the following ditty, which was commonly used in the eighteenth century to teach schoolchildren the units:

> Two mouthfuls are a jigger; two jiggers are a jack; two jacks are a jill; two jills are a cup; two cups are a pint; two pints are a quart; two quarts are a pottle; two pottles are a gallon; two gallons are a pail; two pails are a peck; two pecks are a bushel; two bushels are a strike; two strikes are a coomb; two coombs are a cask; two casks are a barrel; two barrels are a hogshead; two hogsheads are a pipe; two pipes are a tun—and there my story is done![4]

■ **History** ■

The familiar nursery rhyme Jack and Jill was written as a protest against King Charles of England's taxation of jack(pots) of liquor. King Charles not only lost his crown, he lost his head! (He was beheaded.)

Which measures do you recognize? Table 10.2 lists common modern-day units of volume.

TABLE 10.2

Units	Abbreviation	Relationship to other unit(s)
1 teaspoon	tsp	
1 tablespoon	tbsp	3 teaspoons
1 liquid ounce	oz	2 tablespoons
1 cup	c	8 liquid ounces
1 pint	pt	2 cups or 16 ounces
1 quart	qt	2 pints or 32 ounces
1 gallon	gal	4 quarts or 128 ounces

Temperature In 1714, a German instrument maker named Gabriel Fahrenheit made the first mercury thermometer. He designated the lowest temperature he could create in the laboratory as 0° and the normal temperature of the body as 98°. On his scale, the freezing point of water is 32° and the boiling point of water is 212°.

Let us now examine the metric system, with an emphasis on understanding its structure.

---

[4]S. Carl Hirsch, *Meter Means Measure: The Story of the Metric System* (New York: Viking Press, 1973), p. 29.

> **History**
>
> In 1785, Congress passed Thomas Jefferson's proposal for a decimal system of currency.
>
> On July 4, 1790, Thomas Jefferson presented to Congress a report on weights and measures that included two proposals: the first a complete transformation of our system of weights and measures based on powers of 10, and the second a plan for a partial transformation. George Washington supported Jefferson's efforts. However, the Congress decided to adopt neither plan.

The Metric System

The idea of a system of measures based on powers of 10, which is what the **metric system** is, was first proposed in 1670 by Gabriel Mouton, from Lyons, France. Over the next 100 years, many scientists and nonscientists made various proposals for a uniform system of measurement. Before the development of the metric system, there were literally hundreds of systems of measurement in Europe alone. What is particularly important is that they were not uniform, so that a bushel in one location was not the same as a bushel in another location. These differences were often used by the rich to exploit the poor, who obviously resented this. One of the first acts of the French government after the French Revolution in 1789 was to develop a uniform system of measurement so that the rich could not cheat the poor. In 1793, the French Academy of Sciences proposed a new metric system for *all* units of measurement.

The United States has not officially adopted the metric system, and its adoption is not imminent. (Our system is called the **U.S. customary system**.) However, metric units are finding their way into everyday life—for example, some gas stations sell gasoline by the liter, some road markers now give distances in kilometers as well as in miles, and most soda is now sold by the liter. Furthermore, international travel is so common that many Americans encounter the metric system in their travels (even to Canada and Mexico), and most industries that compete in international markets have gone metric, since most of the rest of the world uses the metric system. Thus it is helpful to have a rough idea of the metric system.

There is one feature of the metric system that it is helpful to know before we examine the different units. Once the French had determined the unit, whether it was meter, liter, or gram, they used Greek prefixes for *multiples* of this unit and Latin prefixes for *fractions* of this unit.

Several of these prefixes show up in familiar words; for example, a millennium is 1000 years, a century is 100 years, and a decade is 10 years.

Table 10.3 shows the metric prefixes and their relationship to the basic metric units. Note that in everyday life, we rarely use *deci-*, *deca-*, or *hecto-*.

TABLE 10.3

Prefix	Milli	Centi	Deci	Deca	Hecto	Kilo
Relationship to the basic unit	$\frac{1}{1000}$	$\frac{1}{100}$	$\frac{1}{10}$	10	100	1000
Example	milliliter	centimeter				kilogram

Metric Length

> **Mathematics**
>
> In 1999, NASA lost a $125 million Mars orbiter because one team working on the project was using English units of measurement while another team was using metric units.

The standard unit of length in the metric system is the **meter**, which was defined (most likely after much debate) as one ten-millionth of the length of the line that starts at the equator and goes to the North Pole through Barcelona, Paris, and Dunkirk. In other words, this distance was decreed to be 10 million meters, and the meter was the length that was one ten-millionth of that distance. How they decided upon 10 million and how they knew the exact distance to the North Pole is an interesting story in itself! The most commonly used metric units of length are, from largest to smallest, the *kilometer*, *meter*, *centimeter*, and *millimeter* (see Table 10.4). Most yardsticks and rulers sold in the United States have inches and feet on one side and metric measurements on the other.

Section 10.1 / Systems of Measurement   655

TABLE 10.4

Units	Abbreviation	Relationship to other unit(s)
1 millimeter	mm	
1 centimeter	cm	10 millimeters
1 meter	m	100 centimeters or 1000 millimeters
1 kilometer	km	1000 meters

INVESTIGATION 10.1

Developing Metric Sense

Try to do these problems yourself. Then read on. . . .

A. Insert the decimal point in the proper place.
- The diameter of a penny is 19 centimeters.
- The length of a page of notebook paper is 279 centimeters.
- The common adult height of an elephant is about 39 meters.

B. If the speed limit says 90 kilometers per hour, what is the speed in miles per hour?

DISCUSSION

A. The diameter of a penny is 1.9 centimeters.
- The length of a page of notebook paper is 27.9 centimeters.
- The common adult height of an elephant is about 3.9 meters.

B. We can use dimensional analysis (introduced in Chapter 1 and discussed in Chapter 5) to convert this speed to miles per hour. We can also use reasoning to deduce that we need to divide 90 by 1.6.

$$\frac{90 \text{ kilometers}}{\text{hour}} \times \frac{1 \text{ mile}}{1.6 \text{ kilometers}} \approx \frac{56 \text{ miles}}{\text{hour}}$$

Metric Volume

The metric system was designed so that we use the same units to measure volume (dry) and capacity (liquid). The standard metric unit for volume is the **liter** (see Table 10.5). A liter is approximately 34 ounces, or slightly more than a quart. The liter is also defined to be the volume of a cube whose sides are 10 centimeters. Thus 1 liter is equivalent to 1000 cubic centimeters, abbreviated as cm^3 (see Figure 10.2).

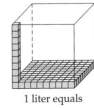

1 cubic centimeter (cm^3)

1 liter equals

FIGURE 10.2

TABLE 10.5

Units	Abbreviation	Relationship to other unit(s)
1 milliliter	ml	
1 liter	l	1000 milliliters

However, a milliliter is also 1/1000 of a liter. Therefore, we have the equivalence between liquid and dry measures: 1 cm^3 = 1 ml. Most canned goods give the amount in milliliters. If a can contains 240 ml, how many liters is that? Do this before reading on. . . .

Because there are 1000 ml in a liter, we must divide by 1000. Therefore, 240 ml = 0.24 liter, about one-fourth of a liter.

The creators of the metric system not only worked to connect liquid and dry measures but also connected volume and mass. They defined a kilogram to be the mass of 1 liter of pure water at a certain temperature.

Metric Mass

The standard of mass in the metric system is the **gram** (see Table 10.6). Technically, the terms *mass* and *weight* refer to different attributes. **Weight** is the force that gravity exerts on an object, and **mass** is the amount of matter that makes up the object. If you went to the moon, you would weigh about one-sixth as much as you do on Earth, but your mass would be the same. Because our planet is much larger than the moon, the force of gravity on your body is much greater on Earth. You may recall hearing that because the force of gravity is less on the moon, we could jump six times as high on the moon as we can on Earth.

TABLE 10.6

Units	Abbreviation	Relationship to other unit(s)
1 milligram	mg	
1 gram	g	1000 milligrams
1 kilogram	kg	1000 grams

Some metric units of mass—kilogram, gram, and microgram—are encountered more often than others. One kilogram is approximately equal to 2.2 pounds. Where do you see metric mass in everyday life in the United States? Think before reading on. . . .

The net weight on most canned goods is generally given in both ounces and grams. This is one of the few places where one might see metric units for mass used in the United States at present. Some scales also have weights in metric units.

Metric Temperature

In 1742, a Swedish astronomer named Anders Celsius proposed a modification in the units of measurement that the Fahrenheit system used. He proposed that the reference points be the freezing point of water (0°) and the boiling point of water (100°) (see Figure 10.3). This *Celsius* system was also called the *centigrade* system (that is, "100 grades").

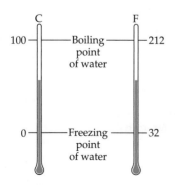

FIGURE 10.3

Time and Angles

As mentioned earlier, the French Academy of Sciences recommended that *all* known measures be based on powers of 10. Thus they recommended changing the calendar so that there would be 10 months in a year, 10 days in a week, and 10 hours in a day. They also proposed that there be 400 degrees in a circle, which would mean that a right angle had 100 degrees. As you know, not all of these proposals survived! When Napoleon came to power, he repealed the law making metric measure compulsory. Although the French continued to use metric units for length, volume, and weight, they (happily) threw out the metric system for time and returned to the more familiar 7-day week, 24-hour day, 60-minute hour, and 60-second minute.

Becoming Comfortable with Metric Measurements

What if you heard that someone is 182 centimeters tall and weighs 80 kilograms? Most Americans can easily visualize 6 feet and 176 pounds, the U.S. customary equivalents of the metric measures given in the first sentence, but are less comfortable with measurements given in metric units. We will address this problem by providing some reference measures and conversion ratios.

The diagrams in Figure 10.4 give approximations for a meter, a centimeter, and a millimeter.

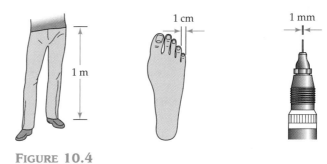

FIGURE 10.4

Here are some common conversions and approximations for other units:

One inch is about 2.5 cm.
One meter is about 39 inches.
One mile is about 1.6 kilometers.
One kilometer is just over 1/2 mile, more closely about 0.6 miles.

One milliliter (ml) is about 1/4 teaspoon of water.
One liter is just over a quart.

One milligram (mg) is about the mass of one strand of hair.
One gram about the mass of 1 raisin or a dollar bill.
One kilogram is about 2.2 pounds.

INVESTIGATION 10.2

Converting Among Units in the Metric System

A. A path is measured as 1.5 km. How many meters is this?
B. A recipe calls for 250 ml. How many liters is this?
C. A sample weighs 700 mg. How many grams is this?
D. Change 435 mm to centimeters.

DISCUSSION

A. Because there are 1000 meters in 1 meter, we multiply 1.5 × 1000 to get 1500 meters. We can also use dimensional analysis (first presented on page 16).

$$1.5 \text{ km} \times \frac{1000 \text{ m}}{1 \text{ km}} = 1500 \text{ m}$$

B. Because there are 1000 milliliters in 1 liter, we divide 250 by 1000 to get 0.25 liter.

$$250 \text{ ml} \times \frac{1 \text{ liter}}{1000 \text{ ml}} = 0.25 \text{ liter}$$

C. Because there are 1000 milligrams in 1 gram, we divide 700 by 1000 to get 0.7 gram.

$$700 \text{ mg} \times \frac{1 \text{ gram}}{1000 \text{ mg}} = 0.7 \text{ gram}$$

D. Because there are 10 millimeters in 1 cm, we divide 435 by 10 to get 43.5 centimeters.

$$435 \text{ mm} \times \frac{1 \text{ cm}}{10 \text{ mm}} = 43.5 \text{ cm}$$

Hint: You can see this on your ruler, which has centimeter and millimeter gradations.

Measurement of Other Quantities

The quantities that we have examined in this section—length, volume, mass, time, and temperature—represent only a small subset of the quantities measured in our world. What other things do we measure? Why do we need to measure them? What are some of the units used to measure those things? What instruments are used? Think about things that we measure, and make a table like Table 10.7 for those things you can think of. Do this before reading on. . . .

TABLE 10.7

What do we measure?	Why?	Units	Instruments
Noise	To study noise levels that can cause hearing damage	Decibels	Audiometer

The following is only a partial list: speed (miles per hour), fuel consumption (miles per gallon), light intensity (candela), electric current (amperes), efficiency (production per unit of time), density (mass per unit volume), and infant mortality (deaths per 1000 live births).

Precision

There is another aspect of measurement that requires some attention. Virtually all measurements are approximations. For example, if you ask for the dimensions of a room, you might be told that it is 14 feet by 11 feet. If you ask for more precision—let's say you are buying a rug—you might be told that it is 14 feet 2 inches by 10 feet 11 inches.

The precision of our measurement depends on the reason why we are measuring. Sometimes we want to measure amounts with much precision, and sometimes less precision is quite acceptable. For example, when selling cloth, clerks generally measure the length roughly and then add an extra couple of inches. However, if you go to a candy store or buy meat, the amount is generally weighed to the nearest tenth of an ounce, and cylinders in cars need to be accurate to the nearest thousandth of an inch. See also Explorations 10.1, 10.2, and 10.3.

Although many people use the words precise and accurate synonymously, the two concepts are not identical. A measurement can be very precise (for example, 34.628 meters) but be inaccurate (if the measurer made a mistake). Similarly, a measurement can be not very precise (for example, $16\frac{1}{2}$ feet) but be very accurate (that is, the actual distance is closer to $16\frac{1}{2}$ feet than to 16 feet or 17 feet).

Greatest possible error We can quantify the amount of precision in our measurement, and we use two terms to do so. First, we can determine the greatest possible error (GPE). If you were to measure the width of a sheet of paper to the nearest millimeter, what would you get? Do so before reading on if you have a ruler close at hand. . . .

It is 21.6 centimeters, or 216 millimeters. That is, my reporting of 216 millimeters means that I believe the true width is closer to 216 millimeters than to 215 millimeters or 217 millimeters. What do you think the greatest possible error would be in this case? That is, if someone were to measure the width of the paper to the nearest tenth of a millimeter, what would be the biggest difference between my measurement and the more precise measurement? Think and then read on. . . .

The greatest possible error would be 0.5 millimeter—that is, one-half the unit that I used to measure.

Relative error The relative error tells us the percent error of the greatest possible error. The following example illustrates this concept. A measurement of 300 centimeters and a measurement of 3 centimeters both have the same greatest possible error, 0.5 centimeter, but 0.5 centimeter is a much bigger part of 3 centimeters than it is of 300 centimeters. What is the relative error in these two cases (3 centimeters and 300 centimeters)? Do this before reading on. . . .

The relative error in the former case is

$$\frac{0.5 \text{ centimeter}}{3 \text{ centimeters}} \times 100 = 17\%$$

The relative error in the latter case is

$$\frac{0.5 \text{ centimeter}}{300 \text{ centimeters}} \times 100 = 0.2\%$$

Measurement as a Function

At a very basic level, measuring tells how much of something there is. From another perspective, measurement is the act of assigning numbers to amounts. The number refers to a specified unit. Do you realize that the operation of measuring an amount is a function? If you do, describe how it fits. If not, go back and review the discussion of functions in Chapter 2. Then write your thoughts before reading on. . . .

In order for an operation to qualify as a function, each element of the input set must be mapped onto exactly one element of the output set. As we found in Chapter 2, the two sets need not be sets of numbers. When we measure, we take the amount to be measured (the input element) and determine the number that corresponds to that amount. Although you need not know that measuring represents a functional relationship between two sets, this is simply one more example of the underlying connectedness and relatedness of so many seemingly different mathematical ideas and topics.

Summary

In this section, we have explored what it means to measure an object, and we have realized that many objects have several attributes that can be measured—for example, weight, surface area, and volume. We have also examined where many of the units came from, both U.S. customary and metric. We have noted that many, if not most, of the numbers we use represent measurements. We have learned that most measurements are approximations and that there is a difference between precision and accuracy. Finally, we considered measurement from a functional perspective.

Exercises 10.1

1. Think about each object below and describe all of the measurable attributes of that object. Next to each property, briefly describe why someone might want to measure that attribute.
   a. a pond
   b. a garage
   c. a house
   d. a carpet
   e. a banana
   f. a tree

2. Select, from the following units, the unit that you would use to express each of the attributes below. Briefly explain your choice.

   Length: millimeters, centimeters, meters, kilometers
   Volume: milliliters, liters
   Mass: milligrams, grams, kilograms

   a. Your height
   b. Your weight
   c. The length of a butterfly
   d. How much fluid there is in a drinking cup
   e. The mass of a fingernail
   f. The volume of a thimble

3. Fill in the blanks:
   a. 500 m = ____ km
   b. 4.5 cm = ____ mm
   c. 670 mL = ____ L
   d. 3.6 L = ____ mL
   e. 450 g = ____ kg
   f. 35 kg = ____ g
   g. 24 mg = ____ g

4. There are about 6 billion people on Earth. If they all lined up and held hands, how long a line would they form?

5. The lengths of most older U.S. swimming pools are multiples of yards, generally 25 yards. However, newer pools are 25 or 50 meters long. Suppose an American swimmer is practicing for a 400-meter race. Eight lengths of the pool is 400 yards, which is "close" to 400 meters. How close is it? That is, how much shorter is 400 yards than 400 meters? First predict the amount, using only mental math. Explain how you got your prediction and how you did the operations in your head. Then determine the amount (to the nearest yard).

6. Alan and Bill are going to have a 400-meter race, but Alan gets a head start because he is slower. Their best times of the season are 82 seconds and 75 seconds. How much (distance) of a head start should Alan get?

7. Let's say your estimate of a length was 100 feet and the actual length was 105 feet. Clearly this would not be "as big" a difference as that between an estimate of 10 feet and an actual length of 15 feet, although both estimates were "off" by 5 feet. Compare the accuracy of the estimates in such a way that we can see from the language that the first estimate was much closer than the other.

8. To determine the relationship between miles and kilometers, if we use the conversion 39.37 inches = 1 meter, we find that 1 kilometer is approximately equal to 0.62 mile. What if we use the conversion 2.54 centimeters = 1 inch? How far apart are the answers (1 kilometer = x miles) obtained by using the different conversions?

9. When examining the relationship between kilometers and miles, we now have two conversion formulas: 1 kilometer = 0.62 mile and 1 mile = 1.6 kilometers. What relationship do 0.62 and 1.6 have?

10. a. How many days would it take you to spend $1 million if you spent $3 per minute?
    b. How high would a stack of 1 million pennies be?
    c. How high would a stack of 1 million quarters be?
    d. How high would a stack of pennies worth 1 million dollars be?
    e. How high would a stack of quarters worth 1 million dollars be?

11. *Classroom Connection* This problem comes from "Figure This!" Every year many birds migrate. The record holder is the Arctic tern, which flies from the Arctic to the Antarctic and back. If the one-way distance is 9000 miles and these birds can fly for 12 hours a day at an average speed of 25 miles per hour, how many days does it take them to make the trip?

    Source: Reprinted with permission from Figure This! (http://www.figurethis.org), copyright 2006 by the National Council of Teachers of Mathematics.

12. Wagon trains in the 1800s took from 4 to 6 weeks to travel from Missouri to California. Although the direct distance is less, the actual mileage for the wagons was 2000 miles. About what average distance did they travel per day?

13. The Biblical Goliath was said to be 6 cubits and 1 span tall (1 Samuel 17:4). How tall do you think he was in feet and inches? Explain both your reasoning and your calculations.

14. Suppose you were King Henry I (1496), who declared that henceforth 1 yard would be equal to the distance from his nose to his forefinger. How long would a yard be, on the basis of *your* body?

15. The accompanying historical table[5] compares the relative lengths of 1 foot in various countries in 1788.
    a. Explain the meaning of the column headed "Parts," as though you were talking to a fifth-grader who didn't understand.
    b. Make a graph for the column headed "Parts."
    c. Describe the variation among the various feet.
    d. What do you think the second column means?
    e. Are the numbers in the two columns proportional? How did you determine your answer to this question?

> A Comparison of the American foot with the feet of other Countries.
>
> The American foot being divided into 1000 parts, or into 12 inches, the feet of several other Countries will be as follow.
>
> | | | Parts | | Inch.lin.points |
> |---|---|---|---|---|
> | America | — | 1000 | — — — | 12 0 0 *dec.* |
> | London | — | 1000 | — — — | 12 0 0 |
> | Antwerp | — | 946 | — — — | 11 4 1, 32 |
> | Bologna | — | 1204 | — — — | 14 5 2, 25 |
> | Bremen | — | 964 | — — — | 11 6 4, 89 |
> | Cologne | — | 954 | — — — | 11 5 2, 25 |
> | Copenhagen | — | 965 | — — — | 11 6 5, 76 |
>
> PIKE'S "ARITHMETICK," 1788
> Early difficulties with weights and measures.

16. Say the United States did go metric.
    a. What would you propose as the size of a standard sheet of paper? Justify your choice.
    b. What would you propose as the two standard sizes for soft drinks to replace 12 ounces and 16 ounces?
    c. Instead of buying a pound of butter, what would we buy?

17. If you filled up your car's gas tank and it took 48.3 liters, approximately how many gallons is that?
    a. Use mental math and estimation to determine approximately how many gallons of gas this amount is equivalent to.
    b. By how much would your conversion be off if you used 1 liter ≈ 1 quart instead of the more precise 1 liter ≈ 34 ounces conversion?

18. The average human heart pumps about 60 milliliters of blood for each beat.
    a. About how many liters of blood does your heart pump each day?
    b. Do you think a reasonable degree of precision for this question would be to report the answer to the nearest 100 liters, 10 liters, 1 liter, or 1 milliliter? Explain your reasoning.
    c. About how many gallons is this?

19. Suppose we measured the passage of time during the day as the Babylonians did. Compare the length of a daytime hour on the longest day of the year to the length of a daytime hour on the shortest day of the year where you live. Part of this question has to do with resourcefulness, an important trait for teachers to develop. Where can you go to get the information needed to answer this question?

20. Which is greater, an increase of 5° Fahrenheit or an increase of 5° Celsius? Justify your answer.

---

[5]Louis Karpinski, *The History of Arithmetic* (New York: Russell & Russell, Inc., 1965), p. 159. Reprinted with permission of Scribner, a Division of Simon & Schuster. From *The History of Arithmetic* by Louis Karpinski. Copyright © 1965 by Russell & Russell. All rights reserved.

21. A train is traveling 60 miles per hour and is about to enter a tunnel that is 1200 feet long. How long will it take the train to pass completely through the tunnel? What additional information do you need in order to solve this problem?

22. a. One way to determine the speed of the current of a river is to measure how long it takes a leaf to pass under a bridge. For example, let's say it takes a leaf 9 seconds to pass under a bridge that is 35 feet wide. What is the speed of the current in miles per hour?
    b. How might we determine the speed of the current if we were in a boat?

23. Two cups are half full. Cup A holds coffee. Cup B holds milk. One teaspoon of coffee is taken from cup A to cup B. Then one teaspoon of the mixture is taken from cup B to cup A. The two cups now have the same amount of liquid. Is there more coffee in cup A than milk in cup B, the same amount of coffee in A as there is milk in B, or less coffee in A than milk in B? Explain your reasoning.

24. Let's say that a 12-ounce cup of coffee is 20 percent milk. If you add 6 ounces of coffee, what is the percentage of milk in the new mixture?

25. a. Determine the following dimensions for your body: height, wrist, neck, leg, arm, foot.
    b. Determine the following ratios: height to foot, height to leg, neck to wrist, height to arm.
    c. Before determining the ratios for the entire class, predict which ratio will have the most variation and which the least variation. Explain your reasoning.
    d. Gather the class data. Decide on a way to determine the variation and to compare the variation.

26. Will the ratio of the heights of an adult and a child be equivalent to the ratio of the lengths of their feet? If it will be, explain why you think so. If not, predict which ratio will be greater and why. Gather some data. Do the data support your predictions? Explain.

27. You will need a scale for these problems.
    a. Buy several bananas. Compare the price per pound that you paid and the price per pound of what you actually eat. If there are enough data, graph the actual price for the entire class. Before you do so, predict the shape of the distribution: normal, uniform, and so on.
    b. Do the same for peanuts.
    c. Buy several oranges of two types, one type with a thick skin and one type with a thin skin. Compare the ratios for both kinds of oranges.

28. Before they wash their faces, most people run the tap until the water is hot. How much water is wasted waiting for the water to get hot?
    a. With a partner, determine a plan for answering this question.
    b. Determine the amount of water (in U.S. customary units) wasted at the place where you live.
    c. Describe the relative precision of your answer. For example, if you say 3 quarts, is your answer to the nearest quart?
    d. Graph the class data.
    e. Let's say the average for the United States is close to the average in this class and that 200 million people in the United States wash their faces every day. How many gallons of water are wasted?

29. You will need a dropper for this problem. Let's say you have a leaky faucet that drips every 6 seconds. How much water is wasted in one day? Explain the assumptions that you made and describe your solution path.

30. Take a sheet of blank paper. Tear it in half. Place one half on top of the other. Tear these two sheets in half. Place one half (two sheets) on top of the other. Assume for the purposes of this problem that you were able to continue this process for a total of 20 times. What would be the height of the stack of paper?

31. Select a trip in your part of the country for which there is not one obvious shortest way.
    a. Determine which way is shortest by using a map. Explain your work.
    b. Have at least one person try more than one way to determine the actual distance and the actual time it took in each case.
    c. Is the shorter way faster?
    d. Describe the possible sources of error in part (a).
    e. Determine the average speed for each route.
    f. On the basis of the data you have, which way would you go, or do you need more data?

32. a. Measure your classroom in meters and then make a scale model of your classroom so that 1 centimeter on your model represents 1 meter.
    b. Measure your classroom in feet and inches, and then make a scale model of your classroom so that 1 inch on your model represents 1 foot.
    c. Are the computations in both problems about the same, or are the computations for the metric-scale map easier? Explain.

33. In *The History of Arithmetic*, Louis Karpinski describes the Babylonian system of measure:

    3 lines = 1 sossus
    10 sossus = 1 palm
    3 palms = 1 small ell (cubit)
    5 palms = 1 large ell
    6 large ells = 30 palms = 1 large seed
    60 palms = 1 gar
    60 gar = 1 ush
    30 ush = 1 kask

    The Babylonian palm was equal to approximately 4 of our inches.

    a. Compare the Babylonian kask to a metric length.

b. Describe the connections between their choice of units and their numeration system.

c. The cubic palm (called a ka) was used as a unit of capacity. Compare the cubic palm to the liter.

d. The weight of 1 ka of water was the unit of weight and was called 1 mina. Thus the units of length and weight were connected. In analyzing the Babylonian system, Louis Karpinski cites the parallels between the ancient Babylonian system and the modern metric system, noting that both systems established a connection between linear and cubic measure and a unit of weight. He further states that the parallel is "one of the most striking [parallels] to be found in the development of scientific ideas."[6] Explain Karpinski's praise for the Babylonian system.

34. Is the relation "the perimeter of" a function? Justify your answer.

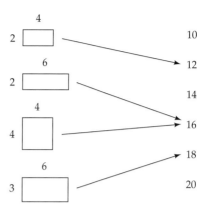

SECTION 10.2 PERIMETER AND AREA

WHAT DO YOU THINK?

- Can two shapes with different perimeters have the same area? Why or why not?
- What does pi mean?

In this section, we will investigate the concepts of perimeter and area, both for geometric figures for which there are formulas and for irregularly shaped objects. It may surprise you that many students do not understand these concepts well. For example, consider the data in Table 10.8 concerning questions reported in *Results from the Fourth Mathematics Assessment*.[7]

TABLE 10.8

		PERCENT CORRECT	
Question		Grade 3	Grade 7
What is the perimeter of the rectangle below?		17	46
What is the area of this rectangle?		20	56
What is the area of this rectangle?		5	46

---

[6]Louis Karpinski, *The History of Arithmetic* (New York: Russell & Russell, Inc., 1965), p. 159.
[7]Mary M. Lindquist, ed., *Results from the Fourth Mathematics Assessment* (Reston, VA: NCTM, 1989), p. 40.

Perimeter

Let us begin with the concept of **perimeter**, essentially the distance around an object. Many practical applications of perimeter involve surrounding an object—for example, fencing a yard or running a baseboard around the base of a room. We will look first at distances around circles.

Circumference and π

When we determine the distances around figures and objects, sometimes the path is not a straight line but rather is a circle. You may remember a formula involving the distance around a circle, or the **circumference**, and that it involves π. You may have constructed a definition of π in Exploration 10.6. Before reading on, take a few moments to think about what π means. That is, the value of π is 3.14 (to two decimal places), but what does π *mean*? For example, where did the number 3.14 come from? How is this number related to the circumference or diameter or radius of a circle (not the formula, but a description of the relationship)? Write your thoughts before reading on. . . .

In one sense, π is a ratio—that is, π is the ratio of the circumference to the diameter of *any* circle. If we could precisely measure the circumference and diameter of any circle and then divide the circumference of the circle by its diameter, we would *always* get π. If we call the circumference C and the diameter d, we have $\pi = C/d$.

Thus we have the formulas

$$C = \pi d \quad \text{and} \quad C = 2\pi r$$

In Figure 10.5, C is the center of the circle, $\overline{AC}$, $\overline{BC}$, and $\overline{DC}$ are radii (*radii* is the plural of *radius*), and $\overline{AD}$ is a diameter.

There is another way to think about π. Imagine placing a string around the circumference of the circle in Figure 10.5 and then straightening out that string. If the diameter of the circle is d, then the length of the string is about $3.14d$. That is, if we unwrap the circumference, its length will always be about 3.14 times the length of the diameter.

Let us now examine one application of these formulas.

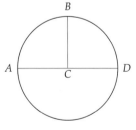

FIGURE 10.5

INVESTIGATION 10.3 What Is the Length of the Arc?

In Figure 10.6, the radius of the circle is 3 centimeters and the measure of angle *BOC* is 42 degrees. How long is arc *BC*? Think about this and then read on. . . .

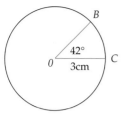

FIGURE 10.6

DISCUSSION

STRATEGY 1: *Make a simpler problem*

What if the angle had been 90 degrees? Can you solve that problem? Think before reading on. . . .

If the angle had been 90 degrees, then the arc would simply have been one-fourth of the circumference of the circle. Do you see why? Can you adapt this strategy to solve this problem now? Try to do so yourself before reading on. . . .

Because 42 does not divide 360 evenly, we don't have an easy fraction. However, we can multiply the circumference of the circle by 42/360.

STRATEGY 2: *Use a proportion*

We could also solve this problem using a proportion. Can you make the proportion? Try to do so before reading on. . . .

Arc $\overarc{BC}$ is a fraction of the circumference of the circle. Similarly, 42 degrees is a fraction of 360 degrees, the number of degrees in a circle. From the ratio construct of fractions, these two ratios are equal. Why is this?

The circumference of the circle is πd, which equals approximately 18.84 inches, so we can set up the following proportion, because the two part: whole ratios are equal:

$$\frac{\text{arc } \overarc{BC}}{18.84 \text{ inches}} = \frac{42°}{360°}$$

When we solve the equation, we find that the length of arc *BC* is about 2.2 inches.

Area

10.6

10.7

Earlier we stated that perimeter questions generally deal with "how much" it takes to *surround* or go around an object. In order to answer perimeter questions, we have to select an appropriate unit, and thus the answer takes the form of how many of those units. Area questions generally deal with "how much" it takes to *cover* an object—for example, how much fertilizer to cover a lawn, how much material to cover a bed. In order to answer area questions, we have to select an appropriate unit, and thus the answer takes the form of how many of those units. However, the units for perimeter and area are not the same. For example, if we have a 20-foot by 10-foot garden, we say that we need 60 feet of fence to surround the garden, but we would say that the area of the garden is 200 *square feet*. The need for units and the difference between units for perimeter, area, and volume are generally not well understood by students and therefore are worth emphasizing more than once.

Many people remember or recognize the basic formulas for determining areas, but few people understand why they work or where they came from. Let us investigate them for a bit. You may have encountered these ideas in Explorations 10.7 and 10.8. Understanding them will pay dividends when we attempt to solve more complex problems. The simplest case for area involves rectangles and squares.

Just as we have found that average is a number that gives us a sense of the middle of a set of data, and standard deviation is a number that gives us a sense of the variation of a set of data, **area** is a number that gives us a sense of the size of something—for example, how much surface that object covers or how much material (cloth, plastic, paint, carpet) it would take to cover that surface. If the geometric figure is a square or a rectangle, we can easily determine the area by multiplying the *base* by the *height*.

Base and height Most students use the more common terms *length* and *width*, so let's take a moment to examine what *base* and *height* mean, because they will become important when we investigate other figures for which the terms *length* and *width* are not appropriate.

Any polygon can be rotated so that its bottom side will be parallel to the bottom of the paper. Thus any side can be taken as the **base** of the polygon. For example, we can rotate the triangle at the left in Figure 10.7 so that any one of its sides is the base. Which one we choose is generally an arbitrary decision. See Figure 10.7.

The height of a polygon is the distance from the side chosen as the base to the point farthest away, measured along a line perpendicular to the base. See Figure 10.7 for an illustration of this notion with triangles.

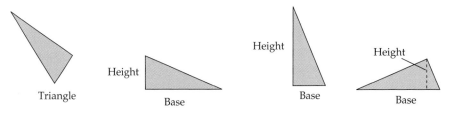

FIGURE 10.7

Thinking about area by looking at rectangles Let us focus on some more subtle aspects of the concept of area.

What does it mean to say that one rectangle (or other figure) covers more area than another? Consider the diagrams in Figure 10.8 of two pieces of plastic sheeting that I use to cover two different woodpiles in my back yard. One piece is 10 feet by 3 feet, and the other is 6 feet by 5 feet.

When I ask young children if the sheets are the same size or if one is bigger, they generally say that the 10-foot by 3-foot sheet is bigger. Older children and adults know that the areas of the sheets are equal, 30 square feet. We can demonstrate this by covering each figure with 30 squares, each of which is 1 foot by 1 foot (see Figure 10.9).

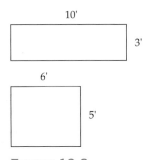

FIGURE 10.8

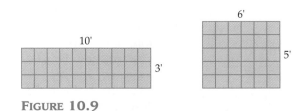

FIGURE 10.9

Although this may seem very simple and straightforward, many people are at least initially stumped when they try to explain what *area* means when the dimensions of a rectangle are not whole numbers—for example, when a rectangle is 3.5 centimeters by 2.5 centimeters. We say that the area of the rectangle is 8.75 square centimeters, because we know that the area of a rectangle is obtained by multiplying its length by its width. Can you explain what that 8.75 means, as if you were talking to a child who understands that in the previous problem, both sheets of plastic have an area of 30 square feet? Do this before reading on. . . .

Figure 10.10 shows one way of explaining what it means to say that the area is 8.75 square centimeters: cutting up the rectangle and reassembling the pieces in a certain way. Do you understand this diagram? Did you explain the meaning of the 8.75 square centimeters in another way that also makes sense? Take some time to see if you can understand the diagrams below before you read my explanation.

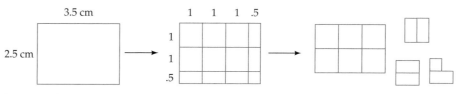

FIGURE 10.10

We can partition the rectangle in a manner similar to the way we used partitioning in the multiplication of fractions and decimals. That is, we have 6 one-centimeter by one-centimeter squares, and we can combine the remaining pieces to make another 2.75 square centimeters (see Figure 10.10).

Area as a function From another perspective, once we specify the unit of measurement, the relationship between any surface and its area is a functional relationship: For every surface, there is a unique number that we associate with its area. Even when the surface is not a rectangle or the dimensions are not whole numbers, the number that represents the area tells us the equivalent number of square units that this object covers.

Let us now turn our attention to understanding some of the more commonly used formulas for determining area.

Understanding the area formula for parallelograms The formulas for many common figures are striking, both in their simplicity and in their connection to one another. It still surprises me that so few people understand why they work. The formula for determining the area of a parallelogram is connected to the formula for determining the area of a rectangle. The diagram at the right in Figure 10.11 illustrates the connection. Can you see it? How would you explain it?

FIGURE 10.11

If we cut the triangle from the parallelogram and reconnect it at the right-hand side, we have transformed the parallelogram into a rectangle. We have not added or subtracted any area, so the areas of the parallelogram and the rectangle must be equal. From Figure 10.11, you can see that the base of the parallelogram is congruent to the base of the rectangle and that the height of the parallelogram is congruent to the height of the rectangle. Because we know that the areas of the two are equal, we can now state that *the area of any parallelogram is equal to the product of its base and its height;* that is, $A = bh$ (see Figure 10.12).

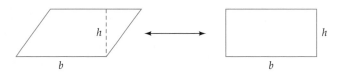

FIGURE 10.12

Understanding the area formula for triangles We will use what we have learned about parallelograms to develop the formula for the area of triangles. Just as we connected the area of any parallelogram to a rectangle, take a few minutes to see if you can adapt this idea to develop formulas for the areas of triangles. Then read on. . . .

Consider triangle *MAN* (see Figure 10.13). If we make a congruent copy of this triangle and move that triangle into place as shown (that is, by rotating it 180 degrees), we form a parallelogram. Do you see how we can use the formula for parallelograms to determine the formula for triangles now? See if you can do it on your own before reading on. . . .

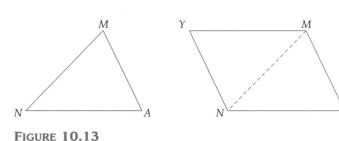

FIGURE 10.13

CLASSROOM CONNECTION

You can see children's thinking about trapezoids in *Measuring Space in One, Two, and Three Dimensions: Casebook*, by D. Schifter, V. Bastable, and S. J. Russell with K. R. Woleck (Parsippany, NJ: Dale Seymour Publications, 2002), pp. 130–135.

The base of the parallelogram and the base of the triangle are identical, and so are the heights.

If the area of the parallelogram = base · height, then (since the area of the two triangles is equal to the area of the parallelogram), *the area of the triangle* = $\frac{1}{2}(base \cdot height)$, or $A = \frac{1}{2}b \cdot h$.

Understanding the area formula for trapezoids Can you now apply what you have learned about the areas of these figures to determine the formula for the area of a trapezoid (see Figure 10.14)? Try to do so before reading on. . . .

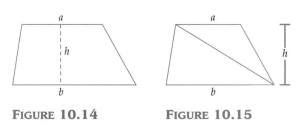

FIGURE 10.14   **FIGURE 10.15**

■ **Mathematics** ■

Did you realize that a trapezoid can be thought of as a triangle with its head cut off? Actually, this is not just a silly thought, because this way of thinking about trapezoids helps us connect the area formula for the trapezoid with the triangle. We could have determined the formula for the area of a triangle by thinking of a triangle as a trapezoid where the top base = 0. In this case, the area is $\frac{1}{2}(b_1 + b_2)h = \frac{1}{2}b_1 h$—that is, $\frac{1}{2}bh$.

There are several ways to derive the formula for the area of a trapezoid. We will discuss two that connect to our preceding discussions. First, we can draw a diagonal of the trapezoid, which cuts (decomposes) it into two triangles. We know how to find the areas of these triangles: The areas are $\frac{1}{2}ah$ and $\frac{1}{2}bh$. The sum of the areas of the two triangles is equal to the area of the trapezoid, so *the area A of the trapezoid is* $\frac{1}{2}ah + \frac{1}{2}bh$, *or* $A = \frac{1}{2}(a + b)h$. See Figure 10.15.

Figure 10.16 shows another method for finding the area of a trapezoid. If we construct a congruent trapezoid and connect it (via a translation and a rotation) to the original trapezoid, we have a parallelogram, and the area of the parallelogram is equal to the product of its base and its height; that is, the area is equal to $(a + b)h$. Because the area of the trapezoid is equal to one-half the area of this parallelogram, the area of the trapezoid $= \frac{1}{2}(a + b)h$.

FIGURE 10.16

INVESTIGATION 10.4

Converting Units of Area

Converting area units is more complicated than converting linear units, and thus an investigation will help you to own this idea. Malik and Lee have decided to pull up their old, shabby carpet and buy new carpet. The room measures 12 feet by 15 feet, and so the area is 180 square feet. However, when they go to the carpet store, they find that the prices are in square yards. How many square yards is their floor?

DISCUSSION

A common answer is 60 square yards—that is, 180 square feet divided by 3, because there are 3 yards in 1 foot. However, that is wrong. Do you see why?

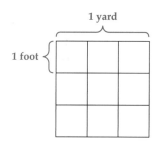

The diagram at the right illustrates why you must divide by 9. Even though there are 3 feet in 1 yard, there are 9 square feet in 1 square yard.

Pythagorean Theorem

Although the relationship between the lengths of the sides of a right triangle had been known long before Pythagoras, it was Pythagoras who proved that *for any right triangle, the sum of the squares of the lengths of the two sides (say, a and b) is equal to the square of the length of the hypotenuse (say, c)*; that is, $a^2 + b^2 = c^2$ [see Figure 10.17(a)].

It is important to note that for the Greeks, this was not an algebraic relationship but rather a geometric relationship [see Figure 10.17(b)]. That is, if you make a square whose sides are congruent to the length of side a, a square whose sides are congruent to the length of side b, and a square whose sides are congruent to the length of the hypotenuse c, then the sum of the areas of the two smaller squares will be equal to the area of the larger square.

At the last count, there were over 370 different proofs of the Pythagorean theorem, including a proof by one of the U.S. presidents, James Garfield. The

670 CHAPTER 10 / Geometry as Measurement

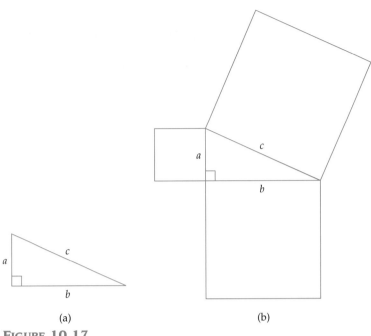

FIGURE 10.17

Pythagorean theorem has many practical applications, such as the one below, where it is used to determine the length of something that we cannot measure directly. The need for the Pythagorean theorem arises both in determining lengths of lines and in determining areas.

INVESTIGATION 10.5

Using the Pythagorean Theorem

We can use the Pythagorean theorem to determine the length of a lake. In Figure 10.18, a right triangle has been created in such a way that the length of the lake is part of the length of one side of the triangle. The lengths of *AB*, *BC*, and *CD* have been measured, because they are on land. The triangle has been carefully made so that all lines are straight lines, so that point *D* lies on side *AC*, and so that angle *C* is 90 degrees. What is the length of the lake? Work on this before reading on. . . .

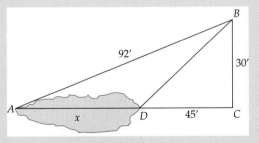

FIGURE 10.18

DISCUSSION

Let us discuss two different strategies and then discuss problem-solving.

Mathematics

The *quadratic formula*, which you may remember from algebra, lets you solve an equation of the form

$$ax^2 + bx + x = 0$$

that is, a quadratic equation. The formula is

$$x = \frac{-b \pm \sqrt{(b^2 - 4ac)}}{2a}$$

STRATEGY 1

Let x be the length of the pond. Using the Pythagorean theorem, we have

$$(x + 45)^2 + 30^2 = 92^2$$

We now have a quadratic equation, which we can solve using the *quadratic formula*:

$$x^2 + 90x + 2025 + 900 = 8464$$
$$x^2 + 90x - 5539 = 0$$
$$x \approx 42 \text{ feet}$$

STRATEGY 2

Looking at the problem from a different perspective, we can find the length of the base of the triangle ($\overline{AC}$) rather easily, and then we can subtract 45 feet from the base to find the length of the lake.

$$AC^2 + BC^2 = AB^2$$
$$AC^2 + 900 = 8464$$
$$AC^2 = 7564$$
$$AC \approx 87 \text{ feet (to the nearest whole foot)}$$

Thus $AD \approx 42$ feet ($87 - 45$).

This investigation illustrates the idea of examining a problem from different perspectives. The second solution is quicker and simpler, although the first solution is not difficult for someone who is adept at algebra. Often in mathematics problems (and in real-world problems too), a problem that at first glance seems very complex and difficult may, when looked at from another perspective, be much simpler than it first appeared.

INVESTIGATION 10.6

10.8

Understanding the Area Formula for Circles

The last area formula that we will consider now is that for the area of a circle. The purpose of this investigation is in the realm of "number sense," helping you to see why $A \approx 3.14r^2$ (and therefore **$A = \pi r^2$**) makes sense. Exploration 10.10 examines another way to make sense of this area formula. Consider the circle in Figure 10.19, whose radius is r inches. In Figure 10.19(b), I have circumscribed a square around the circle. What is the area of the square? Determine this and then read on. . . .

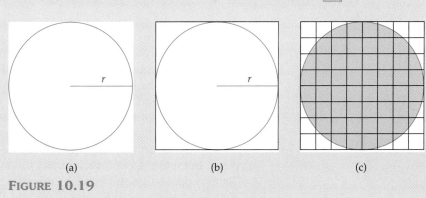

(a)   (b)   (c)

FIGURE 10.19

DISCUSSION

If the radius of the circle is r inches, then the length of each side of the square is $2r$, and thus the area of the square is $(2r)(2r) = 4r^2$.

The area of the circle thus is clearly less than 4 times r^2. We can use our spatial sense to estimate that the circle covers about 3/4 as much space as the square and thus approximate the area of the circle as $3r^2$, or we can place a grid over the figure and determine what fraction of the square is covered by the circle. In this case, we get a more accurate estimate of the area of the circle [see Figure 10.19(c)].

▪ History ▪

Some of the area formulas we have investigated were known by the Egyptians, Babylonians, Chinese, and other peoples thousands of years ago. We know that both the Egyptians and the Babylonians knew not only the formulas for the area of the square and the rectangle, but also the formula for the area of a trapezoid. They knew the formula for finding the area of a right triangle. There were correct and incorrect formulas for the areas of isosceles triangles. For example, there is an example in the Rhind papyrus (from ancient Egypt) of finding the area of an isosceles triangle by finding half the product of the base and the length of one of the equal sides, instead of the altitude. For the circle, the Babylonians developed the formula $A = C^2/12$ and the Egyptians developed the formula $A = \left(\frac{8}{9}d\right)^2$. It will be left as an exercise to see how close their formulas were to the actual formula.

INVESTIGATION 10.7

A 16-Inch Pizza Versus an 8-Inch Pizza

Let's say you are going out to have pizza with several friends. You are thinking of getting one large pizza—let's say its diameter is 16 inches. However, some people are vegetarians, and so you decide to get two little pizzas, each of which is 8 inches in diameter. Suddenly someone asks, "Are we getting the same amount of pizza if we get two little pizzas instead of one large one?" What do you think? Please work on this before reading on. . . .

DISCUSSION

This is generally one of the most amazing investigations in this book because so many people are so surprised. To determine the area, we can use the area formulas: If the large pizza has a diameter of 16 inches, it has a radius of 8 inches. Therefore its area is $\pi(8)^2 = 64\pi \approx 201$ square inches.

If the small pizza has a diameter of 8 inches, it has a radius of 4 inches.

Therefore, its area is $\pi(4)^2 = 16\pi$. Therefore, the area of two small pizzas is 32π, or approximately 100.5 square inches.

Amazing, isn't it? Not only are two 8-inch pizzas not equal to one 16-inch pizza, but it takes *four* 8-inch pizzas to have the same area as *one* 16-inch pizza. Some people still don't believe their eyes. If you find this result amazing, draw a circle to represent the large pizza and then figure out how to place the two smaller pizzas inside the circle.

INVESTIGATION

10.8

CLASSROOM CONNECTION

You can see children's thinking about determining the area of a hand in *Measuring Space in One, Two, and Three Dimensions: Casebook*, by D. Schifter, V. Bastable, and S. J. Russell with K. R. Woleck (Parsippany, NJ: Dale Seymour Publications, 2002), pp. 109–115.

How Big Is the Footprint?

The area of most objects cannot be determined by a formula. For example, biologists and environmentalists want to know the total surface area of the leaves of a tree in order to determine the amount of oxygen the tree produces. What if we wanted to measure the area of another irregularly shaped object, such as the footprint in Figure 10.20?

Take some time to think about how you might determine the area and to try out your ideas. Then read on. . . .

FIGURE 10.20

 10.11

DISCUSSION

There are many, many possible strategies. Several are briefly described below.

STRATEGY 1: Use graph paper

This is one of the simpler strategies. There are two issues worth pursuing: (1) what to do with the partially filled squares, and (2) the advantages and disadvantages of graph paper with smaller squares.

STRATEGY 2: Draw rectangles and triangles

We can partition the footprint into rectangles and triangles. One advantage of this over the previous strategy is that in this case, one rectangle will account for a large portion of the print. One disadvantage is that there is room for error, depending on how the rectangles and triangles are made. This procedure could also become a bit tedious.

STRATEGY 3: Draw trapezoids

We can partition the footprint into trapezoids. This strategy arises from the realization that trapezoids are easily broken into rectangles and triangles; also, this method involves less computation than the previous method. I can approximate the area of the footprint with four trapezoids.

STRATEGY 4: Find the perimeter

We can trace the figure with string to get the perimeter of the print, then make a rectangle with the string and compute the area of the rectangle. There is a flaw in the reasoning behind this strategy. If you do not see this flaw, revisit this problem after doing the next investigation.

STRATEGY 5: Weigh it

We can trace this print onto a piece of thick posterboard and cut it out. We can then cut out another piece of posterboard in the shape of a rectangle. The ratio of the area of the footprint to the area of the rectangle will be equal to the ratio of the weight of the footprint to the weight of the rectangle. Therefore, we can use the following proportion:

$$\frac{\text{Area of footprint}}{\text{Area of rectangle}} = \frac{\text{weight of footprint}}{\text{weight of rectangle}}$$

INVESTIGATION 10.9

Making a Fence with Maximum Area

Joshua has decided to convert part of his back yard into a garden. He knows from his neighbor's experience that he needs to fence in the garden to keep the animals out. At the hardware store, he buys 200 feet of chicken wire. After he gets home and unrolls the wire, he finds to his surprise that there are many different size gardens that he can make with 200 feet of fencing. He decides that he wants to get the largest garden from 200 feet of fencing. If he wants a garden that is rectangular in shape, what are the dimensions of that garden? Work on this before reading on. . . .

DISCUSSION

STRATEGY 1: Use guess–check–revise

TABLE 10.9

Length	Width	Area	Reasoning
80 feet	20 feet	1600 square feet	If the perimeter is 200, then $l + w = 100$. Reduce the length by 10 and increase the width by 10.
70 feet	30 feet	2100 square feet	This has more area, so reduce the length by 10 and increase the width by 10.
60 feet	40 feet	2400 square feet	This has more area, so reduce the length by 10 and increase the width by 10.
50 feet	50 feet	2500 square feet	Why does this feel like it "should" be the biggest garden?

STRATEGY 2: Reason and make a model

A very tactile way to "feel" this investigation is to cut a piece of string and tie the ends together. Now, using your two thumbs and two forefingers, make as skinny a rectangle as you can (see Figure 10.21).

Now slowly move your thumbs and forefingers apart—that is, decrease the length and increase the width. What seems to be happening to the area? Why must it stop increasing when the length and width are equal?

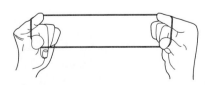

FIGURE 10.21

STRATEGY 3: *Be creative and adventurous*

What if we made the garden in the shape of a circle? Do this before reading on. . . .

Because the circumference of the circle is 200 feet, the radius of the circle will be $r = C/2\pi \approx 200/(2 \cdot 3.14) = 31.8$ feet. Now that we know the radius, we can find the area:

$$A = \pi r^2 \approx 3.14(31.8)^2 = 3175 \text{ square feet}$$

Compare the area of the circular garden to that of the square garden: The circular shape will give Joshua more than 25 percent more area.

Summary

In this section, we examined some measurement ideas and procedures related to perimeter and area. In one sense, perimeter concerns how much is needed to surround an object, and area involves how much is needed to cover an object. We have learned that π can be viewed both as how many times a diameter can wrap around a circle and as the ratio between a circle's circumference and its diameter. We have explored area formulas (and why they work) for triangles, for certain quadrilaterals, and for circles. We have explored the Pythagorean theorem and an application of this theorem in problem-solving. We have moved beyond more routine applications of perimeter and area to examine how areas of irregularly shaped figures are determined. Finally, we have found that the relationship between perimeter and area is not a simple one. For example, if we double the area of a square, we do not double its perimeter. Similarly, two objects can have the same perimeter but different areas, and two objects can have the same area but different perimeters.

EXERCISES 10.2

1. Determine the area and perimeter of the figures below.

   a. Acute scalene triangle

   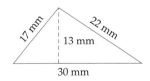

   b. Obtuse scalene triangle

   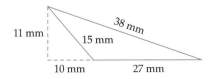

   c. Equilateral triangle

   d. Circle

e. Parallelogram

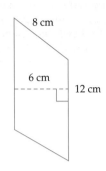

f. Trapezoid

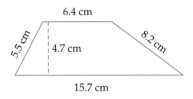

2. Determine the area and perimeter of the figures below.

a.

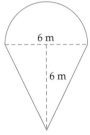

b.

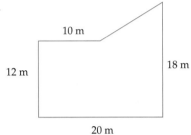

3. Determine the area of each of the following polygons on the Geoboard Dot Paper.

a. 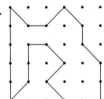 b. 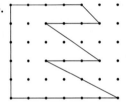 c.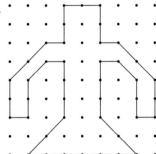

4. Find the area of the triangles below in square centimeters, using a ruler to measure the appropriate lengths.

a. 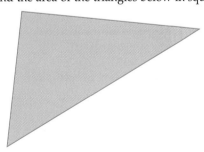 b.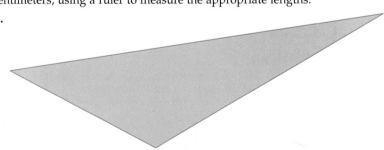

5. Determine the length of the arc $\overarc{AB}$ if the diameter of the circle is 10 feet and the angle is 128 degrees.

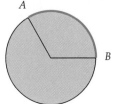

6. Find the length of x, to the nearest tenth of a foot. *Note:* The figure is not drawn to scale.

7. Determine the area of each figure below in two different ways.

   a.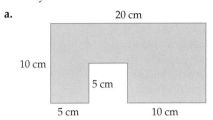

   b.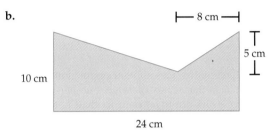

8. Let's say we have a square whose sides all measure 10 inches. Determine to two decimal places the dimensions of the square that has twice the area of this square.

9. How many square inches are there in 1 square foot?

10. Bernice wants to seed her large back yard with grass. Her back yard is in the shape of a rectangle with dimensions 120 feet by 80 feet. The seed costs $3.99 for a 1-pound bag, and each bag covers up to 1050 square feet. How much will the seed cost?

11. One bag of bark mulch will cover 18 square feet. If a landscape architect wants to have a border of plants around three sides of a 6-foot by 10-foot shed, how wide can the border be?

    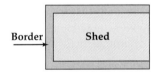

12. How big is an acre? The mathematical answer is 43,560 square feet. The acre was originally an English unit of measurement that described the area that a yoke of oxen could plow in a day. Neither of these measurements gives most people a sense of the size of an acre. Find an area that you have a sense of, such as a football field, a parking lot, or a large supermarket and determine the number of acres in that area.

13. The indoor sod for the 1994 World Cup match in the Silverdome was grown in hexagonal units. Each hexagon was 7.5 feet on a side. How many hexagons were needed to cover the field? The soccer field was about 240 by 360 feet.

14. A farmer has a fence that encloses a square plot with an area of 36 square meters. If the farmer uses this fence to enclose a circular flower garden, what will the area of the garden be?

15. A Little Leaguer asks her mother to make her a home plate for a sandlot baseball game. According to her encyclopedia, an official plate is made from a square by making two 12-inch diagonal cuts as shown below. What is the length of the side of the original square?

16. What percent of the quilt square below is dark blue?

    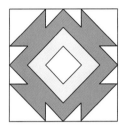

17. An equilateral triangle whose area was 100 square centimeters was reflected to make a congruent equilateral triangle so that the six smaller triangles are all congruent.

    a. What is the area of each of the little triangles?
    b. What is the area of the hexagon in the middle?

18. If each of the circles has a diameter of 10 meters, what is the area of the region inside the four circles?

19. The figure below is a common representation of the famous yin-yang relationship, which originated in China.

    a. Can you explain how the figure is made?
    b. If the radius of the circle is r, what is the length of the curved line in the interior of the circle that separates the white region from the dark region?

20. A circular flower bed is 6 meters in diameter and has a circular sidewalk 1 meter wide around it. Find the area of the sidewalk in square meters.

21. Sixteen circles (of equal area) are cut out of a square sheet of tin whose sides are 40 centimeters. What is the area of wasted tin if the circles are made as large as possible?

22. How many 2-inch by 3-inch by 8-inch bricks will you need to build a (uniformly wide) brick walk with the shape shown in the figure below? (Lay the bricks so that the largest face is up.)

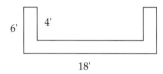

23. a. *Classroom Connection* Suppose a goat was tethered at the middle of a 100-foot-long fence and the length of the rope was 50 feet, as shown in the diagram below. Over how much area could the goat graze? Describe any assumptions you make in order to solve the problem.

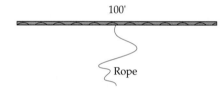

b. What if the point at which the goat was tethered was at the middle of a 50-foot-long fence, as shown in the diagram below? Over how much area could the goat graze? Describe any assumptions you make in order to solve the problem.

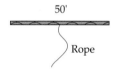

 Classroom Connection You can find a discussion of variations of the goat problem in "Reflections on Mathematics and the Connected Mathematics Project" in the February 1999 issue of *Mathematics Teaching in the Middle School*, pp. 324–330.

24. When I was a child, we visited Sequoia National Park. At that time, there was a redwood that we could actually drive through. Impossible you say? If the circumference of the tree was 83 feet, what was the diameter of the tree?

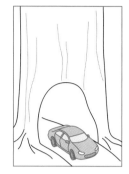

25. Measurement provides another opportunity for children to gain a better sense of large numbers. How many blades of grass are there on a football field? To answer this question, select some small samples of grass on a field and count the number of blades of grass in that sample. Then use those data to determine the number of blades of grass on the field. You can read the experiences of one class that did this project in "Counting Grass" in the September 1999 issue of *Mathematics Teaching in the Middle School*, pp. 7–10.

26. a. Make two rectangles that have the same perimeter but differ in area.

b. Make two rectangles that have the same area but differ in perimeter.

c. Make two parallelograms that have the same perimeter but differ in area.

d. Make two parallelograms that have the same area but differ in perimeter.

27. Archaeologists need to know geometry! Suppose the following piece of a wheel has been found. The archaeologist wants to know how big the actual wheel was. How would you determine this?

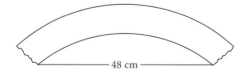

28. *Classroom Connection* This problem comes from "Figure This!" (**http://www.figurethis.org/**). The diagrams below show two kinds of windshield wipers. Which wiper cleans the greater area? Both wipers sweep a portion of a circle, although some thought will be required to figure out the shape that the second wiper covers.

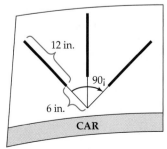 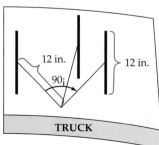

29. a. *Classroom Connection* Cut a 5×5 Geoboard into eight equal regions such that none of the regions is rectangular in shape.

b. Below is one second-grader's solution, which is shown in "Exploring Area with Geoboards" in the October 1997 issue of *Teaching Children Mathematics*, pp. 72–75. Verify that the solution is correct.

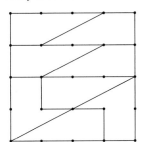

Source: Reprinted with permission from *Teaching Children Mathematics,* copyright © 1997 by the National Council of Teachers of Mathematics.

30. *Classroom Connection* The unit Geoboard consist of a rubber band that encloses the smallest square you can make on a Geoboard.

a. Make a square that has twice the area of the unit square. You can read children's solutions in "Responses to the Patterns in Squares Problem" in the November 2001 issue of *Teaching Children Mathematics*.

b. Make a square that has five times the area of the unit square.

c. Do you think you could make a square that has an area equal to any whole number on Geoboard Dot Paper? Why or why not?

31. Suppose a wire that is stretched tightly around the Earth is elongated by 5 meters and then placed back around the Earth. Could a person crawl under the elongated wire?

32. Let us examine three ways to travel from point *A* to point *D* on the cube below. Assume that the cube is 1 meter on each side.

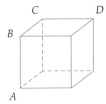

a. What would the distance be if you went from *A* to *B* and then from *B* to *C* and then from *C* to *D*?

b. If you had to travel on the surface of the cube but didn't have to travel on edges, what would be the shortest distance?

c. What if you could go (in space) directly from *A* to *D*? Now what would be the shortest distance?

d. Compare the three distances.

33. How many right triangles with sides 3 inches and 6 inches can be cut from a sheet of paper that is 18 inches by 40 inches?

34. The steps to the entrance of a school are 40 inches high. The school has decided to make the building wheelchair-accessible. If a city ordinance states that the ramp cannot be steeper than 5 degrees, how long must the ramp be? Solve this without using trigonometry.

35. Some people have an intuitive sense of whether things will fit or not. For example, let's say you have bought a circular table top with a diameter of 7 feet and a thickness of 3 inches, and your doorway is 78 inches by 30 inches. First, predict whether or not the table top will fit. Then determine whether or not it will fit.

36. Compare the Babylonian formula for the area of the circle, $A = C^2/12$, and the Egyptian formula, $A = \left(\frac{8}{9}d\right)^2$. Which is more accurate? Justify your choice.

37. *Classroom Connection* *The Librarian Who Measured the Earth* by Kathryn Lasky tells the story of Eratosthenes (New York: Little, Brown, 1994). Around 240 B.C., Eratosthenes, a Greek mathematician, estimated the circumference of the Earth. His estimate was off by only 1 percent! He knew that on a certain day, the sun would be directly overhead at a town called Aswan. However, the sun would not be exactly overhead in Alexandria, which was about 490 miles north of Aswan. Using the figure below, determine how Eratosthenes measured the circumference of the Earth.

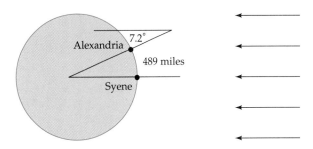

38. On a blank sheet of paper, draw from memory:
- A rectangle the size of a dollar bill
- A rectangle the size of a floppy disk
- A circle the size of a penny
- A circle approximating the top view of a soda can

a. Before you check your estimates, predict the relative accuracy of your estimates, from 1 to 4, with

1 denoting the drawing that you predict will be "closest" to the actual shape.

b. Determine the actual area of each of your shapes and of the actual objects. Describe how you determined which of your measurements was "closest" to the actual measurement.

39. In Investigation 10.8, we discussed several strategies for finding the area of irregularly shaped figures. Describe a situation for which one of the strategies would be preferable to the others. Make a convincing argument that the particular strategy would indeed be more practical and/or more accurate.

40. Describe how you would find the surface area of the top of each of the following:

a.   b.   c.

41. *Classroom Connection* The April 1999 issue of *The Mathematics Teacher* (pp. 294–298) describes a project in which high school students used Pick's theorem to determine the area of leaves, which biologists need to do. The formula is $A = b/2 + I - 1$, where b is the number of boundary points and I is the number of interior points in a figure that has been traced on Geoboard Dot Paper.

a. Use your knowledge of triangles and other shapes to find the area of the figure at the left.

b. Calculate the area of the figure at the left using Pick's theorem.

c. Determine the area of the leaf at the right using Pick's theorem.

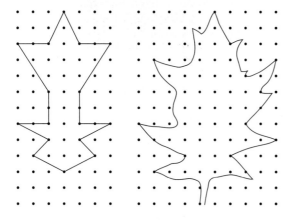

42. Make a photocopy of a map of your state and determine the area of your state.

a. Show and explain your work.

b. Justify the precision in your answer—for example, the choice of unit and the number of decimal places, if any.

c. Describe and explain your degree of confidence in your result.

d. Describe any difficulties you had and how you overcame them.

43. Several years ago the Treasury Department designed a new $100 bill and printed $80 billion in new $100 bills. How large a room would be needed to contain all these $100 bills?

44. One way of buying tennis balls is in a tin that contains three tennis balls. Is the tin taller or bigger around?

45. Draw on Geoboard Dot Paper a polygon whose perimeter is 20 units and whose area is less than 10 square units. Then draw a figure whose perimeter is 20 units and whose area is greater than 20 square units. Describe your solution process. This description should include your reasoning, guesses that didn't work, and what you learned from those guesses.

46. Let's say you have two similar polygons and the ratio of their perimeters is 2:1. What is the ratio of corresponding sides? What is the ratio of their areas? Justify your answer.

47. How is the size of a TV measured? If we are comparing a 15-inch TV and a 30-inch TV, what is twice as big about the bigger TV?

48. If possible, determine the perimeter of the figures below. If there is not enough information, explain why.

a.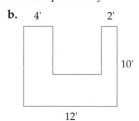

b.

c.

49. If possible, determine the area of the figures below. If there is not enough information, explain why.

a. b.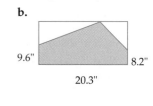

50. For each of the problems below, if there is not enough information given for there to be exactly one answer, explain why and provide one more piece of information that would enable us to determine the solution. If there is enough information, solve the problem.
    a. The hypotenuse of a right isosceles triangle is 10 inches. Find the area.
    b. The perimeter of one rectangle is twice the perimeter of another rectangle. The dimensions of the rectangle with the greater perimeter are 8 inches by 6 inches. Find the dimensions of the smaller rectangle.
    c. The area of a rectangle is 20 centimeters. Find its perimeter.
    d. A farmer is fencing his pasture. The fence posts will be 10 feet apart. The pasture is in the shape of a rectangle whose length is double its width. The farmer needs 54 fence posts. What are the dimensions of the pasture?

51. Anne used a grid to determine the area of an irregularly shaped figure, and she determined that the area was 23 square centimeters. Then she realized that she wanted square inches. How can she convert from square centimeters to square inches? There are 2.5 centimeters in 1 inch.

52. Adam is also confused. He wonders why, if you take a rectangle and multiply the length of each side by 1.5, the area of the new rectangle isn't 1.5 times as big as the area of the old rectangle. How would you explain this to him?

53. If a circle has a radius of 5 inches, then using the formula, we say that the area of the circle is approximately 78.5 square inches. A fellow student has come to you and asked why we use *square inches* to represent the area of the circle. What does 78.5 square inches mean?

54. Cattle ranching is a primary cause of deforestation in Latin America. Since 1960, more than one quarter of all Central American forests have been razed to make pasture for cattle. Just one quarter-pound hamburger imported from Latin America requires the clearing of 6 square yards of rain forest. How much land would be required to provide one hamburger per year for each person in the United States?

55. How much land is used by all the streets and highways in the United States? As of 1990 there were 3,880,000 miles of roadway.

56. The Reverend Robert Walsh served aboard a ship designed to intercept slave ships in 1829. His account of the capture of the *Feloz* reveals the incredible conditions. He reported that 336 male slaves had been crammed into a space that was 3 feet 3 inches high and 40 feet by 21 feet. How much room did each man have?

57. A kitchen gadget called Perfect Pasta Portions enables people to determine the appropriate amount of spaghetti to boil for 1, 2, 3, or 4 portions. If the diameter of the hole for 1 serving is 7/8 inch, what is the diameter of the hole for 4 servings?

58. *Classroom Connection* How many blades of grass are there on a soccer field? Look in the September 1999 issue of *Mathematics Teaching in the Middle School*, pp. 7–10, to see the children's work.

59. *Classroom Connection Counting on Frank* by Rod Clement (Milwaukee: Gareth Stevens Publishing, 1991) examines the kinds of questions that kids ask. Let us consider two here.
    a. How long would it take to fill a bathroom if both faucets are running?
    b. If ten humpback whales would fit in the narrator's house, how big is the narrator's house?

60. *Classroom Connection If You Hopped Like a Frog* by David Schwartz (New York: Scholastic Press, 1999) tells the readers that if you ate like a shrew, you would eat three times your weight every day.
    a. How many hamburgers would that be?
    b. How many boxes of cereal?

61. *Classroom Connection*
    a. What is the greatest number of blank squares you could enclose if you colored in 20 squares on a sheet of graph paper with the condition that each square had to have at least one entire side connected to an entire side of another square?
    b. What if the condition was changed so that two squares could touch just at a vertex? This problem was posted in *Teaching Children Mathematics* in a more child-friendly way using cubes to make a pen to hold animals who took up the space of one cube. The children's solutions are shown in the December 1997 issue, pp. 212–214.

SECTION 10.3 SURFACE AREA AND VOLUME

WHAT DO YOU THINK?

- Is the relationship between surface area and volume a functional relationship? Why or why not?
- If we doubled the volume of a room, would its surface area double too?

In Section 10.2, we examined two attributes of two-dimensional objects: perimeter and area. In this section, we will examine two attributes of three-dimensional objects: surface area and volume. We can think of surface area as the area needed to cover all the faces of a three-dimensional object, and we can think of volume as the amount of space contained within that object.

Understanding the Surface Area of Prisms

Very simply, the *surface area of a prism is equal to the sum of the areas of all of its faces.* Thus, to determine the surface area, we need to be able to determine the dimensions of each face. Let us consider a rectangular prism first. Determine the surface area of the rectangular prism in Figure 10.22 and then read on. It may help first to draw a net of that prism, which is shown at the right.

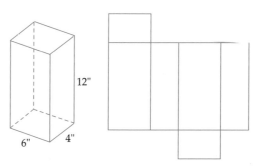

FIGURE 10.22

There are patterns in this prism that make its surface area easier to determine. You can see that the bottom and top bases are congruent; similarly, the front and back faces are congruent, and the right and left faces are congruent. Thus,

Surface area of the prism = $2(24 \text{ in}^2) + 2(72 \text{ in}^2) + 2(48 \text{ in}^2) = 288 \text{ in}^2$

With a concrete example under our belts, let us now investigate the surface area of prisms whose bases are regular polygons. We will use this thinking to develop the general formulas for the surface areas and volumes of some basic three-dimensional geometric figures. Most readers find it necessary to read this material more than once and to read actively, with pencil in hand. After all, I am presenting you a summary of discoveries that took hundreds of years of effort by some of the brightest people of their time!

What do we know about the different surfaces of prisms that would make the task easier? Look at the three diagrams in Figure 10.23, showing three prisms and their nets. What do you see? Think before reading on. . . .

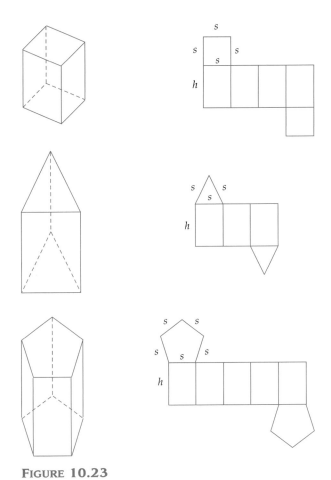

FIGURE 10.23

If you constructed nets in Exploration 10.14, you saw that each prism has a bottom base and a top base that are congruent. Therefore, the areas of the top and bottom bases will be equal. Similarly, the lateral surfaces (sides) of all these prisms are congruent; therefore, their areas will be equal. The number of lateral surfaces is equal to the number of sides of the base. If we let this number be n, we can state a generic formula for *the surface area of all prisms whose bases are regular polygons*:

Surface area = 2 times the area of one base + n times the area of one lateral face

Because the area of the base will not be the same in all cases, we can let B stand for the area of the base. The area of each of the lateral faces is equal to the side s times the height h (Figure 10.23), or $s \cdot h$, so we can now present a general formula:

Surface area of an n-gon prism = $2B + nsh$

We can further simplify the formula by noting that ns is equal to the perimeter p of the base. Thus an alternative representation of this formula (one that will be more useful later) is

Surface area of a prism = $2B + ph$

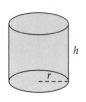

FIGURE 10.24

Understanding the Surface Area of Cylinders

Just as we saw a relationship between cylinders and prisms when we first examined them in Chapter 8, there is a relationship between the formulas for the surface areas of cylinders and prisms. In other words, the formulas for the surface areas of cylinders and prisms are connected. How do you think we might determine the surface area of a cylinder with a circular base (see Figure 10.24)? Even if you can't figure out the whole formula, write down what you do know. Then read on. . . .

DISCUSSION

One of the connections between prisms and cylinders is that both have two congruent bases. Thus, to find the surface area of a cylinder, we have to find the area of one base and multiply that by 2. What about the lateral surface area? How do you think we might find that? If you are not sure, find a cylinder. (Use an empty toilet paper tube or paper towel tube, or wrap a piece of paper around a soup can or soda can.) What is the shape of the lateral face? Do this yourself before reading on. . . .

Yes, it is a rectangle! The height of the rectangle is equal to the height of the cylinder. What about the length of the base of the rectangle? How is it connected to the cylinder's base? Think and then read on. . . .

The length of the base of the rectangle is equal to the circumference of the circle! See Figure 10.25. With this knowledge, see if you can now discover the formula for the surface area of any cylinder in terms of the radius of the cylinder and the height of the cylinder. Then read on. . . .

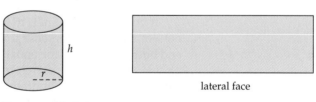

lateral face

FIGURE 10.25

- The area of the base is πr^2, and because we have two bases, the area of the two bases is $2\pi r^2$.
- The area of the rectangle = (the circumference of the circle) times (the height of the cylinder); that is, the area of the rectangle = $2\pi r h$.

Thus,

$$\text{Surface area of a cylinder} = 2\pi r^2 + 2\pi r h$$

Understanding the Surface Area of Pyramids

Whether we have a triangular pyramid, a square pyramid, or whatever, we have a base, and we have as many lateral faces as we have sides of the base. We will develop the formula for pyramids in which the apex is directly above the center of the base so that the lateral faces will be congruent triangles. Can you determine the surface area of the pyramid in Figure 10.26, if s represents the length of each side of the base and l is the slant height of the pyramid—that is, the altitude of each of the lateral faces? Try to do so before reading on. . . .

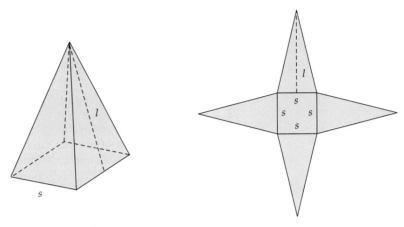

FIGURE 10.26

$$\text{Surface area of the pyramid above} = s^2 + 4\left(\frac{1}{2}sl\right) = s^2 + 2sl$$

In order to determine the general formula, we need to make two substitutions. First, as before, we will let B represent the area of the base of the pyramid. Second, the perimeter of the base of the pyramid in the formula above is equal to $4s$. Because we have a $2s$ in the second part of the formula, we have, in essence, one-half of the perimeter. Thus we will substitute $\frac{1}{2}p$ for $2s$. It will soon become apparent why we want to make this substitution. Substituting B for s^2 and $\frac{1}{2}p$ for $2s$, we have

$$\textbf{Surface area of a pyramid} = B + \frac{1}{2}pl$$

FIGURE 10.27

Understanding the Surface Area of Cones

Just as we saw connections between the surface areas of prisms and cylinders, there are connections between the surface areas of pyramids and cones. Both objects have bases, and both have slant heights (see Figure 10.27). If we model the formula for the surface area of a cone after that of a pyramid, we have

$$\text{Surface area of a pyramid} = B + \frac{1}{2}(\text{perimeter})(\text{slant height})$$

Thus,

$$\text{Surface area of a cone} = B + \frac{1}{2}(\text{perimeter})(\text{slant height})$$

What do we know about B, the area of the base of the cone? What is the equivalent of $\frac{1}{2}p$? What do you think?

For all cones that have a circular base, and so the area of the base B is equal to πr^2. The perimeter (circumference) of a circle is $2\pi r$; thus the value of $\frac{1}{2}p$ for the cone is πr.

Therefore, we can say that

Surface area of a circular cone $= \pi r^2 + \pi r l$

Understanding the Surface Area of Spheres

Understanding the derivation of the formula for the surface area of a sphere is somewhat sophisticated, so we will not go into the derivation here.

Surface area of a sphere $= 4\pi r^2$ *where r is the radius of the sphere*

One interesting connection is that the surface area of the sphere is exactly four times the area of a great circle of the sphere. If you are indeed actively reading this book, you are now thinking that you need to know what a great circle is to understand this relationship. What do you think a great circle is? Think of the planet Earth as a sphere. What might be a great circle on the Earth's surface? Stop and think before reading on. . . .

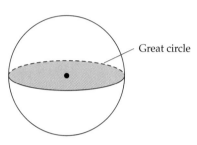

FIGURE 10.28

Imagine passing a plane through the center of a sphere. The intersection (cross section) of that plane and the sphere is a great circle (see Figure 10.28). Thus, any sphere has an infinite number of great circles. On Earth's surface, the equator and all lines of longitude are great circles.

Volume

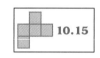

In Section 10.2, we connected the terms *perimeter* and *surround*, and we connected the terms *area* and *cover*. With respect to volume, we can say that **volume** reflects how much it will take to *fill* an object. Exploration 10.16 also helps you to understand basic concepts and formulas related to volume.

Understanding the Volumes of Prisms

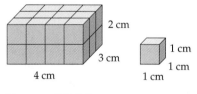

FIGURE 10.29

Let us first develop the formula for the volume of a rectangular prism (more commonly known as a box). Imagine a box whose base is 4 centimeters by 3 centimeters and whose height is 2 centimeters (see Figure 10.29). If we were to fill that box with little cubes, each with dimensions 1 centimeter by 1 centimeter by 1 centimeter—that is, with a volume of 1 cubic centimeter—how many little cubes would it take?

It would take 24 little cubes—that is, $4 \cdot 3 \cdot 2$. Therefore, it makes sense to say that *we can determine the volume of a rectangular prism by multiplying its length, its width, and its height:*

$$V = l \cdot w \cdot h.$$

Volume of an oblique prism What about an oblique prism? What if the two prisms in Figure 10.30 have the same height and the same base? Will they have the same volume, or will their volumes be different? What do you think?

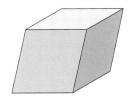

FIGURE 10.30

An approach that is helpful is to think of a stack of paper. We can arrange the stack of paper to look like each of the diagrams in Figure 10.31. However, we still have the same volume of paper. This principle was first articulated by an Italian mathematician, Bonaventura Cavalieri, in the early seventeenth century. Cavalieri proved that two solids with congruent bases and the same height will have the same volume if the following condition is true: Each cross section that is parallel to the plane of the bases has the same area for each solid.

 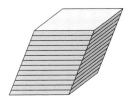

FIGURE 10.31

Thus we can say that the volume of any prism can be determined by multiplying the area of its base by its height.

Understanding the Volumes of Cylinders

What about cylinders? Can you connect what we have learned about determining the volume of prisms to determining the volume of cylinders? Think and then read on. . . .

The following discussion does not constitute a proof of the formula for determining the volume of a cylinder. Rather, it is designed to help you make sense of the formula. Recall the sentence in NCTM Curriculum Standard 3: Students should "believe that mathematics makes sense" (*Curriculum Standards*, p. 29).

In Section 10.2, we used our understanding of the area of a square to shape our understanding of the area of a circle. We will do something similar here. Consider the cylinder at the left in Figure 10.32. We can approximate the volume of the cylinder by breaking it down into a number of prisms—in this case, five prisms, as shown in the middle figure. Recall Investigation 10.8 (How Big Is the Footprint?) and Exploration 10.12 (Irregular Areas). We know that the volume of each of the prisms is Bh. Thus, if we denote the area of the bases of the prisms as B_1, B_2, B_3, and so on, we can say that the volume of the figure in the middle diagram is $B_1 h + B_2 h + B_3 h + B_4 h + B_5 h = (B_1 + B_2 + B_3 + B_4 + B_5)h$. That is, the volume of this object is equal to the product of the height and the total area of the bases. Thus it makes sense to conclude that the volume of a cylinder is equal to the product of the height and the total area of the bases. By breaking the cylinder into smaller and smaller prisms, as in the figure at the right, we can come closer and closer to the actual volume, just as we can closely approximate the area of a circle by covering it with a grid with smaller and smaller squares.

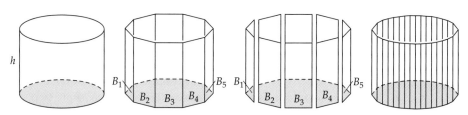

FIGURE 10.32

Finally, just as we applied Cavalieri's principle to state the formula for finding the volume of any prism, not just right prisms, so too we can apply Cavalieri's principle to state the formula for finding the volume of any cylinder, not just right cylinders:

$V = Bh$ where B *represents the area of the base*

Understanding the Volumes of Pyramids

Let's look at pyramids. The formula is difficult for most people to develop, but it becomes much more understandable with concrete models.

We will illustrate the formula using a cube. It so happens that three congruent pyramids will fit inside the cube. Most people need to see this to believe it, and Figure 10.33 provides the necessary information to make three congruent pyramids that can be joined together to make a cube.

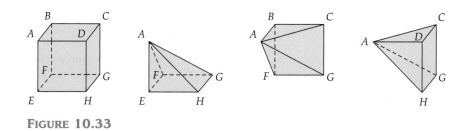

FIGURE 10.33

At this moment, let us examine the implications of this relationship for determining the volume of a pyramid. On the basis of this information, what do you predict will be the formula for the volume of a pyramid? Think before reading on. . . .

If three congruent pyramids can be fit into the cube, and the bases of the pyramids are congruent to the base of the cube, this means that the volume of the pyramid is one-third the volume of the cube and that *the formula for finding the volume of a pyramid is*

$$V = \frac{1}{3}Bh \quad \text{where B is the area of the base (any face) of the cube}$$

Understanding the Volumes of Cones

FIGURE 10.34

What about cones? Knowing the formula for the surface area of a pyramid and thinking of the similarities between pyramids and cones, can you hypothesize a formula for the volume of a cone with radius r, height h, and slant height l (see Figure 10.34)? Try to do so before reading on. . . .

We can deduce that just as the volume of a pyramid is one-third the volume of a prism with the same base and height, the volume of a cone must be one-third the volume of a cylinder with the same base and height:

$$\text{Volume of a cone} = \frac{1}{3}Bh$$

Understanding the Volume of Spheres

As was true of the formula for the surface area of a sphere, the derivation of the formula for the volume of a sphere is rather sophisticated.

$$\text{Volume of a sphere} = \frac{4}{3}\pi r^3 \quad \text{where r is the radius of the sphere}$$

There is an interesting connection between the volume of a sphere and the volume of a cylinder that you can verify if your instructor has a hollow cylinder and sphere of the same radius. The height of the cylinder must be equal to the diameter of the cylinder. Recall that the volume of a cylinder is Bh; if the base of the cylinder is a circle of radius r and the height of the cylinder is $2r$, then the volume of that cylinder is $2\pi r^3$.

First imagine placing the sphere inside the cylinder (Figure 10.35). Next, imagine filling the cylinder with water. Finally, we take out the sphere and put the water back in the cylinder (Figure 10.36). If this experiment is done carefully, the height of the water will be 1/3 the height of the cylinder. It logically follows that the volume of the sphere is 2/3 the volume of the cylinder, so we have

$$\text{Volume of sphere} = \frac{2}{3}(2\pi r^3) = \frac{4}{3}\pi r^3$$

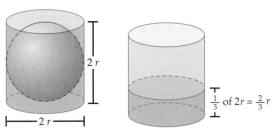

FIGURE 10.35   FIGURE 10.36

Archimedes was the first person to prove the formula for the volume of a sphere.

Outside the Classroom

I hope that in this course, you are coming to realize that mathematics is more than just formulas. With respect to understanding spheres, one of the more interesting mathematical problems is determining how to represent that sphere in two dimensions—for example, making a map of the planet Earth.

If you look at the traditional (Mercator) map of the world (Figure 10.37), you can see that this map misrepresents the relative sizes of the continents. For example, the area of Africa is $1\frac{1}{2}$ times the area of North America. However, North America appears to be much larger than Africa in the Mercator projection. In 1974, the Peters projection (Figure 10.38) was created to represent accurately the relative sizes of Earth's land masses. The mathematical problem here is representing a three-dimensional object in two dimensions. Thus, if you choose to make a map of the planet that is rectangular in shape (there are other possibilities), either you can have the shapes accurate (that is, this is what the continents would look like viewed from space) or you can have the relative sizes accurate, but you cannot have both! This is but one of many examples where the mathematical answer to a problem is "There is no perfect solution." There are many websites where you can explore this map-making issue in more depth, and there are many websites with lesson plans for elementary schools.

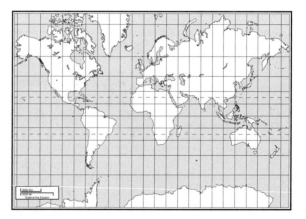

FIGURE 10.37

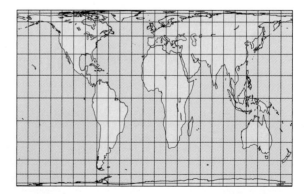

Source: Mercator projection map and Gall-Peters equal-area projection map, from *Which Map Is Best? Projections for World Maps* (1986). Courtesy of American Congress on Surveying and Mapping. Used with permission.

FIGURE 10.38

INVESTIGATION 10.10

Are Their Pictures Misleading?

More and more newspapers, magazines, and brochures that companies put out are using graphs to display data in a way that will catch the reader's eye. The hypothetical example below illustrates the dangers. Let's say that Yummy Soda sold twice as much soda as Good Soda last year. The graph in Figure 10.39 is a pictorial representation of the sales of the two sodas that Yummy included in an ad. Why do you think Good Soda might object to the graph and actually file a lawsuit for unfair advertising? Write down your thoughts before reading on. . . .

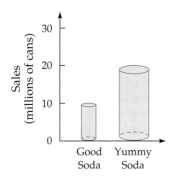

FIGURE 10.39

DISCUSSION

If the reader looks at this as simply a "cute" bar graph, the height of the Yummy bar is twice the height of the Good bar. However, the diameter of the base of the Yummy can is twice the diameter of the base of the Good can, causing the Yummy can to look more than twice as big as the Good can. What is the ratio of the volumes of the two cans? Determine this before reading on. . . .

One solution is to take the actual dimensions:

- The Good Soda can has a radius of 2 millimeters and a height of 10 millimeters.
- The Yummy Soda can has a radius of 4 millimeters and a height of 20 millimeters.

Using the volume formula for cylinders, $V = \pi r^2 h$, we have

- Volume of Good Soda can = 40π
- Volume of Yummy Soda can = 320π

That is, the Yummy Soda can has 8 times the volume of the Good Soda can. *Note:* Do you see why we used π instead of using a decimal approximation? We also could have used the ratios—that is,

Volume of Good Soda can = $\pi r^2 h$

Volume of Yummy Soda can = $\pi(2r)^2(2h) = 8\pi r^2 h$

In either case, we find that the graph gives a visual impression that Yummy Soda is selling much more than twice as much as Good Soda. (It also gives the impression that Yummy cans of soda are bigger!)

Determining the Volumes of Irregularly Shaped Objects

In Section 10.2, we acknowledged that formulas have limited usefulness in real life. Often we need to find the area of an irregularly shaped figure. Similarly, we often need to find the volume of a solid to which the formulas do not apply. How can we do this? Think about this before reading on. . . .

This is actually a very challenging problem that stumped ancient mathematician-scientists for quite some time. One way of addressing this question was developed by one of the greatest mathematicians of all times,

Archimedes, who lived in the city of Syracuse, located in modern Italy, over 2000 years ago.

Archimedes was summoned by the king, who suspected that the goldsmith who had made his crown had not made it of solid gold but had mixed some silver in it. He wanted Archimedes to determine whether the crown was pure gold without having to cut into it.

Poor Archimedes. In those days, one did not disappoint kings. In fact, it was not unusual for disappointed kings to banish or even execute advisors who failed to solve their problems.

Archimedes knew that silver was only about half as dense as gold. (We determine the density of an object by dividing its mass by its volume—for example, grams per cubic centimeter.) Archimedes was aware of the formula volume × density = mass (for example, $cm^3 \times g/cm^3 = g$). Thus, if he could determine the volume of the crown, he would be able to solve the problem. But how was he to find the volume of the crown accurately?

Some time later, possibly that same day, Archimedes was taking a bath when the solution suddenly came to him: He could determine the volume of the crown by submerging it in water, because two objects with the same volume will displace the same amount of water. Legend has it that he became so excited that he ran outside, still naked, shouting "Eureka," which means, in Greek, "I have found it!"

How did this discovery enable him to determine whether the crown was counterfeit?

Now that he knew a way to determine the volume with some precision, he could use this number and the density of gold in the formula given above to determine how much the crown would weigh if it were pure gold. (It turns out that the goldsmith had indeed cheated the king!)

INVESTIGATION 10.11

Finding the Volume of a Hollow Box

Imagine a large object that will be used to hold water or other materials (see Figure 10.40). The outer dimensions are 12 meters by 10 meters by 8 meters. The walls are 1 meter thick, and so is the floor. What is the volume of the container in cubic meters? For example, if we were to make this object from concrete, we would need to know how much concrete to order. Work on this until you determine the volume or become stuck. Then read on. . . .

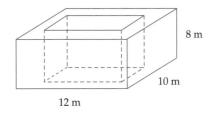

FIGURE 10.40

DISCUSSION

STRATEGY 1: *Break the problem into parts*

Just as a rectangular prism has six surfaces, a prism with no top will have five surfaces. Let us identify them and then determine the dimensions of each of them: bottom, front, right side, left side, back. If you did not think of this strategy, seeing if you can use it to determine the volume will be a useful exercise of your problem-solving tools. If you used a different strategy, do you get the same answer using this strategy? Work on this before reading on. . . .

The volume of the bottom will be

$$12 \text{ m} \cdot 10 \text{ m} \cdot 1 \text{ m} = 120 \text{ m}^3$$

The front and back sides are congruent, and the volume of each will be

12 m · 7 m · 1 m = 84m³

(Do you see why it is not 8 m? If not, make a model of the container.)
The right and left sides are congruent, and the volume of each will be

8 m · 7 m · 1 m = 56 m³

(Do you understand these dimensions? If not, make a model.)

STRATEGY 2: See the problem from a different perspective

For example, determine what the volume would be if the container were not hollow. Then determine the volume of the hollow region. Then subtract! Do you understand this strategy?
Volume of the whole region:

12 m · 10 m · 8 m = 960 m³

Volume of the hollow:

10 m · 8 m · 7 m = 560 m³

What is the difference between the two? Does this answer match the answer from Strategy 1? Exploration 10.16 provides more opportunities to apply your understanding of volume concepts.

Again it is important to note that this discussion illustrated only two of several possible ways to solve this problem.

INVESTIGATION 10.12

CLASSROOM CONNECTION

A similar problem occurs in Challenge 62 of "Figure This!" (**http://www.figurethis.org/**), where students are asked to figure out why a large block of ice will melt more slowly than the same block cut into three cubes.

Surface Area and Volume

In Section 10.2, we discovered that perimeter and area are not related in a simple way; for example, if you double the perimeter, that doesn't necessarily mean that you double the area. Let us investigate the relationship between surface area and volume. As you can imagine, there are many applications of this relationship in fields outside mathematics!

Consider a set of eight small cubes, each of whose dimensions are 1 centimeter by 1 centimeter by 1 centimeter (see Figure 10.41).

FIGURE 10.41

What arrangement of those cubes will have the smallest surface area?
What arrangement of those cubes will have the largest surface area?

Work on these questions before reading on. . . .

DISCUSSION

There are two aspects of this question that are not well defined. First, some students, applying this idea to real-life phenomena such as melting ice, consider the "surface area" to be the amount of surface exposed. Thus they do not count the base. Other students imagine having to paint the whole arrangement, and thus they do count the base. In this investigation, when we say "surface area," we will mean the latter.

The second part of the question that is not well defined is "arrangement." In this investigation, in order for something to count as an arrangement, at least one entire face of each cube will have to cover an entire face of another cube. Thus, simply putting the six cubes down on the table so that they are all disjoined does not count as an "arrangement" as I am defining this problem. With these two aspects of the problem now well defined, solve the problem before reading on. . . .

FIGURE 10.42

Arranging the cubes into a large cube will create the smallest surface area (see Figure 10.42). If you didn't find this, make them into a cube (or draw a diagram) and determine the surface area.

Because all six faces of the cube are congruent, we can multiply the area of one face (4 cm^2) by 6 (the number of faces), to get 24 cm^2.

The largest surface area is obtained by arranging the six cubes in a line (see Figure 10.43). If you didn't find this, please determine the surface area. If you did, you may want to compare your solution path with that of another student. Then read on. . . .

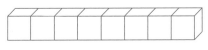

FIGURE 10.43

STRATEGY 1: See this arrangement as a prism with six sides

The surface area of the front, top, back, and bottom is 32 cm^2.
The surface area of the two bases (the two sides in this diagram) is 2 cm^2.
Thus the total surface area is 34 cm^2.

STRATEGY 2: See this arrangement as two end cubes and six cubes in the middle

The surface area of the six cubes in the middle is 24 cm^2.
The surface area of the two cubes on the end is 10 cm^2.
Thus the total surface area is 34 cm^2.

Summary

If this section has been successful, the various formulas for surface area and volume make sense. In order to make sense of these formulas, we used some strategies like the ones we used to make sense of area formulas in Section 10.2 (breaking a figure into parts) and built on previous knowledge. Connections were an important aspect of making sense of these formulas. That is, the formulas for finding the surface area and volume of prisms are fairly straightforward. By seeing how prisms and cylinders are similar, we could then use the prism formulas to understand the cylinder formulas. Similarly, we connected the volume of a pyramid to the volume of a cube in order to understand the formula for the volume of a pyramid. After we developed the formula for finding the surface area of a pyramid, seeing how pyramids and cones are similar enabled us to make sense of the formula for the surface area of a cone. Understanding the formulas and having a good problem-solving toolbox then enabled us to solve nonroutine problems. Finally, we examined the relationship between surface area and volume, realizing as we did so that this relationship, like that between perimeter and area, is not simple.

Exercises 10.3

1. Determine the surface area and volume of each of the following:

   a. ∡ABC is a right angle.
      AB = 10 inches, BC = 12 inches
      AD = 3 feet

   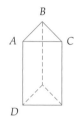

   b. The base is a square.

   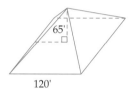

   c. The diameter is 5 feet 6 inches. The height is 11 feet 3 inches.

2. a. Determine the surface area and volume of the watering trough at the left below. The length of the trough is 16 feet, each of the two ends of the trough is an isosceles triangle whose base is 2 feet, and the height is 1 foot.

   b. Determine the surface area of the inside of the room in the center below and the volume of the room. The base of the room is in the shape of an isosceles trapezoid. The longer base is 12 feet 4 inches, and the shorter base is 8 feet 9 inches. Both slant sides are 13 feet. The height of the room is 9 feet.

   c. The base of the pyramid at the right below is a square. The slant height of the pyramid is 18 feet. Determine the surface area and volume.

   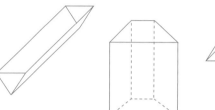

3. a. Determine the surface area and volume of the swimming pool below.

   b. Determine how many gallons of water the pool will hold.

   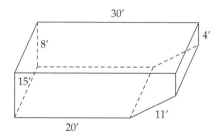

4. Before we had children, my wife and I went camping in a pup tent, shown at the left in the figure that follows. After having two children, we bought a much bigger tent, like the one in the center. Determining the volume of that tent is rather complex, so let us assume that the big tent was actually a square pyramid, like the figure at the right, with a height of 8 feet.

   a. How much bigger is the new tent than the old tent? First predict the answer and explain the reasoning behind your prediction. Then do the computations.

   b. Compare the amounts of material needed to make these tents. First predict the answer and explain the reasoning behind your prediction. Then do the computations.

   c. Using what you have learned about two- and three-dimensional figures in the last three chapters, describe your hypotheses for how you might determine the actual volume of the middle tent. The height of the truncated pyramid is 6 feet, and the height of the small pyramid at the top is 2 feet.

   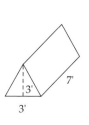
   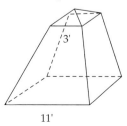
   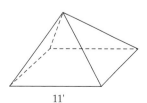

5. The napkin ring pictured is to be resilvered. How many square millimeters of surface area must be covered?

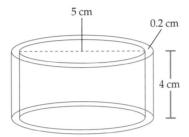

6. This past year the local elementary school has bought 50 boxes of paper, each box containing 10 reams. Each ream contains 500 sheets of paper. There are two shelves in the copy room, each 12 inches wide and 5 feet long. The second shelf is 2 feet above the first shelf and 3 feet below the ceiling. Assuming that the shelves are sturdy, how many sheets of paper can be stored on the shelves? What percent of the yearly purchase is this?

7. The U.S. federal debt, at the time of the writing of this exercise, was $4.8 trillion. One way to get a sense of the enormousness of that number is to imagine how much space 4.8 trillion one-dollar bills would occupy. Determine the space. Then recommend a denomination that would enable a semitrailer to haul that much money.

8. The planned Three Gorges Dam in China will require approximately 950 million cubic feet of concrete. Do something with this number to make it more understandable. For example, it would be equivalent to a block 1 mile long, 1 mile wide, and x feet high.

9. The Great Pyramid was built around 2600 B.C. and is considered to be one of the "Seven Wonders of the World." It was built for Pharaoh Khufu. The pyramid was made from about 2.3 million stone blocks, whose total weight was about 6 million tons! It has been estimated that it took 100,000 workers about 30 years to make it, and we believe they hauled the blocks only during the three months of the year when the Nile was flooded—that is, when the farmers weren't farming. The pyramid has a square base, and each side is 768 feet. The height of the pyramid is 481 feet.
   a. Compare the volume of the pyramid to the volume of the building in which you are taking this course.
   b. Compare the volume of the pyramid to the volume of a building 100 feet high covering a football field (120 yards long and 55 yards wide).
   c. Compare the Great Pyramid to the Transamerica Pyramid in San Francisco. The height and base of the Transamerica Pyramid are 870 feet and 117 feet, respectively.
   d. When I was researching this book, one of the books I read said that the height of each of the triangular faces of the pyramid was 232 meters. Is that figure compatible with the height of the pyramid given here, 481 feet?
   e. Many books have been written about many of the "coincidental" measurements of the pyramid. For example, it is said that the area of one of the triangular faces of the pyramid is equal to the square of the height of the pyramid. Are these two dimensions equal?

10. Joe has eight blocks of ice, each measuring 30 centimeters by 30 centimeters by 30 centimeters.
    a. How should he stack them if he wants them to melt as slowly as possible?
    b. How should he stack them if he wants them to melt as quickly as possible?

11. Let's say a company is manufacturing juice boxes that are rectangular prisms with dimensions 4 inches by 3 inches by 4 inches. The company has a warehouse whose dimensions are 40 feet by 30 feet by 20 feet. If the company makes 200,000 juice boxes, will it be able to store all the juice boxes in the warehouse?

12. Sharon has 150 cassettes and has decided to make a wooden shelf to hold them. The space between cassettes need only be 2 millimeters. Design a shelf that will hold at least 200 cassettes.

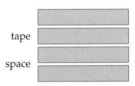

13. Given eight cubes, each with dimensions of 1 centimeter by 1 centimeter by 1 centimeter, can you find arrangements that will have surface areas for every whole number between the minimum and maximum that were found in Investigation 10.12—that is, between 24 square centimeters and 34 square centimeters?

14. One inch is about 2.54 times as long as a centimeter. We could say that it takes just over $2\frac{1}{2}$ centimeters to make an inch.
    a. Predict how many cubic centimeters it would take to fill a container that was 1 cubic inch in volume. Explain your reasoning.
    b. Determine the actual ratio.
    c. Assess your prediction, both in terms of percent error and in terms of reasoning.

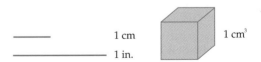

15. Consider two cubes. The smaller cube measures 4 inches on a side, and the larger cube measures 10 inches on a side. What is the ratio of the surface areas of the two cubes? What is the ratio of the volumes of the two cubes?

16. Consider the two similar rectangular prisms below (not drawn to scale). The dimensions of the smaller prism are 3 inches by 2 inches by 4 inches, and the dimensions of

the larger prism are 9 inches by 6 inches by 12 inches. That is, the ratio of their sides is 1:3.

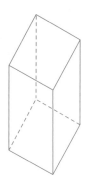

a. Predict the ratio of their surface areas and explain your reasoning. Then determine the actual ratio of the surface areas. If your prediction was off, explain the error in your prediction.

b. Predict the ratio of their volumes and explain your reasoning. Then determine the actual ratio of the volumes. If your prediction was off, explain the error in your prediction.

17. When my children were younger, we bought a set of dominoes. A complete set had every possible combination of numbers from 0 (represented by a blank) to 6. For example:

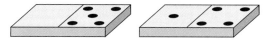

a. How many dominoes are in a complete set?

b. If each domino is 1 inch × 2 inches × 3/8 inch, design a carrying case that would hold all of them.

18. This problem is taken from a book called *Mathematical Investigations: Book One*.[8] I have noticed that "flats" of berries and "flats" of young plants are not cubical in shape but, rather, are rectangular prisms (see the picture below).

Suppose you wanted a flat that would hold 4000 cubic centimeters of strawberries. What would be the dimensions of the prism that would require the least amount of cardboard to make?

19. a. The cake pictured has been cut into three equal pieces. One more person comes. How can you make one cut in the cake so that each of four persons will get the same amount of cake?

b. The cake pictured below has been cut into n equal pieces. One more person comes. How can you make one cut in the cake so that each of $n + 1$ persons will get the same amount of cake?

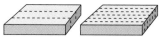

20. Below are three possible floor plans for a new office building. All the plans have almost exactly the same amount of floor space.

a. Analyze the advantages and disadvantages of each of these designs.

b. If each building were to be made of bricks, compare the relative amounts of bricks needed for each.

c. If the outside walls of each building were 10 feet tall and 1 gallon of paint covered approximately 500 square feet, determine the amount of paint that would be needed to paint each building.

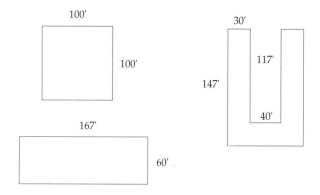

21. Each American produces about 3.5 kilograms of solid waste (garbage) a day. If 1 cubic meter of garbage weighs 90 kilograms, determine the size of the waste produced each day by:

a. The population of your college

b. The people in your state

c. The people in the United States

In each case, describe the assumptions you made in order to answer the question.

22. Many automobile commercials have stressed that the new model being advertised has more carrying capacity.

a. How might you determine the carrying capacity of two similar automobiles?

b. Select two comparable automobiles and determine the carrying capacity of each. Look up the carrying capacity given by the manufacturers. Do you agree with their amounts? If not, do you think one of the figures is erroneous, or do you think the term was defined differently?

---

[8]Randall Souviney, Murray Britt, Salvi Garguilo, and Peter Hughes, *Mathematical Investigations: Book One* (Palo Alto, CA: Dale Seymour Publications, 1988), pp. 10, 14. Reprinted by permission.

23. This problem has been adapted from *Mathematical Investigations: Book One*.[9] When a boat or ship sinks, the Red Cross recommends that people huddle together to stay warm. Let's say four adults huddle together to keep warm. Determine a way to answer the following question: What is the ratio of body surface area exposed to the cold water when floating alone (in a life jacket) to body surface area exposed when huddling?

24. Analyze the accompanying graph for its accuracy. First, predict whether you think it is accurate or not and explain your reasoning. Next do the measuring and calculating and explain your interpretation of the numbers. Then assess your prediction. If the prediction and the reasoning behind it were accurate, great. If either the prediction or the reasoning behind it was not accurate, explain the errors in the prediction.

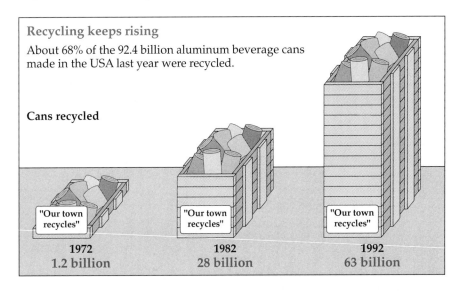

25. Make a net for a soccer ball. Describe your process. For example, how did you determine how many faces the soccer ball had and how to put them together? Describe any discoveries, conjectures, and questions that came from this exploration.

26. Take a standard $8\frac{1}{2}$- by 11-inch sheet of paper and determine how to cut and fold the paper so that you could make a rectangular prism with no top that would contain the greatest volume.

27. Predict how many feet of floor space there are in the building in which you are taking this class. If it is a large building, your instructor may wish to have you determine the floor space in a specific subset of the whole building. Finally, as accurately as you can, determine the actual floor space.

28. How many ping-pong balls would it take to fill a classroom that measures 20 feet by 16 feet by 8 feet?

29. *Classroom Connection* The following questions can be done theoretically with rectangular prisms. However, give them a context, cereal boxes, and you will find these questions and similar ones in many elementary and middle school textbooks.

    a. Can two cereal boxes have same volume but differ in surface area?

    b. Pick a cereal box. Redesign it so that it has the same volume but less surface area.

30. Why do you think most products are not manufactured in the shape that requires the least amount of packaging material?

[9]Randall Souviney, Murray Britt, Salvi Garguilo, and Peter Hughes, *Mathematical Investigations: Book One* (Palo Alto, CA: Dale Seymour Publications, 1988), pp. 10, 14. Reprinted by permission.

CHAPTER SUMMARY

1. To measure an object means to assign a number to some attribute of the object.
2. Two objects do not have to be congruent to have the same measure. For example, two objects can have different shapes but the same volume.
3. In order to measure, we must decide on a unit.
4. All units are conventions, as opposed to absolutes.
5. Virtually all measurements are approximations.
6. On some occasions, we cannot directly measure an attribute and must rely on indirect measurement.
7. Different attributes of an object—perimeter and area, and surface area and volume—are not related in simple ways.

BASIC CONCEPTS

Section 10.1: Systems of Measurement

metric system *654*
U.S. customary system *654*
meter *654*
liter, gram *655*
weight, mass *656*
precise *663*   accurate *663*
greatest possible error (GPE) *663*
relative error *663*

Section 10.2: Perimeter and Area

perimeter *664*, circumference *664*
π *664*   area, base, height *671*
Pythagorean theorem *669*
Understanding the area formulas for parallelograms *667*, triangles *668*, trapezoids *668*, and circles *671*
Areas of irregularly shaped figures *673*
Relationships between perimeter and area *674*

Section 10.3: Surface Area and Volume

Understanding the surface area formulas for prisms *682*, cylinders *684*, pyramids *685*, cones *685*, and spheres *686*
Understanding the volume formulas for prisms *684*, cylinders *687*, pyramids *688*, cones *689*, and spheres *689*.

CHAPTER 10 REVIEW EXERCISES

1. Describe the four-step process of measurement in your own words.
2. Which metric unit would you use to express each of the following? Briefly explain your choice. The choices for length are mm, cm, m, km; the choices for volume are mL, L; the choices for mass are mg, g, kg.
   a. The length of a pencil tip
   b. The volume of a bathtub
   c. The volume of a dose of cough medicine
   d. The mass of an ant
   e. The mass of an elephant
3. Fill in the blanks.
   a. 340 m = ____ cm   b. 0.345 kg = ____ g
   c. 2.75 L = ____ mL
4. a. Describe a situation in which a measurement could be precise but not accurate.
   b. Describe a situation in which a measurement could be accurate but not precise.
5. Describe all the attributes of a puddle that we might measure—for example, the volume of the water in the puddle.
6. Find the exact area of the figure below in two different ways.

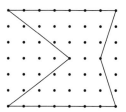

7. If possible, determine the area of the accompanying right triangle. If there is not enough information, explain why.

8. A swimming pool is shaped like the figure below. Each end is a semicircle, and the length and width of the rectangle are 40 feet and 20 feet, respectively. If there were a 1-foot-wide cement border around the pool, what would be the area of the border?

9. A meter trundle wheel is a convenient device for measuring distances along the ground. Every time the wheel makes one complete revolution, it has moved forward 1 meter. If you were to cut this wheel from a square piece of plywood, what would be the dimensions of the piece of wood?

10. Building codes generally require that the ventilating area be not less than 1/150 of the area of the floor area of the attic space being ventilated. If the attic of a house is 50 feet by 25 feet, what is the minimum square inches of ventilation needed?

11. Below is a piece of wire mesh. Determine the dimensions of the square that would contain 1 million holes.

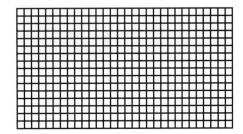

12. If you throw a dart at random and it lands on the dart board, what is the probability of its landing in the shaded region?

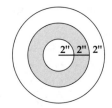

13. Kim decided he wanted to tile his bedroom with pennies. His bedroom measures 10 feet by 10 feet. He knows the area of a penny is πr^2, which is approximately 0.44 square inch. He computed the number of pennies by dividing the area of the room by the area of a penny. Here is his computation: (120 in. · 120 in.)/0.44 sq. in.

   a. However, this is not the right answer. What did he do wrong?
   b. Determine the right answer.

14. Randolph wants to know why you don't find the area of a parallelogram by multiplying the length times the width, as you do for rectangles. Explain why the formula is $A = b \cdot h$ instead of $l \cdot w$.

15. Given that there are 12 inches in 1 foot, why can't we convert 1000 square inches to square feet by dividing by 12?

16. What happens to the area of a circle when you double the radius?

17. What percent of this quilt block is shaded?

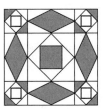

18. The figure below is a map of a pond. Each centimeter on the map represents 10 meters. Determine the area, in square meters, of the pond by covering the pond with a grid and then finding the area of the pond by decomposing the pond into trapezoids. Show your work.

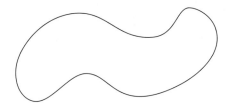

19. Determine the length of the arc $\overarc{AB}$, if the diameter of the circle is 30 cm and $\angle ACB = 123$ degrees.

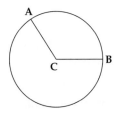

20. The federal government recently passed a law requiring providers of services to have all of their clients sign a privacy form. I filled out a form for

the dentist, orthodontist, doctor, credit cards, and so on. Determine the volume of the space occupied by all of these forms if the average adult filled out 10 of them. Then convert this number to something that will be more readily grasped—for example, the dimensions of a building that would contain all the forms or how many filing cabinets that much paper would fill. Assume that there are 200 million adults in the United States.

21. Each side of the cube is 20 feet.
    a. If someone went from point A to point X (the midpoint of the side) and then from point X to point D, what would be the total distance traveled?
    b. Determine the direct distance between point A and point D.

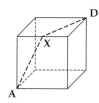

22. Let's say your college has decided to replace the worn cement paths between buildings. The width of each path is 6 feet, the depth of the concrete is 4 inches, and the total length of the new paths is 1200 feet. Concrete is normally ordered by the cubic yard.
    a. How many cubic yards should the company order?
    b. If customers want to allow 10% for spilling and overflow, how much should they order?

23. Determine the surface area (of the sides and roof) and the volume of the house-shaped figure below.

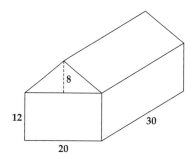

24. How many cubes that measure $\frac{1}{2}$ inch by $\frac{1}{2}$ inch by $\frac{1}{2}$ inch would fit into this container?

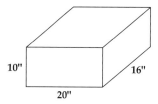

25. Design two different rectangular prisms that have the same surface area. Compare their volumes.

26. A company is manufacturing two kinds of juice boxes. One is in the shape of a rectangular prism with dimensions 10 cm by 6 cm by 4 cm. The other is a cylinder with radius 2.8 cm and height 10 cm. Compare the volumes and surface areas of these two juice boxes.

27. Our family bought a water-saving shower head. We determined that each shower now consumes 4 fewer gallons than before.
    a. If there are four of us and we each take one shower per day, how many gallons do we save per year?
    b. If we had a swimming pool in the form of a rectangular prism, and the pool measured 30 feet by 20 feet by 6 feet, would the amount saved in 1 year equal the volume of the swimming pool?

APPENDIX A

Principles and Standards for School Mathematics[1]

Number and Operations
Instructional programs from prekindergarten through grade 12 should enable all students to—

- understand numbers, ways of representing numbers, relationships among numbers, and number systems;
- understand meanings of operations and how they relate to one another;
- compute fluently and make reasonable estimates.

Algebra
Instructional programs from prekindergarten through grade 12 should enable all students to—

- understand patterns, relations, and functions;
- represent and analyze mathematical situations and structures using algebraic symbols;
- use mathematical models to represent and understand quantitative relationships;
- analyze change in various concepts.

Geometry
Instructional programs from prekindergarten through grade 12 should enable all students to—

- analyze characteristics and properties of two- and three-dimensional geometric shapes and develop mathematical arguments about geometric relationships;

---

[1] http://standards.nctm.org/document/index.htm Reston, VA: National Council of Teachers of Mathematics, 2000. Reprinted with permission from *Principles and Standards for School Mathematics: A Discussion Draft,* copyright 1998, by the National Council of Teachers of Mathematics. All rights reserved.

- specify locations and describe spatial relationships using coordinate geometry and other representational systems;
- apply transformations and use symmetry to analyze mathematical situations;
- use visualization, spatial reasoning, and geometric modeling to solve problems.

Measurement
Instructional programs from prekindergarten through grade 12 should enable all students to—

- understand measurable attributes of objects and the units, systems, and processes of measurement;
- apply appropriate techniques, tools, and formulas to determine measurements.

Data Analysis and Probability
Instructional programs from prekindergarten through grade 12 should enable all students to—

- formulate questions that can be addressed with data and collect, organize, and display relevant data to answer them;
- select and use appropriate statistical methods to analyze data;
- develop and evaluate inferences and predictions that are based on data;
- understand and apply basic concepts of probability.

Problem Solving
Instructional programs from prekindergarten through grade 12 should enable all students to—

- build new mathematical knowledge through problem solving;
- solve problems that arise in mathematics and in other contexts;
- apply and adapt a variety of appropriate strategies to solve problems;
- monitor and reflect on the process of mathematical problem solving.

Reasoning and Proof
Instructional programs from prekindergarten through grade 12 should enable all students to—

- recognize reasoning and proof as fundamental aspects of mathematics;
- make and investigate mathematical conjectures;
- develop and evaluate mathematical arguments and proofs;
- select and use various types of reasoning and methods of proof.

Communication

Instructional programs from prekindergarten through grade 12 should enable all students to—

- organize and consolidate their mathematical thinking through communication;
- communicate their mathematical thinking coherently and clearly to peers, teachers, and others;
- analyze and evaluate the mathematical thinking and strategies of others;
- use the language of mathematics to express mathematical ideas precisely.

Connections

Instructional programs from prekindergarten through grade 12 should enable all students to—

- recognize and use connections among mathematical ideas;
- understand how mathematical ideas interconnect and build on one another to produce a coherent whole;
- recognize and apply mathematics in contexts outside of mathematics.

Representation

Instructional programs from prekindergarten through grade 12 should enable all students to—

- create and use representations to organize, record, and communicate mathematical ideas;
- select, apply, and translate among mathematical representations to solve problems;
- use representations to model and interpret physical, social, and mathematical phenomena.

Answers to Selected Exercises

Chapter 1 Exercises, page 26

1. 8 tricycles and 24 bicycles
4. 500 $2 tickets and 100 $5 tickets
7. Ten 10¢ stamps and ten 5¢ stamps.
10. $704
13. 73,500 apples
15. a. About 3,417,000 miles
17. About 33 days
20. a. One solution is $519 + 327 = 846$.
21. b. (1) Here are three patterns: The second diagonal is the set of odd numbers, which can also be expressed as "the numbers increase by 2 each time"; all the numbers in this triangle are odd numbers; in any row, the two outside numbers are the same, the next two from the end are the same, and so on, all the way to the center.
    (2) Each number in the triangle is the sum of the three numbers above it; for example, $63 = 25 + 25 + 13$.
22. a. □   c. 13   e. 25   g. 2   h. 53
26. 30 pigs and 139 chickens
26. a. One wording: There are some 4s at the beginning, followed by some 8s, and then a 9 at the end. We can be more specific if we talk about the 1st, 2nd, 3rd, etc. product. In this case, the nth product will consist of $(n + 1)$ 4s, (n) 8s, and a 9 at the end.
    c. Answers will vary. You might explore what happens if you increase the first multiplicand but keep the second one at 67; i.e., 667×67, 6667×67, etc. You might explore what happens if there is one more 6 in one multiplicand; i.e., 667×67, 6667×667. You might explore what happens if you try different numbers; e.g., 57×57, 557×557; in this case, the pattern is not as "elegant," but all the numbers do end in 249.
29. a. The first and second numbers increase by 1, the third number increases by 2 more each time, (that is, 2 to 6 is an increase of 4; 6 to 12 is an increase of 6), so the next number will be 20. The number to the right of the equals sign is 1 greater than the number to the left. So the next equation will be $4^2 + 5^2 + 20^2 = 21^2$
30. a. 12 ways
31. The ball bounced 4 times.
33. a. Words: The sum of the first n terms is equal to 1 less than the $(n + 2)$th term.
    Notation: $A_1 + A_2 + A_3 + \cdots + A_n = (A_{n+2}) - 1$
    b. Words: The product of the nth and $(n + 2)$th terms is always 1 more or 1 less than the square of the $(n + 1)$th term.
    Notation: $A_n \times A_{n+2} = (A_{(n+1)})^2 \pm 1$
    $|(A_n \cdot A_{n+2}) - (A_{(n+1)})^2| = 1$
36. 10,440 handshakes

Chapter 1 Exercises, page 54

2. a. If you add two fractions that both have a 2 in the numerator, then the numerator of the product is double the sum of the two denominators, and the denominator of the product is the product of the two denominators.

   In notation: $\dfrac{2}{a} + \dfrac{2}{b} = \dfrac{2(a + b)}{ab}$

5. The deal depends on what your present service is. However, even if your present service costs over 10¢ a minute, it is probably not a good idea. For example, if you make a 1-minute call, you pay 10¢ + $1.95; thus, that minute costs $2.05. Even if you make a 30-minute call, you are charged $3.00 (for the 30 minutes) plus $1.95, which averages to 16.5 cents per minute.

6. a.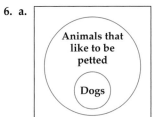

   c.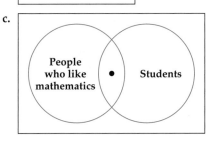

7. **a.** If it is fruit, then it contains sugar.
   c. If it is a cat, then it does not bark when the mail carrier comes to the door.
   e. If you want to drive a car, then you must pass a written test.
8. **a.** Inverse: If you do not exercise three times a week, then you will get sick.
   Converse: If you do not get sick, then you exercise three times a week.
   Contrapositive: If you get sick, then you did not exercise three times a week.
   c. Inverse: If Washington had crossed the Delaware, then we would have won independence from Britain.
   Converse: If we had not won independence from Britain, then Washington would not have crossed the Delaware.
   Contrapositive: If we won independence from Britain, then Washington crossed the Delaware.
9. **a.** Valid   **c.** Invalid   **e.** Invalid
10. **a.** Invalid. Nothing is said about the relationship between professional athletes and rock stars. Although it is possible that some rock stars are professional athletes, it does not follow from the two statements.

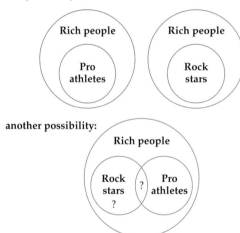

c. Valid

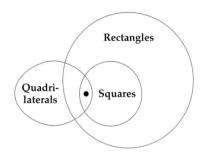

15. **a.** 4, 8, 12, 13, 16, 17, 20, 21, 22, 25, 26, 28, 30, 31, 32, 35, 36, 37, 40, 41, 45, 46, 50, 52, 55, 56, 60, 61, 65, 70, 76, 80, 85, 100
    b. I selected 15: If 1 is one of the four throws, then the remaining three would have to add up to 14, but that is impossible, because it would have to be $10 + 1 + 1 + 1$ or $5 + 5 + 1 + 1 + 1$. So 1 cannot be one of the scores. If 5 is one of the throws, then the remaining three darts would have to add up to 10, but that is impossible, because it would have to be $5 + 1 + 1$. If 10 is one of the throws, then the remaining three darts would have to add up to 5, but that is not possible, because we have only 1's and three 1's cannot add up to 5.
20. **a.** Answers will vary. Here are a few observations and patterns.
    The first three numbers (1, 2, 3) are at the three corners of the triangle. You can get the first three numbers (1, 2, 3) by going clockwise; the next three numbers (4, 5, 6) by going counterclockwise; and the last three numbers (7, 8, 9) by going clockwise.
    You can make triangles from four numbers, and the sum of each of these triangles is 20; e.g., $1 + 5 + 8 + 6$.
26. **a.** 17 bottles
28. There are two pairs of numbers that work: 91 and 64 or 82 and 73.
30. **a.** 5 hours
32. **a.** 1234

Chapter 1 Review Exercises, page 59

1. 316   2. 16
3. $216
4. Nine ways

25¢	10¢	5¢
1	2	1
1	1	3
1	0	5
0	5	0
0	4	2
0	3	4
0	2	6
0	1	8
0	0	10

5. 40 posts
6. two 10¢ stamps and ten 5¢ stamps
7. The product is a four-digit number. The first two digits have a value one less than the number you are multiplying by 99. The second two digits equal what you get if you subtract that number from 100. Using symbols, if we let the number $= ab$, then the first two digits are $(ab - 1)$ and the second two digits are $(100 - ab)$.
8. The letter P
9. **a.** Fill the 9 gallon pail and use it to fill the 4 gallon pail. Empty that pail and fill it again from the 9 gallon pail. You now have 1 gallon left in the 9 gallon pail.
   b. Fill the 4 gallon pail and empty it into the 9 gallon pail. Do it again. Fill the 4 gallon pail a third time and now fill the 9 gallon pail—it takes 1 more gallon to do so. You now have exactly 3 gallons left in the 4 gallon pail.

10. 11 hours
11. 26 packages, 2 ounces
12. a. Three ways: 8, 4, 1, 1; 8, 2, 2, 2; 4, 4, 4, 2
    b. 23, 27, 29, 30, and 31 impossible.
13. There are so many possibilities. Here are several: The magic sum is 34, the sum of the 4 numbers in the center 2 × 2 square is 34; the sum of the four corner numbers is also 34; if you partition the 4 × 4 square into four 2 × 2 squares, the sum of the numbers in each of the 2 × 2 squares is also 34; there are 2 odd and 2 even numbers in each row and column; the two middle rows and/or columns can be switched without affecting the square; the top-left to bottom-right diagonal has a constant difference of 3; if you look at the middle two columns, each pair of numbers are consecutive numbers.
14. The sum of the numbers along the length of the stick equals the number at the end of the stick.
15.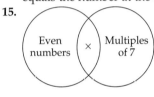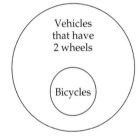

16. a. If it is meat, then it contains protein.
    b. If you want to pass the course, then you must attend class.
17. Inverse: If you don't work hard, then you won't succeed.
    Converse: If you succeeded, then you worked hard.
    Contrapositive: If you didn't succeed, then you didn't work hard.
18. This is invalid; there might be other cameras that do not have disk drives.
19.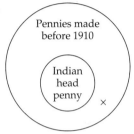

This argument is invalid.

Exercises 2.1, page 76

1. a. $0 \notin \emptyset$ or $0 \notin \{\}$ b. $3 \notin B$
2. a. $D \not\subseteq E$ b. $A \subseteq U$
3. a. {e, l, m, n, t, a, r, y} and $\{x \mid x$ is a letter in the word "elementary"$\}$
   c. {2, 3, 5, 7, 11, 13, 17, 19, 23, 29, 31, 37, 41, 43, 47, 53, 59, 61, 67, 71, 73, 79, 83, 89, 97}; $\{x \mid x$ is a prime number less than 100$\}$
4. b. is not well defined, because the "countries in Europe" are constantly changing and may be defined in various geopolitical terms.
6. b. ⊂   {3} is a subset of the set.
   c. ∈   {1} is an element of this set of sets.
   e. ⊄   {ab} is neither a subset nor an element.
   f. ⊂   The null set is a subset of every set.
7. a. 64
   b. A set with n elements has 2^n subsets.
12. a. 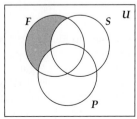 $F \cap \overline{(S \cup P)}$ or $F \cap \overline{S} \cap \overline{P}$

    c. 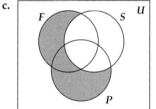   Nonsmokers who either are female or have health problems

    e. $\overline{F} \cap (S \cap P)$
    Males who smoke and have a health problem
14. b. All numbers that don't evenly divide 12, 15, or 20; $\overline{A \cup B \cup C}$ or $\overline{A} \cap \overline{B} \cap \overline{C}$
    d. All numbers except 1 and 3

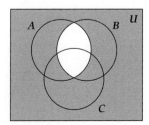

15. a. Students who have at least one cat and at least one dog

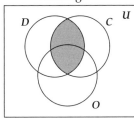

d. $D \cap \overline{C}$

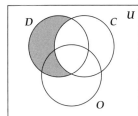

g. $\overline{D \cup C \cup O}$ or $\overline{D} \cap \overline{C} \cap \overline{O}$
Students who have no pets

16. The circles enable us to easily represent visually all the possible subsets. The diagram is not equivalent because there is no region corresponding to elements that are in all three sets.

17.

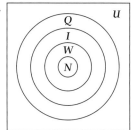

21. b. Technically, this subset represents a lesson in which the students use concrete materials, but not in a lab approach and not in small groups. Pedagogically this doesn't make sense.

Exercises 2.2, page 96

3. a. 7 buses. Each bus will carry a maximum of 34 students and two adults. 7 buses × 34 students per bus = 238 students, so there will be 6 extra seats.

b.

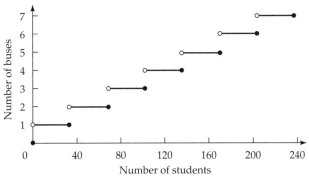

School Buses for Field Trip

5. 164.9 cm, or 1.65 m
8. No
10. a. $T_n^2 - T_{n-1}^2$ **b.** It is a function.
12. It would be less expensive to rent a car for a week. The minimum cost renting by the day is ($45) × (3 days) + ($0.30 per mile) (320 miles round trip) = $231.
13. Three solution paths are presented here:
Solution path 1: Guess–check–revise
Find when the first plan costs $5.

Checks	Charge	Calculations	Reflection/Analysis
0	2.00	2.00	The charge for checks will be $2.
10	3.50	2.00 + 10(0.15)	10 checks added $1.50. Another $1.50 will make $5.
20	5.00	2.00 + 20(0.15)	Got it!

Solution path 2: Represent the problem algebraically
Find when the charges are equal.

Let x = the number of checks written.
Charges in first plan: $2 + 0.15x$
Charges in second plan: 5

$$2 + 0.15x = 5$$
$$0.15x = 3$$
$$x = 20$$

Solution path 3: Represent the problem with a graph
Where the two lines cross is the point where you will be charged the same fee.

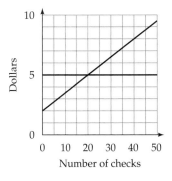

16. a. $3400   **b.** $800   **c.** In $6\frac{1}{4}$ years
19. Two solution paths are shown.

Solution path 1: You can view the parts of the figures as a rectangle plus 1 square sticking out at the top right and 1 square sticking out at the bottom left.

	2nd figure	3rd figure	4th figure	nth figure
Middle rectangle	3 · 1	4 · 2	5 · 3	$(n+1)(n-1)$
Little squares "sticking out"	2	2	2	2

Thus the number of squares in the nth figure is $(n+1)(n-1) + 2$, which simplifies to $n^2 + 1$.

Solution path 2: You can view the parts of the figure as a top and a bottom row and then a middle square.

	2nd figure	3rd figure	4th figure	nth figure
Top and bottom row	$2 + 2$	$3 + 3$	$4 + 4$	$n + n$
Middle square	1	$2 \cdot 2$	$3 \cdot 3$	$(n-1)(n-1)$

Thus the number of squares in the nth figure is $n + n + n^2 - 2n + 1$, which simplifies to $n^2 + 1$.

23. Here we will assume that the area of each face is 1 square unit.
    a. Solution path 1: Make a table and look for patterns. If the stack has 1 cube, the surface area is 6 square units. If the stack has 2 cubes, the surface area is 10 square units. If the stack has 3 cubes, the surface area is 14 square units. In this case, the increase is 4 each time, indicating that this relationship can be represented by a linear equation that has slope 4 and that includes the ordered pair (1, 6). Using what we learned in Algebra 1, we deduce that the equation is $y = 4x + 2$. We can transfer this equation to the question at hand directly: A stack of n cubes will have a surface area of $4n + 2$ square units.

Solution path 2: Break the problem into parts. Each stack will have four equal "sides" (technically, they are faces) and a top and bottom.

	HEIGHT OF THE TOWER			
	1 cube	2 cubes	3 cubes	n cubes
Number of cubes on the sides	4	$4 \cdot 2$	$4 \cdot 3$	$4 \cdot n$
Top + bottom	2	2	2	2

Thus the surface area of a tower of n cubes is $4n + 2$.
25. a. 33 trips
37. a. First graph. The flag is raised at a steady rate.

Exercises 2.3, page 119

2. a. 3031   c. 1666   f. $75,602 (21 \times 60^2 + 0 \times 60 + 2)$
3. a. Egyptian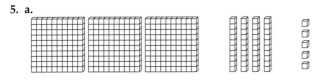
   Roman CCCXII
   Babylonian

   c. Egyptian
   Roman MMMMMM
   Babylonian $(1 \times 60^2 + 40 \times 60 + 0)$
5. a.

   b.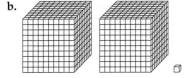

6. a. $300 + 40 + 5$ 7. a. 4859
8. a.
9. a. $7,777_8$
10. a. $500 + 10 + 10 + 5 + 1 + 1 = 527$
    c. H H H △ △ △ ΓII
11. a. 32,570   c.
12. a. $(4 \times 20) + 7 = 87$
    d. $245 = (12 \times 20) + 5$
13. a.
    c. 460,859
14. a. $-400 =$ c. $+80,000 =$ e. $-346,733$
15. a. 40_5 g. 130_5
16. e. 1001_2 g. 113_4
17. b. $(5 \times 6) + (5 \times 1) = 35$
    d. $(2 \times 125) + (1 \times 25) + (4 \times 1) = 279$
18. a. 134_5 c. 1011100_2 g. 112_6
20. $x = 9$
30. b. It has all 6 characteristics because this system is essentially base 6. The places are called cartons, boxes, crates, flats, and pallets. The value of each place is 6 times that of the previous place.
34. b. It has characteristic 2: The value of each place is 10 times the value of the previous place. It "sort of" has characteristic 3, with the modification that each "place" contains two symbols. Some might say that it has characteristics 4 and 5, but the order of the numerals is still a matter of convention—unlike base 10, where changing the order changes the value.
36. a. 26   c. 25   e. 3,400
38. a. About 11.57 days

Chapter 2 Review Exercises, page 123

1. **a.** $\{x \mid x = 10^n, n = 1, 2, 3, \ldots\}$
   b. $\{10, 100, 1000, 10,000, \ldots\}$ or $(10^1, 10^2, 10^3, \ldots)$
2. No, because not all countries are recognized by all other countries.
3. **a.** $\in$ **b.** $\subset$ **c.** $\subset$ **d.** $\not\subset$
4. **a.** $D \not\subset E$ **b.** $0 \notin \{\}$
5. **a.**

U	A	B
	1 3 7 9 11 13 17 19 21 23 27 29	5 15 25	0 10 20

   (outside: 2 4 6 8 12 14 16 18 22 24 26 28 30)

   b. 5, 15, 25
   c. The set of even numbers between 0 and 30.
6. **a.** 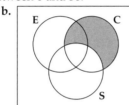 **b.**
7. 50
8. The former means the same elements, and the latter means the same number of elements.
9. **a.**

x	f(x)
0	−2	0 → −2
1	1	1 → 1
2	4	2 → 4
3	7	3 → 7

   b. $f(6) = 16$
10. 28
11. $2(n + 2) + 1 = 2n + 5$
12. $2(2n + 1) + 1 = 4n + 3$
13. **a.** 110 **b.** $n(n + 1)$
14. **a.** 46 **b.** $4n + 6$
15. If he averages 400 or more minutes per month, then use the second plan. Otherwise, use the first plan.
16. Graph (a) is not a match because the temperature of the coffee will not go below room temperature. Graph (b) is not a good match because it implies that the coffee cools at a steady rate. The coffee will cool more rapidly at the beginning. As it gets closer to room temperature, the cooling will get slower. Graph (c) is the best match because the rate of cooling slows down over time.
17.
18. One story: independent is time; dependent is distance. Jackie walked steadily for a period of time and then sat down.
    Another story: independent is time; dependent is money. Pedro worked for several days (the same number of hours each day) and then didn't work for several days.
19.

	Egyptian	Roman	Babylonian
a. 47		XLVII	
b. 95		XCV	
c. 203		CCIII	
d. 3210		MMMCCX	

20. **a.** 410_5 **b.** 1300_5 **c.** 1010_2
21. **a.** 4314_5 **b.** 30034_5 **c.** 1011_2
22. $25 \times 2 + 4 \times 5 + 3 = 73$
23. Because $1000_5 = 125_{10}$, I would you rather have $\$200_{10}$.
24. They both have the value of 3 flats, 2 longs, and 1 singles. Because base 6 flats and longs have greater value than base 5 flats and longs, the two numbers do not have the same value.
25. Because we are dealing with powers. Thus, the value of a base 10 flat is $2 \times 2 = 4$ times the value of a base 5 flat.
26. 16
27. There are many possible responses. Here are three: "One-zero" is the amount obtained when the first place is full.
    It means you have used up all the single digits in your base.
    It is the first two-digit number.
28. The visual representation of the seventh place is a cube.
29. There are several equivalent representations:

    $2000 + 60 + 8$

    $2 \times 1000 + 0 \times 100 + 6 \times 10 + 8 \times 1$ or
    $2 \times 1000 + 6 \times 10 + 8 \times 1$

    $2 \times 10^3 + 0 \times 10^2 + 6 \times 10^1 + 8 \times 10^0$ or
    $2 \times 10^3 + 6 \times 10^1 + 8 \times 10^0$

30. Answers will need to include all six characteristics described in the section.

Exercises 3.1, page 150

3. **a.** $3 + 4 = 7$
   c. $3 \cdot 2 = 6$ (repeated addition)
8. 5 hundreds + 10 tens + 13 ones
   = 5 hundreds + 11 tens + 3 ones
   = 6 hundreds + 1 ten + 3 ones
   = 613
10. **a.** After adding $6 + 8 = 14$, the student wrote the 1 in the ones place, carried the 4, and added it to $3 + 2$.
11. Answers will vary. One possibility for each is given below.
    a. $47 + 53 = 100$ (compatible numbers). Now add 2 more to get 102.
    c. $150 + 15 = 165$
    e. $575 + 125 = 700$. Now add 3 to get 703.
    g. $340 + 16 = 356$
    i. $387 + 24 = 400 + 11 = 411$. Then $411 + 53 = 464$
13. Answers will vary. One possibility for each is given below.
    a. Leading digit: 1173
    $(1000 + 160 = 1160; 1160 + 13 = 1173)$
    c. Leading digit and rounding: 146,000
    Add leading digits: $8 + 3 + 2 = 13$ (representing 130,000). Then round each number to nearest thousand. And add second leading digit: $9 + 3 + 4 = 16$ (representing 16,000). $130,000 + 16,000 = 146,000$.
    e. Leading digit, compatible numbers, and rounding: 2800
    $4 + 3 + 1 + 5 + 9 + 3 = 25$, representing 2500.
    $73 + 34 \approx 100$; $45 + 55 = 100$; $65 + 43 \approx 100$.
    $2500 + 100 + 100 + 100 = 2800$.
    g. Leading digit and rounding: 1,400,000,000
14. **a.** 37   56   74
    c. 24   47   97
    e. 185   153   274
16. **a.** The greatest sum, 1593, is obtained when the 7 and 8 are in the hundreds places, the 3 and 6 are in the tens places, and the 1 and 2 are in the ones places.
18. 372
    168   There are many solutions.
    459
    ___
    999
19. $N = 9, P = 2$
21. **b.** $10,000_5$

Exercises 3.2, page 166

2. **a.** $a - b \neq b - a$, because $b - a = -(a - b)$.
3. **a.** $7 - 5 = 2$
7. **a.** The student subtracted $8 - 6$. Some students automatically subtract the smaller number from the larger number, even if it changes the order of the numbers in the problem.
   b. The student wrote $6 - 8 = 2$, knew that 8 is bigger, and so "borrowed" from the tens place.
   e. The student changed the zero in the ones place to 10 but didn't rename the 7 in the tens place as 6.
10. **b.** 904 $367 + ⑦ = 374$
           -367 $374 + ㉚ = 404$
                      $404 + ⑤⓪⓪ = 904$
    The answer is $500 + 30 + 7 = 537$
13. Answers will vary. One possibility for each is given below.
    a. Add 1 to each number: $88 - 30 = 58$.
    c. $34 + 50 = 84$; the answer is 48.
    e. $206 + 300 = 506$; the answer is 296.
    g. $475 + 25 = 500$; add 125 more; the answer is 150.
    i. Add 1 to each number: $507 - 30$; count backwards by tens: 507, 497, 487, 477.
18. **a.** $876 - 123 = 753$
19. There are many possibilities. Two are shown here: $968 - 734$, $946 - 712$.
23. Answers will vary. One possibility for each is given below.
    a. Leading digit: 2100
    c. 28 to 30 is 2, 30 to 73 is 43; an estimate is 45,000.
    e. 285 to 300 is 15, to 413 is 113 more. $113 + 15 = 128$; estimate is 128,000.
    g. Rounding: about 1,200,000
24. **a.** 475   764   **c.** 382   723

Exercises 3.3, page 186

7. **a.** ㉕   17
           50   8
           100   4 $400 + 25 = 425$
           200   2
           ④⓪⓪ 1
8. **a.** $45^2 - 9^2 = 2025 - 81 = 1944$
13. The 1 being carried represents 10. It is carried so that it can be added to the other values in the tens column.
15. **a.** $587 \times 40 = 23,480$
    c. $587 \times 305 = 179,035$

19. **a.** 8 × 6: The student wrote down the tens digit and carried the ones digit.
    c. 2 × 46: The student wrote the product in the tens and ones places, even though the 2 represents 20.
22. Answers will vary. One possibility for each is given below.
    a. 16 × 16 = 32 × 8 = 64 × 4 = 256
    c. 66 × 10 = 660, so 66 × 5 = 330
    e. 6 × 3 = 18, so 60 × 30 = 1800
    g. 35 × 10 = 350, 35 × 2 = 70; answer is 350 + 70 = 420.
    i. Double 736 twice: 1472, then 2944
25. Answers will vary. One possibility for each is given below.
    a. 40 × 70 = 2800
    c. 60 + 2
       80 + 3
       ──────
       4800 + 200 + 200 = 5200
    e. 30,000 × 50 = 1,500,000
26. Part (b) is wrong.
27. 34 × 65 = 2210
30. About 3600 miles. Answers on cost will vary depending on price of gas and miles per gallon that the car gets.
42. **a.** 530 · 71 = 37,630
43. **a.** 55,999,999,944 Below is one solution path.
       999,999,999 · 56 = (1,000,000,000 − 1)56
       = 56,000,000,000 − 56
45. Alike: The sum (product) of any row, any column, and both diagonals are identical.
    Difference: In a multiplication magic square, it seems that the numbers in the square don't all need to be different.

Exercises 3.4, page 207

8. **a.** The student didn't use zero as a place holder in the quotient to indicate that 2 (in the tens place) divided by 7 is less than 1 (ten).
10. **b.** A reasonable estimate is 4200.
15. Rough: About 500
    Refined: About 670
17. Rough: Just under $50,000
    Refined: About $45,000
18. Rough: About $90
    Refined: About $103
20. About 8 years
21. About $240
25. 25¢   18. 11 cartons
28. 160 tapes   21. 700 rolls
29. **a.** 14 buses
31. 60 cans of juice, assuming 8 oz/glass; $53.40
35. 2,190,000 gizmos
38. Rough: About 8000 days
    Refined: About 7300 days
41. Answers may vary. 3 million crimes per year would mean about 17,000 crimes per school day (dividing by 180 school days per year). If we divide 17,000 by 50 states, that's about 340 crimes per day per state. That still seems high, so the term *crime* must be broad. It must include such things as vandalism and petty theft. Does it mean "crimes reported to police"?
46. **a.** 2)974 = 487
48. $x = 7$ and $y = 13$, or vice versa
54. **a.**

a	b	$a + b$	$a \cdot b$
65	66	131	**4290**
1208	72	1280	86,976

b.

a	b	$a + b$	$a \div b$
19	1	20	19

d.

a	b	$a \cdot b$	$a \div b$
10	2	20	5

55. **a.** 74 · 86 = 6364
    e. 1530 = **31** · 48 + 42
58. **a.** One solution is: 66 + 80 − 3
59. **a.** 14(48 − 32) = 224
60. **a.** [(A + B + C) × D]/100 = E

Chapter 3 Review Exercises, page 212

1. a. $7 + 2$ b. $7 - 4$ c. 2×3 d. $12 \div 4$
2.
   ```
   +--+--+--+--+--+--+--+--+--+--+--+--+--+--+--+--+--+--+--+
   0  1  2  3  4  5  6  7  8  9  10 11 12 13 14 15 16 17 18 19
   ```
3. Responses will vary. This algorithm shows the sum of each place, beginning with the largest place. Once the sum for each place is written down, it is relatively easy to add those numbers mentally to write down the sum.
4. $345 - 97$
5. We have simply changed the representation of the minuend from 8 hundreds to 7 hundreds + 9 tens + 10 ones.
6. Responses will vary. The take-away problems involve actions such as losing, spending, and giving; for example, "Rita had $30 and spent $12 on a CD." The comparison problems involve comparing the sizes of two sets: "Jorge has $30 and needs $42 to buy a video game." or "Kayla has 45 CDs and her friend has 57. How many more does her friend have?
7. Responses will vary. At the heart of the matter is that, regardless of the places involved, 10 of something is being exchanged for 1 of something else.
8. Responses will vary. A common solution path is to see the four partial sums and add them: 6 hundreds, 14 tens, 12 tens, and 28 ones.
9. The digits being multiplied to get the second row are 37×2. However, this represents 37×20, which is 740. We commonly omit the zero and just write 74; however, because its real value is 740, we must place the 7 and the 4 in the correct places.
10. Responses will vary. At the heart of the matter is repeated addition. Essentially, we have 13 groups of 25. We can more easily get the answer by finding 10 groups of 25 (250) and 3 groups of 25 (75) and adding these amounts together.
11. 541×82
12. Responses will vary. One response is to note that a visual representation of the problem involves four partial products, only two of which are shown here. Another response notes that we need 34 groups of 26. 30×20 gives us 20 groups of 30; we need 6 more groups of 30 to give us 26×30; then we need 4 groups of 26 to give us 26×34. Thus, to get the total answer, we need 30×20, 6×30, and 4×26 (which can be broken into 4×20 and 4×6). Thus Pete's method gives us only two of the four products needed.
13. Responses will vary. The heart of the response is to note that if we are adding, when we take 1 from 17 and give it to 29, it is literally 1. However, if we are multiplying, when we take 1 from 17 and give it to 29, we are really taking one group of 29, rather than just 1.
14. Responses will vary. At a concrete level, using the example cited, the origins of 12×20 are (3×4) and (4×5). The origins of 15 and 16 are (3×5) and (4×4). Using the commutative and associative properties, we can show the equivalence of (3×4) and (4×5) and (3×5) and (4×4).
15. 16 and 30
16. Using a calculator, $73932500 \div 97665 \approx 757.00097$. Multiplying 757×97665 gives us 73932405, which we can subtract from 73932500 to get 95. Thus $x = 757$ and $y = 95$.
17. The answer is an infinite sequence beginning with 91 and increasing by 75 each time. That is, 91, 166, 241, 316, and so on.
18. 12
19. $79 \times 9 \times 11 = 7821$
20. In each case, we are determining how much is being used up by the digit in the divisor. That is, when we say 5 goes into 43, 8 times, we are saying that we are using up $80 \times 5 = 400$.
21. Responses will vary. Partitioning problems will generally involve sharing. Repeated-subtraction problems will generally involve having an amount that is being divided into groups that are all the same size.
22. Responses will vary.
23. a. Commutative and associative properties transform the problem into $(36 + 64) + 82$ which equals $100 + 82$.
    b. The distributive property transforms the problem into 20×19.
    c. Representing 1592 as $1600 - 8$, and then using the distributive property of division over subtraction, transforms the problem into $1600/8 - 8/8$.
24.–28. Responses will vary.
29. When we change 638×42 to 638×40, we are decreasing the answer not by 2 but by 638×2, roughly 1200. Thus, when we round up 638, we need to make up that 1200. If we round 638 to 640, we are increasing by 2×42; if we round 638 to 660, we are increasing by 22×42.
30. We have (52 years $\times$ 52 weeks per year $\times$ 3 hours per week)/24 hours per day = 338 days.
31. The car will get approximately 360 miles per tank. Because $400 \times 3 = 1200$, they will need to stop 3 times: at approximately 360, 720, and 1080 miles.
32. a. $500 \times 1200 = 600{,}000$
    b. $18,000
    c. $12,000
33. a. 11330
    b. 2211
    c. 3113
    d. 3214
34. Base 7
35. Base 8
36. Base 4
37. a. $3684 + 4248$
    b. $6003 - 3284$
    c. $a = 4, b = 6$

Exercises 4.1, page 229

3. **a.** 1, 3, 7, 9 or 1, 3, 5, 11
6. **a.** Hint: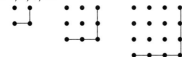
7. **a.** 30 and 15   **b.** 36 and 12
8. **a.** $37 \times 3 \times 1 = 111$; $37 \times 3 \times 2 = 222$;
   $37 \times 3 \times 3 = 333$. The repeated digit in the product is the factor multiplied by 37×3 in the problem.
9. Using expanded form, we can represent a four digit palindrome as:
   $1000a + 100b + 10b + a$, which is equivalent to $1001a + 11b$.
   Since 11 divides 1001, it divides $1001a$ and $11b$. Therefore, it divides their sum.
11. **a.** $4 \mid 76$. Therefore, $4 \mid 1776$, which is therefore a leap year.
    c. 400 does not divide 2100. Therefore, 2100 is not a leap year.
14. **a.** Test divisibility by 3 by adding all the digits:
    $1 + 2 + 3 + 4 + 5 + 6 = 21$, $3 \mid 21$ so $3 \mid 123{,}456$.
    Test divisibility by 4 by examining the last digits. $4 \mid 56$ so $4 \mid 123{,}456$.
    Show divisibility by 6 by showing $2 \mid 123{,}456$ and $3 \mid 123{,}456$.
    Test divisibility by 8 by examining the last three digits $8 \mid 456$ so $8 \mid 123{,}456$.
    To be divisible by 9, a number would have to be divisible by 3, and then the resulting quotient would need to also be divisible by 3. $123{,}456 \div 3 = 41{,}152$, $4 + 1 + 1 + 5 + 2 = 13$ and 3 does not divide 13, so 9 does not divide $123{,}456$.
15. **a.** Divisible by 2, 3, 4, 6, 9, 12
17. **a.** The digit 4 will make the statement true.
    c. The digits 1, 4, and 7 will each make the statement true.
    d. 467,712   **21.** 79
21. You can use a similar processes in number 20:
    The first column is 4, 7, 10, 13, etc.
    The second column is 9, 14, 19, 24, 29, 34, 39, 44, etc.
    The third column is 9, 16, 23, 30, 37, 44, etc.
    You can notice that the ones place of the second column is either 4 or 9. So you can look to see when the ones place of the third column is 4 or 9 and then check for a match in the first column. From another perspective, you will notice a pattern in the third column, that once you find a number that is also in the second column, every fifth number after that will be in the second column. So now you can circle every fifth number past the first match and look to see matches in the first column. The answer is 79.
26. **c.** The last two digits must be 00, 25, 50, 75.
    d. There are several possibilities. One is that the sum of the digits must be an even number. The most elegant is that the number is divisible by 2 iff there are an even number of even digits in the number.
27. **a.** True
    c. False; e.g., $6 \mid (4 \cdot 9)$ but 6×4 and 6×9
    e. True
28. **a.** True   **c.** True
29. **a.** The sum of three consecutive numbers will always be divisible by 3 and by the middle number in the sequence. $x + (x + 1) + (x + 2) = 3x + 3$
31. **a.** Yes   **32. a.** 1
34. **a.**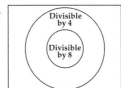

Exercises 4.2, page 239

1. Any prime number has exactly two factors.
2. The first 20 prime numbers are 2, 3, 5, 7, 11, 13, 17, 19, 23, 29, 31, 37, 41, 43, 47, 53, 59, 61, 67, 71.
3. **a.** $2 \times 2 \times 2 \times 2 \times 3$
   b. $5 \times 5 \times 3$
   c. $2 \times 2 \times 23$
   d. $2 \times 2 \times 2 \times 2 \times 3 \times 3$
   e. $2 \times 2 \times 7 \times 7$
   f. $2 \times 2 \times 2 \times 3 \times 3 \times 7$
4. **a.** 61 is prime.
   b. Composite: 5×13
   c. 71 is prime.
   d. 89 is prime.
   e. Prime   **f.** Prime
   g. Composite: 19×23
6. 1951; 1973; 1979; 1987; 1993; 1997; 1999
7. 24
8. Answers may vary: 313, 317, 373
11. **a.** 2520
12. **a.** 1369; factors: 1, 37, 1369
15. **a.** 2, 4, 5, 10, 20, 25, 50
16. **a.** 3, 7, and 1

Exercises 4.3, page 252

1. **a.** 15   **b.** 9   **c.** 1   **d.** 27   **e.** 3
2. **a.** 3
3. **a.** 60   **b.** 150   **c.** 132
4. **a.** 300
6. **a.** 5; 200   **b.** 36; 216   **c.** 84; 1,260
8. 143

10. 224 mm × 224 mm; 28 dominoes
12. a. *a* is a factor of *b*, and *b* is a multiple of *a*.
    c. *a*
13. a. True c. False; GCF(12, 18) = 6 e. True
15. Answers vary. One possibility is 120 cm × 120 cm × 120 cm.
17. a. A 2 by 2 square
    b. A 3 by 3 square
    c. The GCF of *x* and *y*.
18. a. A train that is 12 units long; for example, a train made by joining an orange rod and a red rod
    b. A train that is 24 units long
    c. A train whose length is the LCM of *x* and *y*.
19. a. The purple rod—that is, a rod 4 units long
    b. The yellow rod—that is, a rod 5 units long
    c. A train whose length is the GCF of *x* and *y*.
20. The LCF of any two numbers is 1, by definition. Thus the LCF is not useful. The GCM of any two numbers is infinitely large and therefore not useful.

Chapter 4 Review Exercises, page 254

1. a. Yes, the sum of the digits is divisible by 3.
   b. Yes, the number is divisible by both 2 and 3.
   c. No, the number represented by the last two digits is not divisible by 4.
2. Here are three equivalent statements:
   4 is a factor of 12.
   12 is a multiple of 4.
   12 is divisible by 4.
3. 4320
4. False. Counterexample: 6 divides 24 and 8 divides 24, but 48 does not divide 24.
5. a. 11 is one answer.   b. 107 is one answer.
6. 91, 93, 94, and 95
7. If the sum is odd, then one is even and one is odd. Therefore, their product must be even.
8. It will have a 1 in the units place.

9.

10. a. 2 × 2 × 2 × 3 × 3   b. 2 × 2 × 3 × 3 × 7
11. It is a prime number.
12. 49
13. Both 3 and 5 must divide *m*.
14. a. 8   b. 8   c. 4
15. a. 90   b. 120
16. 54 and 60
17. 54
18. The two numbers must be relatively prime.
19. 192 or 216

Exercises 5.1, page 265

1. a. ⁻218   c. 19   e. 14   g. 7   i. 221
2. a. 238   c. ⁻417   e. 213
3. ⁻35
5. ⁻5° Celsius
7. ⁻5° F
10. a. 20 mph wind speed
    b. 180 mph maximum speed with no wind

12. 19 hours   14. 3760 B.C.
19. a. Always positive.   c. Always positive.
20. a. Always positive.   c. Always positive.
21. 176 pounds

Exercises 5.2, page 281

1. a. 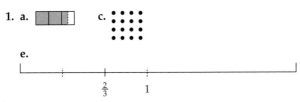   c.

   e.

   g. 7/10 is one possibility.
3. *Note:* Answers will vary. The figure shows one valid representation for each part of 3/5.

11. a. 5/8   c. 9/10   e. 7/9   g. 29/30
15. a. *a/b* > *a/c*   c. *b/c* > *a/c*

16. The new fraction is larger than the original fraction.
19. 1/10 of your blood
21. a. 1/3   b. 1/9
23. Approximately 3/4
25. Approximately 2/3
27. **Religion**   **Fraction**
    Christianity   1/3
    Islam   1/6
    Hinduism   1/8
    Buddhism   1/16
    Judaism   1/300
29. a. 5/16 ≈ 1/3   c. 199/307 ≈ 2/3
31. 1/2 of the plants survived.
32. a. Son 14, father 56; in 7 years, 21 & 63, in 28 years, 42 and 84

b. The number of realistic solutions probably begins with 8/32 and ends with 14/56. My son younger than 8 makes me too young; my son older than 14 would have me pretty old when I would be twice his age!

c. There are many possible answers. One would be: both numbers are even and father is in his 50s.

Exercises 5.3, page 303

1. a. $80\frac{17}{24}$ b. $-2\frac{23}{120}$ c. $23\frac{13}{20}$ d. $191\frac{21}{26}$ e. $-3\frac{7}{12}$
   f. $66\frac{7}{24}$ g. 99 h. 135 i. $13\frac{1}{2}$ j. $\frac{1}{8}$ k. $\frac{3}{8}$
2. a. 13 3/8 d. 19 2/5 g. 44
3. a. Less than 10
   c. Greater than 2
   e. Greater than 20
5. $2/3 - \frac{3}{4}$ is less than zero; $2/3 \times \frac{3}{4}$ is less than either fraction; $2/3 \div \frac{3}{4}$ is less than 1; $2/3 + \frac{3}{4}$ is greater than 1, almost 1 1/2.
7. a.

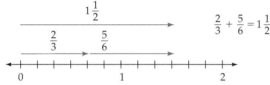

   e.

   $\frac{3}{4} \times 6 = 3 \times 1\frac{1}{2} = 4\frac{1}{2}$

   h.

   We can take $\frac{3}{8}$ out of $\frac{7}{8}$ 2 whole times. What is left is $\frac{1}{3}$ of $\frac{3}{8}$, so we can take $\frac{3}{8}$ out of $\frac{7}{8}$ $2\frac{1}{3}$ times.

8. a. Hexagon + rhombus is equivalent to 8 triangles, so the value of 1 triangle is 1/8.
   c. Hexagon + triangle is equivalent to 7 triangles and trapezoid is equivalent to 3 triangles, so the value of trapezoid is 3/7 of 2 or 6/7.

10. Estimate: 4 ounces
    Exact answer: $4\frac{23}{64}$ ounces
12. Estimate: 72 inches or 75 inches
    Exact answer: $73\frac{1}{2}$ inches
14. Estimate: 90 million people
    Exact answer: $2/3 \times 136{,}800{,}000 = 91{,}200{,}000$
17. Estimate: Maybe yes, maybe no. It will be close.
    Exact answer: Yes, exactly.
18. 120 first-year women.
20. a. $\frac{1}{2}^{10}$ or 1/1024 of the germs remain.
    b. 19,531 germs (theoretically). $1/1024 \times 20{,}000{,}000 = 2 \times 10^7 \div 1024$. A better answer might be to say approximately 1/1000 of 20 million ≈ 20,000 germs.
22. 3/7 are single.
24. 415,220; 1,245,660
26. a. 4 pressings; 6 pressings
28. $\dfrac{bc^2 - bd}{ac}$
32. a. Answers will vary. The pan holds about 120 square inches. What number times $1\frac{1}{2} = 120$? About 80.
37. a. One way: $26 \times 11 \div 12$
38. $\dfrac{7}{8} + \dfrac{1}{9}$
42. $\dfrac{7}{8} \div \dfrac{1}{9}$ if you assumed proper fractions; $\dfrac{8}{2} \div \dfrac{1}{9}$ if you did not.
44. b. They added the denominators.
    e. They simply subtracted the numerators and denominators.
    i. They cross-multiplied, one numerator by the other denominator.
    j. They divided the numerators and divided the denominators.
50. $\dfrac{9\frac{1}{4}}{3\frac{3}{4}} = \dfrac{9 + \frac{1}{4}}{3 + \frac{3}{4}} \neq \dfrac{9}{3} + \dfrac{\frac{1}{4}}{\frac{3}{4}}$. Rather, $\dfrac{9\frac{1}{4}}{3\frac{3}{4}} = \dfrac{9}{3\frac{3}{4}} + \dfrac{\frac{1}{4}}{3\frac{3}{4}}$.

Exercises 5.4, page 331

1. a. 0.075 c. 0.03 e. 1/200
2. a. $4 + \frac{6}{10}, 4\frac{6}{10}$ c. $1 + \frac{2}{10} + \frac{3}{100} + \frac{4}{1000}$; $1\frac{234}{1000}$
3. a. Thirty-two and four hundredths
4. a. 0.067 b. 4060.034
5. a. 2.4 c. 2.40 e. 0.988
6. a. 0.4 c. 0.05
7. a. 0.0084, 0.058, 0.56, 0.6
9. a.

10. b.

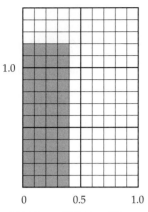

13. $24,440,000 or $24.44 million
14. 102 is how many vials can be completely filled, and .4 is the fraction of another vial that can be filled.
15. a. The right answer is $108. There are several ways to get $1080. One is 24 × 45.
19. 21/60 = 0.35, so it is 7.35 minutes.
21. a. 9/24 of 10 = 3.75, so it's three seventy-five.
  c. 8/10 of 24 is 19.2. 19 hours after midnight is 7 PM and 2/10 of 60 is 12, so it's 7:12 PM.
23. If I go 55 mph, it will take $\frac{30 \text{ miles}}{55 \text{ miles per hour}}$ = 0.55 hours.
If I go 65 mph, it will take 0.46 hours. Thus, it will take 0.09 hours less and 0.09 × 60 = 5.4, so I will save just under 6 minutes.

24. 21,735.3 dominoes per minute, which is 362.3 dominoes per second
27. a. 160 **c.** 0.8 **e.** 0.425 **g.** 0.24
28. a. 1.23456789×10^8 **c.** 5.6×10^{-10}
29. a. 4.182×10^{18} **c.** 1.53×10^{-17}
30. 0.00075 cm
32. a. $1.84467441 \times 10^{19}$
33. 845 items; $1.21
36. a. $1573.15 **b.** $78,657.50
37. Set up a proportion. $70.64
39. a. Option (b) **b.** 30 checks
40. a. 32.7 miles per gallon **b.** $30.62
  c. Her average speed was 405.6 miles/11 hours = 36.9 miles per hour. Also acceptable is 405.6/10.5 = 38.6 miles per hour. This is her average speed while the car was moving.
44. 27.7¢/day
45. a. 2550 dots **b.** 8.415×10^6 dots
47. a. We need to know how long each candle is.
49. At his current job he makes approximately $19,000. Not taking into consideration benefits, the work environment, and other factors, the annual salary of $24,000 seems better.
54. One of the basic rules of place value is only one digit per place.
56. a.
  c. Multiply just like they were whole numbers. 14.23

Chapter 5 Review Exercises, page 335

1. a. ⁻4 **b.** 1/4
2. a. ⁻74 **b.** 19 **c.** ⁻6 **d.** ⁻18
3. ⁻2
4. It means ⁻3 + ⁻3 + ⁻3 + ⁻3 + ⁻3 + ⁻3.
5. The first symbol represents an operation. The second symbol signifies the value of the number. To mix them up is to become sloppy, which is not a good habit to get into. The number sentence given reads, "Negative 3 minus 4 is equal to negative 7."
6. Yes. The whole has been divided into four regions of equivalent value, and three of those four regions are shaded. The horizontal line is extraneous.
7. Because 5/6 = 35/42 and 6/7 = 36/42, you can go to 84ths and see that 71/84 is between the two.
8. 11/15
9.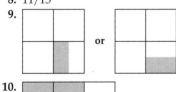
10.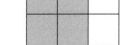
11. • •

   • •
12. 7/10 and 2/3 are both good approximations.

13. a. Both are two parts from 1. Because 11ths are greater than 13ths, 2/11 will be farther from 1, so 13/15 is greater.
  b. 7/12 is greater. Using 1/2 as a benchmark, we can quickly see that 7/12 > 1/2 and 13/28 < 1/2.
14. When we divide 1 by 2 in base 5, we get 0.22222....
15. It means they have the same value.
16. There are many ways to express the reason. One is to say that, by definition, the numerator of a fraction means how many "equal" pieces one has. If the denominators are not identical, then the numerators represent pieces of different size.
17. There are many ways to express why the hypothesis is invalid. One is to liken it to multiplication of two-digit numbers, such as 73 × 21. If you just multiply 70 × 20 and 3 × 1, you are missing the "two longs" regions.
18. We can solve it without the algorithm by finding 7 3/4, 2 1/3 times—that is, $7\frac{3}{4} + 7\frac{3}{4} + \frac{1}{3}(7\frac{3}{4})$. We can still avoid the algorithm by using number sense: 1/3 of 6 3/4 is 2 1/4. Hence we have $7\frac{3}{4} + 7\frac{3}{4} + \frac{1}{3}(6\frac{3}{4} + 1) = 7\frac{3}{4} + 7\frac{3}{4} + 2\frac{1}{4} + \frac{1}{3} = 18\frac{1}{12}$.
19. It will be in region B because the answer is clearly positive, but it will be less than either of the two fractions because of the operator model of fractions.
20. She can make 15 cakes and will have $1\frac{1}{4}$ cups of flour left over.
21. You could do it in either of two ways: $4\frac{3}{4}$ is $\frac{1}{2}$ of $9\frac{1}{2}$, so this problem is more than $\frac{1}{2}$. Or $\frac{1}{2}$ of $8\frac{7}{8}$ would be

Chapter 6   A-17

$4\frac{3.5}{8}$, and because $\frac{3}{4}$ is greater than $\frac{3.5}{8}$, the answer is greater than $\frac{1}{2}$.

22. Answers will vary. Although Liping Ma gives examples using the partitive model, virtually all examples familiar to Americans will use repeated subtraction: There is a whole, and it is being divided into pieces that are the same size.
23. a. 1200 is a reasonable rough estimate, because 2/3 of 18 is 12, and 1754 is close to 1800. Using the distributive property, we can get closer:
    2/3(1800 − 46) = 1200 − 30 = 1170.
24. a. 0.625   b. 0.004   c. 6/25
25. The number of valid answers is infinite. They include 0.44441, 0.44442, and so on.
26. ⁻0.54   0.045   0.454   5/11
27. 23.3. The two nearest tenths are 23.2 and 23.3.
28. $4,306,000,000
29. a. 1.904×10^{18} b. 2.53×10^{-5}
30. a. 0.112 cubic meters. We multiply because we have part of the tank: fraction as operator.
    b. 53 cups. We divide because the problem is repeated subtraction—repeatedly taking away 0.15 liter.
    c. 8 boxes. Same reasoning as in part (b).
    d. $6.06. This can be seen as division as a proportion: 4.85/0.8 = x/1.

Exercises 6.1, page 354

1. 143.75 calories. Answer is an approximation.
3. 165 dentists. Answer is an approximation.
5. a. $750; this is an exact answer.
   c. $977.08; this answer is rounded to two decimal places.
7. 860,000 people. Answer is an approximation (rounded to two significant figures).
9. Approximately 51 feet, 5 inches.
11. Yes; at this rate she will row 2,228.6 meters in 10 minutes, which exceeds her goal.
13. 11 3/4 inches. This is an approximate answer.
15. a. 45 miles. This is an approximate answer.
18. a. 104 km/h   19. $500
21. 40 miles   22. 10 questions
24. The man with five loaves should get 7 cents, and the man with three should get 1 cent.
25. 3.5 feet
26. a. 23 miles per hour
27. a. 37 handshakes per minute, which is one handshake in less than 2 seconds
29. 2520 students
30. $1\frac{1}{2}$ hours more
33. More prison inmates. For a U.S. population of 30,000,000, there are approximately 546,500 physicians and 603,600 prison inmates.
36. 199/325 = 61.2% and 5/8 = 62.5%. These percents are close enough to say that the advertisement is accurate.
39. The person who gets paid twice a month will have a larger paycheck.
40. a. In order to get 40 pledges in 3 hours, they would need 20 pledges by 7:30.
    b. Two lines of reasoning are appropriate: One is that the announcer may have felt that it would turn people off to say that they were behind; the other is that the announcer might realize that they will get more callers from 7:30 to 9 than from 6 to 7:30, and so they would catch up then.
48. a. 36.5, 14.4

Exercises 6.2, page 373

1. a. 18   c. 1,211.8   e. 0.8625   g. 80%
   i. 8.52%   k. 18.75%   m. 46.35   o. 7.192
   q. 32.4   s. 1,185.6
3. 35%   5. 700 patients
7. a. 131%   c. 43%
9. Approximately 6%; 9.6 lb   11. 5%
13. Approximately 2,320,000 (2.32 million) Americans are homeless.
15. Approximately $90,300
17. $4,544.40
19. $440   21. 18%
24. Having the public pick apples will provide more income.
29. a. The average elementary principal makes an average of 59% more than the average teacher, the average junior high principal makes 68% more, and the average senior high principal makes 76% more.
    b. The average junior high principal makes 7% less than the average senior high principal, the average elementary principal makes 13% less, and the average teacher makes 48.1% less.
31. a. 5 out of 1000, or 1 out of 200, children will have a severe reaction.
33. In the United States, inflation caused prices to rise 5%, so a $10 item would cost $10.50. In a country with a 400% inflation rate, a $10 item would cost $50 a year later. The value of the item hasn't changed, but the value of the currency used to purchase the item has diminished.
35. a. $23,946
38. Answers will vary. It is a 20% drop, but only a 0.02 change in blood alcohol content.
40. This interpretation represents an interesting misconception. If we selected 100 students at random, we might or might not find that 8 of them were working full-time. The 8% means that if we take the fraction of students that are working full-time and convert that fraction into a fraction whose denominator is 100, that fraction will be closer to 8/100 than to 7/100 or 9/100.
42. a. 600 square feet of window space
    b. It depends on the size of the windows. If the windows are 5 feet × 4 feet, it would be 30 windows.
43. $121,885.98
47. $2\frac{1}{2}$ hours
51. a. 80 seconds

Chapter 6 Review Exercises, page 378

1. $69.29   2. $1227.50   3. The ratio will increase.
4. They need to add 7 teachers.
5. 559 miles   6. 50 more males
7. The birth rate is a ratio. Thus, to keep the ratio the same, if the population increases, the number of births also needs to increase. If the rate is down, it means that the number of births did not increase at the same rate as the population did.
8. We use rates to enable us to compare "apples and oranges." For example, let's say two cities had the same number of cases of a disease, but the second city had five times as many people as the first. If we get just raw numbers, we don't see the difference. However, if we convert the numbers to rates, the rate for the first city is five times the rate for the second, which tells us that the disease is much more prevalent in the first city.
9. $240,000
10. To keep the same taste, she has to add both juices according to the ratio which, in simplest terms, is 2 cups of grape juice for every 3 cups of orange juice. Thus, if she added 4 cups of grape juice, she would need to add 6 cups of orange juice to keep the ratios the same.
11. $22,373,000   12. 67%
13. 1 hour   14. 4 minutes 12 seconds
15. $1,575,000   16. 66%
17. 4.6% or 5%   18. 30
19. 80%   20. 2.3%
21. I had punched .0055 instead of .055
22. Technically, the answer is $125.64, but it is more likely that the original price was $126. When you take 30% off $126, you get $88.20, which would probably be rounded down to $88.
23. If the original price is greater than $100.
24. a. 75   b. $123/185 \approx 12/18 = 2/3 \approx 67\%$
25. Technically, it comes to $305,300, but the more reasonable answer is $300,000, because the 115% is an average.
26. $3200/8400 = 0.38$ and $20,400/59,000 = 0.35$, so the 1976 purchase took a bigger portion of his income than the 2002 Honda did.
27. 71%
28. 65.9 ppm

Exercises 7.1, page 413

1. **a.** mean 53.3 minutes
   Median is 30 minutes
   Modes are 5 and 30 minutes
   b. They do not tell us anything about the shape of the data—the range, the spread, gaps, clusters, or outliers.
   d.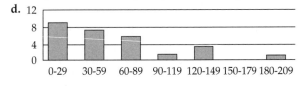
   g.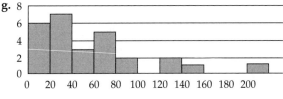
   j.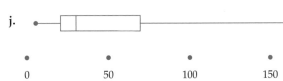

2. **b.** There is no mean or median for these data. The mode is fish.
   c. Technically we could, but it doesn't have much meaning.
   c. Is it "your favorite sport" for participation or as a spectator? We cannot be sure everyone was answering the same question.
   d. The mode is basketball. There is no median or mean.

6. Over 1/3 of our sample had no alcohol in the past week and almost 1/3 had more than 5 drinks, so it is hard to say what the average person looks like in this case. Furthermore, the three "averages" are all quite different: The mode is 0, the median is 2, and the mean is about $4\frac{1}{2}$ drinks if we count the "10–15" as 12.5.

7. **b.** The graph here almost has to be a bar graph. A circle graph is technically valid, but 12 categories make for many slices. Also, because the numbers are small, a change of one person would make the slice much bigger or smaller.

9. Since Willie has an average of 83.4 on the first five exams, we count 83.4 five times in the calculation of his average. Solve the equation $\dfrac{(83.4 \times 5) + x}{6} = 85$ to get $x = 93$.

11. Let x be the number that is removed.
$$\dfrac{4(7) + x}{5} = 6 \Rightarrow 28 + x = 30 \Rightarrow x = 2.$$

12. **a.** The estimated mean becomes a reference point from which all the scores are measured. The mean of the differences tells you how far your estimated mean is from the actual mean.
    b. Answers will vary.

14. It couldn't have been the mean—you don't need too many older people to make a mean of 9 impossible. I don't see how it could be mode—more homeless people are 9 years old than another age?! It's hard to see it as the median—half of all the people are 9 or under—but this is the only one that is even conceivable, assuming that the data are valid. This would mean that most homeless people are families with several children.
15. Answers will vary.
17. a. 0, 2, 2, 3, 4, 5, 5, 6, 9
18. I would use the median. In both cases, more than half of the data are very close to the median. In the first case, four of the five data values are substantially above the mean.
20. a. Mean is 72.75 and median is 77.5
    b. Delete one below and one above the median.
    c. Delete two below the median
    d. Delete two scores whose sum is greater than 146
    e. Delete one score above and below the median; the sum of the scores must be greater than 146.
24. Using raw numbers is like comparing apples to oranges. For example, look at AIDS deaths in the United States. The (raw) number of deaths of whites and blacks is almost identical. However, since there are almost eight times as many whites as blacks, the *rate* for blacks is much higher.
27. c. Answers will vary.

33. a.

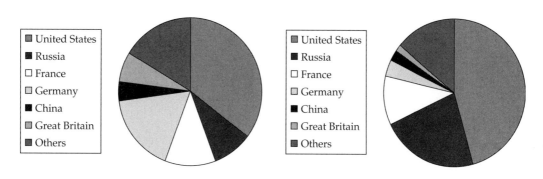

c.

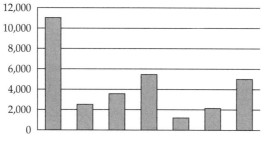
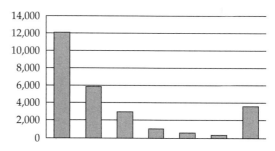

35. Answers will vary.
41. a. In the United States, the number of AIDS deaths rose steadily from 1981 to 1995 when they peaked. The number of deaths dropped sharply for two years and then continued decreasing but at a slower rate. The number increased in 2003.
    b. How do they determine AIDS as the cause of death? I have heard that many early AIDS deaths were actually reported as death from pneumonia or heart failure to hide the fact that the person had died from AIDS.
    c. Line graph is fine—it enables us to see the rate of increase or decrease.
    d. Answers will vary.
    e. Answers will vary.
42. a. Answers will vary. Some possible answers: More than $\frac{1}{2}$ of all AIDS deaths occur between 25 and 45 years of age.
    b. Same as #41.
    c. The circle graph enables us to see percentages. A bar graph would enable us to see the actual numbers.
    d. Answers will vary.
    e. Answers will vary.
45. a. When asked to describe what they wanted students to learn from the lessons that were videotaped, over 1/2 of the German and U.S. teachers' responses fell into a "Skills" category, compared to 1/4 of the Japanese teachers. Conversely, roughly 1/4 of the German and U.S. teachers' reponses fell into a "Thinking" category, compared to almost 3/4 of the Japanese teachers.

c. Side-by-side bar graphs make it easier to compare data.
d. It would be nice if the bars weren't so close together.
48. a. The percentage of drivers involved in fatal crashes with BAC .08 or higher increases as the driver gets older, until the age 25–34 group. Then it declines steadily as drivers get older.
b. The graph is skewed to the right.

Exercises 7.2, page 448

2. a.

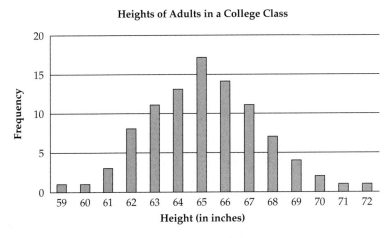

b. Mean: 65.15; standard deviation 2.43. c. 70%
13. The lengths of the rivers range from about 2300 to about 4200 miles. One-fourth of the rivers are longer than 3600 miles in length. The middle 9 rivers are between 2600 and 3600 miles long.
16. 61, 71 One line of reasoning: 5 feet is 6 inches less than the mean, and is 2.4 standard deviations below the mean. Since the curve is higher closer to the mean, we would expect that about 1% of the boys are less than 5 feet. 1% of 1500 boys is 15 boys.
18. The mean will increase by the amount. The standard deviation will not be affected.
20. z-scores, one can approximate the area under the curve. By various means, one can conclude that approximately 35%, or about 700 tires, will wear out before 45,000 miles.
22. Her math score was the highest in terms of standard deviations above the mean. English was 1.33 standard deviations above the mean; Math, 1.33 standard deviations above the mean; and General Info, 1.25 standard deviations above the mean.
23. The mode will have the smallest value, because the frequencies are higher at the low end of the range when a set of data is skewed to the right. The median will have the middle value, because it is affected by the high frequencies at the low end and, to a lesser extent, the presence of outliers at the high end. The mean will have the highest value, because it reflects outlying data at the high end of the range.

29. Answers will vary. Below I describe interesting aspects for the set of data.
a. Did you get a representative sample, including, for example, both rich and poor, both urban and rural, and different parts of the country? For example, if most of your sample is from cities, then my guess is that the numbers with two or more TVs and two or more VCRs are higher than the actual data would be.
b. Do you have a television in your house? (If the answer is no, do not count this person.) If the answer is yes, how many TVs? How many VCRs?
d. One could make two bar graphs to show the data, but a circle graph quickly shows the percentages of each question.
e. I don't see any advantage to the graphs. I could already see that 3/4 of the households that had a TV had more than one TV and that about 1/4 of the households that had a TV had two or more VCRs. *Note:* There is one problem with these data; they imply that every household that has a TV has at least one VCR, because the VCR percentages add to 100%. I don't believe it! I don't believe the data which imply that everyone has a TV and everyone has a VCR. This is a great place to illustrate the notion of a population. In this case, I am pretty sure that population for the survey consists of people who have a TV. That is, of all the people who have a TV, 76% have two or more. Even now the data imply that everyone who has a TV also has at least one VCR.

31. a. Whom did you survey? I would not say that over half of the families I know eat dinner together 5 or more days a week. Does it count if only part of the family is there? Were these data gathered from two-parent families or from one- and two-parent families?

b. How often does your family eat dinner together in an average week during the school year? (I would give them the categories in the table below or I would ask for a specific number. For example, a response of "2 or 3" would create problems in comparing to the data given.)

c.

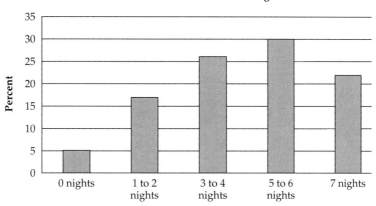

39. Julie is right. Although 148.12 grade points/46 credits = 3.22 GPA, one's total grade points is either a whole number or a mixed number.

Exercises 7.3, page 477

1. a. The probability of picking a black is the same in both cases.
  b. Pick from 6 blacks and 2 whites.
3. a. $11/1105 \approx 0.01$
  b. $27/2197$
  c. The probability of drawing a face card three times in a row has to be less than the probability of getting a face card once.
5. No. The class is not a representative sample of the people who died last year.
7. No, it is not a fair game.
9. Possible results: 0, 1, 2, 3, 4, 5. Most likely difference: 1. Probability that the difference is 1: $10/36 = 5/18$.

11. 1 in 256; $1/256 = 0.0039$, or thirty-nine ten-thousandths.
14. $1/216$
16. Answers will vary. Here is one possibility: One die must have the same number on all its faces; the other die could have 1, 1, 3, 3, 5, 5 or three of one odd number and three of another.
18. Yes, $P(WWR) = P(WRW) = P(RWW) = 1/5$
20. a. $81/256$ **b.** $3/32$
24. Many solutions are possible. If the shape of the spinner remains the same, one possibility is $\$6(1/8) + \$4(1/8) + \$2(1/8) + \$2(1/8) + \$1(1/4) + \$0(1/4) = \$2$.
26. Not fair because the expected value is $108/216 - 125/216$.

Exercises 7.4, page 488

1. 720
3. 5040
5. $1/120$
7. 220
9. If the flavors were scooped in any order, there would be 84 possibilities. If you specified the order of the flavors, there would be 504 possibilities.

11. $\frac{26}{27}$
13. a. 24 **b.** 96
15. $1/17$
17. $1/5040$
19. $_nP_n = n!$
22. a. 11! or 39,916,800 possible words
  b. Answers will vary.

Chapter 7 Review Exercises, page 494

1. a.

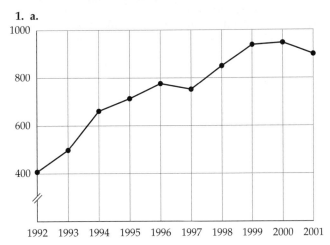

Note that the truncation is acceptable, if noted; the truncation is not necessary.

b. The number of CDs shipped rose steadily between 1992 and 2001, with a dip in 1997 and 2001. The number of units shipped in 2001 was more than double times the number shipped 9 years earlier.

2. a.

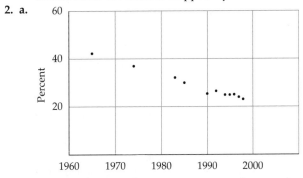

Note that because the years are not evenly spaced, we need to leave blank spots for missing years, and we should not connect the dots, because we can't be sure what the data are for the missing years.

b. The percentage of adults who smoke declined pretty steadily between 1965 and 1990. Since 1990, the percentage has declined only slightly, from 25% to 23%.

3.

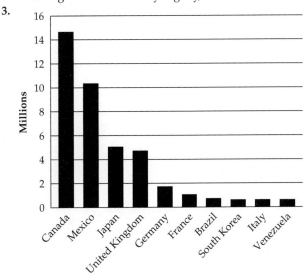

4. a.

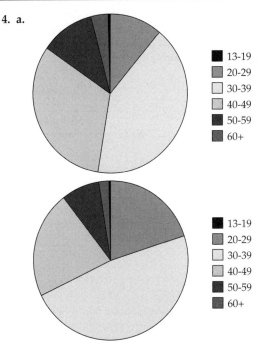

b. Between 1985 and 2000, the ages of AIDS patients increased. An alternative wording: The proportion of AIDS patients in younger age groups decreased, and the proportion of AIDS patients in older age groups increased.

5.–7. Responses will vary.

5. a. The United States and Europe accounted for over half of the world carbon dioxide emissions from the use of fossil fuels in 2000.

b. Were the data obtained from the governments of the countries or from a "neutral" agency?

c. The circle graph is an excellent choice, because it enables us to see the overall proportions. The actual amounts are not as important as the relative amounts.

d. No major problems with the graph. Some might quibble with the coloring or with the labels on the slices being inside in some cases and outside in other cases.

e. If all areas were contributing to the pollution equally, the United States would be responsible for only 5% of the world's total. Our producing almost 25% means that we are polluting at over 5 times a "fair" amount.

6. a. The percentage of cesarean births grew from about 21% to about 24% between 1995 and 2001.

b. My guess is the data came from hospitals, which keep good records, so I would have no concerns about the accuracy of the data.

c. The line graph is an excellent choice—we are looking at changes over time.

d. The graph is truncated, but the graph does not indicate the truncation. The graph makes it appear that the percentage of cesarean births rose dramatically over this time period. Actually, the change from 20.8 to 24.4 is a percent increase of 17%.

e. I really have no idea. I think the graph would be easier for most people to read if it said "Percentage of babies delivered by cesarean section."

7. **a.** It appears that a much higher percentage of lessons have high quality of math content in Japan and Germany than in the United States. Alternatively, the percentage of lessons having a low quality of math content was much greater in the United States than in Germany or Japan.
   b. My questions here would focus on the sample. Were they comparing teachers who were comparable in years of experience, age, etc.?
   c. The choice of graph makes great sense, although three circle graphs would have been fine too, so we could compare the percentages easily.
   d. I have no problems with the graph.
8. **a.**

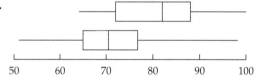

   b. The mean is 23.7; the median is 25.
   c. The number of drops recorded varied from 15 to 30. Over half of the data were between 21 and 27 drops.
9. **a.** 4.1 pounds   **b.** 2.8 pounds
   c. Were the dogs all approximately the same size? For example, comparing the weight loss of a 100-pound dog to the weight loss of a 10-pound dog does not make sense.
10. **a.**

   The box plot gives a quick snapshot. It tells us that the first class had a significantly higher range; that the second class did better overall—it had a higher median; and that $\frac{3}{4}$ of the second class is above 70, compared to only $\frac{1}{2}$ of the first class.
   b.

   The line plot also lets us see the range; it also lets us see the clusters. The first class is relatively spread out; the second class has a cluster in the 80s.
   c. Box: pros—quick snapshot; cons—you don't have all the data.
   Line: pros—you have all the data; you can see the distribution (spread, range, clusters, gaps); cons—you don't have the average.
   d. Means are 72.1 and 80.8; medians are 70.5 and 82.
   e. Ranges are 47 vs. 36.
   f. Standard deviations are 11.9 vs. 9.3.
   g. 71% vs. 70%
11. He needs at least an 83.
12. About 48   **13. a.** 21   **b.** 1.9   **c.** 2
14. **a.** A situation with outliers.
   b. A situation where you would want to know standard deviation; mean and standard deviation go together well; grades.
15. 1/16   **16. a.** 1/2   **b.** 1   **c.** 5/42
17. 9, 1/8   **18.** 1/4   **19.** 1/10, 1/5
20. It's not fair; 3/4 chance of even; 1/4 chance of odd. To make it fair, give 3 points to player A.
21. 120   **22.** $\frac{4}{17}$
23. **a.** 15   **b.** 3   **c.** 5   **24.** 576,000
25. **a.**

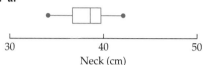

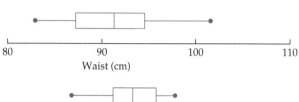

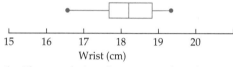

   b. The correlations between neck and wrist is a strong positive relationship. The correlation between neck and waist is a weak positive relationship.

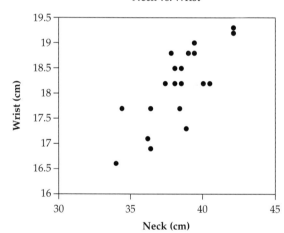

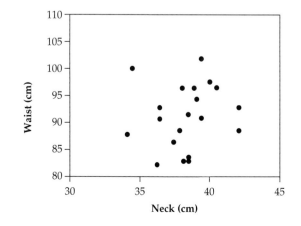

Exercises 8.1, page 518

4. $\overrightarrow{AC}$; $\overrightarrow{BC}$; $\overrightarrow{CA}$; $\overrightarrow{BA}$

6.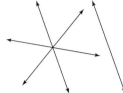

7. a. False; skew lines do not intersect and are not parallel.
   c. True   d. True
   e. False; think of two lines on the plane determined by this sheet of paper and a third line perpendicular to this plane.

8. Several answers are possible. Examples are given.
   a. $\overline{AB}$ and $\overline{DC}$
   c. $\overline{AB}$ and $\overline{AD}$
   e. $\overline{AB}$ and $\overline{EH}$
   g. All intersecting line segments are perpendicular.

10. a. 50°   b. 20°   c. 120°

11. a. 24 times
    c. 150° (and 210°)

13. c. If you divide the space between 90° and 180° in three, you will have 120° and 150°.

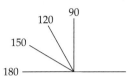

15. a.

    b.   c.

Exercises 8.2, page 555

1. a. 6 sides
   1 pair of parallel sides
   3 right angles
   1 acute angle
   1 obtuse angle
   1 reflex angle
   2 pairs of congruent sides: the two at the top and the 2 that are vertical
   concave figure

2. a.
Square	*Triangle*	*Kite*
4 sides	3 sides	4 sides
convex	convex	concave
4 ≅ sides	No ≅ sides	2 pair ≅ sides
4 right angles	0 right angles	0 right angles
0 acute angles	3 acute angles	3 acute angles

   Also, the top angles of the triangle and kite are ≅. I would say the triangle and kite are most alike, but each pair has similarities.

3. Alike: four sides, same are, same length, both pictured on a grid, four corners, straight sides, both composed of little squares, both are shapes. Different: one is straight, the other slanted; different shapes (they look different); one has more full cubes inside; one is a rectangle the other is not; different perimeters.

4. A pentagon has 5 diagonals, a hexagon has 9 diagonals.

5. a. Squares (two different sizes), rectangle, right isosceles triangle (two different sizes), 9-gon

8. b.

   Impossible on Isometric Dot Paper; can't construct two equal sides adjacent to a 90° angle.

   e.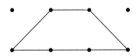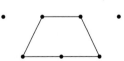

9. a. Concave polygon
   b. Not a polygon
   c. Convex polygon

12. a. Yes. It fits the definition: 2 pair of congruent, adjacent sides.

14. c. Congruent
    Flip the first figure vertically and then horizontally.
    Flip the second figure vertically to show congruence.

15. a. Parallelogram, rhombus, rectangle, square
   c. Square, rhombus
   e. Rectangle, square, isosceles trapezoid
   g. Parallelogram, rhombus, rectangle, square
   i. Rhombus, square
16. Answers will vary. These are some examples.

   a. 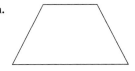   **c.**

   e.

17. a.   Front   Right   Top

   c. The object of this problem is not so much getting one "right" answer as thinking mathematically. Some solution paths are shown. Figures A, B, and C have 3 consecutive right angles.

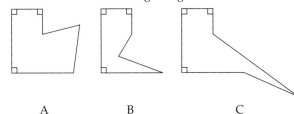

   A   B   C

   Figures D, E, and F have 2 consecutive right angles, a nonright angle, a right angle, and a nonright angle.

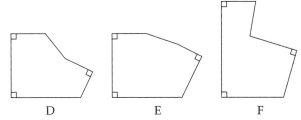

   D   E   F

Figures G and H illustrate a different perspective. Figure G has 5 ≅ sides. Figure H has 3 ≅ sides of one length and 3 ≅ sides of a different length.

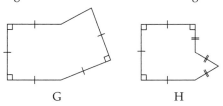

   G   H

18. d.

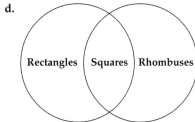

24. Number of diagonals $= \dfrac{n(n-3)}{2}$, where $n =$ number of sides

Hint: There is a pattern in how you make diagonals. For example, if you use a hexagon, you can make 3 diagonals from the first vertex, then 3 from the next vertex, then 2 from the next vertex (the diagonal connecting it to the first vertex is already drawn), and then 1 from the next vertex. That is, the number of diagonals in a hexagon is $3 + 3 + 2 + 1 = 9$.

32. a. 5   **b.** $\sqrt{45} \approx 6.7$
33. a. (2, 7)   **b.** (5, 8)
34. a. The fourth vertex has an x-coordinate of 9.
   c. One solution is (7, 1).
35. a. (0, 0), (6, 0), (6, 6) and (0, 6)
   b. The other two vertices are on the horizontal line going through $(7, -2)$ and are equidistant from $(7, -2)$; for example, $(5, -2)$ and $(9, -2)$.
   c. There are many possibilities: here are two: The coordinates are (10, 0) and (0, 10) or (10, 0) and $(0, -10)$.

Exercises 8.3, page 575

1. Answers will vary. Several correct answers.
   a. △ABC   **b.** A   **c.** $\overline{AB}$
4. $n + 1$ vertices; $n + 1$ faces; $2n$ edges
6. a. Octagonal prism   **c.** Rectangular prism
7. The polyhedron on the right is convex.
10. Yes. If all the edges of the base are of different lengths, then the triangular faces will not be congruent.
12. $n(n - 1)$ diagonals
14. The sides are isosceles triangles.
23. The first one   **24.** The first one

Chapter 8 Review Exercises, page 579

1. Answers will vary.
2. There are multiple answers for each question. One answer is given for each.
   a. ∠ABG and ∠GBF   b. ∠CBG and ∠GBF
   c. ∠ABG and ∠DBE   d. ∠DBE and ∠EBG
3. a. False. Consider two lines that lie on this paper and one that is perpendicular to the paper.
   b. False. They could be skew.
4. If you do not start with some undefined terms, some of the definitions will be circular, similar to dictionary definitions. For example, A is defined in terms of B, and B is defined in terms of C, and C is defined in terms of A.
5. Answers will vary. One option is to say that a triangle is a shape made by three line segments such that the endpoints of each segment touch the endpoints of another segment.
6. 14

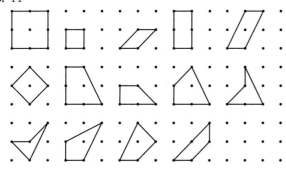

7. Answers will vary. One option is to draw a quadrilateral and one diagonal, thus making two triangles. Knowing that the sum of the angles of a triangle is 180 leads to the conclusion that the sum of the angles of the quadrilateral is 360.
8. They are convex polygons, with four sides, four right angles, opposite sides congruent and parallel, and the diagonals congruent and bisecting each other.
9. a. 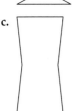 b.
   c.   d. e.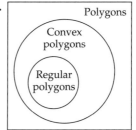

10. The case of five right angles is the most challenging. In a hexagon, there are six angles whose sum is 720. If five of the angles are right angles, then the sixth angle would be 270, which is possible if the hexagon is convex.
11.
The first figure	The second figure
hexagon	hexagon
concave	convex
two sides parallel	three pairs of parallel sides
two pairs of congruent sides	three pairs of congruent sides
4 acute angles	0 acute angles
2 reflex angles	0 reflex angles
0 right angles	2 right angles
0 obtuse angles	4 obtuse angles
3 pairs of congruent angles	3 pairs of congruent angles
1 line of symmetry	2 lines of symmetry
no rotation symmetry	180-degree rotation symmetry

12. Answers will vary. There are many ways to represent the answers. Each vertex can be labeled; figures can be outlined with different colors; figures with the same name can be described to distinguish them—for example, a 2 × 2 square and a 1 × 1 square.
13. a. (1) Kiwis are composed of a pentagon with a line segment protruding from one vertex.
    (2) The protruding segment is perpendicular to the side the segment would intersect if it were extended.
    (3) Kiwis have 2 adjacent right angles, which is equivalent to saying that each Kiwi has 2 parallel sides that meet a third side at 90-degree angles.

    The first not-Kiwi has 6 sides. The second not-Kiwi has the protruding segment not perpendicular to the opposite side. The third not-Kiwi has the line segment inside the pentagon; the fourth not-Kiwi does not have 2 adjacent right angles.
    b. Figure a is not a Kiwi—the protruding segment would not intersect the opposite side if extended. Figure b is a Kiwi.
    Figure c is not a Kiwi—the line segment is partially inside and partially outside the pentagon.
    Figure d is not a Kiwi—it does not have 2 adjacent right angles.
14. Answers will vary. One description: Draw a square and then draw the diagonal that begins at the bottom left corner. Extend the base of the square to the right, about one-third the length of the square. Connect this endpoint to the top right vertex of the square.
15. No, because it implies that there are some rectangles that are not parallelograms.
16. (Polygons ⊃ Convex polygons ⊃ Regular polygons)
17. √29 ≈ 5.4
18. (2.5, 7.5)
19. (2, 3)
20. (6, 6) and (12, 6)
21. 6 vertices, 9 edges, and 5 faces
22. They have a base and an apex.
23. a. Answers will vary. The top, front, and side views will not work here because two different buildings have

the same top, front, and side views. Descriptions can build from the ground up, from one side to another, or from front to back. Also valid is the top view with numbers to indicate the number of cubes in each spot. The key in most cases is to orient the reader correctly.

24.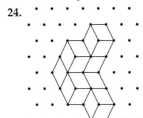

25.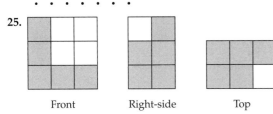
Front   Right-side   Top

26. **a.** From the front view, it will have at least 6 cubes. To get the right-side view, you have to add at least 1 cube; to get the top view, you don't have to add any cubes. Thus, I predict 7 cubes.
   b. It is the mirror image of the right side.
   c.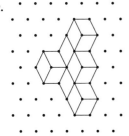

27. There are many possible nets. To be valid, it would have to fold up to make the prism. Two are shown below.

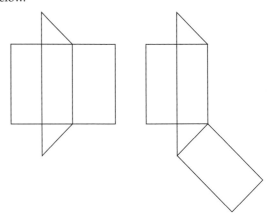

28. Answers will vary. One response: a line segment connecting two nonadjacent vertices.
29. **a.** Trapezoid
    b. Rectangle
30.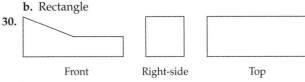
Front   Right-side   Top

Exercises 9.1, page 597

1. Each point of the figure has been moved 2 units to the right and 3 units down.
   Also, the figure has been moved approximately 3.6 units along a vector that makes a 55-degree angle with the x-axis. See diagram.

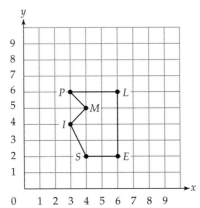

3. See diagram.

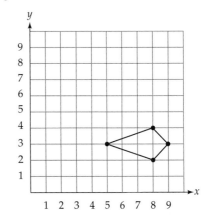

6. **a.** The first and third cases are reflections of each other. The second case is not a reflection because a point and its image are not equidistant from the line of reflection.

8. See diagram.

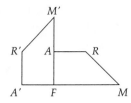

c. See diagram.

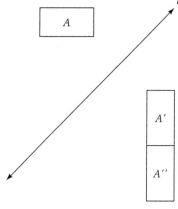

The image from translating and then reflecting does not coincide with the image from reflecting and then translating.

12.

	Point A′	Point B′	Point C′	Point D′
a.	(4, 7)	(6, 7)	(6, 6)	(4, 4)
c.	(4, 3)	(6, 3)	(6, 2)	(4, 0)

17. a. Answers will vary.
b. See diagram.

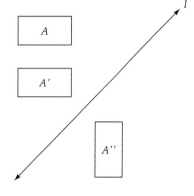

19. a. See diagram.

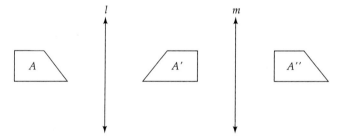

b. Answers will vary.
c. See diagram. A′ is the reflection of figure A across line m. The reflection of figure A′ across line l will not even lie on the paper.

The trapezoid A″ in part (a) is the image produced by reflecting across line l and then across line m. Trapezoid A′ above is the image produced by reflecting across line m. The image produced by reflecting across line m and then line l is off the page (to the left).

21. a. See diagram.

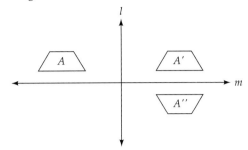

b. Answers will vary.
c. The double reflection produces the same image, no matter what the order.

23. a. Translate $\overline{ST}$ along the vector $\vec{SU}$.
b. Rotate the figure 180 degrees about point M.
c. No; in order for the image of $\overline{UM}$ to be $\overline{MT}$, $\overline{UM}$ would need to be perpendicular to $\overline{SR}$, which would be the case only if STRU were a rhombus.

27. Answers will vary.
a. Baby Blocks: If we see the figure as composed of rhombuses, each rhombus can be mapped onto a neighboring rhombus by a 60-degree rotation. If we see the figures as hexagons (comprising three different-colored rhombuses), each hexagon can be translated onto any other hexagon.

31. a. 9 o'clock

Exercises 9.2, page 630

1. a. 60-, 120-, 180-, 240-, and 300-degree rotation symmetry; no reflection symmetry
c. Horizontal and vertical line symmetry; point symmetry
3. a. Vertical line symmetry
c. No symmetry because of the line segments
4. a. Point symmetry
b. Vertical line symmetry
5. a. Horizontal, vertical, and diagonal line symmetry; 90-, 180-, and 270-degree rotation symmetry
b. 90-, 180-, and 270-degree rotation symmetry
6. a. Each of the right isosceles triangles has one line of symmetry.
The parallelogram has point symmetry.
The square has vertical and horizontal symmetry and two diagonal lines of symmetry; it also has 90-, 180-, and 270-degree rotation symmetry.
8. Answers may vary, according to how the digits are written.
1—no symmetry
2—no symmetry
3—horizontal line symmetry
4—no symmetry
5—no symmetry
6—no symmetry
7—no symmetry
8—horizontal and vertical line symmetry; point symmetry
9—no symmetry
0—horizontal and vertical line symmetry; point symmetry
10. a. False b. True
c. False
16. a. *Note:* All strip patterns, by definition, have translation symmetry. ll; no symmetry.
c. lg; glide reflection symmetry
e. l2; point symmetry
20. a. l2; point symmetry
b. The diamond-shaped unit is translated horizontally; horizontal and vertical line symmetry; point symmetry.
35. a. The pattern has several translation symmetries.
It has vertical and horizontal reflection symmetry.
It has 180-degree rotation symmetry.
It has glide reflection symmetry.
39. a. A and F. Descriptions will vary.
c. A and C (and D and E) are 180-degree rotations of each other about their points of intersection. B, C, and D (and E, F, and G) are 120-degree rotations of each other about the points where the three chevrons in each cluster intersect.
40. a. Tessellates b. Tessellates
c. Tessellates d. Tessellates

Exercises 9.3, page 642

1. a. 13.5 cm
3. Yes; the angles are the same, and the sides of the figure to the right are twice as long as the sides of the smaller triangle.
4. a. No; different isosceles triangles can have different angles.
c. Yes; all squares have four 90-degree angles and proportionate sides.
e. Yes; corresponding sides and angles of congruent polygons are congruent.
6. Yes, in general, but it may be distorted by the angles of the mirrors.
8. c. ≈666.87 ft
10. The length and width of the document are enlarged by 21 percent. The copy is 121 percent the size of the original.

Chapter 9 Review Exercises, page 647

1.

2. The justification will probably involve the observations that corresponding points are the same distance from the line of reflection and that the line segments connecting corresponding points and the line of reflection are vertical.

3. [figure] P 4. P|P P|q / d

5. The only one-step solution is to rotate the figure 90 degrees counterclockwise, through the point that is 3 units to the right and 3 units below the bottom right vertex of the top trapezoid. There are several two-step solutions; for example, translate 6 units down and then rotate 90 degrees counterclockwise.

6. Alike—they both move the object to a different position, keep the figure the same size and the same shape.
Different—the orientation is changed in a reflection but not in a rotation. For example, imagine rotating and reflecting the letter P. To see the rotated P, you could just turn the page until it was "right" again. However, no matter how you turned the page, the reflected P just wouldn't look right—it would feel backwards. The curved part is now facing left instead of right.

7. By including glide reflection, we have the theorem that any figure can be moved to any other spot on a plane in *exactly* one move.

8. a. 1/5 turn symmetry, or 72° rotation symmetry; 5 lines of reflection
   b. 1/4 turn symmetry, or 90° rotation symmetry; 4 lines of reflection
   c. 1/2 turn symmetry, or 180° rotation symmetry; no reflection symmetry

9. There are many possibilities. One valid figure is given for each case.

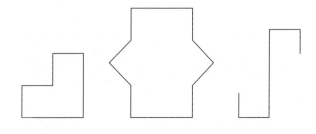

10. Answers will vary.
11. a. It means that there is at least one way to pick up the figure, turn it some amount, and be able to lay it down so it fits back exactly on itself.
    b. It means that there is at least one line that you can draw through the figure such that if you fold the figure along that line, the half of the figure on one side of the line will fit exactly on the other half.
12. The symmetries can be described with notation or visually. The notation is given at the left. On the figures, the reflection symmetries are denoted with lines, the 180-degree rotation symmetries are denoted with points, the translation symmetries are denoted with the translation vector, and the glide symmetries are denoted with the translation vector.

    a. ml
    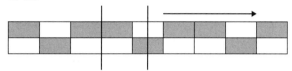

    b. ll
    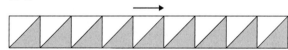

    c. mg
    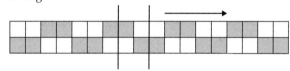

    d. lg

13. a. There is more than one possibility. For example, the upper figure at the right can be seen as the unit. In this case, the means by which it is repeated is a translation to the right. On the other hand, the lower figure can also be seen as the unit. In this case, the means by which it is repeated is also a translation to the right.
    b. In this case, the unit consists of the five chevrons shown at the right. The means by which it is repeated is translation in a diagonal direction.

    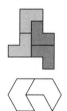

14. a. This figure tessellates. The sum of all angles at every vertex point is 360; this is because all angles are 90° or 270°.
    b. This figure tessellates, because all quadrilaterals tessellate.

c. This figure tessellates, because it is made by beginning with a square, modifying the bottom, and then translating that modification to the top side.
d. This figure tessellates. Because of the symmetry of the hexagon, the four acute angles are congruent and the two reflex angles are congruent. If we label the acute angles a and the reflex angles b, we have that $4a + 2b = 720$, which simplifies to $2a + b = 360$. Thus the sum of the angles at each vertex point is 360.
e. This figure tessellates. The reasoning here is virtually identical to the reasoning for part (d), except that the sum of the angles of a pentagon is 540°.

15. 28 cm
16. Yes. No matter what the size, the ratio between side and hypotenuse is the same.
17.

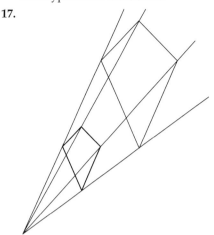

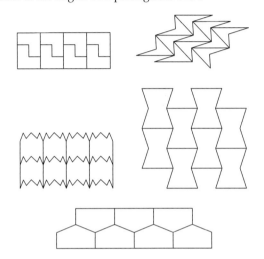

18. (6, 8), (16, 8), and (8, 14). The coordinates of each point of the similar triangle will be double the coordinates of the corresponding points on the original triangle.
19. Answers will vary. For example, any simple, closed curve will suffice for 0; any simple, open curve will suffice for 1, 2, 3. Any simple, closed curve with two straight or curved segments protruding will suffice for 4.

Exercises 10.1, page 660

1. Answers may vary. Following are examples for the given objects.
   a. surface area, volume, amount of pollutants, temperature of the water (at various levels), depth
   c. height, surface area of sides (for painting), surface area of roof (for shingles), surface area of windows, surface area of floors, ratio of area of windows to area of floors (to determine adequacy of ventilation)
2. Answers may vary.
   a. centimeters or meters
   c. millimeters or centimeters
   e. milligrams
3. a. 5.5   c. 0.670   e. 0.450   g. 0.024
6. About 34 meters. If we use their best times, Alan would be 7 seconds behind when Bill crosses the finish line. At Alan's pace (400 meters/82 seconds), he will run about 34 meters in 7 seconds.
8. The difference in conversions is less than an inch.
15. a. The makers of the table essentially set the American/London foot as the unit, with 1000 parts; the other numbers in this column enable us to see the length of other "foots" in relation to the American/London foot. By setting the unit foot as 1000 parts, they were able to avoid decimals.
17. b. 48.3 liters × 34 oz/liter × 1 gallon/128 oz = 12.83 gallons; or, by approximation, 48.3 liters × 1 quart/1 liter × 1 liter/4 quarts = 12.08 gallons. The difference would be about 3/4 gallon.
20. 5° Celsius, because each degree Celsius is equivalent to 1.8 degrees Fahrenheit.
22. a. 35 ft/9 sec × 1 mi/5280 ft × 60 sec/1 min × 60 min/1 hr ≈ 2.65 miles per hour
24. $13\frac{1}{3}$%. 20% of 12 oz = 2.4 oz. 2.4 oz milk/18 oz coffee × 100 = $13\frac{1}{3}$%.
30. About 100 meters! The height depends on the thickness of the paper. Multiply the thickness by the number of sheets of paper; 2^{20} = 1,048,567 sheets of paper. I measured the height of 2000 sheets of paper (4 reams) and got 190 millimeters. This ratio is equivalent to 0.095 millimeter per sheet.
33. a. 1 kask is about 11 kilometers. 1 kask × 30 ush/kask × 60 gar/ush × 60 palm/gar × 4 in/palm × 2.54 cm/in × 1 m/100 cm × 1 km/1000 m = 10.97 km

Exercises 10.2, page 675

1. a. $P = 69$ mm, $A = 195$ mm^2
   b. $P = 80$ mm, $A = 148.5$ mm^2
   c. $P = 6$ cm, $A = \sqrt{3} \approx 1.7$ cm^2
   d. $P \approx 37.68$ cm, $A \approx 113.04$ cm^2
   e. $P = 40$ cm, $A = 72$ cm^2
   f. $P = 35.8$ cm, $A \approx 51.9$ cm^2
3. a. $A = 15$ square units
4. a. Approximately 8.81 square centimeters
5. Approximately 11.2 feet
6. ≈ 7.4 feet
7. Solution paths will vary.
   a. $A = (5 \times 10) + (5 \times 5) + (10 \times 10) = 50 + 25 + 100 = 175$ square centimeters
   $A = (20 \times 10) - (5 \times 5) = 200 - 25 = 175$ square centimeters
8. 14.14 inches on a side. The length of the sides of a square with an area of 200 square inches is $\sqrt{200} = 10\sqrt{2} \approx 10 \times 1.414 \approx 14.14$.
10. $39.90. 9600 square feet/1050 square feet per bag = 9.14 bags. She needs to buy 10 bags.
17. a. ≈ 11.1 square centimeters = 100 square centimeters/9 small triangles.
19. b. πr
21. Approximately 344 square centimeters
23. a. 1250π square feet
32. a. 3 meters
    b. About 2.24 meters
    c. About 1.73 meters
    d. The longest distance is along the edges of the cube; the shortest distance is directly through the cube.
48. a. 10π
49. a. 84 square feet
50. a. 25 square inches   b. Not possible
54. About 732 square miles, $\frac{1}{2}$ the size of Rhode Island
57. $1\frac{3}{4}$ inches
61. a. 16   b. 41

Exercises 10.3, page 695

1. a. Surface area ≈ 1474 square inches
      Volume = 2160 cubic inches
2. a. Surface area ≈ 47 square feet
      Volume = 16 cubic feet
   c. Surface area ≈ 1980 square feet
      Volume = 2985 cubic feet
4. a. Predictions and explanations will vary.
      Volume of new tent ≈ 322.7 cubic feet
      Volume of old tent = 31.5 cubic feet
      The new tent is 291 cubic feet larger than the old tent or has about 10 times the volume!
6. Assuming 1 ream is 2 inches in height, you can get 210 reams, which equals 105,000 sheets and 42% of the yearly purchase.
8. 34 feet
10. a. Stack them into a cube each side being 60 centimeters long.
14. a. Prediction: 15 cubic centimeters
    b. ≈ 16.387 cubic centimeters
16. a. Surface area of small prism : surface area of large prism = 1 : 9
       This is the square of the ratio of their sides.
    b. Volume of small prism : volume of large prism = 1 : 27
       This is the cube of the ratio of their sides.
17. a. 28 dominoes
19. a. The cut should be one-fourth of the way from the end.

Chapter 10 Review Exercises, page 699

1. Answers will vary.
2. a. mm   b. l   c. ml   d. g   e. kg
3. a. 34,000 cm   b. 345 g   c. 2750 ml
4. Answers will vary. Two situations are given here.
   a. Measuring the length of a figure with a ruler but beginning at 1 instead of 0, a mistake commonly made by children.
   b. Reporting the distance between Los Angeles and San Francisco to the nearest 100 miles.
5. Attributes include surface area, perimeter, temperature, depth, and average depth.
6. Here are two ways: (1) Find the area of the rectangle that encloses the figure and subtract the area of the two triangles: $42 - 12 - 3 = 27$ square units. (2) Use the symmetry of the figure—the bottom half is a trapezoid with area 13.5 square units.
7. Area = 6 in^2. Using the Pythagorean theorem, reveals that the length of the other side is 4 in.
8. 146 square feet
9. 31.83 cm by 31.83 cm
10. 1200 square inches
11. 200 cm on a side, which is equivalent to a square that is 2 meters on a side.
12. 0.33, or 33%
13. a. There is empty space between the pennies, because circles do not tessellate.
    b. 25,600 pennies
14. Answers will vary. One response: We can take any parallelogram, slice off a right triangle from one side, and translate that triangle to the opposite side. The parallelogram becomes a rectangle, whose area we know to be $l \times w$. We know first that the rectangle and

parallelogram have the same area, that the length of the rectangle is the same as the base of the parallelogram, and that the width of the rectangle is the same as the height of the parallelogram. Therefore, $l \times w$ (of the rectangle) $= b \times h$ (of the parallelogram).

15. Answers will vary. Two responses: (1) Because 12 inches = 1 foot, 144 square inches = 1 square foot. (2) One could also draw a square that measures 1 foot by 1 foot; thus the area is 1 square foot. When we convert to inches, the same square is 12 inches $\times$ 12 inches = 144 square inches.
16. It quadruples.
17. $37\frac{1}{2}\%$
18. In both cases, the area of the figure will be about 8.1 cm², or 810 mm², which means that the actual area of the pond is 810 square meters.
19. Approximately 32.185 cm
20. 1 ream of paper measures 8.5 inches by 11 inches by 2 inches. Thus all the forms will take up about 433,000 cubic feet. A cube that measures 76 feet on a side would be able to contain all the paper. In more conventional terms, a building 233 feet by 233 feet by 8 feet would be needed to contain all the paper.
21. **a.** 44.7 feet **b.** 34.6 feet
22. **a.** 88.9 cubic yards **b.** 97.8 cubic yards
23. The surface area is approximately 2000 square feet. For the volume, break the figure into a rectangular prism and a triangular prism: 9600 cubic feet.
24. 25,600
25. There are many examples. Two are $15 \times 1 \times 1$, with a surface area of 62 square units and a volume of 15 cubic units; and $5 \times 3 \times 2$, which also has a surface area of 62 square units but a volume of 30 cubic units.
26. Surface area of the prism is 248 cm².
    Surface area of the cylinder is 225 cm².
    Volume of the cylinder is 246 cm³.
    Volume of the prism is 240 cm³.
27. **a.** 5840 gallons
    b. 1 gallon = 0.1337 cubic foot. The pool holds 5840 gallons which converts to 781 cubic feet. Since the swimming pool contains 3600 cubic feet, we would fill about $\frac{1}{5}$ of the pool.

Index

Absolute value, 259–260
Abundant numbers, 238
Accuracy, 650, 659
Active reading, 12
Acute angle, 516
Acute triangle, 531–532
Adaptive beliefs, 2–3
Addends, 130
Adding up strategy, 136, 155, 158, 165
Addition, 128–150
   algorithms, 139–142
   children's understanding of, 83, 128–130
   connections to subtraction, 152, 153, 206, 261
   connections to multiplication, 168, 169, 171, 176–177, 206
   contexts for, 128
   of decimals, 318
   definition of, 130
   estimation strategies, 144–148
   of fractions, 284–286, 289–292
   of integers, 258–259
   mental, 135–136
   model of, 128–131
   of negative numbers, 258–259
   number line model of, 130–131
   in order of operations, 205
   patterns in addition table, 133–134
   pictorial model of, 128–129
   properties of, 83, 131–132
   summary, 149–150
Addition table, 131, 133–134
   patterns in, 133–134
Additive comparisons, 337–338, 444
Additive identity, 132
Additive increase, 366
Additive inverse, 260, 298
Additive numeration system, 105
Adjacent angles, 516
Affirming the hypothesis, 38
Algebra
   elementary teachers and, 79
   as generalized arithmetic, 80, 90–94
   patterns and, 80
   as problem-solving strategy, 10, 259, 275, 302, 345, 350, 362
   as a set of rules and procedures, 81–82
   as study of relationships, 84–91
   as study of structures, 83–84, 298
   summary, 96
   variables in, 81–82
Algorithms, 50, 138, 162
   for addition, 139–142
   alternative, 142, 161, 181, 198
   for division, 194

   for multiplication, 178–181
   for subtraction, 160–161
Al-Khowarizmi, 80, 138, 159
Alphabitia, 108, 131
Altitude
   of triangle, 536
   see also Height
Amperes, 658
Analytic thinking, 38
Angle bisector, 535
Angles, 512–516
   acute, 516
   adjacent, 516
   children's understanding of, 512
   classifying, 516–517
   complementary, 516
   dihedral, 565
   exterior of, 512
   interior of, 512
   naming, 513
   measuring, 513–515
   of polygons, 547–548
   of quadrilaterals, 540–541
   reflex, 516
   right, 516
   of similar figures, 639
   straight, 516
   supplementary, 516
   of triangles, 481
   vertical, 516
Apex
   of cone, 571
   of pyramid, 566
Arc, of circle, 549–550, 664–665
Archimedes, 595, 596–597
Area, 665–675
   basic formulas for, 666–669
   children's understanding of, 663, 668, 673
   of circle, 671–672
   as a function, 667
   of irregular shapes, 673–674
   as a model for multiplication, 170, 178
   of parallelogram, 667
   perimeter and, 663–664
   of rectangle, 666–667
   of square, 666
   summary, 675
   surface area, 681–685
   of trapezoid, 668–669
   of triangle, 668
   units for, 665
Area models
   of fractions, 271–272, 273–275, 285, 287, 293–295, 302
   of multiplication, 169, 170–171
   of multiplying fractions, 293–295

   of probabilities, 470
Argument, 37
Aristotle, 38
Arithmetic sequences, 39
Art
   patterns in, 537, 610–611, 617–618
   transformations in, 582–583, 629
Associative property
   of addition, 132
   division and, 193
   with fractions, 296
   of multiplication, 172
   subtraction and, 154
Attitudes about mathematics, 1–4, 53–54
Attributes
   measured, 650
   of shapes, 524–526
Averages
   calculation of, 389
   as center, 389–392, 399–400
   geometric view of, 393
   as a measure, 391
   three kinds of, 389–393, 399–400
   weighted, 445–447
   see also Mean; Median; Mode
Axis of symmetry, 619

Babylonian mathematics
   areas in, 672
   degrees in circle, 514
   numeration system, 106–108
   time and, 651
Balance point of data, 393, 447
Bar graphs, 384, 388, 427
Base
   of exponential expression, 325
   of numeration system, 109
   of parallelogram, 667–668
   of polygon, 529
   of prism, 565
   of pyramid, 566
   of rectangle, 666
   of square, 666
   of trapezoid, 668–669
   of triangle, 533, 668
Base 5, 115–118
Base 10 numeration system, 109–113, 118
   addition in, 139–142
   advantages of, 110–111
   decimals in, 310–311, 318–320
   division in, 196–198
   geometric representation of, 111–112
   history of, 109–110, 111
   manipulatives for, 111–112, 310–311
   multiplication in, 178–181

I-1

Base 10 numeration system, *(continued)*
  subtraction in, 159–161
Bass, Hyman, 162
Beliefs about mathematics, 1–4, 53–54
Bell curve, *see* Normal distribution
Biconditional statements, 41
Big ideas
  about measurement, 656
Binary operations, 74
Bisector
  angle, 535, 537
  perpendicular, 535–536, 589
Border patterns, 531–534
Borrowing, 164–165
Bound, upper or lower, 182
Box, 565
Box-and-whisker graph, 424–426, 427, 429
Boxplot, 424. *See also* Box-and-whisker graph
Brahmagupta, 257, 268
Break and bridge strategy, 135
Brick patterns, 615–617
Bringing down, 197–198

Calculators
  compound interest with, 372
  dimensional analysis with, 352
  exponents with, 325, 326
  order of operations, 205
  ratios with, 343
Canceling
  of units, 16, 183, 195, 352
  of zeros, 195
  strategy for division, 195
Candela, 658
Capacity, 653, 655
Cardan, 257, 262
Carrying, 164–165
Cartesian coordinate system, 550–551
Cartesian product model
  for classifying triangles, 479
  of multiplication, 171
Cases, 259–260
Cavalieri, Bonaventura, 686, 687
Celsius system, 656
Center
  of circle, 550
  measures of, 389–393, 399–400
  of sphere, 575
Centers of rotation, 590
  inequivalent, 535
Centi-, 654
Centigrade system, 656
Centimeter, 654–655
Centroid, 536
Certain events, 457
Chance, *see* Probability
Change, 4, 337–338
  percent of, 364–365
Checking the solution, 11
Children's understanding
  of addition, 136–138
  of area, 572–573, 575
  of commutative property, 84, 132, 154
  of multiplication, 176–178

of perimeter, 603–614
of shapes, 474, 504–506, 555
of subtraction, 157–159
of units for measurement, 652
Chinese mathematics, 119, 257, 672
Chord of circle, 450
Churn Dash, 523–524
Circle, 549
  arc, 550, 664–665
  area, 671–672
  center, 550
  chord, 550
  circumference, 81, 82, 664
  degrees in, 514
  diameter, 81, 82, 550–551, 664
  great, 685
  radius, 550, 684
Circle graphs, 385, 389, 411, 427
Circumference, 81, 82, 664
Circumcenter, 536
Classes of data, 397
  modal, 399, 427
Classifying
  with sets, 63
  of shapes, 526–527
Closed curves, 526–528
Closed surfaces, 563
Closure property, 132
  of addition, 132
  of congruence transformations, 594
  division and, 193
  of multiplication, 172
  of subtraction, 154
Cluster, of data, 388, 395, 424
Clustering strategy, 145
Codes, 237
Collinear points, 509
Combinations, 67–68, 469, 483–487
Combining, in addition, 128–129
Common denominator, 284, 286, 290. *See also* Least common multiple (LCM)
Common difference, 34
Common ratio, 35
Communication, 44–47
Commutative property
  of addition, 132
  children's understanding of, 84, 132, 154
  division and, 193
  with fractions, 296
  of multiplication, 172
  subtraction and, 154
Comparisons
  additive, 337–338, 444
  of data, 444
  model of subtraction, 152, 153
  multiplicative, 251, 337–338, 444
  with percents, 366
Compatible numbers strategy
  addition, 135, 145, 147
  decimals, 321, 323
  division, 195, 200
  fractions, 290, 292, 300
  multiplication, 175
  subtraction, 156

Compensation strategy
  addition, 135, 137–138
  subtraction, 156
Complement, of a set, 71
Complementary angles, 516
Complementary events, 464
Completeness property, 330
Composite numbers, 232
Composite transformation, 592
Composition, *see* Decomposition; Part-whole relationships
Compound interest, 370, 373
Concave polygon, 545
Concave polyhedron, 563–564
Conceptual understanding, 50
Conclusion
  of argument, 36
  of conditional statement, 36
Concurrent lines, 511
Conditional statements, 36–40
Cones, 564
  surface area, 684
  volume, 688
Congruence, 537, 595
Congruence transformations
  in art, 585–586, 593, 596–597
  combinations of, 592–594
  glide reflections, 592–594, 595
  reflections, 584, 588–589, 591–592
  rotations, 584, 589–590, 591–592
  summary, 597
  symmetry and, 603–604
  tessellations and, 629
  theorems about, 593–594
  translations, 584, 585–588
Connections
  among $+ - \times \div$, 206–207
  among different concepts, 49–50
  among different sets of numbers, 329–330
  among fractions, decimals, and integers, 309, 313
  among fractions, decimals, and percents, 359–360
  among operations, 206
  between algebra and guess-test-revise, 10
  between addition and multiplication, 170, 172
  between addition and subtraction, 165–166
  between "carrying" and "borrowing," 164–165
  between decimals and fractions, 311, 313–314
  between decimals and integers, 309–310, 313
  between fractions and ratios, 270, 341–342
  between GCF and LCM, 248
  between graphs 389, 397
  between guess-test-revise and formula, 10–11
  between integers and other numbers, 256
  between mathematics and real-life, 51

Connections, *(continued)*
  between mathematics and other disciplines, 51
  between models of division, 190–193
  between multiplication and division, 192, 205–260
  between new concepts and old concepts, 49
  between numbers and shapes, 594–595
  between operations, 83
  between probability and sets, 456–457, 463
  between probability and statistics, 380–381
Connecting
  different models of a concept, 50
  conceptual and procedural knowledge, 50
Continuous amounts, 103, 128, 152, 169
Contrapositive, 36–37, 39
Convenience sample, 441
Converse, 36
Convex polygon, 545
Convex polyhedron, 563–564
Coordinate geometry, 550–554
  distance formula, 551–552
  midpoint formula, 553–554
  and similarity, 641–642
Coplanar points, 510
Correlation coefficient, 441
Counterexamples, 38
Counting, 69, 101–103
Counting numbers, *see* Natural numbers
Counting problems, 480–488
  combinations in, 67–68, 482–487
  factorials in, 482
  permutations in, 483–486
Counting systems, *see* Base 10 numeration system; Numeration systems
Cross multiplication, 278
Cross sections, 569
Cube, 565, 567–568
  in base 10 manipulatives, 112
  surface area, 692–693
  volume, 692–693
Cubed, 325
Cubic centimeter, 656
Cuisenaire rods
  addition of fractions, 286
  divisibility, 219
  greatest common factor, 244
  least common multiple, 247
  odd and even numbers, 220
Cuneiform writing, 107
Curved figures, 526–528
Cylinders, 571
  surface area, 683–684
  volume, 687–688

Data
  classes of, 397
  clusters of, 388, 395, 424
  four components, 382–383
  gaps in, 388, 395, 424
  importance in society, 380–382

modal classes of, 399
  uncertainty of sources, 409
  see also Averages; Distributions of data; Graphs
Databases, 67
Deca-, 654
Deci-, 654
Decibels, 658
Decimal numeration system, 111
Decimals, 308–329
  addition of, 318
  converting to fractions, 313–314
  definition of, 308
  division of, 320
  estimation with, 321–324
  history of, 308
  language and, 317–318
  multiplication of, 318–320
  in nonroutine problems, 322–324
  ordering of, 315–316
  repeating, 313
  rounding of, 316
  subtraction of, 318
  summary, 330–331
  usefulness of, 308–309
  zero in, 311–312
  see also Fractions; Rational numbers
Decimal Squares, 312
Decomposition, 141, 250–251
  additive and multiplicative, 250–251
  expanded form, 111, 182, 225, 317, 318
  of shapes, 524, 595
  see also Part-whole relationships
Decrease, percent of, 364–368
Deductive reasoning, 36–41
Deficient numbers, 238
Degrees
  of angles, 513–514, 656–657
  of temperature, 653, 656
Denominator, 268
  common, 284–286
  zero as, 64, 268
Denseness, of rational numbers, 280, 330
Density, 658, 691
Denying the conclusion, 38
Dependent events, 472
Dependent variable, 87
Descartes, René, 90, 257
Diagonal line symmetry, 608–609
Diagonals, of polygon, 540
Diagrams, for problem-solving, 8, 15, 20
Diameter
  of circle, 81, 82, 549–550, 664
  of sphere, 575
Difference, 153
Digits, 111
Dihedral angle, 565
Dimension, 4
Dimensional analysis, 16, 183, 352
Discount, 364
Discrete amounts, 128, 152, 395
Disjoint sets, 72, 464
Dispersion, measures of, 400–402
Distance formula, 551–552
Distortion, geometric, 584
Distributions of data, 400–402

bimodal, 401–402
  on line plot, 388
  normal, 401, 402, 430, 434, 435–436
  skewed, 401–402
  standard deviation of, 431–434
  summary, 402–403
  uniform, 401–402
Distributive property, 172, 176–178, 182
  with fractions, 296
  mental arithmetic with, 182, 195
Divergency constant, 339
Dividend, 192
"Divides," 217
Divisibility, 215–229
  by 2, 223
  by 3, 223–226
  by 4, 226–227
  by 5, 223
  by 6, 227
  by 8, 226–227
  by 9, 226
  by 10, 223
  by 12, 228
  with Cuisenaire rods, 219
  definitions, 217–218
  even numbers, 218–221
  and function language, 221
  of a multiple, 225
  notation, 216–217
  odd numbers, 218–221
  rules for, 223–228
  of a sum, 221–222
  summary, 229
Divisible, 217
Division, 189–201
  algorithm, 194
  algorithms for, 196–198
  contexts for, 189–190
  of decimals, 320
  definition of, 191
  estimation with, 199–201
  of fractions, 296–297
  as a function, 206
  with integers, 264
  mental, 195
  models for, 190–192
  with negative numbers, 264
  number sense with, 201–202
  operation sense with, 203–204
  in order of operations, 205
  properties of, 193
  of rational numbers, 296–297
  with remainders, 194
  by zero, 193, 330
Divisor, 192, 237
  proper, 237
Dodecahedron, 567–568
Domain, of a function, 85, 86
Dot plot, 387
Double and halve strategy, 175

Edge, of polyhedron, 563
Egyptian mathematics
  arithmetic, 193
  areas, 672
  fractions, 268

Egyptian mathematics *(continued)*
   geometry, 498
   numeration system, 104–105
Einstein, Albert, 44
Electric current, 658
Elementary teachers, Chinese, 297–298
Elements, of a set, 64
Ellipsis, 64
Empirical probability, 458
Empty set, 69
Enlarging, 584, 638–641
Equality
   congruence and, 537
   transitive property of, 311
Equal sets, 69
Equals sign, 83
Equally likely outcomes, 457
Equations, 87
Equator, 592
Equilateral triangle, 531–532
   as regular polygon, 546
   symmetry of, 604–605
   tessellation of, 623–624
Equivalent fractions, 10
   strategy for dividing whole numbers, 195
Equivalence, 83
Equivalent sets, 69
Eratosthenes, 233
Error, 659
Escher, M. C., 504, 583, 592–593, 596
Estimation, 143–144
   addition, 144–148
   decimals, 321–324
   division, 199–201
   fractions, 289–292, 299–301
   multiplication, 181–183
   ratios, 346–348
   square roots, 328–329
   subtraction, 163
   see also Precision
Euclid, 236, 507
Euler, Leonard, 39, 82, 233, 263
Euler diagram, 39
Euler line, 537
Euler formula, 569, 653–654
Even numbers, 41–42, 47, 218
Events, 456–457
   complementary, 463
   dependent, 472
   independent, 472
   mutually exclusive, 463
Expanded form, 111
   using, 182, 225, 317, 318
Expected value, 474–476
Experimental probability, 458, 476
Exponents, 112, 324–326
   negative, 325–326
Exterior, of angle, 612
   of a closed curve, 528
Exterior angles, 548

Faces, of polyhedron, 568
Factorial, 482
Factorization, prime, 235–237, 244, 246, 249
Factor-of-change method, 342

Factors, 170, 217
Fahrenheit system, 653
Fair games, 473–475
Fallacy, 38
Fermat, Pierre, 234, 466
Fibonacci numbers, 339
Figure-ground perception, 505
Figurate numbers, 92
Figures, *see* Three-dimensional figures; Two-dimensional figures
Finite sets, 66
Five-number summery, 425
Flat, in base 10 manipulatives, 112
Flips, *see* Reflections
Foot, 652
Formal logic, 36–41. *See also* Proof
Formulas, 10–11, 81–82
Fractal geometry, 611
Fractional dimensions, 4
Fractions, 268
   addition of, 32–33, 284–286
   area models of, 269, 270, 271, 272, 274, 275, 277, 278, 285, 287, 293–295, 302
   contexts of, 269–270
   converting to decimals, 313–314
   definition of, 268
   division of, 296–298
   equivalent, 276–278, 296
   estimation with, 289–292, 299–301
   function concept and, 270, 274
   history of, 268
   improper, 286–288
   as measures, 269–270, 271–272
   mental arithmetic with, 289–292, 299–301
   mixed numbers, 286–288
   multiplication of, 292–295
   in nonroutine problems, 301–302
   ordering of, 279
   percents and, 358, 359, 360
   proper, 286
   properties of operations with, 298
   rational numbers and, 268
   sense, 279
   simplest form, 278–279
   subtraction of, 288
   summary, 280–281, 302–303
   unit vs. whole in, 271, 281
   see also Decimals; Rational numbers; Ratios
Frege, Gottlob, 38
Frequency table, 384
Frequency table, grouped, 384, 397
Function machine, 86
Functions, 84, 85
   area as, 667
   definition of, 85
   domain of, 85
   fractions as, 270, 274
   geometric transformations as, 597
   graphs of, 15, 87, 90–91
   mathematical operations as, 205–206
   measurement as, 659–660
   notation for, 85
   and operations, 127
   range of, 85

   rates as, 344
   variables in, 81–82
Fundamental domain or region, 596
Fundamental Theorem of Arithmetic, 236
Fuzzy sets, 66

Galileo, 466
Gallon, 653
Gambling, 466, 474, 487
Gaps, in data, 388, 395, 424
Gauss, Carl, 17
"Gazinta," 197
GCF (greatest common factor), 241–251
Generalization, 31
Geometric sequences, 34–35
Geometric thinking, 505–506, 522–523
Geometric transformations, *see* Transformations
Geometry
   fractal, 611
   history of, 498–499
   introduction, 498–505
   see also Three-dimensional figures; Two-dimensional figures
Glide reflections, 591–593
   in tessellations, 629
Glide reflection symmetry, 612, 614, 615, 619
GPA (grade point average), 445–447
GPE (greatest possible error), 659
Gram, 656
Graph paper
   for measuring areas, 671–673
Graphs
   bar graphs, 384, 388, 427
   boxplots, 424–426, 427, 429
   circle graphs, 385, 389, 411, 427
   common mistakes, 388, 406, 407
   of functions, 87, 90, 91
   histograms, 395, 398, 401, 413, 427, 429
   interpretation of, 408–412
   line graphs, 405–407, 409–410, 427, 429
   line plots, 387–388, 394
   stem plots, 397, 423
   summary, 412–413
   truncated, 409–410
Great circles, 688
Greatest common factor (GCF), 241–251
Greatest possible error (GPE), 659
Greek mathematics, 215, 220, 235, 506–507
Grope-and-hope, 8
Grouped frequency table, 395
Guess–check–revise, 8

Halve and double strategy, 175
Hand-eye coordination, 505
Hecto-, 654
Height
   defined, 666
   of parallelogram, 576
   of polygons, 666
   of rectangle, 666
   of trapezoid, 667

Height *(continued)*
  of triangle, 666
Heptagon, 529
Hexagon, 529, 546
  tessellation of, 623, 624
Hindu-Arabic numeration system, *see* Base 10 numeration system
Hindu mathematics, fractions in, 268, 296
Hieroglyphics, 104
Histograms, 395, 398, 401, 413, 427, 429
Horizontal line symmetry, 608
Hypotheses, 36

I (set of integers), 64
Icosahedron, 567–568
Identity, algebraic, 81
Identity property
  of addition, 132, 206
  division and, 193
  with fractions, 298–299
  of multiplication, 172
  subtraction and, 154
"If and only if," 41, 67
Iff, 41, 67
If–then statements, 37
Image of a figure, 586
Impossible events, 457
Improper fractions, 287
Incenter, 537
Inch, 652
Increase
  addition as, 128
  additive, 337–338
  multiplicative, 337–338
  percent of, 364–368
Independent events, 472
Independent variable, 87
Indirect proof, 193, 234, 236
Inductive reasoning, 32–34
Infesential statistics, 441–443
Infinite lines, 508
Infinite patterns, 611
Infinite planes, 508
Infinite sets, 66
Input, 85–86
Insurance, 474–476
Integers, 64, 256–265
  addition, 258–259
  defined, 256
  division, 264
  history, 257
  multiplication, 263–264
  properties of, 260
  representations of, 257
  subtraction, 261–262
  summary, 264
Interest, 370–373
Interior, of angle, 513
  of a closed curve, 528
Interquartile range (IQR), 426
Intersecting lines, 498, 511
Intersection, of sets, 71, 72
Intuitive reasoning, 42–43
Invalid arguments, 37–38

Inverse
  additive, 260, 298
  multiplicative, 298
  of a statement, 37–39
IQR (interquartile range), 426
Irrational numbers, 280
Irregular shapes
  areas of, 673–674
  volumes of, 690–691
Isometric drawings, 569
Isosceles triangles, 531–532
  symmetry of, 604–605

Joining, addition as, 128
Jordan curve theorem, 528
Journal, 12, 53

Kilo-, 654
Kilogram, 656
Kilometer, 565, 566, 654–655
Kite, 539
  symmetry of, 606

"Lady or the Tiger," 469–470
Lateral faces, 565
Lattice algorithm, 142, 181
Law of Detachment, 38
Law of Large Numbers, 459
Leading digit strategy
  for addition, 134, 145
  for subtraction, 163
Learning log, 53
Least common multiple (LCM), 245–251
  for adding fractions, 285–286
Length,
  measurement of, 651–652, 654–655
Length models, of fractions, 269, 271, 272, 280, 287, 290, 293, 297, 301
Leonardo of Pisa, 109
Less than, 251
Light intensity, 658
Line graph, 405–407, 409–410, 427, 429
Line plot, 387–388, 394
Lines, 498, 511
  tangent, 549–550
  see also Number lines
Line segments, 510–511
Lines of reflection, 589
Lines of symmetry, 604
  diagonal, 608
  horizontal, 608
  inequivalent, 613
  of patterns, 611–612
  vertical, 608
Liter, 655–656
L numbers, 92–95
Logic, 36–41. *See also* Proof
Long, in base 10 manipulatives, 112
Longitude, 688
Looking back, 11
Lottery, 474, 487–488
Lower bound, 182
Lower quartile, 424
Lowest common denominator, 286. *See also* Least common multiple (LCM)

Ma, Liping, 297
Mahavira, 296
Maladaptive beliefs, 2–3
Manipulatives
  base 10, 112, 135, 137, 140, 157, 160, 173, 179, 196–197, 224, 225, 310–311, 319
  Cuisenaire rods, 219–220, 244, 247, 271, 286
  Decimal Squares, 312
  Geoboards 271, 284
  Mira, 589
  Pattern Blocks, 271, 285
Mappings, 87
Maps, 652, 689
Margin of error, 442
Mass
  density and, 658, 691
  metric units, 656
Mathematical modeling, 88
Mathematical power, 53
Mathematics, basic themes of, 4–5
Matrix, 133
Mean, 389–391, 392
  advantages and disadvantages of, 400, 402
  geometric view of, 393
  standard deviation from, 434–436
  weighted, 445–447
Measurement, 649–693
  of angles, 496–498
  big ideas about, 650
  fractions and, 269, 271, 281
  as a function, 659–660
  of length, 651
  metric system, 654–658
  precision of, 314–315, 650, 658–659
  pseudo-precision of, 312, 361
  statistical analysis and, 441
  systems of, 650–657
  of temperature, 653
  three-step process, 650
  of time, 651, 656–657
  and units, 207
  units for, 650, 652
  of volume, 653, 655
  of weight, 653
Measurement model of division, 190
Measures of dispersion, 400
Measures of the center, 389–393, 399–400. *See also* Averages
Median, 389–391
  advantages and disadvantages of, 401–402
  of triangle, 535
Members of a set, 64
Mental arithmetic
  addition, 135–136
  division, 195
  with fractions, 289–292, 299–301
  multiplication, 175
  subtraction, 155–157
  see also Estimation
Mersenne primes, 234
Meter, 654–655
Metric system, 654–658
Midpoint formula, 553–554

Mile, 652
Milli-, 654
Milligram, 656
Milliliter, 655–656
Millimeter, 654–655
Minuend, 153
Minutes
　of angle, 514
　of time, 651, 657
Mira, 589
Mirror symmetry, *see* Lines of symmetry;
　Planes of symmetry
Misleading advertising, 690
Missing addend model, 152
Missing factor model, 192
Mixed numbers, 287
Modal classes, 398, 427
Mode, 389–391
　advantages and disadvantages of, 400, 402
Models, 23, 88. *See also* Representations
Modus Tollens, 38
"More than," 251, 337–338
Multiple, 170, 217
　least common, 245–250
Multiple representations, 18–20, 23–26, 87, 93–95, 129, 153, 168, 190–192, 271–274, 310–311. *See also* Models
Multiples of 10 strategy, 136, 175
Multiplicand, 170
Multiplication, 168–185
　algorithm for, 178–181
　children's understanding of, 176–178
　contexts for, 169–171
　of decimals, 318–320
　definition of, 170
　estimation with, 181–183
　of fractions, 292–295
　of integers, 263–264
　mental, 175
　models for, 170–171, 178
　with negative numbers, 263–264
　number sense with, 184–185
　in order of operations, 205
　properties of, 172–173
　of rational numbers, 292–295
　in real-life problem, 183–184
　as repeated addition, 169
　summary, 185
　units in, 171–172
Multiplication principle, 467, 472
Multiplication table, 173–175
Multiplicative comparisons, 250–251, 338, 449
Multiplicative identity, 172
Multiplicative increase, 366
Multiplicative inverse, 298
Multiplier, 170
Multistep nonroutine problems, 6
Mutually exclusive events, 463, 472

n! (factorial), 482
N (set of natural numbers), 64
National Council of Teachers
　of Mathematics, *see* NCTM
　standards

Natural numbers, 64, 256, 329
　prime, 232–233
　see also Number theory
NCTM standards, 4–5, Appendix
　algebraic thinking, 80, 84
　area, 664
　communication, 44
　connections, 48
　data interpretation, 382
　estimation, 143
　extending number system, 255
　function concept, 84
　geometry, 505
　making sense of mathematics, 284
　mathematical power, 53, 251
　number theory, 215
　perimeter, 669
　probability, 455
　problem-solving, 6
　process standards, 5
　proportional reasoning, 337
　reasoning and proof, 30–31
　representations, 23
　statistics, 382
Negative exponents, 325–326
Negative numbers, *see* Integers
Negative of a number, 259
Nets, 569, 573
Newton, Isaac, 403
n-gon, 529
Noise measurement, 658
Noncollinear points, 509
Nonroutine multistep problems, 6
Normal distribution, 401, 402, 430, 434, 435–436
Notation, history of
　algebraic, 81
　arithmetic operations, 129
　decimals, 308
　exponents, 324
　negative numbers, 262
　*n*th term, 92
　percent, 358
　see also Symbols
Null set, 69
Number lines
　addition with, 130–131, 135, 137
　decimals on, 314, 315
　fractions with, 269, 271, 272, 280, 287, 290, 293, 297, 301
　division with, 192
　integers with, 257, 259, 260, 261, 264
　multiplication, 169
　subtraction with, 154–158, 162
Numbers, 103
　connections with shapes, 595
　irrational, 280, 327–328
　numerals and, 103
　perfect, 237–238
　prime, 231–236
　real, 329–330
　relative magnitude of, 113–114
　sets of, 64–65, 256, 329
　whole, 64, 256, 329

　see also Integers; Natural numbers;
　　Rational numbers
Number sense, 148
　addition, 148–149
　division, 201–202
　multiplication, 184–185
　subtraction, 164
Number theory, 215–251
　divisibility, 216–229
　greatest common factor, 241–244, 248–250
　importance of, 215
　least common multiple, 245
　prime numbers, 232–237
　summary, 251
Numeral, 103
Numeration systems
　additive, 105
　Babylonian, 106–108
　base of, 109
　base 10, 109–113
　Egyptian, 104–105
　Hindu-Arabic, 109–113, 118
　history of, 101–103
　Mayan, 120
　patterns in, 101–103
　positional, 107
　Roman, 105–106
　see also Base 10 numeration system
Numerator, 268
Numerical sequences, 33–36

Oblique cone, 574
Oblique cylinder, 574
Oblique prism, 565
Obtuse angle, 516
Obtuse triangle, 531–532
Octagon, 546
　tessellation with, 623–624
Octahedron, 567–568
Odd numbers, 41–42, 47, 218
One
　multiplication by, 172, 271
　prime numbers and, 233, 235
One-to-one correspondence, 69, 102
Open curves, 526–528
Operations on sets, 70–72
Operations on shapes, 594
Operation sense, 148, 201–205
Operator context, 270, 294
Opposite of a number, 260
Optical illusions, 504
Ordered pairs, 87
Order of operations, 205
Origin, of number line, 130
Origin, of coordinate system, 550
Ortho center 537
Ounce, 653
Outcomes, 456
Outliers, 388, 395, 424
　defined, 426
Output, 85–86
Owning vs. renting knowledge, 6, 10, 12, 258

Parallel lines, 511
Parallelograms, 539
　areas, 667–678
　symmetry of, 606
Partial sums, 137
Partial products, 180
Partitioning model of division,
　　190–192
　with fractions, 296–297, 303
Part-whole relationships, 180
　addition, 129–130, 165
　division, 191
　fractions, 269–270, 281
　multiplication, 170
　in percent problems, 363
　with ratios, 341–342
　subtraction, 152–153, 165
　see also Decomposition
Pascal, Blaise, 21, 43, 461, 466
Pascal's triangle
　patterns in, 21–22
　in probability problems, 462–463,
　　465
　sets and, 68
Pattern Blocks, 271, 285
Patterns, 16–22
　in addition table, 133–134
　algebra and, 80
　basic characteristics of, 609–612
　in counting, 103–104
　fractal, 611
　geometric, 609–611
　in even and odd numbers, 218–221
　inductive reasoning about, 32–33
　in magic square, 46–47
　in nature, 611
　in numeration systems, 103–104
　in numerical sequences, 33–36
　in problem-solving, 17–20, 92
　strip patterns, 612–615
　symmetry of, 611–612
　tessellations, 623–628
　wallpaper, 615–617
Pencil icon, 3
Pentagons, 529, 546
　tessellation and, 623, 624
Percents, 357–373
　change and, 364–368
　connections with other topics, 363
　decimals and, 360
　fractions and, 358, 359, 360
　greater than 368–370
　history of, 357, 358
　interest, 370–373
　less than one, 368
　number relationships and, 251
　part-whole relationships and, 360–361,
　　363
　proportions and, 357
　rates and, 357
　and scientific notation, 371–372
　summary, 373
　uses of, 357
Perception, spatial, 476–477, 505–506
Perceptual constancy, 505
Perfect numbers, 237–238

Perimeter, 664
　area and, 663–664
　children's understanding of, 663
　circumference of a circle, 81, 82,
　　664–665
　definition of, 664
Permutations, 483–486
Perpendicular bisector, 535, 589
Perpendicular lines, 498, 511
Pi, 81, 82, 664, 671
Piecemeal strategy, 664–665
Pint, 65
Place value, 107
Planes, 508–510
Planes of symmetry, 618–619
Platonic solids, 567
Points, 508
Point symmetry, 606
　in patterns, 614
Polya's four steps, 11, 12
Polygons, 529, 545–548
　angles of, 547–549
　base of, 685
　central angle of, 548
　concave, 545
　convex, 545
　diagonals of, 530
　exterior angles of, 548–549
　height of, 668
　interior angles of, 547
　n-gons, 529 (or "naming," 529)
　polyhedra and, 562–563
　regular, 546
　sides of, 529
　similar, 637–642
　sum of interior angles, 547
　symmetry of, 604–608
　tessellation of, 620–627
　vertices of, 529
　see also Quadrilaterals; Triangles
Polyhedra, 563
　bases, 565
　concave, 564
　convex, 564
　edges, 563
　faces, 563
　inside/outside, 563
　regular, 567–568
　surface of, 563
　surface areas, 698–700, 701
　vertices, 563
　volumes, 685–687
Polyhedron, 563
Populations, 441–443.
　See also Sampling
Positional numeration system, 107
Positive integers, 256
Postulates, 507–508
Pound, 653
Powers, 323–326
　of 10, 105
Precision, 314–315, 650, 659–660
　pseudo-precision, 312, 361
　see also Estimation
Prime factorization, 235
　and tree diagrams, 236

　in exponential form, 244, 246
Prime numbers, 232–237
　codes using, 237
　definition of, 232
　factorization into, 235–236, 244
　infinite number of, 236
　perfect numbers and, 237
　Sieve of Erathosthenes, 233–234
　summary, 238
　testing for, 233–235
Principal, 370
Principal square root, 328
Prisms, 565–566
　surface area, 681–683
　volume, 685–686
Probability, 455–488
　area model of, 470
　basic definitions, 456–457
　combinations and, 483–487
　complementary events, 464
　dependent events, 472
　dice, 466–469
　equally likely, 457–458
　expected value and, 474–476
　experimental, 458–459
　fair games, 472–474
　independent events, 472
　Law of Large Numbers, 459
　lottery, 474, 487–488
　multiplication principle, 467–472
　mutually exclusive events, 463, 472
　outcome, 457
　permutations, 483–487
　real-life importance of, 455
　summary, 488
　theoretical, 458–459
　tree diagram and, 461, 467, 469–470, 471
　true-false tests, 464–465
Problem-solving, 6–16
　algebra for, 9–10, 89, 230, 275, 302
　beliefs about, 13
　diagrams for, 8, 15, 183, 220, 302, 360
　looking back, 11
　NCTM standard, 6
　playfulness in, 19
　Polya's four steps, 11–12
　real-life problems, 6, 14–15, 51–52
　strategies for, 7–16
　tables for, 8, 11, 19–20, 25, 34–35, 37,
　　52, 87, 92
　traditional word problems, 14–15
Procedural understanding, 50, 139, 303
Process standards, 5. *See also* NCTM
　　standards
Product, 170
　partial, 180
Proof, 30–31
　in geometry, 506, 517
　indirect, 193, 234, 236
　in problem-solving, 41–42, 47
Proper divisor, 237
Proper fraction, 286
Proper subset, 66
Proportion, definition of, 339
Proportional reasoning
　fractions and, 340–342

Proportional reasoning *(continued)*
  multiplicative comparisons in, 338
  percents and, 357, 359, 363
  see also Rates; Ratios
Protractor, 514
Pseudo-precision, 312, 361
Pyramids, 566
  surface area, 684
  volume, 687–688
Pythagoreans, 220
Pythagorean theorem, 327, 669–671

Q (set of rational numbers), 64
Quadratic formula, 671
Quadrilaterals, 539–545
  angles of, 540–541
  and attributes, 541–542
  classifying, 544–545
  diagonals of, 540
  symmetry of, 606–607
  tessellation of, 625–626
  and Venn diagrams, 543
  see also specific quadrilaterals
Qualitative graphing, 90–91
Quantity, 4
Quart, 653
Quartiles, 425. *See also* Boxplot
Quotient, 190, 269

Radius
  of circle, 549, 550
  of sphere, 575
Random phenomena, 455
Random sample, 441
  stratified random sample, 442
Range
  of data, 388, 394, 397, 423
  of a function, 85, 86
  interquartile, 426
Rates, 339
  definition of, 339
  as functions, 344
  percent and, 357
  real-life uses of, 339–340, 350–352
  see also Ratios
Rational numbers
  as measure, 269
  as operator, 270
  as quotient, 269
  as ratio, 270
  definition of, 268
  denseness of, 280
  set of, 64, 268
  see also Decimals; Fractions; Ratios
Ratios, 339–342
  definition of, 339
  dimensional analysis with, 352–353
  estimation with, 346–348
  fractions and, 341–342
  in Greek mathematics, 268
  as multiplicative comparisons, 337–338, 353
  rational numbers and, 270
  real-life uses of, 339–340, 348–350
  unit concept and, 338–339

  see also Proportional reasoning; Rates
Rays, 510
Real numbers, 329–330
  properties of, 330
Reasoning, 30–31
  deductive, 36–41
  inductive, 32–34
  intuitive, 42–43
Rectangles, 540, 545
  areas, 666
  symmetry of, 606
Rectangular model, of product, 171, 173, 177, 179–180
Rectangular prisms, 493–494, 565
  surface area, 681–682
  volume, 685–686
Reducing fractions, 278
Reflections (flips), 584, 588–589, 591–592
  in combined transformations, 592–594
  commutative, 594
  properties of, 589
  in tessellations, 621, 629
Reflection symmetry, 612, 613–617
  of patterns, 613–617
  three-dimensional, 618–619
Reflex angle, 516
Regular polygons, 546–547
  tessellation of, 623–624
Regular polyhedra, 567–568
Relationships
  ratios as, 339
  between variables, 81–82, 84–95
Relative error, 659
Relative magnitude, 113–114
Relatively prime, 245
Reliability of data, 408
Remainders, 194
Renting vs. owning knowledge, 10, 11, 363
Repeated addition model, 169
Repeated subtraction model, 190–192
  with fractions, 296–297
Repeating decimals, 313
Representations, 23–26
  connecting geometric and numerical, 111
  exponential, 112
  of functions, 87
  multiple, 24–26
  response rate, 442
  translating among, 73
  see also Models
Rhombus, 540
  symmetry of, 606
Right angle, 516
Right cone, 574
Right cylinder, 574
Right-identity
  for division, 193
  for subtraction, 154
Right prism, 565
Right triangles, 531–532
  Pythagorean theorem, 327, 669–671
Roman mathematics, 105–106, 268
Rotations (turns), 584, 589–590, 591–592

  in combined transformations, 592–594
  in tessellations, 621, 629
  properties of, 590
Rotation symmetry, 612, 613–617
  of patterns, 613–617
  three-dimensional, 618–619
Rounding, 144, 146
  of decimals, 316

Sale price, 364
Sample, 441
Sample size, 442
Sample space, 456
Sampling, 441–443
SAT, 436
Scaffolding algorithm, 199
  for division, 198
Scale factor, 640
Scalene triangle, 531–532
  symmetry and, 604
  tessellation of, 621–622
Scatterplot, 437–440
Scientific notation, 326–327
Seconds
  of angle, 514
  of time, 651, 657
Semiregular tessellation, 624
Sequences, 33–35
Set-builder notation, 64, 65
Set models, of fractions, 271
Sets, 61–75
  classification with, 62–64
  complements of, 71
  descriptions of, 64–66
  disjoint, 72, 464
  elements of, 64
  empty set, 69
  equal, 69
  equivalent, 69
  finite, 66
  importance of, 61
  infinite, 66
  intersection of, 71
  kinds of, 66
  members of, 64
  null set, 69
  of numbers, 64–65
  operations on, 70–72
  subsets of, 64, 67–69
  subtraction of, 72
  summary, 75–76
  union of, 71
  universal, 70
  Venn diagrams of, 70–75
  well-defined, 66–67
  see also Venn diagrams
Shape, 4
Shapes, 523–529
  children's understanding of, 504–506, 522–523
  connections with numbers, 594–595
  see also Three-dimensional figures; Two-dimensional figures
Shrinking
  by operator, 270, 294

Shrinking *(continued)*
    similarity transformation, 637–639
Side
    of angle, 512
    of polygon, 529
Sieve of Eratosthenes, 233–235
Similarity transformations, 637–640
Simple closed surfaces, 563
Simple curves, 526–528
Simple interest, 370
Simplest form, of fraction, 278
Simulations, 459
Skewed distributions, 401–402, 435
Skew lines, 511
Skip counting, 223
Slides, *see* Translations
Slope
    of graph, 90–91
Solid, definition of, 563
Solution paths, 20
Space figure, 563
Spatial thinking, 505–506
Speed, 658
Spheres, 575–576
    center, 575
    surface area, 685
    volume, 688
Spirals, 499, 549
Spread (of data), 397
Spreadsheet
    averaging data with, 385
    compound interest with, 372
    making graph with, 385
Squared, 325
Square feet, 665
Square roots, 328–329
Squares, 539, 544
    areas, 666
    symmetry of, 506
    tessellations with, 623–624
Standard deviation, 431–434
Statements, 36–39
Statistics, 380–413, 423–448. *See also*
    Data
Stem-and-leaf plot, 399, 423
    back-to-back, 423, 429
Stem plot, 397, 423
Step functions, 88
Stevin, Simon, 308
Story problems, 14–16. *See also*
    Problem-solving
Straight angle, 516
Stratified random sample, 442
Stretching
    by operator, 270, 294
String strategy for multiplying, 177
Strip patterns, 612–615
Subsets, 64, 67–69
Subtraction, 152–165
    algorithms for, 159–162
    children's understanding of,
        157–159
    contexts for, 152–153
    of decimals, 318
    definition of, 153
    estimation with, 163

of fractions, 288
of integers, 261–262
mental, 155–157
models of, 153–154
with number lines, 154
with negative numbers, 261–262
in order of operations, 205
pictorial models, 153
properties of, 154
repeated, 190–192, 296–298
of sets, 71
similarities with addition, 165–166
summary, 165
Subtrahend, 153
Sum, 130
    of first 100 numbers, 17–20
Superabundant numbers, 239
Superprime, 239
Supplementary angles, 516
Surface area, 681–685
    cone, 684–685
    cylinder, 683–684
    prism, 681–682
    pyramid, 684
    sphere, 685
    and volume, 692–693
Symbols, *see* Notation
Symmetry, 603–619
    axis of, 619
    breaking, 617–618
    definition of, 604
    glide reflection, 591–593, 594, 612
    imperfect, 611
    importance of, 603–604
    of letters, 608–609
    of patterns, 611–612
    plane of, 619
    point, 606
    of polygons, 607–608
    of quadrilaterals, 606–607
    reflection, 604–609
    rotation, 604–609
    of strip patterns, 612–615
    three-dimensional, 618–619
    translation, 612
    of triangles, 604–605

Tables, 8–9, 11, 19–20, 25, 34–35, 37, 52,
    87, 92
Take-away, 152, 153
Tallies, 101, 105, 106, 107, 110
Tangent line, 549–550
Taxicab language, 586
Teachers, elementary, Chinese,
    297–298
Teflon knowledge, 10
Temperature, 653, 662
Tessellations, 620–629
    congruence transformations and, 629
    definition of, 620
    of regular polygons, 622–624
    semiregular, 624
    of triangles, 621–622
    of trapezoids, 625–626
    vertex points, 626
Tetrahedron, 567–568

Tetris, 499–501
Theoretical probability, 458–459
Thought experiment, 430
Three-dimensional figures, 559–569
    surface area, 681–685
    symmetry of, 618–619
    volume, 685–688
    see also names of specific figures
Time, 651, 656
"Times as much," 337
Toolbox, for problem-solving, 10
Transformations, 582–597, 637–642
    in art, 582–583, 596
    combinations of, 592–594
    congruence transformations,
        585–595
    as functions, 597
    glide reflections, 591–593
    reflections, 584, 588–589, 591–592
    rotations, 584, 589–590, 591–592
    similarity transformations, 584
    summary, 597
    symmetry and, 603
    tessellations and, 620
    theorems about, 594
    topological, 584
    translations, 584, 585–588, 591–592
    see also Symmetry
Transitive property, 311
Translations (slides), 584, 585–588,
    591–592
    in combined transformations,
        592–594
    in tessellations, 621, 629
    properties of, 587–588
    symmetry, 612
Translation symmetry, 612, 613–617
Trapezoids, 539, 544
    area, 668–669
Tree diagram
    for counting problems, 481, 483
    of prime factors, 236
    for probability problems, 461, 467,
        469–471, 472
Trial and error, 7, 8
Triangles, 529
    acute, 531–533
    altitude of, 536
    angles of, 531–532
    areas of, 668
    bases of, 533
    classifying, 531–533
    congruent, 537–539
    equilateral, 513–532, 533, 534, 604, 605,
        608, 623
    heights of, 536
    isosceles, 531–533
    obtuse, 531–533
    properties, 535
    right, 531–533, 669–671
    scalene, 531–533
    similar, 638–640
    symmetry of, 604–605
    tessellations of, 621–624
    and Venn diagrams, 533–535
True-false tests, 464–465

Truncated graph, 410
Turns, *see* Rotations
Two-dimensional figures, 521–555
    classifying, 526–529, 531–533, 544
    congruent, 537–539
    curved, 471, 526–529
    re-creating from memory, 523–524
    see also names of specific figures
Typical representative, 390–391

U (universal set), 70
Uncertainty, 4, 380–381
    of data sources, 409
    see also Data; Probability
Undefined terms, 508
Uniform distribution, 401–402
Union of sets, 71
Unit fractions, 33, 268
Unit pricing, 342–343
Unit-rate method, 343
Units
    for area, 665, 669
    composite, 114
    for measurement, 650, 651–653, 654–655, 656
    in multiplication, 171–172
    and numbers, 114
    of pattern, 609–610
    in ratios, 338–339
    simple, 114
    vs. whole, 271, 281
Unit segment, 130
Universal set, 70
Universe, 70
Upper bound, 182
Upper quartile, 425
U.S. customary system, 659

Valid arguments, 37–38
Validity of data, 408
Van Hiele levels, 522–523, 524, 535
Variables, 81–82
Variation, 394, 400–402, 430
    measurement, 394
    natural, 394
Vector, 587
Venn diagrams, 75
    as a communication tool, 75
    of conditional statements, 39–40
    of divisibility, 228
    of GCF and LCM, 248
    of letter symmetries, 609
    of number sets, 256
    of quadrilaterals, 543, 545
    and relationships between sets, 72–73
    of triangles, 533–534
Vertex
    of angle, 512, 513
    point, 626
    of polygon, 529
    of polyhedron, 563
Vertex point, 626
Vertical angles, 516–517
Vertical line symmetry, 605
Vertices, *see* Vertex
Views, of a building, 569–570
Visual discrimination, 505
Visual memory, 505
Visual perception, 505–506
Volume, 653, 655–656, 685–689
    of cone, 688
    of cylinder, 686–687
    of hollow box, 691–692
    of irregular objects, 690–691
    measurement of, 633, 655–656
    misleading, 690
    of prism, 685–686
    of pyramid, 687–688
    of sphere, 688
    summary, 693
    surface area and, 692–693

W (set of whole numbers), 64
Wallpaper patterns, 615–617
Weight, 653, 656
Weighted averages, 445–447
Well-defined sets, 66–67
What-if programs, 8
"What's my rule?", 85–86
Whiskers, *See* Boxplot
Whole, vs. unit, 271, 280
Whole numbers, 64, 255, 256, 329
Width, 666–667
Word problems, 14–15. *See also* Problem-solving

Yard, 652

Zero, 111
    addition of, 132
    canceling, 195
    in decimals, 311–312
    division by, 193, 330
    as exponent, 324
    invention of, 108, 109, 111
    multiplication by, 173, 299
    need for, 108
    on number line, 130
    number sets and, 329–330
    subtraction of, 154
    two meanings of, 111
Zero property of multiplication, 172